THE ELEMENTS

Name	Symbol	Atomic Number	Atomic Weight*
Actinium	Ac	89	(227)
Aluminum	Al	13	26.981539
Americium	Am	95	(243)
Antimony	Sb	51	121.75
Argon	Ar	18	39.948
Arsenic	As	33	74.92159
Astatine	At	85	(210)
Barium	Ba	56	137.327
Berkelium	Bk	97	(247)
Beryllium	Be	4	9.012182
Bismuth	Bi	83	208.98037
Bohrium	Bh	107	(262)
Boron	B	5	10.811
Bromine	Br	35	79.904
Cadmium	Cd	48	112.411
Calcium	Ca	20	40.078
Californium	Cf	98	(251)
Carbon	C	6	12.011
Cerium	Ce	58	140.115
Cesium	Cs	55	132.90543
Chlorine	Cl	17	35.4527
Chromium	Cr	24	51.9961
Cobalt	Co	27	58.93320
Copper	Cu	29	63.546
Curium	Cm	96	(247)
Dubnium	Db	105	(262)
Dysprosium	Dy	66	162.50
Einsteinium	Es	99	(254)
Erbium	Er	68	167.26
Europium	Eu	63	151.965
Fermium	Fm	100	(257)
Fluorine	F	9	18.9984032
Francium	Fr	87	(223)
Gadolinium	Gd	64	157.25
Gallium	Ga	31	69.723
Germanium	Ge	32	72.61
Gold	Au	79	196.96654
Hafnium	Hf	72	178.49
Hassium	Hs	108	(265)
Helium	He	2	4.002602
Holmium	Ho	67	164.93032
Hydrogen	H	1	1.00794
Indium	In	49	114.82
Iodine	I	53	126.90447
Iridium	Ir	77	192.22
Iron	Fe	26	55.847
Krypton	Kr	36	83.80
Lanthanum	La	57	138.9055
Lawrencium	Lr	103	(260)
Lead	Pb	82	207.2
Lithium	Li	3	6.941
Lutetium	Lu	71	174.967
Magnesium	Mg	12	24.3050
Manganese	Mn	25	54.93805
Meitnerium	Mt	109	(266)
Mendelevium	Md	101	(258)
Mercury	Hg	80	200.59
Molybdenum	Mo	42	95.94
Neodymium	Nd	60	144.24
Neon	Ne	10	20.1797
Neptunium	Np	93	237.05
Nickel	Ni	28	58.69
Niobium	Nb	41	92.90638
Nitrogen	N	7	14.00674
Nobelium	No	102	(259)
Osmium	Os	76	190.2
Oxygen	O	8	15.9994
Palladium	Pd	46	106.42
Phosphorus	P	15	30.973762
Platinum	Pt	78	195.08
Plutonium	Pu	94	(244)
Polonium	Po	84	(209)
Potassium	K	19	39.0983
Praseodymium	Pr	59	140.90765
Promethium	Pm	61	(145)
Protactinium	Pa	91	231.03588
Radium	Ra	88	226.03
Radon	Rn	86	(222)
Rhenium	Re	75	186.207
Rhodium	Rh	45	102.90550
Rubidium	Rb	37	85.4678
Ruthenium	Ru	44	101.07
Rutherfordium	Rf	104	(261)
Samarium	Sm	62	150.36
Scandium	Sc	21	44.955910
Seaborgium	Sg	106	(263)
Selenium	Se	34	78.96
Silicon	Si	14	28.0855
Silver	Ag	47	107.8682
Sodium	Na	11	22.989768
Strontium	Sr	38	87.62
Sulfur	S	16	32.066
Tantalum	Ta	73	180.9479
Technetium	Tc	43	(98)
Tellurium	Te	52	127.60
Terbium	Tb	65	158.92534
Thallium	Tl	81	204.3833
Thorium	Th	90	232.0381
Thulium	Tm	69	168.93421
Tin	Sn	50	118.710
Titanium	Ti	22	47.88
Tungsten	W	74	183.85
Uranium	U	92	238.0289
Vanadium	V	23	50.9415
Xenon	Xe	54	131.29
Ytterbium	Yb	70	173.04
Yttrium	Y	39	88.90585
Zinc	Zn	30	65.39
Zirconium	Zr	40	91.224

* Based on relative atomic mass of $^{12}C = 12$, 1987 IUPAC values. Values in parentheses are the mass numbers of the isotopes with the longest half-life.

World of Chemistry *Essentials*

SECOND EDITION

World of Chemistry
Essentials

SECOND EDITION

Melvin D. Joesten
Professor of Chemistry
Vanderbilt University
Nashville, Tennessee

James L. Wood
Adjunct Professor
David Lipscomb University
Nashville, Tennessee

Mary E. Castellion
Contributing Editor
Norwalk, Connecticut

Several of the Boxed Features entitled *The World of Chemistry* provided by

Isadore Adler†
University of Maryland

Nava Ben-Zvi
Hebrew University of Jerusalem
and Open University of Israel

Saunders Golden Sunburst Series
SAUNDERS COLLEGE PUBLISHING
Harcourt Brace College Publishers

Fort Worth   Philadelphia   San Diego   New York   Orlando   Austin
San Antonio   Toronto   Montreal   London   Sydney   Tokyo

Vice President Publisher: John Vondeling
Vice President Publisher: Emily Barrosse
Product Manager: Pauline Mula
Editorial Coordinator: Ed Dodd
Project Editor: Progressive Publishing Alternatives
Production Manager: Charlene Catlett Squibb
Art Director: Lisa Adamitis
Manager, Photo and Permission: George Semple

Cover Credit: SPL/Photo researchers

Frontispiece: European Space Agency/Science Photo Library/Photo Researchers, Inc.

Printed in the United States of America

World of Chemistry—Essentials, Second Edition
0-03-005888-0

Library of Congress Catalog Card Number 98-5951

Joesten, Melvin D.
World of chemistry: essentials.—2nd ed./Melvin D. Joesten,
James L. Wood; Mary Castellion, contributing editor.
p. cm.—(Saunders golden sunburst series)
Includes index.
ISBN 0-03-005888-0
1. Chemistry. I. Wood, James L., 1926–
II. Castellion, Mary E. III. Title.
QD33.J54 1998
540—dc21 98-5951
CIP

90123456 032 10 9876543

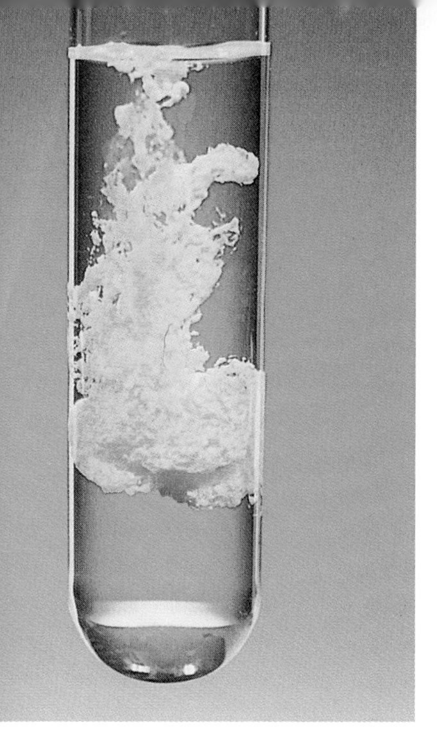
(C. D. Winters)

Contents Overview

(C. D. Winters)

Preface

Approach and Scope

The second edition of *World of Chemistry—Essentials* continues in the tradition of the first edition in providing a text for a one-semester course for nonscience majors. The text is a shortened version of *World of Chemistry, 2/e.* The fundamental approach of the second edition of *World of Chemistry—Essentials* is to teach chemical principles within the framework of real world applications to an even greater extent than was done in *World of Chemistry, 2/e.* Mary E. Castellion, who was a contributing editor to *World of Chemistry, 2/e,* played a significant role in planning, writing, and editing the second edition of *World of Chemistry—Essentials.* The end result is a major revision that both preserves and expands the fundamental approach of the first edition. The comments of reviewers who have used the first edition also influenced our planning and writing of the second edition.

Organization

The philosophical tone for the book is set in Chapter 1. Chapters 2 through 17, while laying the groundwork for intellectual consideration of the effect of chemistry on society, are replete with interesting applications to which liberal arts students readily relate. These include synthetic materials that dramatically alter the human environment; the nutritional basis of healthy living; medicines and drugs; pollution and the conservation of natural resources; and the agricultural production of food for a hungry world population. Basic concepts about atoms, compounds, the periodic table, nuclear changes, chemical bonding, and states of matter are presented in Chapters 2–5. Chemical reactions, acids and bases, oxidation and reduction, are grouped in Chapters 6–8. Organic chemistry and its relevance to energy sources and materials are described in Chapters 9 and 10. Chemical principles of biochemistry are discussed in Chapter 11, and the remaining chapters in the book are topic centered: food chemistry and nutrition (Chapter 12), chemistry and medicine (Chapter 13), useful materials (Chapter 14), water quality and abundance (Chapter 15), air quality (Chapter 16), and feeding the world (Chapter 17).

The following list of specific applications and topics of concern to society illustrates the extent to which applications are integrated with chemical principles: metric system in the U.S. (Chapter 2); scanning tunneling microscope (Chapter 3); medical uses of radioisotopes (Chapter 4); nuclear energy (Chapter 4); problems of radioactive wastes (Chapter 4); recycling metals (Chapter 6); acidic and basic household cleaners (Chapter 7); bleaches, disinfectants, antioxidant vitamins (Chapter 8); new polymer materials, recycling plastics (Chapter 10); shampoos, detergents, and soaps (Chapter 11); hair, creams and lotions (Chapter 11); chiral drugs (Chapter 11); biogenetic engineering (Chapters 11 and 17); new steels, conductors, semiconductors (Chapter 14); solar energy (Chapter 14).

Major Changes in This Revision

Chapter 1 now includes a connection to chemistry-related stories in newspapers and can lead students to discussions of science in the news throughout the course. Chapter 2 introduces the macro-micro aspect of the chemical view of matter, and systematically defines essential terms needed in following chapters. New information on states of matter has been added to chemical bonding in Chapter 5 in response to reviewer and user comments about the need for more emphasis on this topic. The properties of water have been moved from the chemical reactions chapter to Chapter 5.

Updated topics include recycling (Chapters 6 and 10), energy sources (Chapters 4, 9, and 14), global warming and ozone layer depletion (Chapter 16), medicines (Chapter 13), nutrition (Chapter 12), environmental issues (Chapters 15 and 16), and feeding the world population (Chapter 17). The number of chapters has increased by one over the last edition, because of the splitting of Acid–Base Reactions and Oxidation–Reduction Reactions into two separate chapters (Chapters 7 and 8, respectively). The order of the chapters has also slightly changed. Nuclear Reactions (Chapter 4) now comes immediately after the Periodic Table (Chapter 3) rather than after Chemical Reactivity (Chapter 6). The treatment of organic chemistry has also been rearranged to relate the chemical principles more directly to the applications: Chapter 9 discusses Energy and Hydrocarbons; and Chapter 10, Organic Chemicals and Polymers, emphasizes industrial and consumer uses of organic chemicals. In keeping with the theme of stressing the application of chemical principles to the students' everyday lives, the chapter on consumer chemistry has been integrated throughout the text. The topical natures of Chapters 11 through 17 allow for their use individually or in a different sequence. During the revision of every chapter, we have, in addition to updating, selectively diminished coverage of some topics.

New to the Second Edition

A major focus during the revision has been to incorporate into each chapter guides to what is important and tools that will assist students in learning.

Chapter opening questions highlight the most important aspects of the chapter topic.

Essential terms are defined upon their first use. They are boldfaced and definitions are given in the Glossary/Index.

Worked examples within chapters demonstrate how to utilize the simple concepts in this course that require problem solving. Each Example is accompanied by one or more Exercises that are answered in Appendix F, thereby helping students to check their understanding. The Examples and Exercises, together with the Self-Tests featured in the first edition, help students gain confidence about smaller segments of material before they try to answer questions and problems at the end of the chapters.

End of chapter questions and problems are mostly new to this edition of *World of Chemistry—Essentials* and have increased in number. They are divided into two groups, to again meet the varying needs of instructors who do or do not include mathematical problem solving. **Questions for Review and Thought** include some questions that review chapter material and others that require thought about the meaning and application of the chemical concepts. **Problems** are usually mathematical and appear only in appropriate chapters.

The scope of our emphasis on the relationship of chemistry to everyday life has been broadened with three new types of boxed features:

Frontiers in the World of Chemistry describe new developments related to chapter topics, developments that frequently attract attention in the media.

Science and Society essays focus on science-related societal issues that often raise difficult questions and do not have clear-cut answers.

Personal Side boxes highlight the achievements of recent contributors to the science of chemistry.

As was the case in the first edition, **The World of Chemistry** boxes feature interviews from *The World of Chemistry Video Series.*

For today's students, visual impact is approaching equal importance with the written word. With the assistance of the team of creative professionals assembled by Saunders College Publishing, the second edition of *World of Chemistry—Essentials* has a greatly enhanced art and photo program.

Macro-micro visualizations throughout the book, drawn by our outstanding scientific illustrator, George Kelvin, are designed to help students with what is always difficult for them—relating what they can see to the chemist's micro view of materials.

Photos, chosen for their visual appeal, also relate people and everyday circumstances to the chemistry under discussion, while also illustrating chemical concepts and the properties of materials.

Connections to the World of Chemistry videotapes have been expanded by introducing marginal quotations from interviews with scientists, new still photos (identified by the program number so that the tie-in can be discussed), and some new *World of Chemistry* boxes.

Through the services of the **Harcourt Brace Custom Publishing Group,** portions of *World of Chemistry—Essentials, 2/e,* can be packaged according to individual needs. Instructors who wish to augment *World of Chemistry—Essentials, 2/e* with their own material or to make selected chapters available in courses with a different focus than that of the textbook as a whole should contact their local Saunders College Publishing sales representative.

Accompanying Materials to *World of Chemistry—Essentials,* Second Edition

Printed Materials for Instructor and Student

Student Study Guide with Selected Solutions, by Walt Volland of Bellevue Community College, Washington, provides section-by-section lists of main topics, listing of objectives, key terms, and detailed solutions to all the odd-numbered end-of-chapter Questions for Review and Thought. In addition, the *Student Study Guide* includes selected Problems, suggested readings on related topics, "Bridging the Gap" home lab experiments, internet exercises, social issue oriented assignments, crossword puzzles to build vocabulary, and a set of interactive activities designed to connect chemistry to daily life.

Instructor's Resource Manual includes three sections:

1. ***Notes to the Instructor,*** by Walt Volland of Bellevue Community College, includes detailed solutions to the end-of-chapter Questions for Review and Thought and Problems that do not appear in the text and *Student Study Guide with Selected Solutions.*
2. ***Printed Test Bank*** of multiple-choice questions and problems, by James Schreck of Northern Colorado University.
3. ***Correlation Guide I,*** by Cheryl Dembe of Diablo Valley College, California, provides a detailed synopsis of each of the 26 videos from *The World of Chemistry Video Series* with a special section referencing corresponding textbook material.

ExaMaster™ Computerized Test Bank is the software version of the printed test bank. Instructors can create thousands of questions in a multiple-choice format. A command reformats multiple-choice questions into short-answer questions. Instructors can also add or modify existing problems as well as incorporate graphics. ExaMaster™ has gradebook capabilities for recording and graphing students' grades.

Correlation Guide to Overhead Transparencies from World of Chemistry, 2/e comes with the set of 100 full-color acetates that was developed for *World of Chemistry, 2/e.* The acetates come with enlarged labels for easy viewing.

Laboratory Manual to accompany World of Chemistry, 2/e by John Blackburn, John Craig, Paul Langford, and Melvin Joesten includes 45 experiments that are easily incorporated into the course. Includes pre- and post-lab questions, safety considerations, as well as a new, introductory discussion of risk versus benefit.

Instructor's Manual to accompany the Laboratory Manual provides a listing of equipment and reagents needed for each experiment, as well as suggested demonstrations for each pre-lab instruction.

Saunders Student Laboratory Research Notebook can be used as a companion to any of Saunders' general or organic chemistry lab manuals. Available in both long (100 pages) and short (50 pages) versions, this graph notebook features safety instructions, a fold-under cover for space-saving on lab bench, a water-repellent cover, and a periodic table. Pages are numbered, 3-hole punched, and carbonless with perforated duplicates.

Multimedia Materials

The World of Chemistry Video Package. *World of Chemistry—Essentials, 2/e* is presented either as a stand alone course in chemistry for nonscience majors or as an integral part of a comprehensive telecourse package including a series of 26 thirty-minute video programs, a telecourse study guide, telecourse faculty manual, and a telecourse laboratory manual. The video series, entitled *The World of Chemistry,* was developed by the late Dr. Isadore Adler of The University of Maryland, and Dr. Nava Ben-Zvi of Hebrew University of Jerusalem and was sponsored by the Annenberg/CPB Project and corporate sponsors. The video programs feature Nobel laureate and Priestley medalist Roald Hoffmann and provide a comprehensive survey of the field of chemistry and its impact on modern society. The series was produced jointly by the University of Maryland and The Educational Film Center. For information on ordering videocassettes call 1-800-LEARNER.

In addition to the videotapes, JCE:SOFTWARE has produced two videodiscs, *"The World of Chemistry: Selected Demonstrations and Animations I and II"* taken from *The World of Chemistry* videotapes. For information on these videodisks, contact: JCE: Software, John W. Moore and Jon L. Holmes, Department of Chemistry, University of Wisconsin-Madison, Madison, WI 53706.

Saunders Interactive General Chemistry CD-ROM 2.5 with ActivChemistry. This CD-ROM is a powerful multimedia companion for any general chemistry text which gives students the opportunity to be independent thinkers through interactive techniques. Now with the inclusion of *ActivChemistry,* students can actually design and perform simulated lab experiments right on the desktop! Students also can navigate through original animation and graphics, interactive tools, and pop-up definitions, as well as over 100 video clips and chemical experiments enhanced by sound effects and narration. A student workbook accompanies the CD-ROM, and important technical information and suggestions for accessing some of the CD-ROM's unique teaching features are available to professors.

Web site. Be sure to visit us at www.saunderscollege.com for an abundance of teaching aides designed specifically for instructors, such as Power Point™ (Available Fall, 1998), and on-line quizzing.

1999 Chemistry Media Active™. This CD-ROM is a dynamic lecture tool containing imagery from Saunders' 1999 chemistry titles that can be used in conjunction with commercial presentation packages including PowerPoint™, Persuasion™, and Podium™. This CD-ROM works on both Windows and Macintosh platforms.

CalTech Chemistry Animation Project (CAP) is a set of six video units covering the chemical topics of Atomic Orbitals, Valence Shell Electron Pair Repulsion Theory, Crystals and Unit Cells, Molecular Orbitals in Diatomic Molecules, Periodic Trends, and Hybridization and Resonance with unmatched quality and clarity.

Reviewers

We are deeply grateful to all the reviewers who have contributed to the improvement of the manuscript and teaching aids for the second edition. We would especially like to thank Walt Volland, Bellevue Community College, Washington, who provided a thorough review of the edited manuscript, contributed chapter questions, and provided solutions to the chapter questions and problems. We also appreciate the thoughtful comments of all our other reviewers:

John M. Allen *Indiana State University*
Ronald J. Baumgarten *University of Illinois, Chicago*
Joseph Chaiken *Syracuse University*
Thomas C. DeVore *James Madison University*
Marie G. Hankins *University of Southern Indiana*
Mark D. Jackson *Florida Atlantic University*
Robert E. Ludt *Virginia Military Institute*
William H. McMahan *Mississippi State University*
Steve Morris *University of Missouri—Columbia*
Kathleen Richardson *University of Central Florida*
John M. Risley *University of North Carolina at Charlotte*
James Schreck *University of Northern Colorado*
George H. Wahl, Jr. *North Carolina State University*
Kathryn R. Williams *University of Florida*

Reviewers of the first edition included:

Earl C. Alexander *San Diego Mesa College*
Erwin Boschmann *Indiana University—Purdue University at Indianapolis*
Robert C. Byrne *Illinois Valley Community College*
Richard Conway *Shoreline Community College*
Jack Cummins *Metropolitan State College*
Howard D. Dewald *Ohio University*
Ronald Distefano *Northampton Community College*
Alton Hassell *Baylor University*
Chu-Ngi Ho *East Tennessee State University*
Stanley N. Johnson *Orange Coast College*
Jerry L. Mills *Texas Tech University*
Tom Mines *Florissant Valley Community College*
David S. Newman *Bowling Green State University*
Marie Nguyen *Indiana University—Purdue University at Indianapolis*
Robert J. Palma Sr. *Midwestern State University*
James Schreck *University of Northern Colorado*
Berton C. Weberg *Mankato State University*
Donald H. Williams *Hope College*
Bruck Winkler *University of Tampa*
Robert Yolles *DeAnza College*

Acknowledgments

This second edition of *World of Chemistry—Essentials* has been shepherded along from start to finish by Ed Dodd, Editorial Coordinator, of Saunders

College Publishing, and we thank him for his patience, persistence, and good humor. As always, John Vondeling, Vice President/Publisher, has provided support and encouragement. We thank him for his confidence in us.

Because this book is derived from our two-semester text, *World of Chemistry, 2e,* we very much want to acknowledge the fine work of the team that helped us with that earlier project: Beth Rosato, Senior Developmental Editor; Maureen Iannuzzi, Project Editor; Caroline McGowan, Senior Art Director; Charlene Squibb, Senior Production Manager; George Kelvin, science illustrator; Charles Winters, photographer; and Sue Howard, Photo Editor. Their contributions are evident throughout. In addition, a word of thanks to George Semple, Manager, Photos and Permissions, who assisted with photo choices for this edition.

As in all of our previous works, we dedicate this effort to our families and gratefully acknowledge their support and understanding during the preparation of this manuscript.

Melvin D. Joesten
James L. Wood
Nashville, Tennessee

FEBRUARY 1998

(C. D. Winters)

Contents

(C. D. Winters)

(C. D. Winters)

(C. D. Winters)

(C. D. Winters)

(Courtesy of Intel Corporation)

(C. D. Winters)

Index to Boxed Features

(© *Scientific American, Inc./George V. Kelvin)*

CHAPTER 1

Living in a World of Chemistry

1.1 The World of Chemistry

Here you are about to take a chemistry course. What do you already know about chemistry? Maybe you've seen some chemistry demonstrations that produced explosions or dramatic color changes. Many chemists practicing today were first attracted into the profession by just such demonstrations. They wanted to do chemistry themselves—create dramatic changes in materials and understand why they happen.

■ The real meaning of "chemical" is discussed in Chapter 2.

Maybe "chemistry" means "chemicals" to you. And perhaps you think this word should be used only with the adjective "toxic." That belief wouldn't be surprising, since you have probably heard of "toxic chemical spills" or warnings about "toxics" in the environment. Indeed, some chemicals are toxic—very toxic: the arsenic of mystery stories, the poisonous gases of World War I, the chemicals released by the microorganism that grows in badly canned food and causes severe food poisoning.

Learning about chemistry doesn't, however, mean learning only about harmful substances. Nor does it mean learning only about the wonders of our modern world provided by chemistry, such as the medications that cure once-incurable diseases; the synthetic fabrics that are beautiful, durable, and inexpensive; or the colors on the television screen.

Instead, we suggest you come to the subject with a "What's in it for me?" attitude, not from a selfish viewpoint, but as a citizen of the world of chemistry. The first photographs of planet Earth from the Moon provided a forceful reminder to many of us that this planet and the materials on it are finite. The world of chemistry really has but one concern: the materials provided by our planet and what we do with them.

In a global view, some understanding of chemistry will be helpful in dealing with major social issues that lie ahead in the 21st century. The present world population of 5.9 billion is expanding at an ever-increasing rate and is expected to double in the next 50 years. How can everyone be fed, clothed, and housed? How can we prevent spoiling what planet Earth has provided us?

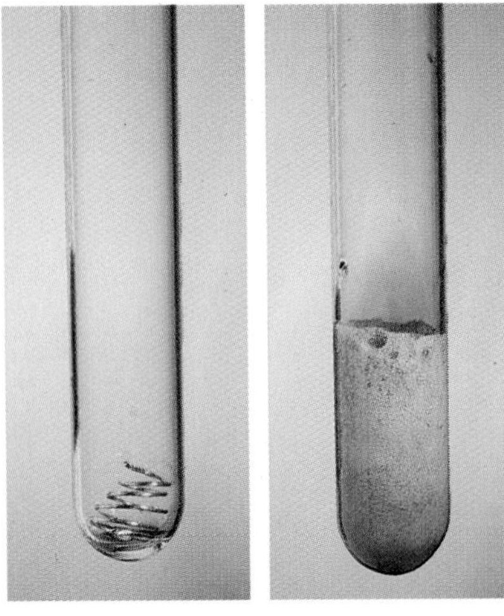

The amazing effect of nitric acid (b) on copper (a). As a young man, Ira Remsen (1846–1927) dropped a copper penny into nitric acid to see what would happen. Remsen, who became an outstanding chemist and teacher, had this reaction: "It resulted in a desire on my part to learn more about that remarkable kind of action. Plainly the only way to learn about it was to see its results, to experiment, to work in a laboratory." *(Larry Cameron)*

■ Population trends and the problem of feeding the world population are discussed in Section 17.1.

How can we reverse some of the damage done when we knew less about the consequences of our activities? How can we prevent doing new damage to our environment? None of these questions can be fully addressed without some serious applied chemistry.

Knowing something about chemistry adds a new dimension to everyday life, too. If an advertisement proclaims that a product "contains no chemicals," what does that tell you about the producers of that product? Is it worth paying twice the price for something that proclaims itself to be "all natural"? (Sometimes it is and sometimes it isn't.) Do you know that you shouldn't mix household ammonia and chlorine bleach because they react to form a very toxic gas?

Lying ahead in the chapters of this book is a look into the chemical view of the world. It is based on observations and facts. A *scientific fact* results from repeated observations that produce the same result every time. (Water boils at 100°C at sea level—that's a scientific fact.) The chemical world is also based on models, theories, and experiments. Scientists use models and theories to organize knowledge and make predictions. The predictions must then be tested by experiments. If experimental results disagree with the predictions, the reason must be explored. Possibly the theory is wrong.

You will also see that the chemical view of the world requires learning to look at materials in two ways. The first is what direct observation provides; the second is the mental image of what scientists have learned about the submicroscopic world that cannot be observed directly.

The nearby drawing of a chemist reading the newspaper is the first of a number of illustrations throughout this book that highlight the relationship between what can easily be seen and the chemists' view of matter at the submicroscopic level. As the chemist in the drawing reads in his newspaper about the latest experiments on DNA (perhaps the cloning of animals), he pictures the spiral twist of the DNA molecule, which controls human heredity. As he reads the latest report on the condition of the ozone layer that protects us from harmful solar radiation, he pictures the three connected atoms of the ozone molecule.

To continue this introduction to the world of chemistry, we want to offer a glimpse of chemistry-related topics that influence everyday events enough to be newsworthy. As you progress through this and subsequent chapters of this book, you will learn more about how such events are related to our submicroscopic picture of matter. The topics discussed in this chapter were chosen from three randomly selected Sunday newspapers when we were writing a longer version of this book. One was a major newspaper from one of the largest cities in the country; the others were average newspapers from cities of about 100,000 people. We chose a few stories from these papers that illustrate chemistry-related topics of general interest and also open the door to some further insight into the science of chemistry. The topics covered remain newsworthy, as illustrated by the updates included in the following discussion.

Science Helps Finger the Suspect
—headline on August 6, 1994

1.2 DNA Fingerprinting, Biochemistry, and Science

First, consider a topic important enough to make the front page of a Sunday newspaper—the beginning of a long feature article about the use of DNA

A mental picture of the newsworthy DNA and ozone molecules.

tests in court. DNA, chemical name *deoxyribonucleic acid,* is present in cells throughout our bodies. The reporter rose to the challenge of explaining the quite complicated chemistry of how DNA analysis provides the equivalent of a fingerprint. Comparison of the DNA analysis of a sample of blood or other biological material left behind at the scene of a crime with that of a sample from a suspect can weigh heavily as evidence for guilt or innocence.

The scientific complexities of DNA testing clearly made it to the front page because of the expected use of such testing in the murder trial of a well-known sports figure. During the ensuing trial many days were filled with intense and detailed descriptions by the prosecution of how DNA fingerprinting works and how its results were interpreted in that particular case. Equally intense efforts were made by the legal defense teams to raise doubts about DNA testing in general, and the methods used in studies for that trial, in particular. (Similar efforts were made to discredit fingerprinting when that was new.)

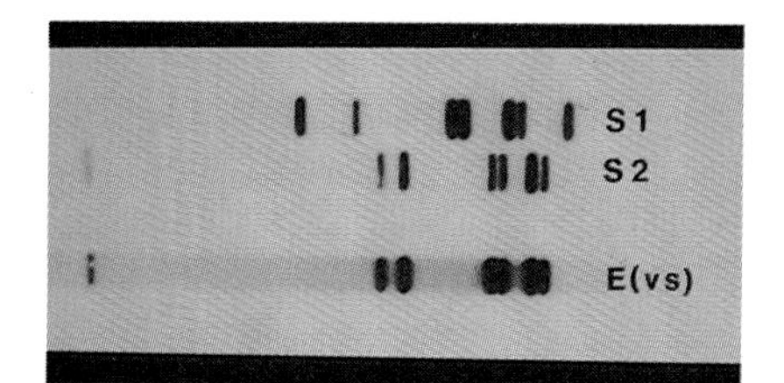

A DNA fingerprint test. The test result shows up as a series of lines that vary with the makeup of an individual's DNA. Comparison shows that the DNA of suspect 2 (S2) matches the evidence blood sample (E(vs)). *(Leonard Lessin/ Peter Arnold, Inc.)*

The DNA test is based on sound and accepted science, and it is being used in thousands of less publicized court cases every year. In the spring of 1996 the National Academy of Sciences announced the results of their review of courtroom use of DNA identification. They concluded that there is no reason to question the reliability of properly analyzed DNA evidence. Already, re-examination of evidence predating the test has freed individuals judged to have been wrongly imprisoned. And the FBI and Scotland Yard report that one third of those accused of rape are cleared of suspicion by DNA tests before the cases go to court.

Concern has been growing, however, over the use of competing scientific "experts" by opposing sides in court cases, which often puts judges in the

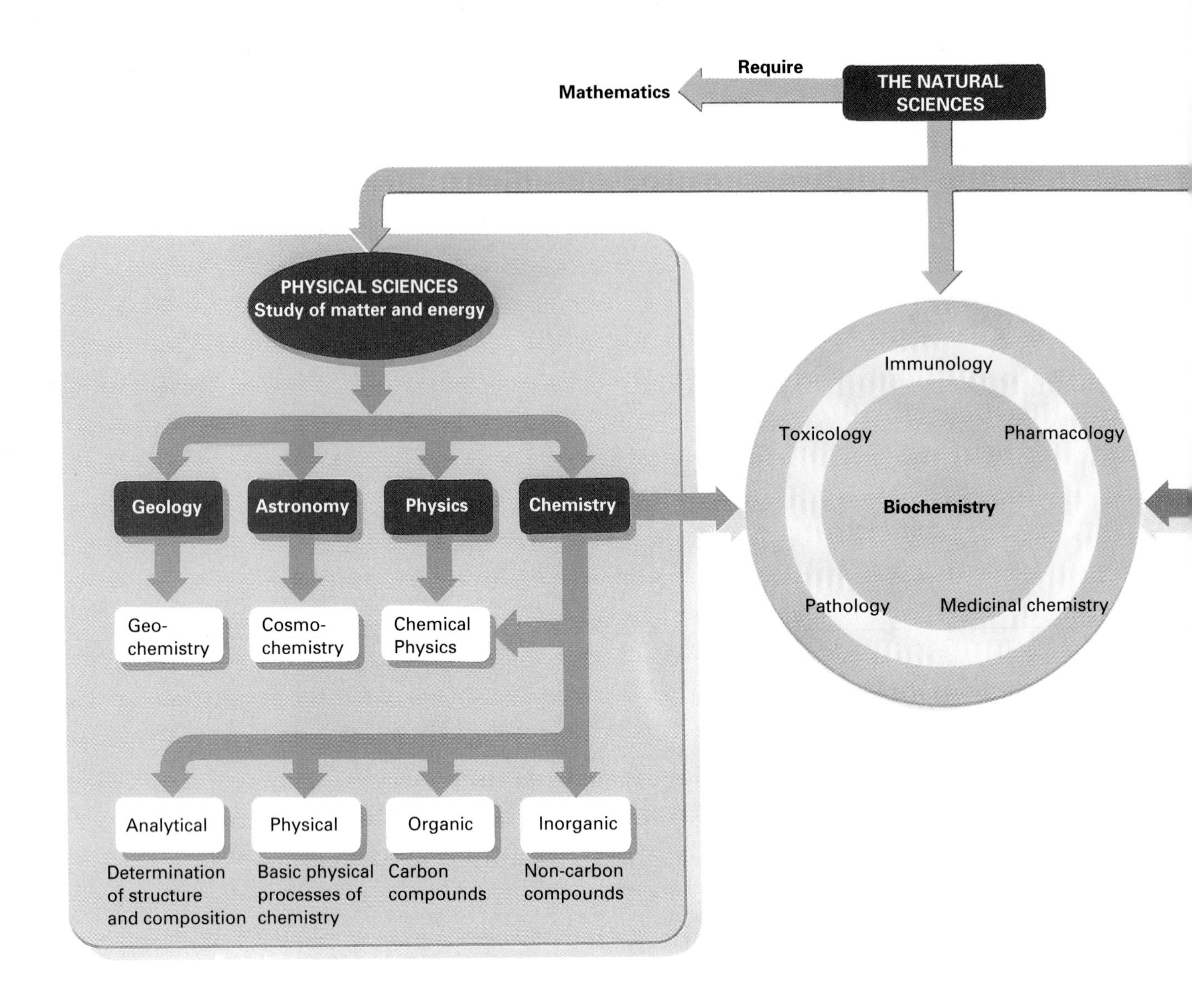

position of ruling on matters about which they know very little. In 1995 the U.S. Supreme Court decided that federal judges, and presumably also state judges, have the power to hold pretrial hearings to validate the competence of experts before allowing them to testify. One goal is to prevent "junk science"—science not widely accepted and not developed with the necessary care—from reaching the courtroom. As a result, in 1996 an effort was begun to educate judges about DNA. Thirty-five judges (of a group expected to grow to 1000 in two years) spent six days studying the subject; they even sequenced their own DNA in the laboratory.

■ Chapter 11 is devoted to biochemistry.

The study of DNA is one aspect of *biochemistry,* the natural science that unites chemistry, a physical science, with the biological sciences. **Chemistry** is often defined as the study of matter and the changes it can undergo. A moment's thought should convince you that this makes an understanding of chemistry fundamental to an understanding of everything in nature. And, of

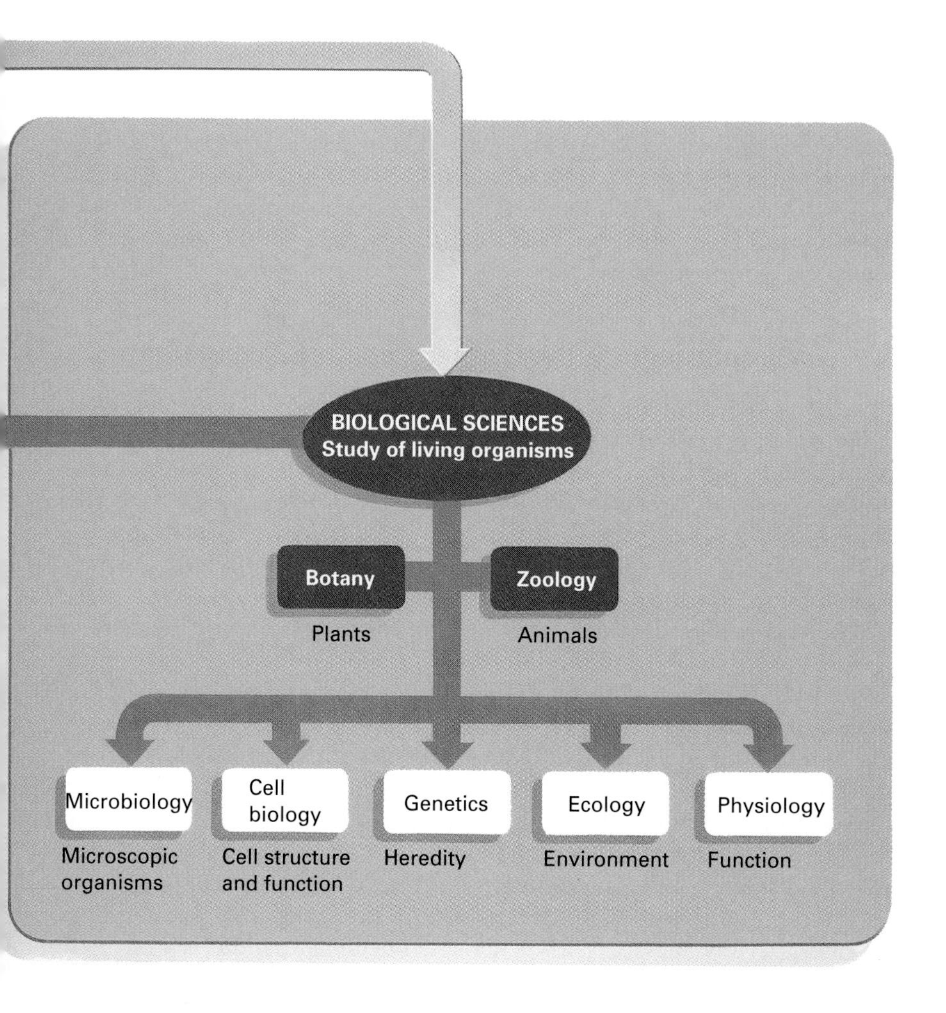

Figure 1.1 The natural sciences, with some of their subdivisions. The number of subdivisions continues to increase as the boundaries between the sciences disappear.

course, with an understanding of natural matter, chemists are able to modify matter and synthesize new kinds of matter.

Historically, the natural sciences have been associated with observations of nature—our physical and biological environment. A traditional classification of the natural sciences is represented in Figure 1.1, with an emphasis on the relationship of chemistry to other sciences. As each science grows more sophisticated, the boundaries become more blurred, however. The dynamic character of science is illustrated by the emergence of new disciplines. There are now individuals who refer to themselves as biophysicists, bioinorganic chemists, geochemists, chemical physicists, and even molecular paleontologists (they use modern chemical methods to analyze ancient artifacts).

The study of DNA is one of the frontier areas of science. Hardly a week passes without the report of some new development, which might be the unraveling of the cause of an inherited disease or the introduction of a new

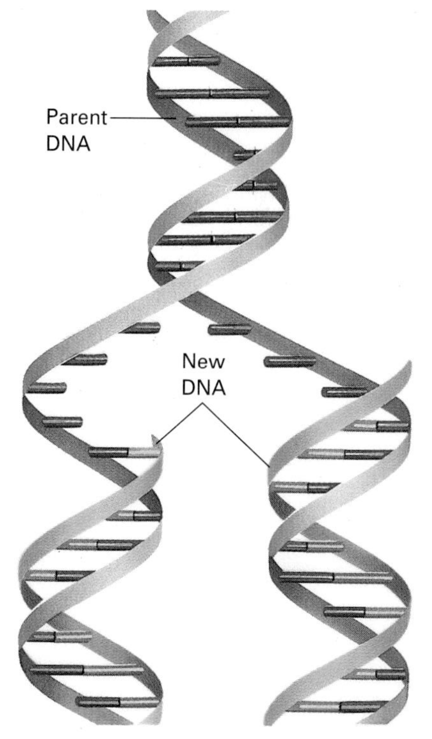

An abstract picture of how DNA unravels and is reproduced. In this way, a parent's DNA is passed to the offspring. (More information about how this happens is given in Section 11.11.)

product of the technology known as *genetic engineering*. Since these developments often relate to human health or behavior, they make the newspapers.

■ Identification of DNA as the basic genetic material in chromosomes was done in 1944 by O. T. Avery, C. M. MacLeod, and M. McCarty. Nine years later James D. Watson and Francis Crick solved the difficult problem of how DNA is constructed (Section 11.11).

About 50 years ago, DNA was identified as the carrier of genetic information from one generation to the next, determining how we will be the same or different from our ancestors. At first, there was a period of intense study of the chemical nature of DNA. The focus has now expanded to how DNA functions in living cells and how it might be manipulated to obtain desired results, such as the cure of inherited diseases or the manufacture of life-saving drugs by genetically engineered bacteria. DNA testing—determination of the exact composition of an individual's DNA—is an analytical tool. Those who use it along with many other kinds of *analytical* chemistry in the investigation of crimes are *forensic* chemists.

Having a Cool Car Is Becoming More Costly

—headline on Sunday, May 15, 1994

1.3 Air-Conditioning, the Ozone Hole, and Technology

Next, we consider a story about something that directly affects more of us than DNA testing—the need to convert automobile air-conditioning systems to the use of a new refrigerant. The background of this need is rooted in some interesting history of chemistry. Refrigeration as we know it has been in use since the late 1800s. The process requires a fluid that absorbs heat as it evaporates, releases heat when it condenses, and can be continuously cycled through evaporation and condensation without breaking down.

In the early 1900s, the fluids used as refrigerants were mostly flammable or toxic. One day, Charles Kettering, then director of research at General Motors, passed along a challenge to Thomas Midgley, a young engineer: "The refrigeration industry needs a new refrigerant if they ever expect to get anywhere." Midgley and a colleague vigorously dug into the problem. They hunted for candidates in the extensive tables of data kept at hand in every chemistry laboratory and surveyed the systematic variations in the properties of the chemical elements. Although fluorine had a reputation for being toxic, they decided that in some cases it might not be. To begin with, they decided to make a new fluorine-containing substance in the laboratory. Because one of the necessary raw materials was a fluorine chemical that was scarce, they quickly purchased the available supply in five small bottles. One of the bottles was used to make the first sample of the new chemical. In Midgley's own words:

> *A guinea pig was placed under a bell jar with it and, much to the surprise of the physician in charge, didn't suddenly gasp and die. In fact, it wasn't even irritated. Our predictions were fulfilled.*

A second sample made from the starting material in a different bottle did, however, kill the guinea pig. Here was a problem to be solved. Some investigation showed that only the first bottle contained pure material. It was a contaminant that had killed the guinea pig, not Midgley's newly prepared chemical.

> *Of five bottles . . . one had really contained good material. We had chosen that one by accident for our first trial. Had we chosen any one of the other four, the animal would have died as expected by everyone else in the world except ourselves. I believe we would have given up what would then have seemed like a "bum hunch."*
>
> *And the moral of this last little story is simply this: You must be lucky as well as have good associates and assistants to succeed in this world of applied chemistry.*

The outcome of this experiment was the widespread presence of refrigerators in our homes. Food spoilage diminished, and after World War II the food processing industry expanded to the production of frozen foods.

So far, this is a success story—the development of chlorofluorocarbons, often referred to as CFCs, as much less hazardous refrigerants. Through the 1950s and 1960s, their use expanded as all kinds of new consumer products were brought to market. The properties of CFCs made them ideal for propellants in aerosol cans and for blowing the tiny holes into materials such as the polyurethane foam used in pillows and furniture cushions. The problem being created did not surface until the 1970s because no one was aware of the fate of these very stable compounds as they were released into the environment.

■ To a chemist, a stable substance is one that is not easily changed into something else. Water is a stable substance.

By the 1980s, it had become clear that CFCs drift unchanged into the stratosphere, where they interact with ozone and destroy it. The result is the "ozone hole" and the potential for damage from the resulting increase in solar radiation that reaches the earth's surface. Herein lies the reason the cost of automobile air-conditioner repair is newsworthy. In 1987 nations that produce CFCs drafted a plan of action designed to reduce CFC use. Later amendments called for complete phaseout of all chemicals that harm the ozone layer and accelerated the schedule for the phaseout. (Chemicals and the ozone layer are further discussed in Chapter 16.) Recent evidence shows that the CFC phaseout may be working. It was reported in 1996 that the effect of CFCs in the atmosphere peaked near the beginning of 1994 and has been declining since then. The decline is slow and must be watched carefully, however.

Our news story explained that most cars made before 1992 used a CFC refrigerant that would not be manufactured after 1994. The strongest candidate for its replacement cannot be used unless the old systems are modified, which can be costly. Economic decisions lie ahead for anyone owning an older car with a leaky air-conditioning system. Sadly, one outcome of the changeover in automobile refrigerants has been the development of a black market in illegal importation of the banned CFCs to the United States.

It is helpful in reading about refrigerants and similar topics to recognize the distinctions among basic science, applied science, and technology. **Basic science**, or basic research, is the pursuit of knowledge about the universe with no short-term practical objectives for application in mind. The biochemists who struggled for years to understand exactly how DNA functions within cells were doing basic science.

The World of Chemistry

Program 1, *The World of Chemistry*

There's still so much to learn. My goodness, the sort of things it takes me years to do, nature does in a matter of seconds. The secrets there still have to be unraveled. Dr. Bertram Fraser-Reid, Duke University

Applied science has the well-defined, short-term goal of solving a specific problem. The search for a better refrigerant by Midgley and his colleagues is an excellent example of applied science: they had a clear-cut, practical goal. To reach the goal, as is done in either basic or applied science, they utilized the recorded observations of earlier studies to make a prediction and then performed experiments to test their prediction.

Technology, also an application of scientific knowledge, is a bit more difficult to define. In essence, it is the sum of the way we apply science in the context of our society, our economic system, and our industry. The first refrigerators and auto air conditioners designed to use CFCs were the products of a new technology. The rapidly expanding number of ways to manipulate DNA to make new medicines or other marketable products is referred to as *biotechnology*.

(a)

(b)

Science and Technology. (a) A basic research laboratory in a pharmaceutical company. (b) People enjoying a laser light show set to rock music. Technology has brought the laser out of the laboratory and into an IMAX theater. *(a, Hank Morgan/Rainbow; b, Courtesy of Audio Visual Imagineering Inc., Orlando, Florida)*

The World of Chemistry

Program 2, *Color*

Now, the way to find new dyes was to get a large number of guys who were competently trained chemists, get them a laboratory, get them the major reagents, set them down and say, Start making things. . . . To make these dyes they had to build up tens of thousands of intermediate compounds. And what you have then is an enormous stable of things that you can use in experiments. And so, as the base broadens, the combinations just become infinite. John K. Smith, science historian, Lehigh University, commenting on the beginnings of the chemical industry in Germany in the late 1800s.

Regardless of the type of scientific discovery, there is a delay between the discovery and its technological application. The incubation intervals for a number of practical applications of various types are given in Table 1.1.

The important point is that technology, like science, is a human activity. Decisions about the uses of technology and priorities for technological developments are made by men and women. How scientific knowledge is used to promote technology depends on those persons who have the authority to make the decisions. Sometimes those persons are all of us—when we go to the polls in a democratic society, we can influence decisions about technology. When we have this chance, it is important to be well enough informed to critically evaluate the societal issues related to the technology.

As the history of CFCs illustrates, science and technology, like social conditions, are constantly evolving. When CFCs were introduced as refrigerants in 1930, they were a great advance for the economy and replaced hazardous materials. In addition, the number of technologically advanced nations and the world population were smaller. People were in the habit of assuming that natural processes would keep the environment healthy, and to a greater extent than today that was true. Moreover, some of the sophisticated instruments that have since revealed ozone depletion were not available then. Only by staying informed can we be ready to adjust to changing times.

TABLE 1.1 ■ Time Needed to Develop Technology for Some Fruitful Ideas

Innovation	Conception	Realization	Incubation Interval (Years)
Antibiotics	1910	1940	30
Nylon	1927	1939	12
Photocopying	1935	1950	15
Photography	1782	1838	56
Roll-on deodorant	1948	1955	7
Videotape recorder	1950	1956	6
X rays in medicine	Dec. 1895	Jan. 1896	0.08
Zipper	1883	1913	30

1.4 Automobile Tires, Hazardous Waste, and Risk

Some risks are pretty obvious, and there can be no doubt that the next chemical story from our randomly chosen newspapers identifies such a risk. Lying in the midst of a woods is the biggest tire dump in New England and possibly in the country. Thirty-three million slowly decomposing tires have been dumped on 14 acres in an inaccessible location near a large population. There is no source of water that could be used to fight a fire. The heat generated as the tires undergo spontaneous chemical transformations, like the heat that sometimes ignites garbage dumps, could ignite the whole pile. According to the news story, a tire fire can burn for months and reach temperatures so high that water would instantly be vaporized.

Neighbors Fear Fire at Massive Tire Dump
—headline on Sunday, August 7, 1994

To imagine the consequences if this pile of tires caught fire, consider what a tire is made of. The essential ingredient is rubber, often not natural rubber from a rubber tree, but a synthetic chemical product. Synthetic rubber is similar in composition to natural rubber or petroleum, another plant-derived material, and burns to produce high temperatures, soot, and gases such as carbon dioxide, nitrogen oxides, hydrogen sulfide (which gives rotten eggs their odor), and sulfur dioxide. The black material in a tire is carbon black (like soot), a filler that provides extra strength and stiffness. Petroleum is also added to rubber to smooth the mixture for processing and to reduce the cost. A long tire fire would continually emit soot and potentially hazardous gases.

■ The reclamation of petroleum from recycled tires can definitely be done if the economics of collecting the tires and processing them is beneficial.

The same tire dump was in the news again in December of 1996. The state has made almost no progress in the now 20-year-long battle to reduce the hazards of the dump and eventually get rid of it. A retroactive state law forbids accumulation of more than 400 tires on private land, and a court order—not complied with—requires removal of a certain number of tires from the dump each year. Also, to make it possible to fight a fire if that nightmare should ever occur, the dump owner has been ordered to build a perimeter road and fire lanes through the pile, which is up to 35 ft high.

It is possible to recycle the materials in tires for use in asphalt or construction materials, and shredded tires can be burned to produce electricity or

THE WORLD OF CHEMISTRY

Discoveries by Accident

Program 2, *Color*

Sometimes important discoveries are the result of what we might call accidents, or serendipity. Use of the term "serendipity" for accidental discoveries was first proposed in 1754 by Horace Walpole after he read a fairy tale titled "The Three Princes of Serendip." Serendip was the ancient name of Ceylon, and the princes, according to Walpole, "were always making discoveries by accident, of things they were not in quest of."

Of course, these accidents have to happen to the right kind of individuals. In science, when an unexpected experimental result turns up, these people look on it as an opportunity. They recognize through a burst of intuition that something needs further exploration. They want to know more about what happened and why it happened.

Little did Perkin know that when he mixed these two substances he would produce a beautiful purple dye.

The history of science is replete with examples of serendipity. A classic example is to be found in the story of W. H. Perkin, as told by science historian John K. Smith of Lehigh University:

He was a brilliant young chemist who, while working in his home laboratory in 1856 in an effort to synthesize badly needed quinine, succeeded instead in creating the dye "mauve." Perkin recognized the importance of his accidental discovery and as a consequence of his efforts succeeded in establishing the beginning of the dye industry. The spinoffs were enormous. There are today as a consequence a large variety of materials such as drugs, explosives, fertilizers which play such an important role in the affairs of society. One of the most important consequences, for example, is aspirin, easily one of the most useful drugs in the history of pharmaceutical chemistry.

What should we do with these tires? *(William McCoy/Rainbow)*

useful heat. However, the stockpiled tires are bad candidates for recycling because their accumulated rocks, dirt, and water hamper the recycling process.

A comprehensive survey of newspapers probably isn't needed to convince you that stories about risks to public health or safety are always newsworthy. Certainly none of us would want to live close to that tire dump, and it is doubtful that any scientist could come up with data to prove that the dump does not present a risk.

Matters are not always so simple, however. How serious a risk is the presence of a pesticide residue on a piece of fruit if that pesticide has the *probability* of causing an estimated one additional cancer death among one million people over the course of a normal 70-year lifetime? There is an equivalent one in one million risk from traveling 10 min by bicycle (Table 1.2). Should the government be regulating how often we ride bicycles?

Risk assessment for individuals involves a consideration of the probability of harm and the severity of the hazard. Risk assessment for society as a whole adds to these concerns the number of people who will be affected, the size of the geographical area involved, and how long-lasting the damage might be.

The science of risk assessment is still evolving, but it is clear that the perceptions of risk by the public and as the result of scientific study are often

TABLE 1.2 ■ Estimates of Risk: Activities with a Probability of One Additional Death per One Million People Exposed to the Risk

Activity	Cause of Death
Smoking 1.4 cigarettes	Cancer, lung disease
Living 2 months with a cigarette smoker	Cancer, lung disease
Eating 100 charcoal-broiled steaks	Cancer
Traveling 6 min by canoe	Accident
Traveling 10 min by bicycle	Accident
Traveling 300 miles by car	Accident
Traveling 1000 miles by jet aircraft	Accident
Drinking Miami tap water for 1 year	Cancer from chloroform
Living 2 months in Denver	Cancer caused by cosmic radiation
Living 5 years at the boundary of a typical nuclear power plant	Cancer

different. For example, statistics show that the risk of injury or death during commercial airplane travel is much lower than that from automobile travel, yet most of us know people who avoid airplane flights out of fear, but who ride in automobiles every day.

What factors influence the public perception of risk? Catastrophic events such as major chemical explosions, oil spills, or nuclear power plant accidents obviously affect public perceptions. People tend to judge involuntary exposure to risk (such as living near a hazardous waste site) as being worse than voluntary risk (such as smoking cigarettes or riding in automobiles). In other words, people rate risks they can control lower than those they cannot control.

■ Some individuals seem to believe that a sufficient amount of public activism and government regulation can guarantee risk-free living. Do you think this is possible?

■ Clean air is discussed in Chapter 16. Clean water is discussed in Chapter 15.

No absolute answer can be provided to the question, "How safe is safe enough?" The determination of acceptable levels of risk requires value judgments that are difficult and complex, involving the consideration of scientific, social, economic, and political factors. Over the years a number of laws designed to protect human health and the environment have been enacted.

Balancing laws balance risks against benefits. The Safe Drinking Water Act, the Toxic Substances Control Act, and the Clean Air Act are laws of this type. The federal Environmental Protection Agency (EPA) is required to balance regulatory costs and benefits in its decision-making activities. Risk assessments are used here. Chemicals are regulated or banned when they pose "unreasonable risks" to or have "adverse effects" on human health or the environment.

Industrial plant in Romania. The black smoke shows that pollutants are being produced from the waste gases being burned. *(Earl Dibble/Photo Researchers)*

Technology-based laws impose technological controls to set standards. For example, parts of the Clean Air Act and the Clean Water Act impose pollution controls based on the best economically available technology or the best practical technology. Such laws assume that complete elimination of the discharge of human and industrial wastes into water or air is not feasible. Controls are imposed to reduce exposure, but true balancing is not attempted. The goal is to provide an "ample margin of safety" to protect the public. As technology

A 1937 portrayal of chemistry: "Chemical industry, upheld by pure science, sustains the production of man's necessities." This illustration appeared in a book entitled *Man in a Chemical World* by A. C. Morrison, published in 1937. *(From a painting by Leon H. Soderston, 1894–1955)*

advances and the cost of the technology decreases, the margin of safety can be adjusted.

Many of our present environmental problems stem from decades of neglect and lack of understanding. The Industrial Revolution brought prosperity, and little thought was given to possible harmful effects of the technology that was providing so many visible benefits. Recognizing this, the government and the chemical industry have initiated serious efforts to solve the problems caused by previous lack of foresight and to evaluate the potential problems of new technology.

Risk management requires value judgments that integrate social, economic, and political issues with risk assessment. Risk assessment is the province of scientists, but determination of the acceptability of the risk is a societal issue. It is up to all of us to weigh the benefits against the risks in an intelligent and competent manner.

Cottonwood Lake near Buena Vista, Colorado. It's up to us to keep such places beautiful. *(Grant Heilman/Grant Heilman Photography)*

1.5 What Is Your Attitude Toward Chemistry?

Before proceeding with this study of chemistry and its relationship to our society, you might want to examine your prejudices (if any) and attitudes about chemistry, science, and technology. Many nonscientists regard science and its various branches as a mystery and think that they cannot possibly comprehend the basic concepts and their relation to societal issues. Many also have what has been christened *chemophobia* (an unreasonable fear of chemicals) and a feeling of hopelessness about the environment. These attitudes are most likely the result of news stories about the harmful effects of technology. Some of these harmful effects are indeed tragic.

What is needed, however, is a full realization of both the benefits and the harmful effects of science and technology. In the analysis of these pluses and minuses, we need to determine why the harmful effects occurred and if the risk can be reduced for future generations. This book will give you the basics in chemistry along with some insight on the role of chemistry in both positive and negative aspects of the world today. We hope this knowledge will afford you a healthier and more satisfying life by allowing you to make wise decisions about personal problems and problems of concern to society as a whole.

CHAPTER 2

The Chemical View of Matter

Practically everything we use has been changed from a natural state of little or no utility to one of very different appearance and much greater utility. Some of these changes are mechanical, such as producing the lumber used to build houses. Many others are chemical, such as the baking of clay to make pottery. Exploring, understanding, and managing the processes by which natural materials can be changed are basic to the science of chemistry. These activities require close examination of the composition and structure of matter, which we begin to do in this chapter.

- What are elements and chemical compounds made of?
- What is the difference between a mixture and a pure substance?
- What is the difference between a chemical and a physical process?
- What is the basic theme of chemistry?
- How are symbols for the elements used in formulas and equations to communicate chemical information?
- Why are standards and measurements essential to chemistry?
- What are the metric units most commonly used in chemistry?

Many times a day you expect things to behave in certain ways and you act on these expectations. If you are cooking spaghetti, you know that you must first boil some water and that after the spaghetti has been in the boiling water for a while it will get soft. If you are pouring gasoline into the fuel tank of your lawn mower, you know that you must protect the gasoline from sparks or it will explode into fire. To cook the spaghetti or safely fill the gas tank, you don't need to think about why these expectations are correct.

Chemistry, however, has taken on the job of understanding why matter behaves as it does. Answering such questions as, "Why does water boil at 100°C?" or "Why does gasoline burn, but water doesn't?" or "Why does spa-

In the chemical view of matter, chemical reactions between molecules in leaves and flowers are made possible by the energy from sunlight. *(Skip Moody/Dembinsky Photo Associates)*

ghetti get soft in boiling water?" requires a different approach than direct observation. We have to look more deeply into the nature of matter than we can actually see, at least under everyday conditions.

Samples of matter large enough to be seen, felt, and handled—and thus large enough for ordinary laboratory experiments—are called **macroscopic** samples. In contrast, **microscopic** samples are so small that they have to be viewed with the aid of a microscope. The structure of matter that really interests chemists, however, is at the **submicroscopic** level. Our senses have limited access into this small world of structure, although new kinds of instruments are beginning to change this condition (Figure 2.1).

In the chemical model of matter, everything around us, matter of every possible kind, is pictured as collections of very small particles. The picture begins with elements and their atoms—*all matter is composed of atoms.*

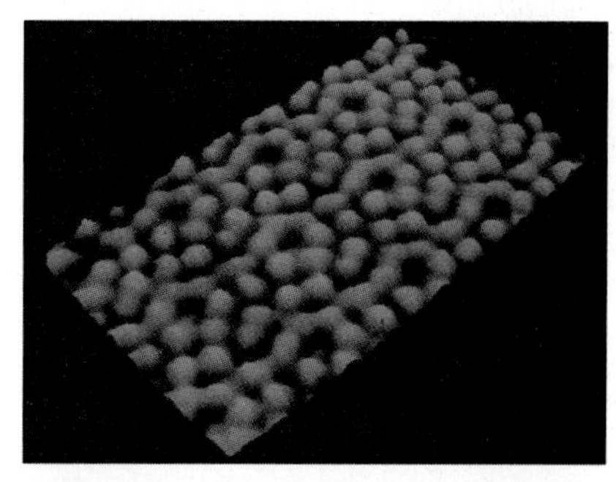

Figure 2.1 Silicon atoms. A scanning tunneling microscope photo shows the regular pattern of atoms on the surface of a piece of silicon. Pictures like this have been possible since the invention of the STM in 1986. (See *The World of Chemistry* box on the scanning tunneling microscope in Chapter 3.) *(Science VV/IBMR/Visuals Unlimited)*

2.1 Elements—The Most Simple Kind of Matter

Most of the materials you handle every day are pretty complex in their structure, and many are mixtures. Only occasionally do you encounter elements not combined with something else. One example is the helium used to fill a birthday balloon. The gas is helium, an element. From the chemical point of view, what does this mean? First, it means that helium cannot be separated or broken down into any other kind of matter that exists independently. Second, it means that the balloon contains the very, very small particles known as

■ The historical development of the chemical model of matter is discussed in Chapter 3.

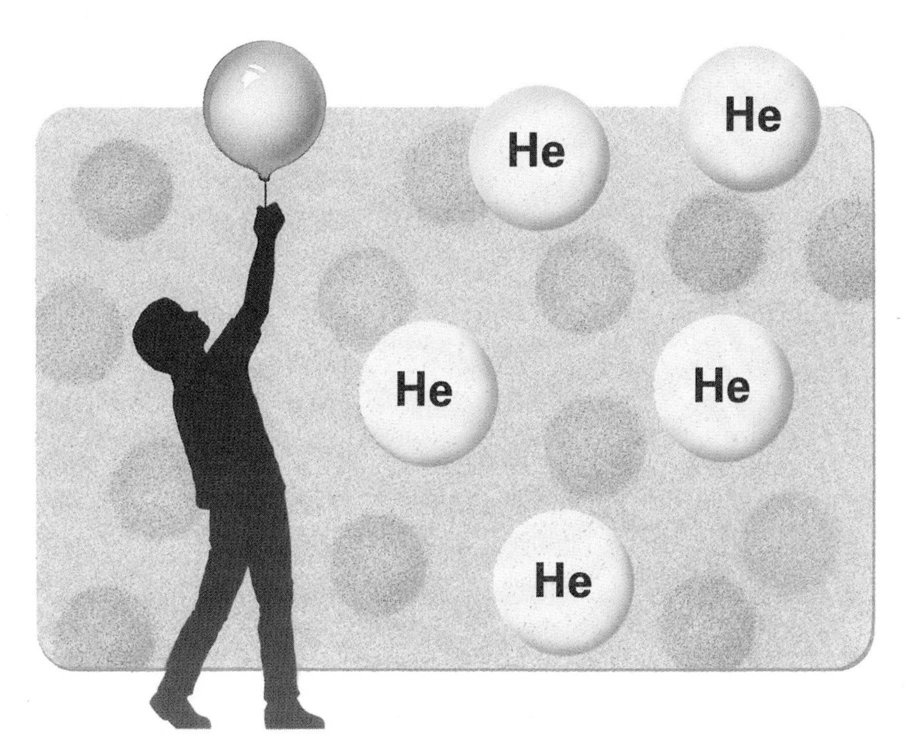

A helium-filled balloon and the helium atoms it contains.

■ **Solids, liquids,** and **gases** are the three common **states of matter**. The fourth state, plasmas, occurs in flames, stars, and the outer atmosphere of the earth.

helium atoms. Third, it means that helium has a unique and consistent set of properties by which it can be identified.

Another element you have probably seen around your house is mercury. It is used to conduct electricity in switches like those in some thermostats and as the liquid in thermometers. Mercury is certainly very different from helium in its properties. It is a liquid, not a gas, and it conducts electricity, which helium cannot do. What mercury and helium have in common is that they are both elements and each is composed of a single kind of atom. The liquid in the mercury thermometer is made of mercury atoms. Because the properties of helium and mercury are so different, we can safely conclude that helium and mercury atoms are not the same (Figure 2.2).

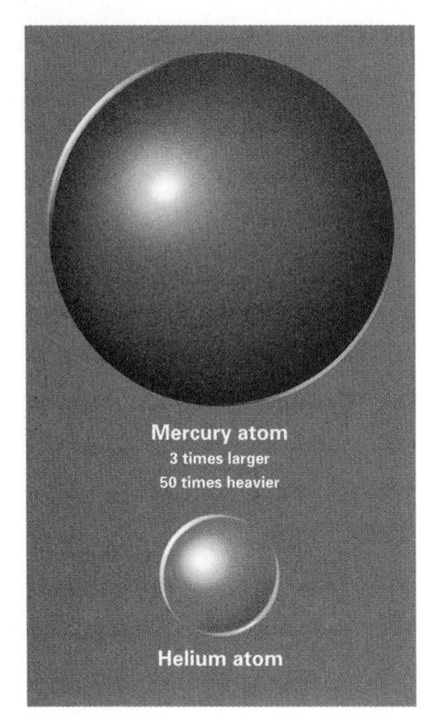

Figure 2.2 The relative sizes of helium and mercury atoms. As you might guess, mercury atoms not only are larger than helium atoms, but also are much heavier (about 50 times heavier).

The chemical model of matter, therefore, starts with the following:

> A **pure substance** is something with a uniform and fixed composition at the submicroscopic level. As you will see, pure substances can be recognized by the unchanging nature of their properties.
>
> An **element** is a pure substance composed of only one kind of atom.
>
> An **atom** is the smallest particle of an element, and the atoms of different elements are different.

2.2 Chemical Compounds—Atoms in Combination

If a pure substance is not an element, it is a chemical compound. To take our most common example, what does it mean that water is a chemical compound? You probably know that water is referred to as "h-2-oh" even if you have never studied chemistry. The "h-2-oh" is a way of reading the notation that chemists use to represent water: H_2O. In this kind of notation, H represents the element hydrogen, and O represents the element oxygen.

Chemical compounds are pure substances made of atoms of different elements combined in definite ways. We know that water is a pure substance because it has the same properties no matter where it comes from. We know that water is a chemical compound because passing energy through it in the form of an electric current causes decomposition to the elements hydrogen and oxygen. Also, under the right conditions, hydrogen and oxygen combine to form water.

■ The symbolism used to represent elements and compounds is explained further in Section 2.7.

Remember that from the chemical point of view all matter consists of tiny particles. In water, each particle is composed of two hydrogen atoms (shown by the subscript number $_2$ in H_2O) and one oxygen atom (shown by the absence of a subscript after O in H_2O), combined in what is called a *water molecule.* Experimentally then, pure substances are classified into two categories: (1) chemical compounds, which can be broken down into simpler pure substances (elements), and (2) elements, which cannot be broken down in this way.

Once elements are combined in compounds, the original, characteristic properties of the elements are replaced by the characteristic properties of the compounds. Consider the difference between table sugar, a white crystalline substance that is soluble in water, and its elements: carbon, which is usually a black powder and is not water soluble; hydrogen, the lightest gas known; and oxygen, the atmospheric gas needed for respiration.

2.3 Mixtures and Pure Substances

Most samples of matter as they occur in nature are mixtures. Some mixtures are obviously **heterogeneous**—their uneven texture is clearly visible, as in the different kinds of crystals in many rocks (Figure 2.3). Other mixtures appear to be **homogeneous** when actually they are not. For example, the air in your room appears homogeneous until a beam of light enters the room and reveals floating dust particles. Blood appears homogeneous as it gushes from a cut in your finger, but a microscope shows that it is quite a heterogeneous mixture (Figure 2.4).

Figure 2.3 A natural heterogeneous mixture. *(Barry L. Runk/Grant Heilman Photography)*

■ **Homogeneous mixtures** are uniform in composition and **heterogeneous mixtures** are not.

■ As science advances, smaller and smaller concentrations of impurities can be detected.

Homogeneous mixtures do exist—we call them **solutions**. No amount of optical magnification will reveal a solution to be heterogeneous, because it is a mixture of particles too small to be seen with ordinary light. In addition, different samples of the *same* homogeneous mixture have the same composition throughout. Examples of solutions are clean air (mostly the elements nitrogen and oxygen); freshly brewed tea; and some brass alloys, which are homogeneous mixtures of the elements copper and zinc. As these examples show, *solutions can be found in the gaseous, liquid, and solid states.*

When a mixture is separated into its components, the components are said to be *purified*. Efforts at separation are usually incomplete in a single step, and repetition of the process is necessary to produce a purer substance. Ultimately, the goal is to arrive at substances that cannot be purified further.

Separation of mixtures is done by taking advantage of the different properties of substances in the mixture. For example, many people want to separate certain substances from their tap water before they drink it. One way to do this is based on the forces of attraction between particles at the submicroscopic level. The water is passed through a material that attracts the undesirable substances and holds them back.

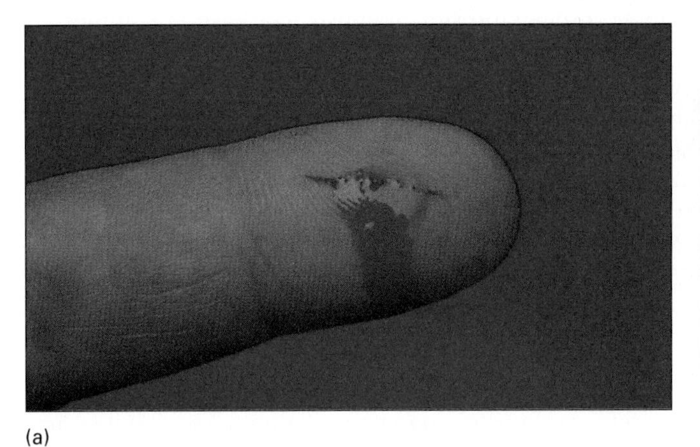

(a)

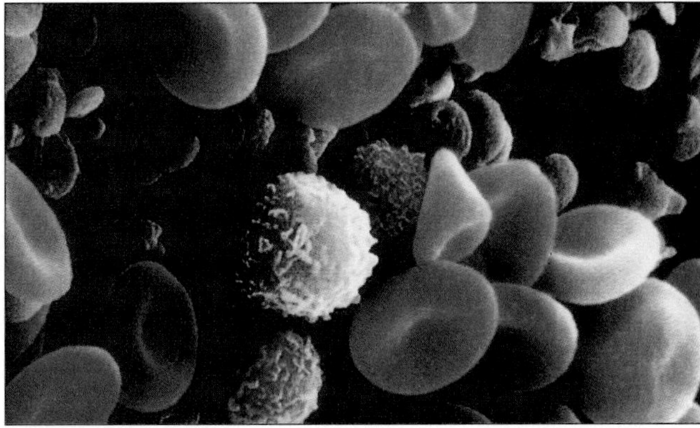

(b)

Figure 2.4 Blood, another heterogeneous mixture. (a) Blood looks homogeneous as it flows from a cut. (b) A microscope shows that it is quite a mixture. Color has been added to this electron microscope picture to show the different kinds of blood cells. The doughnut-shaped ones are red blood cells; the fuzzy ones are immune system cells; the small tan ones are platelets. *(a, Tom Pantages; b, Ken Eward/Science Source/Photo Researchers)*

2.4 Changes in Matter: Is It Physical or Chemical?

Sulfur is yellow; iron is magnetic; water boils at 100°C. These are different properties of matter, but with something in common. We can observe the color of a substance, pick it up with a magnet, or measure its boiling point without changing its chemical identity. Such properties are classified as **physical properties**. In chemistry, the word *physical* is used to refer to processes that do not change chemical identities. Separating sulfur and iron with a magnet is a *physical* separation. Boiling water is a *physical* change—the steam that forms is still water (H_2O).

■ Some physical properties are color, odor, melting point, boiling point, solubility, hardness, density, and state (solid, liquid, or gas).

Whenever possible, the physical properties of substances are pinpointed by making numerical measurements. You might describe lead as heavier than aluminum—if you pick up same-sized pieces of these metals, lead will definitely feel heavier. The difference is better represented though by giving the numerical value of the **density** of each metal, which takes into account the need to compare pieces of equal size, that is, of equal volume. One cubic centimeter (cm^3) of lead weighs 11.3 grams (g) and 1 cubic centimeter of aluminum weighs 2.7 g. Lead is *denser* than aluminum. Stated in the usual way:

■ Don't confuse density and heaviness. A large piece of aluminum could be heavier than a small piece of lead, but the lead is still denser.

■ **Density** is mass per unit volume. The customary measurement units (Section 2.8) for density are grams (g) for mass (weight) and cubic centimeters (cm^3) for the volume of solids or milliliters (mL) for the volume of liquids.

	Densities (at 20°C)
Lead	11.3 g/cm^3
Aluminum	2.7 g/cm^3

By contrast to physical changes such as boiling, there are processes that do result in changes of identity. When gasoline burns, it is converted to a mixture of carbon dioxide, carbon monoxide, and water. Burning in air, a property that gasoline, kerosene, and similar substances have in common, is classified as a **chemical property**.

The word *chemical* is used to describe processes that result in a change in identity. The combination of hydrogen and oxygen to form water is a *chemical change* or **chemical reaction**—a process in which one or more substances (the **reactants**, which can be elements or compounds, or both) are converted to one or more different substances (the **products**, which can also be elements or compounds, or both).

$$\text{Reactants} \longrightarrow \text{products}$$
$$\text{Hydrogen} + \text{oxygen} \longrightarrow \text{water}$$

A chemical reaction produces a new arrangement of atoms. The number and kinds of atoms in the reactants and products remain the same, but the reactants and products are different substances that can be recognized by their different properties. To distinguish between a hydrogen–oxygen mixture and a compound composed of hydrogen and oxygen, we might say that in water, hydrogen and oxygen are *chemically* combined. The meaning is that the hydrogen and oxygen atoms are held together strongly enough to form the individual units we call water molecules. For many compounds, a **molecule** is the smallest unit of the compound; molecules retain the composition of the compound and can have a stable, independent existence.

■ Compounds composed of molecules at the submicroscopic level are referred to as **molecular compounds**.

Some easily observed results of chemical reactions are the rusting of iron, the change of leaf color in the fall, and the formation of carbon dioxide bubbles by an antacid tablet. Often, though not always, the occurrence of a chemical reaction can be detected because there is some observable change (Figure 2.5).

The word "chemical" is often used as a noun. In this sense, every substance in the universe is a *chemical*. You might read of polluted waters that contain chemicals. Used this way, the word really stands for "chemical compound." The chemical compounds in polluted water may be unhealthful or harmful. Anyone who has studied even a little chemistry understands that useful and healthful substances—penicillin, table salt, nylon fabrics, laundry detergents, and all natural herbs and spices—are also made of chemical compounds. So are all plants and animals. Pure water is also a chemical compound.

Some physical changes and almost all chemical reactions are accompanied by changes in energy. Frequently, the energy is taken up or released in the form of heat. Heat must be added for the physical changes of melting or boiling to occur. Heat is released when the chemical reactions of combustion, or burning, take place. But many chemical reactions, such as those of photosynthesis, require energy from their surroundings in order to take place.

Energy is the ability to cause change, or in the formal terms of physics to do work. Energy in storage, waiting to be used, is **potential energy**. There is potential energy in gasoline, known as *chemical energy*—it is released as heat and light when gasoline burns. Chemical energy can also be released in the

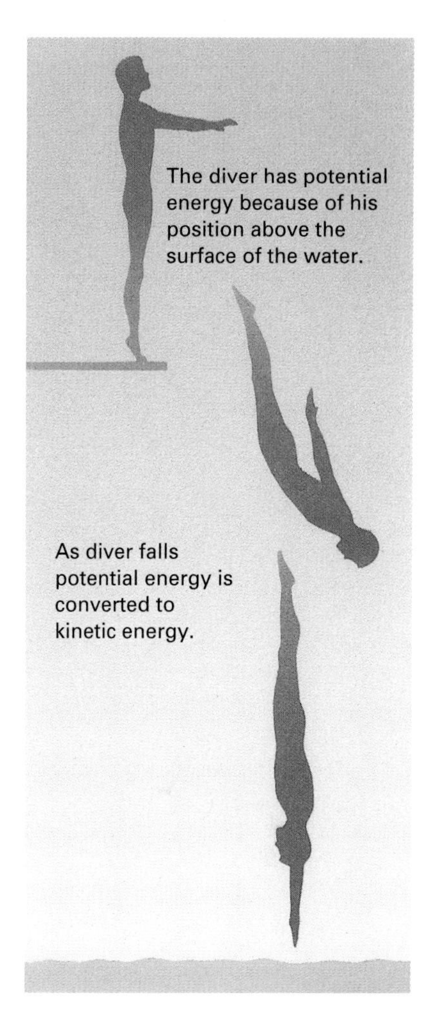

The diver has potential energy because of his position above the surface of the water.
As diver falls potential energy is converted to kinetic energy.

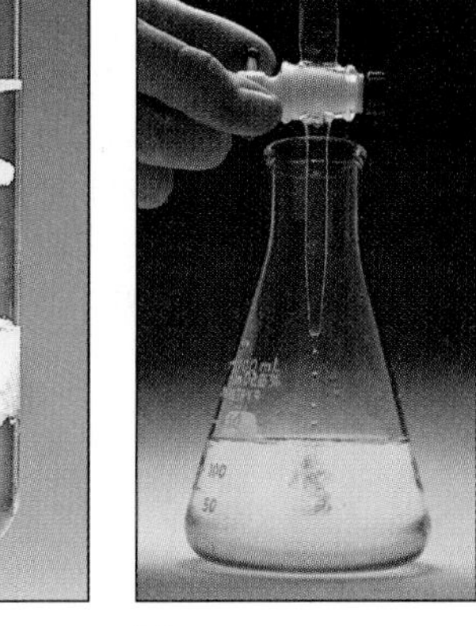

(a) (b) (c) (d)

Figure 2.5 Observable indications of chemical reactions. Chances are that a chemical reaction has occurred when (a) a flame is visible, (b) mixing substances produces a glow, (c) a solid appears when two solutions are mixed, and (d) there is a color change when two substances are mixed. *(C. D. Winters)*

TABLE 2.1 ■ Some Examples of Conversion of Chemical Potential Energy to Other Forms of Energy

Conversion to—	**Electrical Energy**	**Heat**	**Light**	**Mechanical Energy**
Done by—	Batteries Fuel cells	Combustion Digestion of food Many kinds of chemical reactions	Burning candle, logs in fireplace Luminescence in firefly Luminescence from chemical reaction	Rocket Animal muscles Dynamite explosion

form of electrical energy or mechanical energy (Table 2.1). Energy in use, rather than in storage, is **kinetic energy**, the energy associated with motion.

2.5 Classification of Matter

The kinds of matter we have described—elements, compounds, and mixtures—can be classified according to their composition and how they can be separated into other substances, as shown in Figure 2.6.

Heterogeneous samples of matter are all mixtures and can be physically separated into various kinds of homogeneous matter. Homogeneous matter can be a pure substance or a mixture. If it is a mixture, it is described as a solution and has the same composition throughout. Different solutions of the same substances, however, can vary in composition, sometimes over a very wide range. A teacup full of water, for example, might dissolve anywhere from a few grains of sugar to more than one measuring cup of sugar. The water and sugar can be separated by the physical process of evaporating the water.

If the homogeneous matter is a pure substance, it has a fixed composition and must be either an element or a compound. While elements contain only atoms of that element, compounds contain atoms of different elements combined in distinctive ways. At the submicroscopic level, each sugar molecule contains 12 carbon (C) atoms, 22 hydrogen (H) atoms, and 11 oxygen (O) atoms, represented by $C_{12}H_{22}O_{11}$. At the macroscopic level, every 100 g of table sugar contains 42.1 g of carbon, 6.5 g of hydrogen, and 51.4 g of oxygen. Only chemical reactions could produce pure carbon, hydrogen, and oxygen from sugar.

Why Study Pure Substances?

Perhaps by now you are wondering why we should be interested in elements and compounds and their properties. There are three basic reasons.

1. Only by studying pure matter can we understand how to utilize its properties, design new kinds of matter, and make desirable changes in the nature of everyday life. A long time ago, for example, it was known that swallowing extracts from the willow tree could relieve pain, although there were some unpleasant side effects. Once chemists identified the pain-relieving substance in the mixture of extracts, they were

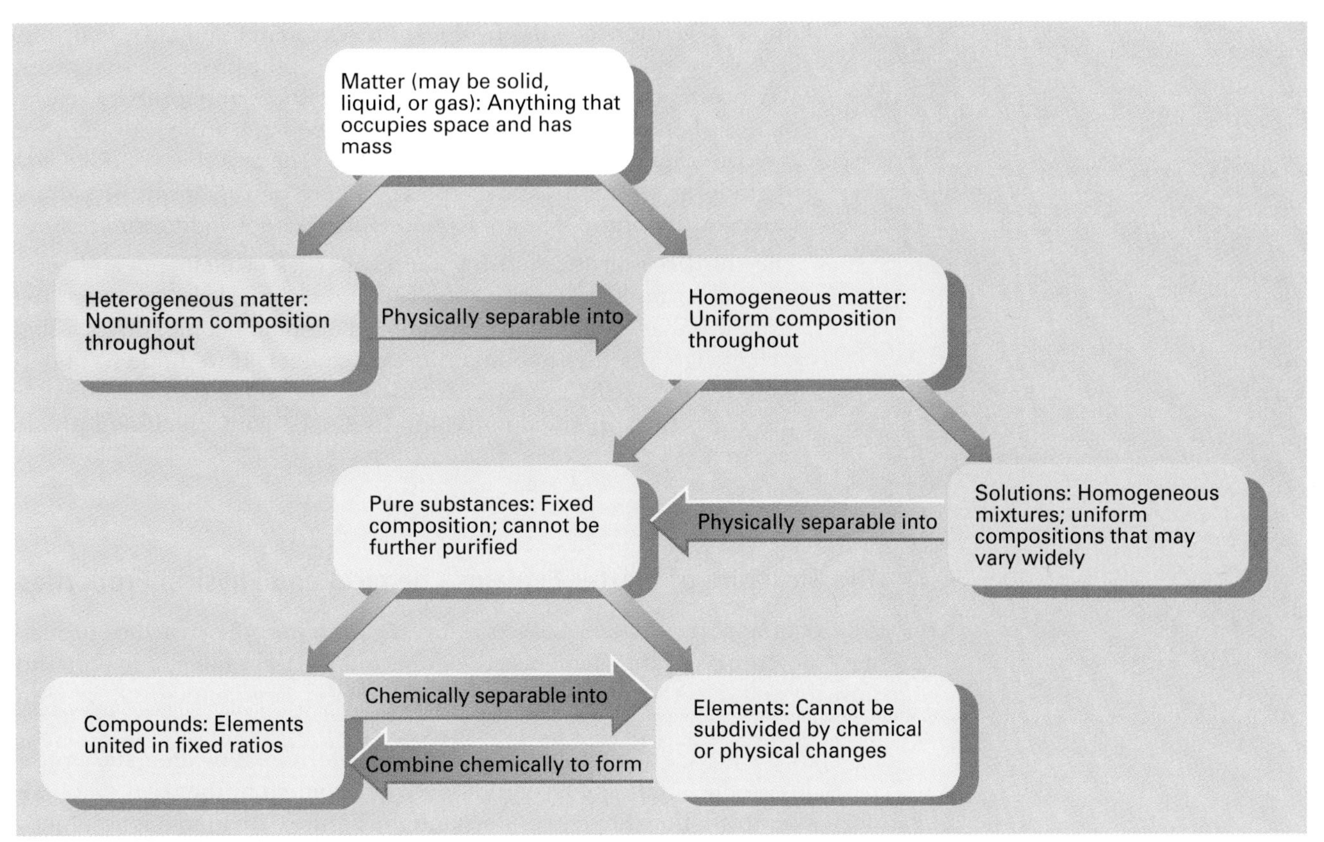

Figure 2.6 A classification of matter. Once elements have combined into compounds, only chemical reactions can separate them.

able to create a molecule that was similar enough to relieve pain, but different enough to have fewer side effects. We know this pure substance as aspirin.

At one time most natural materials could be changed only by physical means. Only a few useful materials, such as pottery and iron, were the products of chemical change. Otherwise, the design of everyday objects was limited to utilizing the properties of natural substances. Today we have evidence all around us that the chemical modification of materials has indeed changed the quality of life. Synthetic fibers, plastics, life-saving drugs, latex paints, new and better fuels, photographic films, audiotapes, and videotapes are but a few of the materials produced by controlled chemical change. It is equally important that understanding the properties of substances allows us to prevent *undesirable* chemical changes, such as the corrosion of metal objects, the spoiling of food, and the effects of hereditary diseases.

■ We shall return to examine the chemistry of many of these modern materials in later chapters.

2. Studying the properties of matter helps us to deal intelligently with our environment, both in everyday life and in the social and political arena. Knowing something about the properties of fertilizers or pesticides helps us decide which ones to use, which ones to avoid, and when their

use is or is not necessary. In the context of community life, many decisions are enlightened by some knowledge of the properties of matter. When you pour waste down a sewer drain, what happens to it? Does it matter whether the waste is motor oil or the water from washing your car? Does your town or city really need to spend millions of dollars to upgrade its sewage treatment plant to keep nitrogen out of natural waters or to remove asbestos from the high school auditorium ceiling?

3. The third reason for studying the properties of elements and compounds is simple curiosity. Chemicals and chemical change are a part of nature that is open to investigation. For some people, this is as big a challenge as a high mountain is to a climber. If we hope to understand matter, the first steps are to discover the simplest forms of matter and study their interactions. Curiosity draws many chemists to basic research.

The Structure of Matter Explains Chemical and Physical Properties

An essential part of basic research is investigation into the **structure of matter**—how atoms of the elements are connected in larger units of matter and how these units are arranged in samples of matter large enough to be seen (Figure 2.7).

The properties of a sample of matter are determined by the nature of its parts, just as the abilities of a computer are determined by the parts that have been assembled. If we hope to understand the nature of matter, it is absolutely necessary that we understand the minute parts and how they are related to each other. Indeed, *the basic theme of this text and of chemistry itself is the relationship between the structure of matter and its properties.* Armed with an understanding of this relationship, chemists are developing an ever-increasing ability to create new substances and predict the properties of these substances.

(a)

(b)

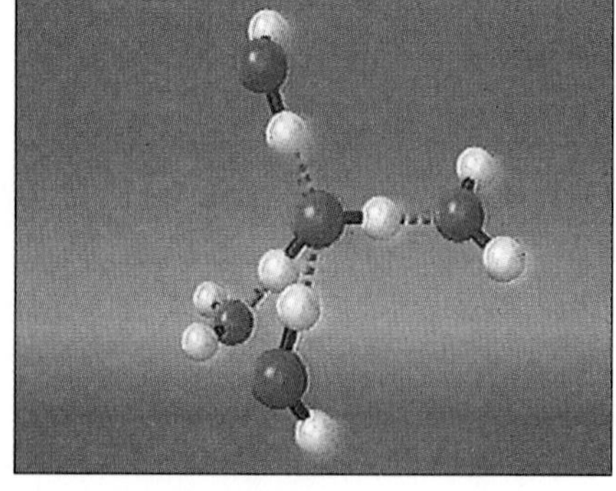
(c)

Figure 2.7 Three ways to look at snow. The beautiful patterns of snowflakes seen under a microscope (b) reflect the submicroscopic, orderly arrangement of water molecules (c). *(a, Grant Heilman; b and c,* The World of Chemistry, Program 12, *Water)*

THE PERSONAL SIDE

Alfred Bernhard Nobel (1833–1896)

Alfred Nobel (The Granger Collection)

To receive a Nobel Prize is looked on by most scientists as the highest possible honor, and the accomplishments recognized by the prize are always among the most significant advances in science. Textbooks usually make note of achievements and individuals worthy of Nobel Prizes as a way of highlighting their importance. It was breakthroughs in the chemistry of explosives that made the Nobel Prizes possible.

Alfred Nobel was a Swedish chemist and engineer, the son of an inventor and industrialist. Together they established a factory to manufacture nitroglycerin, a liquid explosive that is extremely sensitive to light and heat. After two years in operation, there was an explosion at the factory that killed five people, including Alfred Nobel's younger brother. Perhaps motivated by this event, Alfred searched for a way to handle nitroglycerin more safely. He found the answer and in 1867 patented dynamite. By mixing nitroglycerin with a soft, absorbent, nonflammable natural material known as *kieselguhr,* or diatomaceous earth, Nobel had stabilized the explosive power of nitroglycerin. His dynamite could be shipped safely and would explode only when this was desirable. He also invented the blasting caps (containing a mercury compound) needed to set off the dynamite explosion.

Nobel's talent as an entrepreneur combined with his many inventions (he held 355 patents) made him a very rich man. He never married and left his fortune to establish the Nobel Prizes, awarded annually to individuals who "have conferred the greatest benefits on mankind in the fields of physics, chemistry, physiology or medicine, literature and peace."

■ SELF-TEST 2A

1. Which of the following questions would a chemist study?
   (a) Does a grapefruit contain vitamin E?
   (b) What gives a grapefruit its yellow color?
   (c) How many grapefruits does the average tree produce in a season?
2. If a sample of matter contains only one kind of atom it is a (an) ______________ and ______________ be broken down to simpler substances.
3. The two kinds of pure substances are ______________ and ______________.
4. A solution is a ______________ mixture.
5. Solutions may exist in the solid, liquid, or gaseous states. (a) True, (b) False.
6. If you burn the butter in the frying pan, you have carried out a ______________ change because the butter changed in ______________.
7. Which of the following are chemicals? (a) Aspirin, (b) Baking soda, (c) Rubbing alcohol, (d) Vanilla, (e) Soap.

THE WORLD OF CHEMISTRY

Models and Modeling

Program 4, *Modeling the Unseen*

The history of science is filled with examples of models—mental concepts that scientists and investigators use to explain the unseen. Many of these models are triumphs of the human mind and represent extraordinary flights of the imagination. Among the classic examples are models of the atom, the kinetic molecular model of gases as composed of molecules rapidly flying about in mostly empty space, and the biochemists' model of how molecules of DNA unwind to form copies in new generations of plants and animals. Models must agree with facts, be modified to agree with facts, or be abandoned.

A remarkable example of the use of models to explain observed phenomena is found in the observations of Jupiter's moon Io by that astonishing spacecraft Voyager. As explained by Dr. Torrence Johnson, director of the Voyager Project,

Prior to the Jupiter visit, the model of Io, based on terrestrial observations, was of a moon similar to our own in age, composition, etc. The only inconsistency was that the surface was too bright. The model proposed at the time attributed its high reflectivity to the possibility that Io was covered by a layer of ice which had long since evaporated and left behind deposits of highly reflecting salt.

The visit of Voyager showed that this model is invalid. Instead, Io proved to be one of the most active volcanic bodies in the solar system, showing at the time of the Voyager encounter in 1978 at least seven active volcanoes. Enormous quantities of sulfur and sulfur dioxide were being spewed into space. Where does the energy for these volcanoes come from? Why isn't the volcanic material molten rock, as it is in volcanoes on Earth? A new model was needed. It pictures Io with a central core of silicate rock surrounded by the sulfur. The conflicting gravitational pull of Jupiter on one side and another moon on the other side provide the energy to keep subsurface sulfur and sulfur dioxide in a molten state.

Dr. Johnson goes on to explain that

When you build a model like this, you shouldn't get to thinking that the model represents reality. It will always change. In fact, we're not doing our job right if it doesn't change. It will always evolve as we get more information and revise our methods of observation based on the models.

Voyager photograph of Io, one of Jupiter's moons. Note the volcanic activity that appears above the surface at the upper left. The cold moon model of Io had to be changed as a result of these observations.

8. Which of the following mixtures is homogeneous? (a) Sugar and water, (b) Oil and water, (c) Cement, sand, and gravel.
9. Identify by name the reactant(s) shown in the following two equations:
   (a) $\underset{\text{Methane}}{CH_4} + \underset{\text{Oxygen}}{2\,O_2} \rightarrow \underset{\text{Carbon dioxide}}{CO_2} + \underset{\text{Water}}{2\,H_2O}$
   (b) $\underset{\text{Hydrogen peroxide}}{2\,H_2O_2} \rightarrow \underset{\text{Water}}{2\,H_2O} + \underset{\text{Oxygen}}{O_2}$
10. Which of the following are physical properties and which are chemical properties?
    (a) Flammability (b) Density
    (c) Souring of milk
11. Chemical potential energy is converted into light and heat in combustion.
    (a) True, (b) False.

2.6 The Chemical Elements

Iron, gold, sulfur, tin, and a few other substances that occur in the earth's crust were recognized as elements long before the modern meaning of the term developed. Other naturally occurring elements were identified only after long and laborious efforts to separate them from the compounds in which they are found. Although it isn't always easy, chemists can now separate and identify the elements present in the most complex mixtures and compounds.

The names of the elements are listed inside the front cover of this book. About 18 of these elements are not found anywhere in nature, but have only been produced artificially in laboratories, usually in extremely small quantities.

The elements range widely in their properties. As is discussed in Chapter 3 and represented by the arrangement of the elements in the periodic table (inside the front cover of this book), the properties of the elements are largely determined by the composition of their atoms.

Under everyday conditions some elements, including helium, hydrogen, nitrogen, oxygen, and chlorine, are gases. Only two elements—mercury and bromine—are ordinarily liquids. Most of the elements are solids, and of these most are metals. Two distinguishing properties of metals are that they conduct electric current and have a shiny appearance. A second major class of elements, the nonmetals, do not conduct electricity and are not lustrous. The familiar nonmetals include all the elemental gases; and carbon, sulfur, phosphorus, and iodine, which are solids.

■ By "everyday conditions" chemists refer to usual temperatures and pressures at the earth's surface. Changing these conditions can change the state of a substance, for example, the freezing or boiling of water.

Symbols for the Elements

Sometimes people are discouraged about understanding chemistry when they encounter chemical formulas and names: CH_3OH, H_2SO_4, polychlorinated

Three nonmetals: chlorine, a gas; bromine, a liquid; and iodine, a solid. *(Larry Cameron)*

biphenyl, nitric oxide. . . . There is nothing unique about chemistry, however, in needing its own vocabulary. Surely, the symbols and language in the margin are equally mysterious to anyone who has never been a musician or played football. Chemistry, just like music or football, needs special symbols and language. Without them communication would be impossible.

■ Symbols for the elements are the alphabet for the language of chemistry.

Since elements are the building blocks of all matter, symbols for the elements are fundamental to communicating about chemistry. The symbols for the elements are listed next to their names inside the front cover. Some symbols are single letters that are the first letter of the name. Others are two letters from the name, always with the first letter capitalized and the second lowercase. For example, *i*odine is represented by I and *in*dium is represented by In; *n*itrogen is represented by N and *ni*ckel, by Ni; *m*a*g*nesium is represented by Mg and *m*a*n*ganese, by Mn. You can see that it is important to recognize the symbols and know how they differ for different elements.

Eleven metals are represented by symbols not derived from their modern names, but instead from older names in other languages. Most of these metals, listed in Table 2.2, were known in ancient times because they are found free in nature or are easily obtained from their naturally occurring compounds.

Of the known elements, only 17 are nonmetals. Most of the common nonmetals have single letter symbols

H	C	N	O
Hydrogen	Carbon	Nitrogen	Oxygen
P	S	F	I
Phosphorus	Sulfur	Fluorine	Iodine

■ Consult the alphabetical listing inside the front cover to find the symbol from the name of any element, or vice versa.

Earlier we noted that a sample of gaseous helium (also a nonmetal) contains helium atoms. The symbol He is therefore used to represent a single He atom

TABLE 2.2 ■ Elements with Symbols Not Based on Their Modern Names

Modern Name	Symbol	Origin of Symbol
Antimony	Sb	*anti* + *monos:* Greek name, meaning an element not found alone
Copper	Cu	*cuprum:* Latin name, meaning from the island of Cyprus
Gold	Au	*aurum:* Latin name, meaning shining dawn
Iron	Fe	*ferrum:* Latin name for the element
Lead	Pb	*plumbum:* Latin name for the element
Mercury	Hg	*hydrargyrum:* Greek name, meaning liquid silver or quick silver
Potassium	K	*kalium:* Latin name, meaning alkali
Silver	Ag	*argentum:* Latin name for the element
Sodium	Na	*natrium:* Latin name for the element
Tin	Sn	*stannum:* Latin name for the element
Tungsten	W	*wolfram:* Swedish name, meaning devourer of tin (because it interferes with purifying tin)

or a large collection of helium atoms. One of the first milestones in understanding the structure of matter was the discovery that a number of nonmetals are composed not of atoms, but of molecules. A sample of pure hydrogen gas does contain only hydrogen atoms, as expected for an element, but the atoms are joined together in two-atom hydrogen molecules, represented by H_2. There are seven nonmetals that exist under everyday conditions as two-atom molecules, known as **diatomic molecules**:

Element	Symbol	Properties
Hydrogen	H_2	Colorless, odorless, occurs as a very light gas, burns in air
Oxygen	O_2	Colorless, odorless gas, reactive, constituent of air
Nitrogen	N_2	Colorless, odorless, gas, rather unreactive
Chlorine	Cl_2	Greenish-yellow gas, very sharp choking odor, poisonous
Fluorine	F_2	Pale yellow, highly reactive gas
Bromine	Br_2	Dark red liquid, vaporizes readily, very corrosive
Iodine	I_2	Dark purple solid that changes directly to gas when heated gently

2.7 Using Chemical Symbols

Formulas for Chemical Compounds

We have already used a few **chemical formulas**, combinations of the symbols for the elements that represent the stable combinations of atoms in molecules: H_2 for elemental hydrogen, H_2O for water, and $C_{12}H_{22}O_{11}$ for table sugar. The principle is the same no matter how many atoms are combined. The symbols represent the elements and the **subscripts** (like the $_2$ in H_2O) indicate the relative numbers of atoms of each kind. The formulas and properties of a few simple molecular compounds are given in Table 2.3. Such formulas can represent one molecule or a large sample of the compound.

■ Chemical formulas are the words in the language of chemistry.

TABLE 2.3 ■ Some Simple Molecular Compounds

Compound	Formula	Properties
Water	H_2O	Odorless liquid
Carbon monoxide	CO	Odorless, flammable, toxic gas
Carbon dioxide	CO_2	Odorless, nonflammable, suffocating gas
Sulfur dioxide	SO_2	Nonflammable gas, suffocating odor
Ammonia	NH_3	Colorless, nonflammable gas, pungent odor
Methane	CH_4	Odorless, flammable gas
Carbon tetrachloride	CCl_4	Nonflammable, dense liquid

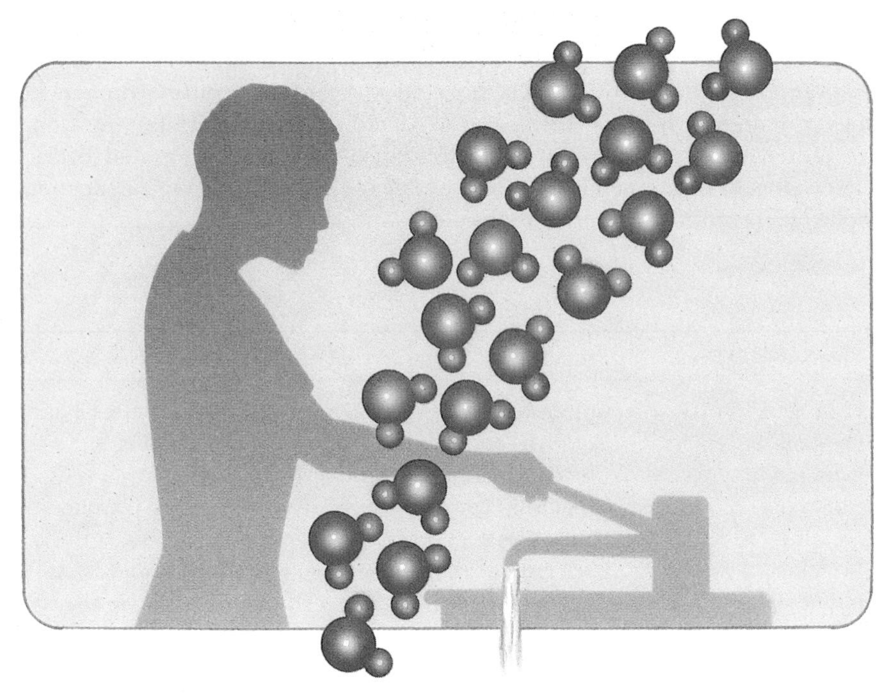

A stream of water and the molecules of which it is composed.

Sometimes a line is drawn between symbols to indicate which atoms are connected in molecules

H—H	H–N(–H)–H (N bonded to three H)	H–C(–H)(–H)–H (C bonded to four H)
A hydrogen molecule	An ammonia molecule	A methane molecule

■ The lines in structural formulas represent chemical bonds, which we'll have much more to say about in Chapter 5.

■ Organic compounds are so numerous and so important that they are the basis for an entire field of chemistry—**organic chemistry** (Chapters 9 and 10). Most compounds produced by living things, most medicines, and the components of most plastics are organic compounds.

Formulas that show the connections in this way are known as **structural formulas**, whereas formulas that give just one symbol for each element present are called **molecular formulas**. The molecular formula for methane is CH_4. Methane and the many millions of compounds that contain carbon combined with hydrogen—and often also with nitrogen, oxygen, phosphorus, or sulfur—are known as **organic compounds**.

Compounds not based on carbon are referred to as **inorganic compounds**. A few that are in everyday use are

	$NaCl$	NH_4NO_3	H_2SO_4	$Mg(OH)_2$
Chemical name:	Sodium chloride	Ammonium nitrate	Sulfuric acid	Magnesium hydroxide
Use:	Table salt	Fertilizer	Battery acid	A laxative

When it is necessary to explore the shapes of molecules beyond what can be shown on a flat piece of paper, chemists resort to physical models or computer-drawn pictures, such as those shown in Figure 2.8.

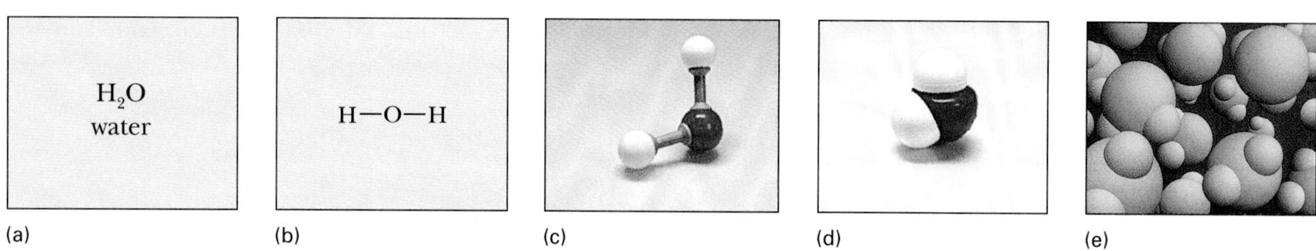

(a) (b) (c) (d) (e)

Figure 2.8 The water molecule, as represented in print (a and b), in physical models that can be handled (c and d), or on a computer screen (e). *(e, Ken Eward/Science Source/Photo Researchers)*

EXAMPLE 2.1 *Chemical Symbols*

Which of the following represent elements and which represent compounds?
(a) KI (b) Co (c) Ag (d) NO (e) Cl_2

SOLUTION

The chemical symbols in (a) and (d) represent compounds—the second letter in an element symbol is never a capital letter; (b) and (c) represent the elements cobalt and silver; and (e) represents the element chlorine, which exists as diatomic molecules (p. 27).

Exercise 2.1

Write the symbols or formulas for (a) lead, (b) phosphorus, (c) a molecule containing one hydrogen atom and one chlorine atom, (d) a molecule containing one aluminum atom and three bromine atoms, and (e) elemental fluorine.

Chemical Equations

■ Chemical equations are the sentences in the language of chemistry.

To concisely represent chemical reactions, symbols and formulas are arranged in **chemical equations**. For example, in words, a chemical reaction could be expressed as "carbon reacts with oxygen to form carbon monoxide," a reaction that happens whenever something containing carbon burns incompletely. Like most solid elements, carbon is represented just by its symbol, C; oxygen must be represented by the molecular formula for its diatomic molecules, O_2; and carbon monoxide molecules, which contain two atoms, one each of carbon and oxygen, are represented by their molecular formula, CO.

Earlier, we pointed out that in chemical reactions atoms are rearranged into different substances, but the number of atoms stays the same. To represent this correctly, chemical equations must be **balanced**—the number of atoms of each kind in the reactants and products must be the same. In this case, one oxygen molecule will combine with two carbon atoms to form two carbon monoxide molecules. All this information is contained in the equation

■ States of matter are shown in equations as (g) for a gas, (ℓ) for a liquid, and (s) for a solid; (aq) is used for a substance dissolved in water (an **aqueous solution**).

$$2\ C(s) + O_2(g) \longrightarrow 2\ CO(g)$$

We have also included here the information that carbon is a solid (s) and that oxygen and carbon monoxide are gases (g). The arrow ($\longrightarrow$) is often read as "yields." The equation then states the following:

At the macro level: Carbon, a solid, plus oxygen gas yields carbon monoxide gas.

At the submicroscopic level: Two atoms of carbon plus one diatomic molecule of oxygen yield two molecules of carbon monoxide.

The number written before a formula in an equation, the **coefficient**, gives the relative amount of the substance involved. The subscripts give the composition of the pure substances. Changing the coefficient changes only the amount of the element or compound involved, whereas changing a subscript would change the identity of the reactant or product. For example, 2 CO represents two molecules of carbon monoxide, whereas CO_2 represents a molecule of carbon dioxide, a very different substance formed in the complete combustion of carbon-containing materials:

$$C(s) + O_2(g) \longrightarrow CO_2(g)$$

EXAMPLE 2.2 *Chemical Equations*

Nitrogen dioxide (NO_2) is a red-brown gas often visible in the haze during a period of air pollution over a city. (a) Interpret in words the information given in the equation for formation of nitrogen dioxide from nitrogen monoxide (NO)

■ NO is commonly known as nitric oxide.

$$2\ NO(g) + O_2(g) \longrightarrow 2\ NO_2(g)$$

(b) Prove that the equation is balanced.

SOLUTION

(a) Nitrogen monoxide, a gas, reacts with oxygen gas to give nitrogen dioxide, also a gas. The coefficients show that two molecules of NO react with one molecule of O_2 to give two molecules of NO_2.

(b) *Reactants:* The coefficient 2 in 2 NO shows the presence of two molecules of NO, which means two N atoms and two O atoms. The O_2 has no coefficient; thus, there is one O_2 molecule, but the subscript shows that the single molecule has two O atoms. Therefore, there are two N atoms and four O atoms on the reactants side. *Products:* The coefficient in 2 NO_2 shows two molecules of NO_2, which gives a total of two N atoms (N has no subscript) and four O atoms (two in each of the two molecules, as shown by the subscript on O). The equation is balanced.

Exercise 2.2

Repeat the question of Example 2.2 for the equation for the formation of hydrogen chloride (HCl):

$$H_2(g) + Cl_2(g) \longrightarrow 2\ HCl(g)$$

■ SELF-TEST 2B

1. Chemists strive to understand the ____________ of matter in order to create new materials.
2. Name two elements that are solids, two that are liquids, and two that are gases.
3. Write the symbols for the six elements you named in question 2.
4. Name the elements combined in the compound K_2HPO_4. How many atoms in total are represented in this chemical formula?
5. Are the following equations balanced?
   (a) $2\ K(s) + Cl_2(g) \rightarrow KCl(s)$
   (b) $2\ Mg(s) + O_2(g) \rightarrow 2\ MgO(s)$
   (c) $SO_3(g) + H_2O(\ell) \rightarrow 2\ H_2SO_4(aq)$
   (d) $NaCl(aq) + AgNO_3(aq) \rightarrow NaNO_3(aq) + AgCl(s)$
6. Consider the chemical equation $CH_4 + 2\ O_2 \rightarrow CO_2 + 2\ H_2O$. Explain what is meant by the symbols:
   (a) O ____________
   (b) $2\ O_2$ ____________
   (c) CH_4 ____________
   (d) $\rightarrow$ ____________
   (e) H_2O ____________
   (f) $2\ H_2O$ ____________
7. A solid element does not conduct electricity. Is it a metal or a nonmetal?
8. Which of the following elements have symbols not based on their modern names? (a) Lead, (b) Carbon, (c) Oxygen, (d) Potassium.

2.8 The Quantitative Side of Science

Several times in the preceding sections we used the numerical results of measurements of the boiling points, masses, or densities of pure substances. These and hundreds of other kinds of measurements are fundamental to chemistry and every other science. The result of a measurement is what we refer to as **quantitative** information—it uses numbers. Weighing yourself is a quantitative experiment. By contrast, there is **qualitative** information that does not deal with numbers. Eating an artichoke if you have never eaten one before is a qualitative experiment.

The result of a measurement is recorded as a number plus a unit. "A boiling point of 52," doesn't mean anything without the unit. Was the temperature measured in Fahrenheit degrees, as is done by the usual household thermometer in the United States, or in Celsius degrees, as would be done in the rest of the world?

■ Everyone making scientific measurements must understand the *significance* of the digits when they are using the results of measurements. This becomes especially important in using electronic calculators. If a 3-cm^3 piece of something weighs 4.52 g, is its density 1.5 g/cm^3, 1.50 g/cm^3, or—as given by a calculator—1.506666667 g/cm^3? Appendix A explains how questions like this are dealt with.

The World of Chemistry

Program 5, *A Matter of State*

It would be inconvenient to measure the components of this circuit board in meters. The millimeter is much more convenient.

The establishment of scientific facts and laws is obviously dependent on accurate observations and measurements. Although measurements can be reported as precisely in one system of measurement as another, there has been an effort since the time of the French Revolution in the late 1700s to have all scientists embrace the same simple system. The hope was and is to facilitate communication in science. The metric system, which was born of this effort, has two advantages. First, it is easy to convert from one unit to another, since smaller and larger units for the same physical quantity differ only by multiples of ten. Consequently, to change millimeters to meters, the decimal point need only be moved three places to the left

$$1 \text{ millimeter} = 0.001 \text{ meter} \quad \text{therefore,} \quad 5.0 \text{ millimeters} = 0.0050 \text{ meter}$$

Compare the decimal shift to the problem of changing inches to yards. Knowing

$$1 \text{ ft} = 12 \text{ inches} \quad 1 \text{ yard} = 3 \text{ ft}$$

we can calculate

$$5.0 \cancel{\text{inches}} \times \frac{1 \cancel{\text{ft}}}{12 \cancel{\text{inches}}} \times \frac{1 \text{ yard}}{3 \cancel{\text{ft}}} = 0.14 \text{ yard}$$

Take a moment to be sure you understand how units are included and canceled in this unit conversion. An excellent way to keep track of what you are doing is to include units in setting up the problem and then to cancel them like numbers. You will find that this method applies to far more than unit conversion or the mathematics of chemistry. Often by including the units with numbers you can figure out what is needed to find a mathematical answer; and when the units cancel to give the desired unit with the answer, chances are that the answer will be the right one (if, of course, your arithmetic is correct). For more information and practice with units and problem solving, you can consult Appendix C at the end of this book.

The second advantage of the modern metric system is that standards for most fundamental units are defined by reproducible phenomena of nature. For example, the metric unit for time—the second—is now defined in terms of a specific number of cycles of radiation from a radioactive cesium atom, a time period believed never to vary.

The World of Chemistry

Program 3, *Measurement: The Foundation of Chemistry*

It doesn't really matter too much what you define a meter to be—you could define it as a very small length like the length of your pen. But unless everyone in the world agrees that that's a meter then you don't have a standard of measurement that's adequate to today's world. Stanley Rasberry, Director, Standard Reference Materials Division, National Institute of Standards and Technology (NIST), formerly the National Bureau of Standards.

The units still in everyday use in the United States have evolved from the English system of measurement, which began with the decrees of various English monarchs. The *yard* started out as the length of the waist sash worn by Saxon kings and obviously varied with the girth of the king. A step toward standardization came when King Henry I decreed that the yard should be the distance from the tip of his nose to the end of his thumb. Standards for the English units in use in the United States today are designated by the National Institute of Standards and Technology (NIST).

Since its introduction in 1790, the metric system has been continually modified and improved. The current version—the International System of Units, abbreviated SI (from the French, Système International d'Unités)—was

TABLE 2.4 ■ Common Prefixes for Multiples and Fractions of SI Units

Prefix	Abbreviation	Meaning	Example
mega-	M	10^6 (1 million)	1 megaton = 1×10^6 tons
kilo-	k	10^3 (1 thousand)	1 kilogram (kg) = 1×10^3 g
deci-	d	10^{-1} (1 tenth)	1 decimeter (dm) = 0.1 m
centi-	c	10^{-2} (1 one-hundredth)	1 centimeter (cm) = 0.01 m
milli-	m	10^{-3} (1 one-thousandth)	1 millimeter (mm) = 0.001 m
micro-	μ*	10^{-6} (1 one-millionth)	1 micrometer (μm) = 1×10^{-6} m
nano-	n	10^{-9} (1 one-billionth)	1 nanometer (nm) = 1×10^{-9} m
pico-†	p	10^{-12} (1 one-trillionth)	1 picometer (pm) = 1×10^{-12} m

* This is the Greek letter mu (pronounced "mew").
† This prefix is pronounced "peako."

■ The prefixes in Table 2.4 are represented by the powers of 10 used in scientific, or exponential, notation for writing large and small numbers. For example, $10^3 = 10 \times 10 \times 10 = 1000$. Appendix B reviews this notation.

adopted by the International Bureau of Weights and Measures in 1960. The SI system defines seven fundamental units from which other units are derived. For example, the fundamental SI unit for length is the meter (abbreviated m), which then gives the SI unit for area as the square meter (m^2) and the SI unit for volume as the cubic meter (m^3). Multiples and fractions of SI units are designated by adding prefixes (Table 2.4). Therefore, when a length unit smaller than the meter is needed, we can choose the *centi*meter (cm) or the *milli*meter (mm); and when a larger length unit is needed, we can use the *kilo*meter (km). For mass units, prefixes are added to gram, such as the kilogram (kg), the milligram (mg), or the microgram (μg, where μ is the Greek letter mu).

EXAMPLE 2.3 *Metric Unit Prefixes*

How many meters are in 2 km?

SOLUTION

The prefix *kilo-* (k) means 1000 times, meaning that 1000 m = 1 km. It is hardly worth the trouble to write anything down in this solution. One just thinks 1000 m for every kilometer as one thinks ten dimes for every one dollar bill. The best way to write this problem, or any more complicated problem, however, is to include the units and cancel them out (like numbers in algebra). This way, if you are left with the correct unit, you have probably done the problem correctly.

$$2\ \cancel{\text{km}} \times \frac{1000\ \text{m}}{1\ \cancel{\text{km}}} = 2000\ \text{m}$$

Exercise 2.3

How much in dollars is 20 *mega*bucks?

A typical laboratory balance.

TABLE 2.5 ■ Some Common Units in Chemistry

Name of Unit	Symbol	Common Equivalent
Meter	m	39.4 inches
Liter	L	1.06 quarts
Gram	g	0.0352 ounce
Degrees Celsius	°C	Water boils at 100°C and freezes at 0°C
Calorie	cal	Energy required to heat 1 g of water 1°C

SCIENCE AND SOCIETY

The Metric System

The United States, Liberia, and Myanmar (formerly Burma) have something in common. We are the only countries still weighing and measuring in pounds, feet, and quarts. In all other countries, kilograms, meters, and liters are in everyday use.

Why have we avoided speaking the same measurement language as the rest of the world? It has long been recognized that communication is best served by a common system of weights and measures. Since 1866, by act of Congress, using the metric system has been legal in the United States. And in 1975 the federal government passed the Metric Conversion Act, which was meant to initiate a gradual changeover. However, the act was based on voluntary change, and very little happened.

In the United States, opposition to change by the public has frequently surfaced as concerns about excessive cost. There are also the usual discomfort with the unfamiliar (attributed by some to widespread math phobia) and the conservatism of those Americans who believe that traditional ways are best.

Suppose you go to the supermarket tonight and find that you have to buy kilos of potatoes, liters of milk, and aluminum foil measured in meters or centimeters. Or suppose that you manufacture kitchen cabinets and will have to redesign them to be measured in meters and centimeters instead of feet and inches. Would you be happy that we are finally joining the rest of the world? Would you be upset over the cost of making new specifications? Would the prospect of additional sales outside the United States make conversion more attractive?

Two ways to convert to the metric system are by retaining the same-sized containers and relabeling or by adopting metric-sized containers. (C. D. Winters)

We already buy soda in 1-L and 2-L bottles, with little inconvenience; and most packaged food labels give both metric and English units. Since 1994 the Federal Trade Commission has also required that products such as detergents, toilet paper, and batteries be labeled with both types of units. And there are reports that most states have met the 1996 deadline for all projects financed by federal highway funds to be designed in metric units.

Nevertheless, a complete changeover to metric units in the United States remains a dim prospect, and the government has not made it a legal mandate despite the potential economic advantages. (For example, the European Union is expected to purchase only metric goods beginning in the year 2000.) Instead in 1997, the NIST launched a "Towards a Metric America" campaign to publicize the value of metrication. Civic leaders, teachers, and manufacturers are being invited to a series of town meetings in major cities throughout the country at which the benefits of metrication and ways to accomplish it are discussed in an open forum.

Because of their convenient sizes, the five units listed in Table 2.5 are commonly used in chemistry and indeed are in everyday use in the rest of the world (see *Science and Society, The Metric System*). We will use the unit symbols listed in Table 2.5 in this text. The volume unit of liters is preferable to the SI-derived unit of cubic meters, which is much too large (1 cubic meter = 1000 liters). (Figure 2.9 gives the relative sizes of some other length units.)

We tend to use the words *mass* and *weight* interchangeably, and that's all right because we understand what we mean. Strictly speaking though *mass* measures the amount of matter in an object and is the same everywhere in the universe, while *weight* measures the force of gravity on that object and is different where the force of gravity is different (e.g., on the moon).

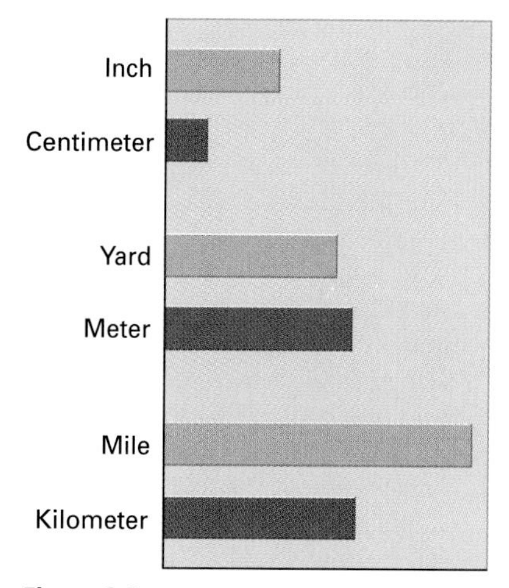

Figure 2.9 Relative sizes of some English and metric length units: 1 inch = 2.54 centimeters; 1 yard = 0.91 meter; 1 mile = 1.61 kilometers.

SELF-TEST 2C

1. The result of a measurement must include both a ____________ and a ____________.
2. The metric system uses ____________ to indicate multiples of ________.
3. The volume unit derived from the centimeter is a ____________.
4. The prefix meaning 1000 times bigger is ____________, and the prefix meaning 0.001, or 1000 times smaller, is ____________.
5. What are the units for the answer to the following calculation?

$$0.500 \text{ ft} \times \frac{12 \text{ inches}}{1 \text{ ft}} \times \frac{1 \text{ m}}{39.4 \text{ inches}} \times \frac{1000 \text{ mm}}{1 \text{ m}} = 152 \_\_\_\_\_\_\_\_$$

6. Which conversion factor should you use to convert miles to kilometers (km)?

   (a) $\frac{1 \text{ mile}}{1.61 \text{ km}}$, (b) $\frac{1.61 \text{ km}}{1 \text{ mile}}$
7. Which of the following experiments is qualitative and which is quantitative?
   (a) Determination of the distance between two atoms in a molecule.
   (b) Determination of the identity of the metal in a piece of wire.

MATCHING SET

____ 1. Produces new substance(s)
____ 2. Air
____ 3. Unchanged by further purification
____ 4. Cannot be reduced to simpler substances
____ 5. SI system
____ 6. Symbol for iron
____ 7. Volume unit
____ 8. Molecule containing three oxygen atoms
____ 9. Carbon monoxide
____ 10. 100 cm
____ 11. SI fundamental unit
____ 12. Deci-
____ 13. Centi-
____ 14. Structural formula
____ 15. Solution
____ 16. Molecular formula

a. Chemical change
b. Element
c. Properties of pure substance
d. Mixture
e. Uses multiples of ten
f. O_3
g. CO
h. Fe
i. 1 m
j. H—S—H
k. Liter
l. One-tenth
m. Meter
n. C_4H_{10}
o. Homogeneous mixture
p. One-hundredth

QUESTIONS FOR REVIEW AND THOUGHT

1. Name as many materials as you can that you have used during the past day and that were not chemically changed from their natural states.
2. Identify the following as physical or chemical changes. Explain your choices.
   (a) Formation of snowflakes
   (b) Rusting of a piece of iron
   (c) Ripening of fruit
   (d) Fashioning a table leg from a piece of wood
   (e) Fermenting grapes
   (f) Boiling a potato
3. Would it be possible for two pure substances to have exactly the same set of properties? Give reasons for your answer.
4. List three physical properties that can be used to identify pure substances. Give a specific example of each property.
5. Describe in your own words what happens when you flick a BIC lighter.
6. Chemical changes can be both useful and destructive to humanity's purposes. Cite a few examples of each kind of change from your own experience. Also give evidence from observation that each is indeed a chemical change and not a physical change.
7. Classify each of the following as a physical property or a chemical property. Explain your answers.
   (a) Density
   (b) Melting temperature
   (c) Decomposition of a substance into two elements on heating
   (d) Electrical conductivity
   (e) The failure of a substance to react with sulfur
   (f) The ignition temperature of a piece of paper
8. Write your last name. How many element symbols can you produce from the letters in your name?
9. Classify each of the following as an element, a compound, or a mixture. Justify each answer.
   (a) Mercury
   (b) Milk
   (c) Pure water
   (d) A piece of lumber
   (e) Ink
   (f) Iced tea
   (g) Pure ice
   (h) Carbon
   (i) Antimony
10. Which of the materials listed in Question 9 can be pure substances?
11. Is it possible for the properties of iron to change? What about the properties of steel, which is an alloy and a homogeneous mixture? Explain your answer.
12. You have a mixture of sand (SiO_2) and salt ($NaCl$). How would you separate these two substances? When you have them as separate substances, how can you prove which is which?
13. Consider the following five elements: nitrogen, sulfur, chlorine, magnesium, and cobalt. By using this text or any other source available at the library, find the major source for these elements and at least one compound that uses the element in combined form.
14. Atrazine is a selective herbicide that has the molecular formula $C_8H_{14}N_5Cl$. This compound is used for season-long weed control in corn, sorghum, and certain other crops. What elements are present in atrazine?
15. Cytoxan, also known as cyclophosphamide, is widely used alone or in combination in the treatment of certain kinds of cancer. It interferes with protein synthesis and in the process kills rapidly replicating cells, particularly malignant ones. Cytoxan has the molecular formula $C_7H_{15}O_2N_2PCl_2$.
    (a) How many atoms are in one molecule of cytoxan?
    (b) What elements are present in cytoxan?
    (c) What is the ratio of hydrogen atoms to nitrogen atoms in cytoxan?
    (d) Would cytoxan be classified as an organic compound?
16. By reading labels, identify a commercial product that contains each of the following compounds:
    (a) Calcium carbonate
    (b) Phosphoric acid
    (c) Water
    (d) Fructose
    (e) Sodium chloride
    (f) Potassium sorbate
    (g) Potassium iodide
    (h) Glycerol
    (i) Aluminum
    (j) Butylated hydroxytoluene (BHT)
17. There are three states of matter—gas, liquid, and solid. Name a material that is a pure substance for each state of matter. Do not use water, oxygen, or salt. Name a material that is a mixture for each state of matter. Do not use air, gasoline, or brass.
18. Given the following sentence, write a chemical equation using chemical symbols that convey the same information: "One nitrogen molecule reacts with three hydrogen molecules to produce two ammonia molecules, each containing one nitrogen and three hydrogen atoms."
19. Is it possible to have a mixture of two elements and also to have a compound of the same two elements? Explain. Can you think of an example?
20. Name four kinds of energy.
21. Describe in words the chemical processes represented by the following equations:
    (a) $2\ Na(s) + Cl_2(g) \rightarrow 2\ NaCl(s)$

(b) $N_2(g) + 3\ Cl_2(g) \rightarrow 2\ NCl_3(g)$ [named as nitrogen trichloride]
(c) $CO_2(g) + H_2O(\ell) \rightarrow H_2CO_3(aq)$ [carbonic acid]
(d) $2\ H_2O_2(aq)$ [hydrogen peroxide] $\rightarrow O_2(g) + 2\ H_2O(\ell)$

22. Prove that each of the equations in Question 21 is balanced.
23. For equations b and d in Question 21, identify the reactant(s) and the product(s).
24. Are the following equations balanced?
    (a) $AgNO_3(aq) + Na_2SO_4(aq) \rightarrow Ag_2SO_4(s) + NaNO_3(aq)$
    (b) $AgNO_3(aq) + HCl(aq) \rightarrow AgCl(s) + HNO_3(aq)$
25. Is the tea in tea bags a pure substance? Use the process of making tea to make an argument for your answer. How would your argument apply to instant tea?
26. Find and list as many pure substances as you can in a kitchen.
27. (a) How many milligrams are there in 1 g?
    (b) How many meters are there in 1 km?
    (c) How many centigrams are there in 1 g?
28. What are the most common units in chemistry for mass, length, and volume?
29. Which of the following quantities is a density?
    (a) 9 cal/g (b) 100 cm/m
    (c) 1.5 g/mL (d) 454 g/lb
30. A cook wants to pour 1.5 L of batter into a 2-quart bowl. Will it fit?

PROBLEMS

1. A rancher needs one acre of grazing land for ten cows. How many acres are needed for 55 cows? Solve this problem (and others where appropriate) by including units. In this case, the answer will have the "units" of acres per cow.
2. There are 200 mg of ibuprofen in an Advil tablet. How many grams is this? How many micrograms?
3. How many meters are in a 10-km race?
4. If a 1-ounce portion of cereal contains 3.0 g of protein, how many milligrams of protein does it contain?
5. Complete the following:
    (a) 4 cm = ________ m
    (b) 0.043 g = ________ mg
    (c) 15.5 m = ________ mm
    (d) 328 mL = ________ L
    (e) 0.98 kg = ________ g
6. An average African gorilla has a mass of 163 kg. What is this mass in grams?
7. An average adult man has a mass of 70 kg. Convert this mass to milligrams.

CHAPTER 3

Atoms and the Periodic Table

Why does an element or compound have the properties it has? Why does one element or compound undergo a change that another element or compound will not undergo? Inanimate matter is the way it is because of the nature of its parts. The use of atoms to represent these parts dates back to about 400 B.C. when the Greek philosopher Leucippus and his student, Democritus (460–370 B.C.), argued for a limit to the divisibility of matter, which was counter to the prevailing view of Greek philosophers that matter is endlessly divisible. Democritus used the Greek word *atomos,* which literally means "uncuttable," to describe the ultimate particles of matter, particles that could not be divided further. However, it wasn't until John Dalton (1766–1844) introduced his atomic theory in 1803 that the importance of using atoms to explain properties of matter was recognized.

Dalton's atomic theory and the development of the periodic table by Mendeleev in 1869 led to the rapid growth of chemistry as a science. In particular, the influence of the location and number of electrons in atoms on the properties of elements has become one of the essential ideas of chemistry. In this chapter we will use the current knowledge about the atom together with the periodic table to serve as the basis for our understanding of the chemical view of matter.

- What are the milestones in the development of atomic theory?
- What is the experimental evidence for the existence of subatomic particles within atoms?
- What are the three basic subatomic particles of the atom?
- What are isotopes?
- Where are electrons in atoms, and how are they arranged?
- How was the periodic table developed?
- Why do elements in the same group in the periodic table have similar chemical properties?

The very different properties of the elements sulfur and bromine are accounted for by the differences between their atoms. *(Larry Cameron)*

- How is the activity of an element related to its position in the periodic table?

3.1 John Dalton's Atomic Theory

John Dalton, drawing from his own quantitative experiments and those of earlier scientists, proposed in 1803 that

1. All matter is made up of indivisible and indestructible particles called atoms.
2. All atoms of a given element are identical, both in mass and in properties. Atoms of different elements have different masses and different properties.
3. Compounds are formed when atoms of different elements combine in the ratio of small whole numbers.
4. Elements and compounds are composed of definite arrangements of atoms, and chemical change occurs when the atomic arrays are rearranged.

The World of Chemistry
Program 6, *The Atom*

The Parthenon is contemporary with early ideas about atoms.

John Dalton's atomic theory was accepted because it could be used to explain several scientific laws that had been established by Dalton and other scientists of the time. Two of these laws are (1) the law of conservation of matter, and (2) the law of definite proportions.

Some years before Dalton proposed his atomic theory, Antoine Lavoisier (1743–1794) had carried out a series of experiments in which the reactants were carefully weighed before a chemical reaction and the products were carefully weighed afterward. He found no change in mass when a reaction occurred, proposed that this was true for every reaction, and called his proposal the **law of conservation of matter**: *Matter is neither lost nor gained during a chemical reaction.* Others verified his results, and the law became accepted. Points (2) and (4) in Dalton's theory imply the same thing. If each kind of atom has a particular characteristic mass, and if there are exactly the same number of each kind of atom before and after a reaction, the masses before and after must also be the same.

Another chemical law known in Dalton's time had been proposed by Joseph Louis Proust (1754–1826) as a result of his analyses of minerals. Proust found that a particular compound, once purified, always contained the same elements in the same ratio by mass. One such study, made by Proust in 1799, involved copper carbonate. Proust discovered that, regardless of how copper carbonate was prepared in the laboratory or how it was isolated from nature, it always contained the same proportions by mass—five parts copper, four parts oxygen, and one part carbon. Careful analyses of this and other compounds led Proust to propose the **law of definite proportions**: *In a compound, the constituent elements are always present in a definite proportion by mass.* For example, pure water, a compound, is always made up of 11.2% hydrogen and 88.8% oxygen by weight. Pure table sugar, another compound, always contains 42.1% carbon, 6.5% hydrogen, and 51.4% oxygen by weight. The source of the pure substance is irrelevant.

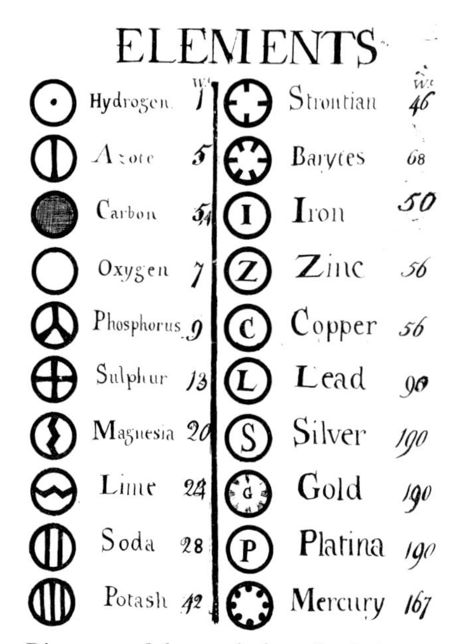

Diagrams of the symbols and relative weights of atoms used by Dalton in his lecture on atomic theory. *(Stock Montage)*

THE PERSONAL SIDE

John Dalton (1766–1844)

John Dalton.
(The Bettmann Archives)

John Dalton, a gentle man and a devout Quaker, gained acclaim because of his work. He made careful measurements, kept detailed records of his research, and expressed them convincingly in his writings. However, he was a poor speaker and was not well received as a lecturer. When Dalton was 66 years old, some of his admirers sought to present him to King William IV. Dalton resisted because he would not wear the court dress. Because he had a graduate degree from Oxford University, the scarlet robes of Oxford were deemed suitable, but a Quaker could not wear scarlet. Dalton, being color blind, saw scarlet as gray; thus, he was presented in scarlet to the court, but in gray to himself. This remarkable man was, in fact, the first to describe color blindness. He began teaching in a Quaker school when only 12 years old, discovered a basic law of physics (the law of partial pressure of gases), and helped found the British Association for the Advancement of Science. He kept more than 200,000 notes on meteorology. Despite his accomplishments he shunned glory and maintained he could never find time for marriage.

3.2 Structure of the Atom

There is now experimental evidence for the existence of more than 60 subatomic particles. However, only three are important to our understanding of the chemical view of matter: **electrons**, **protons**, and **neutrons** (Table 3.1). The mass and charge of electrons and protons were determined with experiments that used electric and magnetic fields. Electrons and protons have the same quantity of charge but different signs. The mass of the positively charged proton is about 1800 times the mass of the negatively charged electron. Although studies in the early 1900s indicated the presence of a third type of particle with no charge but a weight similar to that of the proton, the lack of charge meant that electric and magnetic fields could not be used to detect this type of particle. As a result, the neutron wasn't discovered until 1932 when James Chadwick (1891–1974) detected neutrons with experiments by using emissions from radioactive elements.

TABLE 3.1 ■ Summary of Properties of Electrons, Protons, and Neutrons

	Relative Charge	Relative Mass	Location
Electron	−1	0.00055	Outside the nucleus
Proton	+1	1.00727	Nucleus
Neutron	0	1.00867	Nucleus

Even though protons and electrons had been identified as subatomic particles, the arrangement of these particles within the atom was not known. An atom was thought to be a uniform sphere of protons within which electrons circulated in rings. It wasn't until after the discovery of natural radioactivity that Ernest Rutherford (1871–1937) carried out experiments that led to the idea of a nucleus as a tiny core of the atom.

Natural Radioactivity

Henri Becquerel (1852–1908) discovered natural radioactivity in natural uranium and radium ores in 1896 (Section 4.1). In 1898, Marie Sklodowska Curie (1867–1934)—a student of Becquerel—and her husband Pierre discovered two radioactive elements, radium and polonium. In 1899, Marie Curie suggested that atoms of radioactive substances disintegrate when they emit these unusual rays. She named this phenomenon **radioactivity**. Whether in a compound or uncombined, each radioactive element gives off exactly the same rays; about 25 elements exist only in radioactive forms. Marie Curie's suggestion that atoms disintegrate contradicts Dalton's idea that atoms are indivisible.

Ernest Rutherford (1871–1937) began studying the radiation emitted from radioactive elements soon after experiments by the Curies and others had shown that three types of radiation are spontaneously emitted by radioactive elements (Section 4.1). These rays, referred to as alpha (α), beta (β), and gamma (γ) rays, behave differently when passed between electrically charged plates, as shown in Figure 3.1. Alpha and beta rays are deflected, while gamma rays pass straight through undeflected. This implies that alpha and beta rays are electrically charged particles, because particles with a charge would be attracted or repelled by the charged plates. Even though an alpha particle has an electrical charge (+2) twice as large as that of a beta particle

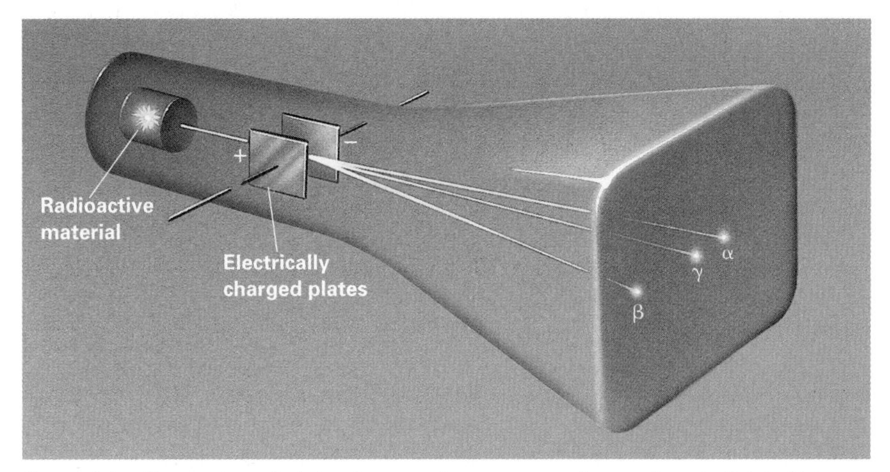

Figure 3.1 Separation of alpha, beta, and gamma rays by an electric field. Alpha rays are deflected toward the negative plate; beta rays are attracted toward the positive plate; and gamma rays are not deflected. Additional studies showed that alpha particles are high-energy helium nuclei, beta particles are high-energy electrons, and gamma rays are high-energy electromagnetic radiation.

■ Beta particles are high-speed electrons.

(− 1), alpha particles are deflected less; hence, alpha particles must be heavier than beta particles. Careful studies by Rutherford showed that **alpha particles** are helium atoms that have lost two electrons (He^{2+}). **Beta particles** were shown to be negatively charged particles identical with the electron. **Gamma rays** have no detectable charge or mass—they behave like light rays.

The Nucleus of the Atom

Rutherford's experiments with alpha particles led him to consider using them in experiments on the structure of the atom. In 1909, he suggested to two of his co-workers that they bombard a piece of gold foil with alpha particles. Hans Geiger (1882–1945), a German physicist, and Ernest Marsden (1889–1970), an undergraduate student, set up the apparatus diagrammed in Figure 3.2 and observed what happened when alpha particles hit the thin gold foil. Most passed straight through, but Geiger and Marsden were amazed to find that a few alpha particles were deflected through large angles, and some came almost straight back. Rutherford later described this unexpected result by saying, "It was about as incredible as if you had fired a 15-inch [artillery] shell at a piece of paper and it came back and hit you."

What allowed most of the alpha particles to pass through the gold foil in a rather straight path? According to Rutherford's interpretation, the atom is mostly *empty space* and therefore offers little resistance to the alpha particles (Figure 3.3).

What caused a few alpha particles to be deflected? According to Rutherford's model of the atom, all of the positive charge and most of the mass of

(text continued on p. 44)

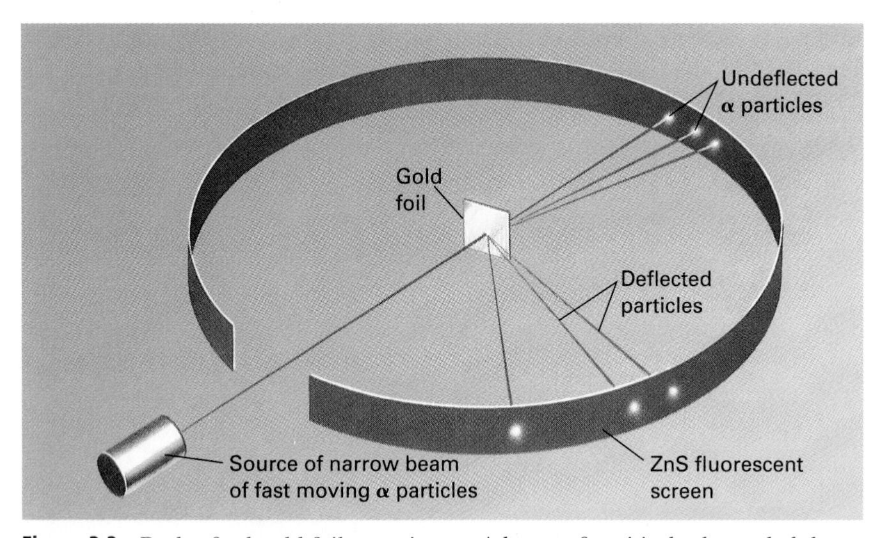

Figure 3.2 Rutherford gold foil experiment. A beam of positively charged alpha particles was directed at a very thin piece of gold foil. A luminescent screen was used to detect particles passing through or deflected by the foil. Most particles passed straight through. Some were deflected slightly, and a few were deflected back toward the source. (Rutherford actually used a movable luminescent screen instead of the circular screen shown.)

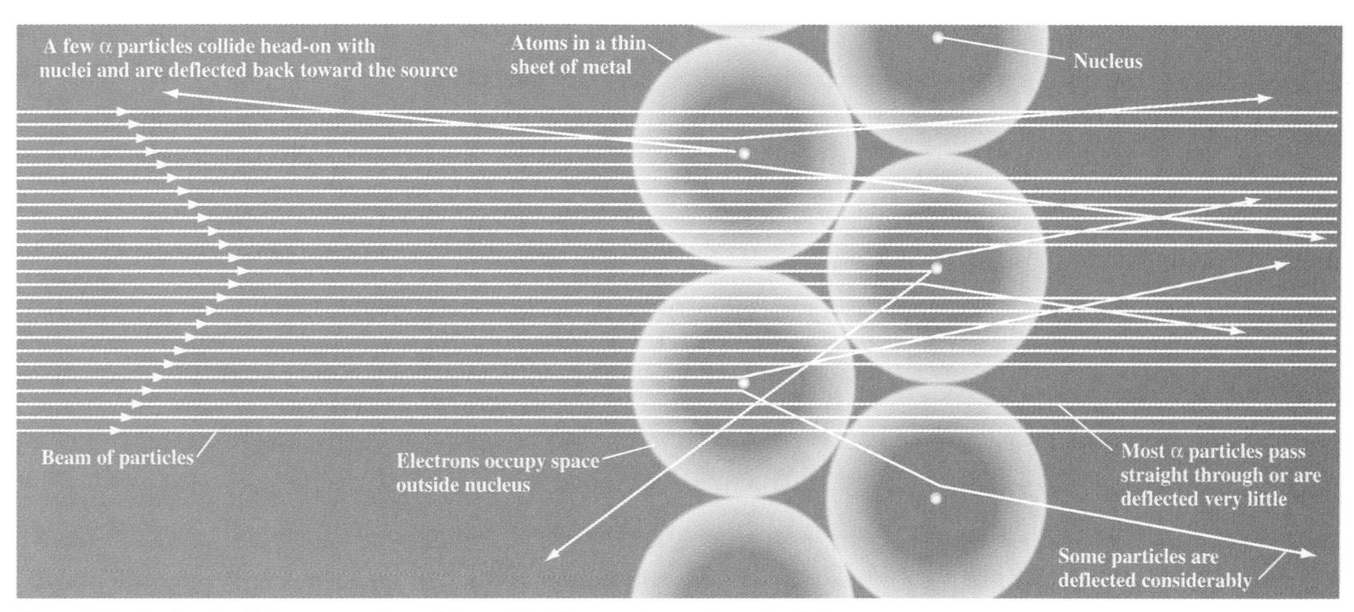

Figure 3.3 Rutherford's interpretation of the gold foil experiment done by Geiger and Marsden. Each circle represents an atom and the dots represent their nuclei. The gold foil was about 1000 atoms thick.

THE PERSONAL SIDE

Ernest Rutherford (1871–1937)

Lord Ernest Rutherford was born in New Zealand in 1871, but went to Cambridge University in England to pursue his Ph.D. in physics in 1895. His original interest was in a phenomenon that we now call radio waves, and he apparently hoped to make his fortune in the field, largely so he could marry his fiancée back in New Zealand. However, his professor at Cambridge, J. J. Thomson, convinced him to work on the newly discovered phenomenon of radioactivity. Rutherford discovered alpha and beta radiation while at Cambridge. In 1899, he moved to McGill University in Canada where he did further experiments to prove that alpha radiation is actually composed of helium nuclei and that beta radiation consists of electrons. For this work he received the Nobel Prize in chemistry in 1908.

Ernest Rutherford (right) *with a colleague in the Cavendish Laboratory at Cambridge University. The sign over Rutherford's head in this photo, taken in 1935, is said to have been directed at the booming voice of Rutherford himself.* (C. E. Wynn-Williams. Courtesy AIP/Emilio Segré Visual Archives)

In 1903, Rutherford and his young wife visited Pierre and Marie Curie in Paris, on the very day that Madame Curie received her doctorate in physics. That evening during a party in the garden of the Curies' home, Pierre Curie brought out a tube coated with a phosphor and containing a large quan-

(continued on next page)

THE PERSONAL SIDE *(continued)*

tity of radioactive radium in solution. The phosphor glowed brilliantly from the radiation given off by the radium. Rutherford later said that the light was so bright he could clearly see Pierre Curie's hands were "in a very inflamed and painful state due to exposure to radium rays."

In 1907, Rutherford moved from Canada to Manchester University in England, and there he performed the experiments that gave us the modern view of the atom. In 1919, he moved to Cambridge and assumed the position formerly held by J. J. Thomson. Not only was Rutherford responsible for very important work in physics and chemistry, but also he guided the work of no fewer than ten future recipients of the Nobel Prize.

the atom must be concentrated in a very small volume at the center of the atom. He named this part of the atom containing most of the mass of the atom and all of the positive charge the **nucleus**. When an alpha particle passes near the nucleus, the positive charge of the nucleus repels the positive charge of the alpha particle; the path of the smaller alpha particle is consequently deflected. The closer an alpha particle comes to a target nucleus, the more it is deflected. Those alpha particles that meet a nucleus head-on bounce back toward the source as a result of the strong positive–positive repulsion, since the alpha particles do not have enough energy to penetrate the nucleus.

Rutherford's calculations, based on the observed deflections, indicate that the nucleus is a very small part of an atom. The diameter of an atom is about 100,000 times greater than the diameter of its nucleus.

Truly, Rutherford's model of the atom was one of the most dramatic interpretations of experimental evidence to come out of this period of significant discoveries.

SELF-TEST 3A

1. Atoms were first proposed (a) by early British philosophers, (b) by early Greek philosophers, (c) in the early 1900s, (d) by John Dalton.
2. The Greek approach to the "discovery" of atoms can best be described as (a) experimentation, (b) philosophy (use of logic), (c) direct observation of atoms, (d) consistent explanation of well-known, established laws of nature.
3. The law of conservation of matter states that matter is neither lost nor ____________ in a ____________ reaction.
4. The composition of sulfur dioxide (SO_2) is 32 parts by mass S and 32 parts by mass O. What is the percentage of sulfur in SO_2?
5. Hydrogen peroxide, H_2O_2, is always 94.12% O. What is the percentage of H in hydrogen peroxide?

6. According to Dalton's atomic theory, a compound has a definite percentage by mass of each element because (a) all atoms of a given element weigh ______________, and (b) all molecules of a given compound contain a definite number and kind of ______________.
7. If the law of definite proportion is true, will the percent mass of silver, Ag, in silver sulfide be the same for all lumps or pieces of Ag_2S? Explain your answer. (a) Yes, (b) No.
8. Natural gas is essentially methane (CH_4). Will methane produced from Texas gas fields have the same composition as methane produced by gas fields in China? Explain your answer. (a) Yes, (b) No.
9. What is the charge on the proton? (a) -1, (b) $+1$, (c) 3.
10. A beta particle is a high-energy (a) neutron, (b) electron, (c) proton, (d) helium nucleus.
11. Ernest Rutherford proposed the modern nuclear model of the atom. (a) True, (b) False.
12. Most of the mass of an atom is concentrated in its (a) nucleus, (b) electrons, (c) protons.
13. The mass of the proton is ______________ times the mass of the electron.

3.3 Modern View of the Atom

Early experiments on the structure of the atom clearly showed that the three primary constituents of atoms are electrons, protons, and neutrons. The nucleus or core of the atom is made up of protons with positive electrical charge and neutrons with no charge. The electrons, with a negative electrical charge, are found in the space around the nucleus (Figure 3.4). For an atom, which has no net electrical charge, *the number of negatively charged electrons around the nucleus equals the number of positively charged protons in the nucleus.*

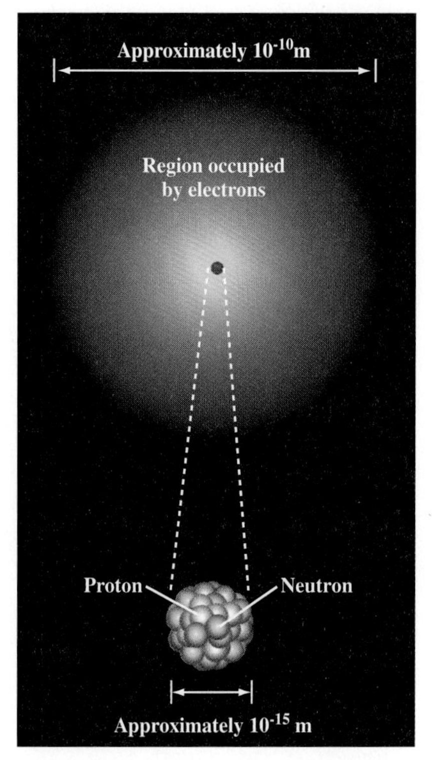

Figure 3.4 Model of atom. All atoms consist of one or more protons (positively charged) and usually at least as many neutrons (no charge) packed into an extremely small nucleus. Electrons (negatively charged) are arranged in a cloud around the nucleus.

Atoms are extremely small, far too small to be seen with even the most powerful optical microscopes. The diameters of most atoms range from 1×10^{-8} cm to 5×10^{-8} cm. For example, the diameter of a carbon atom is 1.5×10^{-8} cm. To visualize how small this is, take a sharp pencil and draw a line 3 cm long (________________). Graphite is made up of carbon atoms, and the 3-cm line of carbon contains 200 million (200,000,000) carbon atoms from end to end and about a million atoms across (the width of the line). If this weren't hard enough to imagine, remember that Rutherford's experiments provided evidence that the diameter of the nucleus is 100,000 times smaller than the diameter of the atom. For example, if an atom were scaled upward in size so that the nucleus were the size of a small marble, the atom would be the size of the Houston Astrodome, and most of the space in between would be empty. Because the nucleus carries most of the mass of the atom in such a small volume, a matchbox full of atomic nuclei would weigh more than 2.5 billion tons. The interior of a collapsed star is made up of nuclear material that is estimated to be nearly this dense.

Scientists have been able to obtain computer-enhanced images of the outer surface of atoms (Figure 3.5 on p. 47) using the scanning tunneling microscope (STM) and the atomic force microscope (AFM).

THE WORLD OF CHEMISTRY

The Scanning Tunneling Microscope

Program 6, *The Atom*

Chemists and physicists had more than ample evidence of the existence of atoms before the invention of the scanning tunneling microscope (STM). They were able to "see" atoms through a large variety of phenomena, but to say that they saw atoms had a special meaning. What they were seeing by such techniques as X-ray diffraction was a manifestation of many atoms and a composite picture created by the scattering of X rays from many planes of a crystal.

Yet chemists and physicists have always dreamed of being able to see individual atoms directly, that is, of being able to produce images with a direct correspondence to the atom's actual position in the sample.

These dreams began to be realized in the 1950s. An early and spectacular effort was reported by Erwin Mueller using a field ion microscope that he invented, which made it possible to image individual atoms on a crystal's surface. The even more remarkable STM not only makes it possible to see individual atoms and how they are arranged on a surface, but also permits the study of atom migration and atomic dislocations on surfaces. The development of the STM is considered an event of such magnitude that its developers, Gerd Binnig and Heinrich Rohrer of IBM's Zurich Research Laboratory in Switzerland, received the Nobel Prize in physics in 1986. The STM is an astonishing device because of its inherent simplicity. It consists of a tungsten needle, hardly more than a single atom wide at the end. When this needle is lowered to within a few atoms' thickness of the surface to be imaged and a small voltage is applied, electrons tunnel; that is, they pass from the tungsten atom into the electron clouds of the atoms on the surface and produce a measurable current. By adjusting the up–down position of the tungsten needle as it moves across the surface, a constant tunneling current is maintained. As this takes place, however, the positions of the atoms are actually measured, giving a picture of the atomic landscape.

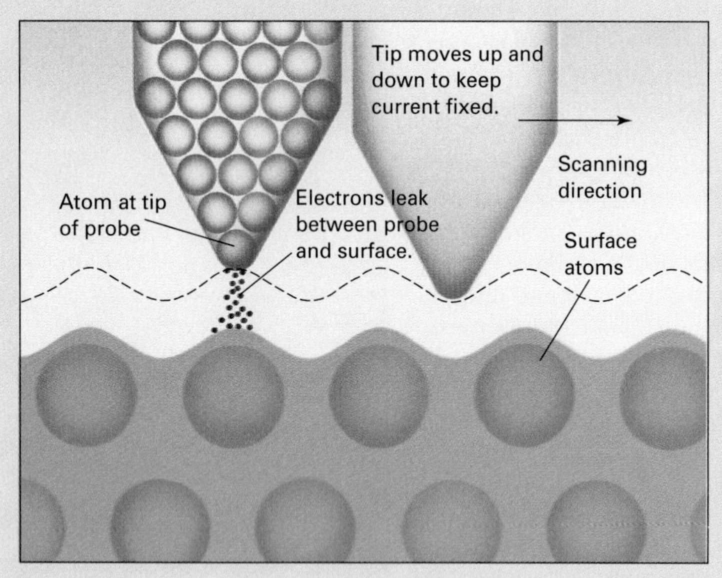

(a)

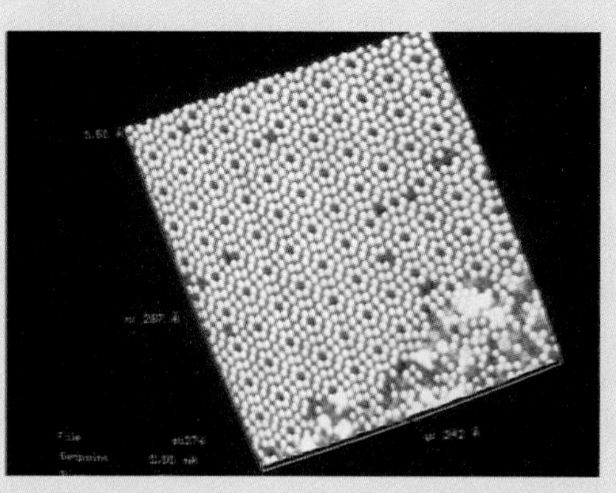

(b)

Scanning tunneling microscope (STM). (a) When an electric current passes through a tungsten needle with a narrow tip (atom's width) into the atoms on the surface of the sample being examined, the electron flow between the tip and the surface changes in relation to the electron clouds around the atoms. By adjusting the position of the needle to maintain a constant current, the positions of the atoms are measured. (b) STM image of atoms of silicon. (b, John Ozcomert/ Michael Trenary)

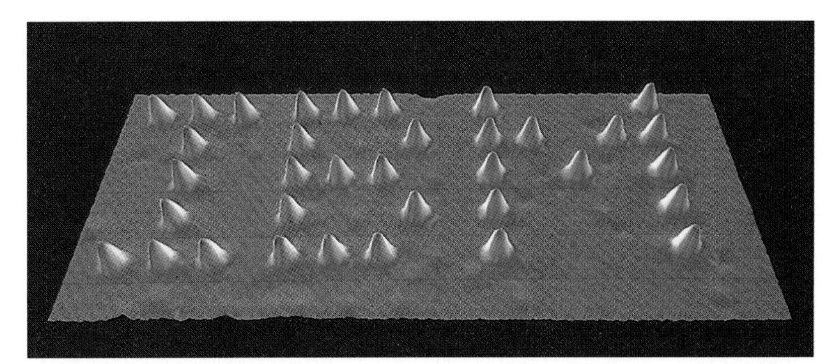

Figure 3.5 IBM spelled out in xenon (Xe) atoms. A few years after invention of the scanning tunneling microscope, scientists discovered that not only could they take pictures of atoms on a surface, but also they could push them around with the tip of the microscope. They created this picture to demonstrate what they could do. The "IBM" is 600 billionths of an inch, or 16.8 nanometers wide. *(IBM Corporation Research Division, Almaden Research Center)*

Atomic Number—Each Element Has a Number

The **atomic number** of an element indicates the number of protons in the nucleus of the atom. All atoms of the same element have the same number of protons in the nucleus. In the periodic table (inside front cover) the atomic number for each element is given above the element's symbol. Beginning with the atomic number 1 for hydrogen, there is a different atomic number for each element. Phosphorus, for example, has an atomic number of 15; thus, the nucleus of a phosphorus atom contains 15 protons.

In a neutral atom the number of protons is equal to the number of electrons; therefore, the atomic number also gives the number of electrons outside the nucleus in an atom of an element.

The World of Chemistry

Program 6, *The Atom*

Well the first time I saw atoms, it was about two or three o'clock in the morning. Of course, I was doing most of my work at night because the instrument itself is extremely sensitive to vibration, so most of our experiments are done at night or on weekends. And it was about two or three o'clock in the morning, and I was staring at this image, and I started to notice this regular pattern appearing, which I knew had to be the positions of the atoms, and it just got better and better, and I was just elated. Eventually, tears came to my eyes because I was so happy, having worked on this day and night for so long. Dr. Robert Hamers, research scientist at IBM.

Mass Number

The **mass number** of a particular atom is the total number of neutrons and protons present in the nucleus of an atom. Because the atomic number gives the number of protons in the nucleus, the difference between the mass number and the atomic number equals the number of neutrons in the nucleus.

A notation frequently used for showing the mass number and atomic number of an atom places subscripts and superscripts to the left of the symbol

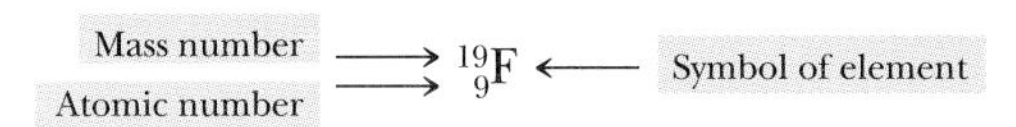

The subscript giving the atomic number is optional because the element symbol tells you what the atomic number must be. For example, the fluorine atom symbolized previously would have the notation $^{19}_{9}F$ or simply ^{19}F. For an atom of fluorine, $^{19}_{9}F$, the number of protons is 9, the number of electrons is also 9, and the number of neutrons is $19 - 9 = 10$.

EXAMPLE 3.1 *Atomic Composition*

How many protons, neutrons, and electrons are in an atom of gold (Au) with a mass number of 197?

SOLUTION

The atomic number of an element gives the number of protons and electrons. If the element is known, its atomic number can be obtained from a periodic table (or the alphabetical list inside the front cover). The atomic number of gold is 79. Gold has 79 protons and 79 electrons. The number of neutrons is obtained by subtracting the atomic number from the mass number.

$$197 - 79 = 118 \text{ neutrons}$$

Exercise 3.1
How many protons, neutrons, and electrons are there in a neutral $^{59}_{28}Ni$ atom?

Isotopes

■ Relative abundances of neon isotopes:

$^{20}_{10}Ne$	90.92%
$^{21}_{10}Ne$	0.26%
$^{22}_{10}Ne$	8.82%

When a natural sample of most elements is analyzed, the element is found to be composed of atoms with different mass numbers. Atoms of the same element having different mass numbers are called **isotopes** of that element.

The element neon is a good example to consider. A natural sample of neon gas is found to be a mixture of three isotopes of neon

$$^{20}_{10}Ne \quad ^{21}_{10}Ne \quad ^{22}_{10}Ne$$

The fundamental difference between isotopes is the different number of neutrons per atom. All atoms of neon have 10 electrons and 10 protons: about 90.92% of the atoms have 10 neutrons, 0.26% have 11, and 8.82% have 12. Because they have different numbers of neutrons, they must have different masses. Note that all the isotopes have the same atomic number. They are all neon.

■ To represent isotopes with words instead of symbols, the mass number is added to the name; for example, neon-20, neon-21, and neon-22.

Over 100 elements are known, yet more than 1000 isotopes have been identified, many of them produced artificially (see Section 4.5). Some elements have many isotopes; tin, for example, has ten natural isotopes. Hydrogen has three isotopes, and two of them are the only known isotopes generally referred to by different names: $^{1}_{1}H$ is commonly called hydrogen, $^{2}_{1}H$ is called deuterium, and $^{3}_{1}H$ is called tritium. Tritium is radioactive.

EXAMPLE 3.2 *Isotopes*

Carbon has seven known isotopes. Three of these have 6, 7, and 8 neutrons, respectively. Write the complete chemical notation for these isotopes, giving mass number, atomic number, and symbol.

SOLUTION

The atomic number of carbon is 6. The mass number for the three isotopes is 6 plus the number of neutrons, which equal 12, 13, and 14, respectively.

$$^{12}_{6}C \quad ^{13}_{6}C \quad ^{14}_{6}C$$

Exercise 3.2

Silver has two isotopes, one with 60 neutrons and the other with 62 neutrons. Give the complete chemical notation for these isotopes.

Atomic Masses and Atomic Weights

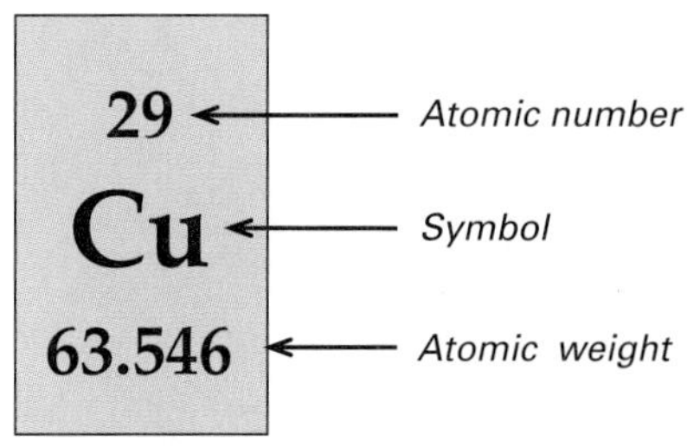

The periodic table entry for copper. (The complete periodic table and its significance are discussed in Section 3.6.)

Although Dalton knew nothing about subatomic particles, he proposed that atoms of different elements have different masses. Eventually, it was found that an oxygen atom is about 16 times heavier than a hydrogen atom. This fact, however, does not tell us the mass of either atom. These are relative masses in the same way that a grapefruit may weigh twice as much as an orange. This information gives neither the mass of the grapefruit nor that of the orange. Nevertheless, if a specific number is assigned as the mass of any particular atom, this fixes the numbers assigned to the masses of all other atoms. The present atomic mass scale, adopted by scientists worldwide in 1961, is based on assigning the mass of a particular isotope of the carbon atom, the carbon-12 isotope, as exactly 12 **atomic mass units (amu)**.

■ The definition of *amu* is 1/12 the mass of the carbon-12 isotope which is exactly 12.000 amu.

The atomic masses are given in the periodic table (inside front cover) below the symbol for the element, as illustrated in the margin for copper. These values are average masses, which take into account the relative abundances of the different isotopes as found in nature. This average is often referred to as the **atomic weight** of the element. For example, boron has two naturally occurring isotopes, $^{10}_{5}B$ and $^{11}_{5}B$, with natural percent abundances of 19.91% and 80.09%. The masses in amu are 10.0129 and 11.0093, respectively. The atomic weight listed in the periodic table is the average mass of a natural sample of atoms, expressed in atomic mass units. The average mass of 10.81 for B is closer to the mass of $^{11}_{5}B$ because of its higher abundance.

■ Chemists usually use the term "atomic weight" of an element instead of "atomic mass" when they are referring to a naturally occurring sample of an element. Although the quantity is more properly called a mass than a weight, the term "atomic weight" is so commonly used that it has become accepted. Atomic weights are established by exacting experiments. The same values are used by all scientists because the relative abundances of the elements are essentially the same everywhere on our planet.

■ SELF-TEST 3B

1. If an atom has an atomic number of ten, then it has ________ protons and ________ electrons. If its mass number is 21, then it has ________ neutrons.
2. In the symbol $^{80}_{35}Br$, the number 35 is the ________, and the number 80 is the ________.
3. Isotopes of an element are atoms that have nuclei with the same number of ________ but different numbers of ________.
4. The negatively charged particles in an atom are ________; the positively charged particles are ________; the neutral particles are ________.
5. In a neutral atom there are equal numbers of ________ and ________.

6. The number of protons per atom is called the ________ number of the element.
7. An atom of arsenic, $^{75}_{33}As$, has ________ electrons, ________ protons, and ________ neutrons.
8. The diameter of the nucleus is ________ smaller than the diameter of the atom.
9. Draw the nuclei of $^{1}_{1}H$, $^{2}_{1}H$, and $^{3}_{1}H$, representing protons and neutrons with circles.
10. The nuclei of all helium atoms contain exactly 2 protons. (a) True, (b) False.
11. All the isotopes of an element have the same relative abundance. (a) True, (b) False.

3.4 Where Are the Electrons in Atoms?

Experiments on the interaction of light with matter provide important information about the energy and location of electrons in atoms.

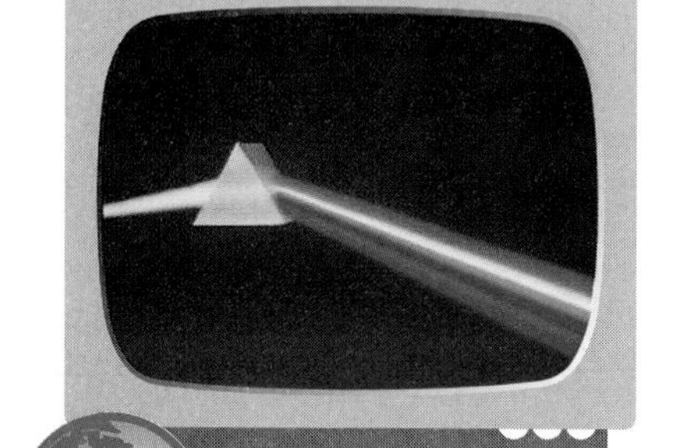

The World of Chemistry

Program 2, *Color*

Spectrum of white light produced by refraction in a glass prism. The different colors blend into one another smoothly.

Continuous and Line Spectra

We are familiar with the spectrum of colors that makes up visible light. The spectrum of white light is a display of separated colors. This type of spectrum is referred to as a **continuous spectrum** and is obtained by passing sunlight or light from an incandescent light bulb through a glass prism. When we see a rainbow, we are looking at a continuous visible spectrum formed when raindrops act as prisms and disperse the sunlight. The different colors of light have different wavelengths. Red light has longer wavelengths than does blue light, but in the continuous spectrum the colors merge from one to another with no break in the spectrum. There are literally millions of colors. White light is a combination of all the colors of different wavelengths.

If a high voltage is applied to an element in the gas phase in a partially evacuated tube, the atoms absorb energy and are said to be "excited." The excited atoms emit light. An example of this is a neon advertising sign, in which excited neon atoms emit orange-red light (Figure 3.6*a*). When light

(a)

(b)

Figure 3.6 Neon. (a) Partially evacuated tube that contains neon gas gives a reddish-orange glow when high voltage is applied. (b) Line emission spectrum of neon is obtained when light from a neon source passes through a prism. *(a, Grant Heilman, Runk/Schoenberger)*

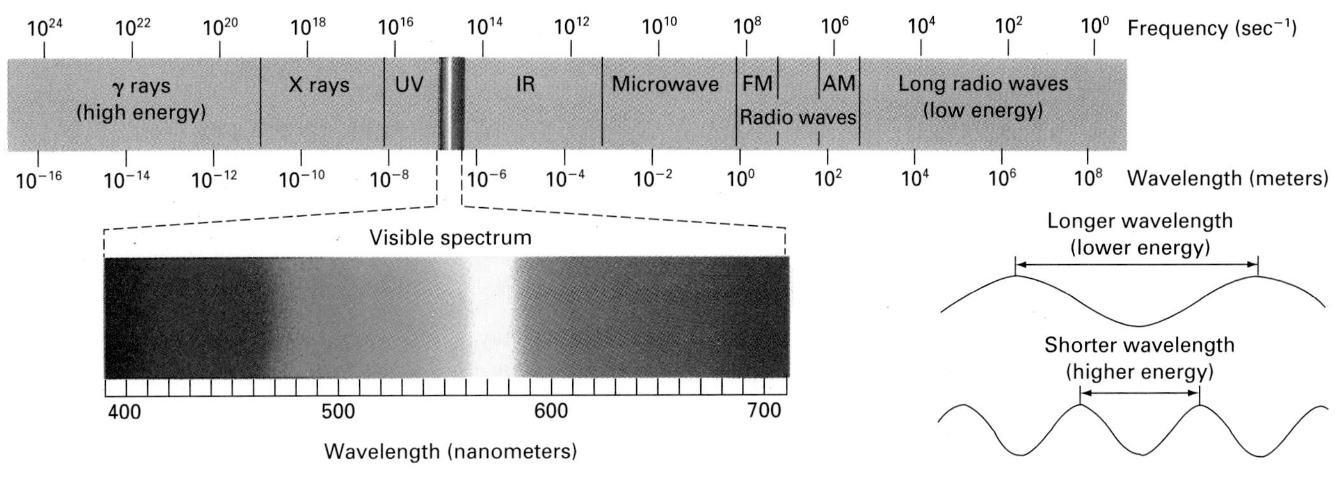

Figure 3.7 The electromagnetic spectrum. Visible light (enlarged section) is but a small part of the entire spectrum. The energy of electromagnetic radiation increases from the radio wave end to the gamma ray end. The frequency of electromagnetic radiation is related to the wavelength by $\nu\lambda = c$ where ν = frequency; λ = wavelength; and c = speed of light, 3.00×10^8 meters (m)/second (s). The higher the frequency, the lower the wavelength and the larger the energy.

from such a source passes through a prism, a different type of spectrum is obtained, one that is not continuous but has characteristic lines at specific wavelengths (Figure 3.6*b*). This type of spectrum is called a line emission spectrum.

Visible light is only a small portion of the electromagnetic spectrum (Figure 3.7). Ultraviolet radiation, the type that leads to sunburn and some forms of skin cancer, has wavelengths shorter than those of visible light; X rays and gamma rays (the latter emitted from radioactive atoms) have even shorter wavelengths. Infrared radiation, the type that is sensed as heat from a fire, has longer wavelengths than visible light. Longer still are the wavelengths of the types of radiation in a microwave oven and in television and radio transmissions. Although the example of a line spectrum shown here is only for the visible region, excited atoms of elements also emit characteristic wavelengths in other regions of the electromagnetic spectrum, as demonstrated by the experiments described in the next section.

Bohr Model of the Atom

In 1913, Niels Bohr introduced his model of the hydrogen atom. He proposed that the single electron of the hydrogen atom could occupy only certain energy levels. He referred to these energy levels as orbits and represented the energy difference between any two adjacent orbits as a single **quantum** of energy. When the hydrogen electron absorbs a quantum of energy, it moves to a higher energy level. When this electron returns to the lower, more stable energy level, the quantum of energy is emitted as a specific wavelength of light (Figure 3.8).

■ Bohr assumed that atoms can exist only in certain energy states.

■ In 1900 Max Planck proposed that energy is not continuous, but comes in discrete "bundles" or "packets" called **quanta**. Something that can have only certain values, with none in between, is referred to as *quantized*.

■ Bohr used the term "orbits," but the modern equivalents of his orbits are called shells or energy levels.

In Bohr's model, each allowed orbit is assigned an integer, n, known as the principal quantum number. The values of n for the orbits range from 1 to infinity. The radii of the circular orbits increase as n increases. The orbit of lowest energy, with $n = 1$, is closest to the nucleus, and the electron of the hydrogen atom is normally in this energy level. Any atom with its electrons in their normal, lowest energy levels is said to be in the **ground state**. Energy must be supplied to move the electron farther away from the nucleus because the positive nucleus and the negative electron attract each other. When the electron of a hydrogen atom occupies an orbit with n greater than 1, the atom has more energy than in its ground state and is said to be in an **excited state**. The excited state is an unstable state, and the extra energy is emitted when the electron returns to the ground state. According to Bohr, the light forming the lines in the bright-line emission spectrum of hydrogen comes from electrons moving toward the nucleus after having first been excited to orbits farther from the nucleus (Figure 3.8). Because the orbits have only certain energies, the emitted light has only certain wavelengths.

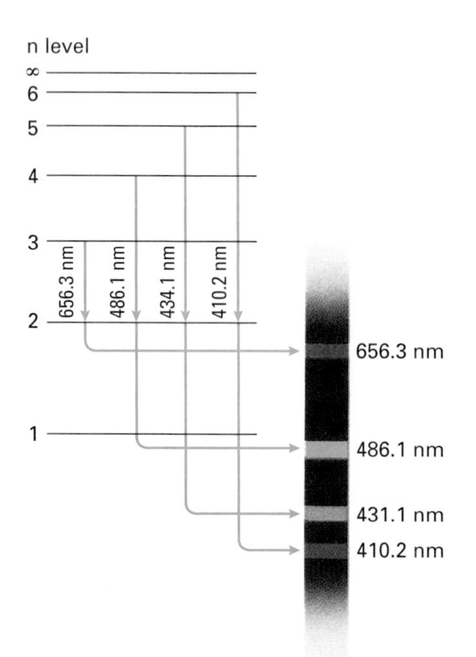

Figure 3.8 Some of the electronic transitions that can occur in an excited H atom. The lines in the visible region result from transitions from levels with values of n greater than 2 to $n = 2$.

With brilliant imagination, Bohr applied a little algebra and some classic mathematical equations of physics to his tiny solar system model of the hydrogen atom. Bohr was able to calculate the wavelengths of the lines in the hydrogen spectrum. By 1900, scientists had measured the wavelengths of lines for hydrogen in the ultraviolet, visible, and infrared regions, and Bohr's calculated values agreed with the measured values. Niels Bohr had tied the unseen (the interior of the atom) with the seen (the observable lines in the

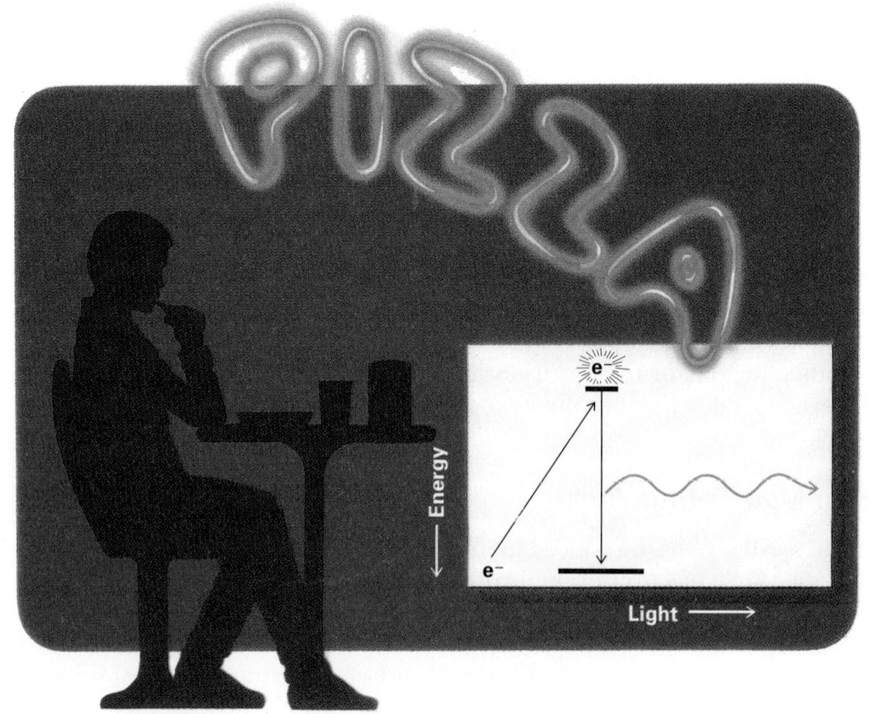

A neon sign and the source of its light—excited electrons losing their extra energy.

THE PERSONAL SIDE

Niels Bohr (1885–1962)

Niels Bohr. (American Institute of Physics)

Niels Bohr was born in Copenhagen, Denmark. He earned a Ph.D. in physics in Copenhagen in 1911 and then went to work first with J. J. Thomson in Cambridge, England, and later with Ernest Rutherford in Manchester, England. It was there that he began to develop the ideas that a few years later led to the publication of his theory of atomic structure and his explanation of atomic spectra. He received the Nobel Prize in 1922 for this work. After being with Rutherford for a very short time, Bohr returned to Copenhagen, where he eventually became the director of the Institute of Theoretical Physics.

Bohr was still in Denmark when Hitler's army suddenly invaded the country in 1940. In 1943, to avoid imprisonment, he escaped to Sweden. There he helped to arrange the escape of nearly every Danish Jew from Hitler's gas chambers. He was later flown to England in a tiny plane, in which he passed into a coma and nearly died from lack of oxygen.

He went on to the United States, where until 1945 he worked with other physicists on the atomic bomb development at Los Alamos, New Mexico. His insistence on sharing the secret of the atomic bomb with other allies, to have international control over nuclear energy, so angered Winston Churchill that he had to be restrained from ordering Bohr's arrest. Bohr worked hard and long on behalf of the development and use of atomic energy for peaceful purposes. For his efforts, he was awarded the first Atoms for Peace Prize in 1957. He died in Copenhagen on November 18, 1962.

hydrogen spectrum)—a fantastic achievement. The concepts of quantum number and energy level are valid for all atoms and molecules.

The Bohr model was accepted almost immediately after its presentation, and Bohr was awarded the Nobel Prize in physics in 1922 for his contribution to the understanding of the hydrogen atom. However, his model gave only approximate agreement with line spectra of atoms having more than one electron. Later models of the atom have been more successful by considering electrons as having both particle and wave characteristics. This led to mathematical treatment of the locations of electrons as probabilities instead of as the precise locations envisioned by Bohr. Thus, in the modern view of the atom, we can picture a space around the nucleus occupied by each electron, somewhat like a cloud of electrical charge of a particular energy. We just don't know where within the cloud a particular electron is at any given instant. Bohr's concept of the main energy levels represented by the quantum number n remains valid, however; and for our purposes, this is all that we need to discuss the location of electrons in atoms.

Level	$2n^2$	Maximum Number of Electrons
Level 1	$2(1)^2$	2
Level 2	$2(2)^2$	8
Level 3	$2(3)^2$	18
Level 4	$2(4)^2$	32
Level 5	$2(5)^2$	50
Level 6	$2(6)^2$	72

Atom Building Using the Bohr Model

Recall that the atomic number is the number of electrons (or protons) per atom of an element. Imagine building atoms by adding one electron to the appropriate energy level as another proton is added to the nucleus. As part of his theory, Bohr proposed that only a fixed number of electrons could be accommodated in any one orbit; and he calculated that this number was given by the formula $2n^2$, where n equals the number of the orbit, or energy level. For the lowest energy level (first orbit), n equals 1, and the maximum number of electrons allowed is $2(1)^2$, or 2. For the second energy level, the maximum number of electrons is $2(2)^2$, or 8. By using $2n^2$, the maximum number of electrons allowed for levels 3, 4, and 5 are 18, 32, and 50, respectively. A general overriding rule to the preceding numbers is that the outside energy level can have no more than eight electrons for a stable atom. Table 3.2 lists the Bohr electron arrangements for the first 20 elements.

Electrons in the highest occupied energy level listed for the elements in Table 3.2 are at the greatest stable distance from the nucleus. These are the most important electrons in the study of chemistry because they are the ones that interact when atoms react with each other. This important observation was first proposed by G. N. Lewis who, independent of Bohr, conceived of the

TABLE 3.2 ■ Electron Arrangements of the First 20 Elements*

Element	Atomic Number	Number of Electrons in Each Energy Level			
		1st	*2nd*	*3rd*	*4th*
Hydrogen (H)	1	1 e			
Helium (He)	2	2 e			
Lithium (Li)	3	2 e	1 e		
Beryllium (Be)	4	2 e	2 e		
Boron (B)	5	2 e	3 e		
Carbon (C)	6	2 e	4 e		
Nitrogen (N)	7	2 e	5 e		
Oxygen (O)	8	2 e	6 e		
Fluorine (F)	9	2 e	7 e		
Neon (Ne)	10	2 e	8 e		
Sodium (Na)	11	2 e	8 e	1 e	
Magnesium (Mg)	12	2 e	8 e	2 e	
Aluminum (Al)	13	2 e	8 e	3 e	
Silicon (Si)	14	2 e	8 e	4 e	
Phosphorus (P)	15	2 e	8 e	5 e	
Sulfur (S)	16	2 e	8 e	6 e	
Chlorine (Cl)	17	2 e	8 e	7 e	
Argon (Ar)	18	2 e	8 e	8 e	
Potassium (K)	19	2 e	8 e	8 e	1 e
Calcium (Ca)	20	2 e	8 e	8 e	2 e

* Valence electrons are shown in color.

idea that electrons in atoms might be arranged in concentric shells, with the nucleus at the center. He proposed that each shell could hold a characteristic number of electrons, and only those electrons in the outermost shell were involved when one atom combined with another to form a chemical compound. These outermost electrons came to be known as **valence electrons.**

For example, look at phosphorus (P). The Bohr arrangement of electrons is 2-8-5. This means that the stable state of the phosphorus atom has electrons in three energy levels. The one closest to the nucleus has two electrons; the second energy level has eight electrons; and the highest energy level has five electrons. The energy level with five electrons is farthest from the nucleus; thus, P has *five valence electrons,* and these are the electrons that are most available for interactions with valence electrons of other atoms in chemical reactions. This interaction results in the formation of bonds, which will be discussed in Chapter 5.

SELF-TEST 3C

1. Which has less energy? (a) Blue light, (b) Red light, (c) Ultraviolet light, (d) Infrared light.
2. Which color of light has the shortest wavelength? (a) Blue, (b) Green, (c) Orange, (d) Red.
3. According to Bohr's theory, light of characteristic wavelength is emitted as an electron drops from an energy level (closer to/farther from) the nucleus to an energy level (closer to/farther from) the nucleus.
4. The maximum number of the electrons in $n = 3$ energy level is ____________.
5. The ground-state Bohr representation for electrons in an atom of K is ____________. K has ____________ valence electron(s).
6. The ground-state Bohr representation of electrons in an atom of Cl is ____________. Cl has ____________ valence electron(s).

3.5 Development of the Periodic Table

So far, in exploring the chemical view of matter, you have seen that everything is made of atoms. Atoms of different elements combined in specific ratios are present in chemical compounds; and all matter consists of elements, compounds, or mixtures of elements or compounds. You have also been introduced to the subatomic composition of atoms—the electrons, protons, and neutrons. The next part of the story is to see how atomic structure and the properties of elements and compounds are related to each other. Fortunately, the chemistry of the elements can be organized and classified in a way that has helped both chemists and nonchemists. The periodic table is the single most important classification system in chemistry because it summarizes, correlates, and predicts a wealth of chemical information. Chemists consult it every day during every possible kind of work. It can be simply a reminder of the symbols and names of the elements, of which elements have similar properties, and of where each element lies on the continuum of atomic numbers. It can also be an inspiration in the search for new compounds or mixtures that will fulfill a specific need. Memorizing the periodic table is no more

Dimitri Mendeleev. *(Oesper Collection in the History of Chemistry, University of Cincinnati)*

necessary than memorizing the map of your home state. Nevertheless, in both cases, it's very helpful to have a general idea of the major features.

On the evening of February 17, 1869, at the University of St. Petersburg in Russia, a 35-year-old professor of general chemistry—Dmitri Ivanovich Mendeleev (1834–1907)—was writing a chapter of his soon-to-be-famous textbook on chemistry. He had the properties of each element written on cards, with a separate card for each element. While he was shuffling the cards trying to gather his thoughts before writing his manuscript, Mendeleev realized that if the elements were arranged in the order of their atomic weights, there was a trend in properties that repeated itself several times. He arranged the elements into groups that had similar properties and used the resulting periodic chart to predict the properties and places in the chart of as yet undiscovered elements.

Thus, the periodic law and table were born, although only 63 elements had been discovered by 1869 (e.g., the noble gases were not discovered until after 1893), and the clarifying concept of the atomic number was not known until 1913. Mendeleev's idea and textbook achieved great success, and he rose to a position of prestige and fame as he continued to teach at the University of St. Petersburg.

Mendeleev aided the discovery of new elements by predicting their properties with remarkable accuracy, and he even suggested the geographical regions in which minerals containing the elements could be found. The properties of a missing element were predicted by consideration of the properties of its neighboring elements in the table. For example, for the element we now know as germanium, which falls below silicon in the modern periodic table (Figure 3.9), Mendeleev predicted a gray element of atomic weight 72 with a density of 5.5 g/cm^3. Germanium, once discovered, proved to be a gray element of atomic weight 72.59 with a density of 5.36 g/cm^3.

The empty spaces in the table and Mendeleev's predictions of the properties of missing elements stimulated a flurry of prospecting for elements in the 1870s and 1880s. As a result, in addition to gallium and germanium, scandium (Sc), samarium (Sm), holmium (Ho), and thulium (Tm) were discovered in 1879; gadolinium (Gd), in 1880; neodymium (Nd) and praseodymium (Pr), in 1885; and dysprosium (Dy), in 1886. Many of these elements are not common even today, yet they are important as ingredients in catalysts and color television screens.

Mendeleev found that a few elements did not fit under other elements with similar chemical properties when arranged according to increasing atomic weight. Eventually, it was found that the *atomic weight is not the property that governs the similarities and differences among the elements.* This was discovered in 1913 by H. G. J. Moseley (1888–1915), a young scientist working with Ernest Rutherford. Moseley found that the wavelengths of X rays emitted by a particular element are related in a precise way to the atomic number of that element. He quickly realized that other atomic properties may be similarly related to atomic number and not, as Mendeleev had believed, to atomic weight.

By building on the work of Mendeleev and others, and by using the concept of the atomic number, we are now able to state the modern **periodic law**: *When elements are arranged in the order of their atomic numbers, their chemical and physical properties show repeatable, or periodic, trends.* Other familiar periodic phenomena include the average daily temperature, which is periodic with time in a temperature climate. A shingle roof has the same pattern over and over and is, therefore, periodic.

Thus, to build up a periodic table according to the periodic law, the elements are lined up in a horizontal row in the order of their atomic numbers, as in Figure 3.9. At an element with similar properties to one already in the row, a new row is started. The columns then contain elements with similar properties. Some chemical and physical properties of the first 20 elements are summarized in Table 3.3. Do you see any trends and similarities among the elements in Table 3.3? For example, lithium (Li) is a soft metal with low density that is very reactive. It combines with chlorine gas to form lithium

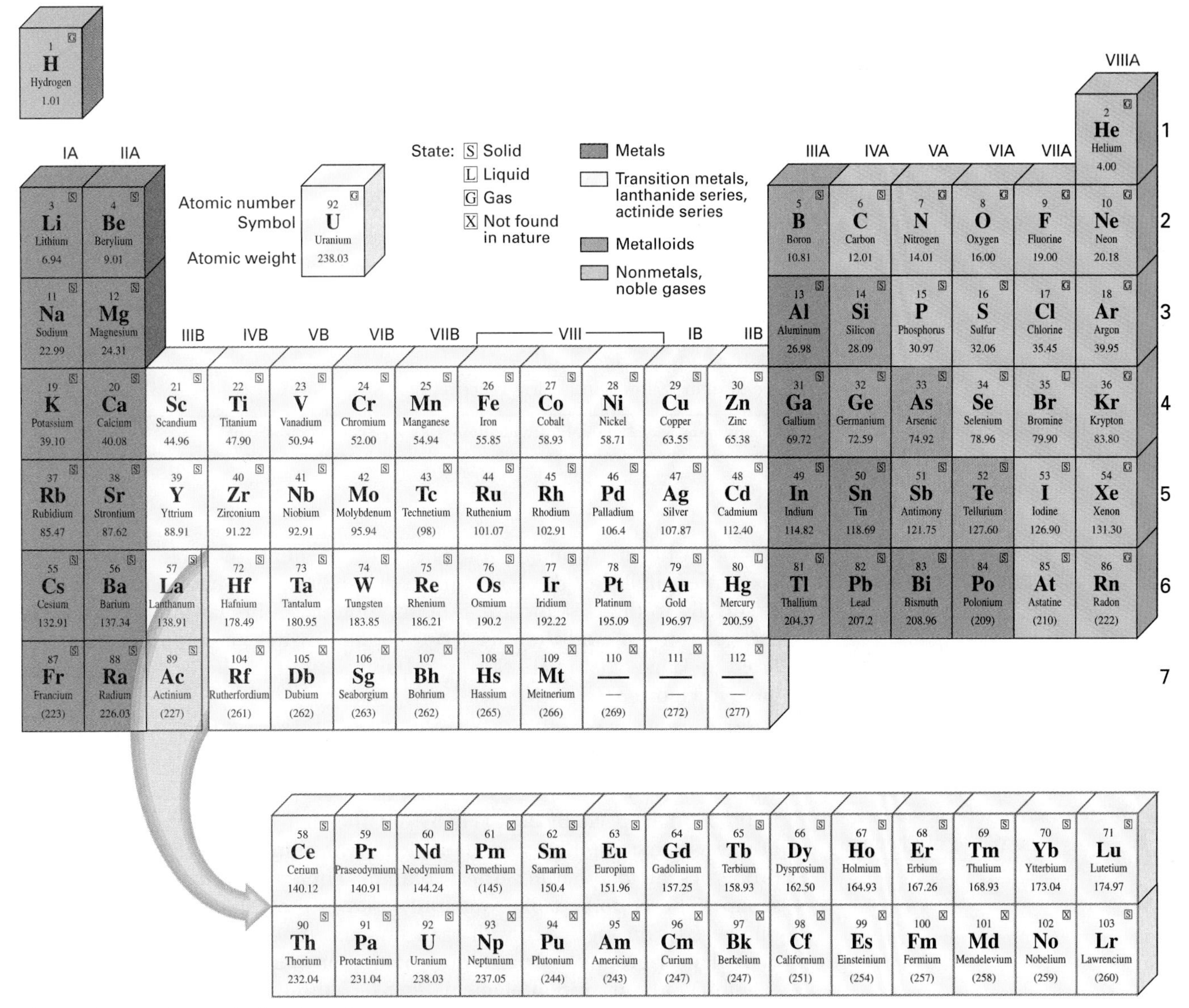

Figure 3.9 Modern periodic table of elements.

TABLE 3.3 ■ Some Properties of the First 20 Elements

Element	Atomic Number	Description	Compound Formation* With Cl (or Na)	With O (or Mg)
Hydrogen (H)	1	Colorless gas, reactive	HCl	H_2O
Helium (He)	2	Colorless gas, unreactive	None	None
Lithium (Li)	3	Soft metal, low density, very reactive	LiCl	Li_2O
Beryllium (Be)	4	Harder metal than Li, low density, less reactive than Li	$BeCl_2$	BeO
Boron (B)	5	Both metallic and nonmetallic, very hard, not very reactive	BCl_3	B_2O_3
Carbon (C)	6	Brittle nonmetal, unreactive at room temperature	CCl_4	CO_2
Nitrogen (N)	7	Colorless gas, nonmetallic, not very reactive	NCl_3	N_2O_5
Oxygen (O)	8	Colorless gas, nonmetallic, reactive	Na_2O, Cl_2O	MgO
Fluorine (F)	9	Greenish-yellow gas, nonmetallic, extremely reactive	NaF, ClF	MgF_2, OF_2
Neon (Ne)	10	Colorless gas, unreactive	None	None
Sodium (Na)	11	Soft metal, low density, very reactive	NaCl	Na_2O
Magnesium (Mg)	12	Harder metal than Na, low density, less reactive than Na	$MgCl_2$	MgO
Aluminum (Al)	13	Metal as hard as Mg, less reactive than Mg	$AlCl_3$	Al_2O_3
Silicon (Si)	14	Brittle nonmetal, not very reactive	$SiCl_4$	SiO_2
Phosphorus (P)	15	Nonmetal, low melting point, white solid, reactive	PCl_3	P_2O_5
Sulfur (S)	16	Yellow solid, nonmetallic, low melting point, moderately reactive	Na_2S, SCl_2	MgS
Chlorine (Cl)	17	Green gas, nonmetallic, extremely reactive	NaCl	$MgCl_2$, Cl_2O
Argon (Ar)	18	Colorless gas, unreactive	None	None
Potassium (K)	19	Soft metal, low density, very reactive	KCl	K_2O
Calcium (Ca)	20	Harder metal than K, low density, less reactive than K	$CaCl_2$	CaO

* The chemical formulas shown are lowest ratios. The molecular formula for $AlCl_3$ is Al_2Cl_6, and that for P_2O_5 is P_4O_{10}.

The World of Chemistry

Program 2, *Color*

An assortment of pure elements.

chloride with the formula LiCl. The other elements in Table 3.3 that have properties similar to those of lithium are sodium (Na) and potassium (K). According to the periodic law, lithium, sodium, and potassium should be in the same group; and they are. Look for similarities among other elements listed in Table 3.3 and check your grouping with that shown in the periodic table in Figure 3.9.

3.6 The Modern Periodic Table

In the modern **periodic table**, Figure 3.9, the elements are arranged in order of their atomic numbers so that elements with similar chemical and physical properties fall together in vertical columns. These vertical columns are called **groups**. The periodic table commonly used in the United States has groups numbered I through VIII, with each Roman numeral followed by a letter A or B. The A groups are the **representative** or **main-group** elements. The B groups are the **transition elements** that link the two areas of representative elements. The **inner transition elements** are the **lanthanide series** and the **actinide series**. They are placed at the bottom of the periodic chart because the similarity of properties within the two series would require their placement between lanthanum and hafnium (lanthanide series) and between actinium and rutherfordium (actinide series).

The horizontal rows are called **periods**. These periods or rows are related to energy levels for electrons in atoms (Figure 3.9). The length of a row is linked to the maximum number of electrons, $2n^2$, that can fit into an energy level. The periods are not equal in size because the maximum number of electrons per energy level increases as the distance of the energy level from the nucleus increases. Periods one through seven have 2, 8, 8, 18, 18, 32, and 23 (incomplete) elements, respectively. Larger periods as the atoms of elements get larger are similar to longer rows and more seats per row in a stadium as you proceed from the field to higher in the stands (Figure 3.10).

■ Chemists from all over the world belong to the International Union of Pure and Applied Chemistry (IUPAC). IUPAC has recommended that groups be labeled 1 through 18 consecutively from left to right. The periodic table on the inside front cover includes both 1 through 18 and A and B group labels for comparison.

Eighty-seven of the elements are **metals** and are found in Groups IA, IIA, parts of Groups IIIA to VIA (red in Figure 3.9), and the B groups (yellow). Characteristic physical properties of metals include malleability (ability to be beaten into thin sheets such as aluminum foil), ductility (ability to be stretched or drawn into wire such as copper), and good conduction of heat and electricity.

Seventeen elements are **nonmetals** (in green), and except for hydrogen they are found in the upper right-hand corner of the periodic table. Hydrogen is shown above Group IA because its atoms have one electron. However, hydrogen is a nonmetal and probably should be in a group by itself, although you may see H in both Group IA and Group VIIA in some periodic tables. Hydrogen forms compounds with formulas similar to those of the Group IA elements, but with vastly different properties. For example, compare NaCl (table salt) with HCl (a strong acid), or compare Na_2O (an active metal oxide) with H_2O (of course, water). Hydrogen also forms compounds similar to those of the Group VIIA elements: NaCl and NaH (sodium hydride), and $CaBr_2$ and CaH_2 (calcium hydride).

■ Hydrogen is the element without a home on the periodic table.

The physical and chemical properties of nonmetals are opposite those of metals. For example, nonmetals are **insulators**; that is, they are extremely poor conductors of heat and electricity.

Figure 3.10 Analogy of periods in periodic table to rows in a football stadium. Larger periods in the periodic table as atoms of elements get larger are similar to longer rows and more seats per row in a stadium as the rows are farther from the playing field. *(Courtesy of Department of Athletics, Vanderbilt University)*

Elements that border the staircase in Figure 3.9 between metals and nonmetals are six **metalloids** (in blue). Their properties are intermediate between those of metals and nonmetals. For example, silicon (Si), germanium (Ge), and arsenic (As) are **semiconductors** and are the elements that form the basic components of computer chips. Semiconductors conduct electricity less than metals, such as silver and copper, but more than insulators, such as sulfur. The six **noble gases** in Group VIIIA have little tendency to undergo chemical reactions. The classifications of metals, nonmetals, and metalloids will enable you to predict the kind of compounds formed between elements.

EXAMPLE 3.3 *Periodic Table*

For the elements with atomic numbers 17, 33, and 82, give the names and symbols and identify the elements as metals, metalloids, or nonmetals.

SOLUTION

Chlorine (Cl) is the element with atomic number 17. It is in Group VIIA. Chlorine and all the other elements in Group VIIA are nonmetals.

Arsenic (As) is the element with atomic number 33. It is in Group VA. Because it lies along the line between metals and nonmetals, arsenic is a metalloid.

Lead (Pb) is the element with atomic number 82. It is in Group IVA. Like other elements at the bottom of Groups IIIA to VIA, lead is a metal.

Exercise 3.3

List the main groups (the A groups) in the periodic table that (a) consist entirely of metals, (b) consist entirely of nonmetals, and (c) include metalloids. Identify the numbers of valence electrons in atoms from groups listed under (a), (b), and (c).

3.7 Periodic Trends

Why do elements in the same group in the periodic table have similar chemical behavior? Why do metals and nonmetals have different properties? G. N. Lewis was seeking answers to these questions during his development of the concept of valence electrons. He assumed that each noble gas atom had a completely filled outermost shell, which he regarded as a stable configuration because of the lack of reactivity of noble gases. He also assumed that the reactivity of other elements was influenced by their numbers of valence electrons.

■ See Section 3.4 for a discussion of valence electrons.

■ The *reactivity* of an element or a compound is its tendency to undergo chemical reactions. Some chemical pollutants, such as DDT, have low reactivity and therefore remain in the environment unchanged for a long time. An element such as potassium that on exposure reacts immediately with water or oxygen in the air is highly reactive.

Lewis Dot Symbols

Lewis used the element's symbol to represent the atomic nucleus together with all but the outermost shell of electrons; he called this the kernel of the

TABLE 3.4 ■ Lewis Dot Symbols for Atoms

IA	IIA	IIIA	IVA	VA	VIA	VIIA	VIIIA
H·							He:
Li·	·Be·	·B·	·C·	·N·	:O·	:F·	:Ne:
Na·	·Mg·	·Al·	·Si·	·P·	:S·	:Cl·	:Ar:
K·	·Ca·						

atom. The valence electrons, which he represented by dots, are then placed around the symbol one at a time until they are used up or until all four sides are occupied; any remaining electron dots are paired with the ones already there. Lewis dot symbols for atoms of the first 20 elements are shown in Table 3.4. Notice that all atoms of elements in a given A group have the same number of valence electrons and that the number of valence electrons equals the group number. All atoms of Group IA elements have one valence electron; Group IIA atoms have two valence electrons, and so forth. The importance of valence electrons in the study of chemistry cannot be overemphasized. The identical number of valence electrons primarily account for the similar properties of elements in the same group. *The chemical view of matter is primarily concerned with what valence electrons are doing in the course of chemical reactions.*

Lewis dot symbols will be used extensively in Chapter 5 in the discussions of bonding.

Atomic Properties

From left to right across each period, metallic character gives way to nonmetallic character (Figure 3.9). The elements with the most metallic character are at the lower left part of the periodic table near cesium (Cs). The elements with the most nonmetallic character are at the upper right portion of the periodic table near fluorine. The six metalloid elements (in blue) that begin with boron and move down like a staircase to astatine (At) roughly separate the metals and the nonmetals.

■ Metallic character means having properties of metals, such as ductility, malleability, and conductivity.

■ Atomic radius is the average distance from the center of the atomic nucleus to the outer surface of the electron cloud in an atom.

Atomic radii show periodicity by increasing with atomic number down the main groups of the periodic table (Figure 3.11). Why do atoms get larger from the top to the bottom of a group? The larger atoms simply have more energy levels inhabited by electrons than do the smaller atoms.

Atomic radii decrease across a period from left to right (Figure 3.11). You may see a paradox in adding electrons and getting smaller atoms, but protons also are being added. The greater nuclear charge pulls electrons in the same energy levels closer to the nucleus and causes contraction of the atom.

We can also use the trends in size of atomic radii to predict trends in reactivity. The valence electrons in larger atoms are farther from the nucleus. The larger the atom, the easier it is to remove the valence electrons because the attractive forces between protons in the nucleus and valence electrons decrease with increasing size of the atom.

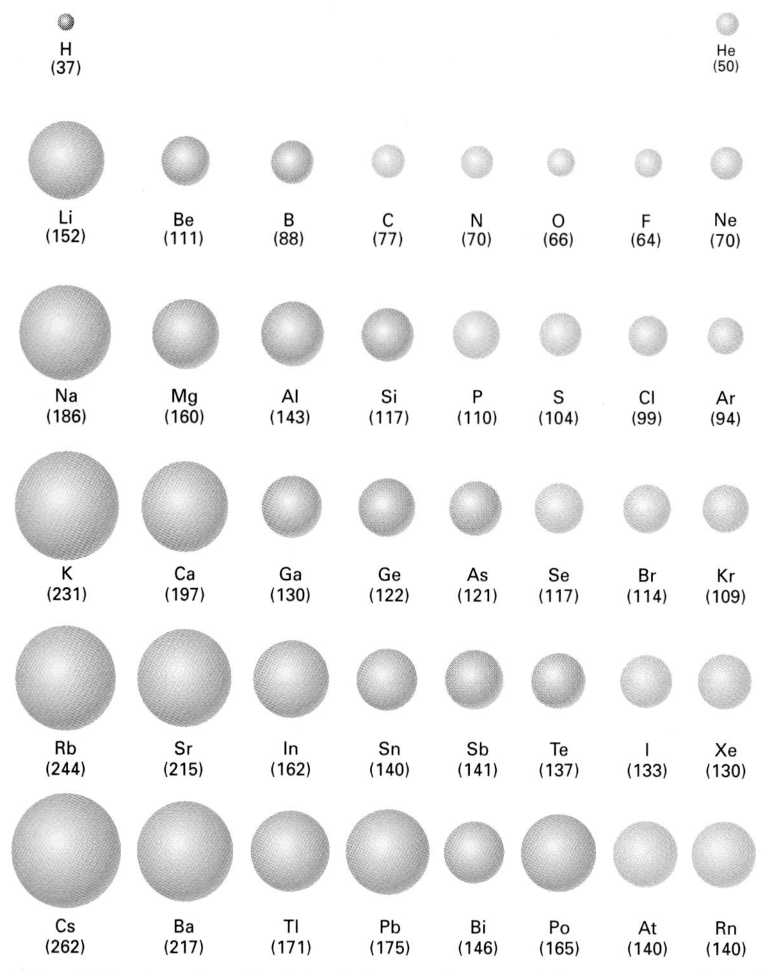

Figure 3.11 Atomic radii of the A Group elements (picometers, pm). A picometer (pm) is 1×10^{-12} m.

■ A general equation for the reaction of water and a metal from Group IA, using M to represent any of the IA metals, is

$$2\ M(s) + 2\ H_2O(\ell) \longrightarrow 2\ MOH(aq) + H_2(g)$$

In ionic compounds, one class of chemical compounds (Section 5.2), atoms have gained and lost electrons to form **ions**, which have positive or negative charges. Metal atoms lose valence electrons to form positive ions. The larger the metal atom, the greater the tendency to lose valence electrons and the more reactive the metal. Therefore, we would predict that the most reactive metal in Figure 3.11 is cesium (Cs), the metal with the largest radius, and this is correct. The result of *increase* in reactivity of metals down a given group in the periodic table is dramatically illustrated by lithium, sodium, and potassium—the first three metals in Group IA. Their atoms increase in size in this order down the group. Each element reacts with water—lithium quietly and smoothly, sodium more vigorously, and potassium much more quickly. The reactions of both sodium and potassium give off enough heat to ignite the hydrogen gas produced by the reaction, but as shown in Figure 3.12, potassium reacts with explosive violence. For elements at the bottom of Group IA, just exposure to moist air produces a vigorous explosion.

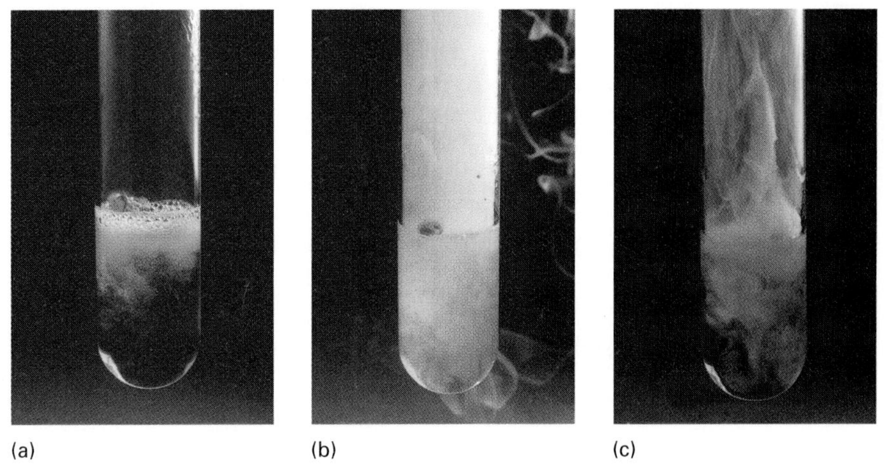

(a) (b) (c)

Figure 3.12 Reaction of alkali metals with water: (a) lithium, (b) sodium, and (c) potassium. *(C. D. Winters)*

Nonmetal atoms gain electrons from metals to form negative ions. The smaller the nonmetal atom, the higher the reactivity of the nonmetal. For example, fluorine atoms are the smallest of Group VIIA elements, and fluorine is the most reactive nonmetal. It reacts with all other elements except three noble gases—helium, neon, and argon. The reaction of Group VIIA elements with hydrogen illustrates how the reactivity of nonmetals *decreases* down the group. Fluorine reacts explosively with hydrogen, but the reaction with hydrogen is less violent for chlorine and is very slow for iodine.

■ A general equation for the reaction of hydrogen with a nonmetal from Group VIIA, using X to represent any of the VIIA nonmetals, is

$$H_2(g) + X_2(g) \longrightarrow 2\ HX(g)$$

Why are there repeatable patterns of properties across the periods in the periodic table? Again, it is because there is a repeatable pattern in atomic structure, and properties depend on atomic structure. Each period begins with one valence electron for atoms of the elements in Group IA. Each period builds up to eight valence electrons, and the period ends. This pattern repeats across periods two through six. As more elements are made by nuclear accelerators (Section 4.6), period seven may be completed someday. When this happens, the periodic table and atomic theory predict the last element in period seven (the final member of the noble gases) will be element number 118. Atoms of element 118 will have eight valence electrons.

EXAMPLE 3.4 *Atomic Radii and Reactivity*

Which element in each pair is more reactive: (a) O or S, (b) Be or Ca, or (c) P or As?

SOLUTION

(a) Oxygen and sulfur are Group VIA nonmetals. Generally, the smaller the atomic radius of a nonmetal, the more reactive the nonmetal is. Because oxygen atoms are smaller than sulfur atoms, oxygen should be more reactive than sulfur. (b) Beryllium (Be) and calcium (Ca) are Group IIA metals. Reactivity of metals in a given group increases down the group as the atomic radius increases. Therefore, calcium should be more reactive than beryllium. (c) Phosphorus and arsenic are Group VA nonmetals; thus, phosphorus, which has a smaller atomic radius, is predicted to be more reactive than arsenic.

Exercise 3.4A

Which element in each pair has the larger radius, that is, which is the larger atom in each pair: (a) Ca or Ba, (b) S or Se, (c) Si or S, or (d) Ga or Br?

Exercise 3.4B

Which element in each pair is more reactive: (a) Mg or Sr, (b) Cl or Br, or (c) Rb or Cs?

3.8 Properties of Main-Group Elements

Elements in a group have similar properties, but not the same properties. Some properties, as already illustrated for atomic radii and reactivity, increase or decrease in a predictable fashion from top to bottom of a periodic group.

Elements in a group generally react with other elements to form similar compounds, a fact accounted for by their identical numbers of valence electrons. For example, the formula for the compound of Li and Cl is LiCl; thus, you can expect there to be a compound of Rb and Cl with the formula RbCl and a compound of Cs and Cl with the formula of CsCl. Likewise, the formula Na_2O is known; therefore, a compound with the formula Na_2S predictably exists, because oxygen and sulfur are in the same group. In general, elements in the same group of the periodic table form some of the same types of compounds. In fact, several main groups have group names because of the distinctive and similar properties of the group elements.

The Group IA elements (Li, Na, K, Rb, Cs, and Fr) are called the **alkali metals**. The name alkali derives from an old word meaning "ashes of burned plants." All the alkali metals are soft enough to be cut with a knife. None are found in nature as free elements, because all combine rapidly and completely with virtually all the nonmetals and, as illustrated in Figure 3.12, with water. These elements form ions with a +1 charge. Francium, the last member of Group IA, is found only in trace amounts in nature, and all its 21 isotopes are naturally radioactive.

Reaction of calcium with water. It is easy to see by comparison with Figure 3.12 that calcium is less reactive than the alkali metals. *(C. D. Winters)*

The Group IIA elements (Be, Mg, Ca, Sr, Ba, and Ra) are called the **alkaline earth metals**. Compared with the alkali metals, the alkaline earth metals are harder, are more dense, and melt at higher temperatures. These elements form ions with a +2 charge. Because the valence electrons of alkaline earths are held more tightly, they are less reactive than their alkali metal neighbors. All the alkaline earth metals react with oxygen to form an oxide MO, where M is the alkaline earth.

$$2\,M(s) + O_2(g) \longrightarrow 2\,MO(s)$$

The **halogens** (F, Cl, Br, I, and As) are Group VIIA elements. In the elemental state, each of these elements exists as diatomic molecules (X_2).

THE WORLD OF CHEMISTRY

Making Glass Stronger

Program 7, *The Periodic Table*

Would you believe that glass was once so rare that it was prized more highly than jewels or gold? Today, household items made of glass are common and inexpensive, which is not surprising because the basic ingredient of glass is sand.

Glassmaking requires that the sand, which is mainly silicon dioxide (SiO_2, also known as *silica*), is mixed with other substances; heated to a temperature high enough to melt the mixture; and then molded, blown, or otherwise formed into the shape it will retain when cooled. The properties of glass, such as its color, melting point, sparkle, and hardness, are determined by the other substances in the mixture, most of which are metal oxides. Everyday glass, for example, contains oxides of two alkali metals, sodium and calcium, plus smaller amounts of magnesium, aluminum, and boron oxides. Chemical laboratory glassware and oven glassware are made of Pyrex, the borosilicate glass developed at Corning Glass Works, which maintains an active glass research program.

Pouring an experimental glass in the laboratory of Corning Glass Works. (James L. Amos/Peter Arnold, Inc.)

As Dr. Gerry Fine, a glass chemist at Corning explains, the periodic table is a key to choosing ingredients to modify the properties of glass.

I cannot imagine working without the periodic table, because scientists are interested in looking for systematic relationships. . . . If we take ordinary glass that contains sodium [ions] and substitute potassium [ions] in the surface—an element that behaves basically in the same way but is slightly larger—we can enhance the strength of that glass and make strong glass. After ordinary window glass has been dipped into molten potassium, the potassium literally stuffs the surface of that glass.

The result is a glass that will not shatter when struck by a steel ball bearing dropped from a height of 20 feet (ft).

Fluorine (F_2) and chlorine (Cl_2) are gases at room temperature, whereas bromine (Br_2) is a liquid and iodine (I_2) is a solid. This illustrates a typical trend within a group—an increase in melting point and boiling point in going down a group. All isotopes of astatine (At) are naturally radioactive and disintegrate quickly. The name halogen comes from a Greek word and means "salt producing." The best known salt containing a halogen is sodium chloride (NaCl), table salt. However, there are many other halogen salts, such as calcium fluoride (CaF_2), a natural source of fluorine; potassium iodide (KI), an additive to table salt that prevents goiter; and silver bromide (AgBr), the active photosensitive component of photographic film.

The **noble gases** (He, Ne, Ar, Kr, Xe, and Rn) are Group VIIIA elements. They are all colorless gases composed of single atoms at room temperature. They are referred to as "noble" because they lack chemical reactivity and generally do not react with "common" elements. Neon is the gas that glows orange-red in tubes of neon lights (Figure 3.13). Other gases and color-tinted tubes are used to give different colors. Radon (Rn) is naturally radioactive. Problems associated with radon indoor pollution are discussed in Section 4.7.

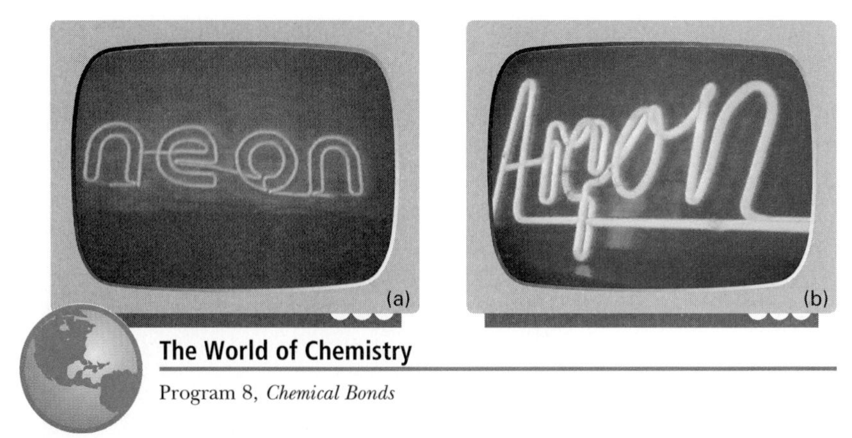

Figure 3.13 Neon and argon signs. Neon glows orange-red and argon glows blue in what are essentially sign-shaped partially evacuated discharge tubes.

■ SELF-TEST 3D

1. How many valence electrons are in each of the following atoms?
   (a) Sodium (Na), Group IA (b) Calcium (Ca), Group IIA
   (c) Boron (B), Group IIIA (d) Aluminum (Al), Group IIIA
   (e) Neon (Ne), Group VIIIA (f) Iodine (I), Group VIIA
2. The periodic table organized by Mendeleev placed elements in order of increasing atomic weight. (a) True, (b) False.
3. The modern periodic table has the elements placed in order of increasing atomic weight. (a) True, (b) False.
4. Metals typically have low numbers of valence electrons. (a) True, (b) False.
5. Nonmetals are typically found in Groups VA through VIIIA. (a) True, (b) False.
6. Which of the following are true about metals?
   (a) They are found only in Group IA and IIA.
   (b) They are good conductors of heat.
   (c) They are good insulators.
   (d) They usually react to form positive ions.
7. Group IIA includes Be, Mg, Ca, Sr, Ba, and Ra. What is the predicted formula for the compound formed between Sr and Cl, if Be and Ba form $BeCl_2$ and $BaCl_2$?
8. Which of the elements in Group IA is the most reactive: Li, Na, K, Rb, or Cs?
9. Which of the following pairs of atoms has the greater number of valence electrons?
   (a) Lithium (Li) or oxygen (O) (b) Calcium (Ca) or sodium (Na)
   (c) Oxygen (O) or fluorine (F) (d) Boron (B) or nitrogen (N)
   (e) Sulfur (S) or arsenic (As) (f) Neon (Ne) or carbon (C)
10. Group IA elements are all metals except for hydrogen. (a) True, (b) False.
11. Isotopes have different numbers of valence electrons. (a) True, (b) False.

12. Which of the following pairs of atoms will be more reactive?
    (a) Lithium (Li) or cesium (Cs)
    (b) Fluorine (F) or bromine (Br)
    (c) Beryllium (Be) or calcium (Ca)
    (d) Oxygen (O) or sulfur (S)
    (e) Lithium (Li) or sodium (Na)
    (f) Neon (Ne) or xenon (Xe)

MATCHING SET I

____ 1. Mass number
____ 2. Unlike electric charges
____ 3. $2n^2$
____ 4. Nucleus
____ 5. Electron
____ 6. ^{22}Ne and ^{20}Ne
____ 7. Atomic number
____ 8. Particles in an H atom
____ 9. Neutron
____ 10. Rutherford
____ 11. Bohr

a. Attract
b. Equal to number of protons in the nucleus
c. Negative particle found outside the nucleus
d. Neutrons plus protons
e. Discovered the nucleus
f. Proton and an electron
g. Developed a theory for representing electrons in energy levels
h. Isotopes
i. Maximum number of electrons in an energy level
j. Uncharged elementary particle
k. Central part of an atom consisting of protons and neutrons

MATCHING SET II

____ 1. Periodic
____ 2. Larger atoms
____ 3. Two valence electrons
____ 4. A noble gas
____ 5. A transition metal
____ 6. A main-group metal
____ 7. Decrease in atomic radius
____ 8. A halogen
____ 9. An inner transition element
____ 10. Valence electrons

a. Electron arrangement 2-8-2
b. Outermost occupied shell
c. Chromium (Cr)
d. Repeated pattern
e. At the bottom of a group
f. Praseodymium (Pr)
g. Eight valence electrons
h. Seven valence electrons
i. Across a period
j. Cesium

QUESTIONS FOR REVIEW AND THOUGHT

1. What is the law of conservation of matter? Give an example of the law in action.
2. State the law of definite proportions. Give an example to illustrate what it means.
3. What kinds of evidence did Dalton have for atoms that the early Greeks (Democritus, Leucippus) did not have?
4. How does Dalton's atomic theory explain these?
    (a) The law of conservation of matter
    (b) The law of definite proportions
5. Describe in detail Rutherford's gold foil experiment under the following headings:
    (a) Experimental setup (b) Observations
    (c) Interpretations
6. What is meant by the term subatomic particles? Give two examples.

7. Give short definitions for the following terms:
   (a) Atomic number (b) Mass number
   (c) Atomic weight (d) Isotope
   (e) Natural abundance (f) Atomic mass unit
8. There are more than 1000 kinds of atoms, each with a different weight, yet there are only 109 elements. How does one explain this in terms of subatomic particles?
9. What do all the atoms of an element have in common?
10. Which of the following pairs are isotopes? Explain your answers.
    (a) ^{50}Ti and ^{50}V (b) ^{12}C and ^{14}C
    (c) ^{40}Ar and ^{40}K
11. A common isotope of Li has a mass of 7. The atomic number of Li is 3. How can this information be used to determine the number of protons and neutrons in the nucleus?
12. The element iodine (I) occurs naturally as a single isotope of atomic mass 127; its atomic number is 53. How many protons and how many neutrons does it have in its nucleus?
13. Which pairs of atoms are isotopes?

	A	B	C	D
Mass number	53	53	52	54
Atomic number	25	24	24	25

14. Complete the following table that contains information about five different atoms:

	Number of Protons	Number of Neutrons	Number of Electrons	Atomic Number	Mass Number
(a)	32	_____	_____	_____	73
(b)	_____	14	14	_____	_____
(c)	_____	_____	_____	28	59
(d)	48	64	_____	_____	_____
(e)	_____	115	_____	77	_____

15. Identify the elements in question 14.
16. What number is most important in identifying an atom?
17. The atomic weight listed in the periodic table for magnesium is 24.305 amu. Someone said that there wasn't a single magnesium atom on the entire earth with a weight of 24.305 amu. Is this statement correct? Why or why not?
18. Complete the following table:

Isotope	Atomic No.	Mass No.	No. of Protons	No. of Neutrons	No. of Electrons
Bromine-81	_____	81	_____	_____	_____
Boron-11	5	_____	_____	_____	_____
^{35}Cl	17	_____	_____	_____	_____
^{52}Cr	_____	52	_____	_____	_____
Ni-60	_____	_____	_____	_____	_____
Sr-90	_____	_____	_____	_____	_____
Lead-206	_____	_____	_____	_____	_____

19. What is constant about a compound? (a) the weight of a sample of the compound, (b) the weight of one of the elements in samples of the compound, (c) the ratio by weight of the elements in the compound
20. Krypton is the name of Superman's home planet and also that of an element. Look up the element krypton and list its symbol, atomic number, atomic weight, and electron arrangement.
21. Write the placement of electrons in their ground-state energy levels according to the Bohr theory for atoms having 6, 10, 13, and 20 electrons.
22. Write the Bohr electron notation for atoms of the elements sodium through argon.
23. How many valence electrons do atoms of each of the elements in question 22 contain?
24. Name five regions of the electromagnetic spectrum that you use most every day. Arrange these regions in order of increasing wavelength.
25. State the periodic law.
26. Give definitions for the following terms:
    (a) Group (b) Period
    (c) Chemical properties (d) Transition element
    (e) Inner transition element
    (f) Representative element
27. How did the discovery of the periodic law lead to the discovery of elements?
28. Describe the relative locations of metals, nonmetals, and metalloids in the periodic table.
29. How do metals differ from nonmetals?
30. Identify each of the following elements as either a metal, nonmetal, or metalloid:
    (a) Nitrogen (b) Arsenic
    (c) Argon (d) Calcium
    (e) Uranium
31. Answer the following questions about the periodic table:
    (a) How many periods are there?
    (b) How many representative groups or families are there?
    (c) How many groups consist of all metals?
    (d) How many groups consist of all nonmetals?
    (e) Is there a period that consists of all metals?
32. What do the electron structures of alkali metals have in common?
33. Pick the following electron arrangements that represent elements in the same chemical family:
    (a) 2, 1, (b) 2, 6, (c) 2, 4, (d) 2, 8, 6, (e) 2, 8, 8, (f) 2, 8, 8, 2, (g) 2, 8, 8, 1, (h) 2, 8, 9, 2.
34. Give the number of valence electrons for each of the following:
    (a) Ba (b) Al
    (c) P (d) Se
    (e) Br (f) K
35. Give the symbol for an element that has
    (a) 3 valence electrons (b) 4 valence electrons
    (c) 7 valence electrons (d) 1 valence electron

36. Draw the Lewis dot symbols for Be, Cl, K, As, and Kr.
37. From their position in the periodic table, predict which will be more metallic:
    (a) Be or B (b) Be or Ca
    (c) As or Ge (d) As or Bi
38. Which atom in the following pairs is more metallic?
    (a) Li or F (b) Li or Cs
    (c) Be or Ba (d) C or Pb
    (e) B or Al (f) Na or Ar
39. What general electron arrangement is conducive to chemical inactivity?
40. Use the information in the periodic table to supply the following:
    (a) The nuclear charge on cadmium (Cd)
    (b) The atomic number of As
    (c) The atomic mass (or mass number) of an isotope of Br having 46 neutrons
    (d) The number of electrons in an atom of Ba
    (e) The number of protons in an isotope of Zn
    (f) The number of protons and neutrons in an isotope of Sr, atomic mass (or mass number) of 88
    (g) An element forming compounds similar to those of Ga
41. Complete the following table:

Atomic Number	Name of Element	Number of Valence Electrons	Period	Metal or Nonmetal
6	______	______	______	______
12	______	______	______	______
17	______	______	______	______
37	______	______	______	______
42	______	______	______	______
54	______	______	______	______

42. Write the symbols of the halogen family in the order of increasing size of their atoms.
43. Why does Cs have larger atoms than Li?
44. How does the atomic radius for a metal atom relate to the reactivity of the metal?
45. How does the atomic radius for a nonmetal atom relate to reactivity?
46. Which atom in the following pairs is more reactive?
    (a) Li or Rb (b) Mg or Ba
    (c) Na or Ar (d) Ne or O
    (e) He or Xe (f) Br or F
47. Why do trends exist in the periodic table?
48. Describe the variation in atomic size (a) across a period, (b) down a group.
49. Rank the following atoms by size, with the largest on the left and the smallest on the right: K, S, Al, P, and Cl.
50. The elements at the bottom of Groups IA, IIA, VIA, VIIA, and VIIIA are all radioactive; thus, less is known about their physical and chemical properties. By using group and period trends described in this chapter, predict which of these five elements—Fr, Ra, Po, At, and Rn—would:
    (a) be the most metallic
    (b) be the most nonmetallic
    (c) have the largest atomic radius
    (d) be the most unreactive
    (e) react most readily with water
51. If element 36 is a noble gas, in what groups would you expect elements 35 and 37 to occur?
52. Oxygen and sulfur are very different elements, in that one is a colorless gas and the other, a yellow crystalline solid. Why then are they both in Group VIA?
53. Give the names and symbols for two elements most like selenium (Se), atomic number 34.

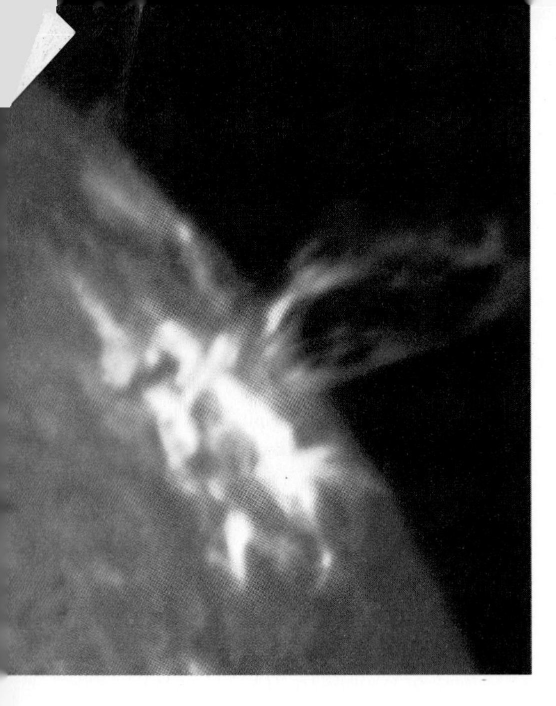

CHAPTER 4

Nuclear Changes

Radioactivity. With what do you associate that word? Hazardous waste? Nuclear power plants? Medical diagnosis? Cancer risks? Nuclear weapons? Cancer cures? All these are appropriate associations with the kind of change that happens only in atomic nuclei. Nuclear changes are very different from ordinary chemical reactions, as you will see. Nuclear changes are usually accompanied by the emission of radiation and can, in some cases, be accompanied by the release of large amounts of energy. In this chapter you will learn something about how and when nuclear changes occur.

You will also see that nuclear changes are pictured in much the same way as chemical changes: reactants going to products. Equations can be written for nuclear changes, and although there are similarities between nuclear and chemical changes, nuclear changes are different in some rather significant ways. Certainly the discovery of nuclear changes has affected our lives, some might say for the worse, but you or one of your friends may be alive today because of an application of what is known about how radioactive isotopes undergo change.

- What are the characteristics of nuclear changes?
- Why do some atoms undergo spontaneous nuclear decay?
- How can an atom of one element be transformed into an atom of another element?
- Why are some radioactive isotopes more dangerous than others?
- What are some of the harmful effects of nuclear reactions and radiation?
- What are some of the useful applications of radioactive isotopes, including energy production?

A large solar flare lasting a few hours releases energy from nuclear reactions that would provide electricity for the United States for about 100,000 years. *(National Solar Observatory/NOAO)*

The nuclear age is often considered to have started either in the late 1800s and early 1900s with the discovery of radioactive elements or in 1945 with the first explosion of an atomic bomb. It is true that early atomic theory said nothing about radioactivity, but that was because radiation cannot be detected directly by our five senses. It took the maturing of the sciences—with such diverse discoveries as how to produce a vacuum, photographic film, electricity, and magnetic fields—to lead to the knowledge that some atoms disintegrate spontaneously and, in the process, produce radiation.

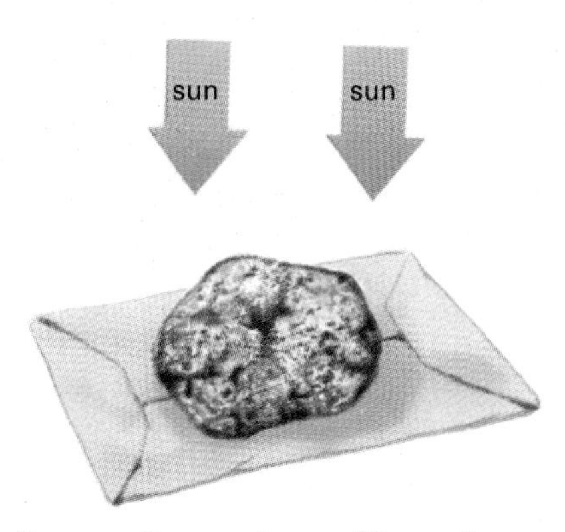

Becquerel's experiment. The goal was to find out if phosphorescence, known to occur when the mineral was exposed to sunlight, included X rays. The presence of X rays would be shown by exposure of the photographic plate.

4.1 The Discovery of Radioactivity

In February 1896, Henri Becquerel was experimenting in France with the relation between the recently discovered X rays and the phosphorescence of certain minerals. X rays had been found to penetrate substances like paper and expose photographic plates. Becquerel had already discovered that phosphorescing uranium minerals also exposed the plates. During several cloudy days some samples were left waiting in a drawer. Out of curiosity, Becquerel developed the plates. He had quite a surprise—the plates were exposed. Obviously, the emission of radiation that penetrated the paper surrounding the plates had nothing to do with phosphorescence but was a property of the mineral. In fact, all uranium compounds, and even the metal itself, exposed photographic plates. Becquerel also discovered that uranium (U) emitted radiation that was capable of causing air molecules to ionize (i.e., to lose electrons and become positively charged particles).

■ Phosphorescent minerals re-emit light after they have been exposed to light.

Before long it was recognized that the radiation from elements like uranium and radium consisted of the three types known as alpha, beta, and gamma rays.

■ As you saw in Section 3.2, identification of alpha rays as helium nuclei (^4_2He) and beta rays as electrons played a significant role in our understanding of the structure of the atom.

In 1899, Ernest Rutherford found that alpha rays could be stopped by thin pieces of paper and had a range of only about 2.5 cm to 8.5 cm in air, while beta rays were capable of penetrating far greater distances in air.

In 1900, Paul Villard identified a third form of natural radiation, gamma (γ) rays. These he discovered were not streams of particles, but instead had the general characteristics of light or X rays. Gamma rays, a high-energy form of electromagnetic radiation (see Figure 3.7), are extremely penetrating; they are capable of passing through more than 22 cm of steel and about 2.5 cm of lead. Figure 4.1 compares the penetrating ability of the three forms of natural radiation.

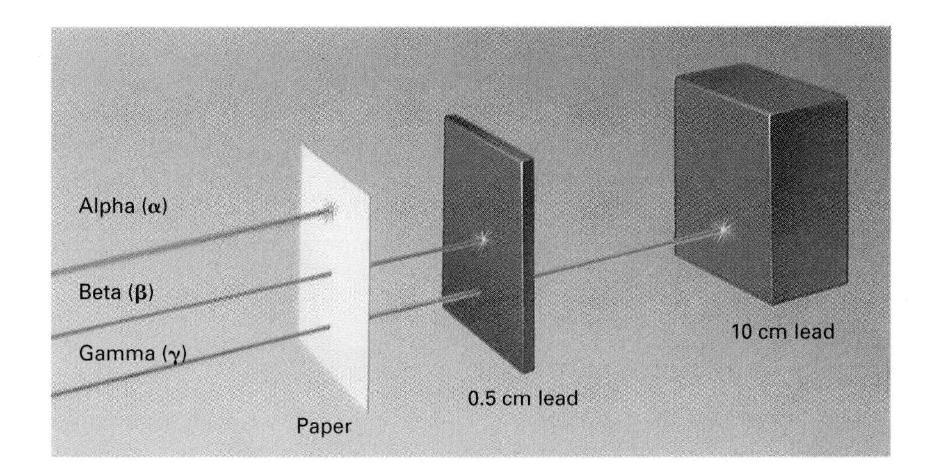

Figure 4.1 The relative penetrating abilities of alpha, beta, and gamma rays. The heavy, highly charged alpha particles are stopped by a piece of paper (or the skin). The lighter, less highly charged beta particles penetrate paper, but are stopped by a 0.5-cm sheet of lead. Because gamma rays have no charge and no mass, they are the most penetrating.

THE PERSONAL SIDE

The Curies

Soon after Becquerel's discovery of uranium's radioactivity, Marie Sklodowska Curie (1867–1934), also working in France, studied the radioactivity of thorium (Th) and began to search systematically for new radioactive elements. She showed that the radioactivity of uranium was an atomic property—that is, its radioactivity was proportional to the amount of the element present and was not related to any particular compound. Her experiments indicated that other radioactive elements were probably also present in certain uranium samples. With painstaking technique, she and her husband Pierre Curie (1859–1906) separated the element radium (Ra) from uranium ore and found that it is more than one million times more radioactive than uranium. In 1903, Marie and Pierre Curie shared the Nobel Prize in physics with Henri Becquerel for their discovery of radioactivity. After Pierre died, Marie continued her research and discovered polonium (Po), which she named after her native Poland. In 1911, she became the first person to win a second Nobel Prize, this one for the discoveries of radium and polonium. In 1921, Marie Curie came to the United States where she was given 1 g of pure radium, purchased with donations from American women interested in her work.

Pierre and Marie Curie with their daughter, Irene. Irene grew up to continue the study of radioactivity with her husband Frédéric Joliot. Together Irene and Frédéric won a Nobel Prize in 1935 for production of the first artificial radioactive isotope. (Stock Montage)

4.2 Nuclear Reactions

After discovery of the natural radioactivity of uranium, thorium, and radium, many other elements were found to have radioactive isotopes. All the elements heavier than bismuth (Bi, atomic number 83) and a few lighter than bismuth have natural radioactivity. While studying radium, Rutherford found that besides emitting alpha particles, radium was also producing radioactive radon gas (Rn). This led Rutherford and one of his students, Frederick Soddy, in 1902 to propose the revolutionary theory that *radioactivity is the result of a natural change of an isotope of one element into an isotope of a different element.* Such a change is a **nuclear reaction**, a process in which an unstable nucleus emits radiation and is converted into a more stable nucleus of a different element. Thus, a nuclear reaction results in a change in atomic number and often a change in mass number as well.

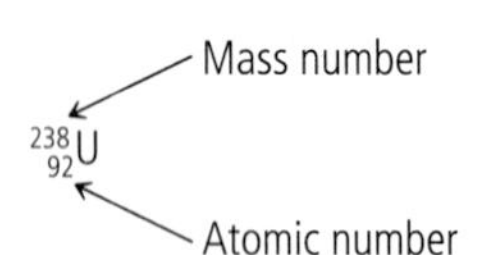

Equations for Nuclear Reactions

In nuclear reactions the total number of nuclear particles, called **nucleons** (protons plus neutrons), remains the same, but the identities of atoms can change. Just as with chemical equations, nuclear equations reflect the fact that matter is conserved. As a result, the sum of the mass numbers of reacting nuclei must equal the sum of the mass numbers of the product nuclei. There must also be nuclear charge balance—the sum of the atomic numbers of the products must equal the sum of the atomic numbers of the reactants.

Consider the equation for the nuclear reaction

$$^{226}_{88}\text{Ra} \longrightarrow {}^{4}_{2}\text{He} + {}^{222}_{86}\text{Rn}$$

Radium-226 Alpha particle Radon-222

The mass number on the left equals the sum of the mass numbers on the right. Similarly, the atomic number on the left equals the sum of the atomic numbers on the right.

When 226 g of radium-226 has completely decayed, 222 g of radon-222 and 4 g of helium will have been formed.

Radium-226 is another way of writing $^{226}_{88}\text{Ra}$.

mass number: 226 = 4 + 222

atomic number: 88 = 2 + 86

The isotope of uranium with atomic mass 238 is also an alpha emitter. When the $^{238}_{92}\text{U}$ nucleus gives off an alpha particle, made up of two protons and two neutrons, four units of atomic mass and two units of atomic charge are lost. The resulting nucleus has a mass of 234 and a nuclear charge of 90, showing that it is an isotope of thorium, which has 90 protons in its nucleus and an atomic number of 90.

$$^{238}_{92}\text{U} \longrightarrow {}^{4}_{2}\text{He} + {}^{234}_{90}\text{Th}$$

Uranium-228 Alpha particle Thorium-234

mass number: 238	=	4	+	234
atomic number: 92	=	2	+	90

Some unstable nuclei are beta emitters. For example, uranium-235 emits a beta particle, which is an electron ($^{0}_{-1}\text{e}$).

$$^{235}_{92}\text{U} \longrightarrow {}^{0}_{-1}\text{e} + {}^{235}_{93}\text{Np}$$

Uranium-235 Beta particle Neptunium-235

mass number: 235	=	0	+	235
atomic number: 92	=	−1	+	93

Knowing that some nuclei emit beta particles leads to a basic question: How can a nucleus containing protons and neutrons emit a beta particle, which is an electron? It has been established that an electron and a proton can combine outside the nucleus to form a neutron. Therefore, the reverse process is proposed to occur inside the nucleus. When a beta particle is emitted from a decaying nucleus, a neutron decomposes, giving up an electron and changing itself into a proton. The ejected electron is the beta particle. The resulting proton remains in the nucleus and increases the atomic number by one.

■ Note that this is a balanced nuclear equation.

$$\text{Beta particle production:} \quad {}^{1}_{0}\text{n} \longrightarrow {}^{0}_{-1}\text{e} + {}^{1}_{1}\text{H}$$

Neutron Electron Proton

Gamma radiation (γ) may or may not be given off simultaneously with alpha or beta rays, depending on the particular nuclear reaction involved. Gamma rays are emitted when the product nucleus must lose some additional energy to become stable. Being electromagnetic radiation, gamma rays have no charge and essentially no mass. The emission of a gamma ray, therefore, cannot alone account for production of a different element.

EXAMPLE 4.1 *Writing an Equation for an Alpha Emission*

Write an equation for alpha emission from a polonium-218 isotope.

SOLUTION

First, write the partial equation and set up a table of mass and atomic number changes under it. The atomic number of polonium is 84.

$${}^{218}_{84}\text{Po} \longrightarrow {}^{4}_{2}\text{He} + ?$$

$$\text{mass number:} \quad 218 \longrightarrow 4 + ?$$

$$\text{atomic number:} \quad 84 \longrightarrow 2 + ?$$

The mass number of the product must be 214 because its mass number plus that of the alpha particle must equal 218, the mass number of the decaying polonium-218 isotope. The atomic number of the product must be 82 because its atomic number plus that of the alpha particle must equal 84, the atomic number of the decaying isotope. Inasmuch as the element with an atomic number of 82 is lead, the product is ${}^{214}_{82}\text{Pb}$.

Exercise 4.1
Write an equation showing the emission of an alpha particle by an isotope of neptunium (${}^{237}_{93}\text{Np}$).

EXAMPLE 4.2 *Writing an Equation for a Beta Emission*

Write an equation for beta emission from a lead-210 nucleus.

SOLUTION

First, write a partial equation that includes what is known: lead-210 is a reactant, and a beta particle is a product. Then, to aid in determining the mass and atomic number changes, set up a table like those used earlier. The atomic number of lead is 82.

$$^{210}_{82}Pb \longrightarrow {}^{0}_{-1}e + ?$$

$$\text{mass number:} \quad 210 \longrightarrow 0 + ?$$

$$\text{atomic number:} \quad 82 \longrightarrow -1 + ?$$

The sum of the mass numbers of the products must equal 210, the mass number of the decaying lead isotope. Since the mass of the beta particle is essentially zero, the mass number of the product nucleus must be 210. The sum of the atomic numbers of the products must also equal the atomic number of lead, 82. So, the atomic number of the product must be 82 [83 + (−1) = 82], which is the atomic number of bismuth (Bi). The product nucleus is $^{210}_{83}Bi$.

Exercise 4.2
Write an equation showing the emission of a beta particle from $^{234}_{91}Pa$.

4.3 The Stability of Atomic Nuclei

Why are some nuclei unstable and radioactive, while others are stable and not radioactive? The fact that there are strong repulsions among all those protons packed inside the nucleus of an atom has something to do with nuclear stability. In addition, the relative numbers of neutrons, which are not charged, play some role. Figure 4.2 shows a plot of the number of protons versus the number of neutrons in the known isotopes from hydrogen ($Z = 1$) to bismuth ($Z = 83$). The nonradioactive (stable) isotopes (black dots) are far fewer in number than the radioactive (unstable) isotopes (red dots).

■ *Z* is the symbol for atomic number.

The stability of nuclei is apparently dependent on the relative numbers of protons and neutrons. The nucleus of the simplest atom, hydrogen, contains only a proton. Its two isotopes, deuterium ($^{2}_{1}H$) and tritium ($^{3}_{1}H$), contain one and two neutrons, respectively. By looking at Figure 4.2, you can see that the band of stable nuclei, the black dots, curves upward toward the neutron axis; this shows that in stable nuclei, the number of neutrons is equal to or greater than the number of protons. From hydrogen to bismuth, except for $^{1}_{1}H$ and $^{3}_{2}He$, *the mass numbers of stable isotopes are always twice as large as the atomic number or even larger.* It appears that the larger numbers of protons in the nuclei of heavier atoms require extra neutrons to gain stability. Any unstable isotope (a red dot in Figure 4.2, on page 76) will decay in such a way that its decay product falls closer to the stable band (the black dots).

■ Perhaps because there are so many unstable (radioactive) isotopes, people sometimes think of "radioactive" when the word "isotope" is mentioned.

Beta emission occurs in isotopes that have *too many neutrons* to be stable. These isotopes appear as red dots above the stable band in Figure 4.2. When beta decay occurs, the conversion of a neutron into a proton and an electron increases the atomic number while lowering the number of neutrons, as was illustrated in Example 4.2, and the new isotope moves toward the stable region.

■ Beta particles result from the conversion of neutrons into protons and electrons

$$^{1}_{0}n \longrightarrow {}^{1}_{1}H + {}^{0}_{-1}e$$

Those isotopes with *too few neutrons* (red dots below the band of stability) decay as well, but in a manner that increases the number of neutrons relative

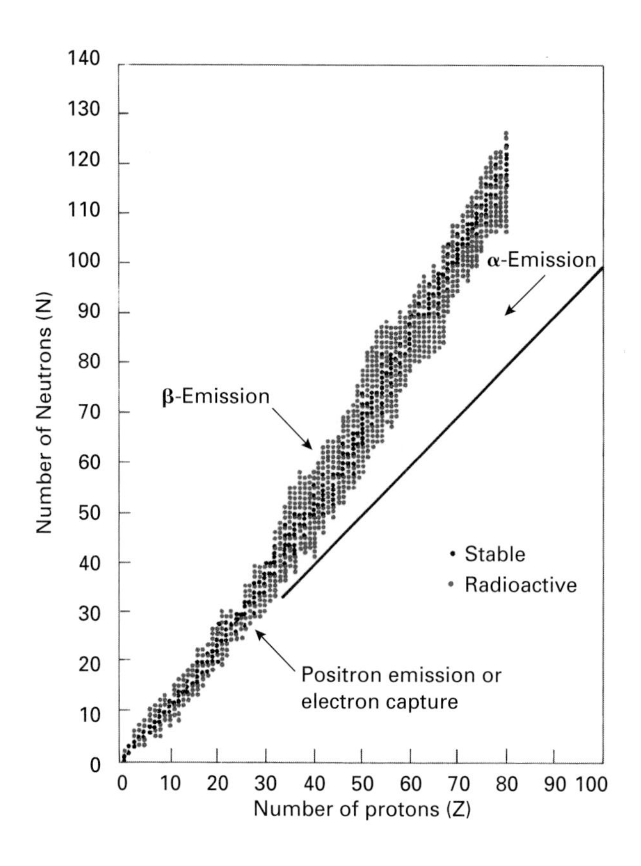

Figure 4.2 The band of nuclear stability, as shown by the black dots for stable isotopes. When radioactive nuclei (red dots) decay, their neutron-to-proton ratio moves closer to the band. Isotopes above the band undergo β-decay (the number of protons increases), isotopes below the band undergo positron emission (the number of protons decreases), and those with more than 83 protons undergo α-decay (the number of protons decreases by 2 and the number of neutrons decreases by 2). *(Redrawn from Oxtoby, Nachtrieb, Freeman:* Chemistry: Science of Change*)*

■ The positron is sometimes called the *antielectron.* The positron is one of a group of *antimatter* particles known to exist. An electron will react with a positron to annihilate each other and produce two high-energy gamma rays.

to the number of protons. One way this can happen is by emission of a type of subatomic particle discovered in 1932, a **positron**—a positively charged electron, ${}^{0}_{+1}e$. For example, the decay of nitrogen-13, an isotope with too few neutrons, is by positron emission.

$${}^{13}_{7}N \longrightarrow {}^{0}_{+1}e + {}^{13}_{6}C$$

The positron results from the decay of a proton.

$$\underset{\text{Proton}}{{}^{1}_{1}H} \longrightarrow \underset{\text{Neutron}}{{}^{1}_{0}n} + \underset{\text{Positron}}{{}^{0}_{+1}e}$$

Because the positron, like the electron, has a mass number of zero, the mass number of the product nucleus is the same as that of the starting nucleus.

All isotopes of the elements beyond bismuth (Z = 83) are unstable. Most of them decay by ejecting an alpha particle. This kind of decay, as illustrated in Example 4.1, decreases the mass number by four and the atomic number by two. The types of radioactive decay we have discussed are summarized in Table 4.1.

TABLE 4.1 ■ Changes in Atomic Number and Mass Number Accompanying Radioactive Decay

Type of Decay	Symbol	Charge	Mass	Change in Atomic Number	Change in Mass Number
Beta	${}_{-1}^{0}e$	−1	0	+1	None
Positron	${}_{+1}^{0}e$	+1	0	−1	None
Alpha	${}_{2}^{4}He$	+2	4	−2	−4
Gamma	${}_{0}^{0}\gamma$	0	0	None	None

4.4 Activity and Rates of Nuclear Disintegrations

The number of radioactive nuclei that disintegrate in a sample per unit time is called its **activity** (this is the "activity" in radioactivity). The activity of a sample containing radioactive isotopes depends on the number of nuclei present and the rate at which they decay. If a sample of matter is "highly radioactive," many atoms are undergoing decay per unit of time. A small number of nuclei decaying at a rapid rate can produce the same activity as a larger number of atoms decaying at a slower rate. Radioactive disintegrations are measured in **curies** (Ci); one Ci is 37 billion disintegrations per second (dps).

■ A more suitable unit is the **microcurie** (μCi), which is 37,000 disintegrations per second (dps). Another unit used to measure radioactive disintegrations is the **becquerel** (Bq), where 1 Bq = 1 dps.

To illustrate the differences in rates of decay of radioactive nuclei, consider first how cobalt-60 is used in medicine to treat cancerous tumors in the human body. When cobalt-60 decays, it produces beta particles as well as gamma rays.

■ Gamma radiation is more damaging to cancer cells because they are duplicating faster than normal cells.

$$ {}_{27}^{60}Co \longrightarrow {}_{28}^{60}Ni + {}_{-1}^{0}e + {}_{0}^{0}\gamma $$

Although the cobalt-60 isotope is radioactive, it is stable enough so that only half of a sample will decay in 5.27 years. A cobalt-60 sample is installed in a well-shielded apparatus that emits a focused beam of gamma rays. Because the half-life of the cobalt-60 radioisotope is fairly long, the sample does not have to be replaced very often. By rotating the radiation source around the patient, the physician can concentrate the rays in the cancerous region being treated, while limiting the radiation somewhat to other parts of the body (Figure 4.3, p. 78).

Copper-64, in the form of copper acetate, is also used in medicine, but as a diagnostic tool rather than for treatment. Also a gamma emitter, copper-64 decays much more rapidly than cobalt-60; half the copper-64 atoms in a sample will decay in 12.9 h. When injected into a patient's blood, the copper compound is carried to the brain and concentrates in any tumorous region that is present. A camera that detects gamma rays is used to locate the tumor. The radiologist knows that the copper-64 nuclei used in diagnosis will have decayed into safer, more stable isotopes in a matter of a few hours or days because of the rapid rate of decay of these nuclei.

■ Further examples of the use of radioisotopes in medicine are given in Section 4.8.

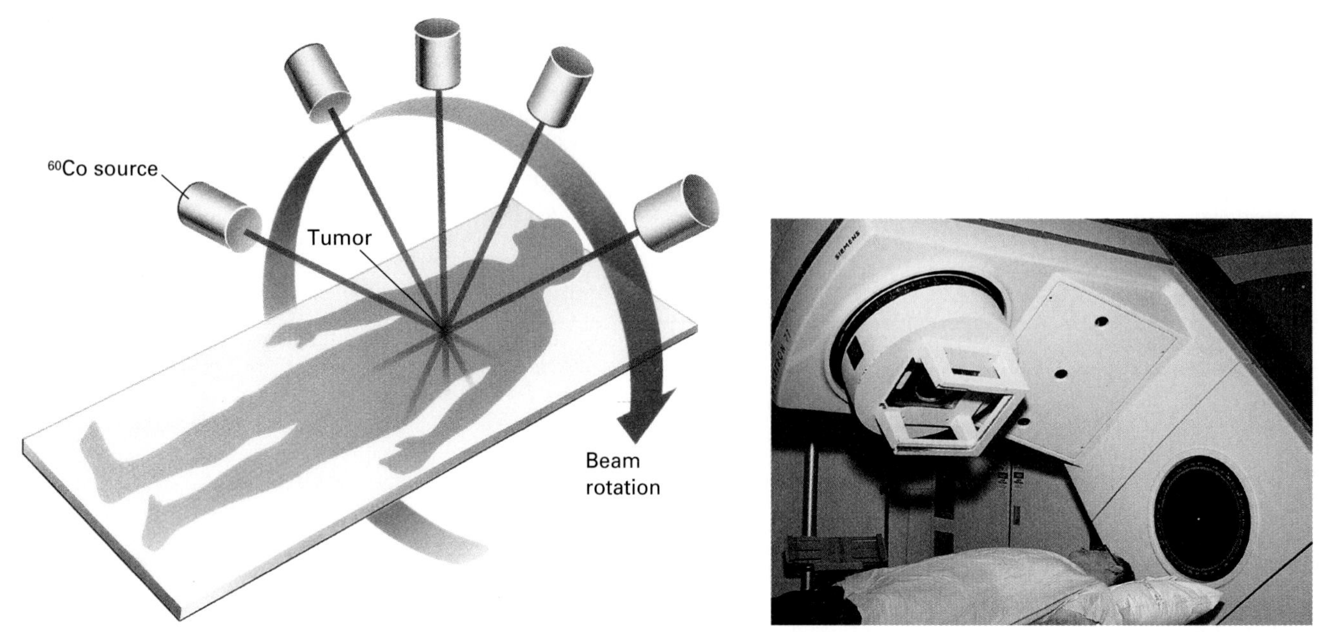

Figure 4.3 Treatment for cancer with gamma radiation from cobalt-60. By adjusting the rotation of the radiation source, the radiation is concentrated where the beams cross at the location of the diseased tissue. *(b, Beverly March; Courtesy of Long Island Jewish Hospital)*

Half-Life

The rate of decay of any radioactive isotope can be represented by its characteristic **half-life**, the period required for one half of the radioactive material originally present to undergo radioactive decay. Short half-lives are the results of high rates of decay, while long half-lives are the results of low rates of decay.

The half-life of an isotope is independent of the amount of radioactive material present and is essentially independent of temperature and the chemical form in which the radioactive atoms are present. Table 4.2 gives the half-lives of some radioactive isotopes. The 12.9 h half-life of $^{64}_{29}Cu$, for example, means that one half of the original amount of copper-64 atoms will remain

■ Mathematically, the fraction of a radioactive isotope remaining after n half-lives is $(\frac{1}{2})^n$. The fraction after two half-lives is $(\frac{1}{2})^2 = \frac{1}{4}$; after three half-lives it is $(\frac{1}{2})^3 = \frac{1}{8}$, and so on.

TABLE 4.2 ■ Half-Lives of Some Radioactive Isotopes

Decay Process	Half-Life
$^{238}_{92}U \longrightarrow ^{234}_{90}Th + ^{4}_{2}He$	4.51×10^9 years
$^{3}_{1}H \longrightarrow ^{3}_{2}He + ^{0}_{-1}e$	12.3 years
$^{14}_{6}C \longrightarrow ^{14}_{7}N + ^{0}_{-1}e$	5730 years
$^{131}_{53}I \longrightarrow ^{131}_{54}Xe + ^{0}_{-1}e$	8.05 days
$^{64}_{29}Cu \longrightarrow ^{64}_{30}Zn + ^{0}_{-1}e$	12.9 h
$^{69}_{30}Zn \longrightarrow ^{69}_{31}Ga + ^{0}_{-1}e$	55 min

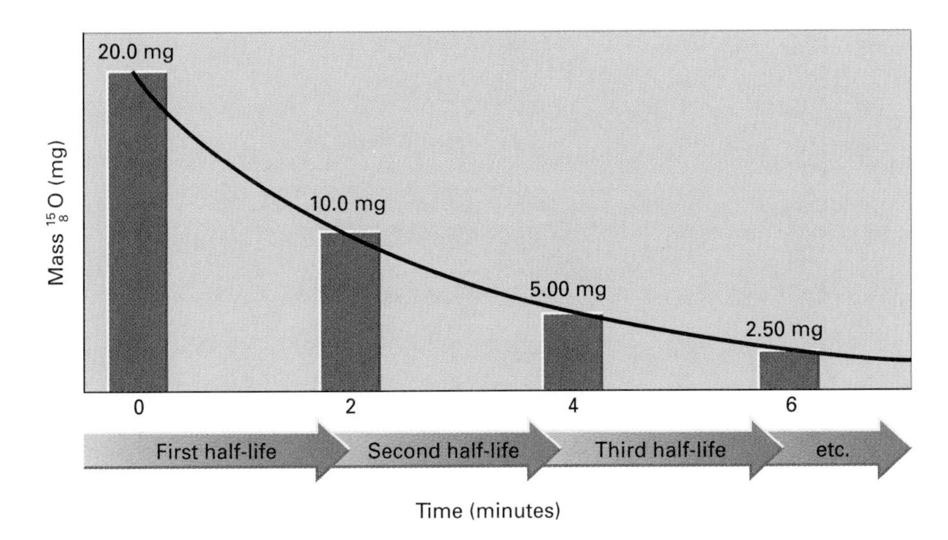

Figure 4.4 Radioactive decay of 20 mg of oxygen-15, which has a half-life of 2.0 min. The plotted data are given in the following table. After each half-life period, the quantity present at the beginning of the period is reduced by one half.

Number of Half-Lives	Fraction of Initial Quantity Remaining	Quantity Remaining (mg)
0	1	20.0 (initial)
1	1/2	10.0
2	1/4	5.00
3	1/8	2.50
4	1/16	1.25
5	1/32	0.625

after 12.9 h. In another 12.9 h, half of the original half, $(\frac{1}{2})^2 = (\frac{1}{4})$, will remain. This process continues indefinitely until virtually all the copper-64 isotopes have decayed. Figure 4.4 illustrates graphically how the concept of half-life works for a radioactive isotope. No matter what its half-life, the fraction of a radioactive isotope remaining will be one half after one half-life, one-fourth after two half-lives, one eighth after three half-lives, and so on.

By focusing attention on the decay products, we also are helped in understanding the concept of half-life. For example, if a sample contains one million copper-64 atoms at some beginning time, 12.9 h later only 500,000 copper-64 atoms would remain. However, there would be 500,000 zinc-64 atoms present that had not been there 12.9 h earlier. After 25.8 h (two half-lives) only 250,000 copper-64 atoms would remain, and there would be 750,000 zinc-64 atoms resulting from the decay of the copper atoms. After many half-lives, almost all the copper atoms will have decayed, and there will be almost one million zinc-64 atoms, which are stable and do not undergo decay.

Natural Radioactive Decay Series

As one would expect for an element with such a long half-life (Table 4.2), relatively large amounts of uranium-238 can be found in certain rocks and mineral deposits. Uranium-238 decays first to thorium-234, which then decays. These first two steps are part of the **uranium series**, which ends with a stable, nonradioactive isotope of lead, lead-206 (Figure 4.5, p. 80). You would not expect to find much of the short-lived isotopes in a sample of rock, and indeed you have to look carefully for them, but they are there. The longer lived isotopes such as uranium-234 and thorium-230 are readily detected. Two other similar natural decay series exist, each of which starts out with a different isotope and proceeds through a different set of radioactive decay products. Most of the naturally occurring radioactive isotopes are members of one of the three decay series.

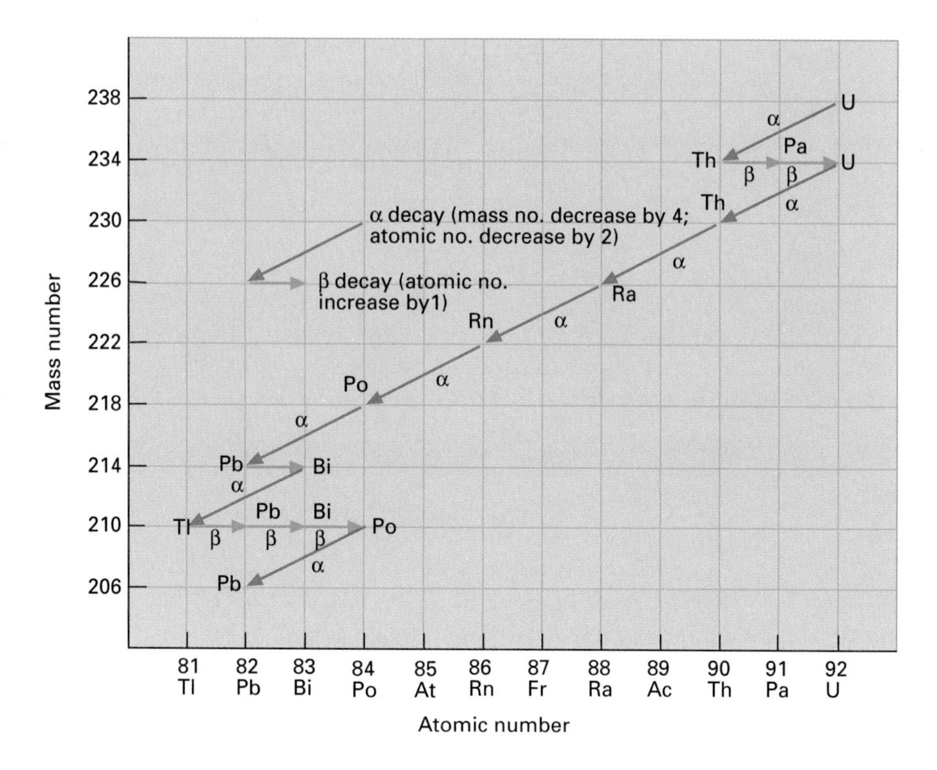

Figure 4.5 The uranium-238 decay series. Radium (Ra) and polonium (Po), the two elements discovered by Marie Curie, are part of this series. Radon (Rn), the radioactive gas of environmental concern, is generated as shown here wherever rocks contain uranium. Lead-206 is not radioactive. The half-lives of the isotopes in this decay series vary considerably. Uranium-238 has a half-life of 4.5 billion years while the half-life of lead-210 is 22 years. Some, like thallium-210, have half-lives of only a few minutes.

■ SELF-TEST 4A

1. The first evidence for radioactivity occurred when photographic films were exposed when placed near samples of uranium. (a) True, (b) False.
2. Which of these rays is the most penetrating radiation? (a) Alpha rays, (b) Beta rays, (c) Gamma rays?
3. The mass number of a nucleus is unchanged after beta particle emission. (a) True, (b) False.
4. The mass number of a uranium-238 nucleus will decrease by two units after alpha emission. (a) True, (b) False.
5. When a $^{216}_{84}Po$ nucleus emits an alpha particle, the nuclear species that results is ____________.
6. Nuclei with atomic numbers greater than 83 are all radioactive. (a) True, (b) False.
7. The half-life of ^{210}Bi is 5 days. If the initial activity was 2 μCi, or 74,000 dps, the activity after 10 days is expected to be ____________.

4.5 Artificial Nuclear Reactions

In 1919, Rutherford was successful in producing the first artificial nuclear change by bombarding nitrogen (N_2) with alpha particles. All the results of the experiment could be explained if one assumed the nuclear reaction to be

$$^{14}_{7}N + ^{4}_{2}He \longrightarrow [^{18}_{9}F] \longrightarrow ^{17}_{8}O + ^{1}_{1}H$$

where $^{18}_{9}F$ is an unstable nucleus that quickly disintegrates to $^{17}_{8}O$ and $^{1}_{1}H$. Both product nuclei are stable. Rutherford had observed an **artificial transmutation**, the conversion of one element into another during a laboratory experiment. Following Rutherford's original experiment, there was considerable interest in discovering new nuclear reactions.

In 1934, Irene Curie Joliot, daughter of Marie and Pierre Curie, and her husband Frédéric Joliot bombarded aluminum (Al) with alpha particles and observed neutrons and a positron. The Joliots discovered that when the flow of alpha particles striking the Al was stopped, the neutron emissions stopped, but the positron emissions continued. They reasoned that the alpha particles reacted with aluminum nuclei to produce phosphorus-30 nuclei, which then decayed to produce positrons.

$$^{27}_{13}Al + ^{4}_{2}He \longrightarrow ^{30}_{15}P + ^{1}_{0}n$$

$$^{30}_{15}P \longrightarrow ^{30}_{14}Si + \underset{\text{Positron}}{^{0}_{+1}e}$$

The second reaction continued because the phosphorus-30 was decaying more slowly than it was being produced. Phosphorus-30 was the *first radioactive isotope to be produced artificially*. Today, more than 1000 other radioactive isotopes have been produced.

■ Alchemists in ancient times tried in vain to transmute one element into another—mainly lead into gold. Nevertheless, they would have been extremely excited at the news of Rutherford's accomplishment.

4.6 Transuranium Elements

In 1940, at the University of California, E. M. McMillan and P. H. Abelson prepared element 93, the synthetic element neptunium (Np). Neptunium was the first of the **transuranium elements**, those with atomic number greater than 92. The neptunium was made by directing a stream of high-energy deuterons ($^{2}_{1}H$) onto a target of uranium-238. The initial reaction was the conversion of uranium-238 to uranium-239

$$^{238}_{92}U + ^{2}_{1}H \longrightarrow ^{239}_{92}U + ^{1}_{1}H$$

Uranium-239 has a half-life of 23.5 min and decays spontaneously to the element neptunium by the emission of beta particles

$$^{239}_{92}U \longrightarrow ^{239}_{93}Np + ^{0}_{-1}e$$

Neptunium is also unstable, with a half-life of 2.33 days; it converts into a second new element, plutonium (Pu)

$$^{239}_{93}Np \longrightarrow ^{239}_{94}Pu + ^{0}_{-1}e$$

Plutonium-239, like neptunium-239, is radioactive, but it has a half-life of 24,100 years. Because of the relative values of the half-lives, very little neptunium could be accumulated, but the plutonium could be obtained in larger quantities. The names of neptunium and plutonium were taken from the mythological names Neptune and Pluto in the same atomic number sequence as the order of the planets Uranus (uranium), Neptune, and Pluto out from the Sun.

■ No one knows for certain how much plutonium has been made—probably in excess of several million kilograms. Most of this amount has been used to make nuclear warheads by the United States, the former Soviet Union, and the other nuclear nations including the United Kingdom, France, China, Israel, and India.

■ Plutonium-239 is important because it is a fissionable isotope (Section 4.10) and can be used for making bombs. In addition, plutonium has the added distinction of being one of the most toxic substances known.

Although Neptune and Pluto are the last of the known planets in the solar system, their namesakes are not the last in the list of elements. The rush of experiments that followed the synthesis of plutonium produced additional elements: americium (Am), curium (Cm), berkelium (Bk), californium (Cf), einsteinium (Es), fermium (Fm), mendelevium (Md), nobelium (No), lawrencium (Lr), and elements 104 through 112. Up to element 101, mendelevium, all the elements can be made by hitting the target nuclei with small particles such as $^{4}_{2}He$ or $^{1}_{0}n$. Beyond element 101, special techniques using heavier particles are required. For example, lawrencium is made by hitting californium-252 with boron nuclei. In this reaction five neutrons are produced

$$^{252}_{98}Cf + ^{10}_{5}B \longrightarrow ^{257}_{103}Lr + 5\,^{1}_{0}n$$

The latest transuranium element to be discovered, as yet unnamed element-112, was prepared by hitting lead-208 nuclei with fast-moving zinc-70 nuclei. This discovery was made in 1996. Perhaps by the time you read this book, element 113 will have been prepared.

$$^{70}_{30}Zn + ^{208}_{82}Pb \longrightarrow ^{277}_{112}? + ^{1}_{0}n$$

The naming of the transuranium elements has produced some interesting problems. Unlike biology, where the right to name new species is granted exclusively to the first person who reports the discovery of that species, chemical elements are named by committee. In several cases, especially for those elements beyond fermium, element 100, working groups of the International Union of Pure and Applied Chemistry (IUPAC) have had to resolve controversies between laboratories that have claimed first discovery of certain elements. In a case involving element 106, American scientists named it seaborgium, after Glen Seaborg, a living scientist, who discovered ten of the transuranium elements (see The World of Chemistry, *A Revision to The Periodic Table,* p. 83). The IUPAC, however, did not wish to use that name, partly because Seaborg is still alive. On hearing this, he joked that he might have to go ahead and die, to be granted that honor. In August 1997, the IUPAC formally accepted the name seaborgium for element 106.

4.7 Radon and Other Sources of Background Radiation

Sources of radiation called background radiation, which we receive during normal daily life fall into two categories; man-made and natural (Figure 4.6). Of these, natural sources produce the bulk of our exposures. When natural sources are coupled with desirable exposures from medical applications, scarcely 3% of the exposures are truly avoidable by society as a whole. To illustrate how unavoidable most exposures to radiation are, consider the element potassium, an element essential to human life that helps regulate water balance in our bodies and is involved in nerve transmissions. A 60-kg (132-lb) person has about 200 g of potassium. It turns out that just over 0.01% of all potassium atoms are potassium-40 atoms, which are radioactive,

THE WORLD OF CHEMISTRY

A Revision to the Periodic Table

Program 7, *The Periodic Table*

Among the most significant contributions to the modern periodic table is that made by Nobel Laureate Glenn Seaborg. Among other things, he demonstrated the importance of maintaining the courage of one's convictions.

Thanks to his insights, it is now very well established that the transuranium elements (atomic numbers greater than 92), a number of which he either discovered or helped to discover during the Manhattan Project, are members of the actinide series. Actinides are the elements following actinium and belonging in a row usually placed at the bottom of the periodic table.

Until Seaborg offered his version of the periodic table, chemists were convinced that Th, Pa, and U belonged in the main body of the table, Th under Hf, Pa under Ta, and U under W. When Seaborg proposed that Th was the beginning of the actinides and that the transuranium elements belonged as a group under the rare earths, some prominent and famous inorganic chemists, many of them Seaborg's friends, tried to discourage him from publishing his idea in the open literature. One very prominent inorganic chemist felt that Seaborg would ruin his scientific reputation. Nevertheless, Seaborg was strongly convinced and persisted. As a result, he properly placed this most important class of elements where they are today. Based on Seaborg's expansion of the periodic table, it was possible to predict accurately the properties of many of the as yet undiscovered transuranium elements. Subsequent preparation in atomic accelerators of these elements proved him right, and it was fitting that he was awarded the Nobel Prize in 1951 for his outstanding work.

Glenn T. Seaborg (1912–) began his college education as a literature major, but changed his major to science in his junior year. He has been co-discoverer of ten of the transuranium elements, which are known only from laboratory synthesis.

with a half-life of 1.25×10^9 years. This means the 60-kg person has roughly 20 mg of radioactive potassium, which disintegrates at a rate determined by the half-life of that isotope. These disintegrations contribute to the background radiation we all receive. In addition, the beta particle from the decay of a potassium-40 disintegration in your neighbor might pass into you to con-

■ Even the altitude above sea level where you live is related to your annual dose of background radiation—the higher the altitude, the greater the cosmic radiation.

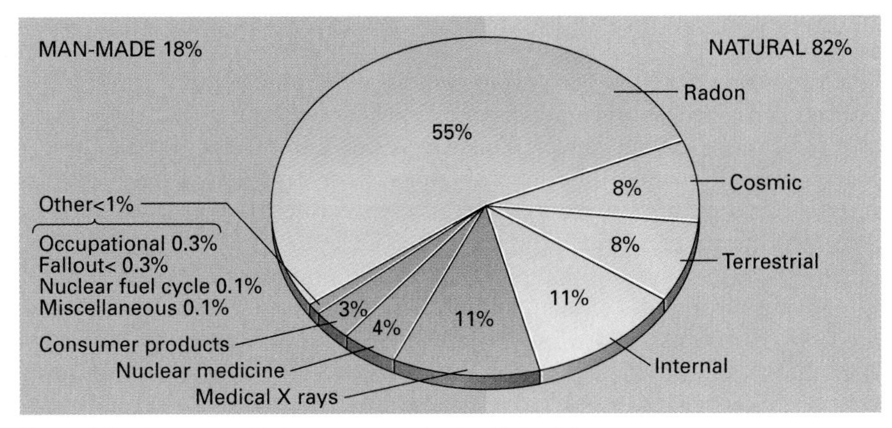

Figure 4.6 Average radiation exposure in the United States.

tribute to your background radiation. Normal background radiation from all sources is 2 or 3 dps.

■ Radon (Rn) is the heaviest member of the noble gas family of elements (He, Ne, Ar, Kr, Xe, and Rn).

Another element that contributes to our background radiation, and hence to our risks of radiation-caused damage, is radon. Radon-222, the most common isotope of radon, is radioactive, with a half-life of 3.82 days. It is a product of the uranium decay series (see Figure 4.5) and results from the alpha decay of radium-226.

$$^{226}_{88}\text{Ra} \longrightarrow \, ^{222}_{86}\text{Rn} + \, ^{4}_{2}\text{He}$$

When radon decays, it produces alpha particles and another short-lived radioisotope, polonium-218.

$$^{222}_{86}\text{Rn} \longrightarrow \, ^{218}_{84}\text{Po} + \, ^{4}_{2}\text{He}$$

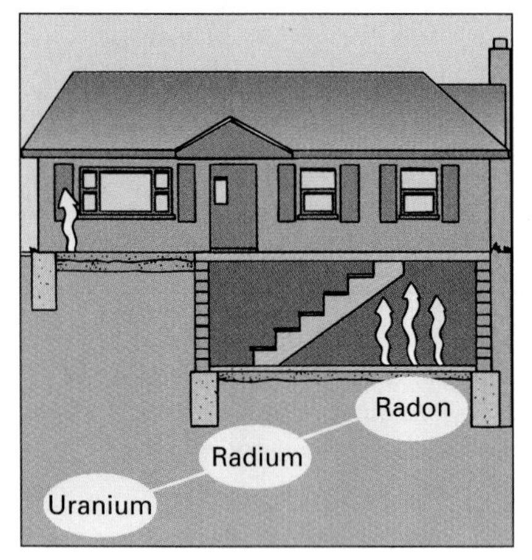

Common route of radon entry to homes—seepage through the foundation.

Polonium-218 (half-life 3.05 min) also decays, producing an alpha particle and lead-214 (half-life 26.8 min).

$$^{218}_{84}\text{Po} \longrightarrow \, ^{214}_{82}\text{Pb} + \, ^{4}_{2}\text{He}$$

These alpha particles from the products of radon decay make breathing air containing radon dangerous.

Radon exists as a gas, and all rocks contain *some* uranium, although in most the amount is small (one to three parts per million). Therefore, radon is constantly being formed in amounts that vary with the type of rock or soil. How much of this radon escapes into the outside air or a building on the surface depends on the porosity and moisture content of the soil and how finely divided it is (escape is easier from a small particle than from a large one). The overall geology of the site is also important.

■ *Daughters* is a term used to describe radioactive decay products. Radon daughters come from the decay of radon.

Because radon is chemically unreactive, radon atoms in the air we breathe are inhaled and exhaled without any chemical change, although some may dissolve in lung fluids. If a radon atom happens to decay within the lungs, however, the nongaseous and radioactive *radon daughters* can remain inside the lungs, where they will continue to decay. Those up to lead-210 have short half-lives and include several alpha emitters. Because alpha particles can travel up to 0.7 mm, the approximate thickness of the epithelial cells of the lung, they can damage delicate lung tissue and create a higher than normal risk of lung cancer.

■ As always, it is wise to keep in mind that all such data are based on statistical studies and risk assessment models. A number of studies of radon risk are still underway.

Miners in deep mines are exposed to far more than average radon levels, and as early as 1950 government agencies began monitoring radon exposures and incidences of lung cancer. Today, it is well known that radon exposure increases one's chance of developing lung cancer. If you smoke, the chances are even greater, since there seems to be a synergistic effect between smoking and radon levels in causing lung cancer.

■ If the radon radiation level is above 4 pCi/L, EPA requires that some form of remediation action be taken.

■ 4 pCi/L is a little less than 1.5 disintegrations every 10 s.

It has been estimated that perhaps 8 million homes in the United States are affected by radon contamination at levels higher than 4 pCi per liter of air (4 pCi/L), which is the "action level" the U.S. Environmental Protection Agency (EPA) has set. Radon has been detected in homes in almost every state.

This EPA action level is important for two reasons. First, levels of radon higher than 4 pCi/L should be reduced because levels above this are judged to lead to unacceptable risks of lung cancer. Second, it is difficult to lower the level of radon in most contaminated homes below 4 pCi/L. This last point is an acceptance of the fact that radiation exposure will always be with us.

■ Outdoor radon concentrations are approximately 0.1 pCi/L to 0.15 pCi/L worldwide.

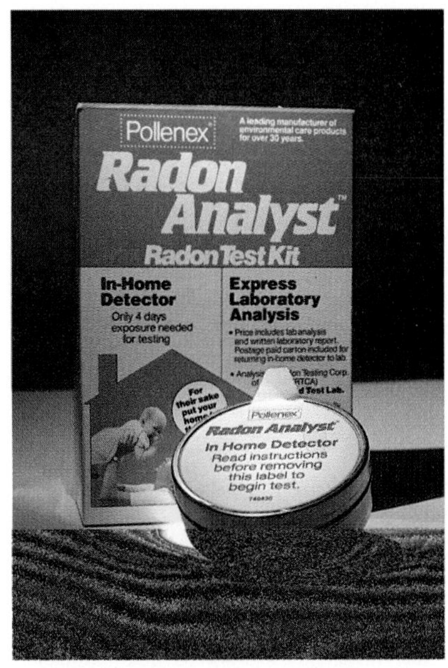

A commercially available kit for testing for radon in the home. *(C. D. Winters)*

■ SELF-TEST 4B

1. When unstable uranium-239 goes through beta decay the product nucleus is ___________.
2. The first transuranium element is ___________.
3. Normal background radiation from all sources produces 2 dps to 3 dps. (a) True, (b) False.
4. Experiments to make elements 104 through 111 require using projectiles bigger than alpha particles or neutrons. (a) True, (b) False.
5. Radon gas comes from the decay of what naturally occurring radioactive element?
6. The EPA action level for radon gas found in homes is ___________.

4.8 Useful Applications of Radioactivity

The damaging aspects of nuclear radiation must always be kept in mind, especially when the possibilities of accidental or unintended exposures are great. However, the radiation from radioisotopes can be put to beneficial use.

Food Irradiation

In some parts of the world, stored-food spoilage may claim up to 50% of the food crop. In the United States, refrigeration, canning, and chemical additives reduce this figure considerably. Still, there are problems with food spoilage and contamination. Food protection costs amount to a sizable fraction of the final cost of food.

Foods may be irradiated to retard the growth of organisms such as bacteria, molds, and yeasts. Food irradiation for this purpose using gamma rays from sources such as cobalt-60 or cesium-137 is common in European countries, Canada, Mexico, and the United States. Such irradiation prolongs shelf life under refrigeration in much the same way that heat pasteurization protects milk. For example, chicken normally has a three-day refrigerated shelf life. After irradiation, chicken may have a three-week refrigerated shelf life.

Foods irradiated at sufficient levels will keep indefinitely when sealed in plastic or aluminum foil packages (Figure 4.7). In 1994, The National Aeronautics and Space Administration (NASA) was granted permission to use irradiated frozen beefsteak on space missions. At present, the Food and Drug Administration (FDA) has approved irradiation of many classes of foods (Table 4.3). It is important to note that in no case is there any chance of irradiated food becoming radioactive. The ongoing debate about the safety of irradiated foods revolves around the possibly harmful nature of would-be "radiolytic products"—products of chemical changes in food caused by the high energy of the radiation. For example, could irradiation of food produce a chemical

Figure 4.7 Cartons of food ready to undergo irradiation. *(Hank Morgan/ Rainbow)*

TABLE 4.3 ■ Irradiated Foodstuffs Approved by U.S. FDA

Food	Purpose
Uncooked beef	Control of food-borne pathogens
Uncooked pork	Control of *Trichinella spiralis*
All fresh foods	Growth and maturation inhibition
All foods	Disinfestation of anthropod pests
Dry enzyme preparations	Microbial disinfection
Dried herbs and spices	Microbial disinfection
Uncooked poultry	Control of food-borne pathogens

Source: U.S. FDA regulations.

■ The FDA approved irradiation of beef in December 1997 in response to concerns due to food poisoning and product recalls of contaminated hamburger meat.

capable of causing genetic damage? To date, no evidence has been found for harmful radiolytic products, but further animal feeding studies are under way. Meanwhile, based on extensive review of current data, the FDA-approved radiation limits have been conservatively set at relatively low levels.

Medical Imaging

Radioisotopes are used in medicine in two distinctly different ways: diagnosis and therapy. In the diagnosis of internal disorders, physicians need information on the locations of the disorders. An appropriate radioisotope is introduced into the patient's body, either alone or combined with some other chemical, and it accumulates at the site of the disorder. There the radioisotope disintegrates and emits its characteristic radiation, which can be detected. Modern medical diagnostic instruments not only determine where the radioisotope is located in the patient's body but also construct an image of the area.

■ The half-life of $^{99m}_{43}Tc$, 6 h, makes it an ideal radioisotope for medical purposes. The more stable isotope $^{99}_{43}Tc$ has a half-life of 2.12×10^5 years. This means it has such a low activity that it will be eliminated from the patient's body before many disintegrations can occur.

Four of the most common diagnostic radioisotopes are listed in Table 4.4. Each produces gamma radiation, which in low doses is less harmful to the tissue than ionizing radiations such as beta or alpha rays. By the use of special carriers, these radioisotopes can be made to accumulate in specific areas of the body. For example, the pyrophosphate ion ($P_4O_7^{4-}$), a simple polyatomic ion, can bond to the technetium-99m radioisotope (the *m* denotes *metastable,* meaning the isotope decays by emitting a gamma ray to form a more stable

TABLE 4.4 ■ Diagnostic Radioisotopes

Radioisotope	Name	Half-Life (Hours)	Uses
^{99m}Tc*	Technetium-99m	6	As TcO_4^- to the thyroid, brain, and kidneys
^{201}Tl	Thallium-201	21.5	To the heart
^{123}I	Iodine-123	13.2	To the thyroid
^{67}Ga	Gallium-67	78.3	To various tumors and abscesses

* The technetium-99m isotope is the one most commonly used for diagnostic purposes. The *m* stands for *metastable,* a term explained in the text.

isotope with the same mass number), and together they accumulate in the skeletal structure where abnormal bone metabolism is taking place. Such investigations often pinpoint bone tumors.

The imaging method is based on the emission of gamma rays from the target organ. As the gamma rays strike a gamma-ray camera, the signal is processed by a computer and displayed as a video image (Figure 4.8).

$$^{99m}_{43}Tc \longrightarrow \, ^{99}_{43}Tc + \, ^{0}_{0}\gamma$$
Gamma ray

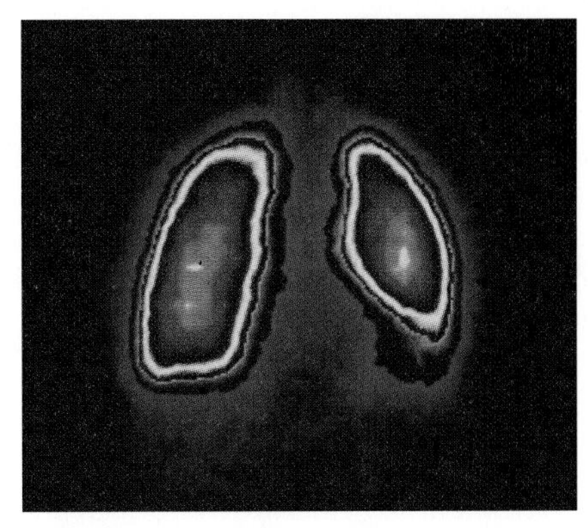

Figure 4.8 Radioisotopes in medical diagnosis. A gamma ray scan of healthy lungs is produced by technetium-99m. *(Jean-Perrin/CNRI/Science Photo Library/Photo Researchers)*

■ SELF-TEST 4C

1. Name two uses of radioactive isotopes.
2. The process of concentrating a radioisotope at a particular site of the body to locate and measure the extent of a disorder is called ______________.
3. In the symbol for the radioisotope technetium-99 m, the *m* stands for (a) middle, (b) mathematical, (c) metastable.
4. With its half-life of approximately 6 h, how much technetium-99m would remain 18 h after injection into a patient?
   (a) One eighth of the original dose
   (b) One half of the original dose
   (c) One sixth of the original dose
   (d) One fourth of the original dose
5. If two radioisotopes available for diagnosis worked equally well and each decayed by giving off gamma rays, but one had a half-life of 13 h and the other had a half-life of 6 h, which one would you recommend?

4.9 Energy from Nuclear Reactions

A vast amount of energy is released when heavy atomic nuclei split—the nuclear **fission** process—and when small atomic nuclei combine to make heavier nuclei—the **fusion** process. In 1938, Otto Hahn, Fritz Strassman, Lise Meitner, and Otto Frisch discovered that $^{235}_{92}U$ is fissionable by neutrons (Figure 4.9). In less than a decade, this discovery led to two important applications of this energy release accompanying fission—the atomic bomb and nuclear power plants.

There is a huge difference between the amount of energy liberated in an ordinary chemical reaction, like the burning of methane in air, and the energy liberated in a nuclear fission reaction. If you compared the energy from the burning of only 16 g of methane with that from the fission of an equivalent amount of uranium-235, the fission reaction would produce almost 25 million times more energy.

The World of Chemistry
Program 6, *The Atom*

Atomic bomb explosion.

Fission Reactions

Nuclear fission can occur when a neutron ($^{1}_{0}n$) enters a heavy nucleus. Certain heavy nuclei with an odd number of neutrons ($^{235}_{92}U$, $^{233}_{92}U$, $^{239}_{94}Pu$) will undergo fission when struck by slow-moving *thermal* neutrons (neutrons with a kinetic energy about the same as that of a gaseous molecule at ordinary temperatures). The splitting of the heavy nucleus produces two smaller nuclei, two or more neutrons (an average of 2.5 neutrons for $^{235}_{92}U$), and much energy. A

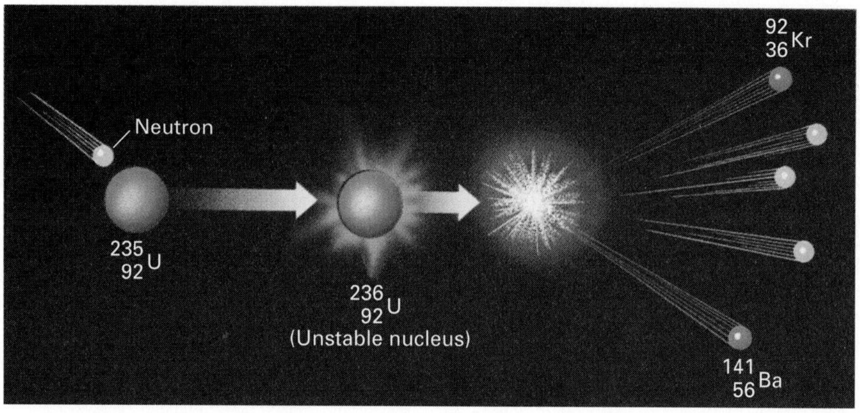

Figure 4.9 Nuclear fission. A slow-moving neutron strikes a uranium-235 atom, which becomes an unstable uranium-236 isotope. This unstable atom then quickly splits into two more-stable atoms of about equal mass. The three neutrons produced here are then available to cause fission in other adjacent uranium-235 atoms.

■ One kilogram of uranium fuel undergoing fission in a nuclear reactor can produce the same amount of energy as the combustion of 3000 tons of coal or 14,000 barrels of oil.

■ The same unstable $^{236}_{92}U$ atom may undergo fission to produce other products. For example

$$^{235}_{92}U + ^{1}_{0}n \longrightarrow ^{236}_{92}U \longrightarrow ^{103}_{42}Mo + ^{131}_{50}Sn + 2\ ^{1}_{0}n + \text{energy}$$

mole of uranium-235 atoms, each undergoing fission, will produce a number of different reaction products. One fission reaction that can take place follows:

$$^{235}_{92}U + ^{1}_{0}n \longrightarrow \underset{\text{Unstable nucleus}}{^{236}_{92}U} \longrightarrow \underset{\text{Typical fission products}}{^{141}_{56}Ba + ^{92}_{36}Kr} + 3\ ^{1}_{0}n + 4.6 \times 10^{9}\ \text{kcal}$$

Note that the same nucleus may split in more than one way. The lighter nuclei produced by the fission reaction are called **fission products**. These fission products, such as $^{141}_{56}Ba$ and $^{92}_{36}Kr$, can also be unstable and emit beta particles ($^{0}_{-1}e$) and gamma rays (γ), and may have long half-lives (see Section 4.4). Eventually, the decay of these fission products leads to stable isotopes.

The neutrons emitted by the fission of one uranium-235 atom can cause the fission of other uranium-235 atoms. For example, the 3 neutrons emitted

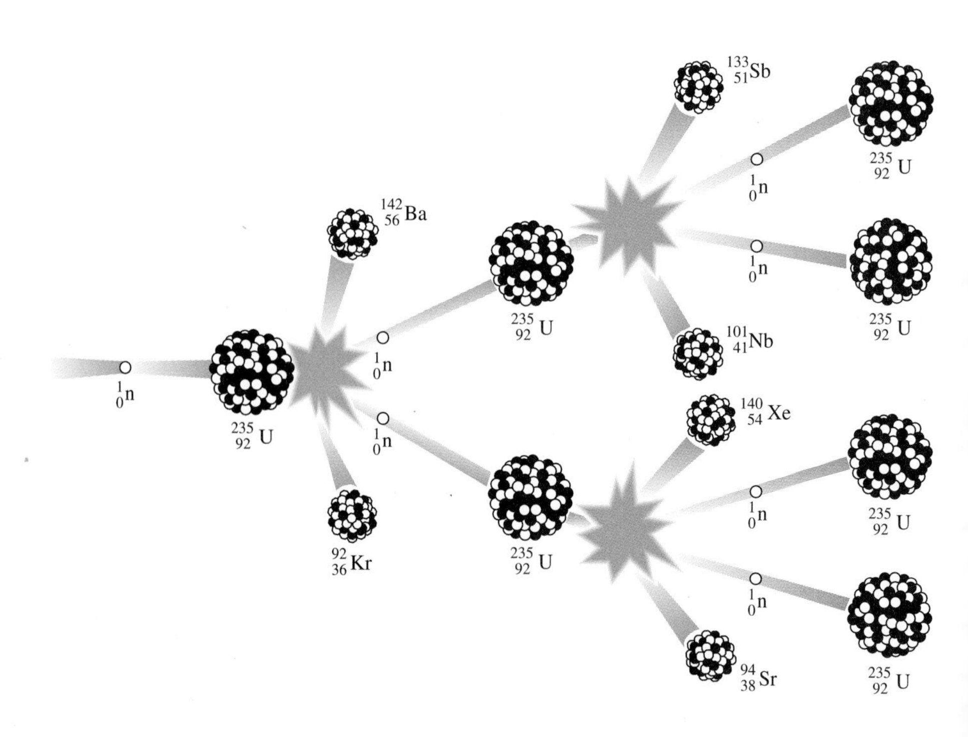

Figure 4.10 Illustration of a self-propagating nuclear chain reaction initiated by the capture of a slow-moving neutron. The fission of uranium-235 produces a number of different products. Thirty-four elements have been detected among uranium fission products, including those shown in this figure. Each fission produces two lighter nuclei, plus two or three neutrons which, in turn, can produce fission in other uranium-235 nuclei.

in the uranium fission (Figure 4.9) could produce fission in three more uranium atoms; the 9 neutrons emitted by those nuclei could produce nine more fissions; the 27 neutrons from these fissions could produce 81 neutrons; the 81 neutrons could produce 243 additional neutrons; those 243 neutrons could produce 729, and so on. This process is called a **chain reaction** (Figure 4.10), and it occurs at a maximum rate when the uranium sample is large enough for most of the neutrons emitted to be captured by other nuclei before passing out of the sample. A sample of fissionable material of sufficient size to self-sustain a chain reaction is termed the **critical mass**. If a critical mass of fissionable material is suddenly brought together, an explosion will occur be-

■ The critical mass for uranium-235 is about 10 kg.

THE PERSONAL SIDE

Lise Meitner (1878–1968)

Lise Meitner (AIP Emilio Segré Visual Archives Herzfeld Collection)

Lise Meitner was born in Vienna, Austria, one of seven children. In 1902, while studying science in Vienna, she became fascinated with the accounts of the discovery of radium by Pierre and Marie Curie and decided to pursue a career studying atomic physics. In 1906, she received her doctorate in physics from the University of Vienna. Two years later she moved to Berlin, where she became associated with Otto Hahn and Fritz Strassman, with whom she discovered atomic fission almost 20 years later. In 1918, together with Hahn, she discovered the element protactinium, which decays to form actinium. While working with Hahn and Strassman, Meitner discovered that something highly unusual was taking place when atoms of uranium-235 were struck by neutrons. In 1938, before she could solve this apparent puzzle, she fled to Sweden to escape the repression of Nazi Germany. After she established herself in Sweden, Otto Hahn announced the discovery that the uranium atom was being split into two approximately equal-sized parts by the neutron and sent Meitner his results for her analysis. Based on the data, she calculated the energy released when a uranium atom is split by the neutron and named the phenomenon nuclear fission. She even reported her findings in the British journal *Nature* in 1939. The uranium nucleus, she wrote, would split into a barium atom and a krypton atom under the influence of a neutron, at the same time producing more than 7.6×10^{-12} cal of energy (equal to 4.6×10^9 kcal/mol of uranium atoms undergoing fission). The enormous energy released by atomic fission was immediately recognized to have military potential, and in the United States the Manhattan Project was begun to create the first atomic bomb. In 1945, when she heard that the first atomic bomb had been dropped on Japan, Dr. Meitner said,

> *I must stress that I myself have not in any way worked on the smashing of the atom with the idea of producing death-dealing weapons. You must not blame us scientists for the use to which war technicians have put our discoveries.*

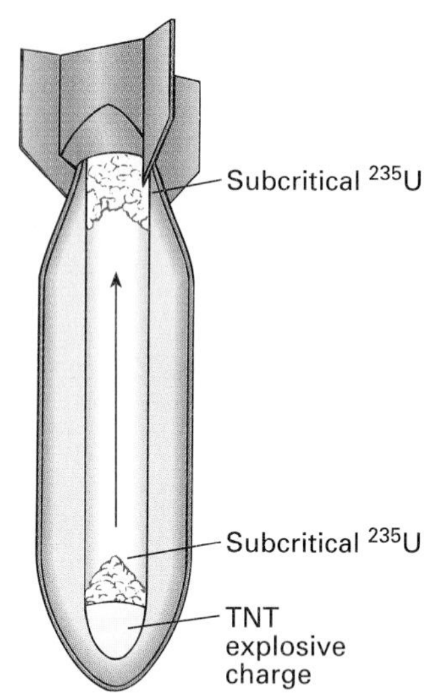

Figure 4.11 Implosion. By carefully shaping a conventional explosive, enough fissionable material can be brought together quickly to produce a massive chain reaction, causing many of the atoms to undergo fission almost simultaneously.

cause the energy released during each fission reaction cannot dissipate rapidly enough.

In a nuclear fission bomb the critical mass is kept separated into several smaller subcritical masses until detonation, at which time the masses are driven together by an explosive device (Figure 4.11). During the split second that the chain reaction occurs, the tremendous energy of billions of fission reactions is liberated, and everything in the immediate vicinity is heated to temperatures of 5 to 10 million kelvins (K) and vaporized. The sudden expansion of hot gases literally pushes aside everything nearby and scatters the radioactive fission fragments over a wide area.

Of the uranium found in nature, only 0.711% is the easily fissionable $^{235}_{92}U$. The other 99.289% is $^{238}_{92}U$, which is not fissionable by thermal neutrons. To make nuclear bombs or nuclear fuel for generation of electricity, the naturally occurring uranium must first be *enriched,* a process that increases the relative proportion of $^{235}_{92}U$ atoms in a sample. It is possible to enrich uranium so that the percentage of $^{235}_{92}U$ is between 2% and 5% by making use of slight differences in the volatility of uranium hexafluoride (UF_6). Of the two kinds of UF_6, $^{238}UF_6$ and $^{235}UF_6$, the molecule containing the lighter isotope is more volatile and will pass through porous barriers faster. After multiple passes, the desired separation is achieved.

■ Uranium enrichment by the movement of gaseous UF_6 molecules through a porous barrier has been replaced by other techniques such as those using laser beams to excite and ionize gaseous uranium −235 atoms, leaving uranium −238 behind as neutral atoms.

The Mass–Energy Relationship

What is the source of the tremendous energy of the fission process? It ultimately comes from the conversion of mass into energy, according to Einstein's famous equation, $E = mc^2$, where E is energy that results from the loss of an amount of mass m, and c is the speed of light (186,000 miles/s, or 3.00×10^8 m/s). If the masses of the products of the fission of a uranium-235 atom by a neutron are compared with the masses of the reactants, it is found that the products have less mass than the reactants. In the case of the fission reaction

$$^{1}_{0}n + ^{235}_{92}U \longrightarrow ^{93}_{37}Rb + ^{141}_{55}Cs + 2\,^{1}_{0}n + \text{energy}$$

the difference in mass is about 0.000214 g/mol of uranium-235. According to Einstein's equation, this mass difference is equivalent to 4.6×10^9 kcal/mol of $^{235}_{92}U$. The mass "lost" as a result of the fission process is the source of the tremendous energy that is released.

■ The binding energy is analogous to the concept of bond energy, in that both are a measure of the energy necessary to separate the whole (nucleus or molecule) into its parts.

The small nucleus of an atom is crowded with neutrons and protons. The very fact that nuclei hold together indicates that there must be some sort of force binding the particles together. This force is called the nuclear **binding energy** and is directly related to the stability of the nucleus. The binding energy depends on the number of particles in the nucleus. For example, if separate neutrons and protons (collectively, these particles are called **nucleons**) are combined to form any particular nucleus, the resulting nucleus always has less mass than the starting nucleons. This mass difference, converted into energy units according to Einstein's equation, is the binding energy and can be expressed for each atom as its *binding energy per nucleon* by dividing the total binding energy for the atom by the number of nucleons in the nucleus.

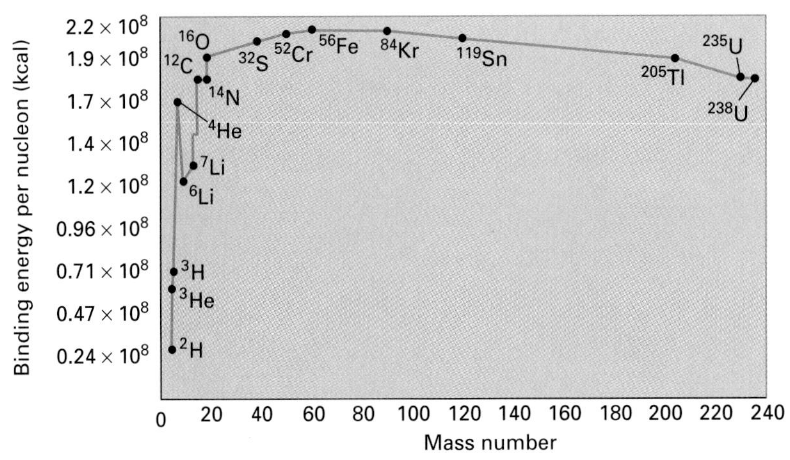

Figure 4.12 Relative stability of nuclei with different mass numbers. Both lighter and heavier isotopes are less stable (less binding energy per nuclear particle) than those isotopes with masses between 50 atomic mass units (amu) and about 65 amu. The most stable isotope is that of iron-56. Light isotopes can be fused together to form more stable atoms (nuclear fusion) while heavier isotopes can be split into more stable, lighter atoms (nuclear fission).

Figure 4.12 shows that light atoms and heavy atoms have lower binding energies per nucleon than atoms with mass numbers between 50 and 65. This means energy can be released when extremely light atoms are combined to make heavier atoms and when extremely heavy atoms are split to make lighter atoms. The greatest nuclear stability (greatest binding energy per nucleon) is at iron-56 ($^{56}_{26}Fe$). This is why iron is the most abundant of the heavier elements in the universe.

■ It takes only about 1 kg of uranium-235 or plutonium-239 undergoing fission to produce the equivalent of about 20,000 tons (20 kilotons) of ordinary explosives such as trinitrotoluene (TNT) or dynamite.

Because of their relative stabilities, most fission products fall into the intermediate range of atomic numbers (Figure 4.12). Therefore, when fission occurs and smaller, more stable nuclei result, these nuclei will contain less mass per nuclear particle. In the process, mass must be changed into energy—the tremendous energy released in a nuclear bomb or, under controlled conditions, in a nuclear power plant.

4.10 Useful Nuclear Energy

Electricity Production

Enrico Fermi (an Italian scientist who had immigrated to the United States in the late 1930s) and others believed that nuclear fission might somehow be made to proceed at a controlled rate. They reasoned that if a way could be found to control the number of thermal neutrons, their concentration could be maintained at a level sufficient to keep the fission process going but not high enough to allow an uncontrolled chain reaction. It would then be possible to drain the heat energy away on a continuing basis to do useful work.

The nuclear power plant at Indian Point, New York. *(Dan McCoy/Rainbow)*

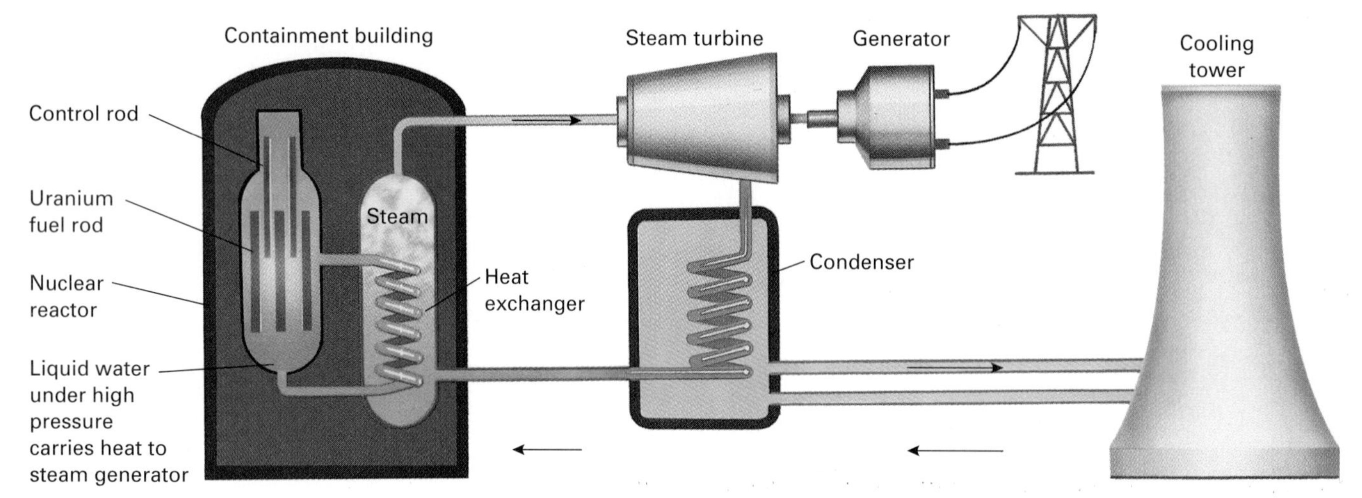

Figure 4.13 A nuclear power plant. In this reactor design (the most commonly used), ordinary water (called light water to differentiate it from D_2O—heavy water—used in some designs) is pressurized and allowed to carry heat energy from the reactor core to a heat exchanger, where high pressure steam is generated. This steam passes through a turbine, which generates electricity. Although simple in concept, safety considerations make the design, testing, and operation of a nuclear power plant a complex and costly operation.

■ Ordinary uranium, which is mostly $^{238}_{92}U$, cannot be used as a fuel in an atomic reactor because of the small concentration of the easily fissionable $^{235}_{92}U$ isotope.

In 1942, working at the University of Chicago, Fermi was successful in building the first atomic reactor.

An atomic reactor has several essential components. The reactor fuel must contain significant concentrations of atoms such as $^{235}_{92}U$, $^{233}_{92}U$, or $^{239}_{94}Pu$ that are fissionable by slow-moving neutrons. Typically, reactor fuel will contain uranium in the form of an oxide, U_3O_8, that has been enriched to contain about 3% or 4% $^{235}_{92}U$. A **moderator** is required to slow the speed of the neutrons produced in the reactions without absorbing them. Graphite and water have been used as moderators. A **neutron-absorber**, such as cadmium or boron steel, is present to provide fine control over the neutron concentration. **Shielding** to protect the workers from dangerous radiation is an absolute necessity. Shielding tends to make reactors heavy and bulky installations. Finally, a **heat-transfer fluid** provides a large and even flow of heat energy away from the reaction center. Water is used as a heat-transfer fluid in many nuclear reactor designs. In addition to water's high heat capacity (Section 5.10), its hydrogen atoms are excellent moderators for neutrons. Conventional technology then allows the heat energy carried by the hot water from the reactor to be used to generate electricity, to power ships, or to operate any device that uses heat energy. A system for the nuclear production of electricity is illustrated in Figure 4.13. Nuclear energy today produces about 100,000 megawatts (MW) of electrical power in 110 power plants throughout the United States. This is about 23% of the electricity used in this country. Some countries produce less energy by nuclear fission than the United States, but in others nuclear energy provides a much larger share of the total energy production (Figure 4.14).

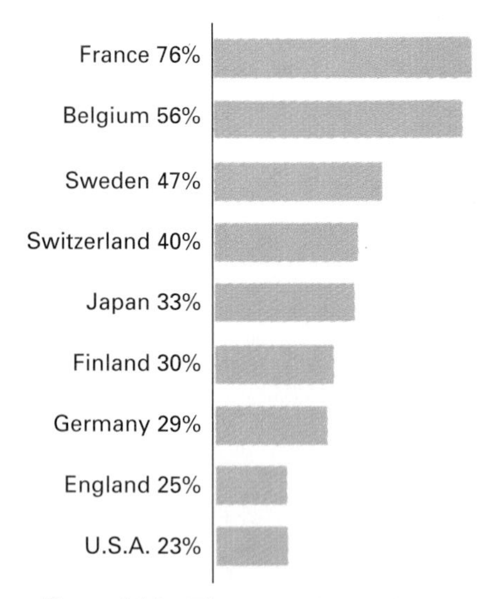

Figure 4.14 The approximate share of electricity generated by nuclear fission in various countries.

There are several extremely vexing problems associated with nuclear energy, any one of which may have adverse effects on its future as an alternative energy source. One problem, which probably is the greatest in the minds of the public, is the risk of a catastrophic accident at a nuclear power facility. Two accidents, of different degrees of seriousness, are generally known to the public. In late March of 1979, an accident occurred at the Three Mile Island power plant near Harrisburg, Pennsylvania. A water pump failed in the reactor and caused a partial reactor core meltdown. Steam vented inside the reactor vessel, and hydrogen gas was produced by the decomposition of the steam at the very high temperatures. This hydrogen gas, with the oxygen that was also produced, caused a risk of a chemical explosion that would have blown apart the safety containment, releasing fission products. In fact, some radioactive gases were vented into the atmosphere, resulting in an increased radiation dosage for people near the plant. There were no deaths directly associated with the Three Mile Island accident. Cleanup of the Three Mile Island facility continues today.

Certainly the most catastrophic nuclear accident occurred on April 26, 1986, at the Chernobyl unit 4 reactor near Kiev, Ukraine. The accident—a core meltdown, explosion, and fire—killed 31 people, hospitalized 500 others, and exposed many thousands of people to potentially harmful radiation. It has been estimated that the radiation from this accident will cause 17,000 extra cancer deaths in the next 70 years or so. So far, an estimated 5000 people have died from causes attributed to the accident. The Chernobyl reactor, built in 1983, had a design quite different from those used elsewhere in the world. Graphite was used exclusively as the moderator. While this design allows a higher efficiency than reactors with other types of moderators, the graphite can be ignited if sufficiently high temperatures are reached. In addition, the Chernobyl reactor lacked an adequate containment vessel to withstand an explosion in the reactor core.

While engineers were running an unauthorized test on the electrical generator, the Chernobyl reactor suddenly increased its power output and before neutron absorbers could be lowered into the core, a meltdown occurred. A vast amount of steam was formed, which together with burning blocks of graphite and radioactive fuel caused the entire reactor roof to blow off. The result was release of a radioactive plume that rose almost 5000 m into the atmosphere, scattering throughout much of Western Europe an estimated 100 million curies of radioisotopes, including the beta particle emitters ^{137}Cs (half-life 30 years) and ^{131}I (half-life 8.05 days). About 4.9 million people in Ukraine, Belarus, and Russia were directly affected by the release. Today the reactor is entombed in more than 300,000 tons of concrete, including an underground concrete liner to protect the groundwater. Indications are that the protective concrete may be slowly developing cracks and leaks.

■ At very high temperatures water will decompose into hydrogen and oxygen

$$2\ H_2O(g) \xrightarrow{\text{High temp}} 2\ H_2(g) + O_2(g)$$

■ In a core meltdown of a nuclear reactor, the failure of cooling would allow temperatures to rise above the melting point of the metal rods containing the uranium fuel (about 1205°C). In the worst scenario, the resulting mass of highly radioactive molten metal would melt through the steel and concrete of the containment vessel beneath it. Once out of the containment vessel, the radioactivity might contaminate groundwater supplies.

■ There are about 10 billion curies (Ci) of radioactive material within an operating nuclear reactor.

■ Cesium is in the same chemical family as sodium and potassium and reacts similarly to these elements. Cesium-137 has a long half-life (30 years) and can easily become a part of the food chain.

Plutonium, A Problem Element

Another problem resulting from nuclear energy production is *unavoidable plutonium production* and its possible diversion to bomb making. A fission reactor makes plutonium because some of the uranium-238 in the fuel rods captures neutrons, first forming $^{239}_{92}U$ (half-life 23.5 min), which decays by beta decay to $^{239}_{93}Np$ (half-life 2.35 days), which in turn forms $^{239}_{94}Pu$ (half-life 24,000 years) by beta decay.

$$^{238}_{92}\text{U} + ^{1}_{0}\text{n} \longrightarrow ^{239}_{92}\text{U}$$

$$^{239}_{92}\text{U} \longrightarrow ^{239}_{93}\text{Np} + ^{0}_{-1}\text{e}$$

$$^{239}_{93}\text{Np} \longrightarrow ^{239}_{94}\text{Pu} + ^{0}_{-1}\text{e}$$

The short half-lives of the intermediates mean that plutonium-239 quickly accumulates in the fuel rods of the reactor.

There are about 400 nuclear power plants worldwide, and these produce about 70 tons of plutonium annually. Already there is an accumulation of 700 tons of plutonium, and assuming that uranium fuel will be used in the future at its current rate, about 2000 more tons of plutonium will be produced by the time the estimated reserves of uranium fuel are used up.

About once every year, a third of the fuel rods in a nuclear reactor are replaced with new ones. In the United States, these fuel rods are currently being stored at the reactor sites while a permanent national storage site is being developed. During the early days of nuclear energy production in this country, nuclear fuel rods were *reprocessed,* allowing unreacted uranium and plutonium to be separated from the fission products and recycled back into new fuel rods. Through the reprocessing of nuclear fuel about 30% of the uranium in any fuel rod assembly could be reused, thus increasing the usefulness of this nonrenewable energy resource. In addition, the plutonium-239 found in the spent fuel rods was also recycled into new reactor fuel. The problem with nuclear fuel reprocessing is that with an additional few simple steps, plutonium of 99% purity can easily be separated from the uranium. This plutonium, mostly the easily fissionable plutonium-239 isotope, can be used to make fission bombs. Although the United States does not currently reprocess nuclear fuel, other countries, such as the United Kingdom, France, and Japan, do. This has led to widespread concern that plutonium from commercial reactors may someday appear in the hands of renegade governments and terrorists in spite of the best efforts to prevent such a thing from happening.

■ The diplomatic flap in 1993–1994 between North Korea and the other nations who signed the Nuclear Nonproliferation Treaty was caused by North Korea's unwillingness to allow inspections of its spent nuclear fuel to see if any of the plutonium it contained had been diverted. In 1962, the United States exploded a nuclear bomb made from plutonium derived from a civilian reactor.

There are other plutonium problems as well. Nuclear disarmament, begun in earnest in 1993, has brought with it the problem of what to do with the tremendous amount of plutonium that was fabricated into nuclear warhead materials. The United States currently dismantles about 1800 nuclear warheads annually. By 2003, about 50 metric tons of weapons-grade plutonium will have accumulated from dismantled warheads. In addition to the plutonium, an even greater amount—about 400 metric tons—of highly enriched uranium will also have been taken from warheads. Similar amounts of both plutonium and uranium will also have been taken from warheads in Russia. The enriched uranium from warhead dismantling will probably find uses in civilian nuclear reactor fuel, but currently no large-scale use of fissionable plutonium is planned. Only Japan and Russia are currently showing interest in using plutonium as a reactor fuel. The United States will probably opt to mix weapons-grade plutonium with other highly radioactive fuel waste from reactors and create glass logs that will weigh about 2 metric tons each. These logs would then be placed in geological repositories. The glass logs would be so highly radioactive that thieves would be unlikely to want to use them as a source of plutonium to make bombs.

■ In 1997, the U.S. Department of Energy (DOE) proposed that in addition to encapsulating waste plutonium in glass logs, some be diverted to use as nuclear fuel in commercial power reactors.

The nuclear wastes from all the U.S. nuclear power plants total about 2000 tons per year, and more than 15,000 tons are now being stored. In 1998, construction is scheduled to begin on an underground repository at Yucca Mountain in Nevada. The rock formation at Yucca Mountain is a dense volcanic ash that is predicted to allow any seepage into the ground-water to travel only about a mile every 3400 to 8300 years. Still, it may not be possible to put all of the stored nuclear power plant wastes at Yucca Mountain because of a "not-in-my-backyard" attitude among residents of the region.

■ Military nuclear wastes present yet another problem. Much more military waste exists, and it is more complex than civilian reactor wastes. Much of it is liquid. About 80 million gallons of high-level waste is currently in storage at several sites in the United States. Other countries with nuclear weapons programs have similar waste-disposal problems.

Fusion Reactions

Nuclear fusion produces about the same amounts of energy as fission on a per-mole basis but fewer radioactive by-products; the products that are radioactive have short half-lives. Hydrogen ($^{1}_{1}H$) has two isotopes, deuterium ($^{2}_{1}H$) and tritium ($^{3}_{1}H$). The nuclei of deuterium and tritium can be fused together at high temperatures to form a helium-4 atom. The energy released is 4.1×10^8 kcal.

$$^{2}_{1}H + ^{3}_{1}H \longrightarrow ^{4}_{2}He + ^{1}_{0}n + 4.1 \times 10^8 \text{ kcal}$$

When very light nuclei, such as those of hydrogen, helium, or lithium, are combined, or *fused*, to form an element of higher atomic number, energy equivalent to the difference between the total mass of the reacting atoms and the smaller total mass of the more stable products is released. This energy, which comes from a decrease in mass, is the source of the energy released by the sun, the stars, and hydrogen bombs. Typical examples of fusion reactions are

$$4\ ^{1}_{1}H \longrightarrow ^{4}_{2}He + 2\ ^{0}_{+1}e + 6.14 \times 10^8 \text{ kcal}$$
$$2\ ^{2}_{1}H \longrightarrow ^{3}_{2}He + ^{1}_{0}n + 7.3 \times 10^7 \text{ kcal}$$
$$2\ ^{2}_{1}H \longrightarrow ^{3}_{1}H + ^{1}_{1}H + 9.2 \times 10^7 \text{ kcal}$$
$$^{3}_{1}H + ^{2}_{1}H \longrightarrow ^{4}_{2}He + ^{1}_{0}n + 4.04 \times 10^8 \text{ kcal}$$

If fusion were to be used as a source of energy here on Earth, a suitable source of fusible atoms (fuel) would be needed. Fortunately, the oceans are a potential source of fantastic amounts of deuterium. There are 1.03×10^{22} atoms of deuterium in a *single liter* of seawater. If all the deuterium atoms in a cubic kilometer of seawater were fused to form heavier atoms, the energy released would be equal to that released from burning 1360 billion barrels of crude oil, and this is approximately the total amount of oil originally present on this planet.

Fusion reactions occur rapidly only when the temperature is of the order of 100 million degrees or more. At these high temperatures, atoms do not exist as such; instead, there is a **plasma** consisting of unbound nuclei and electrons. In this plasma nuclei can combine. The first fusion reactions that scientists were able to create artificially were produced in hydrogen bombs, or thermonuclear bombs. In a thermonuclear bomb, the high temperatures needed to initiate fusion are achieved by using the heat of a fission bomb (atomic bomb).

■ The term *thermonuclear* refers to the extreme temperatures required to cause nuclear fusion to take place.

In one type of hydrogen bomb, lithium deuteride ($^{6}_{3}Li^{2}_{1}H$, a solid salt) is placed around an ordinary $^{235}_{92}U$ or $^{239}_{94}Pu$ fission bomb. The fission reaction is set off in the usual way. A $^{6}_{3}Li$ nucleus absorbs one of the neutrons produced and splits into tritium and helium

$$^{6}_{3}Li + ^{1}_{0}n \longrightarrow ^{3}_{1}H + ^{4}_{2}He$$

The temperature reached by the fission of $^{235}_{92}U$ or $^{239}_{94}Pu$ is sufficiently high to bring about the fusion of tritium and deuterium.

Controlled Nuclear Fusion

Because there is so much available potential fuel in the oceans (as deuterium), controlled fusion seems like a natural candidate as an alternate source of energy. Another attractive feature is the rather limited production of dangerous radioisotopes. Most radioisotopes produced by fusion have short half-lives and therefore are a serious hazard for only a short time.

Three critical requirements must be met for controlled fusion to be a source of energy. First, the temperature must be high enough for fusion to occur. The fusion of deuterium ($^{2}_{1}H$) and tritium ($^{3}_{1}H$) requires a temperature of 100 million degrees or more. Second, the plasma must be confined long enough to release a net output of energy. Third, the energy must be recoverable in some usable form.

Fusion reactions are extremely difficult to control, principally because of the difficulty of holding the hot plasma together long enough for the particles to react. A further problem is reaching the very high temperature and maintaining it long enough to sustain the reaction. Nevertheless, progress is being made. By using a *magnetic bottle,* in which the plasma is held in place between two magnetic fields (Figure 4.15), scientists at the Princeton University plasma physics laboratory have achieved record levels of controlled fusion. In November 1994, fusion reactions in a half-and-half mixture of deuterium and tritium generated 10.7 million watts of power. So far, the longest power burst has lasted 0.20 s and the highest plasma temperature has been 460 million kelvins. Experiments indicate that alpha particles (He nuclei) from the fusion remain inside the plasma, adding their energy to the plasma and heating it up. If

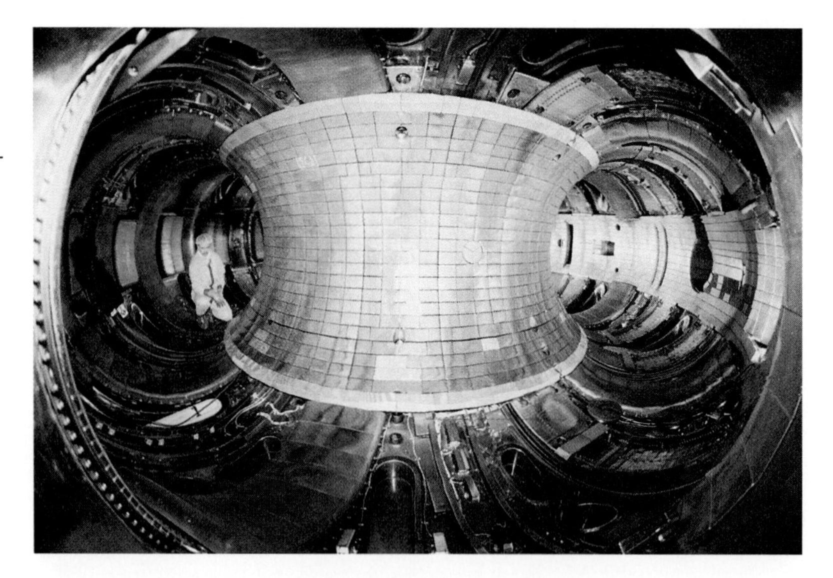

Figure 4.15 Inside the Tokamak fusion test reactor: the chamber where nuclear fusion takes place. Electric current running through vertical and horizontal coils that surround this doughnut-shaped space creates a pair of magnetic fields that interact to hold the plasma in place. The chamber is lined with graphite and graphite-composite tiles that can withstand the high temperatures. *(Princeton Plasma Physics Lab)*

sufficient heating is provided by the alpha particles, the reaction may become self-sustaining because it would need no external source of heat. In this condition, a commercial fusion reactor could produce more energy than it consumes, a condition not yet achieved. Only time will tell if fusion can be controlled to this extent. If it does happen, abundant and low-cost energy may truly become available to everyone.

SELF-TEST 4D

1. Which atom can undergo fission by thermal neutrons? (a) Uranium-238, (b) Uranium-235, (c) Krypton-131.
2. Complete the following nuclear equation

$$^{1}_{0}n + ^{235}_{92}U \longrightarrow [\ \_\_\_\_\_\_\_\_\_\_\_\_\ ] \longrightarrow ^{103}_{42}Mo + ^{131}_{50}Sn + 2\ ^{1}_{0}n$$

3. Complete the following nuclear equation

$$^{1}_{0}n + ^{239}_{94}Pu \longrightarrow ^{104}_{42}Mo + [\ \_\_\_\_\_\_\_\_\_\_\_\_\ ] + 2\ ^{1}_{0}n$$

4. If the masses of the nucleons making up an atom are summed together and compared with the mass of the atom these nucleons can make, the atom will always have a (greater/smaller) mass.
5. What is the most stable element, in terms of binding energy per nucleon? (a) Iron, (b) Hydrogen, (c) Uranium.
6. About what percent of the electricity generated in the United States is from nuclear energy? (a) 100%, (b) 23%, (c) 50%.
7. Plutonium is always produced in nuclear fission reactors. (a) True, (b) False.
8. Which value is closer to the half-life of plutonium-239? (a) 30 days, (b) 24,000 years, (c) 15.9 min.

MATCHING SET

___ 1. Thermal neutrons
___ 2. Most penetrating nuclear radiation
___ 3. Least penetrating nuclear radiation
___ 4. Discovered radioactivity
___ 5. Fissionable atom
___ 6. Co-discovered nuclear fission
___ 7. Same as alpha particle
___ 8. Reactor core meltdown
___ 9. Stable isotope
___ 10. Dangerous to the lungs
___ 11. Source of H that may someday be used as a nuclear fuel
___ 12. Positive electron
___ 13. Useful in radio imaging
___ 14. Nuclear disintegrations decrease by one half
___ 15. Loss from the nucleus causes an increase in atomic number
___ 16. Dangerous by-product from controlled nuclear fission

a. Half-life
b. Plutonium-239
c. He nucleus
d. Lead-206
e. Radon-222
f. The oceans
g. Positron
h. Gamma ray
i. Becquerel
j. Technetium-99m
k. Alpha ray
l. Slow moving
m. Beta particle
n. Uranium-235
o. L. Meitner
p. Chernobyl

QUESTIONS FOR REVIEW AND THOUGHT

1. Which of the following three types of radiation is the most penetrating? (a) Alpha rays (b) Beta rays (c) Gamma rays.
2. Give the type and approximate amount of material required to stop each of the following:
   (a) Alpha particles (b) Beta particles
   (c) Gamma rays
3. Which radioisotope sample is more hazardous, 1 g of ^{238}U with a half-life of 4.5 billion years or 1 g of $^{222}_{86}Rn$ with a half-life of 3 days? Explain your choice.
4. What is the nuclear reaction for the production of a beta particle?
5. What makes Rn-222 a health hazard? What health problem results from Rn-222 exposure?
6. Why must the sum of the mass numbers of the products of a nuclear reaction always equal the sum of the mass numbers of the reactants?
7. What is wrong with the following nuclear equation?

$$^{4}_{2}He + ^{27}_{13}Al \longrightarrow ^{49}_{15}P + ^{1}_{0}n$$

8. What kinds of projectile particles were used for the transmutations to form elements up to atomic number 101?
9. Nobel Laureate Glenn Seaborg proposed the present-day form of the periodic table. What change did he propose for the positions of the rows?
10. What are transuranium elements? How do they differ from the other elements in the periodic table?
11. What is meant by the term "uranium series"?
12. What is the final product of the uranium series?
13. What are the two major applications of radioisotopes in nuclear medicine?
14. What is the reason for irradiating food with gamma rays? What is the effect of the radiation?
15. What are the electrical charges and relative masses for the following?
    (a) Beta particles (b) Alpha particles
    (c) Gamma rays (d) Positrons
    (e) Neutrons
16. Tell why it is necessary to use high-energy accelerators to produce the transuranium radioisotopes from smaller nuclei.
17. Describe what effect gamma rays from cobalt-60 have on cancer cells and why this makes them useful in radiation therapy for cancer.
18. Why does airplane travel increase a person's annual dose of ionizing radiation?
19. Which source of background radiation contributes more to a person's annual dose, weapons test fallout or natural radiation in food, water, and air?
20. *Escherichia coli* contaminated food in fast-food establishments have caused poisonings and deaths in recent years. Contaminated meat was often the source. What effect would food irradiation have on the *E. coli* residues on meat?
21. Hahnium-260 (atomic number 105) was produced by the transmutation of californium-249 (atomic number 98) by bombardment with high-energy $^{15}_{7}N$ projectiles. What other particles resulted from the collisions?
22. What was the significance of exposed photographic films found near samples of uranium during the first experiments with radioactive materials?
23. What was the contribution made by each of the following scientists in the study of radioactivity?
    (a) Becquerel (b) Marie Curie
    (c) Rutherford
24. Describe how the penetrating power of beta rays compares with the penetrating power of gamma rays.
25. What effect does the emission of a beta particle have on the mass number of a nucleus?
26. Describe nuclear fission:
    (a) What is the starting material?
    (b) What is needed to cause fission?
    (c) What are the products of fission?
27. Name three problems that are associated with nuclear energy from fission.
28. What happened at the Chernobyl unit 4 reactor in 1986?
29. (a) What does the term *reprocessing of nuclear fuels* mean?
    (b) What danger is associated with it?
30. What is nuclear fusion?
31. Compare nuclear fission and nuclear fusion as sources of energy. Name the fuels, benefits, problems, and current status.
32. Complete the following nuclear equations:
    (a) $^{1}_{0}n + ^{235}_{92}U \rightarrow ^{142}_{56}Ba + [\_\_\_\_\_\_] + 3\,^{1}_{0}n$
    (b) $^{1}_{0}n + ^{235}_{92}U \rightarrow [\_\_\_\_\_\_] + ^{129}_{50}Sn + 2\,^{1}_{0}n$
    (c) $^{1}_{0}n + ^{239}_{94}Pu \rightarrow [\_\_\_\_\_\_] + ^{123}_{52}Te + 2\,^{1}_{0}n$
33. Balance the following nuclear equations, giving symbols, nuclear charges, and mass numbers:
    (a) $^{64}_{29}Cu \rightarrow \_\_\_\_\_\_ + ^{0}_{-1}e$
    (b) $^{69}_{30}Zn \rightarrow \_\_\_\_\_\_ + ^{69}_{31}Ga$
    (c) $^{131}_{53}I \rightarrow ^{131}_{54}Xe + \_\_\_\_\_\_$

34. Balance the following beta decay equations, giving symbols, nuclear charges, and mass numbers:
    (a) $^{14}_{6}C \rightarrow$ __________ $+ ^{0}_{-1}e$
    (b) $^{210}_{82}Pb \rightarrow$ __________ $+ ^{0}_{-1}e$
35. (a) What is the half-life of plutonium-239?
    (b) How does this add to the dangers of this isotope?
36. In what important way are the isotopes uranium-235 and plutonium-239 similar?
37. Describe how the "magnetic bottle" is used to contain a nuclear fusion reaction.

PROBLEMS

1. How many atoms of tritium will remain after four half-lives if there were initially 300,000 atoms?
2. Radioactive decay of ^{238}U can yield ^{234}Th and an alpha particle. An initial sample contained 200,000 atoms of ^{238}U. How many alpha particles will be produced from this sample after one half-life of 4.5 billion years?

CHAPTER 5

Chemical Bonding and States of Matter

Atoms of elements are only rarely found by themselves in nature, but atoms of the 90 naturally occurring elements are the basic building blocks of all matter. What makes atoms stick together? Valence electrons form the glue, but how? Atoms form a few different types of bonds in combining with other atoms. In this chapter we're going to describe two of them—ionic bonds and covalent bonds. Transfer of valence electrons from an atom of a metal to an atom of a nonmetal produces ionic bonds. Sharing of electrons between atoms of nonmetals produces covalent bonds. These simple ideas about valence electrons are the basis for understanding the bonding in two major classes of chemical compounds, ionic compounds and molecular compounds. In addition, with an understanding of how valence electrons function, we can better understand how the states of matter—gas, liquid, and solid—are formed.

- How are ionic bonds formed?
- How can the periodic table be used to predict the formulas of ionic compounds?
- How are covalent bonds formed?
- Why are some covalent bonds polar?
- How are the states of matter related to the attractions between molecules?
- What is hydrogen bonding, and why is it important?
- How are the properties of gases, liquids, and solids related to the bonding present?

Yellowstone National Park: As hot water evaporates from the blue solution of minerals in a hot spring, steam forms in the air and solid minerals collect around the pool. *(Stan Osolinski/Dembinsky Photo Associates)*

The concept of valence electrons developed by G. N. Lewis is useful for understanding how atoms of different elements interact and why elements in the same group have similar properties (Section 3.7). Lewis assumed that each noble gas atom had a completely filled outermost shell, which he regarded as a stable configuration because of the lack of reactivity of noble gases. Since all noble gases (except He) have eight valence electrons, the observation came to be known as the **octet rule**: *When atoms of elements react, they tend to lose, gain, or share electrons to achieve the same electron arrangement as the noble gas nearest them in the periodic table.* Metals can achieve a noble gas electron arrangement by giving up electrons, and nonmetals can achieve a noble gas electron arrangement by adding or sharing electrons. This is the basis for our discussion of the two major classes of bonding—ionic bonding and covalent bonding.

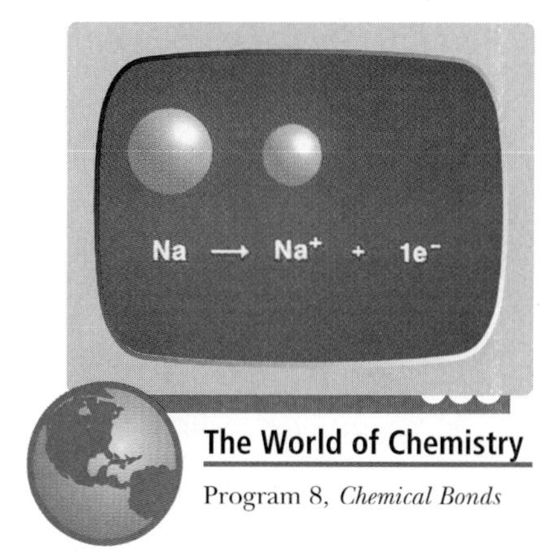

Program 8, *Chemical Bonds*

Loss of an electron to form a sodium cation.

5.1 Ionic Bonds

Sodium chloride (table salt) is always the best starting point for discussing ionic bonding and ionic compounds. We are all familiar with some of its properties because it's in every kitchen and on every dining table. Let's review what you know so far about sodium chloride. It is a white crystalline solid and is representative of a class of compounds known as *salts*. It is composed of two of the most reactive elements—the metal sodium (Na) and the nonmetal chlorine (Cl). Sodium is from Group IA and has a single valence electron, while chlorine is from Group VIIA and therefore has seven valence electrons.

$$\text{Na}\cdot \quad \cdot\ddot{\underset{\cdot\cdot}{\text{Cl}}}:$$

Application of the octet rule to sodium and chlorine atoms shows how they can form a compound. If sodium loses one electron, it will have the same outer electron arrangement as neon, which precedes it by one atomic number in the periodic table. Since the atom now has one less electron than it has protons, it acquires a single positive charge and is converted to what we call a *sodium ion* (Na^+). If chlorine gains one electron, it has the same outer electron arrangement as argon, which immediately follows it in the periodic table. In this case, the neutral atom has been converted to a *chloride ion* (Cl^-), which has a single negative charge because it has one more electron than it does protons.

Program 8, *Chemical Bonds*

Gain of an electron to form a chloride anion.

The reaction of sodium with chlorine to form sodium chloride is therefore fundamentally the transfer of an electron from a metal atom to a nonmetal atom

$$\text{Na}\cdot + \cdot\ddot{\underset{\cdot\cdot}{\text{Cl}}}: \longrightarrow \text{Na}^+ + :\ddot{\underset{\cdot\cdot}{\text{Cl}}}:^-$$

The strong electrostatic attraction between the positive and negative ions is known as the **ionic bond**, and compounds that are held together by ionic bonds are known as **ionic compounds**. Because chemical compounds are overall electrically neutral, sodium chloride must be composed of one sodium ion for every chloride ion. To show this composition, the formula of the compound is written as NaCl. (Note that the ionic charges are not indicated in the formula.) In ionic compounds the simplest ratio of oppositely charged ions that gives an electrically neutral unit is represented in the formula and is called a **formula unit**. The formula unit for sodium chloride is NaCl, or one sodium ion and one chloride ion.

■ To represent the reaction of sodium with chlorine as it actually occurs requires writing Cl_2 for elemental chlorine:

$$2\ Na + Cl_2 \longrightarrow 2\ NaCl$$

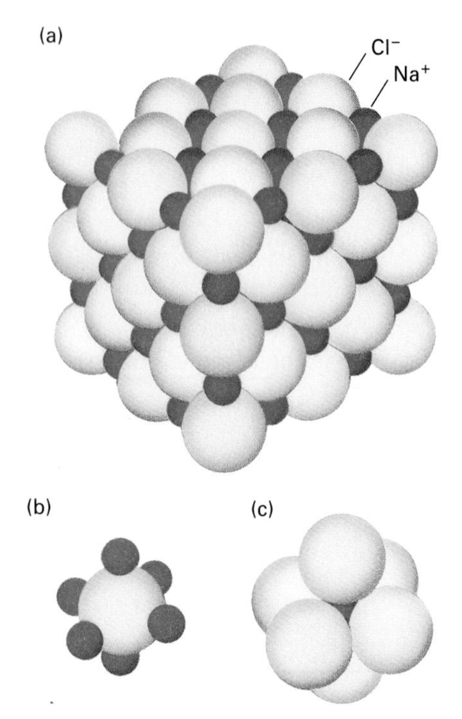

Figure 5.1 Structure of sodium chloride. (a) Model of the three-dimensional sodium chloride crystalline lattice. (b) Each Cl^- ion is surrounded by 6 Na^+ ions. (c) Each Na^+ ion, barely visible as a red area, is surrounded by 6 Cl^- ions.

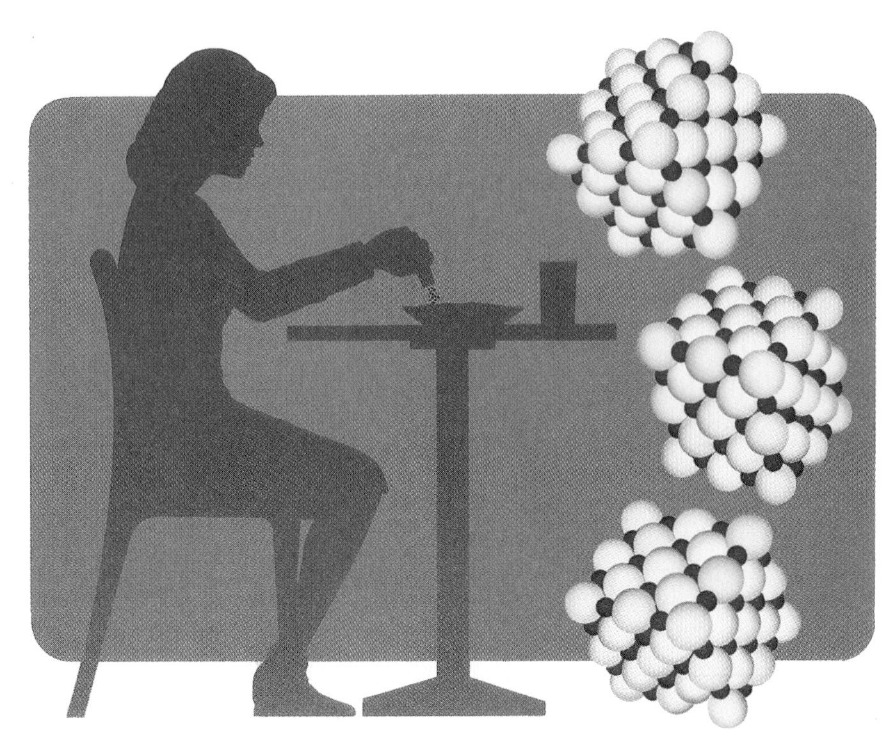

Sodium chloride at the dinner table and the arrangement of its ions.

Ionic crystals are made up of large numbers of formula units to form a regular three-dimensional crystalline lattice. A model of the sodium chloride crystalline lattice is shown in Figure 5.1. Note that each Na^+ ion has six Cl^- ions around it. Similarly, each Cl^- ion has six Na^+ ions around it. In this way, the one-to-one ratio of the singly charged ions is preserved. There is no *unique molecule* in ionic structures; no particular ion is attached exclusively to another ion, but each ion is attracted to all the oppositely charged ions surrounding it.

When atoms become ions, properties are drastically altered. For example, a collection of Br_2 molecules is red, but bromide ions (Br^-) contribute no color to a crystal of a compound such as NaBr. A chunk of sodium atoms (Figure 5.2, *left*) is soft, metallic, and violently reactive with water, but Na^+ ions are stable in water. A large collection of Cl_2 molecules (Figure 5.2, *center*) constitutes a greenish-yellow poisonous gas, but chloride ions (Cl^-) produce no color in compounds and are not poisonous. We have everyday evidence of this because we use NaCl (Figure 5.2, *right*) to season food. When atoms become ions, atoms obviously change their nature.

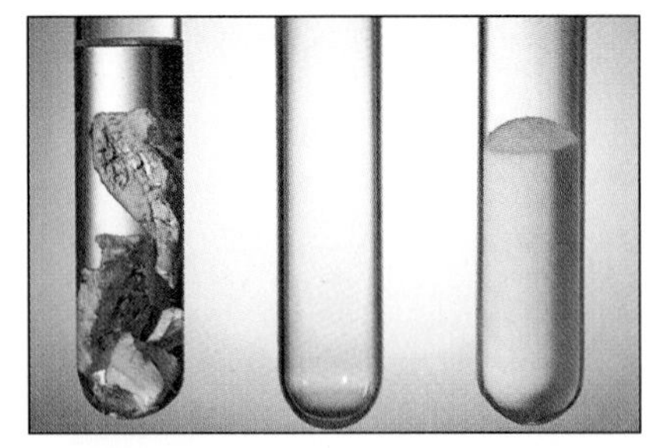

Figure 5.2 Sodium, chlorine, and sodium chloride. *(Larry Cameron)*

5.2 Ionic Compounds

Predicting Formulas

Lewis dot symbols can be used along with the octet rule to predict formulas for ionic compounds. For example, consider the reaction of calcium with oxygen. Calcium is a Group IIA metal and has two valence electrons. According to the octet rule, it will give up two electrons to form an ion with a +2 charge, Ca^{2+}, that has the same configuration as the nearest noble gas (Ar, 2–8–8). Oxygen is a Group VIA element whose atoms have six valence electrons and must gain two electrons to have the same configuration as the nearest noble gas (Ne, 2–8). The oxide ion will have a −2 charge, and the formation of the compound is essentially the transfer of two electrons:

$$\cdot Ca \cdot + \cdot \ddot{O} : \longrightarrow Ca^{2+} + :\ddot{O}:^{2-}$$

The formula of the ionic compound formed by this reaction is CaO, and it is called calcium oxide.

Generally, metals in Groups IA, IIA, and IIIA react with nonmetals in Groups VA, VIA, and VIIA to form ionic compounds. The formulas of thousands of ionic compounds can be predicted by using the periodic table and the octet rule to determine the ions formed by elements from these groups. This procedure can be summarized as follows:

1. Form a positive ion from a metal in an A group by removing the number of electrons equal to the group number.
2. Form a negative ion from a nonmetal in an A group by adding the number of electrons that when added to the group number gives a total of eight.
3. Write the formula unit for the ionic compound that gives the simplest ratio needed to produce an electrically neutral unit.

EXAMPLE 5.1 *Predicting Formulas of Ionic Compounds*

What is the formula of aluminum oxide, the ionic compound formed by the combination of aluminum and oxygen?

SOLUTION

1. Aluminum (Al) is in Group IIIA; so remove three electrons from an Al atom to give the Al^{3+} ion.
2. Oxygen (O) is in Group VIA; so add two electrons to give eight. The resulting ion is O^{2-}.
3. To make this formula electrically neutral: Al^{3+} ion is +3, and two ions would give a total charge of +6; O^{2-} ion is −2, and three ions would give a total charge of −6; +6 and −6 = 0; the formula is Al_2O_3.

Exercise 5.1

What is the formula of calcium fluoride, the ionic compound formed by the combination of calcium (Ca) and fluorine (F)?

Our examples of formula predictions have come from main-group metals because these metals are more predictable in ion formation. Many of the

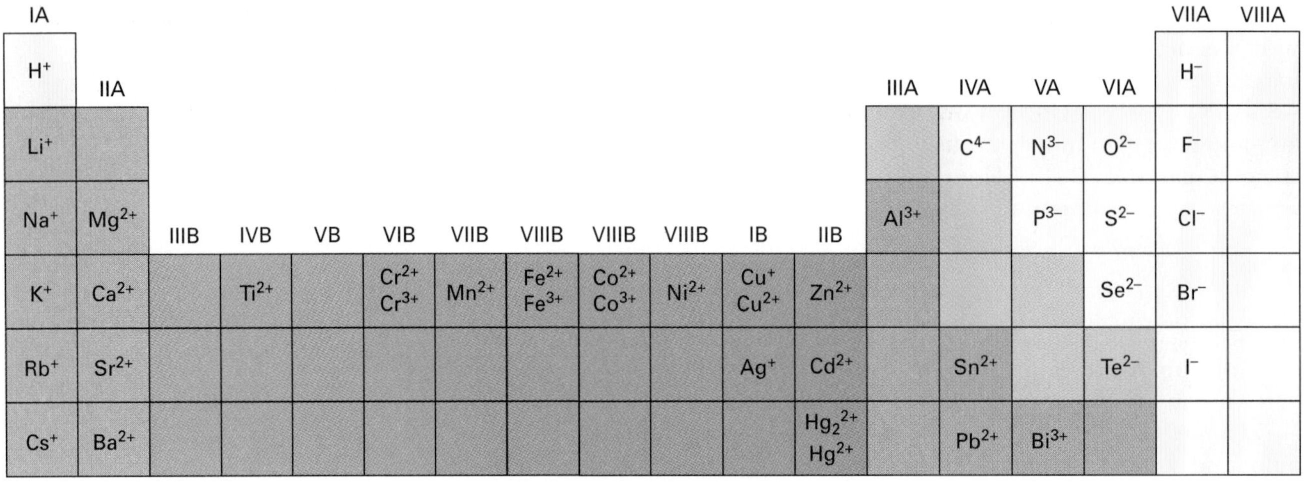

IA	IIA	IIIB	IVB	VB	VIB	VIIB	VIIIB	VIIIB	VIIIB	IB	IIB	IIIA	IVA	VA	VIA	VIIA	VIIIA
H^+																H^-	
Li^+													C^{4-}	N^{3-}	O^{2-}	F^-	
Na^+	Mg^{2+}											Al^{3+}		P^{3-}	S^{2-}	Cl^-	
K^+	Ca^{2+}		Ti^{2+}		Cr^{2+} Cr^{3+}	Mn^{2+}	Fe^{2+} Fe^{3+}	Co^{2+} Co^{3+}	Ni^{2+}	Cu^+ Cu^{2+}	Zn^{2+}				Se^{2-}	Br^-	
Rb^+	Sr^{2+}									Ag^+	Cd^{2+}		Sn^{2+}		Te^{2-}	I^-	
Cs^+	Ba^{2+}										Hg_2^{2+} Hg^{2+}		Pb^{2+}	Bi^{3+}			

Figure 5.3 Common ions. Metals usually form positive ions with a charge given by the group number in the case of the main-group metals (blue). For transition metals (red), the positive charge is variable, and other ions in addition to those illustrated are possible. Nonmetals (yellow) generally form negative ions with a charge equal to eight minus the group number.

■ Positive ions are called **cations**. Negative ions are called **anions**.

transition metals can give up different numbers of electrons to form more than one positive ion and are less predictable using the octet rule. For example, iron forms both Fe^{2+} and Fe^{3+} ions. The stable ions formed by some main-group metals and nonmetals are given in Figure 5.3 along with those of some transition metals. Notice in Figure 5.3 that main-group metals give up electrons according to the octet rule and main-group nonmetals add electrons according to the octet rule.

Naming Binary Ionic Compounds

By starting once again with table salt, we can use the formula and chemical name of this compound, NaCl (sodium chloride) to help remember the rules for naming compounds. The positive ion is named first, followed by the name of the negative ion. The element name is used for the positive ion. If the compound is made up of only one metal and one nonmetal (a **binary compound**), the name of the negative ion ends in *ide*. Examples we have already used besides sodium chloride include calcium oxide (CaO) and aluminum oxide (Al_2O_3). (Note that *binary* means only two elements are present, but the number of atoms in the formula can be more than two.) For transition metals that form ions with different charges, roman numerals are used with the names to indicate the charge. For example, iron(II) ion refers to Fe^{2+}, and iron(III) ion means Fe^{3+}. The name of $FeCl_2$ is iron(II) chloride, and the name of $FeCl_3$ is iron(III) chloride.

EXAMPLE 5.2 *Naming Binary Ionic Compounds*

Name the following ionic compounds: (a) K_2S, (b) $BaBr_2$, (c) Li_2O, (d) Fe_2O_3.

SOLUTION

All the compounds are binary compounds—made up of ions of two elements. The positive ion is named first, followed by the negative ion. The positive ion is named as the element. The negative ion is named by adding *-ide* to the stem of the name of the element. The correct names of the first three are (a) potassium sulfide, (b) barium bromide, and (c) lithium oxide. Compounds of iron require the use of a roman numeral after *iron* to identify its charge. The charge on the iron ion in Fe_2O_3 is determined by calculating the total negative charge (−6 from 3 O^{2-}) and dividing by 2 (because there are two iron ions). This gives a charge of +3; thus, the correct name of Fe_2O_3 is iron(III) oxide.

Exercise 5.2
Name the following compounds: (a) RbCl, (b) Ga_2O_3, (c) $CaBr_2$, and (d) Fe_3N_2.

EXAMPLE 5.3 *Writing Formulas for Binary Ionic Compounds*

Write formulas for the following compounds: (a) cesium bromide, (b) cobalt(III) chloride, (c) barium oxide.

SOLUTION

(a) The correct formula is CsBr since Cs (Group IA) forms Cs^+ and Br (Group VIIA) forms Br^-. (b) The roman numeral III indicates the Co^{3+} ion, and Cl (Group VIIA) forms Cl^-; thus, the correct formula is $CoCl_3$. (c) The correct formula is BaO since Ba (Group IIA) forms Ba^{2+} and O (Group VIA) forms O^{2-}.

Exercise 5.3
What are the formulas of (a) cobalt(II) sulfide, (b) magnesium fluoride, and (c) potassium iodide?

Ionic Compounds with Polyatomic Ions

Atoms of two or more elements can also combine to form a polyatomic ion, a chemically distinct species with an electric charge. Communication in the world of chemistry and understanding many applications require one to know the names, formulas, and charges of the common polyatomic ions listed in Table 5.1.

■ Most common polyatomic ions are negatively charged; ammonium ion (NH_4^+) is the major exception.

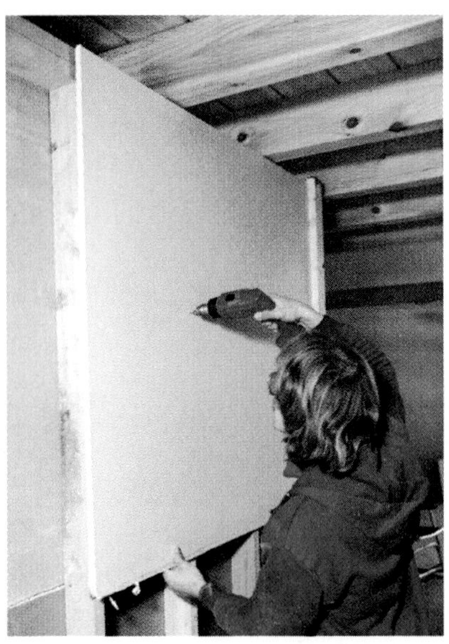

Installation of gypsum wallboard. Uses of gypsum ($CaSO_4 \cdot 2\ H_2O$) include the production of wallboard, portland cement, plaster of paris, and building plasters. There is evidence that the interiors of some of the great pyramids in Egypt were coated with gypsum plaster. *(C. D. Winters)*

TABLE 5.1 ■ Names and Composition of Some Common Polyatomic Ions

Cation (Positive Ion)			
NH_4^+	Ammonium ion		
Anions (Negative Ions)			
OH^-	Hydroxide ion	CO_3^{2-}	Carbonate ion
$CH_3CO_2^-$	Acetate ion	HCO_3^-	Hydrogen carbonate ion (or bicarbonate ion)
NO_2^-	Nitrite ion		
NO_3^-	Nitrate ion	PO_4^{3-}	Phosphate ion
SO_3^{2-}	Sulfite ion	HPO_4^{2-}	Hydrogen phosphate ion
HSO_3^-	Hydrogen sulfite ion	$H_2PO_4^-$	Dihydrogen phosphate ion
SO_4^{2-}	Sulfate ion	ClO^-	Hypochlorite ion
HSO_4^-	Hydrogen sulfate ion	ClO_3^-	Chlorate ion
	(or bisulfate ion)	ClO_4^-	Perchlorate ion
CN^-	Cyanide ion		

The bonding between the atoms within polyatomic ions is just like the bonding in molecular compounds (Section 5.3), but the group of atoms has either more or fewer electrons than protons and therefore has an overall charge. Compounds that contain polyatomic ions are ionic, and their formulas are written by the same procedure as described for binary ionic compounds. The only difference is that the polyatomic ion is enclosed in parentheses when the subscript is larger than one. For example, the formula of aluminum nitrate is $Al(NO_3)_3$. The compounds are also named in the same manner as binary ionic compounds, with the name of the positive ion followed by the name of the negative ion. Examples of some important ionic compounds with polyatomic ions are given in Table 5.2.

EXAMPLE 5.4 *Writing Formulas for Compounds of Polyatomic Ions*

Write the formulas of (a) magnesium sulfate and (b) calcium hydrogen sulfite.

SOLUTION

(a) The sulfate ion is SO_4^{2-} and the charge on the magnesium ion is $+2$; so the formula is $MgSO_4$. No parentheses are needed around the sulfate ion because only one SO_4^{2-} is present.
(b) The hydrogen sulfite ion is HSO_3^- and the charge on the calcium ion is $+2$; thus, the formula is $Ca(HSO_3)_2$.

Exercise 5.4

Write the formulas of (a) magnesium carbonate and (b) sodium dihydrogen phosphate.

TABLE 5.2 ■ Some Commercially Important Ionic Compounds with Polyatomic Ions

Formula	Name (Common Name)	Uses
NH_4NO_3	Ammonium nitrate	Fertilizers and explosives
KNO_3	Potassium nitrate	Gunpowder and matches
$NaOH$	Sodium hydroxide (lye)	Extract Al from ore; prepare soaps, detergents, and rayon; pulp and paper industry
$Mg(OH)_2$	Magnesium hydroxide	Milk of magnesia
Na_2CO_3	Sodium carbonate (washing soda, soda ash)	Water softening, detergents and cleansers, pulp and paper industry, glass and ceramics
$NaHCO_3$	Sodium bicarbonate (baking soda)	Household use, food industry, fire extinguisher
Na_3PO_4	Sodium phosphate	Food additive
$Ca(H_2PO_4)_2$	Calcium dihydrogen phosphate	Fertilizer
$CaSO_4$	Calcium sulfate	Gypsum, drywall (wallboard)
$Al_2(SO_4)_3$	Aluminum sulfate	Water purification

■ SELF-TEST 5A

1. The attractive forces between positive and negative ions in a crystal lattice are called ____________ bonds.
2. Positive ions are formed from neutral atoms by (a) losing electrons (b) gaining electrons.
3. What charge is expected when the following atoms form ions?
   (a) Lithium (Li) (b) Aluminum (Al)
   (c) Sulfur (S) (d) Bromine (Br)
4. Which of the following atoms form positive ions? (a) Potassium (K), (b) Bromine (Br), (c) Nitrogen (N), (d) Sodium (Na).
5. Negative ions are formed from neutral atoms by (a) losing electrons (b) gaining electrons.
6. When nutritionists refer to the importance of low salt intake, they are referring to the compound with the formula ____________ and the name ____________.

5.3 Covalent Bonds

There are a large number of compounds that are not ionic. These compounds are made up of the individual units we know as molecules (Sections 2.2, 2.4). What holds together the atoms in molecules of carbon monoxide (CO), methane (CH_4), water (H_2O), quartz (SiO_2), ammonia (NH_3), carbon tetrachloride (CCl_4), and millions of other compounds in which all the elements are nonmetals? G. N. Lewis proposed that the bonds holding atoms together in molecules consist of one or more pairs of electrons *shared* between the bonded atoms. The attraction of positively charged nuclei for electrons between them pulls the nuclei together. In many molecular compounds, atoms of nonmetals achieve noble gas electron arrangements (octet rule) by sharing electrons. The bond formed between two atoms that share electrons is called a **covalent bond**.

Single Covalent Bonds

■ A single covalent bond is formed when two atoms share a single pair of electrons.

A hydrogen atom has one electron. If it can share its electron with another atom that has an unpaired valence electron, a stable pairing of the two electrons can be achieved, and the H atom can then have the electron structure of helium, a noble gas. This arrangement can be achieved by two H atoms sharing their single electrons. The shared electrons are attracted by the positive nuclei of both atoms.

■ Ionic compounds are almost all solids, but molecular compounds can be gases, liquids, or solids.

Lewis dot symbols are used for the elements combining to form a molecule, and the resulting electron dot representation of the valence electrons in the molecule is called the **Lewis structure**. For example, the Lewis structure for H_2 shows two electrons (two dots) shared between two hydrogen nuclei (two H·)

$$\mathrm{H\cdot + H\cdot \longrightarrow H{:}H \quad or \quad H{-}H}$$

Lewis structure

Since each fluorine atom has one unpaired electron ($\cdot\ddot{\underset{\cdot\cdot}{\mathrm{F}}}{:}$), two fluorine atoms also can share an electron each to form a single covalent bond and an F_2 molecule. Each fluorine atom needs one electron to complete its outer shell. Shared electrons are counted toward the completion of the shells of both atoms.

$$2\,\cdot\ddot{\underset{\cdot\cdot}{\mathrm{F}}}{:} \longrightarrow {:}\ddot{\underset{\cdot\cdot}{\mathrm{F}}}{:}\ddot{\underset{\cdot\cdot}{\mathrm{F}}}{:} \quad \mathrm{or} \quad {:}\ddot{\underset{\cdot\cdot}{\mathrm{F}}}{-}\ddot{\underset{\cdot\cdot}{\mathrm{F}}}{:}$$

Only the shared pair of electrons represented between the two symbols (the two F's) are bonding valence electrons, and these are referred to as a **bonding pair** of valence electrons. The other six unshared pairs of electrons are called **nonbonding pairs** of valence electrons. In Lewis structures the bonding pairs of electrons are usually indicated by lines connecting the atoms they hold together, and nonbonding pairs are usually indicated by dots.

■ Nonbonding pairs of electrons are also called lone pairs.

What about Lewis structures for molecules such as H_2O or NH_3? Oxygen (Group VIA) has the Lewis dot symbol $\cdot\ddot{\underset{\cdot}{\mathrm{O}}}{:}$ and must share two electrons to satisfy the octet rule. This can be accomplished by forming covalent bonds with, for example, two hydrogen atoms:

$$2\,\mathrm{H}\cdot + \cdot\ddot{\underset{\cdot}{\mathrm{O}}}{:} \longrightarrow \mathrm{H}{:}\ddot{\underset{\cdot\cdot}{\mathrm{O}}}{:}\mathrm{H} \quad \mathrm{or} \quad \mathrm{H}{-}\ddot{\underset{\cdot\cdot}{\mathrm{O}}}{-}\mathrm{H}$$

Nitrogen ($\cdot\ddot{\underset{\cdot}{\mathrm{N}}}\cdot$, Group VA) must share three electrons to achieve a noble gas configuration, which can be done by forming covalent bonds with three hydrogen atoms

$$3\,\mathrm{H}\cdot + \cdot\ddot{\underset{\cdot}{\mathrm{N}}}\cdot \longrightarrow \mathrm{H}{:}\ddot{\underset{\underset{\mathrm{H}}{\cdot\cdot}}{\mathrm{N}}}{:}\mathrm{H} \quad \mathrm{or} \quad \mathrm{H}{-}\ddot{\underset{\underset{\mathrm{H}}{|}}{\mathrm{N}}}{-}\mathrm{H}$$

THE PERSONAL SIDE

Gilbert Newton Lewis (1875–1946)

G. N. Lewis was born in Massachusetts but raised in Nebraska. After earning his B.A. and Ph.D. degrees at Harvard University, he began his academic career. In 1912, he was appointed Chairman of the Chemistry Department at the University of California, Berkeley, and he remained there for the rest of his life. Lewis felt that a chemistry department should both teach and advance fundamental chemistry, and he was not only a productive researcher but also a teacher who profoundly affected his students. He developed his concepts about valence electrons and the stability of the noble gas electron configuration (octet rule) to explain periodic trends to students in his introductory chemistry course. Although he was teaching these concepts as early as 1902, he didn't publish them until 1916. Lewis also made major contributions to other areas of chemistry such as isotope studies, and acid–base theory.

Gilbert Newton Lewis (Photo by Frances Simon Courtesy AIP Emilio Segré Visual Archives)

Single Bonds in Hydrocarbons

The simplest large class of compounds is composed of **hydrocarbons**. These are organic compounds containing only carbon and hydrogen. Hydrocarbons that contain only C—C and C—H single bonds are called **alkanes**. Methane (CH_4), the simplest alkane, is the main component of natural gas. Is the formula CH_4 in agreement with what we would predict using Lewis dot symbols, the octet rule, and the Lewis structure? Carbon is in Group IVA; therefore, a carbon atom has four valence electrons and the Lewis dot symbol is $\cdot\dot{\underset{\cdot}{C}}\cdot$. To satisfy the octet rule, a carbon atom needs to gain a share of an additional four electrons. In this case, four hydrogen atoms with one valence electron each are needed; thus, the Lewis structure predicted for methane is

■ All organic compounds can be pictured as derived from hydrocarbons by the addition of various other kinds of atoms or groups of atoms. Organic compounds are so numerous and so important that they are the basis for an entire field of chemistry—organic chemistry (Chapters 9 and 10).

$$\begin{array}{ccc} & H & \\ H & :\ddot{\underset{..}{C}}: & H \\ & H & \end{array} \qquad \text{or} \qquad \begin{array}{ccccc} & & H & & \\ & & | & & \\ H & — & C & — & H \\ & & | & & \\ & & H & & \end{array}$$

An alkane always has four single bonds around each carbon atom. These may be C—C or C—H bonds. For example, the alkane molecule with three carbon atoms, propane, has the following Lewis structure:

■ Structural formulas like the one shown on the left for propane quickly get cumbersome as the number of atoms increase. One way to handle this is to write all of the symbols of the atoms on one line, with the atoms connected together written together. Propane would be $CH_3CH_2CH_3$. Formulas written this way are known as **condensed formulas.**

$$\begin{array}{ccccccccc} & & H & & H & & H & & \\ & & | & & | & & | & & \\ H & — & C & — & C & — & C & — & H \\ & & | & & | & & | & & \\ & & H & & H & & H & & \end{array}$$

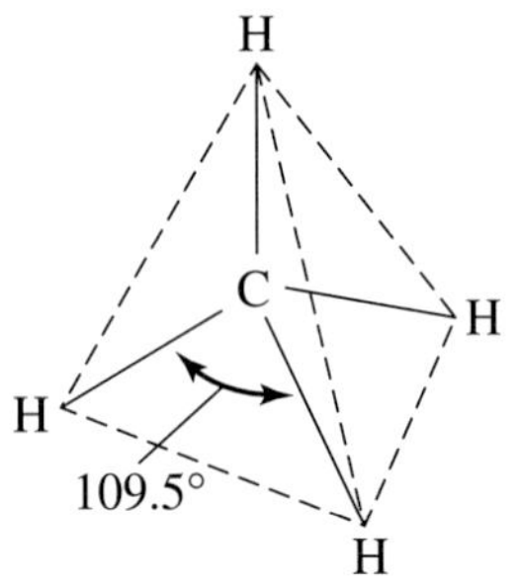

The four faces of a tetrahedron are equilateral triangles, and the angle between any two lines drawn from the center to two corners is 109.5°.

Alkanes are often referred to as **saturated hydrocarbons** because they have the highest possible ratio of hydrogen to carbon atoms bonded in a molecule. The general formula for saturated hydrocarbons is C_nH_{2n+2} where n is the number of carbon atoms.

The simplest hydrocarbon molecule, methane, has a tetrahedral shape, with the carbon atom at the center of the tetrahedron and the four hydrogen atoms at the corners. Every carbon atom in a saturated hydrocarbon molecule has its bonds tetrahedrally arranged. This arrangement allows the hydrogen atoms bonded to carbon atoms to be located as far as possible from one another.

Multiple Covalent Bonds

An atom with fewer than seven valence electrons can form covalent bonds in two ways. The atom may share a single electron with each of several other atoms that can each contribute a single electron. This leads to the single covalent bonds described in the previous sections. The atom can also share two (or three) pairs of electrons with a single other atom. In this case there will be two (or three) bonds between these two atoms.

When two shared pairs of electrons join the same two atoms, the bond is called a **double bond**. For example, carbon dioxide has two carbon–oxygen double bonds, represented by putting a pair of lines between the atom symbols, C═O. This can be predicted by using the octet rule and the Lewis dot symbols.

$$2\,\cdot\ddot{\underset{\cdot}{O}}\!: + \cdot\dot{\underset{\cdot}{C}}\cdot \longrightarrow :\!\ddot{O}\!::\!C\!::\!\ddot{O}\!: \qquad \text{or} \qquad :\!\ddot{O}\!=\!C\!=\!\ddot{O}\!:$$

■ The most common double or triple bonds involve carbon, nitrogen, oxygen, or sulfur atoms.

Each O atom needs two electrons, and C needs four electrons to satisfy the octet rule. To accomplish this, the carbon atom shares two electrons with each of the two oxygen atoms, forming two double bonds. Each double bond consists of two pairs of electrons and counts as part of the octet of each bonded atom. The result is that carbon is surrounded by four pairs of bonding electrons, and each oxygen is surrounded by two pairs of bonding electrons and two pairs of nonbonding electrons.

When two atoms share three pairs of bonding electrons, the result is a **triple bond**. In the N_2 molecule, each nitrogen atom needs to share six valence electrons (or three pairs) with the other to satisfy the octet rule.

$$2\,\cdot\ddot{\underset{\cdot}{N}}\cdot \longrightarrow :\!N\!:::\!N\!: \quad \text{or} \quad :\!N\!\equiv\!N\!:$$

Formation of the strong triple bond in N_2 helps to account for the energy released by many explosives.

Notice that each N atom has eight electrons around it, three bonding pairs and one nonbonding pair. The octet rule is thus satisfied.

Single, double, and triple bonds differ in length and strength. Triple bonds are shorter than double bonds, which in turn are shorter than single bonds. Bond energies normally increase with decreasing bond length. **Bond energy** is defined as the amount of energy required to break a mole of the bonds (the mole is discussed in Section 6.2). Some typical bond lengths and energies are listed in Table 5.3.

FRONTIERS IN THE WORLD OF CHEMISTRY

Fullerenes

The tendency for carbon to form single bonds to itself is well illustrated in the different forms of carbon, known as allotropic forms. Until the mid-1980s, only two allotropic forms of carbon were known—graphite and diamond. However, in 1985 Richard Smalley at Rice University and Harry Kroto of the University of Sussex, England, and their co-workers detected another form of carbon in the soot formed from laser vaporization of graphite. They proposed that the new form of carbon was a C_{60} molecule in the shape of an icosahedron. The C_{60} molecule resembles a hollow soccer ball; the surface is made up of five-membered rings linked to six-membered rings (like those in graphite). The discoverers named the new allotrope *buckminsterfullerene* (or simply *buckyball*) after the innovative American philosopher and engineer R. Buckminster Fuller, who popularized the icosahedral shape by using it in his patented geodesic dome.

Recent research indicates that C_{60} is the first of a family of *fullerenes* that contain an even number of carbon atoms arranged in closed, hollow cages. Others that have been discovered include C_{70}, C_{76}, C_{90}, C_{94}, and even giant fullerenes such as C_{240}, C_{540}, and C_{960}. The original buckyball, C_{60}, though, is the one found in greatest abundance in soot. Some possible uses include lightweight batteries, new lubricants, antitumor therapy for cancer patients (by enclosing a radioactive atom within the cage), and microscopic ballbearings.

Richard Smalley, Harry Kroto, and Robert Curl, Jr. received the 1996 Nobel Prize in chemistry for their discovery of fullerenes.

Structure of fullerene, C_{60}. The soccer ball is a model of the C_{60} structure. The surface of C_{60} is made up of five-membered rings (black rings on soccer ball) and six-membered rings (white rings on soccer ball). Seams of the soccer ball represent covalent bonds between carbon atoms at the intersection of each seam with other seams. (C. D. Winters)

Multiple Covalent Bonds in Hydrocarbons

Ethylene (Figure 5.4) contains a double bond between the carbon atoms and single bonds between the hydrogen atoms and the carbon atoms. Ethylene is the first member of the **alkene** series of hydrocarbons, compounds that have one or more C═C bonds, that is, carbon–carbon double bonds.

TABLE 5.3 ■ Some Bond Lengths and Bond Energies

Bond type	C—C	C═C	C≡C	N—N	N═N	N≡N
Bond length (nm)	0.154	0.134	0.120	0.140	0.124	0.109
Bond energy (kcal/mol)*	83	146	200	40	100	225

* kcal/mol (kilocalories per mole) = thousands of calories necessary to break 6.02×10^{23} bonds (see Section 6.2).

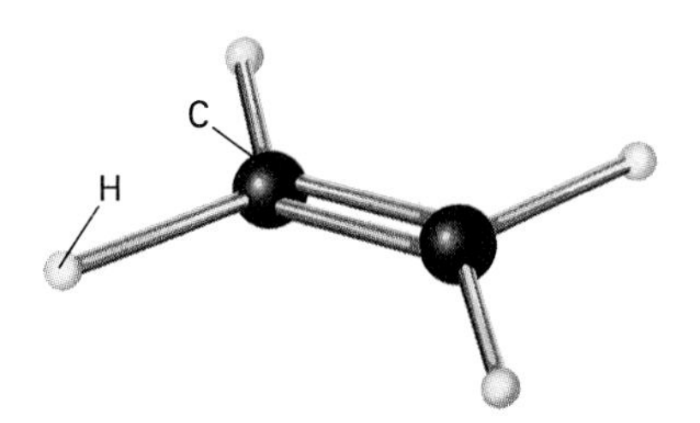

Figure 5.4 Structure of ethylene molecule. All the atoms in this molecule lie in the same plane.

■ The official name of ethylene is ethene.

$$\begin{array}{cc} H & H \\ | & | \\ C & = C \\ | & | \\ H & H \end{array}$$

■ Polyethylene (Section 10.5) is the most widely used polymer. Examples of plastics made from polyethylene include milk bottles, sandwich bags, garbage bags, toys, and molded objects.

More than 48 billion lb of ethylene are produced annually in the United States. About half is used in the manufacture of polyethylene plastics. The structural formula of ethylene illustrates why alkenes are said to be **unsaturated hydrocarbons**; they contain fewer hydrogen atoms than the corresponding alkanes and react with hydrogen to form alkanes.

$$H_2 + \begin{array}{cc} H & H \\ | & | \\ C & = C \\ | & | \\ H & H \end{array} \longrightarrow H-\begin{array}{c} H \\ | \\ C \\ | \\ H \end{array}-\begin{array}{c} H \\ | \\ C \\ | \\ H \end{array}-H$$

The World of Chemistry

Program 9, *Molecular Architecture*

Molecular model of ethylene.

Acetylene, the gas that produces a flame hot enough to cut steel when it is mixed with oxygen and burned, has a carbon–carbon triple bond, $C{\equiv}C$.

$$H-C{\equiv}C-H$$

Names for Binary Molecular Compounds

Hydrogen forms binary compounds with all the nonmetals (except the noble gases). For compounds of oxygen, sulfur, and the halogens, the H atom is generally written first in the formula and is named first using the element name. The other nonmetal is named as if it were a negative ion. For example, HF is hydrogen fluoride and H_2S is hydrogen sulfide.

When there is more than one possible combination of two elements, the number of atoms of a given type in the compound is designated with a prefix such as *mono-, di-, tri-, tetra-,* and so on. Table 5.4 lists the prefixes for up to ten atoms, and some common molecular compounds and their names are given in Table 5.5.

Many molecular compounds were discovered years ago and have names so common they continue to be used. Examples include water (H_2O), ammonia (NH_3), nitric oxide (NO), and nitrous oxide (N_2O).

TABLE 5.4 ■ Prefixes for Number of Atoms in a Compound

Number of Atoms	Prefix	Number of Atoms	Prefix
1	mono-	6	hexa-
2	di-	7	hepta-
3	tri-	8	octa-
4	tetra-	9	nona-
5	penta-	10	deca-

EXAMPLE 5.5 *Naming Molecular Compounds*

What is the name of N_2O_4?

SOLUTION

The prefix for 2 N atoms is *di-* and the prefix for 4 O atoms is *tetra-*; thus, the name of N_2O_4 is dinitrogen tetraoxide.

Exercise 5.5

What is the name of P_4S_3?

TABLE 5.5 ■ Common Molecular Compounds

Compound	Name	Use
CO	Carbon monoxide	Preparation of methanol and other organic chemicals
CO_2	Carbon dioxide	Carbonated beverages, fire extinguisher, inert atmosphere, Dry Ice
NO	Nitrogen monoxide (nitric oxide)	Preparation of nitric acid
NO_2	Nitrogen dioxide	Preparation of nitric acid
N_2O	Dinitrogen oxide (nitrous oxide)	Spray can propellant, anesthetic
SO_2	Sulfur dioxide	Preparation of sulfuric acid, food preservative, metal refining
SO_3	Sulfur trioxide	Preparation of sulfuric acid
CCl_4	Carbon tetrachloride	Solvent
SF_6	Sulfur hexafluoride	Insulator in electric transformers
P_4O_{10}	Tetraphosphorus decaoxide	Preparation of phosphoric acid

■ SELF-TEST 5B

1. The name of SO_2 is ______________________.
2. The formula of sulfur trioxide is ______________________.
3. The name of HBr is ______________________.
4. The formula of dichlorine monoxide is ______________________.
5. The formula of sulfur dichloride is ______________________.
6. The name of $SiCl_4$ is ______________________.
7. How many electrons are shared in (a) a double bond? (b) a triple bond?
8. Which is a strongest bond? (a) C—C, (b) C═C, (c) C≡C.

5.4 Shapes of Molecules

Molecules come in a variety of shapes. If a molecule contains only two atoms, it has the simplest shape; it is linear. Whenever three or more atoms are in a molecule, the number of possible shapes increases. For example, both CO_2 and H_2O have three atoms in their molecules. In CO_2 the central atom is the C atom, and in H_2O the central atom is the O atom. In spite of the similarity in the number of atoms, the CO_2 molecule is linear while the H_2O molecule is bent. The angles between the covalent bonds connecting the central atom and its neighbors vary in molecules with different shapes. In CO_2 the bond angle is 180° while in H_2O the bond angle is 104.5°. This difference results from the different numbers of electron pairs around the central atom and how these electron pairs repel one another.

You have already learned to write Lewis structures for simple molecules. The Lewis structure for CO_2 is

$$:\ddot{O}::C::\ddot{O}:$$

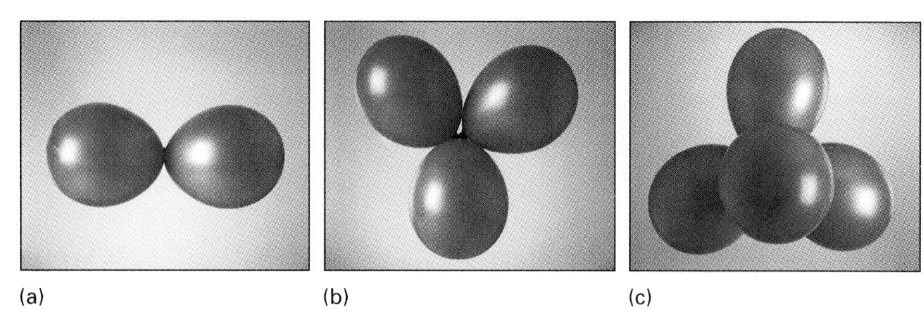

Figure 5.5 Balloon models of geometries predicted for two, three, and four electron pairs. *(Kristen Brochmann/Fundamental Photographs)*

and the Lewis structure for H_2O is

$$\mathrm{H:\ddot{\underset{\cdot\cdot}{O}}:H}$$

Note that there are two electron pairs associated with each double bond in the CO_2 molecule and that these are located between the C and O atoms. There are four electron pairs around the O atom in the water molecule, two of which are located between the O and H atoms. The fact that like charges repel each other tell us that the electron pairs in the CO_2 molecule will tend to get as far from each other as possible. This results in a bond angle of 180°. In the case of the H_2O molecule, the four electron pairs can get as far apart as possible if they are at approximately 109.5° apart. This angle is called the tetrahedral angle and is well known to students of geometry (see p. 110). When some of the electron pairs around the central atom are nonbonding, they can distort the predicted bond angles somewhat. For example, in the water molecule, the H—O—H bond angle is 104.5° instead of the tetrahedral angle. While this small difference is important, it is not as important as the fact that the water molecule is bent and that it contains two nonbonding pairs of electrons on the O atom.

■ By placing all four electron pairs in the same plane a 90° angle results between them. Since this angle is less than the tetrahedral angle, the H—O—H angle in the H_2O molecule is not 90°.

What is the arrangement of only three electron pairs about the central atom? They would be arranged in a plane at 120° apart. This arrangement, along with those for two electron pairs and four electron pairs about the central atom, can account for the shapes of a large number of both simple and complex molecules (Figure 5.5). With these ideas, we can now turn our attention to the subject of polar bonds and polar molecules.

5.5 Polar and Nonpolar Bonding

In a molecule like H_2 or F_2, where both atoms are alike, there is equal sharing of the electron pair. Where two unlike atoms are bonded, however, the sharing of the electron pair is unequal and results in a shift of electric charge toward one partner. The more nonmetallic an element is, the more that element attracts electrons.

When two atoms are bonded covalently and their abilities to attract electrons are the same, there is an equal sharing of the bonding electrons, and the bond is a **nonpolar** covalent bond. The bonds in H_2, F_2, and NCl_3 (N and Cl have equal abilities to attract electrons) are nonpolar.

Two atoms with different abilities to attract electrons bonded covalently form a **polar** covalent bond. The bonds in HF, NO, SO_2, H_2O, CCl_4, and BeF_2 are polar. In a molecule of HF, for example, the bonding pair of electrons is drawn more toward the fluorine atom and away from the hydrogen atom (Figure 5.6). The unequal sharing of electrons makes the fluorine end of the molecule more negative than the hydrogen end.

Polar bonds fall between the extremes of nonpolar covalent bonds and ionic bonds. In a covalent bond between two atoms of the same element, there is no charge separation; that is, the negative charge of the electrons is evenly distributed over the bond. In ionic bonds there is complete separation of the charges, and in polar bonds the separation falls somewhere in between.

Polar bonds in molecules can result in the molecule itself being polar when the shape of the molecule allows a permanent separation of charge. A linear diatomic molecule containing a polar bond, such as HF, is polar because there is a separation between the positive end of the molecule and the negative end (Figure 5.6*b*). Some linear molecules can themselves be nonpolar because the separated changes cancel out each other. The carbon dioxide (CO_2) molecule is an example. The oxygen atoms attract electrons to themselves, leaving the carbon atom with a partial positive charge, but because the oxygen atoms are opposite one another, the molecule itself is not polar.

$$\mathrm{O}\overset{\longleftarrow\!\!\!+}{=}\mathrm{C}\overset{+\!\!\!\longrightarrow}{=}\mathrm{O}$$

The ↔ indicate bond polarity. The arrow points in the direction where the partial negative charge is found.

The water molecule, on the other hand, is an example of a polar molecule. The shape of the water molecule is bent; thus, the partial charges of the polar bonds between the hydrogen atoms and the oxygen atom do not cancel out one another.

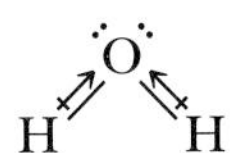

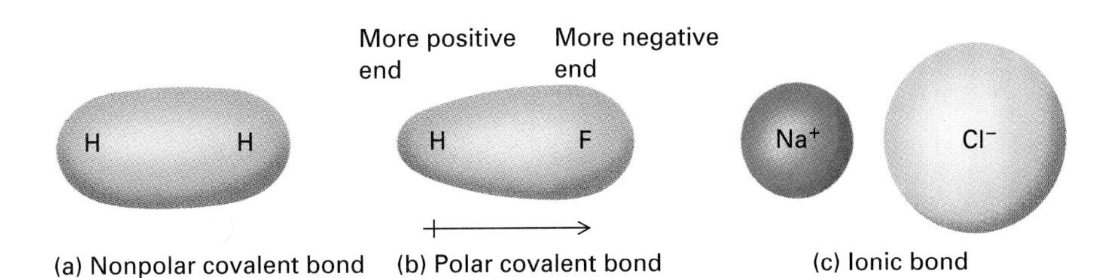

Figure 5.6 Nonpolar, polar, and ionic bonds. (a) In a nonpolar molecule such as H_2, the valence electron density is equally shared by both atoms. (b) In a polar molecule like HF, the valence electron density is shifted toward the fluorine atom. An arrow is used to show the direction of molecule polarity, with the arrowhead pointing toward the negative end of the molecule and the plus sign at the positive end of the molecule. (c) In ionic compounds such as NaCl, the valence electron or electrons of the metal are transferred completely to the nonmetal to give ions.

5.6 Properties of Molecular and Ionic Compounds Compared

The general properties of ionic and molecular compounds are summarized in Table 5.6. All these properties can be interpreted using the chemical view of matter. Ionic compounds form hard, brittle, crystalline solids with high melting points. Their crystalline nature is explained by the regular arrangement of positive and negative ions needed to place each ion in contact with those of the opposite charge. Their hardness is accounted for by the strong electrostatic forces of attraction that hold the ions in place. The differences in melting and boiling points (high for ionic compounds and low for molecular compounds) are largely accounted for by the greater strength of the electrostatic attraction between ions than between molecules. Melting requires adding enough energy to overcome these forces.

Most highly polar molecules are still more weakly attracted to each other than positive or negative ions. Whether a molecular compound is a gas, liquid, or solid is a function of the strength of the attractions between molecules. Only when there are huge molecules, with attractions between many polar regions in the molecule, as in polymers (Figure 10.12), do we find very hard and strong molecular substances.

Another major difference between ionic and molecular compounds is in their ability to conduct electricity, an ability that depends on the presence of mobile carriers of positive or negative charge. A solid ionic compound cannot conduct electricity because the ions are held in fixed positions. When melted or dissolved in water, the situation is different—the ions are free to move and current can flow, as demonstrated in Figure 5.7. Positive ions move toward the negative electrode and negative ions move toward the positive electrode. A compound that conducts electricity under these conditions is referred to as an **electrolyte**. Ionic compounds, to whatever extent they dissolve, are elec-

TABLE 5.6 ■ Properties of Ionic and Molecular Compounds

Ionic Compounds	Molecular Compounds
Examples: $NaCl$, CaF_2	*Examples:* CH_4, CO_2, NH_3, $CH_3CH_2CH_3$
Many are formed by combination of reactive metals with reactive nonmetals	Many are formed by combination of nonmetals with other nonmetals or with less reactive metals
Crystalline solids	Gases, liquids, and solids
Hard and brittle	Solids are brittle and weak, or soft and waxy
High melting points	Low melting points
High boiling points (700°C to 3500°C)	Low boiling points (−250°C to 600°C)
Good conductors of electricity when molten; poor conductors of heat and electricity when solid	Poor conductors of heat and electricity
Many are soluble in water	Many are insoluble in water but soluble in organic solvents

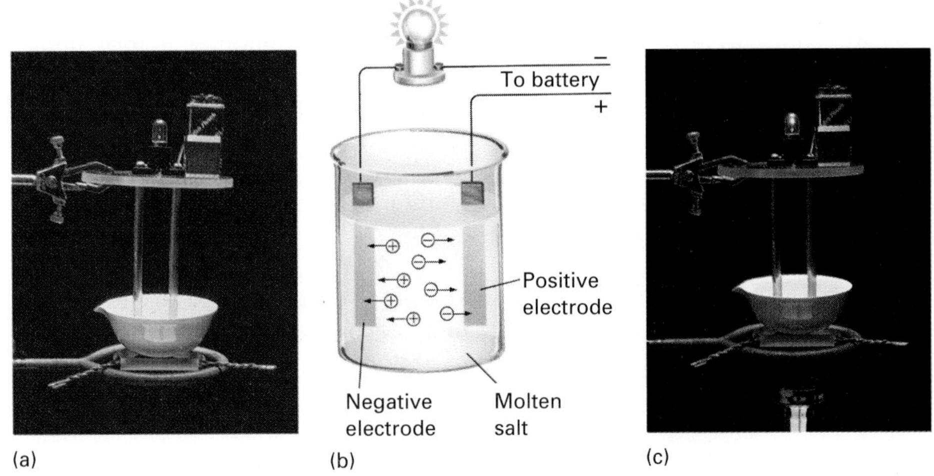

Figure 5.7 Conductivity. (a) Solid salts do not conduct electricity. (b) Molten salts do conduct electricity. Ions in a melted salt are free to move and migrate to the electrodes dipping into the melt. (c) The lighted bulb shows that the electric circuit is complete. *(c, C. D. Winters)*

trolytes in water solution because the ions separate from the crystal and can move about in the solution. Molecules, whether in pure compounds or dissolved in water or any other liquid, have no overall charge and therefore do not carry current; they are referred to as **nonelectrolytes**.

SELF-TEST 5C

1. (a) An example of a molecule with covalent bonds in which the electrons are equally shared between the atoms is ____________. (b) One where electrons are unequally shared is ____________.
2. What are the bond angles in CH_4?
3. Which is a more polar bond?
   (a) H—H or H—F
   (b) C—H or C—O
   (c) H—N or C—N
4. Which is a polar molecule? (a) H_2O, (b) H_2, (c) O_2, (d) CCl_4.
5. Which is a nonpolar molecule with polar covalent bonds? (a) H_2O, (b) H_2, (c) O_2, (d) CCl_4.

5.7 Intermolecular Forces

Although atoms within molecules are held together by strong chemical bonds, the attractive forces between two separate molecules are much weaker. For example, only about 1% of the energy required to break one of the C—H bonds in a methane (CH_4) molecule is required to pull two methane molecules away from one another. These small forces between molecules are called **intermolecular forces**. Even though these attractions are small, they give sam-

■ Methane (CH_4) is the major component of natural gas.

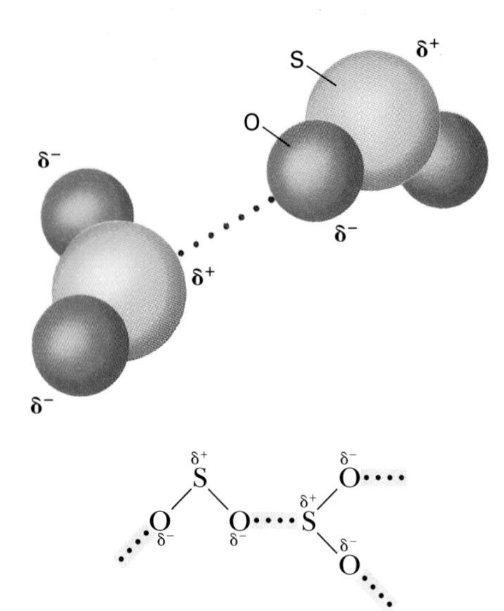

Figure 5.8 Polar sulfur dioxide molecules attracted to one another (oxygen atoms attract electrons more than do sulfur atoms).

ples of matter exceedingly important properties. For example, intermolecular attractions are responsible for the fact that all gases condense to form liquids and solids. Both pressure and temperature play a role in the liquefaction of a gas. When the pressure is high, the volume decreases and the molecules of a gas are close together. This condition allows the effect of attractive forces to be appreciable (see Section 5.8).

For polar molecules (Section 5.5), intermolecular forces act between the positive end of one polar molecule and the negative end of an adjacent polar molecule. Molecules of SO_2, (Figure 5.8), are polar, with the partially negative region of one molecule (represented by δ^-) being attracted to the partially positive region (represented by δ^+) of an adjacent molecule.

Even nonpolar molecules can have momentary unequal distribution of their electrons, which results in weak intermolecular forces. For example, why would nitrogen—composed of nonpolar molecules—liquefy or why would carbon dioxide—also composed of nonpolar molecules—form a solid? When the molecules get close enough, one molecule will cause an uneven distribution of charge in its neighbor. The two molecules become momentarily polar and are attracted to one another. These induced attractive forces are weak but become more pronounced when the molecules are larger and contain more electrons.

Hydrogen Bonding

An especially strong intermolecular force called **hydrogen bonding** acts between molecules in which a hydrogen atom is covalently bonded to a strongly electron-attracting atom with nonbonding electron pairs. The electron-attracting atom may be fluorine, oxygen, or nitrogen. The hydrogen bond is the attraction between such a hydrogen atom with a partial positive charge on one molecule and a small, electron-attracting atom (F, O, or N) of another molecule that has a partial negative charge. The greater the electron attraction of the atom connected to H, the greater the partial positive charge on the H, hence, the stronger the hydrogen bond is between it and a partially negative atom on another molecule. Hydrogen bonds are typically shown as dotted lines between the atoms, and partial charges are shown as δ^+ and δ^-.

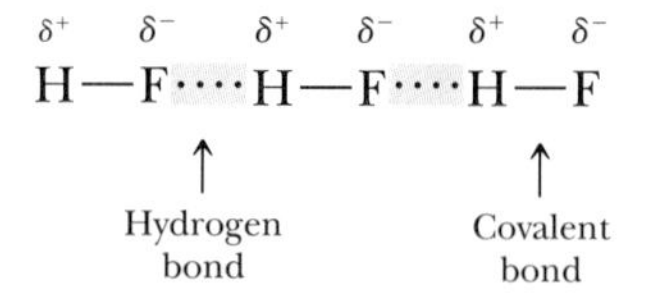

Water provides the most common example of hydrogen bonding. Hydrogen compounds of oxygen's neighbors and family members in the periodic table are all gases at room temperature. However, water is a liquid at room temperature, and this indicates a strong degree of intermolecular attraction. Figure 5.9 illustrates that the boiling point of H_2O is about 200°C higher than would be predicted if hydrogen bonding were not present.

Since each hydrogen atom can form a hydrogen bond to an oxygen atom in another water molecule and since each oxygen atom has two nonbonding electron pairs, each water molecule can form a maximum of four hydrogen

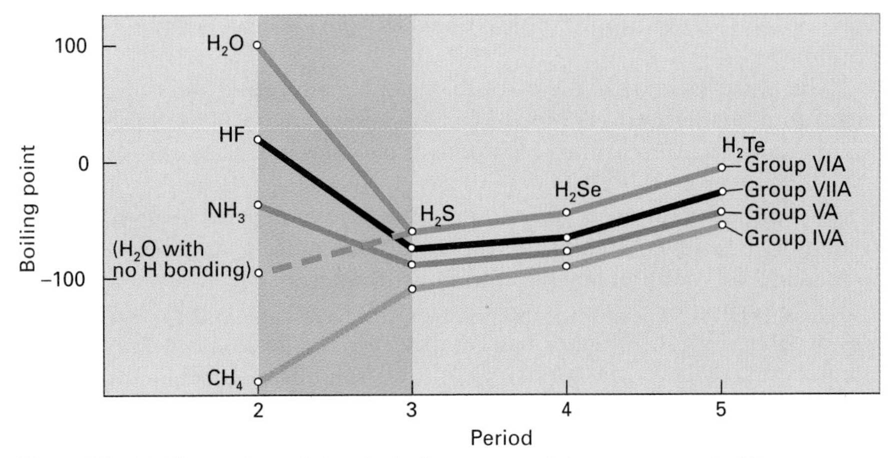

Figure 5.9 Boiling points of simple hydrogen-containing compounds. Lines connect molecules in which hydrogen combines with atoms from the same periodic table group. As shown for water, the point at which the compounds would boil if there were no hydrogen bonding is found by following the straight parts of the lines on the right down to the left.

bonds to four other water molecules (Figure 5.10). The result is a tetrahedral cluster of water molecules around the central water molecule.

Although hydrogen bonds are much weaker than ordinary covalent bonds, they play a key role in the chemistry of life. Later chapters in this text will discuss hydrogen bonding.

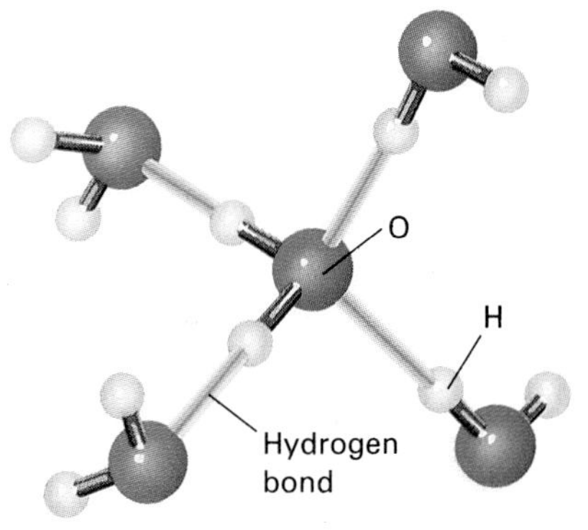

Figure 5.10 The four hydrogen bonds between one water molecule and its neighbors.

5.8 The States of Matter

By now you know enough about bonding to understand the states of matter: gas, liquid, and solid. Matter in any state is composed of atoms, molecules, or ions in constant motion. In gases (Figure 5.11*a*), the particles are in rapid motion relative to one another. In addition, they are, on average, far apart for one another. In liquids (Figure 5.11*b*), the particles are in motion, but they are so close together that they are touching one another. In solids (Figure 5.11*c*), the particles are also touching one another and are in motion, but

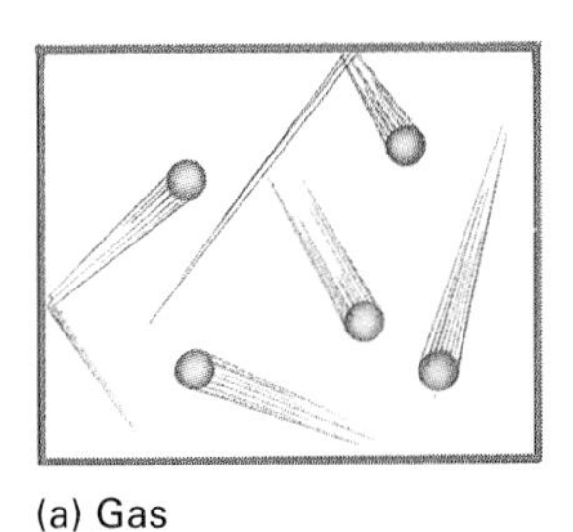

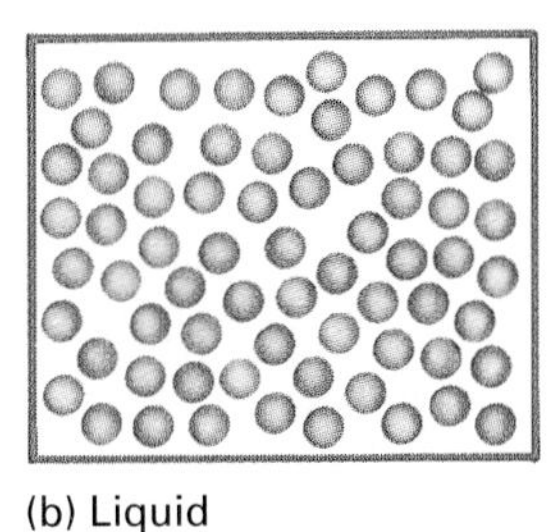

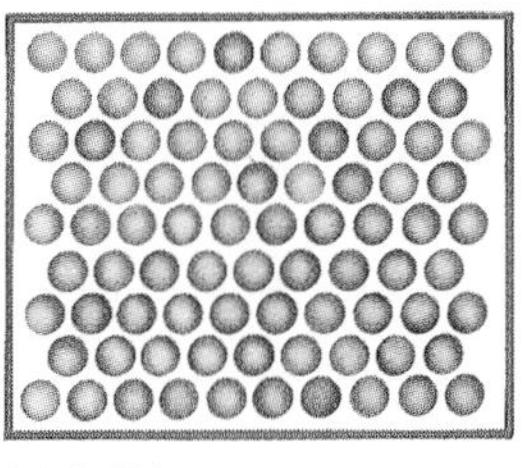

Figure 5.11 The three states of matter. The particles represented by the circles can be atoms, molecules, or ions (in liquids and solids). (a) In a gas, particles are very far apart and move rapidly in straight lines. (b) In a liquid, the particles move about at random, alone or in clusters. (c) In a solid, the particles or ions are in fixed positions and can only vibrate in place.

they do not move appreciable distances, but instead vibrate in fixed positions.

The particles in a sample of gas move at high speeds when they are at or near room temperature. For example, the molecules found in the air in your room are moving, on average, at about 1000 miles per hour and they travel about 400 times their own diameter between collisions with another molecule. If you cool a gas to a very low temperature, the molecules move much less rapidly. On further cooling, the molecules are moving so slowly that they attract one another and begin to form tiny droplets of liquid, a phenomenon called **condensation**. If that liquefied sample of gas is cooled even further, it will become a solid. All gases behave this way. For those gases consisting of nonpolar molecules, the condensation and solidification temperatures are quite low. However, for those gases whose molecules are polar, the temperatures are much higher. In the table in the margin, the condensation and solidification temperatures for SO_2, which is a polar molecule, and H_2O, which can engage in hydrogen bonding, are given for comparison.

■ Condensation and solidification temperatures for some gases at 1 atm pressure follow:

	Condenses (°C)	Solidifies (°C)
He	−269	—
Ar	−185.9	−189
O_2	−183.0	−218.4
N_2	−195.8	−209.8
Cl_2	−34.6	−101.0
SO_2	−10.1	−72.7
H_2O	100	0

In contrast to solids, whose particles are in rigid arrangements, liquids and gases have the property of being **fluid**—that is, they flow—because their atoms, ions, or molecules are not as strongly attracted to each other as they are in solids. Not being confined to specific locations, the particles in a liquid can move past one another. For most substances, the particles are a little farther apart in a liquid than in the corresponding solid, so that the volume occupied by a given mass of the liquid is a little larger than the volume occupied by the same mass of the solid. This means the liquid is less dense than the solid, and the solid form of a sample of matter sinks in its liquid (Figure 5.12). There is a rather important exception to this rule: solid water floats on

(a)

(b)

Figure 5.12 Water (H_2O) and benzene (C_6H_6). On the left, ice floats on water. On the right, solid benzene (melting point 5.5°C) sinks in liquid benzene. *(C. D. Winters)*

liquid water. The importance of this property of water is discussed in Section 5.10.

For solids, liquids, or gases, the higher the temperature, the faster the particles move. A solid melts when its temperature is raised to the point at which the particles vibrate fast enough and far enough to get away from the attraction of their neighbors and move out of their regularly spaced positions. As the temperature goes higher, the particles move even faster, until finally they escape their neighbors and become independent; the substance becomes a gas.

5.9 Gases and How We Use Them

Gases surround us in our atmosphere. We breath a mixture of nitrogen, oxygen, and other gases. Every breath we take carries with it the oxygen gas we need to burn the foods we eat. When we exhale, the carbon dioxide gas that is produced by the food-burning processes in our cells leaves our bodies. This is one of the ways we rid ourselves of waste materials.

Interestingly, all gases possess a set of common properties. At constant temperature, all gases expand when the surrounding pressure decreases and contract when the pressure increases. At constant pressure, all gases expand with increasing temperature and contract with decreasing temperature. All gases have the ability to mix in any proportion with other gases. These properties are explained by the fact that the gas particles are far apart, move fast, and have very little chance to interact. Their molecular properties, such as size, number of electrons, and shape, have virtually no effect on the properties of the gas as a whole.

A sample of gas confined inside a container exerts a **pressure**, which is caused by the individual particles of the gas sample striking the walls of the container. The earth's atmosphere, a mixture of gases, exhibits a pressure that is dependent on the altitude relative to sea level and temperature.

Another general property of gases is **compressibility**. All gases can be compressed by applying a pressure on a confined sample. This property of all gases, known as *Boyle's law* after Robert Boyle who discovered it in 1661, is explained by the great distances between gas particles—applying pressure only confines the particles in a smaller space. Of course, a gas will always expand if the pressure is reduced. If a sample of gas is released into deep space, where the pressure is effectively zero, the molecules would begin randomly moving away from one another and eventually would become attracted to some nearby star system.

Compression and expansion of a sample of gas. *(Tom Pantages)*

Perhaps one of the more interesting properties of all gases is that of **miscibility**—the ability to mix in all proportions with other gases. The miscibility of gases is explained by the great distances between gas molecules. In effect, "there is always room for some more molecules."

To illustrate how gases mix with one another, consider what happens when a person wearing a strong perfume enters a room. The smell of the perfume is noticed immediately by those close by and eventually by everyone in the room. The perfume contains **volatile**, meaning easily vaporized, compounds that gradually mix with the other gases in the room's atmosphere. Even if there were no apparent movement of the air in the room, the smell of perfume would eventually reach everywhere in the room. This mixing of two or more gases due to random molecular motion is called **gaseous diffusion**. Given time, the molecules of one component in a gas mixture will thoroughly and completely mix with all the other components to form a homogeneous mixture.

■ **Vaporization** is the movement of molecules from the liquid state to the gaseous state.

■ SELF-TEST 5D

1. Arrange the states of matter (liquid, gas, solid) by increasing order of the particles.
2. As temperature increases, molecular motion (decreases/increases).
3. Name the two states of matter in which the particles are very close to each other, on average.
4. Which two states of matter can be described as fluid?
5. Which state of matter is compressible? Which two states of matter are noncompressible?
6. As the temperature of a sample of gas decreases, the volume ______________ at constant pressure.
7. All gases mix with one another in all proportions. (a) True, (b) False.
8. Which kind of intermolecular force is stronger—hydrogen bonding or attractions caused by shifting electrons in molecules?

5.10 Water

There would be no life as we know it on the earth without water and its unique properties. Certainly there are other media on the earth and in the universe wherein much chemistry occurs. However, on the earth the chemistry in water solutions and the chemistry of water dominate. Water plays an important role as a reactant, a product, or a solvent in most of the chemical reactions in our environment.

Some Properties of Water

1. ***Water is a liquid at room temperature as a direct consequence of hydrogen bonding between adjacent water molecules.*** Pure water is a liquid between 0°C and 100°C.
2. ***The density of solid water (ice) is less than that of liquid water.*** Put another way, water expands when it freezes. If ice were a normal solid, it would be denser than liquid water, and lakes would freeze from the bottom up. This would have disastrous consequences for marine life, which could not survive in areas with winter seasons. The application of pressure causes ice to melt. This is a consequence of the structure of ice. The pressure causes ice to change to a form with a smaller volume, and since liquid water occupies a smaller volume than does ice, the ice converts to a liquid.
3. ***Water has a relatively high heat capacity.*** Water can absorb large quantities of heat without large changes in temperature, because the added heat can break hydrogen bonds instead of increasing the temperature.

■ **Heat capacity** is defined as the amount of heat required to raise the temperature of a sample of matter of a given size by 1°C. The heat capacity of water is 1 calorie per gram (cal/g).

For comparison, the heat capacity of water is about ten times that of copper or iron. Water's heat capacity accounts for the moderating influence of lakes and oceans on the climate. Huge bodies of water absorb heat from the Sun and release the heat at night or in cooler seasons. The Earth would have extreme temperature variations if it were not for this property of water. By contrast, the temperatures on the surface of the Moon and the planet Mercury vary by hundreds of degrees through the light and dark cycles.

4. ***Water has a high heat of vaporization.*** For a liquid to vaporize, heat is required. The **heat of vaporization** of a liquid is a measure of the intermolecular attractions holding the molecules together in the liquid. Water has one of the highest heats of vaporization. A consequence of this high heat of vaporization is the cooling effect that occurs when water evaporates from moist skin. Evaporating water molecules take with them a considerable amount of energy, which was needed to overcome the attractions between the leaving molecules and those remaining behind.

■ **Heat of vaporization** is defined as the heat required to vaporize a given quantity of liquid at its boiling point. The heat of vaporization of water at 100°C is 540 cal/g.

5. ***Water has a high surface tension.*** Unlike gases, liquids have surface properties, and these are extremely important in the overall behavior of many liquids. Molecules beneath the surface of the liquid are completely surrounded by other molecules and experience forces in all directions due to intermolecular attractions. By contrast, molecules at the surface are attracted only by molecules below or beside them (Figure 5.13). This unevenness of attractive forces at the surface of the liquid causes the surface to contract, making it act like a skin. The

■ Benzene, chloroform, ethyl alcohol, and octane—all organic compounds that are liquids at room temperature—have surface tensions about one third as strong as that of water.

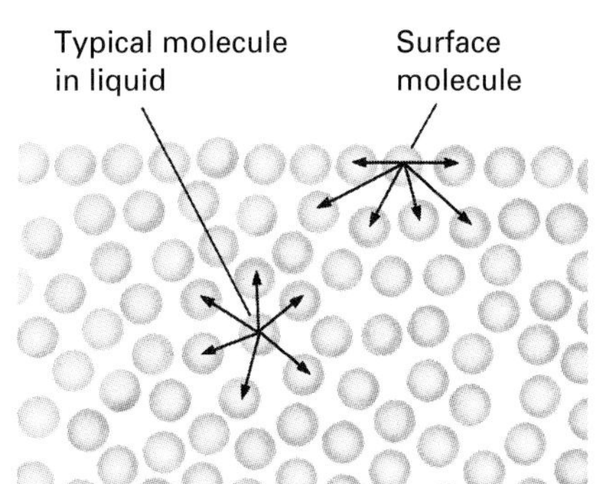

(a)

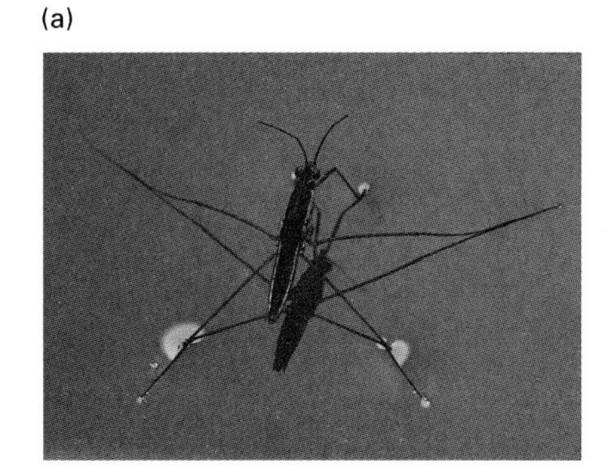

(b)

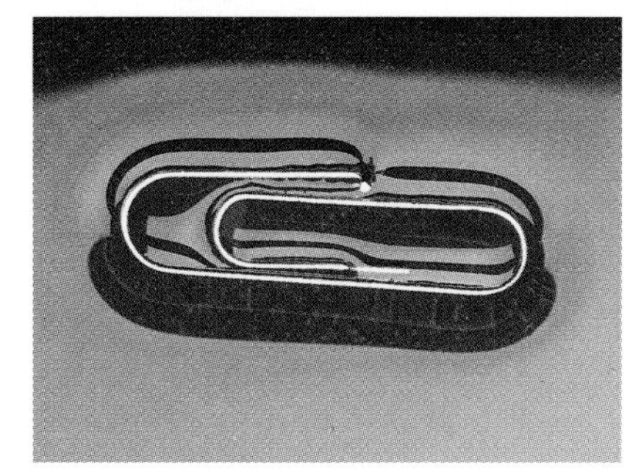

(c)

(d)

Figure 5.13 Surface tension. (a) The surface is strengthened by intermolecular forces attracting surface molecules. (b) The water strider, a lightweight insect that does not provide enough force per unit area to break through the surface tension. Note that the strider does not walk on the sharp ends of its "toes." (c) With care, a paper clip can be placed so that it won't sink in the water. (d) On a dirty car, this wouldn't happen. The dirt would overcome the surface tension of the water droplets. *(b: Runk/Schoenberger/Grant Heilman Photography; c: C. D. Winters; d: Richard Megna/Fundamental Photographs)*

Gypsum sand at White Sands National Monument in New Mexico. The sand has formed as the result of weathering, erosion, and the water solubility of gypsum ($CaSO_4 \cdot 2\ H_2O$). *(Jon Mark Stewart/ Biological Photo Service)*

energy required to overcome the "toughness" of this liquid skin is called the **surface tension**, and it is higher for liquids that have strong intermolecular attractions. Water's surface tension is high compared with those of most other liquids because of the extensive hydrogen bonding that holds water molecules to each other.

6. ***Water is an excellent solvent, often referred to as the universal solvent.*** Because it is such a good solvent, water from natural sources is not pure water but is instead a solution of substances dissolved by contact with water.

Some More About Properties of Pure Liquids

In a liquid the molecules are close enough together that, unlike a gas, a liquid is only slightly compressible. However, the molecules remain mobile enough that the liquid flows. Because they are difficult to compress and their molecules are moving in all directions, confined liquids can transmit applied pressure equally in all directions. This property is used in the *hydraulic fluids* that operate automotive brakes and airplane wing surfaces, tail flaps, and rudders.

Every liquid has a **vapor pressure**, which is the pressure in the gaseous state of those molecules that have escaped from the liquid at a given temperature. As you would expect, the vapor pressure of a liquid increases with increasing temperature because more molecules escape the liquid (vaporize) and enter the gaseous state. The higher the temperature, the greater the volatility, because a larger fraction of the molecules have sufficient energy to overcome the attractive forces at the liquid's surface. Our everyday experiences such as heating water on a stove or spilling a liquid on a hot pavement in the summer tell us that raising the temperature of the liquid makes evaporation take place more readily. Conversely, at lower temperatures, the volatility of a liquid will be lower. The same amount of water spilled on the pave-

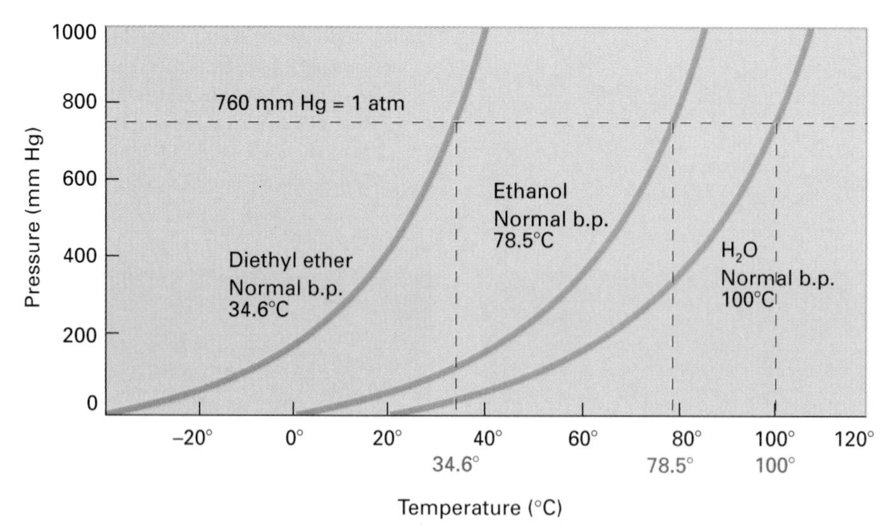

Figure 5.14 Vapor pressures. The curves show how boiling points change with pressure. The normal boiling point of a liquid, shown here in red, is where the curves reach 760 mm Hg. The higher its vapor pressure at a given temperature, the more volatile the liquid.

ment on a winter day will remain there much longer. Figure 5.14 shows how the vapor pressures increase for three common liquids as the temperature is increased. The three compounds have different volatilities due to the different strengths of their intermolecular attractions, but in all cases vapor pressure increases with increasing temperature. This is true for all liquids.

The temperature at which the vapor pressure equals the atmospheric pressure is the **boiling point** of the liquid. If the atmospheric pressure is 1 atm, the boiling temperature is called the **normal boiling point**.

5.11 Solutions

Solutions, as explained in Section 2.3, are homogeneous mixtures that can be in the gaseous, liquid, or solid states. Most commonly, however, we encounter liquid solutions. How many liquid solutions are familiar to you? How about sugar or salt dissolved in water, oil paints dissolved in turpentine, or grease dissolved in gasoline? In each of these solutions, the substance present in the greater amount, the liquid, is the **solvent**; and the substance dissolved in the liquid, the one present in a smaller amount, is the **solute**. For example, in a glass of tea, water is the solvent; and sugar, lemon juice, and the components extracted from the tea leaves that impart taste and color are solutes.

A substance's **solubility** is defined as the quantity of solute that will dissolve in a given amount of solvent at a given temperature. Solubility is determined by the strengths of the forces of attraction between solvent molecules, between solute atoms, molecules, or ions, and between solute and solvent. The forces acting between the solvent and solute particles must be greater than those within the solute for the solute to dissolve. In other words, the solute and solvent must like each other more than they like themselves. If the forces between a liquid solvent and a solute are strong enough, then the solute will dissolve (Figure 5.15).

Although some solutes are so much like the solvent they are dissolving in that they dissolve in all proportions (a property called **miscibility**, like that of gases), there are limits to the solubility of most solutes in a given solvent. When a quantity of solvent has dissolved all the solute it can, the solution is said to be **saturated**. If more solute can be dissolved in the solution, the solution is said to be **unsaturated**. Our everyday experiences illustrate this kind of behavior in solutions. For example, if a spoonful of sugar is added to a glass of iced tea, it quickly dissolves with a little stirring, which hastens the mixing process. The resulting solution is unsaturated (it can dissolve more sugar); and if you like your iced tea a little sweeter, you just add another spoonful of sugar and stir. However, if sugar is continually added to the solution, a point is reached when no more appears to dissolve and the added sugar simply sinks to the bottom of the glass. At this point the solution has become saturated in the dissolved sugar.

If almost none of the solute dissolves in a solvent, then it is said to be **insoluble** in that solvent. Metals, except those that react with water, are insoluble in water. Oily substances are also insoluble in water, as is water in oily substances. The low solubility of oily substances in water is caused by a lack of attraction between the highly polar water molecules and the nonpolar molecules of an oily substance. The water solubility of a nonpolar molecule can be increased if a polar part can be added. Conversely, the water solubility of a

(a)

(b)

Figure 5.15 Solubilities. (a) Polar water, with a bit of nonpolar iodine (I_2) dissolved in it, floats on top of nonpolar carbon tetrachloride (CCl_4), with which it is immiscible. (b) Nonpolar iodine is much more soluble in nonpolar carbon tetrachloride than in water. Therefore, shaking the mixture in (a) causes the iodine molecules to migrate into the carbon tetrachloride, where they produce a purple color.

■ Air, a mixture of gases, is a solution. Sterling silver, a mixture of silver and copper metals, is also a solution.

molecule decreases as the nonpolar portion of the molecule increases. The simpler alcohols are very water soluble because the —OH group is polar and forms strong hydrogen bonds with water. As the hydrocarbon chain lengthens, the influence of the —OH group decreases, and with it, the water solubility of the molecule.

The effect of temperature on the solubility of solutes in various solvents is difficult to predict. Experience with everyday solutions such as sugar in water would lead us to predict that increasing the temperature of the solvent will cause more solute to dissolve. Although this is true for sugar and most other nonionic solutes, it is not true for all ionic compounds. In fact, the solubility of table salt in water is about the same at all temperatures.

Solubility of Oxygen in Water at Various Temperatures*

Temperature (°C)	Solubility of O_2 (g O_2/L H_2O)
0	0.0141
10	0.0109
20	0.0092
25	0.0083
30	0.0077
35	0.0070
40	0.0065

* These data are for water in contact with air at 1 atm pressure.

Gases dissolve in liquids to an extent dependent on the similarity of the gas molecules and the solvent molecules. Polar gas molecules dissolve to a greater extent in polar solvents than nonpolar molecules do. Pressure also affects gas solubility. At higher pressure more gas dissolves in a given volume of liquid than at lower pressure. When the pressure is lowered, gas will be evolved from a gas-in-liquid solution. The behavior of a carbonated beverage when the cap is removed is a common illustration of this principle.

Temperature has a greater effect on gas solubility in liquids than it has on solids or liquids dissolving in liquids. Without exception, lower temperatures cause more gas to dissolve in a given volume of liquid. Higher temperatures cause less gas to dissolve. As every fisherman knows, fish prefer deeper water in the summer months. This is because more oxygen dissolves in colder water, and the water temperature is generally cooler at greater depths.

5.12 Solids

In contrast to gases and liquids, in which molecules are in continual random motion, the movement of atoms, molecules, or ions in solids is restricted to vibration and sometimes rotation around an average position. This leads to an orderly array of particles and to properties different from those of liquids or gases.

■ Bucky balls (see p. 111), another form of carbon, might someday find use as a lubricant. Looking at their structure, it is easy to see why.

Because the particles that make up a solid are very close together, solids are difficult to compress. In this respect solids are like liquids, the particles of which are also very close together. Unlike liquids, however, solids are rigid, so they cannot transmit pressure in all directions. Solids have definite shapes, occupy fixed volumes, and have varying degrees of hardness. Hardness depends on the kinds of bonds that hold the particles of the solid together. Graphite is one of the softest solids known; diamond is one of the hardest. Graphite is used as a lubricant. At the atomic level, graphite consists of layered sheets that contain carbon atoms. The attractive forces between the sheets are very weak; as a result one sheet can slide along another and be removed easily from the rest. In diamond, on the other hand, each carbon atom is strongly bonded to four neighbors, and each of those neighbors is strongly bonded to three more carbon atoms, and so on throughout the solid. Because of the number and strength of the bonds holding each carbon atom to its neighbors, diamond is so hard that it can scratch or cut almost any other solid. The cutting and abrasive uses of diamonds are far more important commercially than their gemstone uses.

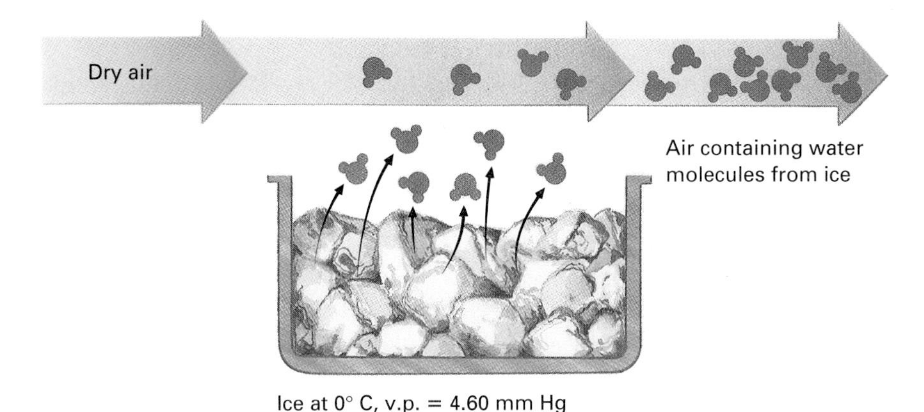

Figure 5.16 Why ice cubes shrink in the freezer—sublimation. Even a solid like ice has a vapor pressure caused by its molecules in the vapor state. Dry air will sweep these molecules away.

Properties of Solids

When a solid is heated to a temperature at which molecular motions are violent enough to partially overcome the interparticle forces, the orderliness of the solid's structure collapses, and the solid melts. The temperature at which melting occurs is the **melting point** of the solid. Melting requires energy to overcome the attractions between the particles in a solid lattice. In the reverse of melting, **solidification** or **crystallization**, energy is evolved. Melting points of solids depend on the kinds of forces holding the particles together in the solid, because these forces must be overcome for a solid to become a liquid. Table 5.7 gives some melting points for several types of solids.

Molecules can also escape directly from the solid to enter the gaseous state by a process known as **sublimation**. A common substance that sublimes at normal atmospheric pressure is solid carbon dioxide, which has a vapor pressure of 1 atm at −78°C. Solid CO_2 is known by its trade name, Dry Ice, because it is cold like ice (although much colder) and produces no liquid residue because it does not melt. Ice can sublime or melt. Have you ever noticed that ice and snow slowly disappear even if the temperature never gets above freezing? The reason is that ice sublimes readily in dry air (Figure 5.16). Given enough air passing over it, a sample of ice will sublime away, even at temperatures well below its melting point, leaving no trace behind. This is what happens in a frost-free refrigerator. A current of dry air periodically blows across any ice formed in the freezer compartment, taking away water vapor (and hence the ice) without having to warm the freezer compartment to melt the ice.

TABLE 5.7 ■ Melting Points of Some Solids

Solid	Melting Point (°C)
Molecular Solids: Nonpolar Molecules	
H_2	−259
O_2	−218.4
N_2	−209.8
F_2	−219.6
Cl_2	−101.0
Molecular Solids: Polar Molecules	
HCl	−115
HBr	−88
H_2S	−86
HF	−83.1
NH_3	−77.7
SO_2	−72.7
HI	−51
H_2O	0
Ionic Solids	
NaI	662
NaBr	747
$CaCl_2$	782
NaCl	800
Al_2O_3	>2072
MgO	2852

Strongly hydrogen bonded molecules shown in color.

■ SELF-TEST 5E

1. Most solids will not float in their liquids. (a) True, (b) False.
2. The density of solid water is less than that of liquid water. (a) True, (b) False.

3. In a solution, the substance dissolved is called the ____________, and the substance doing the dissolving is called the ____________.
4. Carbon tetrachloride is a nonpolar compound. In which solvent would you expect it to dissolve? (a) Water (a polar solvent), (b) Corn oil (a nonpolar solvent).
5. More of a gas will dissolve in warm water than in cold water. (a) True, (b) False.
6. Increasing the pressure of a gas over a liquid (increases/decreases) its solubility in the liquid.
7. Ionic solids generally have melting points higher than those of nonpolar molecular solids. (a) True, (b) False.
8. The opposite of crystallization is called ____________.
9. The process of molecules escaping the surface of a solid and going into the vapor state is called ____________.

MATCHING SET

___ 1. Ionic bonds
___ 2. Covalent bonds
___ 3. Single covalent bond
___ 4. Double covalent bond
___ 5. Triple covalent bond
___ 6. NaCl
___ 7. Molecule
___ 8. NH_3
___ 9. Mixable in all proportions
___ 10. Made of carbon atoms all covalently bonded to four other carbon atoms
___ 11. Causes water's high boiling point
___ 12. Solvent that will dissolve most salts
___ 13. Causes more gas to dissolve in a liquid
___ 14. Number of neighbors around a water molecule in ice
___ 15. Going from solid to gas
___ 16. Going from gas to liquid
___ 17. Phosphate ion

a. An electrically neutral arrangement of covalently bonded atoms
b. Positive ions attracted to negative ions
c. Polyatomic ion
d. Bond with four shared electrons
e. Bond with two shared electrons
f. Molecule
g. Ionic compound
h. Bond with six shared electrons
i. Shared electrons
j. Hydrogen bonding
k. Any two gases
l. Condensation
m. Sublimation
n. Four
o. Water
p. Pressure
q. Diamond

QUESTIONS FOR REVIEW AND THOUGHT

1. Give definitions for the following terms:
   (a) Cation (b) Anion
   (c) Octet rule (d) Formula unit
2. Give definitions for the following terms:
   (a) Nonmetal (b) Binary compound
3. Give definitions for the following terms:
   (a) Shared pair (b) Double bond
   (c) Triple bond (d) Unshared pair
   (e) Single bond (f) Multiple bond
4. Give definitions for the following terms:
   (a) Covalent bond (b) Polyatomic ion
   (c) Ionic bond (d) Binary compound
5. Give definitions for the following terms:
   (a) Nonpolar bond (b) Polar bond
6. What is the octet rule?
7. Describe what each of the following terms means:
   (a) Hydrocarbon
   (b) Saturated hydrocarbon

(c) Unsaturated hydrocarbon
(d) Alkenes

8. Is Ca^{3+} a possible ion under normal chemical conditions? Why or why not?
9. Predict the ions that would be formed by:
   (a) Br (b) Al
   (c) Na (d) Ba
   (e) Ca (f) Ga
   (g) I (h) S
   (i) All Group IA metals
   (j) All Group VIIA nonmetals
10. An ion has 12 protons, 13 neutrons, and 10 electrons. What is its charge? Consult the periodic table and write the symbol of the ion.
11. Write the formula and name of the ionic compounds formed from atoms of each of the following pairs of elements:
    (a) Al and I (b) Sr and Cl
    (c) Ca and N (d) K and S
    (e) Al and S (f) Li and N
12. What holds ionic solids together?
13. Go to your local grocery store and see if you can find at least ten different products that have an ionic compound as a component. Don't use NaCl; it is too common. Try to find as many different ionic compounds in use as you can.
14. Name the following compounds:
    (a) $CaSO_4$ (b) Na_3PO_4
    (c) $NaHCO_3$ (d) K_2HPO_4
    (e) $NaNO_2$ (f) $Cu(NO_3)_2$
15. Write correct formulas for the ionic compounds you expect to be formed when the following pairs of elements react:
    (a) Li and Te (b) Mg and Br
    (c) Ga and S
16. Describe the difference between an ionic bond and a covalent bond.
17. Predict the type of bond formed between each of the following pairs of elements:
    (a) Sodium and sulfur
    (b) Nitrogen and bromine
    (c) Calcium and oxygen
    (d) Phosphorus and iodine
    (e) Carbon and oxygen
18. Complete the following table by writing the predicted formulas for each pair of elements:

	F	O	Cl	S	Br	Se
Na	NaF					
K				K_2S		
B		B_2O_3				
Al						
Ga			$GaCl_3$			
C					CBr_4	
Si						$SiSe_2$

19. Name the following compounds:
    (a) NO (b) SO_3
    (c) N_2O (d) NO_2
20. Draw Lewis structures for the following:
    (a) CO (b) SiF_4
    (c) C_2H_4 (d) H_2S
    (e) C_2H_2 (f) C_2H_6
    (g) OH^- (h) NF_3
21. Summarize the differences between ionic, polar covalent, and nonpolar covalent bonding.
22. Which of the following compounds has the most polar bonds? (a) H—F, (b) H—Cl, (c) H—Br.
23. Which are the more polar bonds in the following molecules?
    (a) Chloroethane (CH_3CH_2Cl) (b) Freon 12 (CCl_2F_2)

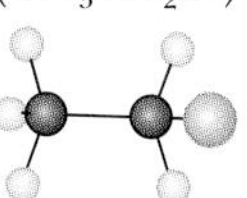

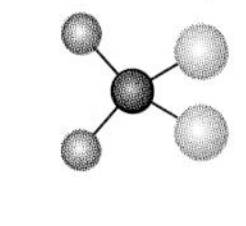

24. Which of the following molecules is polar and which is nonpolar? Explain.
    (a) Acetone (CH_3COCH_3), a common solvent

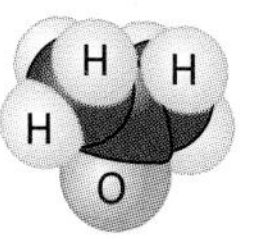

(b) Butane ($CH_3CH_2CH_2CH_3$), a common fuel

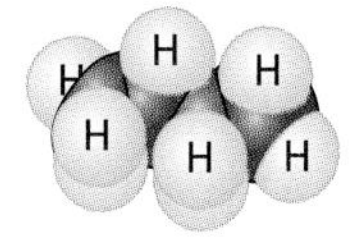

(c) Ammonia (NH_3)

25. Which of the following molecules is (are) not polar? For each polar molecule, which is the negative and which is the positive end of the molecule? (a) CO, (b) GeH_4, (c) BCl_3, (d) HF.
26. The structural formula for ethanol, the alcohol in alcoholic beverages, is

```
    H   H
    |   |
H — C — C — O — H
    |   |
    H   H
```

Give the total number of: (a) valence electrons, (b) single bonds, (c) bonding pairs of electrons. How many extra pairs of electrons are left? What are these called, and where should they be placed in the structural formula?

27. Describe the following states of matter: (a) Gas, (b) Liquid, (c) Solid.

28. In which state of matter are the particles the greatest average distance apart? (a) Liquid, (b) Solid, (c) Gas.
29. In which state of matter are the particles in fixed positions? (a) Liquid, (b) Solid, (c) Gas.
30. Which state of matter can be described as "particles close together and in constant, random motion"? (a) Liquid, (b) Solid, (c) Gas.
31. Explain why the pressure of the atmosphere decreases with increasing altitude.
32. What happens to the pressure in an automobile tire in cold weather? Explain.
33. Explain why molecules of a perfume can be detected a few feet away from the person wearing the perfume.
34. Draw a structure showing four water molecules bonded to a central water molecule by means of hydrogen bonding. Indicate all the hydrogen bonds by drawing arrows to them.
35. Name two properties of water that are unusual because of the presence of hydrogen bonding between adjacent water molecules.
36. Which of the following compounds would you expect to exhibit hydrogen bonding? Explain your answer. (a) CH_3OH, (b) NH_3, (c) SO_2, (d) CO_2, (e) CH_4, (f) HF, (g) CH_3OCH_3.
37. Whenever a liquid evaporates, heat is required. Use this statement to explain why you get chilled when you come out of a swimming pool on a windy day.
38. Why are gases compressible while liquids and solids are not?
39. Based on Figure 5.9, approximately what would be the boiling point of ammonia if there were no hydrogen bonding between the molecules?
40. What causes surface tension in liquids? Name a compound that has a high surface tension.
41. Why is a gas more soluble in a solvent when a higher pressure is applied? Give an example of where you see this behavior of gases dissolving in liquids.
42. Thermal pollution is a name given to industrial discharges of warm waters that contain sufficient heat to warm the water into which they are discharged. Why is thermal pollution harmful to fish?
43. Explain how a frost-free refrigerator defrosts itself.

An oil fire in Kuwait. In an act of environmental sabotage, the oil wells were set on fire in 1991 at the end of the Gulf War. It took months to extinguish the fires, providing a dramatic demonstration of the power of favorable chemical reactions. *(AP/Wide World Photos)*

CHAPTER 6

Chemical Reactivity: Chemicals in Action

Biologists, physicians, chemists, psychologists, and sometimes even sociologists and economics professors depend on experiments to test their theories. Chemists have one advantage in such experiments. Under *identical conditions*, pure chemicals always react with each other in the same way. Sometimes it is difficult, but with effort identical conditions can be achieved. Those who study living things or social interactions can never be sure. Are two plants, two laboratory rats, or two social groups ever identical?

In this chapter the conditions that influence the outcome of chemical reactions are explored.

- What is the meaning of a balanced equation?
- Why are the mole and the molar mass essential concepts?
- In what ways can reaction rates be influenced?
- What is happening in a chemical reaction that has come to equilibrium?
- What two kinds of changes influence the favorability of chemical reactions?
- What are the first and second laws of thermodynamics?
- What are the major issues in recycling metals and other materials?

When journalists investigate a story for page one in the newspaper, they want to answer six questions:

Who? What? When? Where? How? Why?

To fully investigate a chemical reaction requires answering a similar list of questions:

What? How much? How fast? How far? Why?

Some scientists spend a lifetime seeking answers to these questions about a single complex reaction. Others devote themselves to answering one of these questions about many chemical reactions.

In Chapter 2 we introduced chemical reactions and the information needed to answer the "What?" question. What are the reactants and products? Hydrogen and oxygen, the reactants, combine to form water, the product.

$$\underset{\text{Hydrogen}}{2\ H_2(g)} + \underset{\text{Oxygen}}{O_2(g)} \longrightarrow \underset{\text{Water}}{2\ H_2O(\ell)}$$

Now we're going to pick up the story of chemical reactions and pursue the meaning of the other questions listed in the introduction to this chapter.

6.1 Balanced Chemical Equations and What They Tell Us

■ The law of conservation of matter: *Matter is neither lost nor gained in chemical reactions* (Section 3.1). The only known exception to this law is in nuclear reactions, which occur only with radioactive isotopes or under the special conditions of synthetic nuclear reactions (Section 4.2). The conservation law (so far) has always been reliable for chemical changes other than nuclear reactions.

Chemical equations are the best way we have to represent what happens in chemical reactions at the submicroscopic level that we cannot see. An equation will not be faithful to reality if the chemical formulas are wrong or if the equation is not balanced.

In balancing chemical equations, we are applying the law of conservation of matter—the atoms in the reactants must all be there in the products. To be sure an equation is balanced requires counting up the atoms of each kind in the reactants and products (Section 2.7). Doing this, of course, requires knowing the identity and correct formulas of the reactants and products. An equation can *never* be balanced by changing the subscript in a chemical formula. This changes the identity of the compound. Only coefficients can be changed to achieve balance.

The products of the complete burning, or combustion, of any hydrocarbon are always carbon dioxide and water. Thus, for the burning of methane (CH_4), the major ingredient in natural gas, the *unbalanced* equation is

$$\underset{\text{1 C atom}}{CH_4(g)} + O_2(g) \longrightarrow \underset{\text{1 C atom}}{CO_2(g)} + H_2O(g)$$

The C atoms are balanced here—one C on each side of the arrow. But the O and H atoms are not balanced. Often, it is easiest to first balance atoms that appear in only one formula on each side. Balancing H requires a coefficient of 2 in front of H_2O

$$\underset{\text{4 H atoms}}{CH_4(g)} + O_2(g) \longrightarrow CO_2(g) + \underset{\text{4 H atoms}}{2\ H_2O(g)}$$

Now the O atoms must be balanced. With four of them on the right, providing two O_2 molecules on the left finishes the job

$$CH_4(g) + 2\,O_2(g) \longrightarrow CO_2(g) + 2\,H_2O(g)$$

4 O atoms   4 O atoms

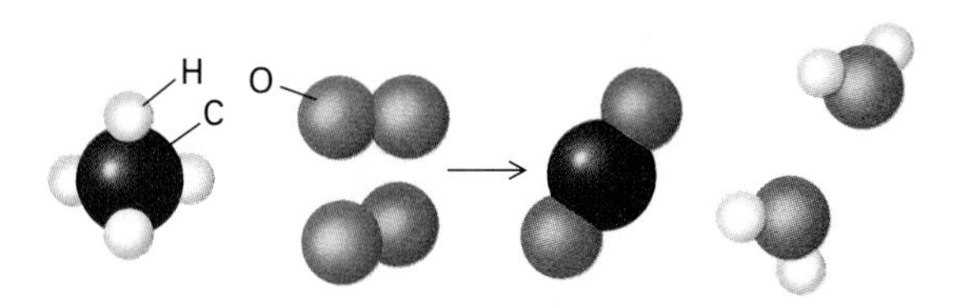

Note in these equations the difference between *subscripts* (e.g., the subscript $_4$ in CH_4), which relate to the need for correct formulas, and **coefficients** (e.g., the 2 with O_2 as a reactant), which relate to the need for a balanced equation.

EXAMPLE 6.1 *Equation Balancing*

Balance the following equation for the reaction of hydrofluoric acid [$HF(aq)$] with glass, which can be represented as calcium silicate ($CaSiO_3$). Decorative glass is etched using this reaction

$$CaSiO_3(s) + HF(aq) \longrightarrow CaF_2(s) + SiF_4(g) + H_2O(\ell)$$

■ The states of the reactants and products are indicated by
(g) for a gas
(s) for a solid
(ℓ) for a liquid
(aq) for something dissolved in water

SOLUTION

The Ca and Si atoms are balanced. To balance the three O atoms on the left requires three H_2O molecules on the right. There must then be six H atoms on the left. Putting in both of these coefficients gives

$$CaSiO_3(s) + 6\ HF(aq) \longrightarrow CaF_2(s) + SiF_4(g) + 3\ H_2O(\ell)$$

There are now six F atoms on each side of the equation, and it is fully balanced.

On the left: 1 Ca 1 Si 6 H 6 F 3 O
On the right: 1 Ca 1 Si 6 H 6 F 3 O

Exercise 6.1A

Balance the following equation for the preparation of aluminum chloride, which is an ingredient in some antiperspirants.

$$Al(s) + Cl_2(g) \longrightarrow AlCl_3(s)$$

Exercise 6.1B

Are the following equations balanced?
(a) $CaO(s) + H_2O(\ell) \rightarrow Ca(OH)_2$
(b) $SiO_2(s) + C(s) \rightarrow Si(s) + CO(g)$

6.2 The Mighty Mole and the "How Much?" Question

Moles and Molar Masses

■ You might want to look back at the description of atomic weights in Section 3.3.

Eventually, anyone who wants to carry out a chemical reaction must figure out how much of the reactants must be combined to make the desired amount of product. Somehow a connection must be made between atoms, molecules, and ions at the submicroscopic scale and weighable amounts of chemicals. Here is where balanced equations are essential. The relative atomic weight scale and balanced chemical equations together make it possible to answer "How much?" questions.

To understand the situation, consider the equation for hydrogen burning in chlorine to form hydrogen chloride:

$$H_2(g) + Cl_2(g) \longrightarrow 2\ HCl(g)$$

Counting by weighing. It is much easier to buy 5 lb of nails than to count them individually. *(C. D. Winters)*

The equation shows that one molecule of hydrogen and one molecule of chlorine combine to form two molecules of hydrogen chloride. Is this information of any help in figuring out, for instance, how much hydrogen and chlorine would be needed to make 100 g of hydrogen chloride? The essential problem is that molecules are very small, so small that it is impossible to count them one by one.

The solution to the problem is counting by weighing. The atomic weights given for each element in the periodic table are relative weights. The weight of one average neon atom is 20. atomic mass units (amu), and the weight of one calcium atom is 40. amu, both relative to an atomic weight of exactly 12 amu for carbon-12. Translating these numbers to masses big enough to measure means that 20. g of neon, 12 g of carbon-12, and 40. g of calcium all contain the same number of atoms. This type of relationship is at the heart of the quantitative use of chemical equations.

■ The mole as a unit is symbolized by mol, e.g., 2.5 mol.

The chemists' counting unit is named the **mole**, and it is defined as equal to the number of atoms in exactly 12 g of carbon-12. The mole is used in the same way as a dozen. In the grocery store you may need one dozen apples. In the laboratory, you may need 1 mol of carbon. You can count out 12 apples, but to get 1 mol of carbon, you have to weigh out 12 g of carbon.

■ If you ever see a car with a bumper sticker that reads, "***Everyone has Avogadro's number***," you can assume that the car probably belongs to a chemist.

The actual number of individual atoms, molecules, or ions in 1 mol is known from experiments. Just as a dozen apples is 12 apples, a mole of atoms is about 602,000,000,000,000,000,000,000, or 6.02×10^{23} atoms. How big is this number? A mole of sand would cover a city the size of Los Angeles to a depth of 600 m. The number of atoms, molecules, ions, or anything else in a mole, 6.02×10^{23}, is known as **Avogadro's number**.

The mass in grams of 1 mol of a substance is referred to as its **molar mass**. The molar mass of an element (except for those that exist as two-atom molecules, e.g., H_2) is the mass in grams numerically equal to the atomic weight of the element (Figure 6.1). The molar mass of helium, the lightest noble gas, is 4.0 g. What about a heavy element like lead? Checking the periodic table shows that the molar mass of lead is 207 g.

■ We don't need all the digits for atomic weights given in the periodic table inside the front cover, so we will round them off.

Since the mole is a counting unit, every balanced chemical equation can

Figure 6.1 One-mole quantities of some elements. The cylinders (*left to right*) hold mercury (201 g), lead (207 g), and copper (64 g). The two Erlenmeyer flasks hold sulfur (*left*, 32 g) and magnesium (*right*, 24 g). All rest on 1 mol of aluminum in the form of foil (27 g). *(C. D. Winters)*

be interpreted in terms of moles—moles of atoms, moles of molecules, moles of ions, or anything else. For example,

$$\underset{1\ H_2\ \text{molecule}}{H_2(g)} + \underset{1\ Cl_2\ \text{molecule}}{Cl_2(g)} \longrightarrow \underset{2\ HCl\ \text{molecules}}{2\ HCl(g)}$$

also means

$$\underset{1\ \text{mol of}\ H_2}{H_2(g)} + \underset{1\ \text{mol of}\ Cl_2}{Cl_2(g)} \longrightarrow \underset{2\ \text{mol of}\ HCl}{2\ HCl(g)}$$

Now, what is needed to interpret this equation in terms of masses of the reactants and products, which happen to be molecular substances? No problem. The molar mass of any pure substance is found by adding up the molar masses of the atoms shown in its formula. One mole of hydrogen molecules (H_2) has a molar mass of 2×1.0 g/mol H = 2.0 g/mol H_2. One mole of chlorine molecules has a molar mass of 2×35.5 g/mol Cl = 71.0 g/mol Cl_2. How about hydrogen chloride, HCl? It has a molar mass of 36.5 g.

■ Sometimes the term *molecular weight* is used instead of molar mass. Both terms refer to the relative mass of a substance according to the atomic weight scale. Strictly speaking, a molecular weight is for one molecule and would be in atomic mass units. We need use only molar masses.

H molar mass	1.0 g
Cl molar mass	35.5 g
	36.5 g = molar mass of HCl

EXAMPLE 6.2 *Moles and Masses*

What is the equivalent in moles of 3.7 g of water, roughly the amount in a teaspoonful?

SOLUTION

The molar mass of water, H_2O, is 18 g. Molar mass provides the conversion factors that connect mass and numbers of moles. Therefore

$$3.7 \text{ g } H_2O \times \frac{1 \text{ mol } H_2O}{18 \text{ g } H_2O} = 0.21 \text{ mol}$$

Exercise 6.2

What is the mass equivalent to 50. mol of barium nitrate, $Ba(NO_3)_2$, which gives a green color to fireworks?

Masses of Reactants and Products

■ Rarely can 100% of reactants actually be converted to products. One goal of industrial preparation of chemicals is to come as close to 100% as is possible and practical.

The mole provides us with the means to answer the "How much?" question about any chemical reaction. If all the reactants are converted to product, 2.0 g of H_2 (1 mol) should react with 71 g of Cl_2 (1 mol) to produce 73 g of HCl (2 mol).

$H_2(g)$	+	$Cl_2(g)$	$\longrightarrow$	$2\,HCl(g)$
1 H_2 molecule		1 Cl_2 molecule		2 HCl molecules
1 mol of H_2		1 mol of Cl_2		2 mol of HCl
2.0 g of H_2		71 g of Cl_2		2 mol × 36.5 g/mol = 73 g of HCl

Have you noticed that these masses adhere to the law of conservation of matter? The 2.0 g + 71 g of reactants produce 73 g of product.

Copper face found in a tomb in Peru. As time passes, green copper sulfide and black copper oxide form on a copper surface as the result of reactions with hydrogen sulfide and oxygen in the atmosphere. *(Heinz Plenge/Peter Arnold, Inc.)*

EXAMPLE 6.3 *Information about Masses from a Chemical Equation*

Interpret the equation for a reaction that produces copper from copper ore in terms of moles, molar masses, and masses of reactants and products.

$$Cu_2S(s) + 2\,Cu_2O(s) \xrightarrow{\text{Heat}} 6\,Cu(s) + SO_2(g)$$

Copper(I) sulfide, Copper(I) oxide, Copper, Sulfur dioxide

SOLUTION

The equation in terms of moles shows by the coefficients that 1 mol of copper(I) sulfide reacts with 2 mol of copper(I) oxide to produce 6 mol of metallic copper and 1 mol of sulfur dioxide. The molar masses of the reactants

THE PERSONAL SIDE

Amadeo Avogadro (1776–1856)

Amadeo Avogadro. (The World of Chemistry, Program 11, *The Mole*)

Today, Avogadro's name is most often associated with "his" number. The mole and the modern definition of the number of particles in it, however, were not directly his invention. His fame lies in a single simple statement, made in 1811: ". . . *the number of integral molecules in gases is always the same for equal volumes.*" This concept opened the door to understanding atomic weight and the formulas for chemical compounds.

Avogadro was a quiet man who was totally devoted to his studies. He was born into a prominent Italian family of lawyers and, as was surely expected of him, completed his training as a lawyer and entered government service. At age 24 he turned his back on his heritage and devoted the rest of his life to science. He became a professor of physics and mathematics and was content to pursue his studies alone, never attending scientific meetings nor seeking out colleagues with whom to exchange ideas. He published very little, but the intensity of his studies is shown in the 75 volumes of handwritten notes left behind when he died at age 80.

Perhaps because of Avogadro's private nature, the true value of his realization about gas volumes was not recognized until after his death, when another Italian chemist (Stanislao Cannizzarro) brought it to the attention of the scientific community. By comparing the masses of equal volumes of gases at identical temperature and pressure, the relative weights of the molecules could be compared. And by examining the volumes of gases that combined in reactions, it was possible to deduce the formulas. Before this, it was not recognized that a hydrogen molecule consisted of two hydrogen atoms rather than one, or that water molecules consist of two hydrogen atoms and one oxygen atom, rather than just one hydrogen atom and one oxygen atom.

and products are found from the molar masses of the elements combined in each reactant and product. Using the molar masses for copper (Cu, 64 g/mol), sulfur (S, 32 g/mol), and oxygen (O, 16 g/mol) gives for Cu_2S, 160 g/mol[(2 × 64 g) + 32 g]; for SO_2, 64 g/mol [32 g + (2 × 16 g)]; and for Cu_2O, 144 g/mol [(2 × 64 g) + 16 g]. Thus, the equation gives the following information about reactants and products:

$Cu_2S(g)$ +	$2\ Cu_2O(g)$	$\longrightarrow$	$6\ Cu(s)$	+ $SO_2(g)$
1 mol Cu_2S	2 mol Cu_2O		6 mol Cu	1 mol SO_2
160 g Cu_2S	2 × 144 g Cu_2O/mol = 288 g Cu_2O		6 × 64 g Cu/mol = 384 g Cu	64 g SO_2

The World of Chemistry

Program 11, *The Mole*

So using the mole concept, we're able—in the laboratory, in the real world of chemistry—to measure out numbers of molecules that we need for our chemical reactions.
John Massingill, research and development chemist, Dow Chemical, Freeport, Texas.

Exercise 6.3

Interpret the equation for making methanol in terms of moles, molar masses, and masses of reactants and products.

$$CO(g) + 2\,H_2(g) \longrightarrow CH_3OH(\ell)$$

SELF-TEST 6A

1. Balancing chemical equations is an application of the ____________.
2. The ____________ is used by chemists the same way the ________ is used by an egg farmer.
3. Balance the following equations:
   (a) ______ $Si(s)$ + ______ $Cl_2(g) \rightarrow SiCl_4(g)$
   (b) ______ $Al(s)$ + ______ $O_2(g) \rightarrow$ ______ $Al_2O_3(s)$
   (c) ______ $(NH_4)_2CO_3(aq)$ + ______ $Cu(NO_3)_2(aq) \rightarrow$
   ______ $CuCO_3(s)$ + ______ $NH_4NO_3(aq)$
4. The molar mass of iron is the mass in grams numerically the same as the ____________.
5. In photosynthesis, carbon dioxide combines with water to form oxygen and the simple sugar glucose ($C_6H_{12}O_6$).
   (a) Balance the equation

   ____________ $CO_2(g)$ + ______ $H_2O(\ell) \longrightarrow$
   ____________ $C_6H_{12}O_6(aq)$ + ______ $O_2(g)$

   (b) How many molecules of CO_2 are needed to produce one molecule of glucose?
   (c) How many moles of CO_2 are needed to produce 1 mol of glucose?
   (d) What is the molar mass of glucose?
   (e) What mass in grams of CO_2 is needed to make 1 mol of glucose?
6. Which is the molar mass of nitrogen gas (N_2)? (a) 7 g, (b) 14 g, (c) 28 g.

Potassium plus water, $2\,K(s) + 2\,H_2O(\ell) \rightarrow 2\,KOH(aq) + H_2(g)$, is a naturally fast reaction.

6.3 Rates and Reaction Pathways: The "How Fast?" Question

Reaction Pathways

The chemical reactions you may have seen as demonstrations are usually fast. The color change, bubbles of gas, or explosion happens right away as visible proof that a reaction has occurred. Many reactions are also naturally slow, however. At everyday conditions of temperature and pressure, for example, the conversion of carbon monoxide to carbon dioxide is slow.

$$2\,CO(g) + O_2(g) \longrightarrow 2\,CO_2(g)$$

It is sometimes unfortunate that this reaction isn't fast. Breathing too high a concentration of carbon monoxide is fatal.

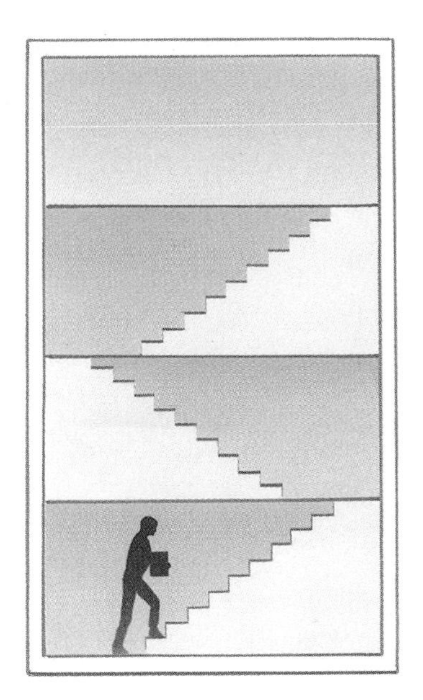

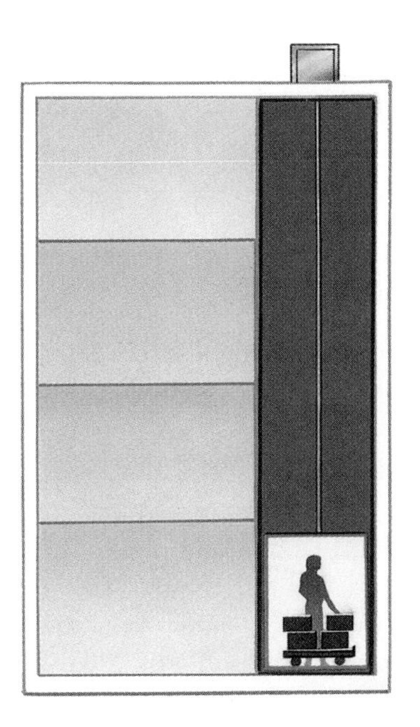

Which pathway has a higher rate of books moved per hour?

Out of curiosity and also for practical reasons, it is interesting to discover what makes reactions fast or slow. Ideally, to do this a chemist would like to watch the pathway of each atom from its position in the reactants to its position in the products.

What is the connection between the rate of a process and its pathway? Suppose that there are several hundred books in a storeroom on the first floor and that you have been hired to move them to the new third-floor library. Depending on the conditions, there are a variety of possible pathways. One pathway might require the following steps: (1) put ten books (the maximum number you can lift) into a carton, (2) carry the carton up one flight of stairs to the second floor, (3) rest a bit, (4) carry the carton up another flight of stairs to the third floor, (5) empty the carton, and (6) return for another load. Another pathway might involve a different series of steps: (1) fill four cartons with books, (2) pile the cartons onto a dolly, (3) push the dolly onto the elevator, (4) ride to the third floor, (5) push the dolly off the elevator, (6) empty the four cartons, and (7) return for another load. The second pathway would probably be faster than the first. To compare them quantitatively, the rate for each pathway could be measured in books moved per hour.

The details of chemical reaction pathways can be extremely complex, and they are very hard to study. With the aid of computers, sophisticated electronics, and clever new techniques, however, advances in this area of observation are being announced regularly.

By using your imagination instead, what might be seen in a simple one-step chemical reaction between two different gases? The gas molecules are flying about at random. Now and then, two molecules head for a collision. As they get closer together, repulsion builds up between their negatively charged electrons. If their kinetic energies are not great enough to overcome this repulsion, the molecules just veer away from each other. No chemical reaction

(a) Unsuccessful collision

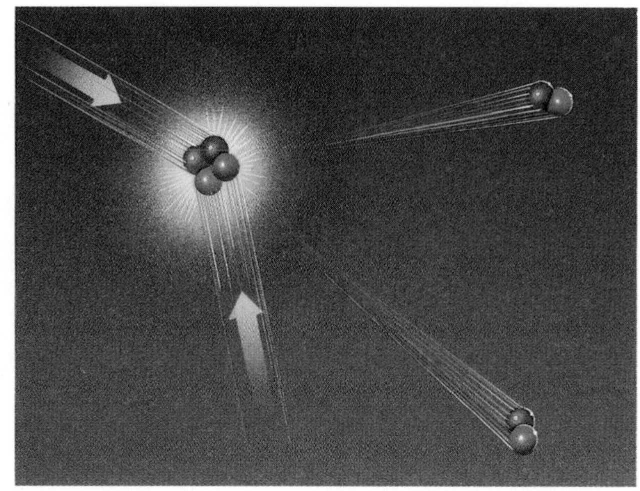

(b) Successful collision

Figure 6.2 Collisions and chemical reactions. Some collisions are successful and some are not. The difference is determined to a great extent by the kinetic energy of the particles.

occurs (Figure 6.2*a*). But if the kinetic energy of the approaching molecules is great enough to drive them together in spite of the repulsion, then they collide. At this instant a chemical reaction might occur. If some of the original bonds break so that new bonds can form, the collision is successful and yields the product (Figure 6.2*b*).

A rapid chemical reaction and the collisions causing it.

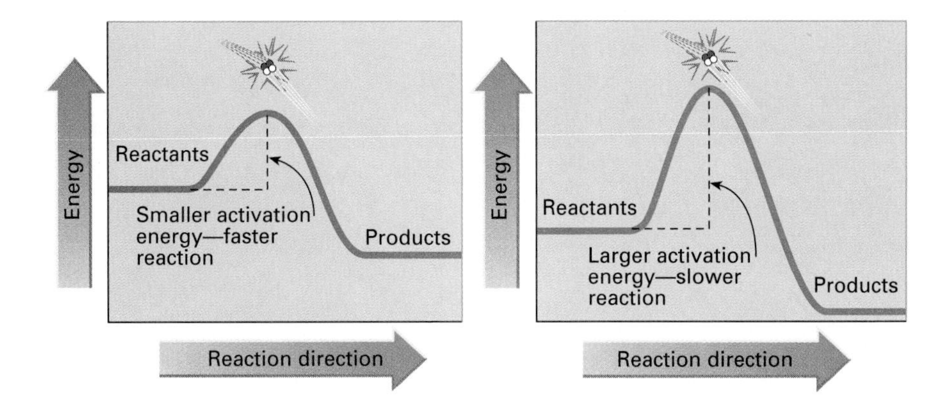

Figure 6.3 Energy pathways of chemical reactions. The height of the activation energy hill determines the rate of the reaction.

An Energy Hill to Climb

Reaction rates are usually expressed as the amount of reactant converted to product in a specific unit of time. For fast reactions the time unit might be seconds; for slow reactions it might be days. The number of successful collisions per second, minute, hour, or day determines the reaction rate.

■ In addition to sufficient energy, successful collisions also require that the reacting parts of molecules or ions physically connect with each other.

For each reaction, there is a distinctive quantity of energy needed for successful collisions, known as the **activation energy**. A high activation energy is like a steep mountain that must be climbed to reach the valley on the other side. In a given collection of reactant particles, only a small number have enough energy to get over a very high activation energy hill in a given time; thus, the reaction is slow. In the opposite condition, many reactant particles can get over a low activation energy hill; therefore, the reaction is fast (Figure 6.3).

Some reactions do not take place at all unless enough energy is supplied to get things started by pushing a few reactants over the activation energy barrier. After that, as in the explosion of a hydrogen–oxygen mixture or the combustion of a fuel, the reaction provides enough energy to keep itself going.

Controlling Reaction Rates

To control the rate of a reaction requires either increasing the population of reactants with enough energy to get over the activation energy barrier or lowering the barrier. Three strategies are available: (1) adjust the temperature, (2) adjust the concentration, or (3) add a catalyst.

1. ***Effect of temperature on reaction rate.*** At higher temperatures, molecules (on average) move faster, so that more of them have enough energy to get over the activation energy hill and react (Figure 6.4). We make use of this principle in cooking by raising the temperature to speed up roasting a piece of meat and in preserving foods by lowering the temperature to slow down the reactions that spoil the food.
2. ***Effect of concentration on reaction rate.*** The quantity of a substance in a given quantity of a mixture is its *concentration*. For example, a solution might have a concentration of 5 g of sodium chloride in 1 L of water

Figure 6.4 Temperature and reaction rate. When a light stick is bent, an inner glass tube breaks, allowing reactants to mix. The result is a chemical reaction that releases energy as light. You can see by the dimmer light in the ice water (right) that the colder temperature slows down the reaction. *(Richard Megna/Fundamental Photographs)*

An alligator crossing a highway. Alligators are cold-blooded animals—the rate of their metabolism depends on the temperature. This alligator crawled onto a major highway during an evening when there was a sharp temperature drop. The resulting slow-down in reaction rate put him to sleep in the middle of the road, where considerate police directed traffic around him. Finally, at midday the temperature rose and he walked off without any coaxing. *(AP/Wide World Photos)*

(5 g/L, or 5 grams per liter). A solution of 10 g of sodium chloride in 1 L of water (10 g/L) has a higher concentration.

Increasing the concentration of reactants increases reaction rates. The more reactant atoms, molecules, or ions, the more frequent the collisions that have enough energy to be successful. There is, for example, a dramatic increase in the rate of combustion in pure oxygen compared with that in air, which is about 20% oxygen.

A similar effect occurs if a reactant is very finely divided, which essentially increases concentration by increasing the surface area at which reactions can occur. We don't think of flour as an explosive substance, but a spark can set off the explosion of flour suspended in the air. Dust explosions are a hazard in grain storage and coal mines.

3. ***Effect of a catalyst on reaction rate.*** Sometimes a reaction speeds up dramatically when a substance other than the reactants is added to the mixture. You may have a bottle containing a dilute solution of hydrogen peroxide on your bathroom shelf—hydrogen peroxide is often used as an antiseptic. When stored in a brown or opaque bottle, it decomposes very slowly to water and oxygen ($2\ H_2O_2 \rightarrow 2\ H_2O + O_2$). Have you noticed that when you put hydrogen peroxide on a cut, it bubbles vigorously? There is a substance in blood (known as *catalase*) that speeds up the decomposition of H_2O_2 (Figure 6.5).

Figure 6.5 Catalyzed decomposition of hydrogen peroxide ($2\ H_2O_2(aq) \rightarrow 2\ H_2O(\ell) + O_2(g)$). The liver is well supplied with an enzyme for this reaction. *(Larry Cameron)*

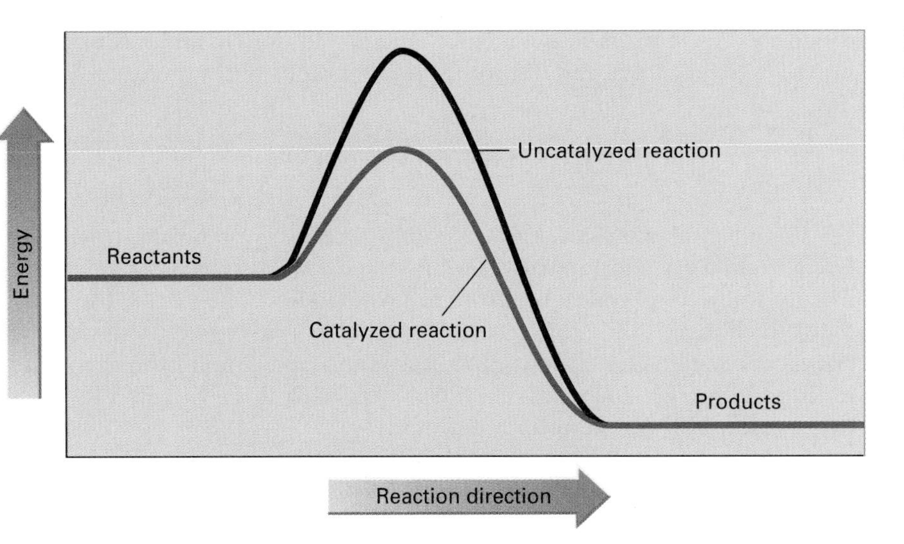

Figure 6.6 Effect of a catalyst. All a catalyst does is lower the activation energy. Because more reactants have enough energy to collide successfully, the reaction speeds up.

Such substances that increase the rate of a chemical reaction without being changed themselves are known as **catalysts**. In the presence of a catalyst, an alternate pathway with a *lower* activation energy is made available (Figure 6.6). More collisions are successful because less energy is required for success. What makes catalysts so practical is that many times they can be recovered after the reaction is over and used again and again. Many industrial processes rely on rare and expensive metals as catalysts, making recovery of the catalyst an economic necessity.

Living things are even more dependent on catalysts than the chemical industry. Engineers can manipulate temperature and concentrations in an industrial process to control reaction rates. Our bodies can't. If body temperature varies far from 37°C or if the concentrations of chemicals in body fluids vary much from the normal, we are in serious trouble. Catalysis is the major strategy available for controlling biochemical reactions. The amazing constancy of our internal chemistry is maintained by biological catalysts known as *enzymes* (Section 11.8). There is a good reason why blood and the liver are well supplied with a fast-acting catalyst for the decomposition of hydrogen peroxide. Because it is highly reactive, hydrogen peroxide must be destroyed before it has the chance to damage essential substances in its surroundings.

The World of Chemistry

Program 13, *The Driving Forces*

Food—it's the study of messy chemistry. In a food which has so many different organic compounds and inorganic compounds, there are lots and lots of different reactions that could have caused spoilage. What we can do is narrow them down to several classes. One is reactions that are enzyme-catalyzed.
Theodore Labuza, Food Chemist

6.4 Chemical Equilibrium and the "How Far?" Question

In a closed container partly filled with water the air over the liquid soon becomes mixed with water vapor. Once the air holds all the water vapor it can, some of the water vapor condenses. Eventually, the evaporation and condensation of the water establish a **dynamic equilibrium**—a state of balance between exactly opposite changes occurring at the same rate. To in-

Figure 6.7 A system at equilibrium. Water has reached equilibrium with its vapor in this Ecosphere. Equilibrium also has been reached by the food and waste products of the inhabitants—a carefully balanced community of plants, shrimp, and a hundred or so kinds of microorganisms. *(Photo courtesy of Ecosphere Associates, Ltd. of Tucson, Arizona, who are responsible for developing this first successful sealed community)*

dicate this equilibrium in a chemical equation, a double arrow is placed between the symbols for water in the liquid and vapor states (Figure 6.7):

$$H_2O(\ell) \underset{\text{Condensation}}{\overset{\text{Evaporation}}{\rightleftharpoons}} H_2O(g)$$

Chemical reactions establish the same kind of equilibrium. In **chemical equilibrium** a chemical reaction and its reverse are occurring at equal rates. Theoretically, all chemical reactions are **reversible**—able to take place in either direction and therefore to come to equilibrium. Studies of many, many chemical reactions have shown that reactions reach equilibrium at a characteristic and predictable point. Understanding what this is provides an answer to the "How far?" question for a reaction.

Some reactions go virtually to what we call *completion*—the conversion of such a large quantity of the reactants to products that what is unconverted is not noticeable and is unimportant. The combination of hydrogen and oxygen to form water is a reaction of this kind. Once a spark has gotten the first few molecules over the activation energy hill, the reaction continues rapidly and explosively until one or both reactants are used up. A slow reaction can also go to completion. As time passes, an iron nail exposed to the atmosphere continues to gradually rust away until only the rust remains.

You may get the idea that all chemical reactions go to completion when you watch a hydrogen–oxygen explosion or watch a piece of wood burn in the fireplace. Chemicals do not always react to form products with the complete extinction of reactants, however. Whenever the point is reached at which the forward reaction is proceeding at the same rate as the reverse reaction, equilibrium is established and the amounts of reactants and products remain unchanged. But because equilibrium is a *dynamic* condition, the forward and reverse reactions are still happening so that each reactant or product is replaced as soon as it is consumed.

For equilibrium to be reached, it is important that none of the reactants or products can escape. An example is provided by the conversion of limestone (calcium carbonate) to lime (calcium oxide). By heating limestone in open pits, early U.S. settlers used this reaction to produce lime for mortar.

$$\underset{\substack{\text{Calcium carbonate}\\ \text{(limestone)}}}{CaCO_3(s)} \xrightarrow{\text{Heat}} \underset{\substack{\text{Calcium oxide}\\ \text{(lime)}}}{CaO(s)} + \underset{\text{Carbon dioxide}}{CO_2(g)}$$

Because the CO_2 gas escapes from the open pit, the reaction keeps going until all of the calcium carbonate in the limestone is converted to lime. If instead some dry limestone is sealed in a closed container and heated, the result is different. As soon as some CO_2 accumulates in the container, the reverse reaction starts to occur. Once the concentration of CO_2 reaches the appropriate point for the conditions, equilibrium is established.

$$CaCO_3(s) \rightleftharpoons CaO(s) + CO_2(g)$$

If there aren't any changes in the pressure or the temperature, the forward and reverse reactions continue to take place at the same rates, and the concentration of CO_2 and the amounts of the solid $CaCO_3$ and CaO in the container remain the same.

Many reactions in which all reactants and products are dissolved in water occur in nature or are carried out by chemists. How far such a reaction goes before equilibrium is reached is always a matter of interest. Acids are a very important class of water-soluble chemical compounds about which we'll have more to say in the next chapter. Acetic acid provides a good example of a reaction that reaches equilibrium with just a small amount of reactant converted to product.

$$\underset{\text{Acetic acid}}{CH_3COOH(aq)} + \underset{\text{Water}}{H_2O(\ell)} \rightleftharpoons \underset{\text{Acetate ion}}{CH_3COO^-(aq)} + \underset{\text{Hydronium ion}}{H_3O^+(aq)}$$

Since the hydronium ion (H_3O^+), which is present in all acid solutions (Section 7.1), is what makes vinegar sour, it's a good thing that this reaction doesn't go to completion. If it did, vinegar would be of little use on a salad.

■ Vinegar is 5% acetic acid by mass. Of this amount, only 0.5% undergoes this reaction.

For an illustration of the importance of constant conditions to maintaining equilibrium, we can return to the decomposition of limestone that has reached equilibrium in a closed container. If the container is opened briefly to let some of the CO_2 out and then sealed again, the forward reaction will outpace the reverse reaction, and the CO_2 concentration will increase until equilibrium is again established. This type of change illustrates a very impor-

THE WORLD OF CHEMISTRY

A Catalyst in Action

Program 14, *Molecules in Action*

Because of their remarkable chemical properties, catalysts are used throughout the industry. Dr. Norman Hochgraf, vice president of Exxon Chemical, shares with us his viewpoint:

Within the petroleum industry, where most of the feedstocks for chemicals come from, catalysts are used to produce motor gasoline, heating oil, and feedstocks for chemicals. And within the chemical industry, catalysts are used not only to purify those feedstocks, but they're also used to polymerize the feedstocks, to make plastics, to make rubbers, and to make synthetic fibers. This industry not only wouldn't be profitable without catalysts, but in a sense, it wouldn't exist. Almost everything we do is the result of catalytic activity, which has been carefully designed, carefully selected, and built into our commercial operations.

Let's look at Eastman Kodak's group of plants in Kingsport, Tennessee. Synthetic products—fabrics, plastics, films, and aspirin—are all made there. These products are all made using acetic anhydride, which is also manufactured at the same location. Yet without the rare South African metal, rhodium, there would be no plant at all. Eastman Kodak used to make acetic anhydride from oil. With the rise in oil prices in the early 1970s, the company began looking for alternative inexpensive materials. Coal was the obvious choice. First, it is gasified, producing hydrogen and carbon monoxide. At a later stage, carbon monoxide is reacted with methyl acetate to produce acetic anhydride. This crucial step requires rhodium as a catalyst.

If rhodium is so expensive, how can it be profitable to use the catalyst in such a large-scale reaction? The answer lies in the fact that catalysts are not used up in reactions. Each catalyst molecule may react with thousands and thousands of molecules of reactant. Thus, this catalyst needs to be present only in tiny amounts. In addition, as the mixture is drawn off, the catalyst is carefully separated from the acetic anhydride and recycled back into the reactor. Very little rhodium is lost.

Without the development of a viable catalyst system for producing acetic anhydride from methyl acetate and carbon monoxide, the chemicals-from-coal complex at Kingsport, Tennessee, would not have been built.

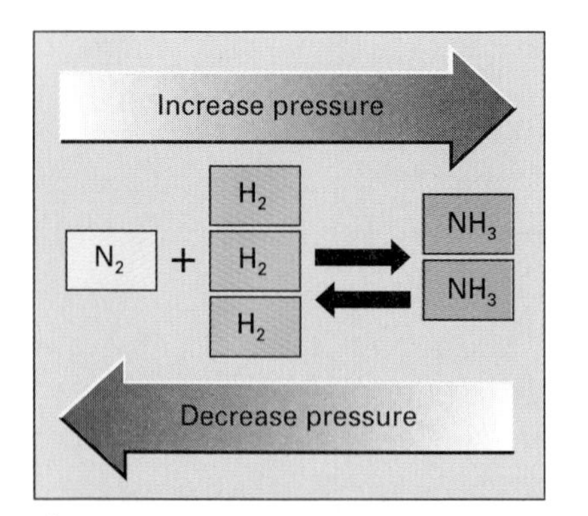

Figure 6.8 Ammonia synthesis. In a demonstration of Le Chatelier's principle, pressure shifts equilibrium of reaction with different amounts of gaseous reactants and products. Higher pressure shifts the reaction toward smaller amounts, and therefore smaller volumes, of gas. To increase the amounts of ammonia produced, the synthesis is done under pressure.

tant principle that applies to all systems at equilibrium: *If a stress is applied to a system at equilibrium, the system will adjust to relieve the stress.* Known as **Le Chatelier's principle** for the French scientist, Henri Le Chatelier, who first stated it in 1884, this principle means that whatever the disruption, the reaction will shift in the direction that re-establishes equilibrium. For a chemical reaction, the stresses might be adding or taking away a reactant or product, changing the temperature, or, in some cases, changing the pressure.

As another example, consider a reaction that takes place entirely in the gaseous state, the synthesis of ammonia. Because of the need for ammonia in the production of fertilizers, this reaction is of great commercial significance (Section 17.4).

$$N_2(g) + 3\ H_2(g) \rightleftharpoons 2\ NH_3(g)$$

There are more reactant gas molecules than product gas molecules, which means that changing the pressure will stress the reaction (Figure 6.8). If the pressure is increased, the equilibrium will shift in the direction that decreases the pressure. To do this, the number of gas molecules must be decreased, which for ammonia synthesis means the forward reaction will be favored, and more ammonia will form.

■ By a combination of experiments and calculations, chemists can find the amounts of reactants and products present at equilibrium for a given reaction.

■ SELF-TEST 6B

1. The rate of a chemical reaction might be expressed as the amount of ____________ converted to ____________ per second.
2. If colliding molecules do not have enough ____________, they cannot react with each other.
3. A reaction can be speeded up by increasing either the ____________ of a reactant or the ____________.
4. A substance that speeds up a reaction without being a reactant or product is called a ____________.
5. The ____________ is like a hill that must be climbed to get a reaction going.
6. Are the amounts of reactants and products always equal to each other at equilibrium?
7. Are the amounts of reactants and products present at equilibrium always the same for the same reaction under all conditions?
8. At equilibrium, are the rates of the forward and reverse reactions equal?
9. Changing the concentration of a reactant to change the equilibrium concentrations of the products is an application of ____________.

Rust on the George Washington Bridge. Rusting is slow, but favorable and inevitable. *(C. D. Winters)*

6.5 The Driving Forces and the "Why?" Question

Why does one reaction produce more product at equilibrium than another? Why do some chemicals react when they are mixed, while others have absolutely no tendency to react unless the conditions are changed? In general, **why** is one chemical reaction favorable and another not favorable?

To examine the driving forces that account for such differences requires looking at two kinds of change—changes in energy as heat and changes in the amount of order that accompany chemical reactions. We experience such changes every day. Turn on the gas stove, and energy is released as heat. Let the refrigerator absorb heat from a mixture of cream and flavorings to produce ice cream. Create major disorder on your desk while rushing to finish a project. Invest the energy needed to restore order to your desk.

Energy Change as a Driving Force

First, it is important to note that how fast a chemical reaction proceeds and the amount of energy associated with it are in no way connected. Our earlier analogy about moving books illustrates this principle. No matter what pathway is chosen and no matter how long the move takes, the total change in potential energy of the books is the same. In terms of a reaction, the difference between the energy stored in the reactants and products is the same, no matter how high or low the activation energy barrier.

■ The potential energy change of books moved up two stories depends only on their mass and the vertical distance they were moved.

The natural direction of chemical reactions is much like the natural direction of more familiar changes in everyday life. We all know that water going over a waterfall is favorable. Likewise, water going back up the waterfall is not favorable. To move water from the bottom of the waterfall back to the top would require energy and work—electricity could drive a mechanical pump that could move the water.

Water held back behind a dam has stored energy—potential energy that can be converted to kinetic energy if the gates in the dam are opened. Like the water behind a dam, chemical compounds have potential energy; it is stored in chemical bonds. In many reactions, this energy is released as heat. The reaction of potassium permanganate and glycerine (an organic compound that can burn) is a good example. When the reactants are mixed, the mixture bursts into flame (Figure 6.9). Any reaction that releases heat is described as **exothermic**, and most exothermic reactions, once they get going, are favorable—no input of energy is required to keep them going.

■ A **favorable reaction** has a natural tendency to happen, like water running downhill. An **unfavorable reaction** (the reverse of a favorable reaction) can only be made to occur by the expenditure of energy.

The opposite condition is found in an **endothermic** reaction, one that requires energy to take place. Most endothermic reactions, such as the decomposition of limestone in an open pit, don't happen at all without a continuous supply of energy.

So far, you've seen evidence for one answer to the "Why?" question. Like water going over a waterfall, chemical reactions are favorable when they release energy and the products thus have less potential energy than the reactants. Usually the energy is released as heat, although there are chemical reactions that generate light and, under the right conditions, electric current. A few reactions, however, are favorable but endothermic—they absorb heat from their surroundings and keep going without any outside influence. Here is evidence that there must be another driving force for change.

The World of Chemistry

Program 13, *The Driving Forces*

Figure 6.9 A favorable exothermic reaction. The favorable driving forces combine to make the reaction of potassium permanganate and glycerine dramatically exothermic.

Entropy Change as a Driving Force

Perhaps you have seen a demonstration of the endothermic but favorable reaction of barium hydroxide and ammonium thiocyanate. This reaction absorbs so much heat from its surroundings that it can freeze water in contact with the reaction flask (Figure 6.10)

$$Ba(OH)_2 \cdot 8\ H_2O(s) + 2\ NH_4SCN(s) \longrightarrow Ba(SCN)_2(aq) + 2\ NH_3(aq) + 10\ H_2O(\ell)$$

The reactants are both crystalline solids and, as such, their components are held together in a repetitive, orderly arrangement. Look at the products—there is liquid water, gaseous ammonia that mostly dissolves in the water, and an ionic compound ($Ba(SCN)_2$) that dissolves in the water to give separate ions (Ba^{2+} and 2 SCN^-). What a mixture! And this is the clue to the driving force. The system has moved from an ordered state to a disordered state.

■ Chemical reactions, like more familiar changes, follow the path of least resistance until they reach a lower energy condition.

The melting of ice is another change that naturally proceeds from order to disorder. You should have no trouble identifying this tendency on a personal scale. It is effortless to create a random mixture of possessions in your room, and it certainly takes energy to put them back in order.

■ Some ionic compounds incorporate water molecules in their crystals. To indicate this, the formula is written with a dot: $Ba(OH)_2 \cdot 8\ H_2O(s)$ represents a crystalline solid composed of 2 OH^- ions and 8 H_2O molecules for every Ba^{2+} ion.

Physical scientists have given the name **entropy** to the disorder of matter, and it can be measured. Gases have a higher entropy than liquids. Liquids have a higher entropy than crystalline solids. Large molecules often have higher entropy than small ones because their atoms can rotate around the bonds in many different ways. The more molecules or the more different kinds of molecules there are in a mixture, the higher the entropy of the mixture.

■ For some reactions, changing the temperature can change the favorability. For example, carbon and water will not react at all at room temperature, but at high temperatures this reaction is the basis for an industrial process for making methanol and other organic compounds (Section 9.8).

Entropy is the second factor that determines the answer to the "Why?" question. Reactions are favorable when they result in a *decrease* in energy *and* an *increase* in disorder. When one of these changes is favorable but the other is not, the greater effect controls the favorability of the reaction.

What happens when a process is unfavorable because it requires energy and creates order? Such a process is not forbidden by nature—it can be made to happen when energy is supplied from some other process. Ordering pro-

(a)

(b)

Figure 6.10 A favorable but endothermic reaction. The reaction of barium hydroxide and ammonium thiocyanate is one of the uncommon examples of a favorable reaction that absorbs heat from its surroundings. *(C. D. Winters)*

cesses must be driven by favorable, disordering processes. The result is always that the net disorder of the universe is increased when an unfavorable process is driven by a favorable one.

The First Law of Thermodynamics

The first and second laws of thermodynamics summarize the universal conditions for changes in energy and entropy (Table 6.1). Because they have such broad application and meaning, not just in science, they are often referred to familiarly as "the first law" and "the second law." Here is a formal statement of the **first law of thermodynamics**, sometimes known as the **law of conservation of energy**:

Energy can be converted from one form to another but cannot be destroyed or created.

When gasoline burns in an auto engine, *all* the energy released could be accounted for if the resulting mechanical energy, the friction of moving parts, the energy that leaves the car in the exhaust, the energy converted to electrical and then chemical potential energy in the battery, and the increase in temperature of the engine and everything surrounding it could be measured.

■ **Thermodynamics** is the movement of energy.

Looked at another way, the first law means that *the total quantity of energy in the universe is constant.* The sun and the energy stored in chemicals on the earth are what we have to use—that is all. Creation of new energy is not possible. All we can do is change it from one form to another. The first law shows up in conversation whenever someone says, "Oh well, you can't get something for nothing."

TABLE 6.1 ■ Some Statements of the First and Second Laws of Thermodynamics

The First Law

The energy of the universe is constant.

Energy can be converted from one form to another, but cannot be destroyed or created.

You can't get something for nothing.

There's no such thing as a free lunch.

The Second Law

The total entropy of the universe is constantly increasing.

Entropy is time's arrow.

The state of maximum entropy is the most stable state for an isolated system.

Energy is conserved in quantity but not in quality.

Every system that is left to itself will, on the average, change toward a condition of maximum probability.

Things are getting more screwed up every day.

You can't break even.

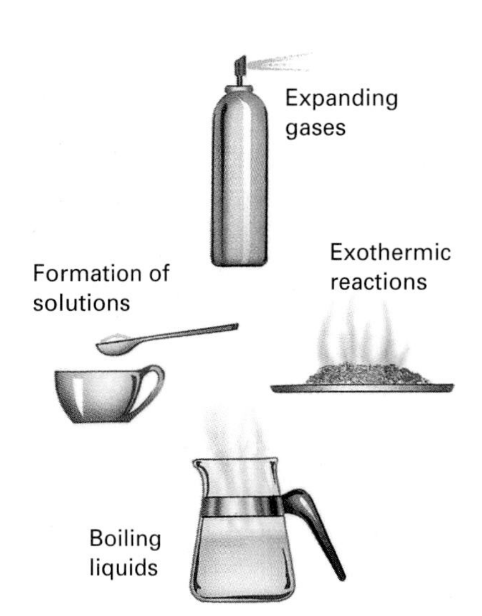

Natural processes that increase entropy.

THE WORLD OF CHEMISTRY

Energy, Entropy, and Industrial Design

Program 13, *The Driving Forces*

Whether a reaction will go or not depends on the balance of energy and entropy of the reactant and the product. We can trade off one of these against the other.

In industry, reactions have to work. Both energy and entropy effects determine that. If a reaction does work—if new molecules can be created—then the industrial design is given to the engineers, and they focus on energy and materials.

Let's look at an example at the Union Carbide plant in West Virginia. Chemicals are made for 500 different products: detergents, adhesives, plastic wraps, car seats, paints, and waxes. Probe the panoply of pipes and towers and you find more than a flow of materials resulting from the scores of chemical reactions. There's a flow of energy, and the engineers must conserve this valuable commodity. Usually a reaction is exothermic, giving off heat. Plant designers want to reuse this heat to drive other reactions, to minimize the waste of energy in the whole plant. Thus, through some of the pipes, they'll transport steam; and out in the plant, steam is piped from point to point, reaction to reaction. The basic raw materials coming into the plant are coal or petroleum, both high in energy. First, ethane gas is produced, and then by selective addition of oxygen to ethane we get a variety of industrial chemicals. The plant is constructed so that each product in the chain of reactions has a successively lower level of energy.

Directing chemicals and heat from one reaction to another requires the maze of pipes visible in chemical manufacturing plants.

The famous insight of Albert Einstein in 1900 when he recognized that matter and energy are interconvertible created an extension of the first law:

■ The production of energy by nuclear fission and fusion takes advantage of Einstein's extension of the law (Section 4.9).

The total amount of matter and energy in the universe is constant.

The Second Law of Thermodynamics

Even more ways have been found to express the **second law of thermodynamics** than the first (Table 6.1). Each statement reflects the observation that although the energy of the universe is constant, once energy is converted to entropy it is never again available for useful purposes.

In chemistry textbooks, the usual statement of the second law is

The total entropy of the universe is constantly increasing.

Visualizing body heat. Thermogram photos show heat in reds and yellow, illustrating that the body heats up with exercise. *(Dan McCoy/Rainbow)*

The universal truth of this law may be hard to accept at first. The formation of the stars and planets, the formation of continents and oceans, the formation of crystalline mineral deposits, and the growth of plants and animals are all ordering processes. Life itself is a constant struggle against entropy. The essential connection is that every ordering process in one small corner of the universe creates disorder somewhere else in the universe. As we are eating, breathing, and producing new biomolecules, we are emitting disordered waste products and contributing disorder to the atmosphere by giving off heat.

In fact, *every* time energy is generated and used to do work, some of the energy is converted to heat. Consider the release of energy by burning coal, petroleum, or wood. The principal products of combustion, carbon dioxide and water, will not burn and release more energy. The energy from the reactants is dispersed as heat into the random molecular motion of the surroundings, where it is not available to do more work. In the burning process, matter and energy are conserved. However, the products and the energy converted to heat are much less useful than the reactants and their stored energy. Observations of this type are the basis of one of the alternative statements of the second law:

Energy is conserved in quantity but not quality.

The implications of the second law are wide ranging. Our economy is based on extracting raw materials from our surroundings, using energy to process them and, in marketplace terms, "adding value" to the raw materials. The second law reminds us that no matter how carefully a process is designed, energy is lost at each manufacturing step.

6.6 Recycling: New Metal for Old

Recycling of metals, paper, and plastic has come to many communities in the United States and is on the rise as awareness of our dwindling natural resources increases. The goal is to see 30% of the total municipal solid waste recycled by the year 2000. (In 1996, we were at about 24%.)

How many of these items could have been repaired or manufactured to last longer? *(Courtesy of the Institute of Scrap Recycling Industries, Inc.)*

There is a long way to go, however, before we can congratulate ourselves too heartily on reducing the quantity of material that is simply dumped somewhere. Between 1960 and 1990 the generation of municipal solid waste almost doubled, while the U.S. population increased by only one third. In other words, each of us produced more garbage. The best remedy remains to generate less waste, rather than to expend time, energy, and money in collecting, processing, and recycling it.

Also, recycling alone is not the best answer. Consider what the laws of thermodynamics mean for recycled materials. Recycling counteracts the natural direction of increasing entropy and can only be done with the expenditure of energy in collecting, transporting, and remanufacturing the waste to produce newly useful materials. Remember the second law—some of the energy used at each step is lost forever to entropy.

A different and more direct approach to conserving resources is to increase the useful lifetime of our household materials and objects. Another is to diminish excess use of materials, even those that are recyclable. The three R's of waste prevention, in order of their importance, are

■ Do any of you remember the time when milk and soda pop were sold in glass bottles that were returned, washed, and used again? Shall we make this the law?

Reduce ***the amount of waste as much as possible.***
Reuse ***products as much as possible.***
Recycle ***materials as much as possible.***

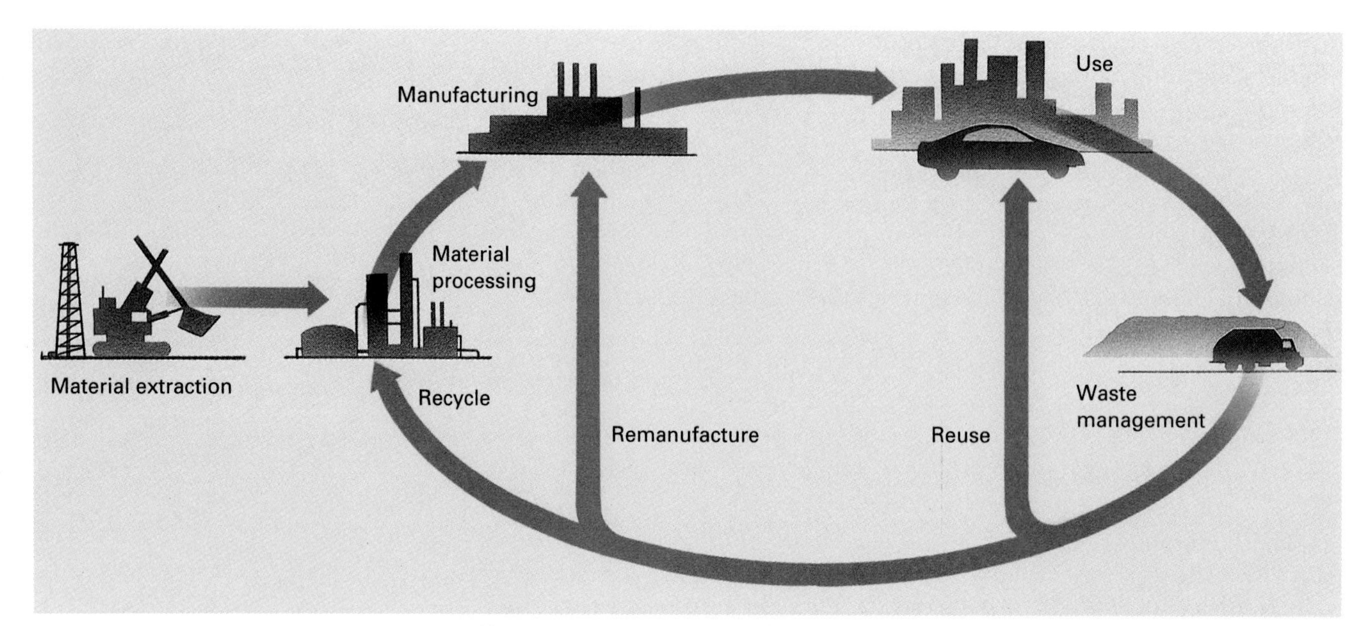

Figure 6.11 Stages in the production, use, and disposal of manufactured products. At each stage energy is used; there is a possibility of air or water pollution; and there is generation of waste. Therefore, there is also the possibility for conservation and recycling at each stage.

To get the whole picture of the amount of waste we generate and its impact on the environment, it's important to keep in mind that municipal waste is only a small fraction of the total. Waste is generated at each stage of a product's life cycle, meaning, of course, that there are opportunities for waste reduction at each stage (Figure 6.11).

Factors in Metal Recycling

- Market value of the metal
- Quantity of the metal in use
- Life span of products containing the metal
- Cost of collection and transport of scrap
- Cost of reprocessing the waste metal
- Cost of disposal of nonrecycled scrap
- Environmental impact of nonrecovery of the metal

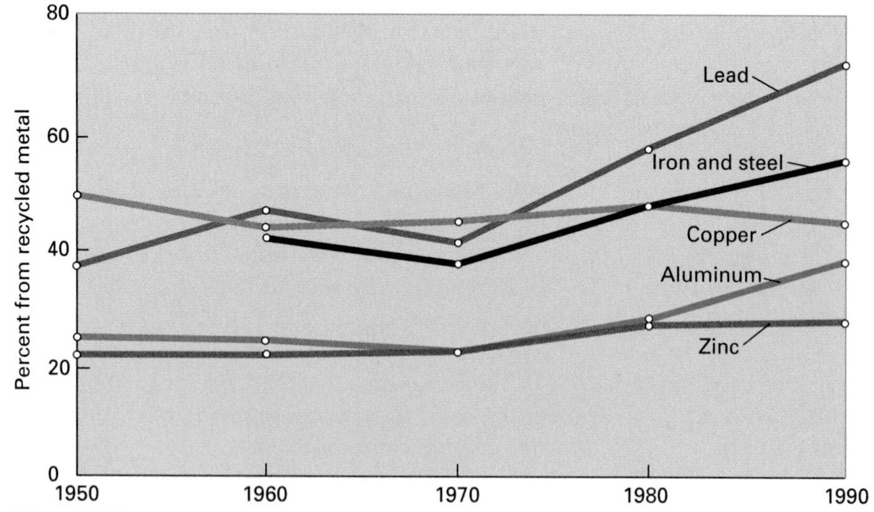

Figure 6.12 The percentage of the annual total consumption of metals from recycled materials. The total of recycled metals includes both production scrap and metal recycled after use in a consumer application. *(Data from U.S. Bureau of Mines)*

FRONTIERS IN THE WORLD OF CHEMISTRY

Green Design

Most products on the market today have been designed either without concern for their impact on the environment or with concern only for their waste-management stage (Figure 6.11). Consumer interest in environmentally friendly products is causing changes. The more desirable approach of considering environmental impact at all stages in a product's life cycle has been christened "green design."

The goals of green design are summarized by the U.S. Office of Technology Assessment as shown in the diagram. Chemistry has a role to play in choices at every stage. Currently, for example, the plastic parts of cars are not recycled; they are shredded along with the metals, but sent to landfills. Applying the green design concept might require formulation by chemists of new plastics with the properties required by automotive parts *and* the ability to be recycled. The plastic parts of cars would then be labeled according to type, and cars would be designed so that the plastic parts could be removed easily at the scrap yard before the metals are shredded for recycling. Economics would still, of course, be a major governing factor in whether this type of change is actually made.

Another approach to green design is providing products that can be remanufactured, that is, restored to new quality and reused, giving them a longer life before they enter the waste stream. Yet another is reducing the mass and volume of material in a product, most noticeably in avoiding overpackaging.

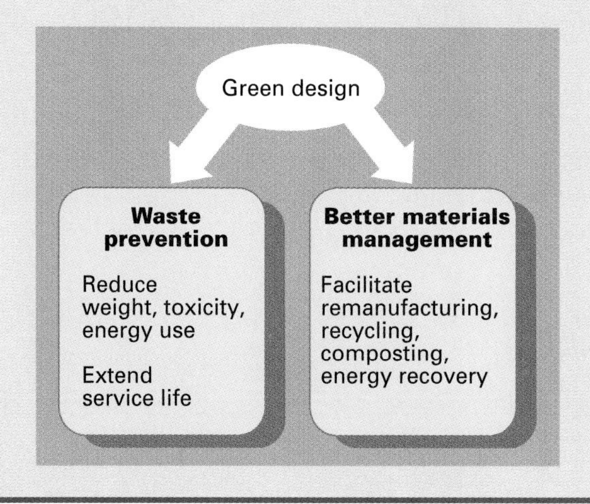

It is important to understand that decision making in green design, like that in risk management, must constantly balance positive and negative outcomes. Many people are happy to purchase their own telephones instead of paying a monthly rental, as was the case before the disbanding of the nationwide telephone monopoly. However, most telephones then were designed to be remanufactured numerous times before they were retired. Today, several million phones are discarded each year after use by only one consumer. Many of these phones have broken down and were not designed to be repaired.

In another example, perhaps you have received a notice from your electric utility company urging the use of fluorescent light bulbs that consume less energy than incandescent bulbs—the plus side of their use. Currently, some fluorescent bulbs contain enough mercury to be designated as hazardous waste, which is the minus side. However, green design is coming to fluorescent bulbs; the amount of mercury in each bulb is gradually being decreased. It's easy to see that green design will be a frontier of chemistry for a long time to come.

Metals are not incinerated, but go either to waste dumps or to recycling plants. Many factors determine the extent to which a metal is recycled. The influence of these factors can be seen in the percentage recycled of the five metals produced in largest volume each year (Figure 6.12). According to the U.S. Bureau of Mines, these metals combined account for more than 99% of the quantity and about 92% of the market value of metals recycled.

Lead is the most recycled metal, for several reasons: it is too toxic to go to landfills; the major use is in automobile batteries, which have a predictable

■ Plastics recycling is discussed in Section 10.7.

■ Do you think the deposit paid on beverage cans and the resulting motivation for scavengers to collect them has contributed to their greater entry into the recycling stream? It has certainly helped to clean up public areas in some parts of the country.

life span; and used batteries are collected at legally designated locations in most states. Iron and steel are second in percentage recycled, most of which is recycled within the industry rather than from consumer products. Iron and steel are used in vastly greater quantities than all other metals combined, resulting in a huge pool of scrap to be dealt with. Not recovering it would be a financial as well as an environmental burden. Also, all steel is potentially recyclable without separation of pure metals from the mixture.

The significant increase in aluminum recycling illustrated in Figure 6.12 has been motivated by the rising cost of the energy needed to make new aluminum, the public's concern about waste management, and the resulting increase in recycling of beverage cans.

SELF-TEST 6C

1. In an exothermic reaction, heat is ____________.
2. In an endothermic reaction, heat is ____________.
3. Some favorable chemical reactions start up as soon as the reactants are mixed, but others do not start till energy is added. (a) True, (b) False.
4. The products of an exothermic reaction store less ____________ than the reactants.
5. Which of the following is a system with higher entropy? (a) A new deck of cards as it comes from the box, (b) A deck of cards after the cards are shuffled.
6. When water freezes, its entropy ____________.
7. The two driving forces for favorable chemical change are a decrease in ____________ and an increase in ____________.
8. "You can't get something for nothing," is one way of stating the ____________.
9. "You can't ever break even," is one way of stating the ____________.
10. In recycling, it is hard to avoid using ____________ and increasing ____________.

MATCHING SET

____	1. Number of atoms shown by the formula of $(NH_4)_3PO_4$	a. Speeds up a reaction
____	2. Mass of 1 mol of H_2O_2	b. 48 g
____	3. Adding a catalyst	c. Same number of atoms of each element on both sides of the arrow
____	4. Cannot be destroyed	d. 34 g
____	5. Mass of 1 mol of titanium	e. Slows down a reaction
____	6. Balanced chemical equation	f. Dynamic equilibrium
____	7. Lowering the temperature	g. Formation of gas that escapes
____	8. Forward and reverse reactions proceeding at equal rates	h. 20
____	9. Drives a reaction to completion	i. 15
____	10. Decreases chemical potential energy	j. An exothermic reaction
____	11. Number of moles of N_2 needed to produce 30 mol of ammonia, $N_2 + 3\ H_2 \rightarrow 2\ NH_3$	k. Energy

QUESTIONS FOR REVIEW AND THOUGHT

1. Look at the following balanced chemical equation for burning ethanol

$$CH_3CH_2OH(g) + 3\ O_2(g) \longrightarrow 2\ CO_2(g) + 3\ H_2O(g)$$

   (a) How many hydrogen atoms are on the product side of the equation? How many are on the reactant side of the equation?
   (b) How many oxygen atoms are on the product side of the equation? How many are on the reactant side of the equation?
   (c) What are the balancing *coefficients* in the reaction?
   (d) Explain, in your own words, how this balanced equation obeys the law of conservation of matter.
2. Equations can only be balanced by adjusting the coefficients of the reactants and products, while the subscripts within the formulas of the reactants and products cannot be changed and still keep the original sense of the equation. Explain why this is so.
3. Balance the following chemical equations:
   (a) $Al + Cl_2 \rightarrow AlCl_3$
   (b) $Mg + N_2 \rightarrow Mg_3N_2$
   (c) $NO + O_2 \rightarrow NO_2$
   (d) $SO_2 + O_2 \rightarrow SO_3$
   (e) $H_2 + N_2 \rightarrow NH_3$
4. Balance the following chemical equations:
   (a) $CH_3OH(\ell) + O_2(g) \rightarrow CO_2(g) + H_2O(g)$
   (b) $C_3H_8(\ell) + O_2(g) \rightarrow CO_2(g) + H_2O(g)$
   (c) $C_6H_6(\ell) + O_2(g) \rightarrow CO_2(g) + H_2O(g)$
   (d) $C_3H_8O(\ell) + O_2(g) \rightarrow CO_2(g) + H_2O(g)$
5. Balance the following equations:
   (a) $Ba(s) + H_2O(\ell) \rightarrow Ba(OH)_2(aq) + H_2(g)$
   (b) $Fe(s) + H_2O(\ell) \rightarrow Fe_3O_4(s) + H_2(g)$
   (c) $Na(s) + H_2O(\ell) \rightarrow NaOH(aq) + H_2(g)$
   (d) $Li(s) + H_2O(\ell) \rightarrow LiOH(aq) + H_2(g)$
6. Balance the following equations:
   (a) $HBr(aq) + KOH(aq) \rightarrow KBr(aq) + H_2O(\ell)$
   (b) $H_2S(g) + NaOH(aq) \rightarrow Na_2S(aq) + H_2O(\ell)$
   (c) $HNO_3(aq) + Ca(OH)_2(aq) \rightarrow Ca(NO_3)_2(aq) + H_2O(\ell)$
   (d) $HCl(aq) + Al(OH)_3(s) \rightarrow AlCl_3(aq) + H_2O(\ell)$
7. Balance the following equations:
   (a) $Sn(s) + HBr(aq) \rightarrow SnBr_2(aq) + H_2(g)$
   (b) $Mg(s) + HCl(aq) \rightarrow MgCl_2(aq) + H_2(g)$
   (c) $Ca(s) + H_2O(\ell) \rightarrow Ca(OH)_2(aq) + H_2(g)$
   (d) $Zn(s) + HNO_3(aq) \rightarrow Zn(NO_3)_2(aq) + H_2(g)$
   (e) $Cs(s) + H_2O(\ell) \rightarrow CsOH(aq) + H_2(g)$
8. Give two interpretations of the number 2 in front of HCl in the following balanced equation:

$$H_2(g) + Cl_2(g) \longrightarrow 2\ HCl(g)$$

9. Identify the atom used as the basis of the atomic weight scale.
10. What do 35.5 g of chlorine (Cl) and 12.011 g of carbon (C) have in common?
11. What do 103.5 g of lead (Pb) and 6.006 g of carbon (C) have in common?
12. Look up the unit called the *gross.* How many pairs of jeans are in 4 gross of jeans? In what way is the gross unit like Avogadro's number?
13. What is the numerical value of Avogadro's number?
14. What is meant by the term *molar mass?*
15. What is the definition for the mole?
16. Interpret the following equation for the complete combustion of octane in terms of moles of reactants and products, and their molar masses:

$$2\ C_8H_{18}(\ell) + 25\ O_2(g) \longrightarrow 16\ CO_2(g) + 18\ H_2O(\ell)$$

17. Give at least one reason why some chemical reactions are fast while others are slow. Use your own analogy to explain the reason.
18. What influence does temperature usually have on the rate of a chemical reaction? Explain why this is so.
19. What effect does freezing food have on reaction rates? Why is freezing used for preservation of food, tissue samples, and biological samples?
20. Explain the term *activation energy* as it applies to chemical reactions.
21. How does the magnitude of the activation energy influence the rate of a chemical reaction?
22. What is the effect of a catalyst on the activation energy of a chemical reaction?
23. Explain why hydrogen and oxygen can remain mixed at room temperature without any noticeable reaction, yet they will combine explosively to form water if the mixture is ignited with a tiny spark.
24. Describe how each of the following changes will affect the rate of a reaction:
   (a) Increase in temperature
   (b) Increase in concentration
   (c) Introduction of a catalyst
25. Paper burns fairly rapidly in air. How would you expect paper to burn in pure oxygen? Explain.
26. What is the general name for biological catalysts?
27. What is meant by the term *reversible* when describing chemical reactions?
28. What is meant by the term *dynamic equilibrium?*
29. If you heat limestone in a closed vessel, explain which way you would expect the equilibrium to shift if
   (a) you added more CO_2
   (b) you allowed some of the CO_2 to escape from the vessel

$$CaCO_3(s) \rightleftharpoons CaO(s) + CO_2(g)$$

30. Give a statement of Le Chatelier's principle.
31. What is meant when it is said that the reaction shifts in favor of the products of the reaction?
32. What is meant when it is said that the reaction shifts in favor of the reactants?
33. The reaction between HCl and NaOH is described as going to completion. Explain what this means in terms of how much of the reactants remain unreacted.
34. The reaction between N_2 and H_2 to produce ammonia (NH_3) is described as an equilibrium reaction

$$N_2(g) + 3\,H_2(g) \rightleftharpoons 2\,NH_3(g)$$

After the reaction reaches equilibrium, what is present in the reaction mixture?
35. Define the following terms:
   (a) Potential energy (b) Exothermic
   (c) Endothermic (d) Entropy
   (e) Favorable reaction
36. State the first law of thermodynamics. What does it mean?
37. State the second law of thermodynamics. What does it mean?
38. Does entropy increase or decrease in the following reactions? Explain your answer.
   (a) Nitrogen reacting with oxygen to form nitrogen dioxide

$$N_2(g) + 2\,O_2(g) \longrightarrow 2\,NO_2(g)$$

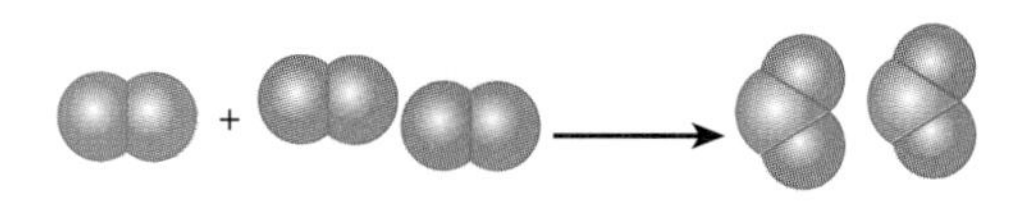

   (b) Acetylene reacting with oxygen to form carbon dioxide and water

$$2\,HC{\equiv}CH(g) + 5\,O_2(g) \longrightarrow 4\,CO_2(g) + 2\,H_2O(g)$$

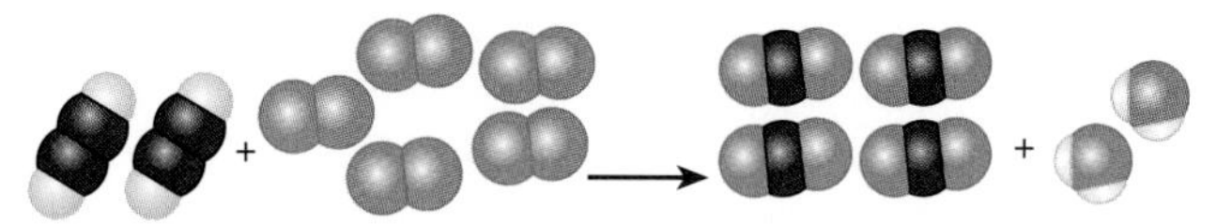

39. Burning methane in air gives off heat. Is this reaction endothermic or exothermic?
40. Give three examples of how your everyday world tends to become disordered. Are these favorable processes? Explain.
41. Photosynthesis involves taking the small molecules CO_2 and H_2O and making larger, more complex molecules like glucose ($C_6H_{12}O_6$). Is this a favorable process? Explain. What is the source of energy for photosynthesis?
42. What does photosynthesis have in common with building a house? Draw as many similarities as you can including ones between the building materials, the finished products, and the energy involved.
43. Does recycling of waste overcome the second law, which predicts the continuous increase in entropy?
44. What is the difference between *quality* of energy and *quantity* of energy?

PROBLEMS

1. Calculate the molar mass of the following:
   (a) H_2O (b) I_2
   (c) KOH (d) NH_3
   (e) CO_2 (f) CO
2. Calculate the molar mass of the following:
   (a) $C_6H_{12}O_6$ (b) H_2SO_4
   (c) Na_2HPO_4 (d) $Ca(NO_3)_2$
   (e) $C_{57}H_{104}O_6$
3. Which will have the larger mass? (a) 1 mol of Cu, (b) 1 mol of Pb, (c) 1 mol of Na.
4. Which will have the larger mass? (a) 0.5 mol of CO_2, (b) 2 mol of Li, (c) 12 mol of H_2.
5. Based on the balanced equation

$$Cu(s) + Cl_2(g) \longrightarrow CuCl_2(s)$$

   (a) How many moles of copper will be required to react with 0.5 mol of chlorine?
   (b) How many moles of $CuCl_2$ can be formed from 1.5 mol of copper?
6. Balance the equation for the fermentation of glucose and then answer the following questions

$$C_6H_{12}O_6(aq) \longrightarrow CH_3CH_2OH(aq) + CO_2$$

   (a) In this reaction, how many moles of ethanol (CH_3CH_2OH) can be prepared from 6 mol of glucose?
   (b) If, during the fermentation process, 10.5 mol of carbon dioxide is found to be produced, how many moles of ethanol will have been produced?

CHAPTER 7

Acid–Base Reactions

Chemists today follow in the tradition started centuries ago by individuals who closely observed their surroundings and wanted to understand what they saw. In this chapter, we'll introduce you to a class of chemical reactions that have been known for a long time because they are common in our natural surroundings—acid–base reactions. These reactions influence our personal health, the health of our environment, and the health of our economy. A small deviation in the delicate balance of acid–base chemistry in our bodies can be life threatening. Sulfuric acid is so important to industrial production that the amount sold each year is taken as an indicator of the state of a nation's economy. Furthermore, acids and bases are common ingredients in chemicals that we use every day in our homes.

- What are the chemical properties of acids and bases?
- What happens chemically when neutralization occurs?
- How is pH defined and why is it useful in describing acid and base solutions?
- What are buffers, and why are buffers in the body so important to our health?
- What are some everyday uses of acids and bases?

7.1 Acids and Bases: Chemical Opposites

Whether hydrangeas are pink or blue is determined by the acidity of the soil. *(Tony Stone Images)*

Everyone should know a few practical things about acids and bases, which we usually encounter as solutions in water. They can be harmful. The harm can range from the stinging sensation when you accidentally squirt lemon juice (citric acid) into your eye, to the severe and persistent burns of the skin that result if you spill battery acid (sulfuric acid) or lye (sodium hydroxide, the active ingredient in Drano and some oven cleaners) on your skin and do not flush it immediately with lots of water. Acidic and basic solutions can also eat holes in clothing, very quickly if they are strong solutions. The first lesson, then, is that strongly acidic or basic substances must be handled with care.

The potential for harm from strong acids and bases, however, does not mean all acids and bases are to be avoided. The dilute solutions of acids and bases that are in everyday use would be sorely missed—orange juice, vinegar, soda pop, household ammonia, and most of our soaps and detergents, to name just a few.

It turns out that all acidic and basic water solutions, whatever their sources or applications might be, have some chemical properties in common. Acids and bases are closely related classes of chemical compounds that are highly reactive, both with other substances and with each other.

The word "acid" comes from the Latin *acidus* meaning sour or tart because in water solutions, acids have a sour or tart taste. Lemons, grapefruit, and limes taste sour because they contain citric acid and ascorbic acid (Vitamin C). Vinegar is sour because it contains acetic acid. Another common property of acids is their ability to change the color of compounds known as **acid–base indicators**. For example, acids change the color of litmus indicator from blue to pink (Figure 7.1).

■ You should never taste anything in a laboratory.

The properties that acids have in common, listed in Table 7.1, result from the ability of acids to release a hydrogen ion (H^+) in a water solution (an *aqueous solution*, symbolized by *aq*). The ionization of an acid is therefore often represented by an equation like the following for hydrochloric acid:

$$HCl(aq) \longrightarrow H^+(aq) + Cl^-(aq)$$

Hydrogen ions, because of their small size and resulting concentration of positive charge in a small area, however, do not remain independent in water solution. Instead, they bond to water molecules to give **hydronium ions** (H_3O^+). Therefore, it is more nearly correct, and as you will see more useful,

Figure 7.1 Litmus paper, our oldest acid–base indicator. The color change of organic dyes like litmus indicates whether they have reacted with an acid or a base. (a) An acid turns litmus from blue to pink, and (b) a base turns litmus from pink to blue.

TABLE 7.1 ■ Properties of Acids and Bases

Acids	*Bases*
Sour taste	Bitter taste
Provide H^+ ions	Provide OH^- ions
React with active metals to give hydrogen	Slippery feeling
Change colors of indicators (e.g., litmus turns from blue to red)	Change colors of indicators (e.g., litmus turns from red to blue)
Produce CO_2 when added to limestone	Neutralize acids
Neutralize bases	
Some Acidic Substances	***Some Basic Substances***
Vinegar	Household ammonia
Tomatoes	Baking soda
Citrus fruits	Soap
Carbonated beverages	Detergents
Black coffee	Milk of magnesia
Gastric fluid	Oven cleaners
Vitamin C	Lye
Aspirin	Drain cleaners

to write an equation for the ionization of hydrochloric acid (and other acids) with the hydronium ion as a product.

$$HCl(aq) + H_2O(\ell) \longrightarrow H_3O^+(aq) + Cl^-(aq)$$

Hydrochloric acid, $HCl(aq)$, which is formed when hydrogen chloride gas dissolves in water, contains hydronium ions and chloride ions (the anions from the acid). A solution that contains H_3O^+ ions and has the properties common to acids is referred to as an **acidic solution**.

Bases also have a number of properties in common (Table 7.1). Water solutions of bases are slippery or soapy to the touch and change litmus from red to blue, which is the reverse of the change caused by acids. The classic properties by which bases are recognized are caused by the presence in water solution of hydroxide ions (OH^-). The hydroxides of sodium and potassium (NaOH and KOH) and of calcium and magnesium ($Ca(OH)_2$ and $Mg(OH)_2$) are among the most common bases used in industry and chemical laboratories. These are ionic compounds that yield hydroxide ions when they dissolve. For example, sodium hydroxide, also known as *lye* or *caustic soda*, dissolves as represented by the equation

$$NaOH(s) \xrightarrow{\text{Water}} Na^+(aq) + OH^-(aq)$$

A solution that contains OH^- ions and has the properties common to bases, is described as a **basic**, or **alkaline**, **solution**.

Cleaning products that contain bases.
(C. D. Winters)

Neutralization Reactions

A water solution cannot be both acidic and basic. An *acidic solution* contains a higher concentration of H_3O^+ ions than OH^- ions, and a *basic solution* contains a higher concentration of OH^- ions than H_3O^+ ions. When the solution of an acid is mixed with the solution of a base, the hydronium ions from the acid react with the hydroxide ions from the base to produce water

$$H_3O^+(aq) + OH^-(aq) \longrightarrow 2\,H_2O(\ell)$$

With bases such as potassium hydroxide and acids such as hydrochloric acid, the second product is a salt, which in this case is potassium chloride (KCl). A **salt** is a compound composed of the positive metal ion from a base and the negative ion from an acid.

$$\underset{\text{Base}}{KOH(aq)} + \underset{\text{Acid}}{HCl(aq)} \longrightarrow \underset{\text{Salt}}{KCl(aq)} + \underset{\text{Water}}{H_2O(\ell)}$$

In this reaction, KOH, HCl, and KCl are all water-soluble compounds that yield ions in solution, while water is a molecular compound.

When exactly equivalent amounts of acid and base react so that all of the H_3O^+ and OH^- ions are used up in forming water, the result is a **neutralization reaction**. The reaction of sodium hydroxide with sulfuric acid, like the reaction of potassium hydroxide with hydrochloric acid, is a neutralization.

$$\underset{\text{Base}}{2\,NaOH(aq)} + \underset{\text{Acid}}{H_2SO_4(aq)} \longrightarrow \underset{\text{Salt}}{Na_2SO_4(aq)} + \underset{\text{Water}}{2\,H_2O(\ell)}$$

In neutralization reactions, acidic and basic properties are eliminated.

Acid–Base Definitions

Our formal definitions of acids and bases take into account the role of the hydrogen ion in acid–base reactions. An **acid** is a molecule or ion able to donate a hydrogen ion to a base. A **base** is a molecule or ion able to accept a hydrogen ion from an acid. The result of these definitions is that every reaction in which a hydrogen ion is exchanged between reactants is an **acid–base reaction**. The following structural formulas show how a hydrogen ion is exchanged between water (written here as H—O—H) and hydrogen chloride gas in the formation of a hydrochloric acid solution. This is an acid–base reaction in which water is the base and HCl is the acid.

$$\underset{\substack{\text{Base} \\ (H^+\text{ acceptor})}}{H{-}O{-}H} + \underset{\substack{\text{Acid} \\ (H^+\text{ donor})}}{H{-}Cl} \longrightarrow H{-}\underset{\substack{| \\ H}}{O}{-}H^+ + Cl^-$$

■ When a hydrogen atom loses an electron, the remaining charged particle is a proton. Thus, acids and bases are also sometimes described as proton donors and acceptors.

In the reaction between sodium hydroxide and the hydronium ion in a hydrochloric acid solution, OH^- is the base and H_3O^+ is the acid

$$\underset{\substack{\text{Base} \\ (\text{H}^+ \text{ acceptor})}}{\text{O—H}^-} + \underset{\substack{\text{Acid} \\ (\text{H}^+ \text{ donor})}}{\text{H—}\underset{|}{\overset{}{\text{O}}}\text{—H}^+ \text{(H)}} \longrightarrow \text{H—O—H} + \text{H—O—H}$$

The picture of acids and bases as hydrogen ion donors and acceptors explains not only the properties of the acids and bases we have described thus far, but also the basicity of ammonia (NH_3) and ions such as the carbonate ion (CO_3^{2-}). The neutral ammonia molecule can accept a hydrogen ion from a water molecule to produce a solution that is basic because it contains hydroxide ions. In this reaction, water donates the hydrogen ion and therefore is the acid

$$\underset{\substack{\text{Base} \\ (\text{H}^+ \text{ acceptor})}}{NH_3(aq)} + \underset{\substack{\text{Acid} \\ (\text{H}^+ \text{ donor})}}{H_2O(\ell)} \rightleftharpoons NH_4^+(aq) + OH^-(aq)$$

Although they contain no hydroxide ions, carbonate salts such as sodium carbonate (Na_2CO_3, washing soda) and potassium carbonate (K_2CO_3) also dissolve in water to give basic solutions. How does a carbonate salt produce hydroxide ions in water? Like ammonia, the carbonate ion accepts a hydrogen ion from a water molecule, which leaves behind a hydroxide ion.

$$\underset{\substack{\text{Base} \\ (\text{H}^+ \text{ acceptor})}}{CO_3^{2-}(aq)} + \underset{\substack{\text{Acid} \\ (\text{H}^+ \text{ donor})}}{H_2O(\ell)} \rightleftharpoons HCO_3^-(aq) + OH^-(aq)$$

■ The description of basic substances as *alkaline* derives from the Arabic word *al-qali*, meaning "plant ashes." Potassium carbonate, commonly known as potash, is found in ashes from wood fires. Long ago it was discovered that this compound dissolves in water to yield a solution that feels slippery, tastes bitter, and reacts with acids.

The equations in this section illustrate a very important point. In the reaction with hydrogen chloride, water reacts as a base. In the reactions with ammonia molecules and the carbonate ion, water reacts as an acid. Since all water molecules are the same, water must be able to react as an acid or a base depending on whether it reacts with a base or an acid. This property of water plays an important role in our water-based world. Whenever a substance that can accept or donate a hydrogen ion dissolves in water, the result is a solution with basic or acidic properties. In removing stubborn dirt (Section 7.5), this can work to our advantage. In the production of acid rain (Section 16.9), this causes problems.

7.2 The Strengths of Acids and Bases

What is the difference between a strong acid such as hydrochloric acid, sold in hardware stores (as *muriatic acid*) to clean brick and concrete, and a weak acid such as the acetic acid in vinegar? The strength of an acid or base is determined by the *extent* of ionization in aqueous solution. The greater the ionization, the stronger the acid or base.

■ By "extent of ionization" we refer to the point at which ionization of an acid reaches equilibrium and the ion concentrations become constant (Section 6.4).

Common acids like sulfuric acid, hydrochloric acid, and nitric acid react completely with water to give hydronium ions and anions. Therefore, like ionic compounds (Section 5.2), they are electrolytes; and because they are entirely converted to ions in solution, they are *strong electrolytes*. The solutions

TABLE 7.2 ■ Common Acids

Name	Formula	Strength	Use and Occurrence
Sulfuric acid*	H_2SO_4	Strong	Cleaning steel; car batteries; making plastics, dyes, fertilizers
Hydrochloric acid	HCl	Strong	Cleaning metals and brick mortar
Nitric acid	HNO_3	Strong	Making fertilizers, explosives, plastics
Phosphoric acid	H_3PO_4	Moderate	Making fertilizers, detergents, food additives
Acetic acid	CH_3COOH	Weak	Vinegar
Propionic acid	CH_3CH_2COOH	Weak	Swiss cheese
Citric acid	$HOC(COOH)(CH_2COOH)_2$	Weak	Fruit
Carbonic acid	H_2CO_3	Weak	Carbonated beverages
Boric acid	H_3BO_3	Weak	Eye drops, mild antiseptic

* Sulfuric acid is a polyprotic acid, meaning it has more than one acidic hydrogen. It first ionizes completely to give HSO_4^- (hydrogen sulfate ion), which is a weak acid and partially ionizes to give SO_4^{2-} (sulfate ion).

of these **strong acids** have high concentrations of hydronium ions and are very acidic (Table 7.2). The equation for the reaction of a strong acid such as nitric acid with water is written with a single arrow, which indicates that the reaction goes to completion.

$$HNO_3(aq) + H_2O(\ell) \longrightarrow H_3O^+(\ell) + NO_3^-(aq)$$

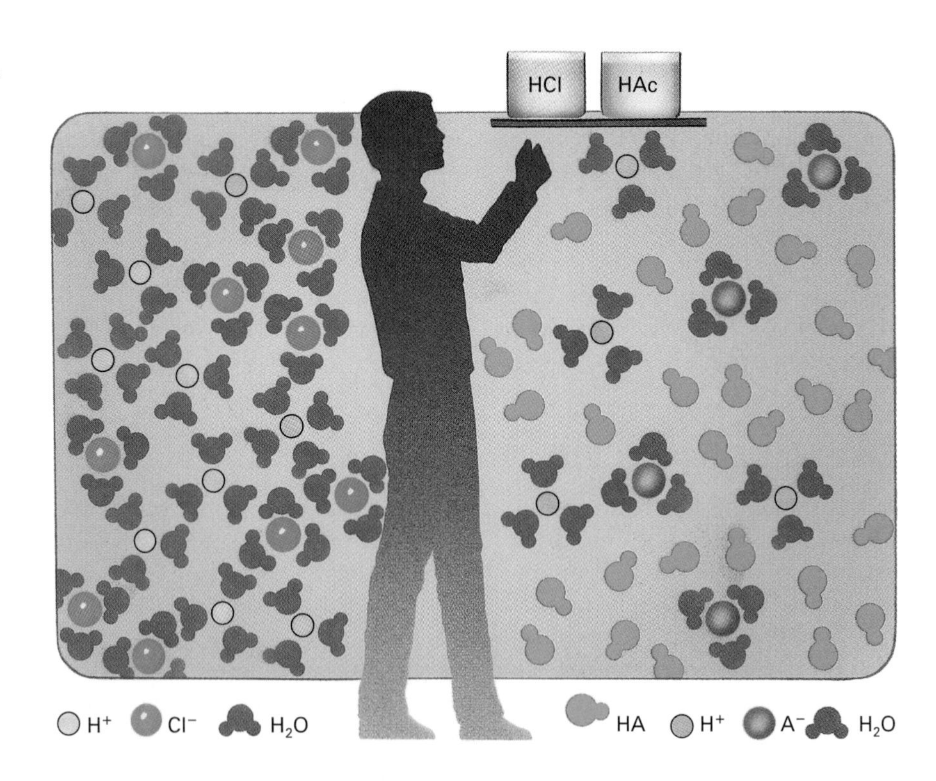

A strong acid (hydrochloric acid, HCl), a weak acid (acetic acid, HA), and the difference between their solutions at the molecular level.

Many acids establish an equilibrium with water at a point where not all the acid molecules have been converted to ions. These are the **weak acids**—they are only slightly ionized in aqueous solution because an equilibrium is established (Section 6.4). Acetic acid is a typical weak acid:

$$CH_3COOH(aq) + H_2O(\ell) \rightleftharpoons H_3O^+(aq) + CH_3COO^-(aq)$$

The CH_3COOH molecules undergo ionization while H_3O^+ and CH_3COO^- ions simultaneously recombine to give CH_3COOH molecules and water. Since only a few percent of the acetic acid molecules are ionized at any given time, aqueous solutions of acetic acid contain mostly acetic acid molecules with a few hydronium ions and acetate anions, and are only weakly acidic.

■ Although there are several H atoms in acetic acid, only the one bonded to oxygen in the —COOH group reacts with a base. This condition is typical of organic compounds that contain the —COOH group.

As previously explained, the metal hydroxides are all ionic compounds and therefore **strong bases** because when dissolved in water, they are completely converted to ions (to the extent that they are soluble). Ammonia is our most common **weak base**, a base that establishes an equilibrium with water to produce a solution with relatively few ammonium and hydroxide ions.

$$NH_3(g) + H_2O(\ell) \rightleftharpoons NH_4^+(aq) + OH^-(aq)$$

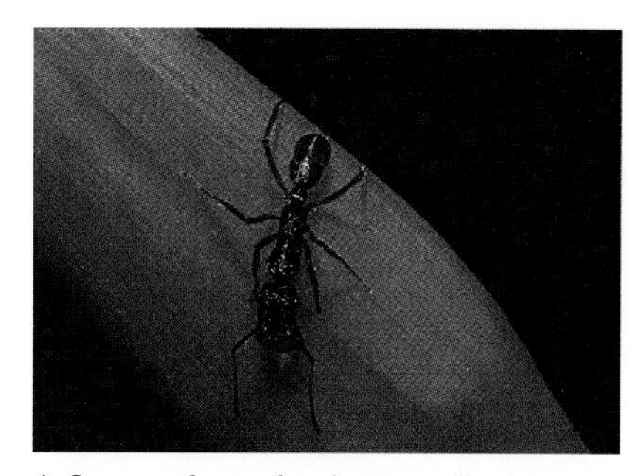

A fire ant from the Amazon Basin on a ginger plant. The sting of these and of other ants contains the simplest organic acid, formic acid (HCOOH). In the Amazon, fire ant colonies can be so large that streams become polluted by their formic acid. *(© 1990 M & A Doolittle/Rainbow)*

The anions of weak acids are also bases (Table 7.3). Look at the reverse of the reaction of acetic acid with water. The acetate ion (CH_3COO^-) accepts a hydrogen ion from the acidic hydronium ion and is therefore reacting as a base. You will see that this property of anions plays an important role in solutions, such as blood, in which the acid concentration must remain constant (Section 7.4). In both blood and the many household uses of sodium bicarbonate, the basic nature of the bicarbonate ion (HCO_3^-) is put to use in controlling acid concentration.

$$HCO_3^-(aq) + H_2O(\ell) \rightleftharpoons H_2CO_3(aq) + OH^-(aq)$$

Because it is not stable, the carbonic acid (H_2CO_3) formed in this reaction, breaks down to form carbon dioxide gas.

$$H_2CO_3(aq) \rightleftharpoons CO_2(g) + H_2O(\ell)$$

TABLE 7.3 ■ Common Bases

Name	Formula	Strength	Use
Sodium hydroxide	$NaOH$	Strong	Drain cleaner; producing aluminum, rayon, soaps, detergents
Potassium hydroxide	KOH	Strong	Producing soaps, detergents, fertilizers
Calcium hydroxide	$Ca(OH)_2$	Strong	Producing bleaching powder, paper and pulp, softening water
Ammonia	NH_3	Weak	Producing fertilizer, explosives, plastics, insecticides, detergents
Sodium bicarbonate	$NaHCO_3$	Weak	Antacid
Sodium carbonate	Na_2CO_3	Weak	Detergents, glassmaking

■ SELF-TEST 7A

1. An acid is a hydrogen ion ____________ and a base is a hydrogen ion ____________.
2. To produce a desirable sour taste, an ____________ is added to many carbonated beverages.
3. Soap solutions are slippery and taste bitter, which indicates that soap is a(n) (basic/acidic) substance.
4. Complete the following equation for the reaction of an acid with a base:

$$H_3O^+(aq) + \_\_\_\_\_\_\_\_\_\_ \longrightarrow 2\ H_2O(\ell).$$

5. Identify each of the following chemical compounds as an acid, a base, or a salt: (a) Na_2SO_4, (b) $HF(aq)$, (c) KCl, (d) KOH.
6. Ammonia is a ____________ base because it establishes an ____________ with water in which ____________ are formed.
7. Because it has a small size and a high charge, H^+ exists in water as the ion known as the ____________ ion, which has the formula ____________.
8. The strong acid of greatest economic importance is ____________.
9. Complete the following equation:

$$\_\_\_\_\_\_\_\_\_\_ + KOH(aq) \longrightarrow KCl(aq) + \_\_\_\_\_\_\_\_\_\_.$$

10. Some acids are described as ____________ acids because they establish equilibrium with their ions in solution instead of being completely ionized.

7.3 Molarity and the pH Scale

Measuring the acidity of soil with a pH meter. The proper acid–base balance is essential to the health of these plants. *(© Leonard Lessin/Peter Arnold, Inc.)*

Water, you have seen, is capable of acting as either a hydrogen ion donor or a hydrogen ion acceptor, that is, as an acid or a base, depending on the properties of the substance with which it reacts. Water can also act as an acid or a base toward itself, although the reaction occurs to only a very small extent. About 1 of every 550,000,000 water molecules is ionized at any given time.

$$H_2O + H_2O \rightleftharpoons H_3O^+ + OH^-$$

Pure water is **neutral** because it contains equal numbers of hydronium ions and hydroxide ions.

The self-ionization of water provides the basis for a convenient method for expressing numerically just how acidic or basic any water solution is. You may have seen this quantity in use for a consumer product, for example, on the label of a "pH-balanced" shampoo. (The label refers to a controlled acidity of the solution, so that it is less likely to damage hair.) To understand what pH is requires first understanding how to express the **concentration of a solution**, which is the quantity of a solute dissolved in a specific quantity of solvent or solution. We might give the concentration of a solution of sodium hydroxide as 4.0 g/L, that is 4.0 g of NaOH per liter of solution. In chemistry, however, because of its relation to quantities of chemicals in reactions (Sections 6.2), concentration based on the mole is preferred.

Molarity

Concentrations of solutions in chemistry are usually expressed as **molarity**, which is the number of moles of solute per liter of solution. For example, a 1 molar (1 M) solution contains 1 mol of solute per liter of solution. If the solute is sodium hydroxide, a 1.00 M solution contains 1 mol, or 40.0 g, of NaOH per liter of solution. To find the molarity of a solution requires knowing the mass of dissolved solute, the volume of the solution, and the molar mass of the solute. For example, 4.0 g of NaOH is 0.10 mol (4.0 g/40.0 g/mol). Thus, a solution with 4.0 g of NaOH dissolved in 1 L of solution is a 0.10 M NaOH solution.

EXAMPLE 7.1 *Molarity*

What is the molarity of a solution that contains 10.0 g of HCl (molar mass, 36.5 g/mol) per liter of solution?

SOLUTION

First, we find the number of moles of HCl equivalent to 10.0 g of HCl

$$10.0 \text{ g HCl} \times \frac{1 \text{ mol HCl}}{36.5 \text{ g HCl}} = 0.274 \text{ mol HCl}$$

■ Finding the number of digits to include in the answers to mathematical problems is reviewed in Appendix A.

With 0.274 mol dissolved per liter of solution, the molarity is 0.274 M in HCl.

Exercise 7.1

How many grams of NaOH have been dissolved per liter of solution to make a 4.0 M NaOH solution?

The pH Scale

In pure water (and all neutral solutions) the concentrations of hydronium ions and hydroxide ions are the same. Expressed as molarity, their concentrations are 1.0×10^{-7} M at 25°C.

The product of the molarity of the hydronium ions and hydroxide ions in pure water is $(1.0 \times 10^{-7})(1.0 \times 10^{-7}) = 1.0 \times 10^{-14}$. *The value* 1.0×10^{-14} *is important to an understanding of all aqueous solutions because it is a constant that is always the product of the molar concentration of* H_3O^+ *and* OH^- *in the solution.* Using square brackets as is customary to represent the molarity of a substance, this relationship is written as

■ Scientific notation, in which large and small numbers are represented by a number between 1 and multiplied by 10 with an exponent, is reviewed in Appendix B.

$$[H_3O^+][OH^-] = 1.0 \times 10^{-14}$$

If acid is added to pure water, the concentration of H_3O^+ will be greater than 1.0×10^{-7}, and the concentration of OH^- will be less than 1.0×10^{-7}. However, the product of the two must equal 1.0×10^{-14}. This relationship is the basis for calculating the concentration of one of the two ions, hydronium or hydroxide, when the other one is known.

Consider 0.10 M NaOH. Because sodium hydroxide is a strong base that is completely ionized, the molar concentration of hydroxide ions in this solution is also 0.10 M. To compensate, the concentration of hydronium ions must be significantly less than the pure water value of 1.0×10^{-7} M. Using the preceding equation, the concentration of hydronium ions in a 0.10 M NaOH solution is found to be 1.0×10^{-13} M. In the following calculation, the value of 0.10 M is expressed in scientific notation as 1.0×10^{-1} M

$$(1.0 \times 10^{-1})[H_3O^+] = 1.0 \times 10^{-14}$$

$$[H_3O^+] = \frac{1.0 \times 10^{-14}}{1.0 \times 10^{-1}} = 1.0 \times 10^{-13}\text{ M}$$

An equivalent calculation for a 0.10 M HCl solution, which contains 0.10 M H_3O^+, would show that its concentration of OH^- ions is 1.0×10^{-13} M.

The difference in H_3O^+ concentration between 0.1 M HCl and 0.1 M NaOH is one trillion times because each change in the exponent is a power of ten, and the difference between 10^{-1} $[H_3O^+]$ and 10^{-13} $[H_3O^+]$ is 10^{12}, or one trillion. The Danish biochemist S. P. L. Sørensen proposed in 1909 that these exponents be used as a measure of acidity. He devised a scale that would be useful in his work of testing the acidity of Danish beer. Sørensen's scale came to be known as the pH scale, from the French *pouvoir hydrogene,* which means hydrogen power. pH is defined as *the negative logarithm of the hydronium ion concentration.*

■ With a calculator that includes base-10 logarithms, the pH is easily found from any H_3O^+ concentration by finding its log; the H_3O^+ concentration can be determined from any pH by finding the value of 10^{-pH}.

$$pH = -\log [H_3O^+]$$

To find the pH of a solution, write the concentration of the hydronium ion as a power of 10 and use the exponent without the negative sign. For example, a 0.01 M solution of HCl has a hydronium ion concentration of 1×10^{-2}, and the pH is 2.0.

The larger the concentration of hydronium ion in a solution, the more acidic a solution is. Therefore, as a solution gets more acidic, for example, as $[H_3O^+]$ goes from 1×10^{-3} (or 0.001) to 1×10^{-1} (or 0.1), the pH becomes a smaller number—in this case it goes from pH 3 to pH 1. *The smaller the pH value, the more acidic the solution.*

In a neutral solution, the pH is 7 (for a hydronium ion concentration of 1×10^{-7}). As the hydronium ion concentration becomes smaller than this, a solution becomes basic. Therefore, in basic solutions the pH is larger than 7. *The larger the pH value, the more basic the solution.*

if $pH < 7.0$, solution is acidic

if $pH = 7.0$, solution is neutral

if $pH > 7.0$, solution is basic

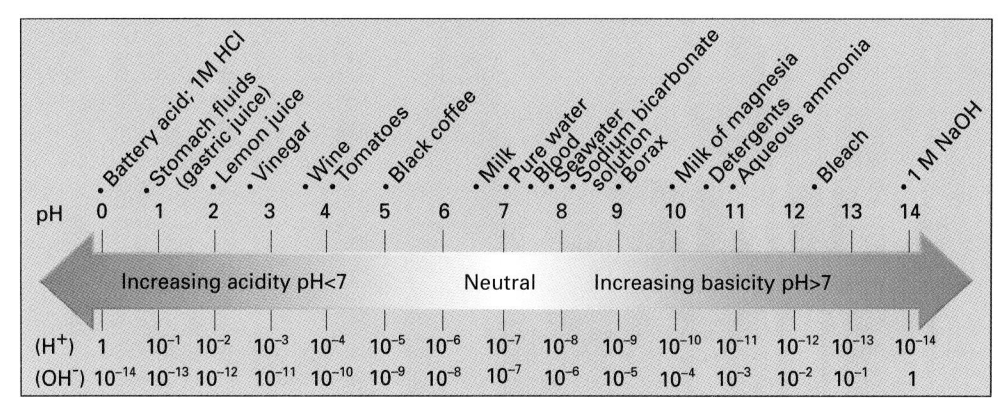

Figure 7.2 Relationship of pH to the concentration of hydrogen ions (H^+ or H_3O^+) and hydroxide ions (OH^-) in water at 25°C. The pH values for some common substances are included in the diagram.

The relationship between pH and the hydronium ion concentration, plus the pH of some common solutions, is illustrated in Figure 7.2.

Because weak acids are only slightly ionized, the pH of a weak acid solution is not given by the concentration of the acid, but is dependent on its extent of ionization. For example, in a 1.0 M HCl solution, the hydronium ion concentration is 1.0 M and the pH is 0, but in a 1.0 M acetic acid solution, the hydronium ion concentration is only 4.3×10^{-3} M and the pH is 2.4. Since many common substances have hydronium ion concentrations in the range of 0.1 M to 10^{-14} M, the pH scale provides a more convenient way of expressing low acid concentrations.

■ Remember that as the hydroxide ion concentration becomes larger, the hydronium ion must become smaller because of the constant relationship between their values, $[H_3O^+][OH^-] = 1 \times 10^{-14}$.

EXAMPLE 7.2 *Finding pH*

Many soft drinks contain carbonic acid and phosphoric acid. If the hydronium ion concentration of a cola beverage is 1×10^{-3}, what is the pH?

SOLUTION

pH is defined as the negative of the logarithm of hydronium ion concentration, $pH = -\log [H_3O^+]$. For a concentration of 1×10^{-3}, the exponent -3 shows that the pH is 3.

Exercise 7.2

Most tomatoes are acidic and have $[H_3O^+] = 1 \times 10^{-4}$ M. What is the pH of these tomatoes? Are they more or less acidic than the cola drink of Example 7.2?

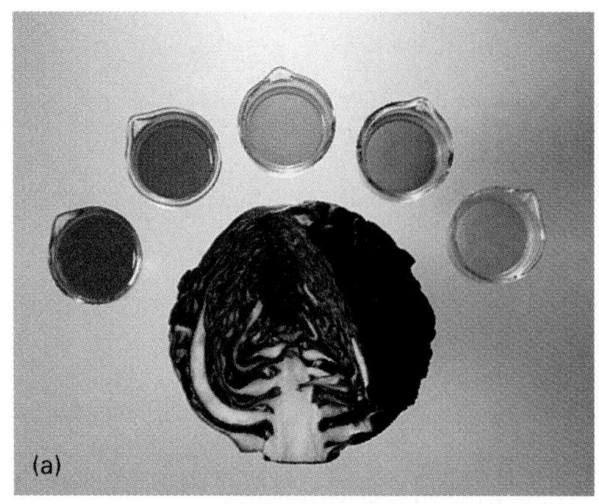

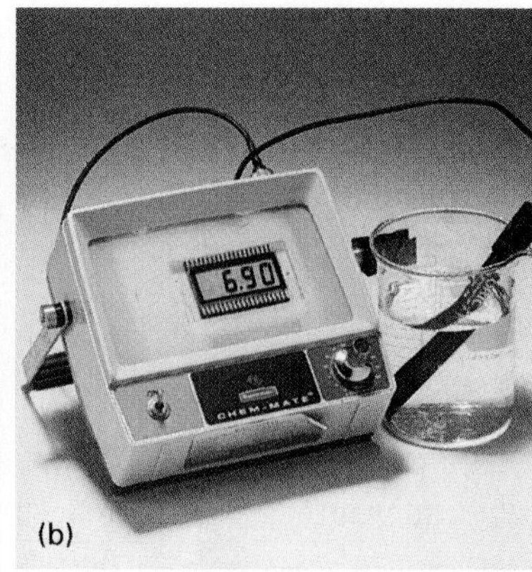

Finding pH. To find the pH of a solution, it can be tested with an acid–base indicator, for example, (a) the natural indicator in a red cabbage solution, which from left to right has the colors shown for solutions of pH 1, 4, 7, 10, and 13. (b) For a more accurate determination, an instrument known as a pH meter is used. *(a, Charles Steele; b, Leon Lewandowski)*

7.4 Acid–Base Buffers

An acid–base buffer is like a shock absorber—something to prevent a disturbance while retaining the original conditions or structure. The control of pH requires maintaining a steady concentration of H_3O^+ even when sudden "shocks" of acid or base are added. **Buffer solutions** contain a base that can react with an acid and an acid that can react with a base so that the pH of a solution remains close to its original value. An acid–base combination suitable for controlling pH is known as a **buffer**. You see the term on bottles of "buffered" aspirin, meaning that the tablets contain some sort of base to offset the acidity of aspirin. The principal benefit is that such tablets may dissolve faster, and thereby go to work faster. The buffering does not, however, actually buffer the highly acidic environment of the stomach, which has a pH of 1.5 to 2.

Buffers are very important to many industrial and natural processes. In fact, the control of the pH of your blood is essential to your health. The pH of the blood is about 7.40, and your good health depends on the ability of buffers to maintain the pH within a narrow range. If the pH falls below 7.35, a condition known as *acidosis* occurs; increasing pH above 7.45 leads to *alkalosis*. Both these conditions can be life threatening.

How does a buffer solution maintain its pH at a nearly constant value? Not only must the acid in the buffer react with added base and the base must react with added acid, but it is also necessary that the acid and base components of a buffer solution not react with each other. To meet these conditions, buffers usually are mixtures of a weak acid and its weakly basic anion (e.g., acetic acid and acetate ion, CH_3COOH and CH_3COO^-) or a weak base and its weakly acidic cation (e.g., ammonia and ammonium ion, NH_3 and NH_4^+). In a solution that contains an acetic acid–acetate ion buffer, added base will react with the acetic acid

$$CH_3COOH(aq) + OH^-(aq) \rightleftharpoons CH_3COO^-(aq) + H_2O(\ell)$$

and added acid will react with the acetate ion

$$CH_3COO^-(aq) + H_3O^+(\ell) \rightleftharpoons CH_3COOH(aq) + H_2O(\ell)$$

Notice that in these reactions no new substances are produced. The products are always components of the buffer.

Carbonic acid, which forms when carbon dioxide dissolves in water, and bicarbonate ion form one of several buffer pairs that keep the pH of blood within the necessary safe range. Their buffering reactions with base and acid are as follows

$$H_2CO_3(aq) + OH^-(aq) \rightleftharpoons HCO_3^-(aq) + H_2O(\ell)$$

$$HCO_3^-(aq) + H_3O^+(aq) \rightleftharpoons H_2CO_3(aq) + H_2O(\ell)$$

7.5 Corrosive Cleaners

■ A *corrosive* substance has the ability to weaken or destroy materials by chemical reactions, often acid–base or oxidation–reduction reactions (Chapter 8).

There are some places in the home where really tough cleaning jobs exist. For these jobs, cleaners are formulated with extremes in pH, which allow the acidity or alkalinity of the cleaner to quickly attack the unwanted dirt, grease, or stain (Figure 7.3). Toilet-bowl cleaners usually contain hydrochloric acid, which can dissolve most mineral scale (mostly carbonates) and iron stains. Other acids, such as phosphoric acid and oxalic acid, are also used in these products. The pH of toilet-bowl cleaners is usually below 2, and because they contain strong acids, they can be quite harmful to skin and eyes on contact. They should be handled with extreme caution, and rubber gloves should be worn when using them.

On the other end of the pH scale are drain cleaners and oven cleaners, which have a pH of 12 or higher. These formulations almost always contain the strong base sodium hydroxide (NaOH). Drains usually clog as a result of oils, grease, and hair caught on rough edges inside the drainpipes. As the foreign matter builds up it becomes more tightly packed, until the flow of water from the drain practically stops. The only way to get rid of the material clogging the drain is to either dismantle the drain plumbing (often very difficult), use a plumber's "snake," or dissolve the material. Bases like sodium hydroxide are very good at dissolving drain clogs because they can cause rapid breaking of bonds in oils and greases of animal and vegetable origin. Once the bonds are broken, smaller, more soluble molecules are formed that can be washed down the drain. Hair and other proteins are also broken down by sodium hydroxide, so that a mat of grease and hair in a clogged drain will quickly succumb to the action of the strong base. Numerous products containing sodium hydroxide are available, including solid NaOH pellets, flakes, and concentrated solutions. The solutions offer the easiest way to apply the drain cleaner, but if they become diluted, their effectiveness is diminished. Some solid drain cleaners also contain pieces of aluminum metal, which react

Figure 7.3 Corrosive household cleaners. Reading the labels and handling these chemicals with care are essential. *(C. D. Winters)*

with aqueous sodium hydroxide to form hydrogen gas (caution—flammable) that helps agitate the mixture and hastens the unclogging process

$$2\ \mathrm{Al}(s) + 2\ \mathrm{NaOH}(aq) + 6\ \mathrm{H_2O}(\ell) \longrightarrow 2\ \mathrm{Na^+}(aq) + 2\ \mathrm{Al(OH)_4^-}(aq) + 3\ \mathrm{H_2}(g)$$

Oven cleaners also contain strong bases such as sodium hydroxide. Many oven cleaners use aerosol sprays to distribute the cleaner on the inner surface of the oven. Care must be taken not to breath air containing these aerosols because strong bases are quite corrosive to nasal tissue, bronchial tubes, and lungs.

7.6 Heartburn: Why Reach for an Antacid?

The walls of a human stomach contain thousands of cells that secrete hydrochloric acid, the main purposes of which are to kill microorganisms and to aid in digestion. The stomach's inner lining is not harmed by the presence of hydrochloric acid with such a low pH (1.5–2.0), since the mucosa, the inner lining of the stomach, is replaced at the rate of about a half million cells per minute.

The uncomfortable condition known as "heartburn" occurs when the acid contents of the stomach back up into the esophagus and cause a burning sensation in the chest and throat. Among the causes are overeating; spicy, acidic, or fatty foods; and some medications, such as aspirin.

■ Until recently, it was believed that excess stomach acid caused stomach ulcers. It has now been proved that the cause is instead a bacterial infection and that ulcers can be successfully treated with antibiotics.

Antacids are bases used to neutralize the acid that causes heartburn. The most common antacid ingredients are magnesium and aluminum hydroxides, and bicarbonate or carbonate salts (Table 7.4). Baking soda (sodium bicarbonate) was used to relieve indigestion before many of the other commercial products became available. The bicarbonate ion, a basic anion of a weak acid,

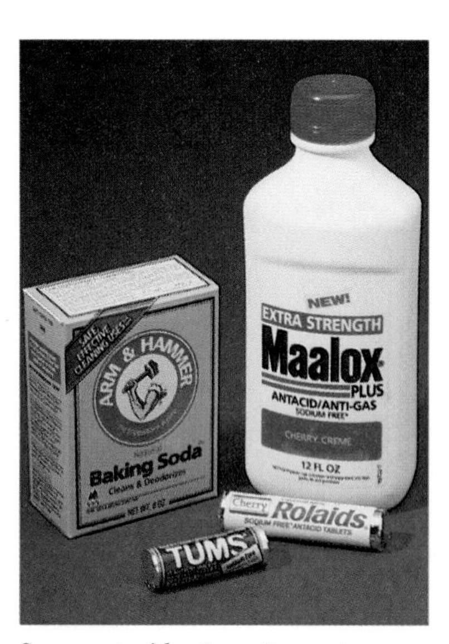

Some antacids. *(Larry Cameron)*

TABLE 7.4 ■ Some Common Antacids

Compound	Formula	Examples of Commercial Products
Magnesium hydroxide	$Mg(OH)_2$	Phillips' Milk of Magnesia
Calcium carbonate	$CaCO_3$	Tums, Titralac
Sodium bicarbonate	$NaHCO_3$	Alka-Seltzer, baking soda
Aluminum hydroxide	$Al(OH)_3$	Amphojel
Aluminum hydroxide and magnesium hydroxide		Maalox, Mylanta, Di-Gel tablets
Aluminum hydroxide, magnesium hydroxide, and magnesium carbonate	$MgCO_3$	Di-Gel liquid
Dihydroxyaluminum sodium carbonate	$NaAl(OH)_2CO_3$	Rolaids
Calcium carbonate and magnesium hydroxide		Sodium-free Rolaids

reacts with the hydronium ion from hydrochloric acid to form carbonic acid, which decomposes to give carbon dioxide and water.

$$HCO_3^-(aq) + H_3O^+(aq) \longrightarrow H_2CO_3(aq) + H_2O(\ell)$$
$$H_2CO_3(aq) \longrightarrow CO_2(g) + H_2O(\ell)$$

Alka-Seltzer contains sodium bicarbonate, potassium bicarbonate, citric acid, and aspirin. The fizz of the Alka-Seltzer tablet in water is carbon dioxide gas given off by the reaction of the citric acid with the bicarbonates to give carbonic acid (the preceding reaction).

Milk of magnesia is a suspension of magnesium hydroxide (which is not very soluble) in water. Magnesium hydroxide acts as an antacid in small doses, but in large doses it is a laxative. Calcium carbonate is the active ingredient in Tums. The carbonate ion, a base, neutralizes hydronium ion.

$$CO_3^{2-}(aq) + 2\ H_3O^+(aq) \longrightarrow H_2CO_3(aq) + 2\ H_2O(\ell)$$
$$H_2CO_3(aq) \longrightarrow CO_2(g) + 2\ H_2O(\ell)$$

The World of Chemistry

Program 16, *The Proton in Chemistry*

Even the sight and smell of appetizing food is enough to make the stomach produce more acid. Dr. Paul Maton, National Institutes of Health

Although small amounts of calcium carbonate are safe, regular use can cause constipation. Aluminum hydroxide, the active ingredient in Amphojel, can also cause constipation in large doses. Because calcium carbonate and aluminum hydroxide can cause constipation, antacids such as Maalox and Mylanta contain aluminum hydroxide mixed with magnesium hydroxide to counteract the constipating effects of the former with the laxative action of the latter.

Another class of medications to combat heartburn has recently become available by being reclassified from prescription-only drugs to over-the-counter drugs. Instead of neutralizing stomach acid, these new drugs (for example, Tagamet and Pepcid) prevent its secretion.

SELF-TEST 7B

1. Which is more acidic, a pH of 6 or a pH of 2?
2. High pH means (a) high hydronium ion concentration, (b) low hydronium ion concentration.
3. Low pH means (a) high hydronium ion concentration, (b) low hydronium ion concentration.
4. In the buffer system H_2CO_3/HCO_3^-, which component neutralizes added base? Which component neutralizes added acid?
5. What is the pH of 1.0×10^{-3} M hydrochloric acid?
6. What is the pH of 1.0×10^{-3} M sodium hydroxide?
7. The pH of household ammonia is 11. What is the hydronium ion concentration? What is the hydroxide ion concentration?
8. The pH of a sodium hydroxide solution is 10. What is the hydroxide ion concentration?
9. What is the $[H_3O^+]$ for a 0.01 M HCl solution?
10. Most toilet-bowl cleaners are strongly ______________ and most oven cleaners are strongly ______________.

11. Which of the following compounds can function as an antacid in the treatment of heartburn? (a) Sodium bicarbonate, (b) $MgCl_2$, (c) $MgCO_3$, (d) $Al(OH)_3$, (e) Aluminum chloride.

MATCHING SET

____	1. M	a. 7
____	2. pH of pure water	b. Molarity, moles of solute per liter of solution
____	3. Acidic pH	c. Hydrogen ion donor
____	4. Strong acid	d. Maintains pH
____	5. Strong base	e. Hydrogen ion acceptor
____	6. Acid definition	f. 10
____	7. Base definition	g. CH_3COOH, acetic acid
____	8. Weak acid	h. NaOH
____	9. Buffer	i. H_2SO_4
____	10. Basic pH	j. 3

QUESTIONS FOR REVIEW AND THOUGHT

1. What do the following terms mean?
   (a) Acidic solution (b) Basic solution
   (c) Neutral solution
2. What is a neutralization reaction?
3. What is a salt?
4. What do the following terms mean?
   (a) H^+ ion donor (b) H^+ ion acceptor
   (c) Hydronium ion (d) Hydroxide ion
5. What do the following terms mean?
   (a) Extent of ionization (b) Strong acid
   (c) Strong base (d) Weak acid
   (e) Weak base
6. Identify four common substances that are acidic.
7. Name four common substances that are basic.
8. Label each of the following substances as acidic or basic:
   (a) Vinegar (b) Citrus fruits
   (c) Aspirin (d) Lye
   (e) Black coffee (f) Milk of magnesia
   (g) Detergents (h) Household ammonia
   (i) Vitamin C
9. Label each of the following substances as acidic or basic:
   (a) Gastric fluid (b) Tomatoes
   (c) Oven cleaners (d) Soap
   (e) Carbonated beverages (f) Baking soda
10. Indicate which of the following is a property of an acid and which is a property of a base:
    (a) Sour taste (b) Bitter taste
    (c) Slippery feeling
    (d) Change color of red litmus to blue
    (e) Change color of blue litmus to red
11. What do the following terms mean?
    (a) Acid–base indicator (b) Molarity
    (c) Concentration
12. What is the definition of pH?
13. What is a buffer?
14. What is an antacid?
15. Write the balanced equations for the following neutralization reactions:
    (a) Acetic acid, $CH_3COOH(aq)$, with potassium hydroxide, $KOH(aq)$
    (b) Sulfuric acid, $H_2SO_4(aq)$, with calcium hydroxide, $Ca(OH)_2(aq)$
    (c) Sulfuric acid, $H_2SO_4(aq)$, with sodium hydroxide, $NaOH(aq)$
16. Balance the following equations:
    (a) $HBr(aq) + Ca(OH)_2(aq) \rightarrow$ ____________
    (b) $HNO_3(aq) + Al(OH)_3(aq) \rightarrow$ ____________
17. What reaction occurs when an antacid like Tums, $CaCO_3(aq)$, reacts with stomach acid, $HCl(aq)$?
18. Which is more basic: a solution with pH of 2 or a solution with pH of 10? Explain.
19. Which is more acidic: a solution of black coffee with a pH of 5.0 or milk with a pH of 6.5? Explain.
20. Which of the following are acidic and which are basic?
    (a) Vinegar, pH = 3.0
    (b) Baking soda solution, pH = 8.5
    (c) Beer, pH = 4.2
    (d) Lye solution, pH = 14
    (e) Soft drink, pH = 4.5
    (f) Milk of magnesia, pH = 10.4

21. Which is more acidic: gastric juice with a pH of 1 or tomato juice with a pH of 4?
22. Which is more basic: a soap solution with pH of 10 or a household ammonia solution with pH of 12?
23. Which of the following pairs of compounds would be useful for making a buffer solution? Explain.
    (a) $NaOH(aq)$ and $HCl(aq)$
    (b) CH_3COOH and $NaCH_3COO$
24. A buffer can be made using $NaHCO_3(aq)$ and $H_2CO_3(aq)$. Explain how this combination resists changes in pH when small amounts of acid or base are added.
25. Two solutions contain 1% acid. Solution A has a pH of 4.6 and solution B has a pH of 1.1. Which solution contains the stronger acid?
26. Moist baking soda is often put on acid burns. Why? Write an equation for the reaction assuming the acid is hydrochloric acid (HCl).

PROBLEMS

1. What is the molarity of solutions containing the following masses of solute dissolved on 1.0 L of solution?
    (a) 5.0 g of HCl (b) 40.0 g of NaOH
    (c) 58.5 g of NaCl (d) 1.0 g of NaCl
2. How many grams of NaCl have been dissolved per liter of solution to make a 1.5 M solution of this salt?
3. How many grams of sulfuric acid have been dissolved per liter of solution to make a 0.10 M solution of this acid?
4. What is the pH for each of the following solutions?
    (a) 1.0×10^{-2} M HCl (b) 0.001 M HNO_3
    (c) 0.1 M NaOH (d) 0.10 M HBr
5. What is the pH for each of the following solutions?
    (a) 1.0×10^{-3} M NaOH (b) 1.0×10^{-3} M HCl
    (c) 0.01 M KOH (d) Neutral water at 25°C
6. What is the molarity of H_3O^+ for each of the following solutions?
    (a) Solution with a pH of 1.0
    (b) Solution with a pH of 0.0
    (c) Solution with a pH of 5.0
    (d) Solution with a pH of 3.0

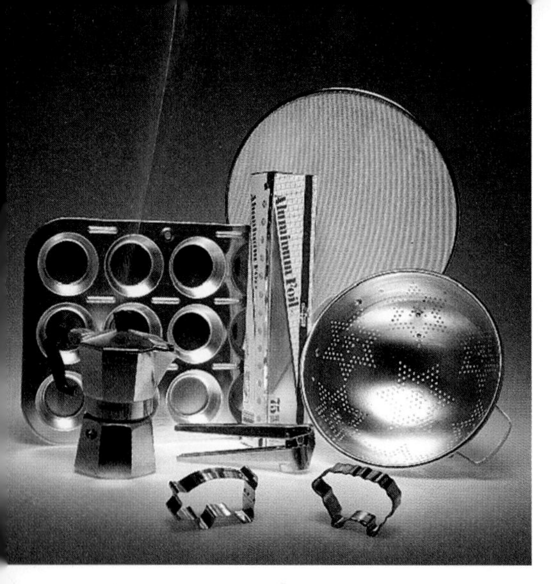

CHAPTER 8

Oxidation–Reduction Reactions

Like the acid–base reactions discussed in Chapter 7, oxidation–reduction reactions are common in our surroundings. As you might guess from the name, oxygen is a reactant in many of these reactions. Most importantly, we rely on the reactions of oxygen for energy, whether it is by burning fossil fuels to heat our homes and generate electricity, or by oxidizing glucose to provide our biochemical energy. We have also come to rely on "redox" reactions in another way—the controlled reactions that occur in batteries. Imagine what a day would be like without the chemical energy provided by the batteries in your car, your calculator, your portable radio or tape player, and all the other battery-driven conveniences that you take for granted.

- What are oxidation and reduction?
- How can you recognize oxidation–reduction reactions?
- What is necessary to release the energy stored in chemical compounds as electric current?
- What is the difference between the chemical reactions in batteries and those used in electrolysis, and their applications?
- What is corrosion and what are some ways to control it?

8.1 Oxidation and Reduction

"Oxidation" got its name from the chemical changes that occur when oxygen (O_2) combines with other elements or with compounds. Many reactive metals are mined as **oxides** (compounds of oxygen with another element) that were formed by reaction of the metals with oxygen in the air. For example, the major aluminum ore, bauxite, contains aluminum oxide (Al_2O_3); and hematite, a principal iron ore, contains the iron(III) oxide (Fe_2O_3). When materials containing carbon, nitrogen, or sulfur burn in air, the products are carbon

Aluminum is made available for our common household items by oxidation–reduction reactions driven by electricity. (C. D. Winters)

oxides (CO_2, CO), nitrogen oxides (NO, N_2O, and others), and sulfur oxides (SO_2, SO_3).

A substance that has combined with oxygen is described as having been *oxidized,* and the reaction is classified as *oxidation.* All **combustion** reactions are oxidations. In the combustion of the methane in natural gas, for example, the carbon is oxidized to carbon dioxide and the hydrogen is oxidized to water.

$$CH_4(g) + 2\ O_2(g) \longrightarrow CO_2(g) + 2\ H_2O(g)$$

The gradual rusting away of iron objects begins with the oxidation of iron.

$$4\ Fe(s) + 3\ O_2(g) \longrightarrow 2\ Fe_2O_3(s)$$

In the blast furnaces that produce iron from iron ore, one of the important chemical changes is the reaction of the iron oxide in hematite with carbon monoxide.

■ **Combustion** is the reaction of a substance with the oxygen in air, accompanied by the emission of heat and light.

$$Fe_2O_3(s) + 3\ CO(s) \longrightarrow 2\ Fe(\ell) + 3\ CO_2(g)$$

If the addition of oxygen to iron is oxidation, how is the removal of oxygen from iron described? It is *reduction,* a term sometimes used to mean "to bring something back." To metallurgists hundreds of years ago, reduction meant bringing a metal back from its ore. In today's chemical sense, the iron oxide has been reduced by the removal of oxygen.

Notice that in the reaction of iron oxide with carbon monoxide, oxygen has added to the carbon monoxide to produce carbon dioxide—the carbon monoxide has been oxidized. Here is a fundamental concept of chemistry—*oxidation and reduction always occur together.* If one reactant is oxidized, another must be reduced.

Over time, the definitions of oxidation and reduction have been extended. With a more modern definition, you can better see why oxidation and reduction always occur together. An atom or ion is said to be oxidized when it loses electrons. Consider the reaction of sodium, a metal, with bromine, a nonmetal (Figure 8.1):

$$2\ Na(s) + Br_2(\ell) \longrightarrow 2\ NaBr(s)$$

The product is sodium bromide, a white, crystalline solid that is an ionic compound composed of equal numbers of Na^+ and Br^- ions.

In the reaction of sodium with bromine, sodium atoms lose electrons to give sodium ions. **Oxidation** is the loss of electrons, and sodium has been oxidized. Except when electrons are flowing through a wire from a negative region to a positive one, they are held to positively charged nuclei. Because we know that matter is neutral, we must ask where the electrons lost in oxidation go. The reaction of sodium with bromine occurs because as sodium atoms lose electrons, bromine atoms gain electrons.

Reduction is the gain of electrons, and in the conversion of bromine to bromide ions in this reaction, bromine has been reduced. Reactions in which one reactant is oxidized and another is reduced are known as **oxidation–**

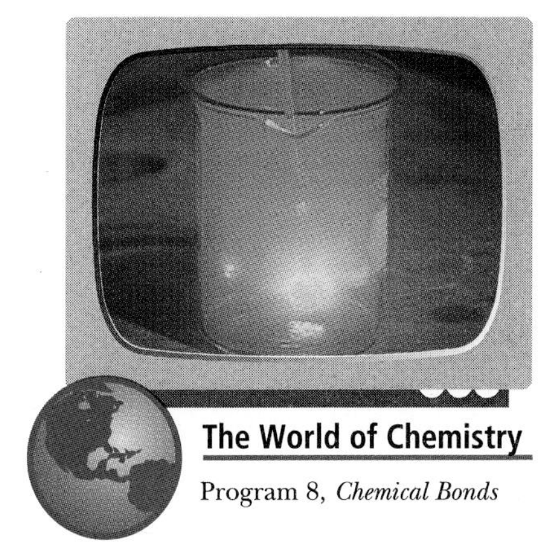

Figure 8.1 Oxidation of sodium by bromine. The reaction between metallic sodium and bromine releases a large amount of energy. The product is sodium bromide (NaBr) a white crystalline ionic compound.

■ In oxidation the charge of an atom or ion can increase, and in reduction the charge of an atom or ion can decrease (it's *reduced*).

reduction reactions, usually referred to as **redox reactions.** For example, every reaction in which a metallic element combines with a nonmetallic element is a redox reaction in which the metal is oxidized and the nonmetal is reduced.

EXAMPLE 8.1 *Recognizing Oxidation and Reduction*

For the reaction of copper with oxygen to give an ionic oxide,

$$2\ Cu(s) + O_2(g) \longrightarrow 2\ CuO(s)$$

identify which reactant is oxidized and which is reduced.

SOLUTION

The reaction of Cu with O_2 can be recognized as a redox reaction because it is the addition of oxygen to a reactant and also because it is the combination of a metal and a nonmetal. Since the product is an ionic compound, the electron gain and loss is determined by the charges on the ions. The oxide CuO must be composed of Cu^{2+} and O^{2-} ions. Therefore, the copper metal has been oxidized by the loss of two electrons from each atom, and the oxygen has been reduced by the gain of two electrons by each atom.

Exercise 8.1
For the reaction of the very active metal lithium with oxygen

$$4\ Li(s) + O_2(g) \longrightarrow 2\ Li_2O(s)$$

identify what is oxidized and what is reduced.

8.2 Oxidizing Agents: They Bleach and They Disinfect

You've seen that whenever oxygen adds to another reactant, that reactant is *oxidized.* The oxygen in such a reaction is the **oxidizing agent**, a reactant that causes an oxidation by gaining electrons. The halogens are also oxidizing agents, for example

$$2\ Al(s) + 3\ Cl_2(g) \longrightarrow 2\ AlCl_3(s)$$
$$Sn(s) + F_2(g) \longrightarrow SnF_2(\ell)$$

Chlorine-containing tablets being added to swimming pool filter. *(Yoav Levy/Phototake NYC)*

In each of these reactions, the neutral halogen molecule has been reduced by the addition of electrons to give -1 ions. Here is another condition common to all redox reactions: *the oxidizing agent is reduced.* Some common oxidizing agents are listed in Table 8.1.

Somewhere around home, most of us have one or more oxidizing agents. The most common household oxidizing agent is sodium hypochlorite (NaOCl), a water-soluble ionic compound. In the laundry, NaOCl is a **bleaching agent**—it removes unwanted color by oxidizing colored chemical com-

TABLE 8.1 ■ Some Oxidizing and Reducing Agents

Name	Formula	Uses
	Oxidizing Agents	
Lead dioxide	PbO_2	Automobile batteries
Manganese dioxide	MnO_2	Batteries
Sodium hypochlorite solution	$NaOCl(aq)$	Laundry bleach, disinfection
Oxygen	O_2	Metabolism of foods, burning fuels
Ozone	O_3	Water purification, hazardous chemical destruction
Chlorine	Cl_2	Drinking water and wastewater purification, chemical synthesis
Hydrogen peroxide	H_2O_2	Bleaching, antiseptic
	Reducing Agents	
Hydrogen	H_2	Fuel, chemical synthesis
Sulfur dioxide	SO_2	Chemical synthesis
Carbon	C	Iron production
Zinc	Zn	Batteries

pounds to produce whiter clean clothes. If you spill an NaOCl solution (e.g., Clorox or Javex) on your jeans, its effectiveness as a bleach is immediately obvious as a white spot appears. Substances have color because their molecular structures absorb portions of the visible spectrum of light. Many of the substances that contribute unwanted color to textiles or paper are organic compounds with long chains of alternating single and double carbon–carbon bonds (Section 5.3). The color is created by the absorption of some of the light by electrons in these bonds. A bleach disrupts the alternating pattern by breaking bonds or converting double bonds to single bonds. The result is loss of the ability to absorb visible light. A very strong bleaching solution can also break bonds in fabric to the extent that the fabric develops thin spots and holes.

Sodium hypochlorite is also a *disinfectant.* When you use a liquid bleach solution to wash the bathroom floor or the kitchen counter, you are taking advantage of both the disinfecting and bleaching properties of this chemical. By oxidizing molecules in the outer surfaces of bacteria, a disinfectant is able to disrupt the structure of the cells and kill them.

Another familiar oxidizing agent is hydrogen peroxide (H_2O_2), and it too is a bleaching agent. At times, its most popular use has been in bleaching hair to produce "peroxide blonde." A dilute solution (usually 3%) of hydrogen peroxide in water is useful in the medicine cabinet because it is an *antiseptic,* a substance that can be used to cleanse a wound to prevent bacterial infection. An antiseptic acts in the same manner as a disinfectant, but, unlike a disinfectant, is mild enough for use on human tissue without doing damage.

■ Concentrated hydrogen peroxide solutions and pure hydrogen peroxide are extremely hazardous substances.

THE PERSONAL SIDE

Richard Feynman and the* Challenger *Explosion

Richard Feynman (1919–1988), winner of the 1965 Nobel Prize in physics, was one of the most original thinkers of this century. He wrote, "Scientific knowledge is a body of statements of varying degrees of certainty—some most unsure, some nearly sure, but *none* absolutely certain. Now, we scientists are used to this, that it is possible to live and *not* know."

Richard Feynman demonstrating that a cold O-ring of the type used in the space shuttle Challenger *does not retain its flexibility when cold and thus would allow gases to escape from the rocket.* (Marilyn K. Yee/NYT Pictures)

Feynman gave the United States a lesson in how science works when he used a simple experiment to uncover the reason for the disastrous explosion of the space shuttle *Challenger*. On launch day, January 28, 1986, the weather was unusually cold in Florida—the temperature was 29°F. A few moments after launch, the world watched in horror as the shuttle and its rockets exploded in a gigantic fireball, killing all the astronauts aboard.

A commission was appointed to investigate the cause of the explosion. It was Feynman who reasoned that due to the cold temperature, rubber O-rings used to seal joints in the solid-fuel booster rockets had not expanded properly. This failure allowed hot flames from the booster rocket to burn through the hydrogen fuel tank. The result was the violent redox reaction that occurs when hydrogen and oxygen combine to form water. To prove his point, Feynman performed a dramatic—but very simple—experiment. During a public hearing, Feynman held a sample rubber O-ring like the one in the rocket engine tightly in a clamp and immersed it in a glass of ice water. Everyone watching his experiment could see that the rubber did not spring back to its original shape. The poor low-temperature characteristics of the rubber O-ring had doomed the *Challenger* and its crew.

Chlorine itself is an oxidizing agent of vital importance in ensuring the purity of drinking water and water that is returned to natural waterways. Treatment of drinking water and waste water is described in Section 15.11.

8.3 Reducing Agents: For Metallurgy and Good Health

When a metal reacts with another substance, the metal usually acts as a **reducing agent**, a reactant that causes a reduction by giving up electrons. In reactions with oxygen or the halogens, for example, the metal is the reducing agent and the oxygen or halogen is reduced.

$$2\ Cu(s) + O_2(g) \longrightarrow 2\ CuO$$
$$2\ Al(s) + 3\ Cl_2(g) \longrightarrow 2\ AlCl_3(s)$$

The metals have lost electrons to form the positive ions in copper(II) oxide and aluminum chloride. In these, and all redox reactions, *the reducing agent is oxidized.* Some common reducing agents are included in Table 8.1.

Reducing agents are crucial in the production of metals from their ores because most metals occur in nature only in chemical compounds. Carbon, usually in the form of coke, is a valuable metallurgical reducing agent.

$$ZnO(s) + C(s) \longrightarrow Zn(g) + CO(g)$$

In the production of zinc, this reaction is carried out at such a high temperature that the zinc is formed as a gas that is condensed for further purification.

Hydrogen gas, like the metals, is a reducing agent in virtually all of its reactions. Hydrogen can, for example, reduce the copper in copper(II) oxide:

$$CuO(s) + H_2(g) \longrightarrow Cu(s) + H_2O(\ell)$$

The addition of hydrogen to another compound is also classified as a reduction. In a reaction that may be of increasing importance in the production of methanol as an alternative fuel (Section 9.9), carbon monoxide is reduced to methanol by the addition of hydrogen.

$$CO(g) + 2\ H_2(g) \longrightarrow CH_3OH(\ell)$$

Much has appeared in the popular media about the value to human health of consuming fruits and vegetables or vitamin supplements that contain *antioxidants.* Drugstores now offer a variety of "antioxidant vitamins," supplements that usually contain vitamin E, vitamin C, and vitamin A or beta-carotene (which converts to vitamin A in the body).

An antioxidant is, in chemical definition terms, a reducing agent. It prevents oxidation by reducing a potential oxidizing agent. The oxidizing agents of concern in the body are atoms or molecular fragments known as **free radicals**. A free radical contains an unpaired electron and does not stay around for long without grabbing another electron from a nearby molecule (Figure 8.2, p. 180). If this molecule happens to have an important function, then that function will be disrupted. For example, if a free radical connects with the part of a DNA molecule that governs cell division (Section 11.11), the result might be abnormal cell division—cancer.

Free radicals arise in the body from normal biochemical reactions and are also produced in the presence of toxic substances from cigarette smoke, polluted air, and other sources. They are quickly deactivated by picking up an electron to form a less-reactive molecule. Thus, the antioxidants are molecules that in one way or another are able to donate an electron to a free radical before it can do any damage.

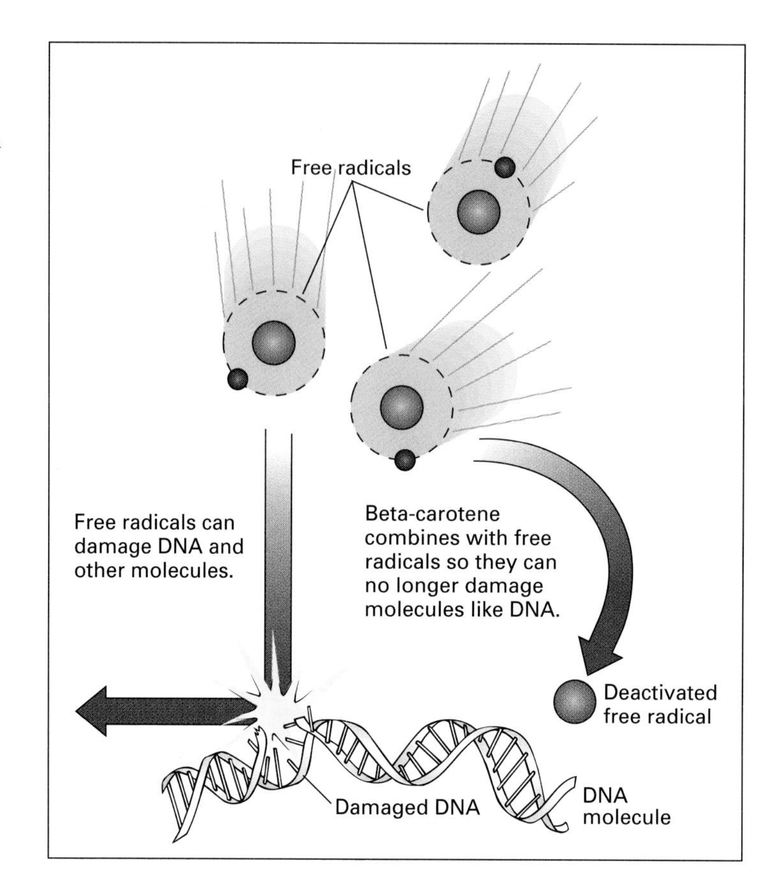

Figure 8.2 Action of an antioxidant. Once a free radical has formed, it will react as soon as it has a chance with something that will remove or pair with the unpaired electron. By intercepting free radicals, beta-carotene and other antioxidant vitamins prevent them from damaging DNA or other crucial biomolecules.

To summarize, we have illustrated three ways to recognize oxidation and reduction:

Oxidation	Reduction
Addition of oxygen	Loss of oxygen
Loss of electrons	Addition of electrons
Loss of hydrogen	Addition of hydrogen

EXAMPLE 8.2 *Oxidizing and Reducing Agents*

For each of the following redox reactions, identify the oxidizing agent and the reducing agent:

(a) $2\ CO(g) + O_2(g) \rightarrow 2\ CO_2(g)$

(b) $CuO(s) + H_2(g) \rightarrow Cu(s) + H_2O(\ell)$

SOLUTION

(a) Oxygen is the oxidizing agent and is added to carbon monoxide. The carbon monoxide is the reducing agent.
(b) Copper oxide is the oxidizing agent as shown by its loss of oxygen. Hydrogen is the reducing agent and is oxidized by the addition of oxygen.

Exercise 8.2
Indicate whether the reactant is being reduced or oxidized in the reactions represented by the following *incomplete* equations:
(a) $Cu(s) \rightarrow CuO(s)$
(b) $CH_3C{\equiv}N(g) \rightarrow CH_3CH_2{-}NH_2(g)$
(c) $SnO(s) \rightarrow Sn(s) + H_2O(g)$

SELF-TEST 8A

1. Which is more oxidized, (a) CO or (b) CO_2?
2. A chemical that causes reduction to take place is called a(n) ________.
3. The conversion of Na to Na^+ is an oxidation reaction. (a) True, (b) False.
4. When coke (a form of carbon) reacts with iron ore in a blast furnace, the coke is the (a) oxidation product, (b) reducing agent, (c) oxidizing agent.
5. When nitrogen reacts with hydrogen to form ammonia (NH_3), the nitrogen is said to be (a) oxidized, (b) reduced.
6. A chemical that causes oxidation to occur is a(n) ________.
7. Which is the oxidized form of zinc, (a) Zn^{2+} ion or (b) Zn atom?
8. In the reaction $2\ Li(s) + S(s) \rightarrow Li_2S(s)$, sulfur is the oxidizing agent. (a) True, (b) False.
9. Which of these two compounds is more reduced than the other: (a) $HC{\equiv}N$ or (b) CH_3NH_2?
10. In chemical definition terms, an antioxidant is a(n) ________.
11. The freeing of a metal from its ore usually requires a chemical reactant that is a(n) ________.
12. A disinfectant frequently is a chemical classified as a(n) ________.

8.4 Batteries

A device that produces an electric current from a chemical reaction is called an **electrochemical cell**. Strictly speaking, a **battery** is a series of electrochemical cells. We will, however, stick with the everyday use of "battery" to refer to any device that converts chemical energy to electrical energy.

To function, a battery takes advantage of the relative ease with which metals lose electrons. Consider the reaction that begins to occur as soon as a piece of zinc is placed in a solution of copper ions (Cu^{2+}). The blue color of the copper ions in solution fades as metallic copper is deposited on the surface

THE WORLD OF CHEMISTRY

The Pacemaker Story

Program 15, *The Busy Electron*

Sometimes an advance in science can come from an unlikely source. Several years ago, the inventor Wilson Greatbatch had an outrageous dream to prolong human life. His story is fascinating:

I quit all my jobs, and with two thousand dollars I went out in the barn in the back of my house and built 50 pacemakers in two years.

I started making the rounds of all the doctors in Buffalo who were working in this field, and I got consistently negative results. The answer I got was, well, these people all die in a year, you can't do much for them, why don't you work on my project, you know.

When I first approached Dr. Shardack with the idea of the pacemaker, he alone thought that it really had a future. He looked at me sort of funny, and he walked up and down the room a couple of times. He said, "You know—if you can do that—you can save a thousand lives a year."

In 1958, a medical team implanted the first heart pacemaker, but for the next few years there was one major problem.

After the first ten years, we were still only getting one or two years out of pacemakers, two years on average, and the failure mechanism was always the battery. It didn't just run down, it failed. The human body is a very hostile environment; it's worse than space; it's worse than the bottom of the sea. You're trying to run things in a warm salt-water environment. The first pacemakers could not be hermetically sealed, and the battery just didn't do the job. Well, after ten years, the battery emerged as the primary mode of failure, and so we started looking around for new power sources. We looked at nuclear sources, we looked at biological sources, of letting the body make its own electricity, we looked at rechargeable batteries, and we looked at improved mercury batteries. And we finally wound up with this lithium battery. It really revolutionized the pacemaker business. The doctors have told me that the introduction of the lithium battery was more significant than the invention of the pacemaker in the first place.

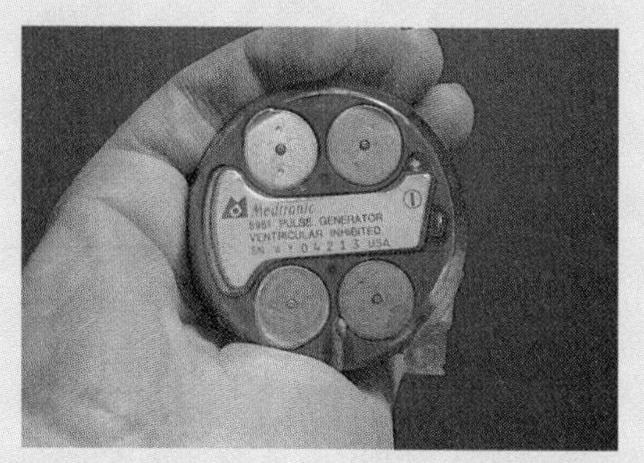

A cardiac pacemaker. (© Yoav Levy/Phototake NYC)

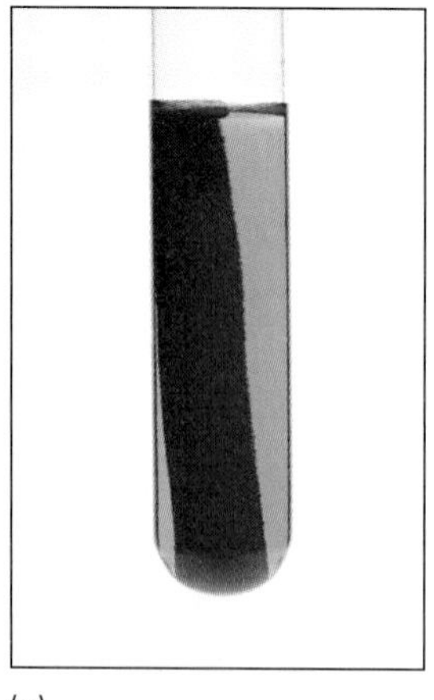

(a)

(b)

(c)

Figure 8.3 Oxidation of zinc by copper ions. (a) A piece of zinc is immersed in a solution containing Cu^{2+} ions, which give the solution a blue color. (b) After a few minutes the blue color begins to fade and copper builds up on the remaining zinc. (c) After about an hour, the solution is almost colorless, indicating that most of the Cu^{2+} ions have been reduced to copper atoms, which have formed metallic copper on what is left of the zinc. The Zn^{2+} ions from oxidation of the zinc are colorless. *(C. D. Winters)*

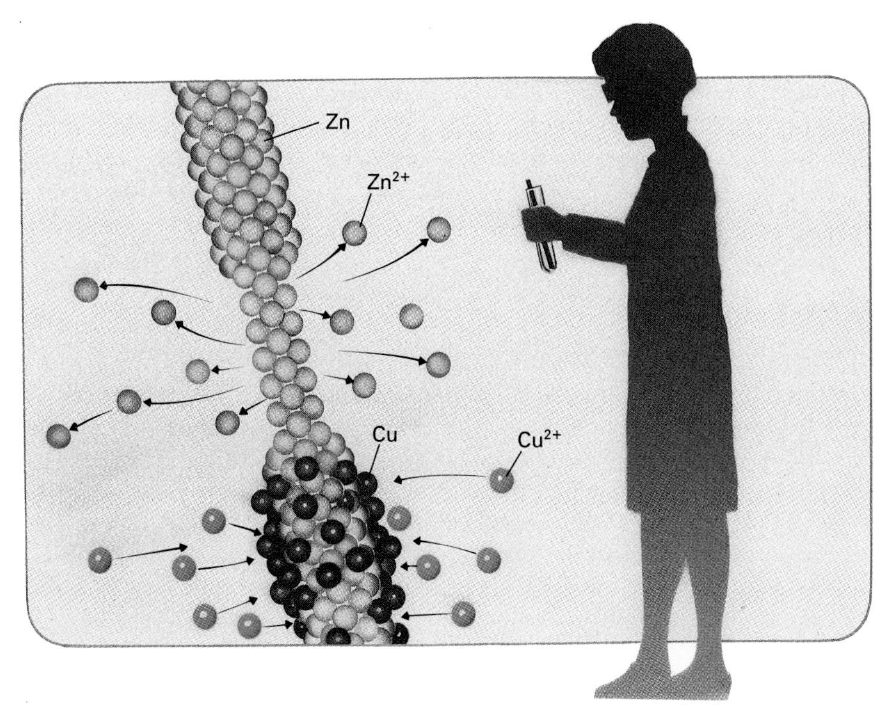

A laboratory experiment with a zinc rod immersed in a solution of copper ions and the action at the atomic level.

of the zinc (Figure 8.3). The copper ions are obviously being reduced—adding electrons and being converted to copper metal.

Since oxidation must accompany reduction, what is being oxidized? That is, what is losing electrons? Careful observation shows that the zinc strip is gradually being consumed in the reaction. The zinc is being oxidized to zinc ions in solution, which are colorless.

The reduction of copper ions by zinc can be thought of as a competition for the available electrons. From observing the reaction in Figure 8.3, it appears that the Zn atoms give up their electrons to the Cu^{2+} ions in the reaction

$$\underset{}{Zn(s)} + \underset{\text{Blue}}{Cu^{2+}(aq)} \longrightarrow \underset{\text{Colorless}}{Zn^{2+}(aq)} + Cu(s)$$

If instead, Cu atoms more easily gave up their electrons to Zn^{2+} ions, the favorable reaction would be the *reverse* of the preceding one. This reaction is unfavorable, however, and does not occur on its own. Apparently, zinc is a stronger reducing agent than copper because it gives up its electrons more easily.

The electrons transferred between a metal that is a good reducing agent and the ion of another metal can provide the electron flow—the *current*—in a battery. A battery is essentially a favorable oxidation–reduction reaction occurring inside a container that has two **electrodes**. At one of these electrodes oxidation takes place and electrons flow out of the cell (it is marked

Throw-away batteries, some with their positive electrode connections marked. *(Dr. Jeremy Burgess/SPL/Photo Researchers)*

■ To remember the difference between an anode and a cathode note the first letters of these words: **o**xidation occurs at the **a**node (both vowels); **r**eduction takes place at the **c**athode (both consonants).

with a − sign). This electrode is the **anode**. At the other electrode, reduction takes place as electrons flow into the cell (it is marked with a + sign). This electrode is the **cathode**. When all the reactants inside the battery are used up and it is impossible to convert the reaction products back to their original form, the battery is "dead" and must be discarded. On the other hand, if the reactants can be converted back to their original form, the battery can be used again.

Throw-Away Batteries

The general arrangement of the parts in an electrochemical cell is diagrammed in Figure 8.4. The reaction between zinc and copper ions discussed at the beginning of this section provides an example of how such a cell works. The zinc is separated from the copper ion solution and the two are connected so that the reaction proceeds as the electrons are transferred through the connecting wire as shown in Figure 8.5. A *salt bridge* allows ions to flow from one electrode chamber to the other. Such a connection is necessary because Zn^{2+} ions are produced at the anode and negative ions must flow into that chamber to prevent positive charge from building up and stopping the reaction. In the other chamber, Cu^{2+} ions are being used up, and positive ions must move into that one for the same reason.

The electrons flow from the Zn anode through the connecting wire and then flow into the copper cathode where reduction of Cu^{2+} ions occurs. The products of the reaction in this simple battery cannot easily be converted to their original form. For this to happen, the copper deposited on the copper electrode would somehow have to be dissolved back into solution, and the Zn^{2+} ions would have to be converted back into the zinc metal strip. In other

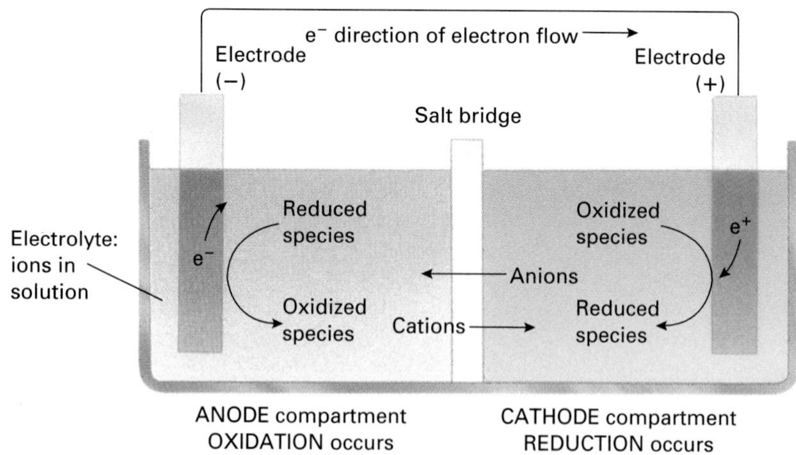

Figure 8.4 Components of an electrochemical cell. Oxidation (loss of electrons) occurs in the anode compartment and electrons flow out of the cell through the external circuit. Electrons re-enter the cell at the cathode, and reduction (gain of electrons) takes place in the cathode compartment. A salt bridge (or other connection) allows ions to flow between the two compartments to maintain charge balance.

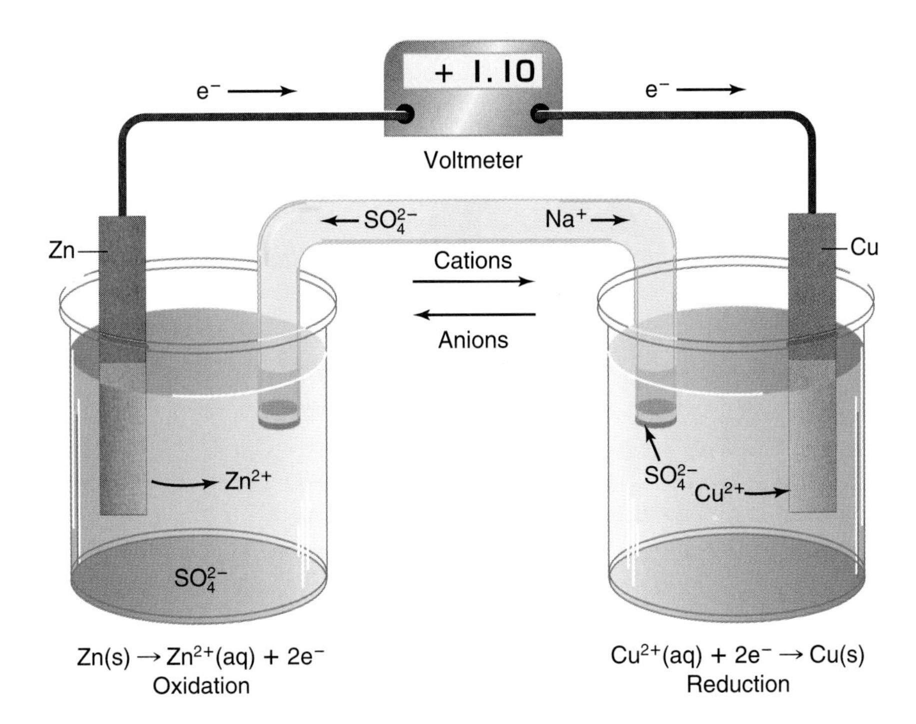

Figure 8.5 A simple electrochemical cell. The cell consists of a zinc electrode in a solution containing Zn^{2+} ions *(left),* a copper electrode in a solution containing Cu^{2+} ions *(right),* and a salt bridge through which ions flow. When the electrodes are connected by a conducting circuit, electrons flow from the zinc electrode, where zinc is oxidized, to the copper electrode, where copper is reduced. The overall reaction in this cell is $Zn(s) + Cu^{2+}(aq) \rightarrow Zn^{2+}(aq) + Cu(s)$.

words, because it is not easily reversible, this reaction is best used in a *throwaway* battery, one that cannot be recharged. Batteries of this type are called **primary batteries**.

One of the most common primary batteries today is the *alkaline battery.* Zinc in the presence of potassium hydroxide (KOH, the alkaline substance) serves as the anode. The zinc is separated from the other chemicals (Figure 8.6) by a porous paper, which serves as the salt bridge. The cathode is made of graphite (carbon) combined with manganese dioxide (MnO_2), which is the oxidizing agent. The oxidation at the anode is conversion of Zn to Zn^{2+} (in solid ZnO) and the reduction at the cathode is conversion of Mn^{4+} (in MnO_2) to Mn^{3+} (in Mn_2O_3) so that the overall cell reaction is

$$Zn(s) + 2\,MnO_2(s) \longrightarrow ZnO(s) + Mn_2O_3(s)$$

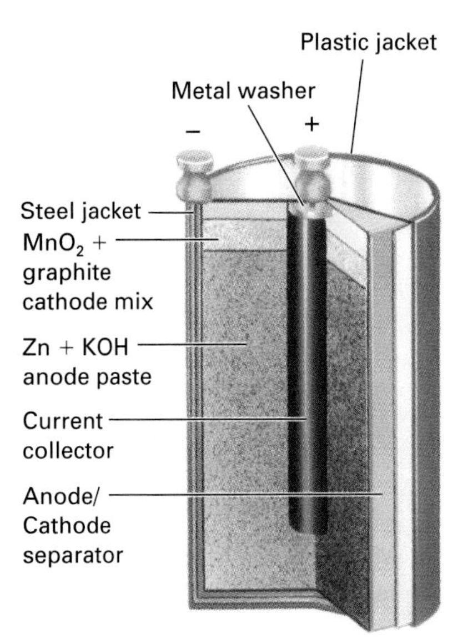

Figure 8.6 An alkaline battery. This type of throw-away battery provides a constant voltage of 1.54 V throughout its useful life.

Reusable Batteries

Some batteries allow the oxidation–reduction reactions at the electrodes to be reversed by the addition of energy, so that the battery can be recharged. Batteries of this type are called **secondary batteries**. Under favorable conditions, secondary batteries may be discharged and recharged many times over.

The lead–acid automobile battery is the most familiar secondary battery. As this battery is discharged, metallic lead is oxidized to lead sulfate at the anode, and the Pb^{4+} in lead dioxide (PbO_2) is reduced at the cathode to the Pb^{2+} in lead sulfate ($PbSO_4$). The reaction takes place in the

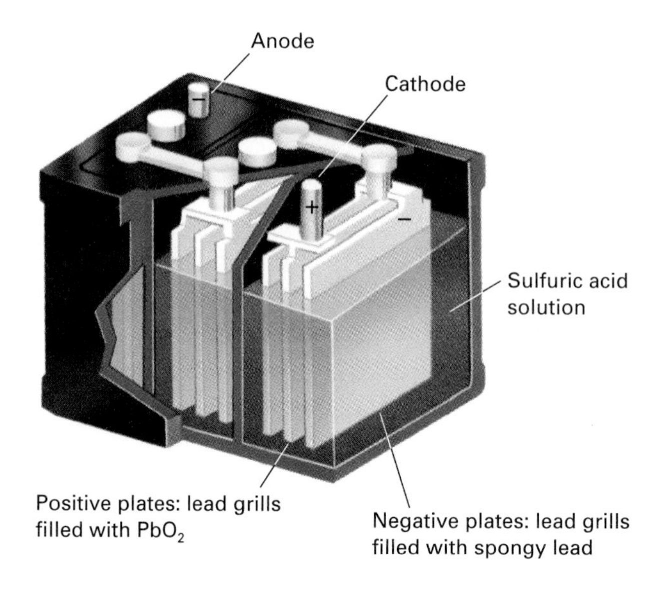

Figure 8.7 A lead–acid battery. Most lead–acid batteries have a useful life of three years or less. The lead in these batteries can be a health hazard and thus there are stringent environmental requirements concerning their disposal.

presence of sulfuric acid (battery acid) and the equation for the overall reaction is

$$Pb(s) + PbO_2(s) + 2\ H_2SO_4(aq) \longrightarrow 2\ PbSO_4(s) + 2\ H_2O(\ell)$$

The lead–acid battery (Figure 8.7) is reusable because the lead sulfate formed at both electrodes is insoluble and stays on the electrode surface. Then when the battery needs recharging, the lead sulfate is available for the reverse reaction.

Recharging a secondary battery requires reversing the direction of electron flow through the battery, which can be accomplished by a generator or an alternator. When recharging occurs, the reactions at the two electrodes are reversed

$$\text{Discharge in a battery: chemical energy} \longrightarrow \text{electrical energy}$$

$$\text{Recharging a battery: electrical energy} \longrightarrow \text{chemical energy}$$

An assortment of batteries. In front of the lead–acid automobile battery are, from left to right, three types of rechargeable nickel–cadmium batteries, three types of not rechargeable alkaline batteries, and a zinc–graphite dry cell (also not rechargeable). *(© 1990 Richard Megna/Fundamental Photographs)*

Normal recharging of an automobile lead–acid battery occurs during driving. The voltage regulator senses the output from the alternator, and when the alternator voltage exceeds that of the battery, electrical energy is added back into the battery and the battery is recharged. During the recharging cycle in a lead–acid battery, some water is reduced to hydrogen at the cathode and some water is oxidized to oxygen at the anode. The result is a potentially explosive mixture of hydrogen and oxygen at the top of the battery. Under normal driving conditions, automobile batteries do not explode; however, internal short circuits can produce explosions in older batteries, and batteries must always be protected from sparks.

All in all, the lead–acid battery is relatively inexpensive, reliable, and simple, and has an adequate life. Its high weight is its major fault. Newer secondary batteries have found use in some applications such as electronics, but none of these newer batteries can perform like the lead–acid battery does for its cost.

Fuel Cells

Fuel cells, like batteries, have an electrode where oxidation takes place and an electrode where reduction takes place. Unlike batteries, which use energy stored in chemicals in the electrode compartments, fuel cells produce energy from reactants that continuously flow into the compartments while the chemical reaction products flow out of them.

A successful application of fuel cells has been in the space program on board the Gemini, Apollo, and space shuttle missions. In the space shuttle fuel cells, hydrogen reacts with oxygen to produce water and energy.

$$2\ H_2(g) + O_2(g) \longrightarrow 2\ H_2O(\ell) + \text{energy}$$

As we have mentioned, if a mixture of hydrogen and oxygen is sparked, the energy is released suddenly in the form of a violent explosion. In a fuel cell (Figure 8.8), the reaction takes place in a controlled manner, with the electrons lost by the hydrogen molecules flowing out of the fuel cell at the anode

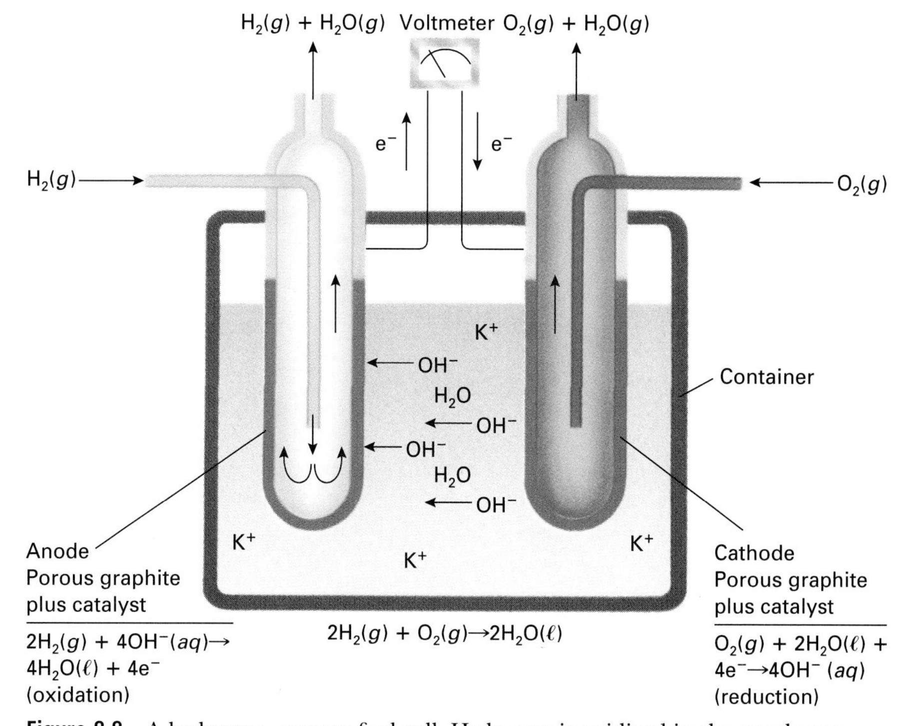

Figure 8.8 A hydrogen–oxygen fuel cell. Hydrogen is oxidized in the anode compartment. Electrons flow out of the cell, provide current to an external circuit, and flow back into the cathode compartment, where oxygen is reduced. The water produced can be purified and used for drinking.

SCIENCE AND SOCIETY

Electric Automobiles: Moving off the Drawing Boards

In an effort to decrease air pollution, California has mandated that by 2003 10% of all vehicles sold in the state must have zero emissions of air pollutants such as carbon monoxide, oxides of nitrogen, and unburned hydrocarbons. Electric, battery-operated vehicles are necessary to meet these conditions. Based on recent sales figures, 200,000 of them should be on the road by 2003. The state had to retreat from an earlier goal of 40,000 electric vehicles to be sold in 1998 due to delays in availability of the cars and public acceptance.

A turning point in the production of battery-powered automobiles was reached in 1996, however, with the introduction of the General Motors EV1, the first production model electric vehicle from a major manufacturer. The EV-1 is powered by a bank of advanced design lead–acid batteries and is reported to have a range of 70 to 90 miles before a 2-h recharge (for 60% of capacity) is needed. In its first five months on the market, GM leased 176 of the cars to drivers who were first carefully interviewed to be sure they understood what they were getting. (The EV-1 was not yet available for sale.)

Despite the slow acceptance by consumers, other major manufacturers are joining GM in the California marketplace. The newer vehicles make a big step forward in being powered not by old-style lead–acid batteries, but by a new nickel–metal hydride (NiMH) battery, which is lighter in weight and promises a longer driving range and quicker recharge (60% in 15 min). In early 1997, a prototype car with a nickel–metal hydride battery was reported to have traveled 375 miles on a single charge. Unlike conventional batteries, the anode in this battery is an alloy (a mixture of metals) that includes vanadium, titanium, and chromium (as well as nickel).

The General Motors EV-1, the first production model electric car from a major auto manufacturer. (Courtesy General Motors Corporation)

While battery-driven automobiles do not emit pollutants, they do, of course, depend for recharging on electricity mostly generated by fossil fuel combustion. Assuming oil is used to generate the electricity, an electric vehicle will travel about 1100 miles on a barrel of oil, while a gasoline-powered vehicle can go only 670 miles per barrel. Perhaps electric vehicles can have some impact on the rate at which petroleum supplies are being depleted. If you live in California, or some other state hard-hit by urban air pollution caused by cars, chances are you may have an electric automobile in your future. Would you like to see your state mandate the use of electric cars?

and back in again at the cathode, where oxygen is reduced. A bonus to the use of this type of fuel cell aboard a spacecraft is that the water can be used for drinking. Also, these cells deliver the same power that batteries weighing ten times as much would provide. On a typical seven-day mission, the shuttle fuel cells consume 1500 lb of hydrogen and generate 190 gal of water.

Developments in the use of fuel cells for both stationary energy production and transportation energy production marked several milestones in the late 1990s. For example, a test stack of 400 fuel cells in Australia provided 3 to 5 kilowatts of electricity during a 200 hour test run. It is expected that this technology can be used to provide electricity to rural communities or industrial facilities and can also be integrated into existing power grids.

Fuel cells may ultimately be a better nonpolluting source of energy for automobiles than electric batteries (see *Science and Society* box). Battery-driven vehicles must plug in to conventionally generated electric current to be recharged, while a fuel cell requires no petroleum combustion. In the fall of 1997, two major manufacturers (Daimler-Benz and Toyota) exhibited fuel-cell driven automobiles that produce hydrogen from methanol, thereby freeing the vehicles from on-board hydrogen storage tanks. Work is also under way on engines that produce the hydrogen for fuel cells from gasoline and on hybrid vehicles that carry conventional batteries to augment the power from fuel cells.

8.5 Electrolysis: Chemical Reactions Caused by Electron Flow

Chemical reactions that are unfavorable can be forced to proceed by the input of energy, as in the recharging of a secondary battery. **Electrolysis reactions** are oxidation–reduction reactions driven by electrical energy from an external power supply. Where electrons flow into the cell, the electrode becomes negatively charged, positive ions in solution migrate toward that electrode, and reduction takes place. In this type of cell (Figure 8.9), the electrodes need not be in separate compartments.

■ In electrolysis electrical energy produces chemical change. In batteries chemical change produces electrical energy.

The metal in some ores is so resistant to reduction that few reducing agents are strong enough to cause the reaction, and electrolysis is a good alternative way to provide enough energy for the reaction. Most importantly, aluminum is a metal of this type. Aluminum, in the form of Al^{3+} ions, is the third most abundant element in the earth's crust (7.4%). The quantity and commercial value of aluminum used in the United States each year is exceeded only by that of iron and steel.

■ The original top of the Washington Monument was aluminum, made in 1884 by the sodium-reduction method.

From the time of its discovery in 1825 until near the turn of the century, aluminum was made by reducing $AlCl_3$ with a more active metal (potassium or sodium) but only at a very high cost. Even though there was commercial production of aluminum by 1854, aluminum was considered to be a precious metal—as gold and platinum are today—and one of its early uses was for jewelry. Napoleon III saw the possibilities of aluminum for military use, however, and commissioned studies on improving its production. The French had a ready source of aluminum-containing ore, bauxite, named for the French town of Les Baux. In 1886, a 23-year-old Frenchman, Paul Heroult, conceived the electrochemical method that is still in use today. In an interesting coincidence, an American, Charles Hall, who was 22 at the time, announced his invention of the identical process in the same year. Hence, the commercial process is now known as the Hall–Heroult process.

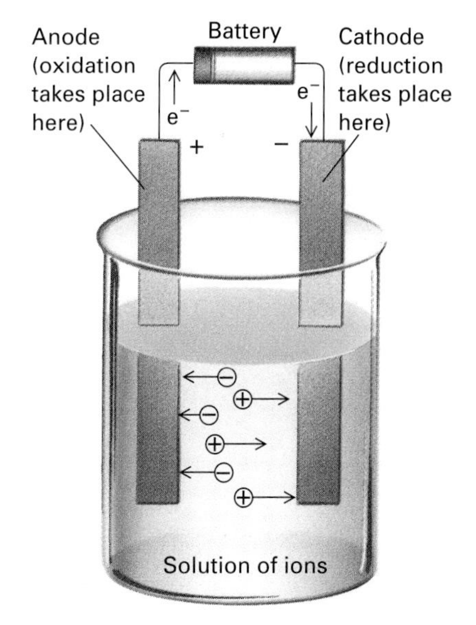

Figure 8.9 An electrolysis cell. For electrolysis, the electrodes need not be in separate compartments. Electrons enter the cell at the cathode, where reduction occurs. Ions flow through the electrolyte solution to maintain charge balance. At the anode, where oxidation occurs, electrons leave the cell.

In the Hall–Heroult process, purified aluminum oxide from bauxite is dissolved in a molten ionic compound (cryolite, Na_3AlF_6). The mixture is electrolyzed in a cell with carbon anodes and a carbon cell lining that serves as the cathode (Figure 8.10).

Because aluminum is such an active metal, separating it from its oxide requires a lot of energy. This, combined with the energy needed to maintain the molten cryolite bath, makes aluminum production highly energy-intensive. One reason for the success of aluminum recycling is the large saving in energy cost—making aluminum beverage cans from recycled aluminum

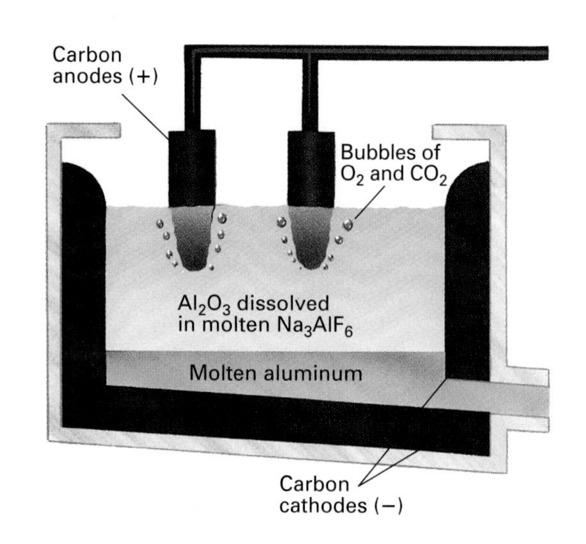

Figure 8.10 Aluminum production by electrolysis. At the cathode, aluminum ions are reduced to aluminum metal. The anode reaction is production of oxygen gas, which reacts with the carbon anodes. (The cell reaction is $2\ Al_2O_3 + 3\ C \rightarrow 4\ Al + 3\ CO_2$.) Molten aluminum is denser than the molten salt mixture in the cell and collects at the bottom.

■ Aluminum usage: 35% of U.S. aluminum production is used in containers and packaging; 20% each, in transportation and construction; and the rest, in electrical applications, durable consumer goods, machinery and equipment, and other miscellaneous uses.

requires only 5% of the energy used in making the cans from new aluminum, and the process is about 20% cheaper overall.

Another practical application of electrolysis is *electroplating*—the coating of one metal, which is made the cathode in an electrolysis cell, with another metal, which is made the anode. For example, dull iron or steel surfaces are electroplated with chromium to provide a mirror-like finish and also to protect against corrosion.

8.6 Corrosion: Unwanted Oxidation–Reduction

In the United States alone, more than $10 billion is lost each year to **corrosion**—the unwanted oxidation of metals during exposure to the environment. Much of this corrosion is the rusting of iron and steel, although other metals may corrode as well. The oxidizing agent causing all this unwanted corrosion is usually oxygen. Iron is most severely affected by corrosion because rust does not adhere strongly to the metal's surface. The continuing loss of surface iron as rust forms and then flakes off eventually causes structural weakness.

The driving forces behind metal corrosion are the activity of the metal as a reducing agent and the strength of the oxidizing agent. Whenever a strong reducing agent (the metal) and a strong oxidizing agent (such as oxygen) are in contact, a reaction between the two substances is likely. Factors governing the rates of chemical reaction, such as temperature and concentration (Section 6.3), affect the rate of corrosion as well. Consider the corrosion of an iron spike (Figure 8.11). There are tiny microcrystals of loosely bound iron atoms on the surface of the metal that can easily be oxidized.

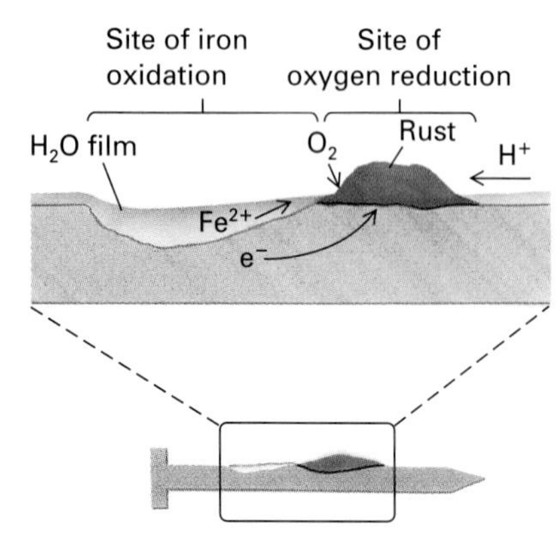

Figure 8.11 Corrosion of iron. The site of iron oxidation may be different than the site of oxygen reduction because iron is a conductor of electricity and electrons can move through it from one site to another.

$$Fe(s) \longrightarrow Fe^{2+}(aq) + 2\ e^-$$

Because iron is a good conductor of electricity, the electrons produced by this oxidation can migrate through the metal to a point where they can reduce something. The fact that iron is a conductor of electricity is important because corrosion would come to an abrupt halt as a result of a buildup of excessive negative charge if the electrons were not conducted away. One location on the surface of the iron where electrons can be used is any tiny drop of water containing dissolved oxygen. Here the oxygen gains electrons, forming hydroxide ions.

$$O_2(g) + 2\,H_2O(\ell) + 4\,e^- \longrightarrow 4\,OH^-(aq)$$

The Fe^{2+} ions are further oxidized to Fe^{3+} ions, which react with OH^- ions to form the hydrated iron oxide known as rust.

$$2\,Fe^{3+}(aq) + 6\,OH^-(aq) \longrightarrow \underset{\text{Rust}}{Fe_2O_3 \cdot 3H_2O(s)}$$

Corrosion has made this bridge unsafe. *(Tom Pantages)*

The rate of rusting is enhanced by salts, which dissolve in the water on the surface of the iron. The hydroxide ions and iron ions migrate more easily in the ion solutions produced by the presence of the dissolved salts. Automobiles rust more quickly when exposed to road salts in wintry climates. If road salts are used in your driving area, it's a good idea after snowy seasons to wash the undersides of automobiles to remove the accumulated salts.

For rust to form, three reactants are necessary. These are iron, oxygen, and water. Rusting can be prevented by protective coatings such as paint, grease, oil, enamel, or a corrosion-resistant metal such as chromium. Most of these coatings keep out moisture. Some of the metals that are more active than iron form adherent oxide coatings when they corrode. Coating iron with these metals provides corrosion protection. One of these active metals is zinc. When the zinc coating of a galvanized object is exposed to air and water, a thin film of zinc oxide forms that protects the zinc from further oxidation. If the zinc coating should get scratched so that iron is exposed to the air, zinc will quickly reduce any Fe^{2+} ions formed because zinc is more active than iron in giving up electrons.

■ SELF-TEST 8B

1. A battery contains two electrodes, named the ____________ and the ____________.
2. A battery is dead when ____________.
3. It is necessary for ____________ to flow between the electrode compartments in a battery so that ____________ balance is maintained.
4. The conversion of metallic zinc to zinc ions (Zn^{2+}) is ____________ and will occur at the ____________ in an electrochemical cell.
5. The conversion of copper ions (Cu^{2+}) to metallic copper is ____________ and will occur at the ____________ in an electrochemical cell.
6. In recharging a secondary battery, ____________ energy is converted to ____________ energy.
7. The reduction of bauxite ore by the process of ____________ produces ____________ metal.

8. Why is rusting of iron an especially destructive form of corrosion?
9. Name the three reactants needed for rusting of iron.

MATCHING SET

___	1. Chemicals in automobile lead storage battery	a. PbO_2, Pb, and H_2SO_4
___	2. Reduction of Cu^{2+} ions produces this	b. Hydrogen
___	3. Oxidizing agent used to purify drinking water	c. Gain of oxygen
___	4. Oxidation	d. Reduction
___	5. Reduction	e. SO_2
___	6. Gain of hydrogen	f. NO_2
___	7. Oxidized form of sulfur	g. Loss of oxygen
___	8. Oxidized form of nitrogen	h. Lead–acid battery
___	9. Allows ions to pass from one electrode compartment in a battery to the other	i. Chlorine
___	10. Fuel used in some fuel cells	j. Copper metal
___	11. Battery used in most automobiles	k. Salt bridge

QUESTIONS FOR REVIEW AND THOUGHT

1. Define the following terms:
   (a) Oxidation in terms of electrons
   (b) Reduction in terms of electrons
   (c) Oxidation in terms of oxygen
   (d) Reduction in terms of hydrogen
   (e) Oxidizing agent
   (f) Reducing agent
2. Define the following terms:
   (a) Anode (b) Primary battery
   (c) Corrosion (d) Cathode
   (e) Secondary battery
3. In each of the following pairs, which substance is more highly oxidized?
   (a) CO or CO_2 (b) NO_2 or NO
   (c) SO_3 or SO_2 (d) CrO or CrO_3
   (e) CaO or Ca (f) N_2 or NH_3
4. Name two common uses of the oxidizing properties of chlorine (Cl_2).
5. What is the most common oxide of the element hydrogen?
6. What is the difference between oxidation and combustion?
7. What oxide of carbon is the product of incomplete combustion?
8. A potassium atom can form a K^+ ion. When it does, the potassium ion is said to be (a) oxidized or (b) reduced.
9. Beside each of the following reactions, indicate whether oxidation or reduction is occurring to the underlined element:
   (a) $2\ \underline{H}_2(g) + O_2(g) \rightarrow 2\ H_2O(\ell)$
   (b) $2\ \underline{Cu}(s) + O_2(g) \rightarrow 2\ CuO(s)$
   (c) $2\ \underline{Sn}O(s) \rightarrow 2\ Sn(s) + O_2(g)$
   (d) $\underline{Fe}_2O_3(s) + 3\ C(s) \rightarrow 2\ Fe(s) + 3\ CO(g)$
   (e) $\underline{N}_2(g) + 3\ H_2(g) \rightarrow 2\ NH_3(g)$
   (f) $2\ \underline{S}(s) + 3\ O_2(g) \rightarrow 2\ SO_3(g)$
10. The atmosphere of Earth can be described as an oxidizing atmosphere. Explain.
11. The atmosphere of Jupiter contains considerable hydrogen and methane. Would you describe that atmosphere as an oxidizing atmosphere or a reducing atmosphere?
12. The main source of magnesium is seawater, which contains Mg^{2+} ions. Is this magnesium in an oxidized or reduced form?
13. When zinc metal reacts with an acid, Zn^{2+} ions form. Is the zinc metal oxidized or reduced?
14. In each of the following reactions, tell which substance is oxidized and which is reduced, and then name the oxidizing agent and the reducing agent in each reaction:
   (a) $2\ Al(s) + 3\ Cl_2(g) \rightarrow 2\ AlCl_3(s)$
   (b) $S(s) + O_2(g) \rightarrow SO_2(g)$
   (c) $CuO(s) + H_2(g) \rightarrow Cu(s) + H_2O(g)$
   (d) $C_2H_4(g) + H_2(g) \rightarrow C_2H_6(g)$
   (e) $N_2(g) + 2\ O_2(g) \rightarrow 2\ NO_2(g)$
   (f) $Fe_2O_3(s) + 3\ C(s) \rightarrow 2\ Fe(s) + 3\ CO(g)$

15. Why would rusting of automobiles be less of a problem in Arizona than in Chicago?
16. Would you expect a piece of iron to rust on the surface of the moon? Explain.
17. Oxidation always occurs at the anode of an electrochemical cell. (a) True, (b) False.
18. Describe a simple battery, naming three essential parts.
19. How is a fuel cell similar to a battery? How is it different?
20. Besides electricity, what do the fuel cells used on the space shuttle produce?
21. A brand new battery has a mass of 249.6 g. After it has been fully discharged its mass is still 249.6 g. Explain this.
22. Think of ways electricity might be distributed to consumers of electric automobiles. Compare all the methods you have thought of with the methods used to distribute hydrocarbon fuels (gasoline). List as many benefits and problems as you can for each.

CHAPTER 9

Energy and Hydrocarbons

The average use of energy per individual is near the highest point in the history of the world. In the United States alone, with only 5% of the world's population, we consume 24% of the daily supply of energy. The combustion of the fossil fuels (coal, petroleum, and natural gas) provides 85% of all the energy used in the world. At current usage rates, proven world reserves of coal, natural gas, and petroleum are estimated to last 230, 60, and 40 years, respectively.

Hydrocarbons are the principal component of fossil fuels. Natural gas is primarily methane, crude petroleum is a complex mixture of thousands of hydrocarbons, and coal is an even more complex mixture of hydrocarbons. Many of the fuels we use, such as gasoline and jet fuel, are obtained from petroleum.

Fossil fuels are also the major source of hydrocarbons that are used to make thousands of consumer products. This chapter describes the chemistry of hydrocarbons and their importance to our energy needs, and Chapter 10 emphasizes the industrial uses of hydrocarbons. Alcohols and ethers are part of the energy discussion because of the need to improve emissions and reduce pollution.

- What are fuels?
- How do fuels produce energy?
- What are the major classes of hydrocarbons?
- What different types of isomers are possible for hydrocarbons, and why are they important?
- How is petroleum refined?
- How is high-octane gasoline produced?
- What are oxygenated gasolines, and why are they used?
- Why are methanol and ethanol receiving attention as alternate fuels?

A thermogram showing in bright colors the heat energy escaping from a house that, like almost all of our houses, is heated by burning hydrocarbons. *(Daedalus Enterprises, Inc./Peter Arnold, Inc.)*

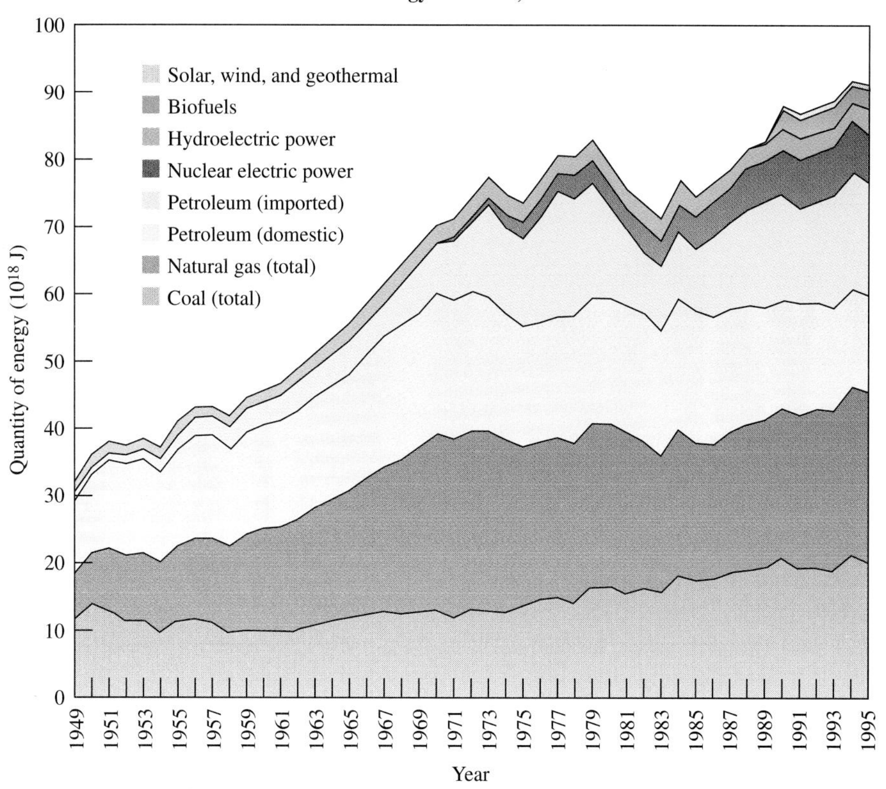

Figure 9.1 Energy resources of the United States. Use of energy resources in the United States is plotted from 1949 to 1995. (An energy resource is a naturally occurring fuel, such as petroleum, or a continuous supply, such as sunlight.) In 1949, coal and petroleum were almost equally important, with natural gas third. Today petroleum and natural gas are used in greater quantity than coal, and more than half of the petroleum is imported. Nuclear electric power did not exist in 1949 but contributes significantly today, while hydroelectric electricity generation has grown slightly. *(Data from Annual Energy Review, 1995, U.S. Department of Energy, Energy Information Administration. http://www.eia.doe.gov/emeu/aer/aergs/aer2.html.)*

9.1 Energy from Fuels

In our homes and our industries, we obtain most of our energy by burning fossil fuels (Figure 9.1). Why does burning fuels provide energy? Fuels are reduced forms of matter that burn readily in the presence of oxygen, and combustion is an oxidation reaction (Section 8.1) that produces heat. The burning of methane, the principal component of natural gas, can be used to illustrate how the combustion of fuels produces energy.

What determines whether energy is released or absorbed in a chemical reaction? For a chemical reaction to occur requires breaking bonds in reactants and forming bonds in products. If the bonds in the reactants are stronger than those formed in the products, energy must be added for the reaction to occur. If the bonds in reactants are weaker than those formed in products, energy will be released. For example, energy is released when methane burns because it takes less energy to break the bonds in the reactants, methane and oxygen, than is produced when the products carbon dioxide and water are formed. The average bond energies for bonds in methane, oxygen, carbon dioxide, and water (Table 9.1) can be used to calculate the heat of combustion of methane.

$$\mathrm{H{-}\overset{\displaystyle H}{\underset{\displaystyle H}{\overset{|}{\underset{|}{C}}}}{-}H} + 2\,\mathrm{O{=}O} \longrightarrow \mathrm{O{=}C{=}O} + 2\,\mathrm{H{-}O{-}H}$$

■ Although the International System of Units or Système International (SI) unit of energy is the joule, the more familiar unit of heat is the calorie. A calorie is the amount of heat required to raise the temperature of 1 g of water 1°C. One calorie equals 4.18 joules (J). 1000 calories (cal) = 1 kilocalorie (kcal). For example, you use 140 kcal/h walking and 80 kcal/h even when you are asleep.

TABLE 9.1 ■ Average Bond Energies

Bond	Energy (kcal/mol)
C—H	99
O=O	118
C=O	192
O—H	111

■ In chemical reactions, heat can be released, as in the combustion of fuels, or absorbed, as in decomposition. For example, decomposing water into hydrogen and oxygen requires heat. The amount of heat released or absorbed in every case depends on the differences in bond strengths of the reactants and products.

■ Bonding in the alkane, alkene, and alkyne classes of hydrocarbons (compounds of hydrogen and carbon) was discussed in Section 5.3.

TABLE 9.2 ■ Energy Values of Some Fuels

Substance	Heat (kcal/g)
Hydrogen	34.0
Natural gas (mostly methane)	11.7
Gasoline	11.5
Anthracite coal	7.4
Wood (pine)	4.3

■ Alkanes are referred to as saturated hydrocarbons because they contain the highest ratio of hydrogen to carbon possible.

Rice fields in the Philippines. Decay of organic matter in rice fields is estimated to make up one fifth of all methane emitted each year due to human activities. *(Kjell B. Sandved/Visuals Unlimited).*

It takes a total of 632 kcal to break all the bonds in the reactants (breaking 2 mol of O═O bonds and 4 mol of C—H bonds is 2(118 kcal) + 4(99 kcal) = 632 kcal. The making of the bonds in the products produces 828 kcal (formation of 2 mol of C═O bonds and 4 mol of O—H bonds is 2(192 kcal) + 4(111 kcal) = 828. This gives a net release of 196 kcal/mol of methane (828 kcal − 632 kcal), which closely agrees with the experimental value of 192 kcal/mol.

The amounts of energy per gram obtained by burning some common fuels are given in Table 9.2. Any hydrocarbon or hydrocarbon mixture can be classified as a fuel, and fossil fuels are complex mixtures of hydrocarbons. Hence, an understanding of hydrocarbon chemistry is central to the study of fossil fuels.

There are four classes of hydrocarbons: the **alkanes**, which contain carbon–carbon single bonds (C—C); the **alkenes**, which contain one or more carbon–carbon double bonds (C═C); the **alkynes**, which contain one or more carbon–carbon triple bonds (C≡C); and the **aromatics**, which consist of benzene, benzene derivatives, and fused benzene rings.

9.2 Alkanes: Backbone of Organic Chemistry

The simplest alkane is methane (CH_4) the principal component of natural gas. Alkanes are saturated hydrocarbons (Section 5.3) with the general formula C_nH_{2n+2} (Table 9.3). Notice that all hydrocarbon formulas are traditionally written with the C atom first, followed by the H atom; and that all alkanes have *-ane* as the suffix in their name. When $n = 1$ to 4, the first part of the name is something of historical origin; these are common names that we just have to remember. When $n = 5$ or more, the Greek prefixes (Table 9.3) tell how many carbon atoms are present. For example, the compound with six carbons is called hexane.

The tetrahedral structure of CH_4 has been discussed (Section 5.3), but it is important to recognize that *every* carbon in an alkane has a tetrahedral environment because all carbon atoms in saturated hydrocarbons have four single bonds. The tetrahedral environment of the carbon atoms in alkanes is difficult to draw in two dimensions.

To save time and space, chains of carbon atoms are usually represented with straight lines as in Figures 9.2*a*, 9.3*a*, and 9.4*a*. However, keep in mind that these drawings are not an accurate representation of the tetrahedral bond angles, which are 109.5°.

Our ability to understand these tetrahedral structures is helped by the use of models. Two types of models are generally used—the "ball-and-stick"

Figure 9.2 The structure of methane, as represented by (a) its structural formula, (b) a ball-and-stick model, and (c) a space-filling model. *(c, C. D. Winters)*

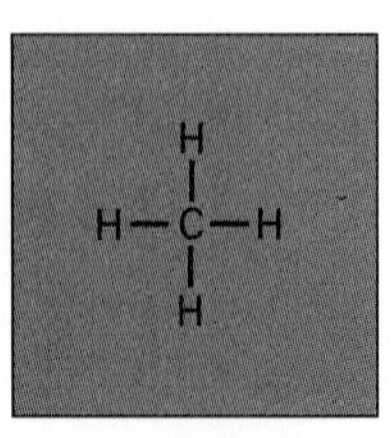

(a)

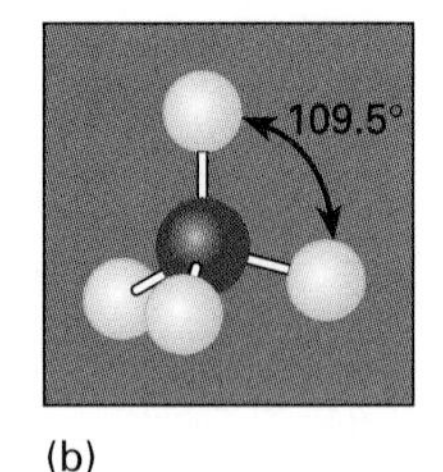

(b)

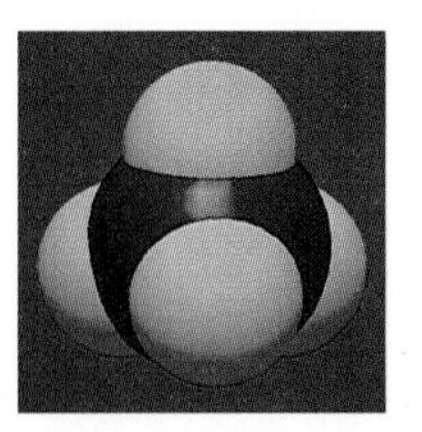

(c)

TABLE 9.3 ■ The First Eight Straight-Chain Saturated Hydrocarbons

Name	Formula	Boiling Point, °C*	Structural Formula	Use
Methane	CH_4	−162	H H—C—H H	Principal component in natural gas
Ethane	C_2H_6	−88.5	H H H—C—C—H H H	Minor component in natural gas
Propane	C_3H_8	−42	H H H H—C—C—C—H H H H	Bottled gas for fuel
Butane	C_4H_{10}	0	H H H H H—C—C—C—C—H H H H H	
Pentane	C_5H_{12}	36	H H H H H H—C—C—C—C—C—H H H H H H	Some of the components of gasoline
Hexane	C_6H_{14}	69	H H H H H H H—C—C—C—C—C—C—H H H H H H H	
Heptane	C_7H_{16}	98	H H H H H H H H—C—C—C—C—C—C—C—H H H H H H H H	
Octane	C_8H_{18}	126	H H H H H H H H H—C—C—C—C—C—C—C—C—H H H H H H H H H	

* Notice the gradual increase in boiling point as the molecular weight increases. Fractional distillation of petroleum is possible because of these differences (Section 9.6).

model and the "space-filling" model. In a ball-and-stick model, balls represent the atoms and short pieces of wood or plastic represent bonds. For example, the ball-and-stick model for methane (Figure 9.2*b*) has a black ball representing carbon, with holes at the correct angles connected by sticks to four white balls representing hydrogen atoms. The space-filling model (Figure 9.2*c*) is a more realistic representation because it depicts both the relative sizes of the atoms and their spatial orientation in the molecule. This is done by scaling the pieces in the model according to the experimental values of atom sizes. The pieces are held together by links that are not visible when the model is assembled.

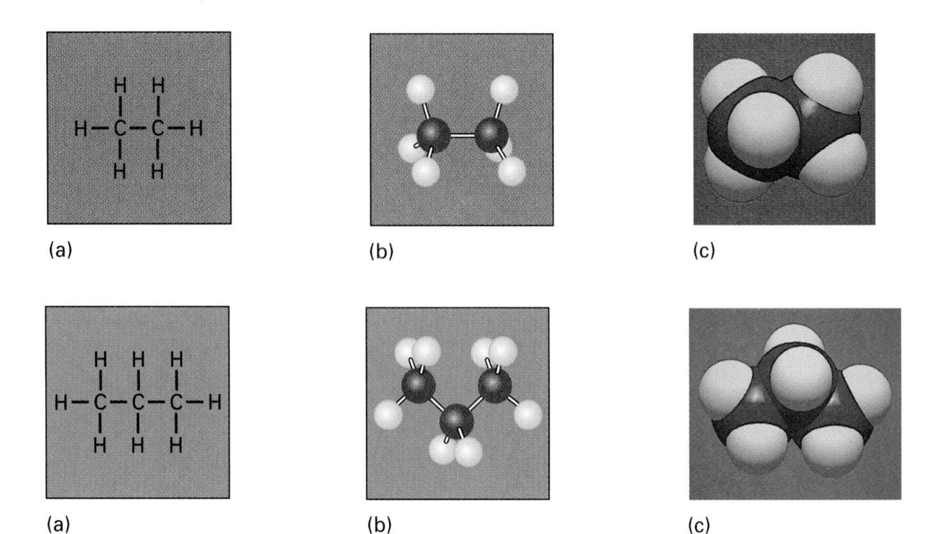

Figure 9.3 The structure of ethane, as represented by (a) its structural formula, (b) a ball-and-stick model, and (c) a space-filling model. *(c, C. D. Winters)*

Figure 9.4 The structure of propane, as represented by (a) its structural formula, (b) a ball-and-stick model, and (c) a space-filling model. *(c, C. D. Winters)*

Illustrations of ball-and-stick models, such as those shown in Figures 9.2*b*, 9.3*b*, and 9.4*b*, will be used extensively in the discussion of hydrocarbons and hydrocarbon derivatives to help you visualize the molecular geometry of molecules. For example, notice that the three carbon atoms in the ball-and-stick model of propane in Figure 9.4*b* do not lie in a straight line because of the tetrahedral geometry about each carbon atom. This illustrates why straight-chain drawings, such as the one for propane in Figure 9.4*a*, are not accurate representations of the tetrahedral H—C—H and C—C—C bond angles.

■ Historically, straight-chain hydrocarbons were referred to as *normal* hydrocarbons, and *n*- was used as a prefix in the name of straight-chain hydrocarbons such as butane (*n*-butane). The current practice is not to use *n*-. If a name is given without indicating that the compound is a branched chain, assume the compound is a straight-chain hydrocarbon.

Straight- and Branched-Chain Isomers of Alkanes

The first three alkanes, CH_4, C_2H_6, and C_3H_8, each have only one possible structural arrangement. However, two structural arrangements are possible for C_4H_{10}—a straight-chain arrangement and a branched-chain arrangement.

Structural formulas:

```
                                    H
                                    |
    H  H  H  H               H   H—C—H   H
    |  |  |  |               |      |      |
 H—C—C—C—C—H             H—C——————C——————C—H
    |  |  |  |               |      |      |
    H  H  H  H               H      H      H
```

The condensed formulas and properties of the two compounds, butane and methylpropane, are as follows:

Condensed formulas:	$CH_3CH_2CH_2CH_3$ Butane	$CH_3CH(CH_3)CH_3$ Methylpropane (isobutane)
Melting point	−138.3°C	−160°C
Boiling point (1 atm)	0.5°C	−12°C
Density (at 20°C)	0.579 g/mL	0.557 g/mL

These molecules have different properties even though they have the same number of atoms in the molecule. Ball-and-stick models of the two structures are shown in Figure 9.5.

Two or more compounds with the same molecular formula but different arrangements of atoms are called **isomers**. Isomers differ in one or more physical or chemical properties such as boiling point, color, solubility, reactivity, and density. Several different types of isomerism are possible for organic compounds. **Branched-chain** and **straight-chain** isomers are examples of **structural isomers** that differ in the order in which the atoms are bonded together. Structural isomerism can be compared to the results you might expect from a child building many different structures with the same collection of building blocks and using all the blocks in each structure.

The branched-chain isomer for C_4H_{10} methylpropane, a common component of bottled gas, has a "methyl" group ($—CH_3$) attached to the central carbon atom. This is the simplest example of the fragments of alkanes known as **alkyl groups**. In this case, removal of H from methane gives a **methyl** group.

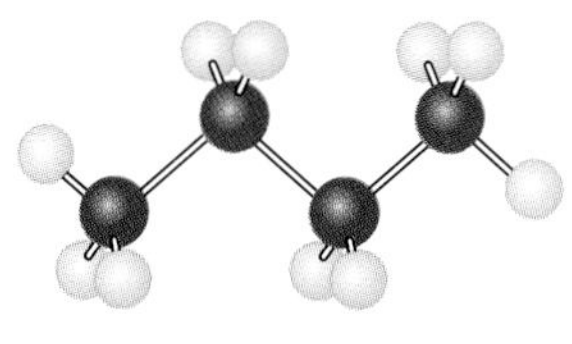

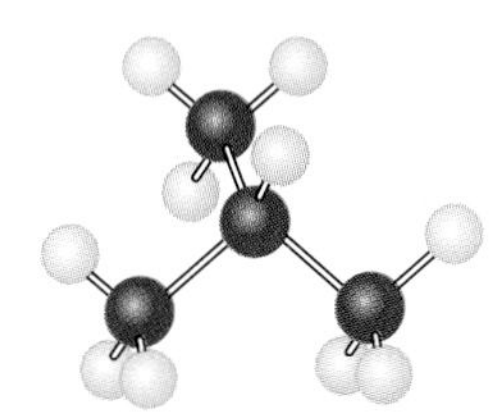

Figure 9.5 Ball-and-stick models of butane, a four-carbon straight-chain hydrocarbon, and methylpropane, a four-carbon branched-chain hydrocarbon.

$$\text{H}—\overset{\text{H}}{\underset{\text{H}}{\overset{|}{\underset{|}{\text{C}}}}}—\text{H} \xrightarrow{-\text{H}} \text{H}—\overset{\text{H}}{\underset{\text{H}}{\overset{|}{\underset{|}{\text{C}}}}}— \quad \text{or} \quad —\text{CH}_3 \left(\begin{array}{c}\text{also written}\\ \text{as CH}_3—\end{array}\right)$$

Removal of an H from ethane gives an **ethyl** group

$$\text{H}—\overset{\text{H}}{\underset{\text{H}}{\overset{|}{\underset{|}{\text{C}}}}}—\overset{\text{H}}{\underset{\text{H}}{\overset{|}{\underset{|}{\text{C}}}}}—\text{H} \xrightarrow{-\text{H}} \text{H}—\overset{\text{H}}{\underset{\text{H}}{\overset{|}{\underset{|}{\text{C}}}}}—\overset{\text{H}}{\underset{\text{H}}{\overset{|}{\underset{|}{\text{C}}}}}— \quad \text{or} \quad —\text{C}_2\text{H}_5 \left(\begin{array}{c}\text{also written}\\ \text{as CH}_3\text{CH}_2—\end{array}\right)$$

Notice that more than one alkyl group is possible when an H atom is removed from C_3H_8

$$\text{H}—\overset{\text{H}}{\underset{\text{H}}{\overset{|}{\underset{|}{\text{C}}}}}—\overset{\text{H}}{\underset{\text{H}}{\overset{|}{\underset{|}{\text{C}}}}}—\overset{\text{H}}{\underset{\text{H}}{\overset{|}{\underset{|}{\text{C}}}}}—\text{H}$$

−H from either end carbon:

$$\text{H}—\overset{\text{H}}{\underset{\text{H}}{\overset{|}{\underset{|}{\text{C}}}}}—\overset{\text{H}}{\underset{\text{H}}{\overset{|}{\underset{|}{\text{C}}}}}—\overset{\text{H}}{\underset{\text{H}}{\overset{|}{\underset{|}{\text{C}}}}}— \quad \text{or} \quad \text{CH}_3\text{CH}_2\text{CH}_2—$$

Propyl

−H from middle carbon:

$$\text{H}—\overset{\text{H}}{\underset{\text{H}}{\overset{|}{\underset{|}{\text{C}}}}}—\overset{\text{H}}{\underset{|}{\text{C}}}—\overset{\text{H}}{\underset{\text{H}}{\overset{|}{\underset{|}{\text{C}}}}}—\text{H} \quad \text{or} \quad (\text{CH}_3)_2\text{CH}—$$

Isopropyl

Alkyl groups are named by dropping "-ane" from the parent alkane and adding "-yl." Theoretically, any alkane can be converted to an alkyl group. Some of the more common examples of alkyl groups are given in Table 9.4.

TABLE 9.4 ■ Some Common Alkyl Groups

Name	Condensed Structural Representation
Methyl	$CH_3—$
Ethyl	$CH_3CH_2—$ or $C_2H_5—$
Propyl	$CH_3CH_2CH_2—$ or $C_3H_7—$
Isopropyl	$CH_3CH(CH_3)—$ or $(CH_3)_2CH—$
Butyl	$CH_3CH_2CH_2CH_2—$ or $C_4H_9—$
t-Butyl*	$CH_3C(CH_3)_2—$ or $(CH_3)_3C—$

* *t* stands for *tertiary*, sometimes abbreviated *tert*, which means that the central C atom is bonded to three other C atoms.

Many structural arrangements will be possible. *(Eamonn McNulty/SPL/Photo Researchers)*

The number of structural isomers predicted for C_6H_{14}, C_7H_{16}, and C_8H_{18} is 5, 9, and 18, respectively. Every predicted isomer, *and no more*, has been isolated and identified for these hydrocarbons. The large number of structural isomers illustrates the complexity and variety organic chemistry can have even for simple hydrocarbons.

Naming Branched-Chain Alkanes

Many alkanes and other organic compounds have both common names and systematic names. Why are both common and systematic names used? Usually the common name came first and is widely known. Many consumer products are labeled with the common name, and when only a few isomers are possible, the common name adequately identifies the product for the consumer. For example, "isobutane," the common name for methylpropane, is sufficient because there is only one branched-chain isomer possible for C_4H_{10}. However, a system of common names quickly fails when several structural isomers are possible.

You have probably heard of the "octane rating" of gasoline. Octane (C_8H_{18}) has 18 possible isomers. One of these isomers, 2,2,4-trimethylpentane,

$$\overset{1}{CH_3}—\overset{2}{C}(CH_3)_2—\overset{3}{CH_2}—\overset{4}{CH}(CH_3)—\overset{5}{CH_3}$$

2,2,4-Trimethylpentane

■ Rules for naming organic compounds are given in Appendix D.

is used as a standard in assigning octane ratings of various gasolines. In this case, a common name such as isooctane would not provide enough information about which isomer was actually being used as the standard. However, the systematic name provides complete information. The "pentane" part, which means a straight five-carbon chain, identifies the longest chain in the molecule. The numbers "2,2,4-" indicate the locations of the three groups attached to the pentane chain, and "tri" is used as a prefix for "methyl" to indicate that all three groups are methyl groups.

■ The numbers in an organic compound name are locators. They give the address of a group along the spine of the molecule.

EXAMPLE 9.1 *Structural Isomers*

Three isomers are possible for the isomeric pentanes (C_5H_{12}). Draw condensed formulas for these isomers.

SOLUTION

A good plan to follow in drawing all possible isomers—no more and no less—is to start with the straight-chain isomer and then remove one methyl group at a time, placing that methyl group on the remaining chain and checking all possibilities before removing another methyl group. In this case, start with the straight-chain five-carbon pentane as the first isomer.

■ You can simplify constructing possible isomers if you follow a systematic process.

$$CH_3CH_2CH_2CH_2CH_3$$

Condensed formula: pentane

Removing one methyl and placing it on the second carbon gives a second isomer, 2-methylbutane. Convince yourself that this is the only possible one with one methyl attached to a four-carbon chain, since putting the methyl group on the next C gives the same isomer. The third possible isomer is obtained by removing a second methyl and placing it on the second C to give 2,2-dimethylpropane.

$$\begin{array}{c} CH_3 \\ | \\ CH_3CHCH_2CH_3 \end{array} \qquad \begin{array}{c} CH_3 \\ | \\ CH_3CCH_3 \\ | \\ CH_3 \end{array}$$

Condensed formula: 2-methylbutane   Condensed formula: 2,2,-dimethylpropane

Exercise 9.1

Draw the condensed structural formulas for the following compounds: (a) 2-methylpentane, (b) 3-methylpentane, (c) 2,2-dimethylbutane, and (d) 2,3-dimethylbutane.

9.3 Alkenes and Alkynes: Reactive Cousins of Alkanes

Petroleum contains alkenes, and their presence in gasoline raises the octane rating (Section 9.7). Alkenes for use in commercial applications are also obtained from petroleum by a cracking process (Section 9.7). Ethene, best

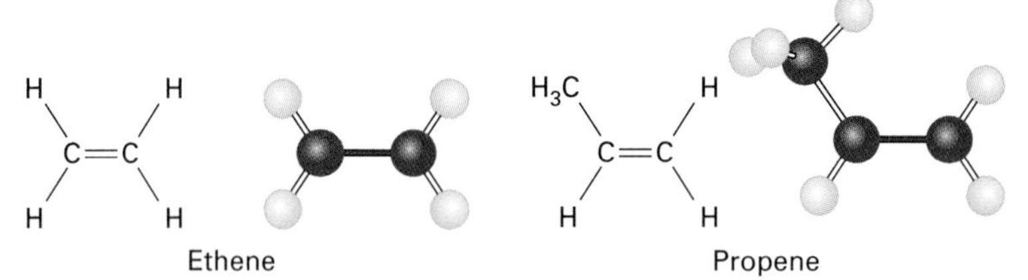

Figure 9.6 The two smallest alkenes: ethene, commonly known as ethylene, and propene, commonly known as propylene.

■ The "-ene" suffix is used for hydrocarbons with one or more double bonds.

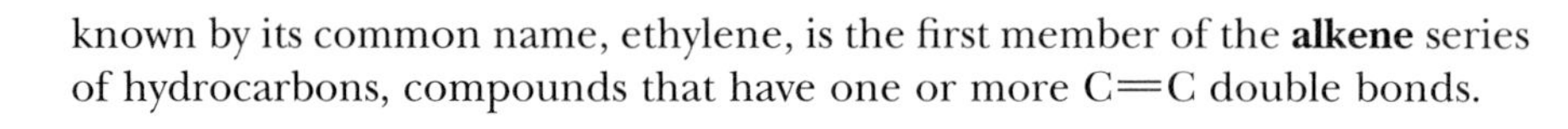

known by its common name, ethylene, is the first member of the **alkene** series of hydrocarbons, compounds that have one or more C=C double bonds.

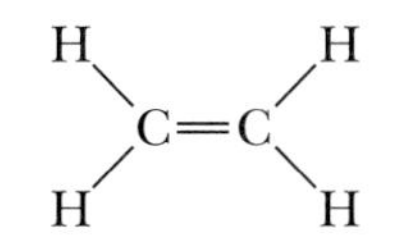

Green tomatoes. On their way to market these tomatoes may be ripened by exposure to ethylene gas. *(Jan Halaska/Photo Researchers)*

More ethylene is manufactured each year than any other organic chemical. In 1996, ethylene was the top organic chemical produced in the United States, with the manufacture of more than 24 million tons. Much of the ethylene is used in the production of polyethylene (Section 10.5).

Ethylene is also found in plants, where it is a hormone that controls seedling growth and regulates fruit ripening. The discovery of this property led to the use of ethylene by food processors for ripening fruits and vegetables after harvest.

The general formula for alkenes with one double bond is C_nH_{2n}. The second member of the alkene series is propene (propylene). Propylene is manufactured in large quantities for use in the production of polypropylene (Section 10.5). Ball-and-stick models for ethene and propene are shown in Figure 9.6. Unlike alkanes, which undergo few chemical reactions easily (except for combustion), alkenes are quite reactive. The site of the chemical change is usually the double bond. This reactivity is essential to making many kinds of plastics, as discussed in Section 10.5. Also, addition of hydrogen to the double-bond carbon atoms converts unsaturated fats and oils, which are liquids, to the more solid consistency needed for margarine (Section 11.3).

■ Propylene was number 6 in manufactured chemical production in the United States in 1996.

Structural Isomers of Alkenes

In the alkene series, the possibility of locating the double bond between two different carbon atoms creates additional structural isomers. Ethene and propene have only one possible location for the double bond. However, the next alkene in the series, butene, has two possible locations for the double bond.

```
   H  H  H  H              H  H  H  H
   |  |  |  |              |  |  |  |
H—C1=C2—C3—C4—H         H—C1—C2=C3—C4—H
         |  |              |        |
         H  H              H        H
     1-Butene                2-Butene
```

When groups such as methyl or ethyl are attached to carbon atoms in an alkene, the longest hydrocarbon chain is numbered from the end that will give the double bond the lowest number, and then numbers are assigned to the attached group. For example, in the following compound the longest chain has seven carbons (heptene); the double bond is between C2 and C3 (2-heptene); and the three (tri-) methyl groups are on the third, fourth, and sixth carbons (3,4,6-trimethyl-).

```
      H  H  CH3  CH3  H  CH3  H
      |  |   |    |   |   |   |
   H—C1—C2=C3———C4———C5—C6———C7—H
      |              |   |   |   |
      H              H   H   H   H
```

Hence, the name is 3,4,6-trimethyl-2-heptene.

EXAMPLE 9.2 *Drawing Alkenes*

Draw the structure of 2,3-dimethyl-2-pentene.

SOLUTION

First, draw the parent alkene and put the double bond in the correct location. In this case, 2-pentene is the parent.

```
     |   |   |   |   |
   —C1—C2=C3—C4—C5—
     |           |   |
```

Then place the alkyl groups on the appropriate carbons. In this case, there are two methyl groups, one on C2 and one on C3. Remember that numbering the double bond takes precedence. Also check your drawing to make sure you don't have more than four bonds per carbon.

```
      H  CH3   CH3   H  H
      |   |     |    |  |
   H—C—C=====C———C—C—H
      |              |  |
      H              H  H
```

Exercise 9.2
Draw the structure of 3-methyl-1-butene.

Stereoisomerism: *Cis* and *Trans* Isomers in Alkenes

Some alkenes can also have ***cis*** and ***trans*** **isomers**, one of two forms of **stereo-isomerism.** *Here the isomers have the same molecular formulas and the same atom-to-atom bonding sequences, but the atoms differ in their arrangement in space.* The other form of stereoisomerism, optical isomerism, is discussed in Section 11.1.

An important difference between alkanes and alkenes is the degree of flexibility of the carbon–carbon bonds in the molecules. Rotation around single carbon–carbon bonds in alkanes occurs readily at room temperature, but the carbon–carbon double bond in alkenes is strong enough to prevent free rotation about the bond. Consider ethene (C_2H_4). Its six atoms lie in the same plane, with bond angles of approximately 120°.

$$\begin{matrix} H & & H \\ & C{=}C & \\ H & & H \end{matrix}$$

If two methyl groups replace two hydrogen atoms, one on each carbon atom of ethene ($H_2C{=}CH_2$), the result is 2-butene ($CH_3CH{=}CHCH_3$). Experimental evidence confirms the existence of two compounds with the same set of bonds. The difference in the two compounds is in the location in space of the two methyl groups: the ***cis*** isomer has two methyl groups on the same side in the plane of the double bond and the ***trans*** isomer has two methyl groups on opposite sides of the double bond. Note that the properties of the *cis* and *trans* isomers of 2-butene are quite different.

■ If free rotation occurred around a carbon–carbon double bond, these two compounds would be the same.

■ The health risks of ***trans*** fatty acids are discussed in Section 11.3.

$$\begin{matrix} H & & H \\ & C{=}C & \\ H_3C & & CH_3 \end{matrix} \qquad \begin{matrix} H & & CH_3 \\ & C{=}C & \\ H_3C & & H \end{matrix}$$

	***cis*-2-Butene**	***trans*-2-Butene**
Melting point	−138.9°C	−105.5°C
Boiling point (1 atm)	3.7°C	0.9°C
Density (at 20°C)	0.621 g/mL	0.604 g/mL

The third possible isomer, 1-butene (a structural isomer of the *cis* and *trans* isomers), does not have *cis* and *trans* structures. Since one carbon atom has two identical groups (H atoms), its properties are different from those of the 2-butene isomers.

■ Many other *cis* and *trans* isomers are possible. For example, *cis*-1,2-dichloroethene and *trans*-1,2-dichloroethene are possibilities when one hydrogen atom on each carbon atom of ethene is replaced with a chlorine atom.

$$\begin{matrix} H & & H \\ & C{=}C & \\ H & & CH_2CH_3 \end{matrix}$$

	1-Butene
Melting point	−185.3°C
Boiling point (1 atm)	−6.3°C
Density (at 20°C)	0.595 g/mL

FRONTIERS IN THE WORLD OF CHEMISTRY

Organic Metals

Organic compounds are generally good insulators, while metals conduct electricity. However, researchers have been successful in making organic compounds that are conductors. Acetylene can be polymerized in the presence of a catalyst to polyacetylene, a typical plastic that does not conduct electricity.

$$2n\,\text{H—C}\equiv\text{C—H} \xrightarrow{\text{Catalyst}} \left(\begin{matrix}\text{H} & & \text{H} & & \text{H} & & \text{H} \\ | & & | & & | & & | \\ \text{C} & = & \text{C} & — & \text{C} & = & \text{C}\end{matrix}\right)_n$$

This polymer appears as a black powder in the usual laboratory preparation and received little attention prior to 1970. In that year a Korean university student, having trouble understanding his Japanese instructor, Hidek Shirakawa, prepared the polymer using an excessive amount of the catalyst. The result was a silver film that looked more like a metal than anything else. Furthermore, the polyacetylene film conducted electricity, which was a first for plastic materials.

The conductance of the shiny polyacetylene film can be explained in terms of very long polymer molecules that are lined up in a crystalline structure. Electric charges are more readily passed along the alternating double bonds in the polyacetylene molecules that are ordered in one direction. Evidence of this is the observation that conductance is greater along the aligned chains than perpendicular to the chains. The black-powder form is an insulator because the long chains of polyacetylene are jumbled in a random fashion, which prevents conductance.

In 1975, at the University of Pennsylvania, Alan MacDiarmid began a systematic study of this new form of polyacetylene. Adding small amounts of iodine (similar to the doping of semiconductors, Section 14.4) during the preparation of polyacetylene increases the electric conductivity of the plastic 10^{12} times, or a trillionfold. This rivals the conductivity of metals.

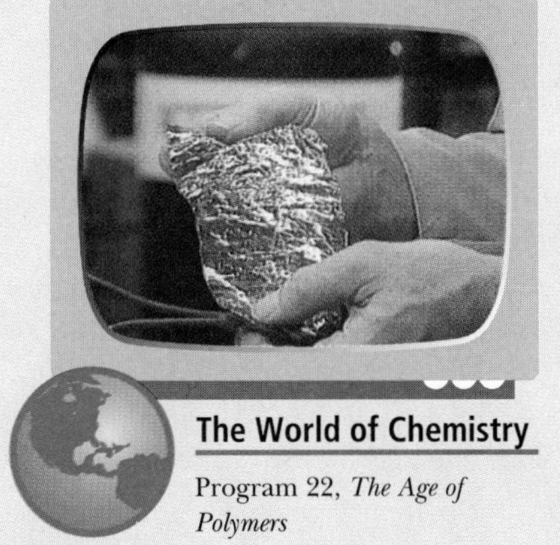

The World of Chemistry

Program 22, *The Age of Polymers*

A piece of electrically conducting polyacetylene film.

References R. B. Kaner and A. G. MacDiarmid: Plastics that conduct electricity. *Scientific American*, Vol. 258, pp. 106–111, February 1988.

R. Dagani: Organic metals, *Chemical and Engineering News*, pp. 8–9, August 31, 1992.

Cis–trans isomerism in alkenes is possible only when both of the double-bond carbon atoms have two different groups.

The **alkynes** have one or more triple bonds (—C≡C—) per molecule and have the general formula C_nH_{2n-2}. The simplest one is ethyne, commonly called acetylene (C_2H_2) (Section 5.3). The naming of alkynes is similar to that of alkenes, with the lowest number possible being used for locating the triple bond.

$$\begin{matrix} & \text{H} & & & & & \text{CH}_3 & & \text{H} & \\ & | & & & & & | & & | & \\ \text{H—} & \underset{1}{\text{C}} & \overset{180°}{—} & \underset{2}{\text{C}} & \equiv & \underset{3}{\text{C}} \overset{180°}{—} & \underset{4}{\text{C}} & — & \underset{5}{\text{C}} & \text{—H} \\ & | & & & & & | & & | & \\ & \text{H} & & & & & \text{H} & & \text{H} & \end{matrix}$$

The 180° bond angles around the triple bond make the C—C≡C—C section of the molecule linear.

The name of the preceding compound is 4-methyl-2-pentyne. As with alkenes, changing the location of the multiple bond produces an isomer. For example, 4-methyl-1-pentyne is a different compound than 4-methyl-2-pentyne.

$$\begin{array}{ccccccccccc} & & & & \mathrm{H} & & \mathrm{CH_3} & & \mathrm{H} & & \\ & & & & | & & | & & | & & \\ \mathrm{H}- & \underset{1}{\mathrm{C}} & \equiv & \underset{2}{\mathrm{C}} & -\underset{3}{\mathrm{C}}- & & \underset{4}{\mathrm{C}} & - & \underset{5}{\mathrm{C}} & -\mathrm{H} \\ & & & & | & & | & & | & & \\ & & & & \mathrm{H} & & \mathrm{H} & & \mathrm{H} & & \end{array}$$

However, *cis* and *trans* isomers are not possible for alkynes because the geometry around the triple-bond carbon atoms is linear.

■ SELF-TEST 9A

1. Which fossil fuel furnishes the most heat energy per gram? (a) Coal, (b) Petroleum, or (c) Natural gas.
2. Which fuel furnishes the most heat energy per gram? (a) Natural gas, (b) Hydrogen, or (c) Coal.
3. All combustions of fossil fuels give off energy. (a) True, (b) False.
4. Hydrocarbons react with ______________ to produce CO_2 and ______________.
5. Each carbon in a saturated hydrocarbon has ______________ geometry.
6. ______________ is the first member of the alkene series.
7. ______________ is the first member of the alkyne series.
8. The formula for the ethyl group is ______________.
9. Butane and 2-methylpropane are examples of ______________ isomers.
10. The number-one organic chemical produced in the United States is ______________.
11. The rigidity of the carbon–carbon double bond allows for the possibility of ______________ isomers.

9.4 The Cyclic Hydrocarbons

Hydrocarbons can form rings as well as straight chains and branched chains. Two important classes of cyclic hydrocarbons found in petroleum and coal are the cycloalkanes and the aromatics.

Cycloalkanes

Cycloalkanes are saturated hydrocarbons with ring structures. The simplest cycloalkane is cyclopropane, a highly strained ring compound.

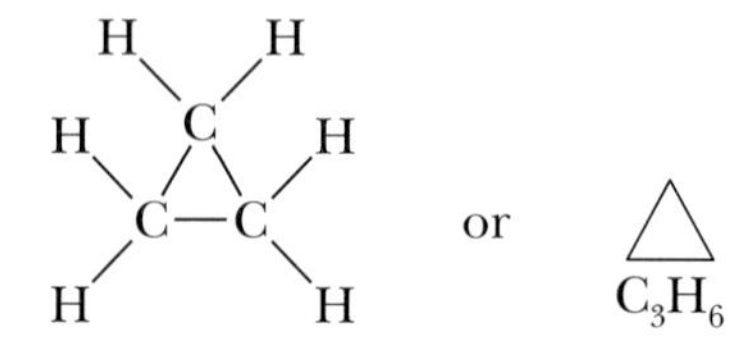

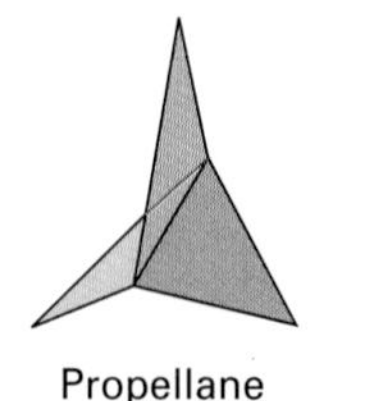
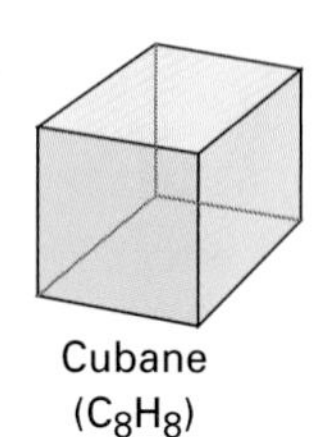

Propellane (C_6H_8) Cubane (C_8H_8)

Two small, strained-ring compounds with appropriate names.

The ring is strained because of the 60° angles in the ring; angles larger than 90° show a much greater stability. Cyclopropane, a volatile, flammable gas (bp −32.7°C), is a rapidly acting anesthetic. A cyclopropane–oxygen mixture is useful in surgery on babies, small children, and "bad risk" patients because of its rapid action and the rapid recovery of the patient. Helium gas is added to the cyclopropane–oxygen mixture to reduce the danger of explosion in the operating room.

The cycloalkanes are commonly represented by polygons in which each corner represents a carbon atom and two hydrogen atoms and the lines represent C—C bonds. The C—H bonds are not shown, but are understood. Other common cycloalkanes include cyclobutane (C_4H_8), cyclopentane (C_5H_{10}), and cyclohexane (C_6H_{12}). These are represented as

Aromatic Compounds

Hydrocarbons containing one or more benzene rings (Figure 9.7) are called **aromatic compounds**. The word "aromatic" was derived from "aroma," which describes the rather strong and often pleasant odor of these compounds. Benzene and most other aromatic compounds, however, are toxic and often **carcinogenic.**

■ Carcinogens are cancer-causing agents. The type of cancer caused may vary from one carcinogen to another. Benzene causes a form of leukemia, and benzopyrene causes skin cancer and lung cancer.

The main structural feature, which is responsible for the distinctive chemical properties of the aromatic compounds, is the six-carbon benzene ring. Figure 9.7*a* illustrates "smearing" of the bonding electrons above and below the plane of the ring. In other words, all the carbon–carbon bonds are equivalent and benzene is a planar molecule. The "smearing" of electrons around

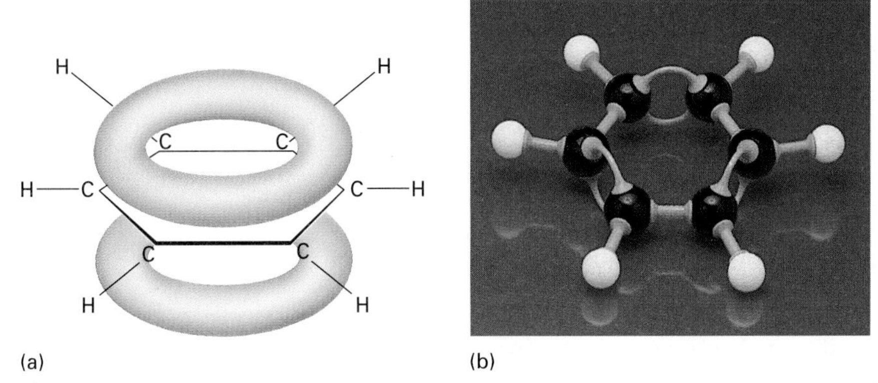

Figure 9.7 Benzene, the smallest aromatic compound. (a) The equal distribution of bonding electrons around the ring can be represented by electron clouds above and below the plane of the ring. (b) Another way to represent the bonding electrons of benzene is as alternating double and single bonds, shown here in a ball-and-stick model. Because all bonds in benzene are the same, (a) is a more correct representation. *(b, C. D. Winters)*

the ring makes benzene and other aromatic compounds less reactive than alkenes. Benzene can be represented as

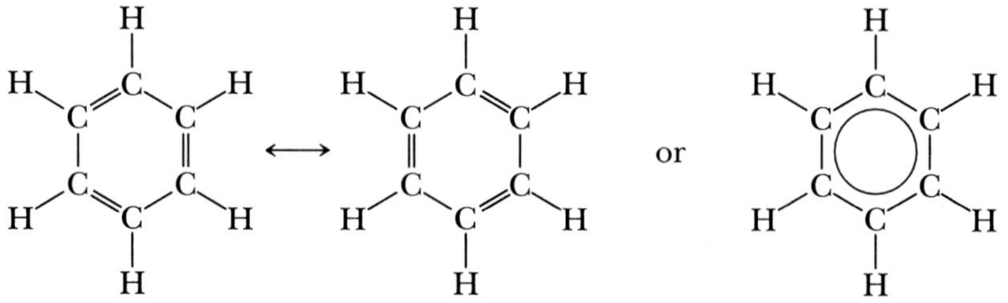

H H H
H C H H C H H C H
C C C C C C
↔ or
C C C C C C
H C H H C H H C H
H H H

where the circle represents the evenly distributed, smeared electrons.

When hydrogen and carbon atoms are not shown, benzene is represented by a circle in a hexagon. Each corner in the hexagon represents one carbon atom and one hydrogen atom. Remember that this symbol stands for C_6H_6

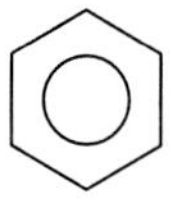

and a hexagon without a circle stands for cyclohexane (C_6H_{12}).

Derivatives of Benzene

Benzene and many of its derivatives are among the most widely used chemicals (see Figure 10.2) because of their use in the manufacture of plastics, detergents, pesticides, drugs, and other organic chemicals. Several important derivatives are monosubstituted benzenes, with one atom or group replacing one of the hydrogen atoms. For example, substitution of a methyl group for one of the hydrogen atoms in benzene gives methylbenzene, usually called toluene, a common solvent. Ethylbenzene, another common chemical, contains an ethyl group substituted for one of the hydrogen atoms of benzene.

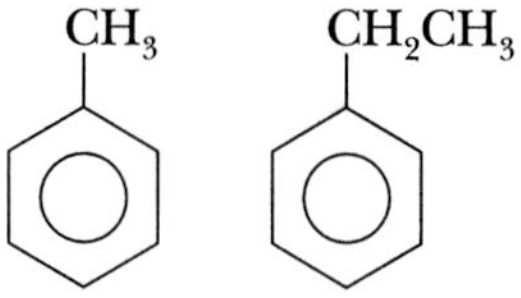

CH_3 CH_2CH_3

Toluene Ethylbenzene

Structural Isomers of Aromatic Compounds

■ Benzene, toluene, and xylenes are important because they raise the octane rating of gasoline (Section 9.6).

Since the benzene molecule has a planar structure, structural isomers are possible when two or more groups are substituted for hydrogen atoms on the benzene ring.

Three isomers are possible if two groups are substituted for two hydrogen atoms on the benzene ring. The prefixes *ortho-*, *meta-*, and *para-* or numbers are used to distinguish among the isomers. When the name of the compound is written, usually only the first letter of one of these terms is given. For example, when the two groups are methyl groups, the three isomers are commonly known as *o*-xylene, *m*-xylene, and *p*-xylene.

1,2-dimethylbenzene (*ortho*-xylene) mp −25°C | 1,3-dimethylbenzene (*meta*-xylene) mp −47.9°C | 1,4-dimethylbenzene (*para*-xylene) mp 13.3°C

(The xylenes are used in making dyes, insecticides, and drugs)

Another type of aromatic compound has two or more benzene rings sharing ring edges. Examples include naphthalene, anthracene, benzopyrene, and phenanthrene.

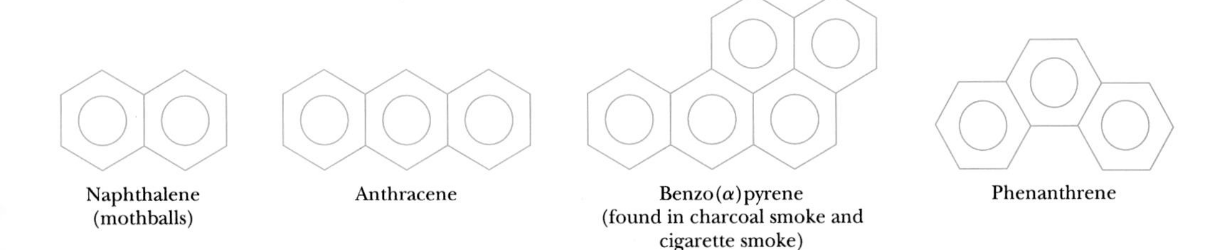

Naphthalene (mothballs) | Anthracene | Benzo(α)pyrene (found in charcoal smoke and cigarette smoke) | Phenanthrene

Many organic compounds found in nature are cyclic hydrocarbons that include both aromatic rings and cycloalkane or cycloalkene rings fused together. Steroids (Section 11.3) are good examples. Chemists who isolate organic compounds from plants and develop methods for making them in the laboratory are called natural product chemists. For example, Percy Julian was the first chemist to synthesize hydrocortisone, a steroid; and physostigmine, a compound useful in the treatment of glaucoma.

One source of aromatic compounds. Both the smoke and the char on the meat contains polycyclic aromatic compounds. *(Hank Morgan/Rainbow)*

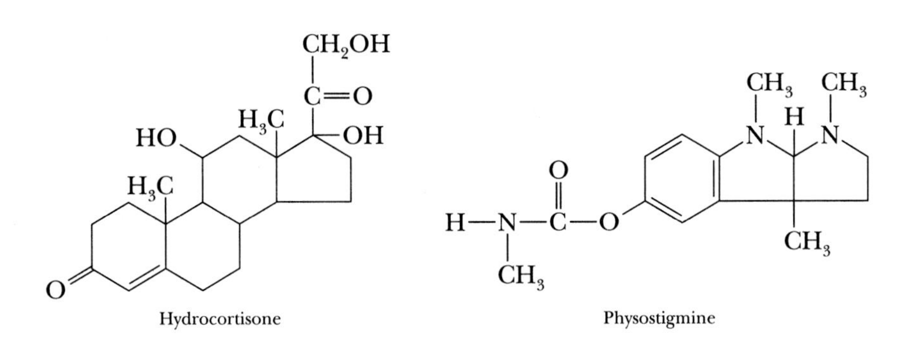

Hydrocortisone | Physostigmine

9.5 Alcohols: Oxygen Comes on Board

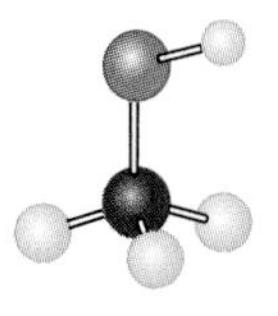

Methanol, CH_3OH

Alcohols

Several alcohols are currently being used as fuels and fuel additives. These include methanol, ethanol, and 2-methyl-2-propanol (commonly known as *tertiary*-butyl alcohol). Alcohols, one of several classes of organic compounds

THE PERSONAL SIDE

Percy Lavon Julian (1899–1975)

Percy Julian's list of achievements reads like that of others who have made it to the top in their professions: a doctorate in chemistry in Vienna in 1931 quickly followed, back home in the United States, by outstanding achievements as a researcher and university professor; 18 years as Director of Research in an industry where he led the way in bringing to market valuable products from soybeans; and the founding of his own research institute, the Julian Laboratories. To grasp the measure of the man, add to this brief outline dozens of scientific publications, over 100 patents granted, numerous academic honors, and positions of responsibility in many civic and humanitarian organizations.

Percy Julian (Chemical Heritage Foundation)

But there were some differences from a successful career path along the way. After completing the eighth grade he had to leave his home in Montgomery, Alabama, for further studies—no more public education was available there for a black man. He enrolled as a "subfreshman" at DePauw University in Indiana. On his first day, a white student welcomed him with a handshake. Julian later related his reaction: "In the shake of a hand my life was changed, I soon learned to smile and act like I believed they all liked me, whether they wanted to or not."

Early in his career, other challenges had to be met. As a successful businessman, he and his family were the first black residents of an upscale suburb of Chicago. There, on Thanksgiving Day in 1950, his home was attacked by arsonists. The Julian family stayed on to become respected and welcome members of the community.

An organic chemist, Julian built his career around the study of chemicals of plant origin, many of them of medicinal value. The synthesis of a complicated natural molecule is a major goal in such work. Julian was first to achieve synthesis of hydrocortisone, now available in every drugstore because of its value in treating allergic skin reactions. He originated the production of soybean protein and the isolation from soybean oil of compounds from which the first synthetic sex hormone (progesterone) could be made.

Julian's talents were evident early when he and a colleague devised a series of nine chemical reactions that produced a compound identical with natural *physostigmine.* Originally isolated from the Calabar bean from Nigeria, physostigmine had already proved valuable for treating glaucoma by lowering fluid pressure in the eye.

■ A **functional group** is an atom or group of atoms within a molecule that gives the substance a characteristic chemical behavior. The —OH group is the **alcohol** functional group. Additional classes of functional groups are discussed in Chapter 10.

that have a characteristic **functional group**, contain one or more —OH groups and have the general formula ROH where R is an alkyl group.

Methanol, (CH_3OH), also called methyl alcohol, can be prepared from a mixture of carbon monoxide and hydrogen known as **synthesis gas**. High pressure, high temperature, and a catalyst are used to increase the yield.

$$\underset{\text{Coal}}{C(s)} + \underset{\text{Steam}}{H_2O(g)} \longrightarrow \underset{\text{Synthesis gas}}{CO(g) + H_2(g)}$$

$$CO(g) + 2\ H_2(g) \xrightarrow[300°C]{ZnO,\ Cr_2O_3} CH_3OH(g)$$

An old method of producing methanol involved heating a hardwood such as beech, hickory, maple, or birch in the absence of air. For this reason, methanol is sometimes called *wood alcohol.* Methanol is highly toxic. Drinking as little as 30 mL can cause death, and smaller amounts (10–15 mL) cause blindness.

Ethanol, (C_2H_5OH), also called ethyl alcohol or grain alcohol, can be obtained by the fermentation of carbohydrates (starch, sugars).

■ The International Union of Pure and Applied Chemistry (IUPAC) names of alcohols include the name of the hydrocarbon to which the alcohol corresponds and indicate the number of carbon atoms; the suffix *-ol* denotes an alcohol. Common names use the name of the alkyl group (represented as R in ROH) attached to —OH. For example, methyl alcohol, ethyl alcohol, and *tertiary*-butyl alcohol are the common names for methanol, ethanol, and 2-methyl-2-propanol.

$$\underset{\text{Glucose}}{C_6H_{12}O_6} \xrightarrow{\text{Yeast}} \underset{\text{Ethanol}}{2\ C_2H_5OH} + 2\ CO_2$$

The yeast contains enzymes that are catalysts for the fermentation process. A mixture of 95% ethanol and 5% water can be recovered from the fermentation products by distillation. Ethanol is the active ingredient of alcoholic beverages. Ethanol is receiving increased attention for use as an alternative fuel and as a fuel additive for oxygenated fuels. At present, most ethanol is used in a blend of 90% gasoline and 10% ethanol (first introduced in the 1970s and known as *gasohol*).

Ethers

Ethers have the general formula R—O—R′ where R and R′ stand for alkyl groups (Table 9.4), which may be the same or different. Methyl-*tertiary*-butyl ether (MTBE) is the most important commercial ether because of its use in oxygenated and reformulated gasolines.

$$CH_3—O—\overset{\displaystyle CH_3}{\underset{\displaystyle CH_3}{\overset{|}{\underset{|}{C}}}}—CH_3$$

MTBE

Before the development of MTBE as an octane enhancer, diethyl ether ($C_2H_5OC_2H_5$), an organic solvent, and methyl propyl ether ($CH_3OCH_2CH_2CH_3$), an anesthetic known as neothyl, were the most common ethers.

■ SELF-TEST 9B

1. The difference between cyclohexane and benzene is the number of ____________ atoms.

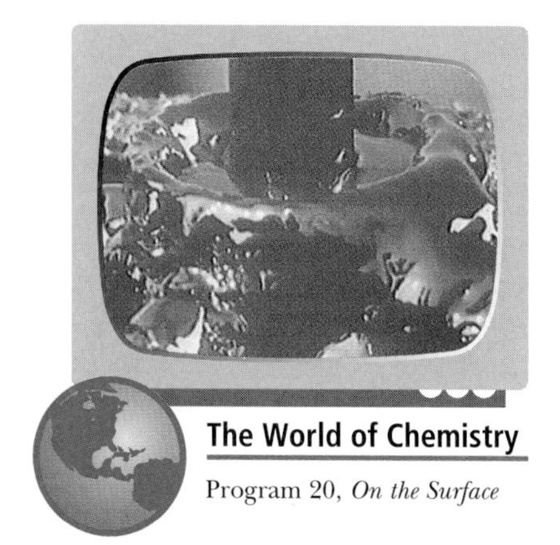

The World of Chemistry

Program 20, *On the Surface*

Pumping crude oil. The yield from a well is improved by adding soaplike molecules, which make the oil flow more freely.

2. How many atoms does the symbol represent?
3. Synthesis gas is a mixture of ________ and ________.
4. ________ structural isomers are possible for trichlorobenzene.
5. ________ structural isomers are possible for dimethylbenzene.
6. Fermentation of carbohydrates yields (a) ethanol, (b) methanol.
7. Ethers are a class of compounds that contain the ________ linkage.

9.6 Petroleum

Crude petroleum is a complex mixture of thousands of hydrocarbon compounds, and the actual composition of petroleum varies with the location in which it is found. For example, Pennsylvania crude oils are primarily straight-chain hydrocarbons, whereas California crude oil is composed of a larger portion of aromatic hydrocarbons.

■ There are 42 gal of oil per barrel.

How long will petroleum be viable as a source of energy and starting materials for consumer products? This is difficult to estimate because of continued revisions of recoverable crude oil resources. For example, estimates of global petroleum reserves increased 43% between 1984 and 1994, primarily because of re-evaluation of oil reserves in the Middle East, where more than 65% of the world's oil resources are located. Counterbalancing this increase in oil reserves is the substantial increase in energy demand expected during the next 20 to 30 years from developing countries in Asia and Latin America. Current projections are for oil production to peak between 2010 and 2025. Oil production would continue for several decades after this; however, the increasing cost and lower availability will favor increased use of natural gas, coal, and alternative sources, such as wind energy and solar energy, particularly electricity obtained through the use of photovoltaic cells.

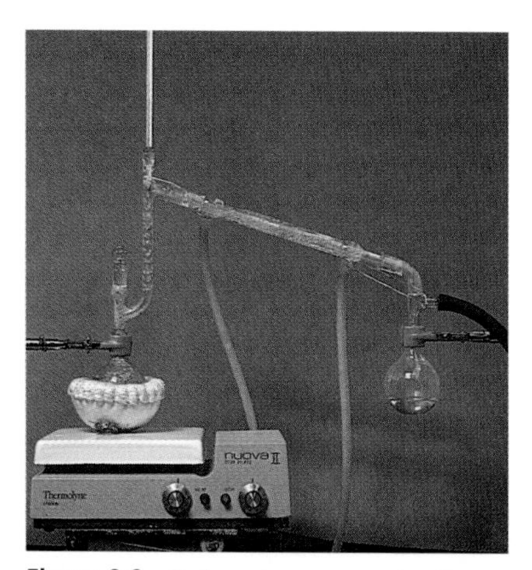

Figure 9.8 Laboratory apparatus for fractional distillation. Vapor from the boiling liquid rises in the vertical column, which has many indentations. As it rises, the vapor is repeatedly condensed and then re-evaporated as it is heated by rising vapors. With each evaporation, the vapor becomes richer in the lowest boiling component of the mixture. The vapor flows down the cooled condenser, and fractions with different boiling points are collected at the end of the condenser. *(J. W. Morgenthaler)*

Petroleum Refining

The refining of petroleum begins with the separation of fractions according to boiling point ranges by a process called **fractional distillation**. The difference between simple distillation and fractional distillation is the degree of separation achieved for the mixture being distilled. For example, water that contains dissolved solids or other liquids can be purified by simple distillation. The impure solution is heated to boiling; the water vapor is condensed and collected in a separate container (Figure 9.8). Since petroleum contains thousands of hydrocarbons, separation of the pure compounds is not feasible or even necessary. The products obtained from distillation of petroleum are still mixtures of hundreds of hydrocarbons; so they are called **petroleum fractions**.

Figure 9.9 illustrates a fractional distillation tower used in the petroleum refining process. The crude oil is first heated to about 400°C to produce a hot vapor and liquid mixture that enters the fractionating tower. The vapor rises and condenses at various points along the tower. The lower boiling petroleum fractions (those that are more volatile) will remain in the vapor stage longer

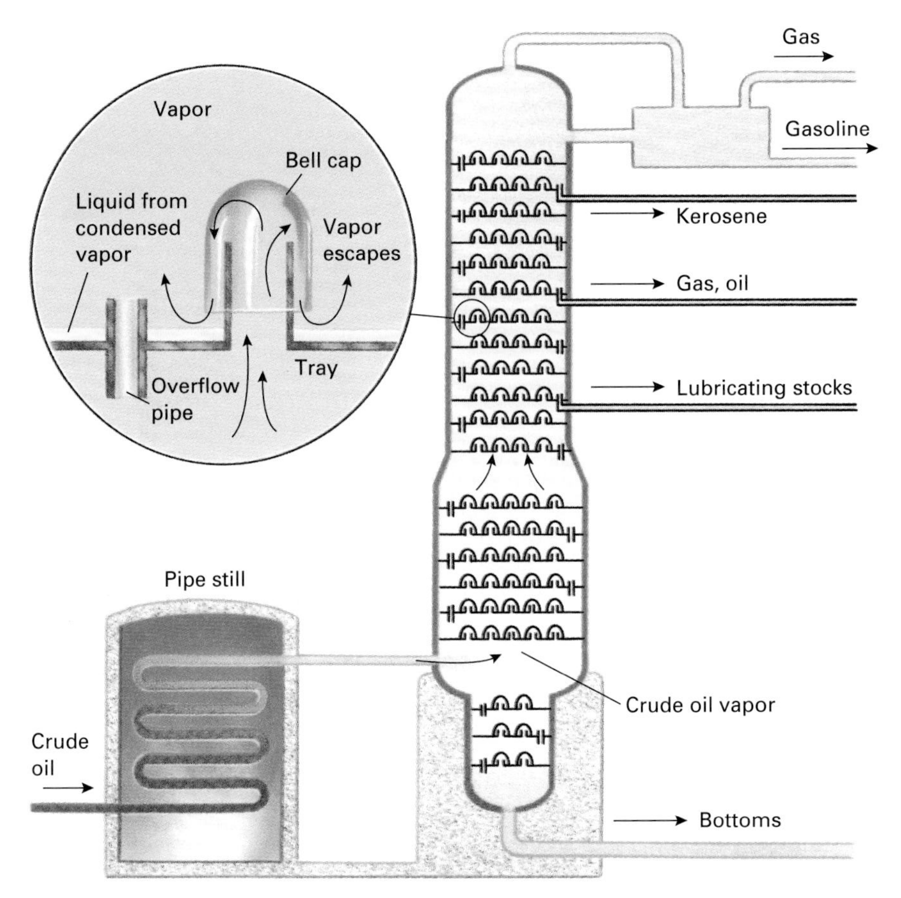

Figure 9.9 Petroleum fractionation. Crude oil is first heated to 400°C in the pipe still. The vapors then enter the fractionation tower. As they rise in the tower, the vapors cool down and condense so that different fractions can be drawn off at different heights. This shows how the rising vapor is repeatedly condensed and collected at the numerous bell caps.

than the higher boiling fractions. These differences in boiling point ranges allow the separation of fractions. Some of the gases do not condense and are drawn off the top of the tower, while the unvaporized residual oil is collected at the bottom of the tower. Typical products of the fractional distillation of petroleum are listed in Table 9.5.

■ Petroleum fractions are mixtures of hundreds of hydrocarbons with boiling points in a certain range.

TABLE 9.5 ■ Hydrocarbon Fractions from Petroleum

Fraction	Size Range of Molecules	Boiling Point Range (°C)	Uses
Gas	C_1–C_4	0–30	Gas fuels
Straight-run gasoline	C_5–C_{12}	30–200	Motor fuel
Kerosene	C_{12}–C_{16}	180–300	Jet fuel, diesel oil
Gas–oil	C_{16}–C_{18}	Over 300	Diesel fuel, cracking stock
Lubricants	C_{18}–C_{20}	Over 350	Lubricating oil, cracking stock
Paraffin wax	C_{20}–C_{40}	Low-melting solids	Candles, wax paper
Asphalt	Above C_{40}	Gummy residues	Road asphalt, roofing tar

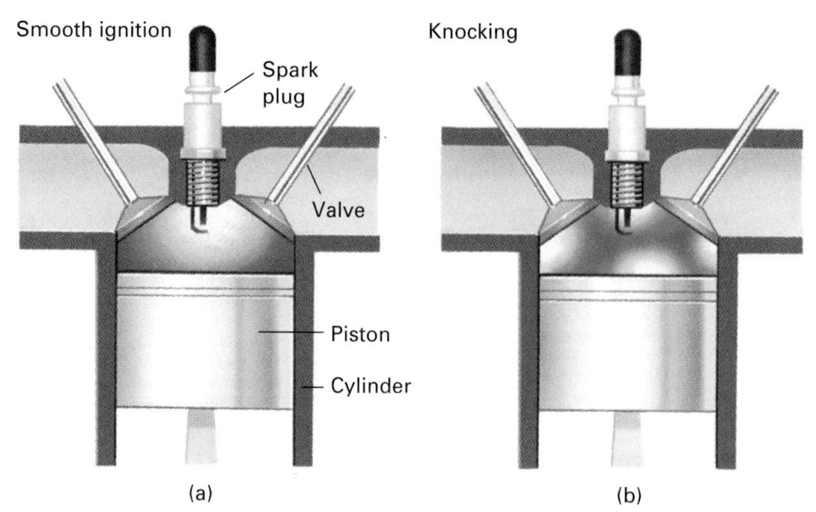

Octane rating relates to smoothness of ignition.

Octane Rating

The "straight-run" gasoline fraction obtained from the fractional distillation of petroleum has an octane rating of only 55 and needs additional refinement because it contains primarily straight-chain hydrocarbons that burn too rapidly to be suitable for use as a fuel in internal combustion engines. Rapid burning causes uncontrolled explosion of the fuel as evidenced by a "knocking" or "pinging" sound in the engine. This reduces engine power and may damage the engine.

Burning properties of hydrocarbons depend on their structure. Isooctane (2,2,4-trimethylpentane) is the standard used to assign octane ratings. The octane rating is an arbitrary scale for rating the relative knocking properties of gasolines, and it is based on the operation of a standard test engine. Heptane knocks considerably and is assigned an octane rating of 0, while 2,2,4-trimethylpentane burns smoothly and is assigned an octane rating of 100. The octane rating of a gasoline is determined by first using the gasoline in a standard engine and recording its knocking properties. The test results are then compared with the behavior of mixtures of heptane and isooctane, and the percentage of isooctane in the mixture with identical knocking properties is called the **octane rating** of the gasoline. Thus, if a gasoline has the same knocking characteristics as a mixture of 13% heptane and 87% isooctane, it is assigned an octane rating of 87. This corresponds to regular unleaded gasoline. Other higher grades of gasoline available at gas stations have octane ratings of 89 (regular plus) and 92 (premium).

Typical octane ratings for gasoline available at gas stations. *(C. D. Winters)*

The octane rating of a gasoline can be increased either by increasing the percentage of branched-chain and aromatic hydrocarbon fractions or by adding octane enhancers (or a combination of both). Since the octane rating scale was established, fuels superior to isooctane have been developed, so that the scale has been extended well above 100. Table 9.6 lists octane ratings for some hydrocarbons and octane enhancers.

TABLE 9.6 ■ Octane Numbers of Some Hydrocarbons and Gasoline Additives

Name	Octane Number
Octane	−20
Heptane	0
Pentane	62
1-Pentene	91
2,2,4-Trimethylpentane (isooctane)	100
Benzene	106
Methanol	107
Ethanol	108
Tertiary-butyl alcohol	113
Methyl *tertiary*-butyl ether (MTBE)	116
Para-xylene	116
Toluene	118

Catalytic Re-Forming

The **catalytic re-forming process** is used to increase the octane rating of straight-run gasoline by converting straight-chain hydrocarbons to branched-chain hydrocarbons and aromatics. This is accomplished by using certain catalysts, such as finely divided platinum on a support of Al_2O_3.

$$\underset{\substack{\text{Pentane} \\ \text{(octane rating 62)}}}{CH_3CH_2CH_2CH_2CH_3} \xrightarrow{\text{Catalyst}} \underset{\substack{\text{2-Methylbutane} \\ \text{(octane rating 94)}}}{CH_3\overset{\displaystyle CH_3 \atop |}{C}HCH_2CH_3}$$

In this process, straight-chain hydrocarbons with low octane numbers can be re-formed into their branched-chain isomers, which have higher octane numbers. Catalytic re-forming is also used to produce aromatic hydrocarbons such as benzenes, toluene, and xylenes by using different catalysts and petroleum mixtures. For example, when the vapors of straight-run gasoline, kerosene, and light oil fractions are passed over a copper catalyst at 650°C, a high percentage of the original material is converted into a mixture of aromatic hydrocarbons, from which benzene, toluene, xylenes, and similar compounds may be separated by fractional distillation. This process can be represented by the equation for converting hexane into benzene.

■ Review the discussion of catalysts in Section 6.3.

$$\underset{\substack{\text{Hexane} \\ \text{(octane rating 25)}}}{CH_3CH_2CH_2CH_2CH_2CH_3} \xrightarrow{\text{Catalyst}} \underset{\substack{\text{Benzene} \\ \text{(octane rating 106)}}}{C_6H_6} + 4\ H_2$$

The catalytic re-forming process is also a major source of hydrogen gas.

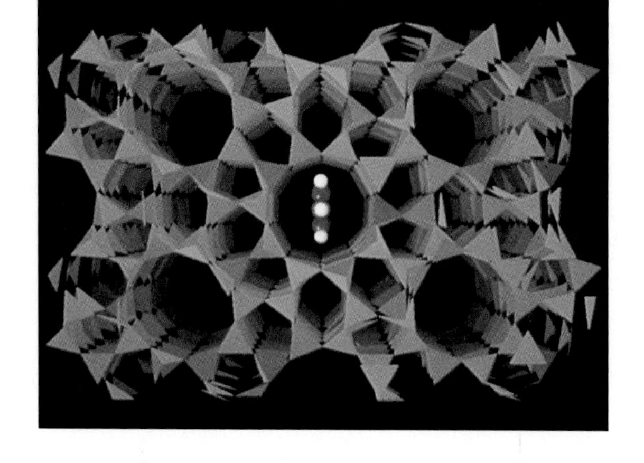

Figure 9.10 The framework of a zeolite catalyst, a network of aluminum, silicon, and oxygen atoms. The dimensions of the zeolite channels are similar to the sizes of the hydrocarbon molecules in gasoline. During the cracking process, the larger molecules of high-boiling petroleum fractions are converted to small hydrocarbons in the gasoline fraction. A model of the benzene molecule is shown in the center of the catalyst. *(Mobil Corporation)*

■ Cracking breaks larger molecules into smaller ones.

Catalytic Cracking

Part of the petroleum refinement process involves adjusting the percentage of each hydrocarbon fraction to match commercial demand. For example, the demand for gasoline is higher than that for kerosene. As a result, chemical reactions convert the larger kerosene-fraction molecules into molecules in the gasoline range in a process called "cracking." The **catalytic cracking process** uses a zeolite catalyst (Figure 9.10) and involves heating saturated hydrocarbons under pressure in the absence of air (Figure 9.11). The hydrocarbons break into shorter chain hydrocarbons—both alkanes and alkenes, some of which will be in the gasoline range.

$$\underset{\text{An alkane}}{C_{16}H_{34}} \xrightarrow[\text{Heat}]{\text{Pressure}} \underset{\substack{\text{An alkane} \\ \text{in the gasoline range}}}{C_8H_{18}} + \underset{\text{An alkene}}{C_8H_{16}}$$

The World of Chemistry

Program 22, *The Age of Polymers*

Figure 9.11 Catalytic cracking unit at a petroleum refinery.

Since alkenes have a higher octane rating than alkanes, the catalytic cracking process also increases the octane rating of the mixture. Catalytic cracking is also important for the production of alkenes used as starting materials in the organic chemical industry.

■ Approximately one third of the compounds in refined gasoline are aromatic compounds.

Octane Enhancers

The octane number of a given blend of gasoline can also be increased by adding antiknock agents, or octane enhancers. Prior to 1975, the most widely used antiknock agent was tetraethyllead, $(C_2H_5)_4Pb$. The addition of 3 g of $(C_2H_5)_4Pb$ per gallon increases the octane rating by 10 to 15. Before the Environmental Protection Agency (EPA) required reductions in lead content, both regular and premium gasoline contained an average of 3 g of $(C_2H_5)_4Pb$ or tetramethyllead, $(CH_3)_4Pb$, per gallon. However, in the Clean Air Act of 1970 Congress required that 1975-model cars emit no more than 10% of the carbon monoxide and hydrocarbons emitted by 1970 models. The platinum-based catalytic converter chosen to reduce emissions of carbon monoxide and hydrocarbons required lead-free gasolines, since lead deactivates the platinum catalyst by coating its surface. For this reason, new automobiles manufactured since 1975 have been required to use lead-free gasoline to protect the catalytic converter.

■ As little as two tankfuls of leaded gasoline can destroy the activity of a catalytic converter.

Since tetraethyllead can no longer be used, other octane enhancers are being added to gasoline to increase the octane rating. These include benzene, toluene, xylenes, 2-methyl-2-propanol, MTBE, methanol, and ethanol. The most popular octane enhancer is MTBE.

■ A side benefit of the removal of lead from gasolines has been a decrease of emissions of this toxic element into the environment.

Oxygenated and Reformulated Gasolines

The 1990 amendments to the Clean Air Act require cities with excessive levels of ozone and carbon monoxide pollution to use oxygenated and reformulated gasolines to reduce hydrocarbon and toxic compound emissions (Section 16.7). **Oxygenated gasolines** are blends of gasoline with organic compounds that contain oxygen, such as MTBE, methanol, ethanol, and 2-methyl-2-propanol (*tertiary*-butyl alcohol). The oxygenated gasolines can be produced either by blending in additives such as MTBE at the refinery or by adding ethanol or methanol at the distribution terminals.

Reformulated gasoline is gasoline whose composition has been changed to reduce the percentage of unsaturated hydrocarbons, aromatics, volatile components, and sulfur; and to add oxygenated additives such as MTBE. This requires significant changes in the refining process, which makes reformulated gasoline more expensive to produce.

■ The difference between oxygenated gasoline and reformulated gasoline is in the refining process. Oxygenated gasoline is produced by adding oxygenated organic compounds to refined gasoline. Reformulated gasoline requires changes in the refining process to alter the percentage composition of the different types of hydrocarbons, particularly olefins and aromatics.

Oxygenated gasolines are required to be used during the four winter months in cities that have serious carbon monoxide pollution (Section 16.10). All gasolines sold in the 41 cities listed in Table 9.7 must contain enough oxygenated organic compounds to provide an average of 2.7% oxygen by weight. Oxygenated gasolines ignite more easily and burn more cleanly, which reduces the need for the fuel-rich operating conditions otherwise required for ignition in winter, and this reduces carbon monoxide emissions. However, oxygenated fuels yield less energy per gram. Use of oxygenated gasolines is estimated to reduce carbon monoxide emissions by 17%.

■ Sulfur in gasoline coats the catalytic converter and reduces its ability to catalyze full combustion of the fuel. This causes an increase in carbon monoxide emissions.

TABLE 9.7 ■ Cities Using Oxygenated Gasolines During Winter Months

Alaska: Anchorage, Fairbanks	*Maryland:* Baltimore
Arizona: Phoenix, Tucson	*Minnesota–Wisconsin:* Minneapolis–St. Paul
California: Chico, Fresno, Los Angeles–Anaheim–Riverside, Modesto, Sacramento, San Diego, San Francisco–Oakland–San Jose, Stockton	*Montana:* Missolula
Colorado: Colorado Springs, Denver–Boulder, Fort Collins–Loveland	*Nevada:* Las Vegas, Reno
Connecticut: Hartford–New Britain–Middletown	*New Mexico:* Albuquerque
New York: New York metropolitan area, including northern New Jersey; Syracuse	*Ohio:* Cleveland–Akron–Lorain
North Carolina: Greensboro–Winston Salem–High Point, Raleigh–Durham	*Oregon-Washington:* Grant's Pass, Klamath County, Medford, Portland–Vancouver, Seattle–Tacoma, Spokane
	Philadelphia–Trenton–Wilmington (Delaware) metropolitan area
	Texas: El Paso
	Utah: Provo–Orem
	Washington, D.C. metropolitan area

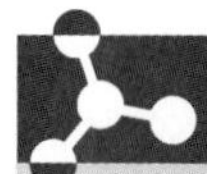

SCIENCE AND SOCIETY

Reformulated Gasoline Regulations

A ruling issued by the Environmental Protection Agency on June 30, 1994 requires that 30% of the oxygenated organic compounds (called oxygenates) used in reformulated gasoline come from renewable sources. Since most gasoline producers have been using MTBE—which is made from methanol—to meet the 1990 Clean Air Act, the new mandate would require more use of ethanol, a renewable resource, and ethyl *tert*-butyl ether (ETBE), which is made from ethanol. Proponents of the mandate argue that the use of ethanol will reduce reliance on oil imports, cut farmers' reliance on federal farm subsidies, and increase the use of renewable resources.

In response to the EPA ruling, the American Petroleum Institute and the National Petroleum Refiners Association filed suit in the U.S. Court of Appeals for the District of Columbia calling for the court to set aside the EPA ruling and institute a stay to prevent the agency from implementing the mandate. The suit accuses EPA of violating the regulatory process by ruling in favor of ethanol producers and farmers who grow the corn used to produce ethanol. The suit also alleges that EPA is ignoring environmental and economic factors related to the use of ethanol as an oxygenate in reformulated fuels. On April 28, 1995, the U.S. Court of Appeals approved the petition by issuing a stay prohibiting the EPA from requiring the use of renewable oxygenates in reformulated gasolines.

Some gasoline containing ethanol is already available. (Roy D. Farris/Visuals Unlimited)

■ The nine cities with the most serious ozone pollution are Baltimore, Chicago, Hartford, Houston, Los Angeles, Milwaukee, New York, Philadelphia, and San Diego.

Nine cities with the most serious ozone pollution are required by the 1990 regulations to use reformulated gasolines, and another 87 cities that are not meeting the ozone air-quality standards can choose to use them.

9.7 Natural Gas

Natural gas is a mixture of gases trapped with petroleum in the earth's crust and is recoverable from oil wells or gas wells where the gases have migrated through the rock. The natural gas found in North America is a mixture of C_1 to C_4 alkanes—methane (60–90%), ethane (5–9%), propane (3–18%), and butane (1–2%)—with a number of other gases, such as CO_2, N_2, H_2S, and the noble gases present in varying amounts. In Europe and Japan, the natural gas is essentially all methane.

Natural gas is the fastest-growing energy source in the United States, and U.S. production of natural gas supplies 17% more energy than does U.S.-produced oil. About half of the homes in the United States are heated by natural gas, followed by electricity (18.5%), fuel oil (14.9%), wood (4.8%), and liquefied gas such as butane and propane (4.6%). Coal and kerosene come in at a low 0.5%, and the percentage of homes using solar heating is even lower. However, the United States has only about 5% of the known world reserves of natural gas, which at the present rate of use is enough to last until the year 2050.

Natural gas is also being used as a vehicle fuel, and worldwide there are about 700,000 vehicles powered by natural gas. Although the number of natural-gas-powered vehicles in the United States (10,000) is much smaller than in countries such as Italy (300,000) and New Zealand (100,000), California and several other states are encouraging the use of natural-gas vehicles to help meet new air-quality regulations. Vehicles powered by natural gas emit minimal amounts of carbon monoxide, hydrocarbons, and particulates; and the price of natural gas is about one third that of gasoline. The main disadvantages of natural-gas vehicles include the need for a cylindrical pressurized gas tank and the lack of service stations that sell compressed natural gas.

Although most natural gas is used as an energy source, it is also an important source of raw materials for the organic chemical industry. (Figure 10.1 shows the uses of alkanes obtained from natural gas and petroleum.)

9.8 Coal

Coal is a mixture of hydrocarbons with a relatively small but variable amount of sulfur. By way of contrast with petroleum, coal has more fused rings of carbon atoms, and the organic structure of coal is much more complicated.

About 88% of our annual coal production is burned to produce electricity. Only 1% is used for residential and commercial heating. Although the use of coal is on the rise, its use as a heating fuel declined because it is a relatively dirty fuel, bulky to handle, and a major cause of air pollution (because of its sulfur content). The dangers of deep coal mining and the environmental disruption caused by strip mining contributed to the decline in the use of coal.

Given our great dependence on coal for the production of electricity and our smaller but still significant dependence on coal for the production of industrial chemicals, just how much coal do we have and how long is it likely to last? World coal reserves are vast relative to supplies of the other fossil fuels. The known world reserves are estimated to be about 1024 billion tons, of which about 29% is in the United States. How much coal has been used and how long coal is expected to last are summarized in Figure 9.12.

Coal can be converted into a combustible gas (coal gasification) or a liquid fuel (coal liquefaction). In each case environmental problems can be averted, but at additional costs per energy unit obtained from these fuels.

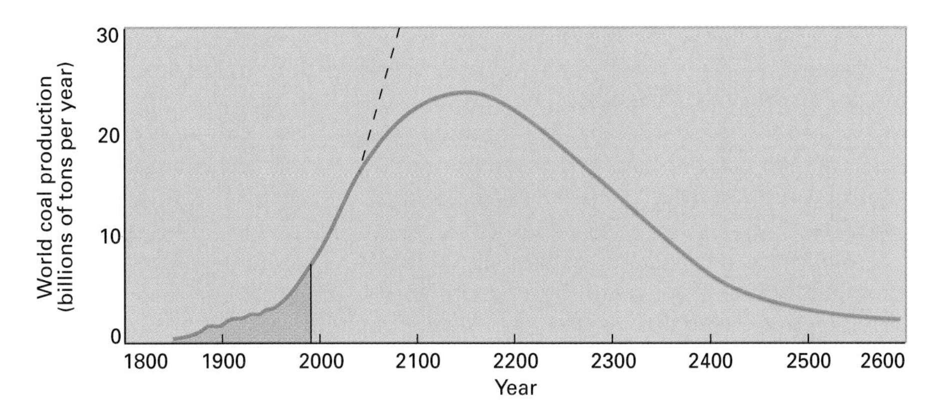

Figure 9.12 Coal resources. The coal mined to date (shaded area) is only a small fraction of the recoverable coal. The rate of increase in coal consumption (dashed line) is 4% per year. It is obvious that such a rise cannot continue.

Coal Gasification

The formation of synthesis gas is an example of a reaction in which the products have higher bond energies than the reactants, as indicated by the absorption of 31 kcal in the reaction.

When coal is pulverized and treated with superheated steam, a mixture of CO and H_2 (synthesis gas) is obtained in a process known as **coal gasification.**

$$C + H_2O + 31 \text{ kcal} \longrightarrow CO + H_2$$

Synthesis gas is used both as a fuel and as a starting material for the production of organic chemicals and gasoline.

The carbon atoms in coal, methane, and other hydrocarbons end up as CO_2 molecules, which contribute to global warming (see Section 16.12).

In a newer coal gasification process, methane is the end product. Crushed coal is mixed with an aqueous catalyst; the mixture is dried, and CO and H_2 are added. The resulting mixture is then heated to 700°C to produce methane and carbon dioxide. The overall reaction is

$$2\,C + 2\,H_2O + 2 \text{ kcal} \longrightarrow CH_4 + CO_2$$

Although the overall reaction is endothermic, the combustion of the methane produced releases 192 kcal/mole; thus, the process is an energy-efficient way to obtain methane, an environmentally clean fuel.

Coal Liquefaction

Liquid fuels are made from coal by reacting the coal with hydrogen gas under high pressure in the presence of catalysts (hydrogenating the coal). The process produces hydrocarbons like those in petroleum. The resulting crude oil type of material can be fractionally distilled to give fuel oil, gasoline, and hydrocarbons used in the manufacture of plastics, medicines, and other commodities. About 5.5 barrels of liquid are produced for each ton of coal. At the present time, the cost of a barrel of liquid from coal liquefaction is about double that of a barrel of crude oil. However, as petroleum supplies diminish and the cost of crude oil increases, coal liquefaction will become economically feasible.

9.9 Methanol as a Fuel

Methanol is being considered as a replacement for gasoline, especially in urban areas that have extremely high levels of air pollution caused by motor vehicles. For example, Southern California has been testing methanol-powered cars since 1981. About half the cars use 100% methanol (M100), while the other half are flexible-fueled vehicles (FFVs) that use either M85, a blend of 85% methanol and 15% gasoline, or gasoline. Although methanol fuels M85 and M100 have received more attention than corresponding ethanol fuels E85 and E100, FFVs are being built to test the use of both M85 and E85.

What are the advantages and disadvantages of switching to methanol-powered vehicles? Methanol burns more cleanly than gasoline, and levels of troublesome pollutants such as carbon monoxide, unreacted hydrocarbons, nitrogen oxides, and ozone are reduced. However, there is concern about the higher exhaust emissions of carcinogenic formaldehyde from methanol-powered vehicles. Since the number of methanol-powered vehicles is limited, it is still difficult to assess the extent to which these formaldehyde emissions will contribute to the total aldehyde levels from other sources.

The technology for methanol-powered vehicles has existed for many years, particularly for racing cars that burn methanol because of its high octane rating of 100. However, methanol only has about one half the energy content of gasoline, which would require fuel tanks to be twice as large to give the same distance per tankful. This is partially compensated for by the fact that methanol costs about half as much as gasoline, so the price per mile would be competitive. Because methanol burns with a colorless flame, something needs to be added (a small amount of gasoline, for example) to methanol so that it can be seen when it burns. Another disadvantage is the tendency for methanol to corrode regular steel, so it will be necessary to use stainless steel for the fuel system or have a methanol-resistant coating. Until sufficient numbers of methanol-powered vehicles are on the road, cars equipped to run on *either* methanol or gasoline will be necessary because of the lack of service stations selling methanol. As the problems of distribution and storage are solved, better engineered methanol-fueled engines will be designed and produced, which will lead to more efficient utilization of methanol as a fuel.

■ Cars at the Indianapolis 500 are powered by methanol.

Another option is to use methanol to make gasoline. Mobil Oil Company has developed a methanol-to-gasoline process that is currently not competitive with refined gasoline prices in the United States, but is competitive in those regions of the world, such as New Zealand, where the price of gasoline is much higher. In fact, the production of 92-octane gasoline from methanol is taking place in New Zealand.

$$2\ CH_3OH \xrightarrow[\text{Catalyst}]{} \underset{\text{Dimethyl ether}}{(CH_3)_2O} + H_2O$$

$$2\ (CH_3)_2O \xrightarrow[\text{Catalyst}]{} \underset{\text{Ethylene}}{2\ C_2H_4} + 2\ H_2O$$

$$C_2H_4 \xrightarrow[\text{Catalyst}]{} \underset{\text{Gasoline}}{\text{Hydrocarbon mixture in the } C_5-C_{12} \text{ range}}$$

The New Zealand plant is currently producing 14,000 barrels per day of gasoline with an octane rating of 92 to 94. This is about one third the amount of gasoline used in New Zealand.

Plant in New Zealand that converts natural gas to methanol, which is then converted to gasoline. *(Mobil Corporation)*

■ SELF-TEST 9C

1. The fractions of petroleum are separated by ____________.
2. The principal component in natural gas is ____________.
3. The octane enhancer used most by gasoline producers at the present time is ____________.
4. The ____________ process is used to produce branched-chain and aromatic hydrocarbons from straight-chain hydrocarbons.
5. The ____________ process is used in refining petroleum to convert molecules in the higher boiling fractions to molecules in the gasoline fraction.

6. Which of the following hydrocarbons would be expected to have the highest octane rating?
   (a) $CH_3CH_2CH_2CH_2CH_2CH_2CH_3$

   (b) $CH_3CH_2CH(CH_3)CH_2CH_2CH_3$

   (c) $CH_3-C(CH_3)_2-CH(CH_3)-CH_3$

MATCHING SET

____	1. Hydrocarbon	a. Benzene
____	2. Alkane	b. R—OH
____	3. Alkyl group	c. Major component of natural gas
____	4. Alkene	d. Contains only C and H
____	5. Alkyne	e. R—O—R′
____	6. Aromatic hydrocarbon	f. C_nH_{2n+2}
____	7. Methane	g. Mixture of CO and H_2
____	8. Alcohol	h. Contains C=C bond
____	9. Ether	i. Hydrocarbon that is missing an H atom
____	10. Synthesis gas	j. Contains C≡C bond

QUESTIONS FOR REVIEW AND THOUGHT

1. What is the definition of a fossil fuel?
2. What are the three major fossil fuels?
3. What is a hydrocarbon?
4. What is the primary component of natural gas?
5. What is the heat of combustion?
6. Give definitions for the following terms.
   (a) Alkane (b) Alkene
   (c) Alkyne (d) Aromatic
7. Give definitions for the following terms.
   (a) Isomer
   (b) Straight-chain hydrocarbon
   (c) Branched-chain hydrocarbon
8. What is an alkyl group? Give the formulas for the methyl and ethyl alkyl groups.
9. Saturated hydrocarbons are so named because they have the maximum amount of hydrogen present for a given amount of carbon. The saturated hydrocarbons have the general formula C_nH_{2n+2} where n is a whole number. What are the names and formulas of the first four members of this series of compounds?
10. What is the structural formula for 1-pentene?
11. Give the structural formulas for the following:
    (a) 2-Methylpentane
    (b) 4,4-Dimethyl-5-ethyloctane
    (c) 2-Methyl-2-hexene
12. Draw as many different isomers as you can that have the formula C_5H_{12}.
13. Name the isomers in Question 12.
14. Draw the condensed structural formulas of all possible structural isomers for C_6H_{14} and name them.
15. Draw the structures of all possible isomers that are dimethylbenzenes.
16. Draw the structure of 2,3,3-trimethyl-1-pentene.
17. Why does 2-butene have *cis* and *trans* isomers but 1-butene doesn't?
18. Draw the *cis* and *trans* isomers of 1,2-dichloroethene.
19. Explain how fractional distillation is used in the refinement of petroleum.
20. What is "straight-run" gasoline?
21. List three gasoline additives that increase the octane rating of gasoline.
22. What is gasohol?

23. Explain how synthesis gas and methane can be obtained from coal, and write equations that represent these processes.
24. What is meant by the following terms?
    (a) Catalytic re-forming (b) Catalytic cracking
    (c) Octane rating
25. Explain the Mobil Oil process for converting methanol to gasoline.
26. What types of hydrocarbons have high octane ratings?
27. What factors are likely to lead to an increased demand for methanol in the next decade?
28. How is the octane rating of a refined gasoline determined? What are oxygenated gasolines?
29. Why do oxygenated gasolines cause less pollution than regular gasolines?
30. What is the difference between oxygenated gasolines and reformulated gasolines?
31. How is coal gasified?
32. What are the advantages and disadvantages of using methanol as an alternative fuel for vehicles?
33. What do the terms M100, E100, M85, and E85 refer to when describing alternate fuels? What is an FFV?
34. The oxygenation of gasolines can be done using ethanol or ETBE. This would make use of grains such as corn to provide the ethanol. Give two advantages and two disadvantages of using agricultural land to produce fuel for machines instead of food for people.
35. Label each of the following as an alkane, alkene, aromatic, or alkyne.

    (a) Methane (CH_4) 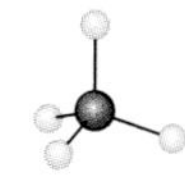

    (b) Benzene (C_6H_6) 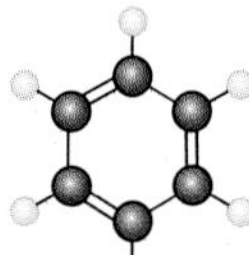

    (c) 1-Butene ($CH_2CHCH_2CH_3$) 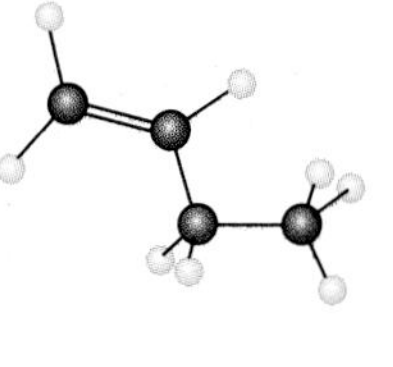

    (d) Acetylene (CHCH) 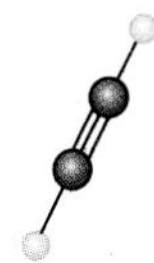

36. The following show the formulas and ball-and-stick models for various compounds. Label each of the following as an alkane, alkene, aromatic, alkyne, alcohol, or ether.

    (a) Propane ($CH_3CH_2CH_3$) 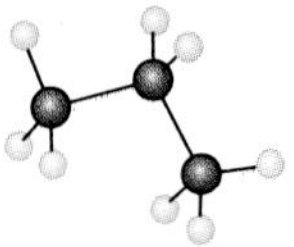

    (b) 1-Butyne ($CHCCH_2CH_3$) 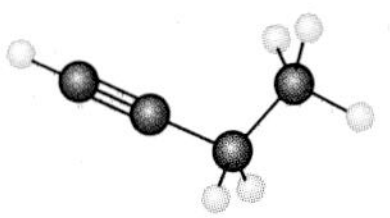

    (c) Diethyl ether ($CH_3CH_2OCH_2CH_3$) 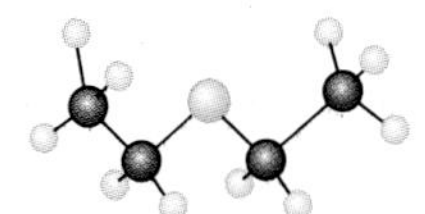

    (d) Ethanol (CH_3CH_2OH) 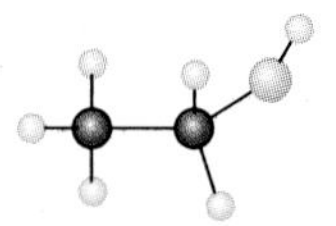

    (e) Ethylbenzene ($CH_3CH_2C_6H_5$) 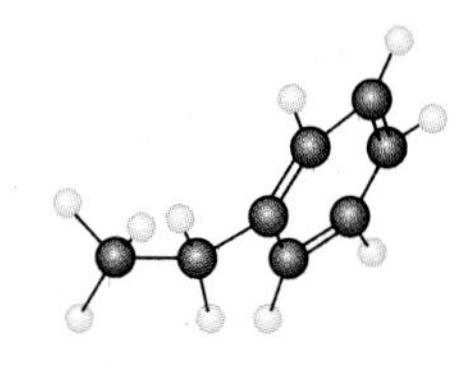

CHAPTER 10

Organic Chemicals and Polymers

Fossil fuels are not only our major source of energy, but they are also the major source of the hydrocarbons that are used to make thousands of consumer products. About 6% of the petroleum refined today is the starting material for the synthesis of organic chemicals of commercial importance. These chemicals are essential to making plastics, synthetic rubber, synthetic fibers, fertilizers, and thousands of other consumer products. For this reason the organic chemical industry is often referred to as the **petrochemical** industry.

A few of the major classes of organic compounds and some of their reactions, especially those used in making polymers, are described in this chapter, with the goal of introducing you to a major segment of chemistry and the chemical industry.

- Why are there so many organic compounds?
- What are the characteristic functional groups?
- What are some common addition and condensation polymers?
- Will coal become a major source of chemicals?
- What types of plastics are being recycled?

Carbon compounds hold the key to life on Earth. Consider what the world would be like if all carbon compounds were removed; the result would be much like the barren surface of the Moon. If carbon compounds were removed from the human body, there would be nothing left except water and a small residue of minerals. The same would be true for all living things. Carbon compounds are also an integral part of our lifestyle. Fossil fuels, foods, and most drugs are made of carbon compounds. Since we live in an age of plastics and synthetic fibers, our clothes, appliances, and most other consumer goods contain a significant portion of carbon compounds.

Over 11 million of the more than 13 million known compounds are carbon compounds, and a separate branch of chemistry, **organic chemistry**, is

Pumping oil—this is where many everyday materials start the journey to our households. *(Bill Ross/Tony Stone Images)*

TABLE 10.1 ■ Top Ten Chemicals Produced in the United States in 1996*

Rank	Name	How Made	End Uses
1.	Sulfuric acid	Burning sulfur to SO_2, oxidation of SO_2 to SO_3, reaction with water; also recovered from metal smelting	Fertilizers, petroleum refining, manufacture of metals and chemicals
2.	Ethylene	Cracking hydrocarbons from oil and natural gas	Plastics, antifreeze production, fibers and solvents
3.	Ammonia	Catalytic reaction of nitrogen, air, and hydrogen	Fertilizers, plastics, fibers, and resins
4.	Phosphoric acid	Reaction of sulfuric acid with phosphate rock; burning of elemental phosphorus and dissolution in water	Fertilizers, detergents, and water-treating compounds
5.	Chlorine	Electrolysis of NaCl, recovery from HCl users	Chemical production, plastics, solvents, pulp and paper
6.	Propylene	Cracking oil and oil products	Plastics, fibers, solvents
7.	Sodium hydroxide	Electrolysis of NaCl solution	Chemicals, pulp and paper, aluminum, textiles, oil refining
8.	Ammonium nitrate	Reaction of ammonia and nitric acid	Explosives, fertilizers
9.	Urea	Reaction of NH_3 and CO_2 under pressure	Fertilizers, animal feeds, adhesives and plastics
10.	Styrene	Dehydrogenation of ethylbenzene	Polymers, rubber, polyesters

* Data from *Chemical and Engineering News,* pp. 41, 42, June 23, 1997. Organic compounds are highlighted.

devoted to the study of them. Why are there so many organic compounds? The discussion of hydrocarbons and their structural and geometric isomers in Chapter 9 indicates two reasons: (1) the ability of thousands of carbon atoms to be linked in sequence with stable carbon–carbon bonds in a single molecule and (2) the occurrence of isomers. A third reason will be discussed further in this chapter: the variety of functional groups that bond to carbon atoms.

The economic importance of the organic chemical industry can be seen by looking at the list of the top ten chemicals produced in the United States. Of the top ten listed in Table 10.1, four are organic chemicals.

■ Catalytic cracking was described in Section 9.6.

10.1 Organic Chemicals

Many of the organic chemicals used in the chemical industry are obtained from fossil fuels. For example, ethylene, propylene, butylene, and acetylene are obtained by catalytic cracking of natural gas or petroleum; Figure 10.1 summarizes the uses of these as starting materials. Petroleum and coal tar, obtained by heating coal at high temperatures in the absence of air, are the primary sources of aromatic compounds used in the chemical industry (Figure 10.2). Distilling coal tar yields the aromatic compounds listed in Table 10.2.

■ Heating coal at high temperatures in the absence of air produces a mixture of coke, coal tar, and coal gas. The process, called pyrolysis, is represented by

$$\text{Coal} \longrightarrow \text{coke} + \text{coal tar} + \text{coal gas}$$

One ton of bituminous (soft) coal yields about 1500 lb of coke, 8 gal of coal tar, and 10,000 cubic feet (ft^3) of coal gas. **Coal gas** is a mixture of H_2, CH_4, CO, C_2H_6, NH_3, CO_2, H_2S, and other gases. At one time coal gas was used as a fuel.

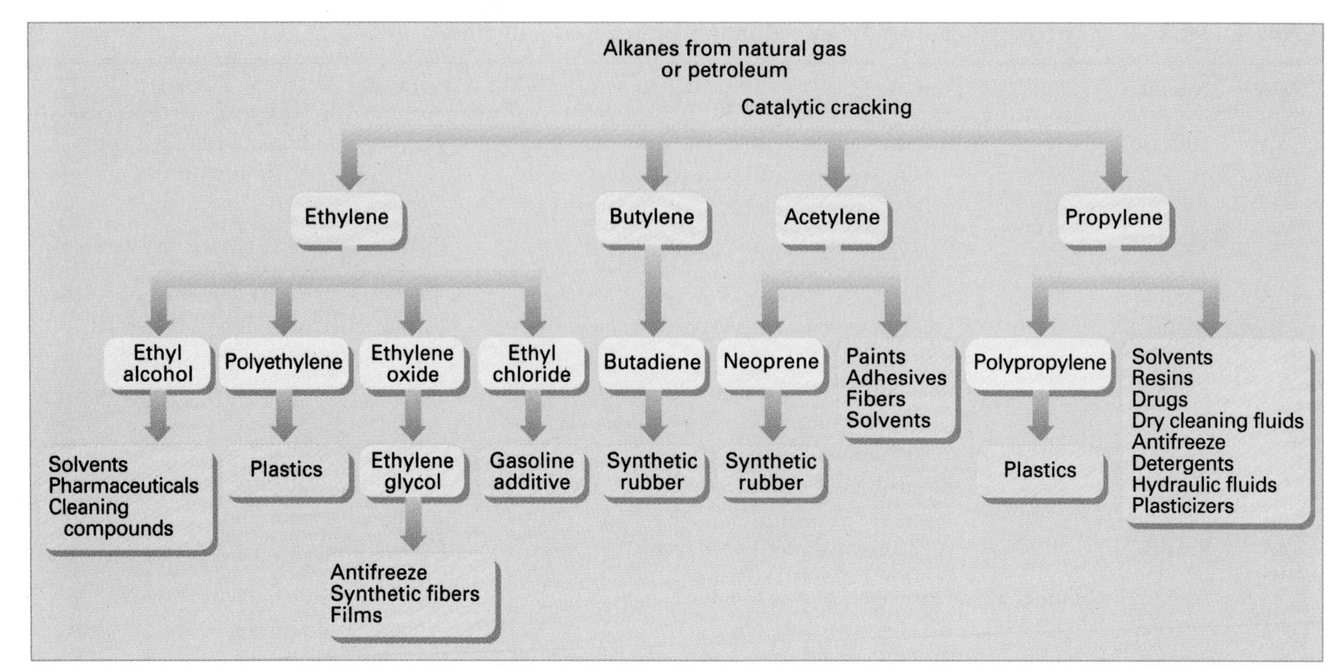

Figure 10.1 Hydrocarbons from petroleum or natural gas as raw materials. Catalytic cracking produces ethylene, butylene, acetylene, and propylene, which are converted into other chemical raw materials and many kinds of consumer products.

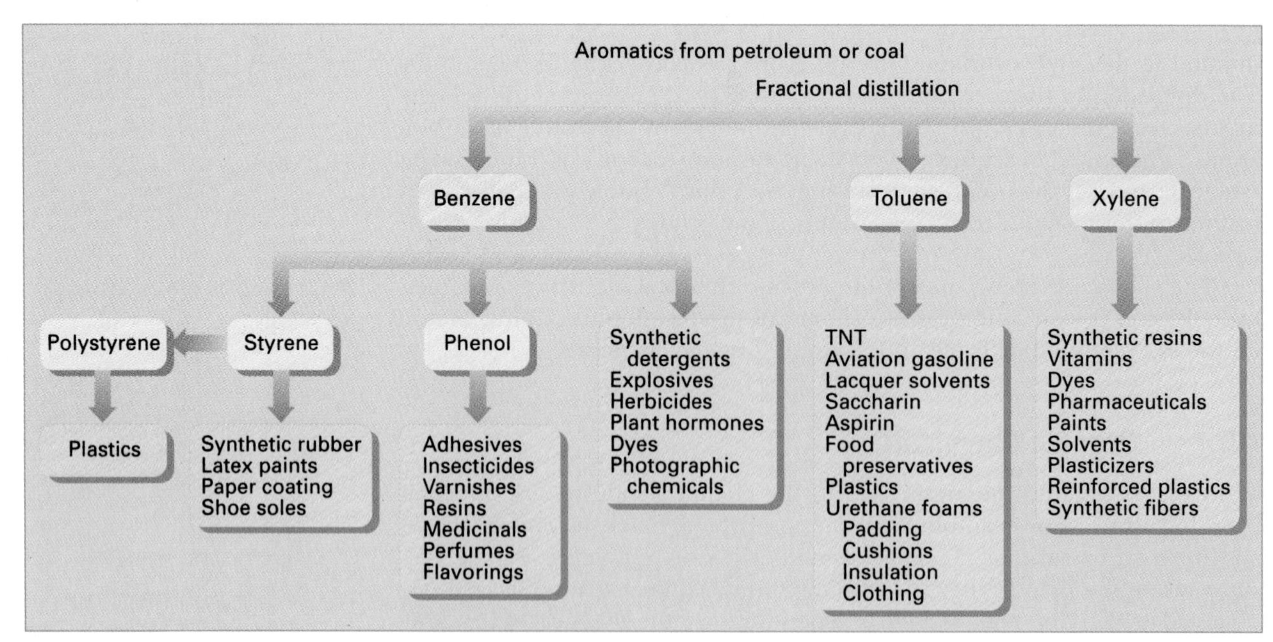

Figure 10.2 Aromatic compounds from petroleum or coal and their uses as raw materials.

TABLE 10.2 ■ Fractions from Distillation of Coal Tar

Boiling Range (°C)	Name	Tar, Mass %	Primary Constituents
Below 200	Light oil	5	Benzene, toluene, xylenes
200–250	Middle oil (carbolic oil)	17	Naphthalene, phenol, pyridine
250–300	Heavy oil (creosote oil)	7	Naphthalenes and methylnaphthalenes, cresols
300–350	Green oil	9	Anthracene, phenanthrene
Residue		62	Pitch or tar

Organic chemicals were once obtained only from plants, animals, and fossil fuels; and these are still direct sources of hydrocarbons (Figure 10.1 and 10.2) and many other important chemicals, such as sucrose from sugarcane and ethanol from fermented grain mash. However, the development of organic chemistry has led to cheaper methods for the synthesis of both naturally occurring substances and new substances.

The millions of organic compounds include classes of compounds that are obtained by replacing hydrogen atoms of hydrocarbons with atoms or groups of atoms known as **functional groups**. The functional groups for alcohols and ethers were discussed in Section 9.5. Alcohols and their oxidation products—aldehydes, ketones, and carboxylic acids—are among the most useful functional group classes of compounds.

■ A table of functional groups and further information on the naming of organic compounds are given in Appendix D.

10.2 Alcohols and Their Oxidation Products

Alcohols contain one or more —OH groups bonded to carbon atoms and are a major class of organic compounds. The importance of methanol, ethanol, and 2-methyl-2-propanol as fuels and fuel additives was described in Sections 9.6 and 9.9. Additional uses of these and other commercially important alcohols are listed in Table 10.3 (p. 228). Alcohols are classified according to the number of carbon atoms bonded to the —C—OH carbon as primary (one other C atom), secondary (two other C atoms), and tertiary (three other C atoms). The reactivities of these classes of alcohols are different.

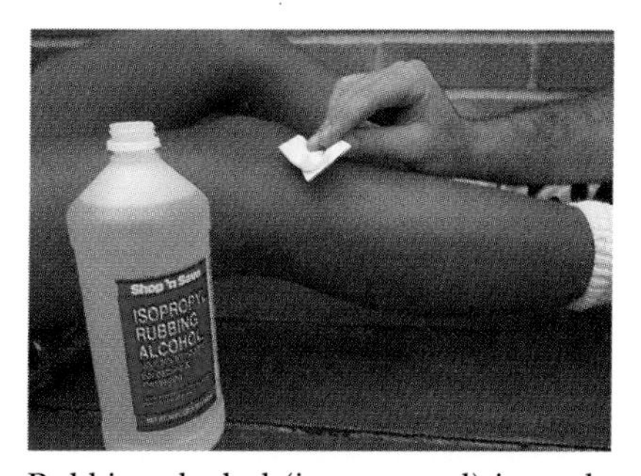
Rubbing alcohol (isopropanol) is used by athletic assistants to clean out cuts and scrapes. *(C. D. Winters)*

Primary	Secondary	Tertiary
R—C(H)(H)—OH	R′—C(R)(H)—OH	R′—C(R)(R″)—OH

■ An "R" is used to represent any kind of alkyl group. The use of R, R′, and R″ indicates that all R groups are different.

Ethanol and 1-propanol are primary alcohols. 2-Propanol, or isopropyl alcohol, is familiar to us as rubbing alcohol, a 70% solution of 2-propanol sold in drugstores and grocery stores. 2-Propanol is a secondary alcohol and is one of the two structural isomers of an alcohol with three carbon atoms.

TABLE 10.3 ■ Some Important Alcohols

Condensed Formula	Boiling Point (°C)	Systematic Name	Common Name	Use
CH_3OH	65.0	Methanol	Methyl alcohol	Fuel, gasoline additive, making formaldehyde
CH_3CH_2OH	78.5	Ethanol	Ethyl alcohol	Beverages, gasoline additive, solvent
$CH_3CH_2CH_2OH$	97.4	1-Propanol	Propyl alcohol	Industrial solvent
$CH_3CH(OH)CH_3$	82.4	2-Propanol	Isopropyl alcohol	Rubbing alcohol
$CH_2(OH)CH_2(OH)$	198	1,2-Ethanediol	Ethylene glycol	Antifreeze
$CH_2(OH)CH(OH)CH_2(OH)$	290	1,2,3-Propanetriol	Glycerol (glycerin)	Moisturizer in foods and cosmetics

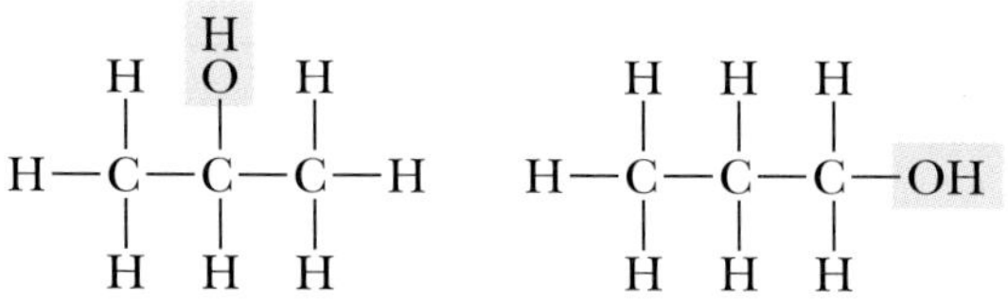

$$H-\underset{H}{\overset{H}{C}}-\underset{H}{\overset{OH}{C}}-\underset{H}{\overset{H}{C}}-H$$

2-Propanol

$$H-\underset{H}{\overset{H}{C}}-\underset{H}{\overset{H}{C}}-\underset{H}{\overset{H}{C}}-OH$$

1-Propanol

For an alcohol with four carbon atoms, there are four structural isomers, including 2-methyl-2-propanol (*tertiary*-butyl alcohol), whose use as a gasoline additive was described in Section 9.6.

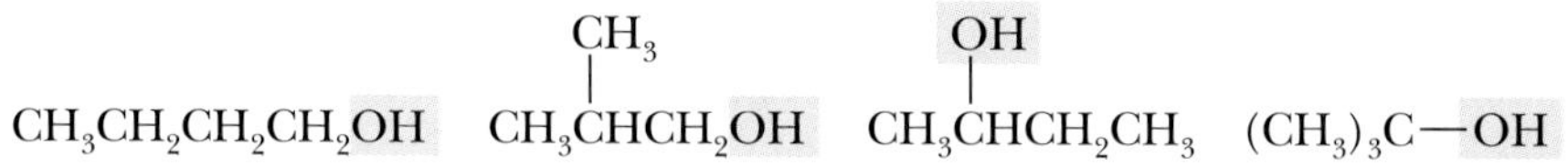

$CH_3CH_2CH_2CH_2OH$ — 1-Butanol, A primary alcohol

$CH_3CH(CH_3)CH_2OH$ — 2-Methyl-1-propanol, A primary alcohol

$CH_3CH(OH)CH_2CH_3$ — 2-Butanol, A secondary alcohol

$(CH_3)_3C-OH$ — 2-Methyl-2-propanol, A tertiary alcohol

Alcohols can serve as the starting substances for the preparation of many other types of organic compounds. Oxidation of alcohols may yield

aldehydes, $R-\overset{O}{\overset{\|}{C}}-H$, **ketones**, $R-\overset{O}{\overset{\|}{C}}-R'$, or **carboxylic acids**, $R-\overset{O}{\overset{\|}{C}}-OH$,

■ Oxidation of organic compounds is usually the addition of oxygen or the removal of hydrogen, and reduction is usually the removal of oxygen or the addition of hydrogen (see Sections 8.1 and 8.2).

depending on the alcohol used and the extent of the oxidation. If the starting compound is a primary alcohol, the first oxidation product is an aldehyde and the second oxidation product is a carboxylic acid. For example, the oxi-

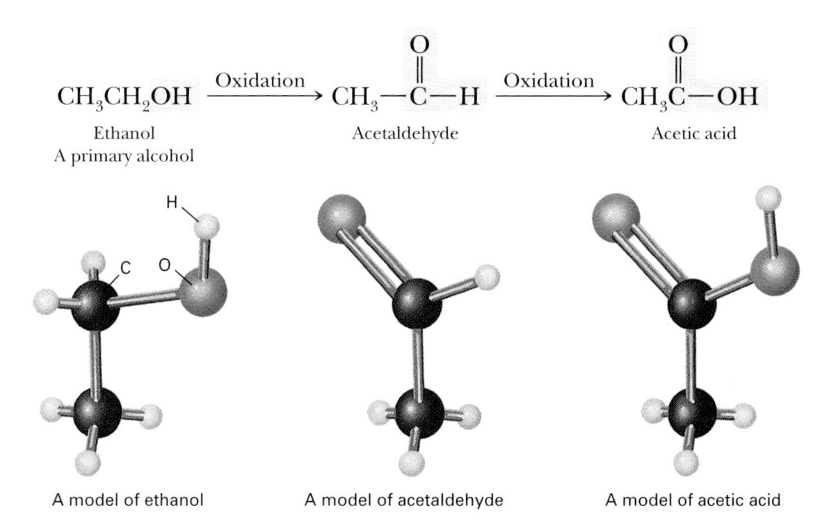

■ The oxidation of ethanol to acetaldehyde is an example of how the reactivity of primary alcohols differs from the reactivity of secondary alcohols.

dation of ethanol, a primary alcohol, can be used to make acetaldehyde and acetic acid.

Formaldehyde (HCHO), the simplest aldehyde, is obtained by oxidation of methanol. It is a gas at room temperature but is often used as a 40% aqueous solution called formalin. Although formaldehyde is an important starting material for polymers (Section 10.5), it presents a number of health hazards because of its toxicity and carcinogenicity. Formaldehyde is also an air pollutant (Section 15.3), being produced in trace amounts in the incomplete combustion of fossil fuels. One of the concerns about methanol as a fuel (Section 9.9) is the potential for increased levels of formaldehyde in urban areas because of the production of formaldehyde from incomplete combustion of methanol.

Secondary alcohols are oxidized to give ketones. The oxidation of 2-propanol (isopropyl alcohol) gives acetone, a ketone widely used as an organic solvent.

$$\underset{\text{2-Propanol}}{CH_3CH(OH)CH_3} \xrightarrow{\text{Oxidation}} \underset{\text{Acetone}}{CH_3C(=O)CH_3}$$

Hydrogen Bonding in Alcohols

■ See Section 5.7 for a discussion of hydrogen bonding.

The physical properties of alcohols provide an example of the effects of hydrogen bonding between molecules in liquids. This is illustrated by the boiling points in Table 10.3. Hydrogen bonding explains why methanol (molar mass 32 g) is a liquid, whereas propane (molar mass 44 g), which has a higher molar mass but no hydrogen bonding, is a gas at room temperature. The boiling point of methanol is lower than that of water because methanol has only one O—H hydrogen through which it can hydrogen bond.

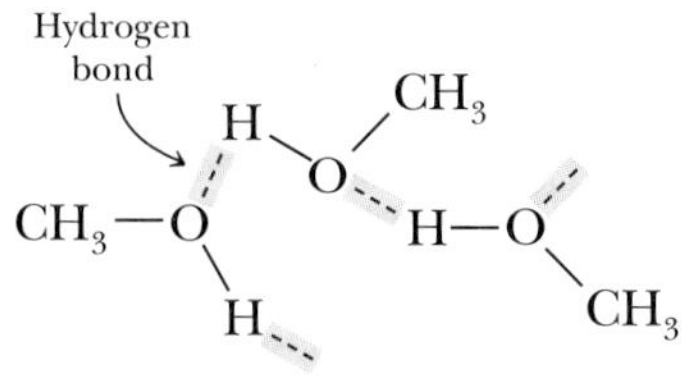

The higher boiling point of ethylene glycol can be attributed to the presence of two —OH groups per molecule. Glycerol, with three —OH groups, has an even higher boiling point.

The alcohols listed in Table 10.3 are also very soluble in water because of hydrogen bonding between water molecules and the —OH group in alcohol molecules. However, as the length of the hydrocarbon chain increases, the resulting alcohols are less soluble because the nonpolar hydrocarbon portion has greater influence on the solubility than the hydrogen bonding by the —OH group.

Ethanol

Ethanol is the "alcohol" of alcoholic beverages and is prepared for this purpose by fermentation (Section 9.5) of carbohydrates (starch, sugars) from a wide variety of plant sources (Table 10.4). The growth of yeast is inhibited at alcohol concentrations higher than about 12%, and fermentation comes to a stop. Beverages with a higher ethanol concentration are prepared either by distillation or by fortification with ethanol that has been obtained by the distillation of another fermentation product. The maximum concentration of ethanol that can be obtained by distillation of ethanol–water mixtures is 95%. The "proof" of an alcoholic beverage is twice the volume percent of ethanol; 80-proof vodka, for example, contains 40% ethanol by volume; 95% ethanol is 190 proof.

■ Proof = 2 × volume percent.

Although ethanol is not as toxic as methanol (Section 9.5), 1 pint of pure ethanol, rapidly ingested, would kill most people. Ethanol is a depressant, and the effects of different blood levels of ethanol are shown in Table 10.5. Rapid consumption of two 1-ounce "shots" of 90-proof whiskey or of two 12-ounce beers can cause one's blood alcohol level to reach 0.05%. Ethanol is quickly absorbed into the bloodstream and metabolized by enzymes produced in the

TABLE 10.4 ■ Common Alcoholic Beverages

Name	Source of Fermented Carbohydrate	Amount of Ethyl Alcohol	Proof
Beer	Barley, wheat	5%	10
Wine	Grapes or other fruit	14% maximum, unless fortified	20–28
Brandy	Distilled wine	40–45%	80–90
Whiskey	Barley, rye, corn, etc.	45–55%	90–110
Rum	Molasses	~45%	90
Vodka	Potatoes	40–50%	80–100

TABLE 10.5 ■ Effects of Alcohol Blood Level*

Number of Drinks‡	% Alcohol by Volume	Effect
1 to 4	0.05–0.15	Lack of coordination, altered judgment, exaggerated emotions
5 to 10	0.15–0.20	Intoxication (slurred speech, altered perception and equilibrium)
10 to 15	0.30–0.40	Unconsciousness, coma
Over 15	0.50	Possible death

* In many states a person with a blood alcohol level of 0.10% or higher is legally intoxicated.
‡ 1 drink = 12 oz of 4.5% beer
4 oz of 14% wine
1 to 1½ oz of 45% alcohol liquor

liver. The rate of detoxification is about 1 ounce of pure ethanol per hour. Ethanol is oxidized to acetaldehyde, which is further oxidized to acetic acid; eventually, CO_2 and H_2O are produced and eliminated through the lungs and kidneys.

$$\underset{\text{Ethanol}}{\mathrm{CH_3CH_2OH}} \xrightarrow[\text{enzymes}]{\text{Liver}} \underset{\text{Acetaldehyde}}{\mathrm{CH_3CHO}}$$

■ Many consumer products like Listerine™ and Nyquil™ contain ethanol.

The federal tax on alcoholic beverages is about $20 per gallon. Since the cost of producing ethanol is only about $1 per gallon, ethanol intended for industrial use must be *denatured* to avoid the beverage tax. **Denatured alcohol** contains small amounts of a toxic substance, such as methanol or gasoline, that cannot be removed easily by chemical or physical means.

Apart from being used in the alcoholic beverage industry, ethanol is used widely in solvents, in the preparation of many other organic compounds, and as a gasoline additive (Section 9.5). For many years industrial ethanol was also made by fermentation. However, in the last several decades, it has become cheaper to make the ethanol from petroleum by-products, specifically by the catalyzed addition of water to ethylene. More than 1 billion pounds of ethanol are produced each year by this process.

A small structure difference makes a big difference in properties. Antifreeze contains ethylene glycol, which is composed of two carbon atoms with an —OH group on each one, and is extremely poisonous. Glycerol, which is composed of three carbon atoms with an —OH group on each one, is nonpoisonous, sweet, syrupy, and holds moisture to the skin. It is used in soap and also as a food additive (Section 12.9). *(C. D. Winters)*

Ethylene Glycol and Glycerol

More than one alcohol group can be present in a single molecule. Ethylene glycol, a di-alcohol (Table 10.3) is used in permanent antifreeze and in the synthesis of polymers.

Glycerol, a tri-alcohol (Table 10.3), is a by-product from the manufacture of soaps. Because of its moisture-holding properties, glycerol has many uses in foods and tobacco as a digestible and nontoxic humectant (which gathers and holds moisture), and in the manufacture of drugs and cosmetics. It is also used in the manufacture of nitroglycerin and numerous other chemicals. Perhaps the most important compounds of glycerol are its natural esters, which are the fats and oils found in plants and animals (Section 11.3).

SELF-TEST 10A

1. What is the R group of 1-propanol, $CH_3CH_2CH_2OH$?
2. Coal tar is the principal source of what major class of hydrocarbons?
3. Ethanol is quickly absorbed into the bloodstream and oxidized to ____________ in the liver.
4. 2-Propanol is commonly known as ____________.
5. The primary component of antifreeze is ____________.
6. Gin that is 84 proof contains what percentage of ethanol?
7. Ethanol intended for industrial use is ____________ by the addition of small amounts of a toxic substance.
8. Glycerol has ____________ alcohol groups.
9. Fats and oils are esters of the alcohol ____________.
10. ____________ is the formula of the alcohol oxidized to formaldehyde.

10.3 Carboxylic Acids and Esters

Carboxylic Acids

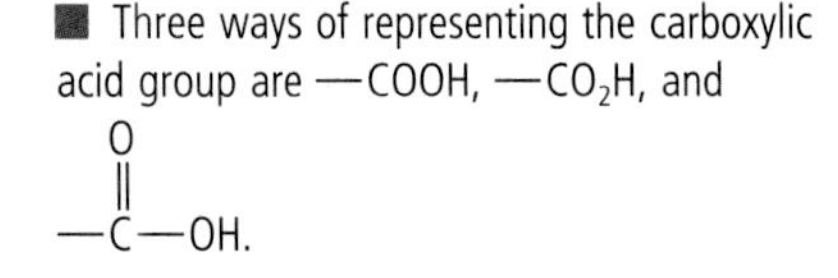

Three ways of representing the carboxylic acid group are —COOH, —CO₂H, and
O
||
—C—OH.

Carboxylic acids contain the —COOH functional group and are prepared by the oxidation of alcohols or aldehydes. These reactions occur quite easily, as evidenced by the souring of wine to form vinegar, which is the oxidation of ethanol to acetic acid in the presence of oxygen from the air.

Carboxylic acids are polar and readily form hydrogen bonds. This hydro-

TABLE 10.6 ■ Some Simple Carboxylic Acids

Structure	Odor	Common Name	Systematic Name	Boiling Point (°C)
$HC(=O)OH$	Sharp	Formic acid	Methanoic acid	101
$CH_3C(=O)OH$	Vinegar	Acetic acid	Ethanoic acid	118
$CH_3CH_2C(=O)OH$	Swiss cheese	Propionic acid	Propanoic acid	141
$CH_3(CH_2)_2C(=O)OH$	Rancid butter	Butyric acid	Butanoic acid	163
$CH_3(CH_2)_3C(=O)OH$	Manure	Valeric acid	Pentanoic acid	187

FRONTIERS IN THE WORLD OF CHEMISTRY

Chemical Peels

For many years, several carboxylic acids have been used by dermatologists and cosmetic surgeons to remove the *stratum corneum,* the top layer of dead skin cells, on people with skin that has become aged or weathered by wind and sun. The treatment feels much like a sunburn. A 35% solution of trichloroacetic acid (Cl_3CCOOH), a moderately strong acid, is used for a "medium" peel. A 50% to 70% solution of glycolic acid, one of several α-hydroxy acids (AHAs) currently used for this purpose, gives a milder "superficial" peel because it is a much weaker acid than trichloroacetic acid. The technique for both the medium and superficial peel is the same. A solution of the acid is placed on the skin for a few minutes to essentially burn off the top layer of dead skin cells, get rid of some surface damage, and accelerate growth of new skin cells.

The market for AHAs is growing rapidly, primarily because AHAs are beginning to be used in a wide variety of hand and body creams, and not just for facial treatment. AHAs used for this purpose include glycolic, citric, lactic, and tartaric acids (Table 10.7). The percentage of acids in these over-the-counter products is lower than in the preparations used by dermatologists for skin peels. AHAs are also used to remove brown spots and precancerous lesions.

Reference E. M. Kirschner: *Chemical and Engineering News,* p. 19, March 3, 1997.

gen bonding results in relatively high boiling points for the acids, even higher than those of alcohols of comparable molar mass. For example, formic acid (46 g/mol) has a boiling point of 101°C, whereas ethanol (46 g/mol) has a boiling point of only 78°C.

All carboxylic acids are weak acids (see Section 7.2) and react with bases to form salts.

$$\underset{\text{Acetic acid}}{CH_3\overset{\overset{\displaystyle O}{\|}}{C}{-}OH(aq)} + NaOH(aq) \longrightarrow \underset{\text{Sodium acetate}}{CH_3\overset{\overset{\displaystyle O}{\|}}{C}O^-Na^+(aq)} + H_2O(\ell)$$

A number of carboxylic acids are found in nature and have been known for many years. As a result, some of the familiar carboxylic acids are almost always referred to by their common names (Table 10.6).

Acetic acid, the acid found in about 5% concentration in vinegar, is produced in large quantities for use in the manufacture of cellulose acetate, a polymer used in the manufacture of photographic film base, synthetic fibers, plastics, and other products (Section 10.5).

Three other acids produced in large quantity have two carboxylic acid groups and are known as dicarboxylic acids.

$$\underset{\text{Adipic acid}}{HO\overset{\overset{\displaystyle O}{\|}}{C}(CH_2)_4\overset{\overset{\displaystyle O}{\|}}{C}OH} \qquad \underset{\text{Terephthalic acid}}{HO{-}\overset{\overset{\displaystyle O}{\|}}{C}{-}C_6H_4{-}\overset{\overset{\displaystyle O}{\|}}{C}{-}OH} \qquad \underset{\text{Phthalic acid}}{C_6H_4(\overset{\overset{\displaystyle O}{\|}}{C}{-}OH)_2}$$

Spinach and rhubarb contain oxalic acid (Table 10.7). Individuals prone to kidney stones composed of highly insoluble calcium oxalate must limit their intake of foods containing oxalic acid. *(C. D. Winters)*

TABLE 10.7 ■ Some Naturally Occurring Carboxylic Acids

Name	Structure	Natural Source
Glycolic acid	$HO—CH_2—COOH$	Sugarcane
Citric acid	$HOOC—CH_2—C(OH)(COOH)—CH_2—COOH$	Citrus fruits
Lactic acid	$CH_3—CH(OH)—COOH$	Sour milk
Oleic acid	$CH_3(CH_2)_7—CH{=}CH—(CH_2)_7—COOH$	Vegetable oils
Oxalic acid	$HOOC—COOH$	Rhubarb, spinach, cabbage, tomatoes
Stearic acid	$CH_3(CH_2)_{16}—COOH$	Animal fats
Tartaric acid	$HOOC—CH(OH)—CH(OH)—COOH$	Grape juice, wine

These three acids are used to manufacture polymers (Section 10.5). There are other acids, however, whose names are familiar to you (Table 10.7), since they occur in nature.

Esters

■ Esters contain the $—\overset{O}{\overset{\|}{C}}—OR$ functional group, also represented as $—CO_2R$ or $—COOR$.

Carboxylic acids react with alcohols in the presence of strong acids to produce **esters**, which contain the —COOR functional group. In an ester the —OH of the carboxylic acid is replaced by the OR group from the alcohol. For example, when ethanol is mixed with acetic acid in the presence of sulfuric acid, ethyl acetate is formed. This reaction is a dehydration in which sulfuric acid acts as a catalyst and dehydrator.

$$\underset{\text{Acetic acid}}{CH_3\overset{O}{\overset{\|}{C}}—OH} + \underset{\text{Ethanol}}{H—OCH_2CH_3} \xrightarrow{H_2SO_4} \underset{\text{Ethyl acetate}}{CH_3\overset{O}{\overset{\|}{C}}—OCH_2CH_3} + H_2O$$

Some household products containing ethyl acetate, which is an excellent solvent with a pleasant odor. *(C. D. Winters)*

Names of esters are derived from the names of the alcohol and the acid used to prepare the ester. The alkyl group from the alcohol is named first followed by the name of the acid changed to end in *-ate.* For example, ethyl acetate is the name of the ester prepared from the reaction of ethanol and acetic acid. Ethyl acetate is a common solvent for lacquers and plastics and is often used as fingernail polish remover.

Unlike the acids from which they are derived, esters often have pleasant odors (Table 10.8). The characteristic odors and flavors of many flowers and fruits are due to the presence of natural esters. For example, the odor and flavor of bananas is primarily due to the ester 3-methylbutyl acetate (also known as isoamyl acetate).

TABLE 10.8 ■ Some Acids, Alcohols, and Their Esters

Acid	Alcohol	Ester	Odor of Ester
CH_3COOH Acetic acid	$CH_3CH(CH_3)CH_2CH_2OH$ 3-Methyl-1-butanol	$CH_3C(=O)OCH_2CH_2CH(CH_3)CH_3$ 3-Methylbutyl acetate	Banana
$CH_3CH_2CH_2CH_2COOH$ Pentanoic acid	$CH_3CH(CH_3)CH_2CH_2OH$ 3-Methyl-1-butanol	$CH_3CH_2CH_2CH_2C(=O)OCH_2CH_2CH(CH_3)CH_3$ 3-Methylbutyl pentanoate	Apple
$CH_3CH_2CH_2COOH$ Butanoic acid	$CH_3CH_2CH_2CH_2OH$ 1-Butanol	$CH_3CH_2CH_2C(=O)OCH_2CH_2CH_2CH_3$ Butyl butanoate	Pineapple
$CH_3CH_2CH_2COOH$ Butanoic acid	$C_6H_5—CH_2OH$ Benzyl alcohol	$CH_3CH_2CH_2C(=O)OCH_2—C_6H_5$ Benzyl butanoate	Rose

Food and beverage manufacturers often use mixtures of esters as food additives. The ingredient label of a brand of imitation banana extract reads "water, alcohol (40%), isoamyl acetate and other esters, orange oil and other essential oils, and FD&C Yellow #5." Except for the water, these are all organic compounds.

10.4 Organic Chemicals from Coal

Although petroleum is now the source of over 90% of the organic chemicals used to synthesize consumer products, the projected depletion of the U.S. petroleum reserves (Section 9.1) is focusing more attention on coal.

An example of the use of coal for the industrial synthesis of organic chemicals is an Eastman Kodak process that produces acetic anhydride used to make cellulose acetate. The first complete "chemicals from coal" plant in the United States, built by Eastman Kodak in Kingsport, Tennessee, started production in 1983. Figure 10.3*a* is a schematic drawing of the various components of the plant, which is pictured in Figure 10.3*b*. The basic reactions are to produce synthesis gas from coal, to use the synthesis gas to make methanol, and to use the methanol in the synthesis of acetic anhydride.

■ Acetic anhydride reacts with water to give acetic acid

$$\underset{\text{Acetic anhydride}}{CH_3—\overset{O}{\overset{\|}{C}}—O—\overset{O}{\overset{\|}{C}}—CH_3} + H_2O \longrightarrow 2\ CH_3—\overset{O}{\overset{\|}{C}}—OH$$

Within the complex pictured in Figure 10.3*b* are nine separate plants: four are related to the gasification of coal (Section 9.8); two are for synthesis gas preparation; and three are for the synthesis of methanol, methyl acetate, and acetic anhydride. Acetic anhydride is needed to make cellulose acetate, a polymer used in the manufacture of photographic film base, synthetic fibers,

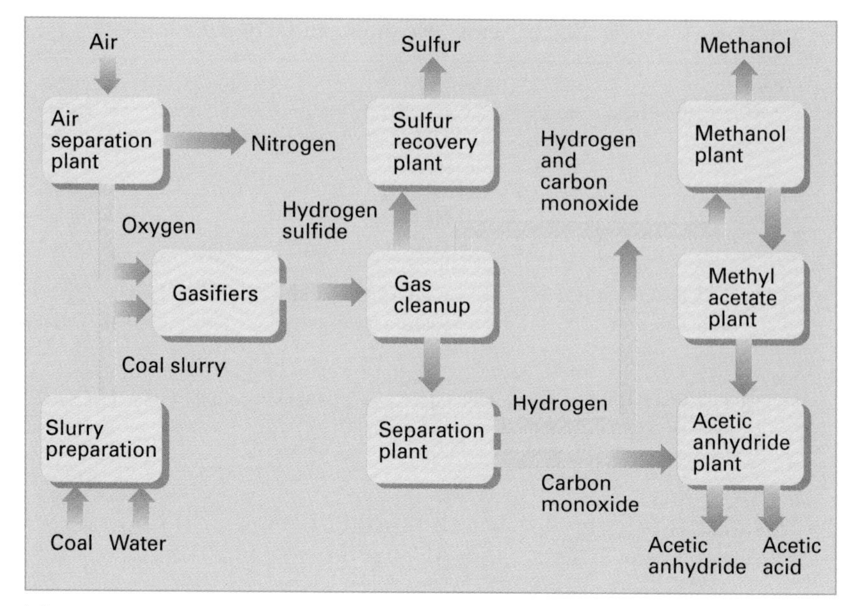

(a)

(b)

Figure 10.3 Chemicals from coal. (a) Flow diagram showing the production of methanol, methyl acetate, and acetic anhydride from coal. Some of the acetic acid produced in the final step is recycled into the preparation of methyl acetate (see equation in text). (b) The Eastman Kodak chemicals-from-coal facility in Kingsport, Tennessee. Numbers in the photograph represent different parts of the plant: 1, coal unloading; 2, coal silos; 3, steam plant; 4, slurry preparation; 5, coal gasification plant; 6, gas cleanup and separation; 7, sulfur recovery plant; 8, gas flare stack; 9, chemical storage; 10, methanol plant; 11, methyl acetate plant; 12, acetic anhydride plant. *(b, Courtesy of Tennessee Eastman)*

plastics, and other products. The main chemical reactions used in the process are

1. $$\underset{\text{Coal}}{C} + \underset{\text{Steam}}{H_2O} \longrightarrow \underset{\text{Synthesis gas}}{CO + H_2}$$

2. $$CO + 2\,H_2 \longrightarrow \underset{\text{Methanol}}{CH_3OH}$$

3. $$CH_3OH + \underset{\text{Acetic acid}}{CH_3\overset{\overset{\displaystyle O}{\|}}{C}OH} \longrightarrow \underset{\text{Methyl acetate}}{CH_3\overset{\overset{\displaystyle O}{\|}}{C}OCH_3} + H_2O$$

4. $$CH_3\overset{\overset{\displaystyle O}{\|}}{C}OCH_3 + CO \longrightarrow \underset{\text{Acetic anhydride}}{CH_3{-}\overset{\overset{\displaystyle O}{\|}}{C}{-}O{-}\overset{\overset{\displaystyle O}{\|}}{C}{-}CH_3}$$

An acid anhydride has the general formula

$$R{-}\overset{\overset{\displaystyle O}{\|}}{C}{-}O{-}\overset{\overset{\displaystyle O}{\|}}{C}{-}R$$

About 900 tons per day of high-sulfur coal from nearby Appalachian coal mines are ground up and mixed with water to form a slurry of 55% to 65% by weight of coal in water. The slurry is fed into two gasifiers to make synthesis gas. To produce the same amounts of these chemicals by conventional means would require the equivalent of 2 million barrels of oil per year. This is an example of how the use of U.S.-produced coal can lower our dependence on imported oil.

The plant design uses the latest control technologies to protect the environment. For example, the sulfur recovery unit converts the hydrogen sulfide gas that was removed during the gasification of coal into free sulfur. This process removes more than 99% of the sulfur from the coal, and this sulfur is then sold to chemical companies for other uses.

SELF-TEST 10B

1. The simplest aldehyde is ____________.
2. The organic acid found in vinegar is ____________.
3. Ethyl acetate can be prepared by reacting ____________ with ____________.
4. The name of the ester obtained by reacting methanol with acetic acid is ____________.
5. Citric acid found in citrus fruits has ____________ carboxylic acid groups.
6. The simplest ketone is ____________.
7. The characteristic odors and flavors of many fruits are caused by (a) carboxylic acids, (b) esters, (c) alcohols.

Figure 10.4 Just a few everyday items made from synthetic polymers. *(C. D. Winters)*

10.5 Synthetic Organic Polymers

It is impossible for us to get through a day without using a dozen or more synthetic organic **polymers**. The word *polymer* means "many parts" (Greek *poly*, meaning "many," and *meros*, meaning "parts"). Polymers are giant molecules with molar masses ranging from thousands to millions. Our clothes are polymers; our food is packaged in polymers; our appliances and cars contain a number of polymer components (Figure 10.4).

Both natural and synthetic polymers are known. Examples of natural polymers include proteins, nucleic acids, starch, cellulose, and rubber. Natural polymers are discussed in Chapter 11.

Approximately 80% of the organic chemical industry is devoted to the production of synthetic polymers. The prominence of synthetic polymers in consumer products is indicated by the fact that about half of the top 50 chemicals produced in the United States are used in the production of plastics, fibers, and rubbers. The average production of synthetic polymers in the United States exceeds 200 lb per person annually. Many synthetic organic polymers are **plastics** of one sort or another. All plastics are polymers, but not all polymers are plastics. Examples of items often made of plastics include dishes and cups, containers, telephones, plastic bags for packaging and wastes, plastic pipes and fittings, automobile steering wheels and seat covers, and cabinets for appliances, radios, and television sets. In fact, such plastic items, along with textile fibers and synthetic rubbers, are so widely used that they are commonly taken for granted.

■ A plastic is a substance that will flow under heat and pressure and hence is capable of being molded into various shapes. All plastics are polymers, but not all polymers are plastics.

Some of our most useful polymer chemistry has resulted from copying giant molecules in nature. Rayon is remanufactured cellulose (discussed in Section 11.2); synthetic rubber is copied from natural latex rubber. As useful as these polymers may be, however, polymer chemistry is not restricted to nature's models. Polystyrene, nylon, and Dacron are a few examples of synthetic molecules that do not have exact duplicates in nature.

Polymers are made by chemically joining together many small molecules into one giant molecule, or macromolecule. The small molecules used to synthesize polymers are called **monomers**. Synthetic polymers can be classified as **addition polymers**, made by monomer units directly joined together; or **condensation polymers**, made by monomer units combining so that a small molecule, usually water, is split out between them.

■ A *macromolecule* is a molecule with a very high molar mass.

Addition Polymers

Polyethylene

The World of Chemistry

Program 1, *The World of Chemistry*

Most of the chemical industry in the United States today is based on petroleum. You really have a huge network from the raw materials which are primarily petroleum to the refinery, which is primarily the place in which you get starting materials. And those starting materials are then used to make petrochemicals. The petrochemicals are then used to make all kinds of things. Mary Good, Allied Signal Corporation (currently Under Secretary for Technology, U.S. Department of Commerce).

The monomer for addition polymers normally contains one or more double bonds. The simplest monomer of this group is ethylene ($CH_2{=}CH_2$). When ethylene is heated to between 100°C and 250°C at a pressure of 1000 atmospheres (atm) to 3000 atm in the presence of a catalyst, polymers are formed with molar masses of up to several million. A reaction of ethylene usually begins with breaking of one of the bonds in the carbon–carbon double bond, so an unpaired electron, a reactive site, remains at each end of the molecule. This step, the **initiation** of the polymerization, can be accomplished with initiator chemicals such as organic peroxides that are unstable and easily break apart into free radicals, which have unpaired electrons (Section 8.3). The free radicals react readily with molecules containing carbon–carbon double bonds to produce new free radicals. This eliminates the double bond.

■ An organic peroxide, RO—OR′, produces free radicals, RO·, each with an unpaired electron.

$$\mathrm{H_2C{=}CH_2} \xrightarrow[\cdot\,\mathrm{OR}]{\text{Free radical}} \cdot\mathrm{CH_2{-}CH_2{-}OR}$$

The growth of the polyethylene chain then begins as the unpaired electron bonds to a double bond electron in an unreacted ethylene molecule. This leaves another unpaired electron to bond with yet another ethylene molecule. For example,

$$\cdot\mathrm{CH_2{-}CH_2}\cdot + \cdot\mathrm{CH_2{-}CH_2{-}OR} \longrightarrow \cdot\mathrm{CH_2{-}CH_2{-}CH_2{-}CH_2{-}OR} \xrightarrow{n\mathrm{CH_2{=}CH_2}} \left(\!\!-\mathrm{CH_2{-}CH_2}-\!\!\right)_n$$

Polyethylene
n ranges from 1000 to 50,000

In the process, the unsaturated hydrocarbon monomer, ethylene, is changed to a saturated hydrocarbon polymer, **polyethylene**.

Polyethylene is the world's most widely used polymer. More than 24 million tons were produced in 1996 in the United States alone. What are some of the reasons for this popularity? The wide range of properties of polyethylene leads to many uses.

Polyethylenes formed under various pressures and catalytic conditions have different molecular structures and hence different physical properties. For example, chromium oxide as a catalyst yields almost exclusively the linear polyethylene shown at the end of this paragraph—a polymer with no branches on the carbon chain. The zigzag structure represents the shape of the chain more closely because of the tetrahedral arrangement of bonds around each carbon in the saturated polyethylene chain. Long, linear chains of polyethylene can pack closely together, and give a material with high density (0.97 g/mL) and high molar mass, referred to as high-density polyethylene (HDPE). This material is hard, tough, and rigid. The plastic milk bottle is a good example of an application of HDPE. Other HDPE containers are shown in Figure 10.5.

The World of Chemistry

Program 22, *The Age of Polymers*

Many places where we use metals today, particularly in structural applications, will be taken over by polymers in the future. . . . A prime reason for this is that polymers allow an almost infinite variation in structure, and hence, almost infinite variation in their properties. Whereas we now know almost all there is to know about metals and glasses—obviously not everything, but almost—there are an almost infinite variety of polymers that we can conceive. We can fine tune their structures and fine tune their properties given enough time.
Howard E. Simmons, Vice President, DuPont Company.

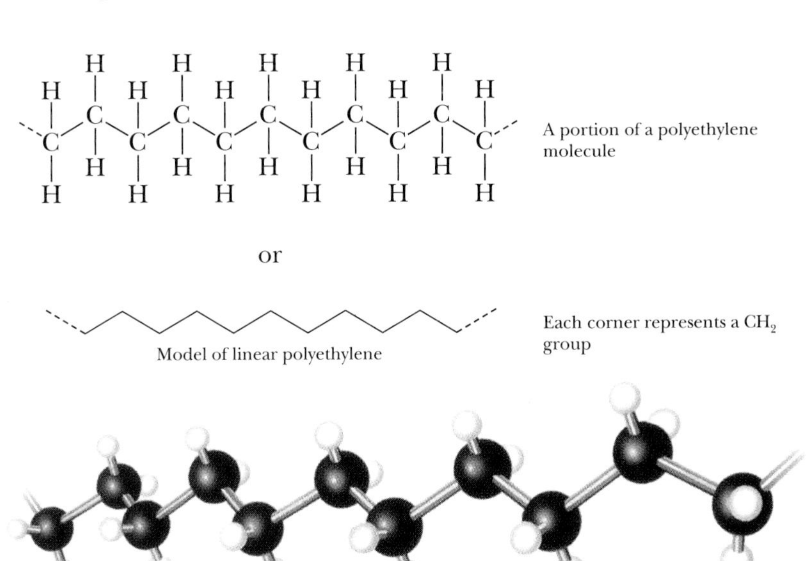

Figure 10.5 Some bottles made of HDPE, one of the most commonly recycled kinds of polymeric materials. *(C. D. Winters)*

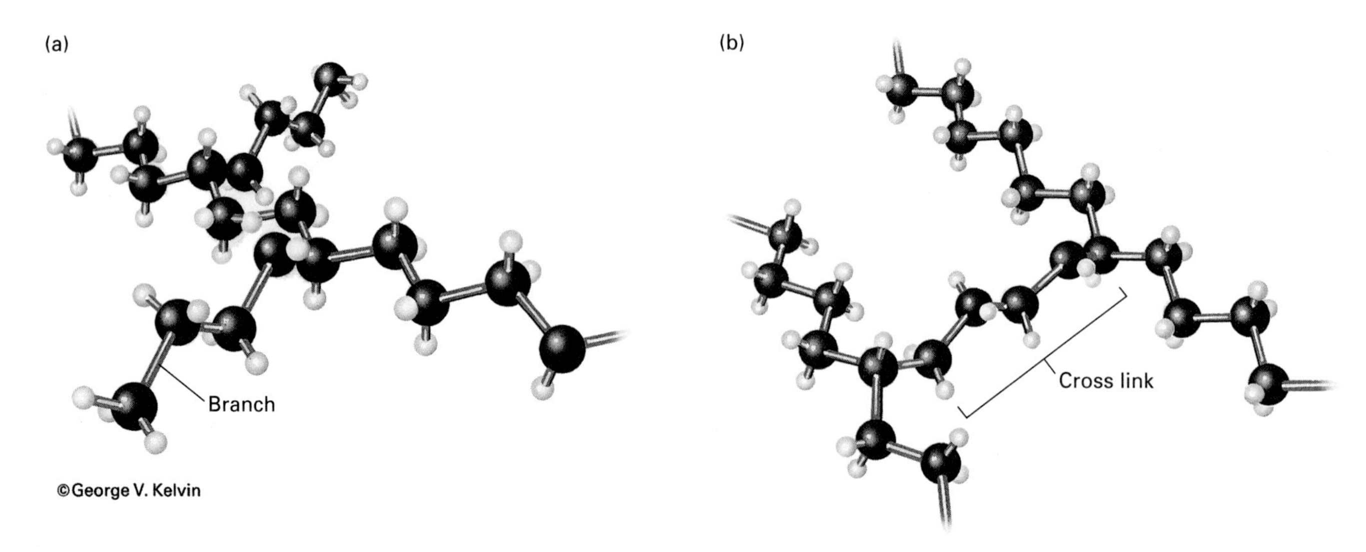

Figure 10.6 Models of (a) branched and (b) cross-linked polyethylene.

If ethylene is heated to 230°C at a pressure of 200 atm, free radicals attack the polyethylene chain at random positions, causing irregular branching (Figure 10.6*a*). Branched chains of polyethylene cannot pack closely together, so the resulting material has a lower density (0.92 g/mL) and is called low-density polyethylene (LDPE). This material is soft and flexible (Figure 10.7). Sandwich bags are made from LDPE. Other conditions can lead to cross-linked polyethylene, in which branches connect long chains to each other (Figure 10.6*b*). If the linear chains of polyethylene are treated in a way that causes cross-links between chains to form cross-linked polyethylene (CLPE),

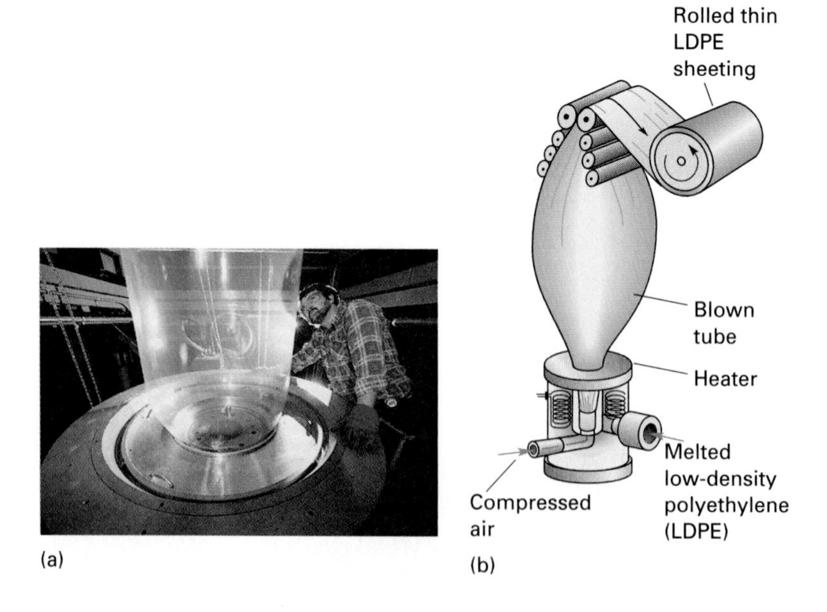

Figure 10.7 Making LDPE film. (a) The blown tube of polyethylene film. (b) Diagram showing how the tube is blown and the film collected in a roll. The process is similar to blowing a bubble gum bubble. *(a, Gary Gladstone/The Image Bank)*

a very tough form of polyethylene is produced. The plastic caps on soft drink bottles are made from CLPE.

Polymers of Ethylene Derivatives

Many different kinds of addition polymers are made from monomers in which one or more of the hydrogen atoms in ethylene have been replaced with either halogen atoms or a variety of organic groups. If the formation of polyethylene is represented as

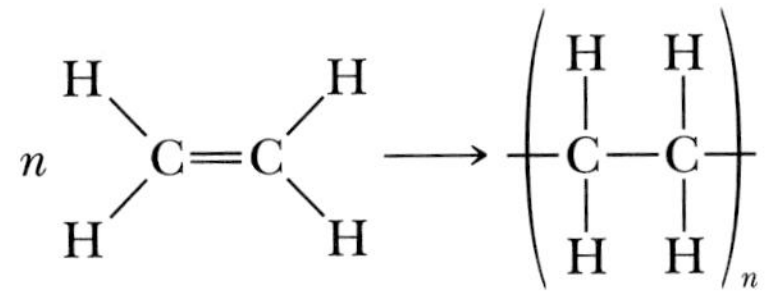

$$n\ H_2C{=}CH_2 \longrightarrow \left(-CH_2-CH_2- \right)_n$$

then the general reaction

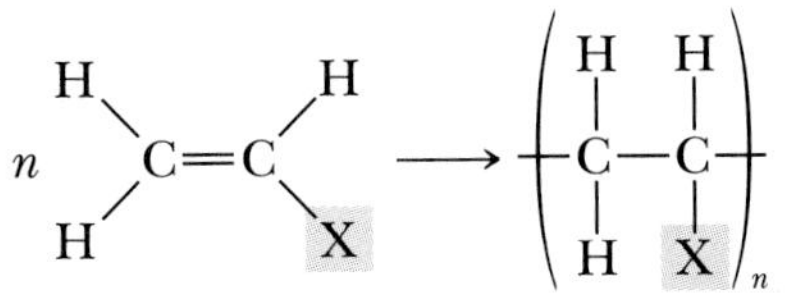

$$n\ H_2C{=}CHX \longrightarrow \left(-CH_2-CHX- \right)_n$$

can be used to represent a number of other important addition polymers, where X is Cl, F, or an organic group (Table 10.9 on p. 242).

For example, the monomer for making **polystyrene** is styrene, and n is about 5700.

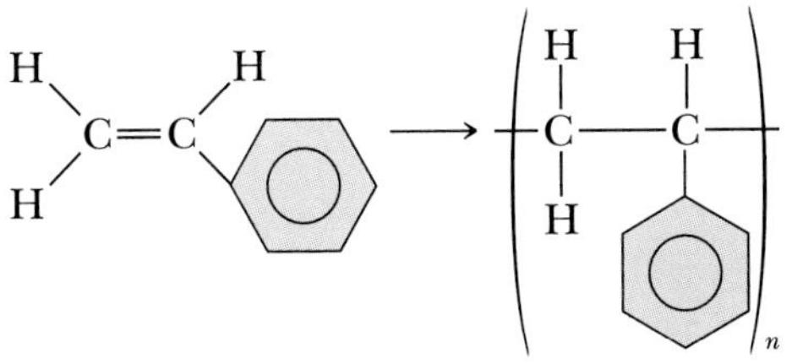

$$H_2C{=}CH(C_6H_5) \longrightarrow \left(-CH_2-CH(C_6H_5)- \right)_n$$

Styrene Polystyrene

Polystyrene is a clear, hard, colorless solid at room temperature that can be molded easily at 250°C. Styrene is tenth on the list of top ten chemicals, primarily because of its use in making polystyrene. More than 6 million tons of polystyrene are produced in the United States each year to make food containers, toys, electrical parts, insulating panels, appliance and furniture components, and many other items. The variation in properties shown by polystyrene products is typical of synthetic polymers. For example, a clear polystyrene drinking glass that is brittle and breaks into sharp pieces somewhat like glass is much different from the polystyrene coffee cup that is soft and pliable.

A major use of polystyrene is in the production of Styrofoam by "expansion molding." In this process, polystyrene beads are placed in a mold and heated with steam or hot air. The beads, 0.25 mm to 1.5 mm in diameter,

TABLE 10.9 ■ Ethylene Derivatives that Undergo Addition Polymerization

Formula	Monomer Common Name	Polymer Name (Trade Names)	Uses	U.S. Polymer Production (Tons/Year)
$H_2C=CH_2$	Ethylene	Polyethylene (Polythene)	Squeeze bottles, bags, films, toys and molded objects, electrical insulation	24 million
$H_2C=CH-CH_3$	Propylene	Polypropylene (Vectra, Herculon)	Bottles, films, indoor–outdoor carpets	12 million
$H_2C=CH-Cl$	Vinyl chloride	Poly(vinyl chloride) (PVC)	Floor tile, raincoats, pipe	5 million
$H_2C=CH-CN$	Acrylonitrile	Polyacrylonitrile (Orlon, Acrilan)	Rugs, fabrics	1 million
$H_2C=CH-C_6H_5$	Styrene	Polystyrene (Styrene, Styrofoam, Styron)	Food and drink coolers, building material insulation	6 million
$H_2C=CH-O-C(=O)-CH_3$	Vinyl acetate	Poly(vinyl acetate) (PVA)	Latex paint, adhesives, textile coatings	500,000
$H_2C=C(CH_3)-C(=O)-O-CH_3$	Methyl methacrylate	Poly(methyl methacrylate) (Plexiglas, Lucite)	High-quality transparent objects, latex paints, contact lenses	450,000
$F_2C=CF_2$	Tetrafluoroethylene	Polytetrafluoroethylene (Teflon)	Gaskets, insulation, bearings, pan coatings	7000

contain 4% to 7% by weight of a low-boiling liquid such as pentane. The steam causes the low-boiling liquid to vaporize; this expands the beads, and as the foamed particles expand, they are molded in the shape of the mold cavity. Styrofoam is used for egg cartons, meat trays, coffee cups, and packing material (Figure 10.8).

Polypropylene, used in indoor–outdoor carpeting, bottles, fabrics, and battery cases, is made from propylene.

$$n\ \text{H}_2\text{C}{=}\text{CH}(\text{CH}_3) \longrightarrow \left[\!-\text{CH}_2-\text{CH}(\text{CH}_3)-\!\right]_n$$

Propylene — Polypropylene

Poly(vinyl chloride) (PVC), used in floor tiles, garden hoses, plumbing pipes, and trash bags has a chlorine atom substituted for one of the hydrogen atoms in ethylene.

$$n\ \text{H}_2\text{C}{=}\text{CHCl} \longrightarrow \left[\!-\text{CH}_2-\text{CHCl}-\!\right]_n$$

Vinyl chloride — Poly(vinyl chloride)

Although the representation

$$\left[\!-\text{CH}_2-\text{CH}_2-\!\right]_n$$

saves space, keep in mind how large the polymer molecules are. Generally n is 500 to 50,000, and this gives molecules with molar masses ranging from 10,000 to several million. The molecules that make up a given polymer sample are of different lengths and thus are not all of the same molar mass. As a result, only the average molar masses can be determined.

In summary, the numerous variations in substituents, length, branching, and cross-linking make it possible to produce a variety of properties for each type of addition polymer. Chemists and chemical engineers can fine-tune the properties of the polymer to match desired properties. Appropriate selection of monomer and reaction conditions accounts for the widespread and growing use of these giant molecules.

Figure 10.8 Some familiar applications of polystyrene foam, which is an excellent thermal insulator. *(C. D. Winters)*

Wigs, among the many items made from acrylic polymers. *(Tom Pantages)*

EXAMPLE 10.1 *Addition Polymers*

Draw the structural formula of the repeating unit for the following addition polymers: (a) polypropylene, (b) poly(vinyl acetate), and (c) poly(vinyl alcohol).

SOLUTION

The names show that the monomers for these polymers are propylene ($CH_2{=}CHCH_3$), vinyl acetate ($CH_2{=}CHOOCCH_3$), and vinyl alcohol ($CH_2{=}CHOH$). The repeating units in the polymers therefore have the same structures, but without the double bonds.

$$\text{(a)}\ \left(-\overset{\displaystyle H}{\underset{\displaystyle H}{C}}-\overset{\displaystyle H}{\underset{\displaystyle CH_3}{C}}- \right)_n \qquad \text{(b)}\ \left(-\overset{\displaystyle H}{\underset{\displaystyle H}{C}}-\overset{\displaystyle H}{\underset{\displaystyle O-C(=O)-CH_3}{C}}- \right)_n \qquad \text{(c)}\ \left(-\overset{\displaystyle H}{\underset{\displaystyle H}{C}}-\overset{\displaystyle H}{\underset{\displaystyle OH}{C}}- \right)_n$$

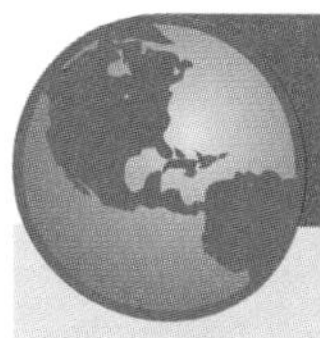

THE WORLD OF CHEMISTRY

Discovery of a Catalyst for Polyacrylonitrile Production

Program 20, *On the Surface*

Oil companies invest considerable sums in equipment and human effort to develop new catalysts to make fuels and chemicals from petroleum. Research and development in this area is a multimillion-dollar gamble. There is no guarantee the money spent will produce anything useful. But if it does, the payoff can be enormous. Just one catalyst breakthrough made more than half a billion dollars for Standard Oil of Ohio, now part of BP America. In the late 1950s, SOHIO researchers came up with a new catalytic process that soon dominated all others in the production of polyacrylonitrile, the polymer used to make textiles, tires, and car bumpers. Oddly enough, SOHIO researchers weren't even trying to produce acrylonitrile at first. They simply wanted to make a metal oxide catalyst to convert waste propane gas from petroleum refining into something more valuable. As Dr. Jeanette Grasselli said,

The theory, the hypothesis at the time, was that we could take the oxygen from the catalyst and insert it into the propane, a relatively unreactive molecule. So this was a tough technical objective. And, in turn, we wanted to generate or take the catalyst back to its original oxidized form by using oxygen from the air.

But the theory didn't hold up. Propane was too stable to react, and the catalyst particles broke down in service. Management gave the research team three more months to show results, so they made some changes. They replaced propane with a more reactive refinery gas, propylene. They made their catalyst from different metal oxides, and they added ammonia to promote, or speed up, the reaction.

To their surprise, ammonia reacted. Rather than just encouraging the reaction to go faster, as a promoter, ammonia reacted and became part of the reaction sequence, and acrylonitrile was made in one step. The researchers had struck pay dirt. Their new catalyst, combined with ammonia, had made a valuable product from a cheap gas. Management quickly saw the value of the new process and wasted no time building a plant to use it. As Dr. Grasselli recalls,

Jeanette Grasselli

In 1960, when our plant came on stream, acrylonitrile was selling for 28 cents a pound. We were making it for 14 cents a pound. And we shut down every other commercial process. Today, 90% of the world's acrylonitrile is manufactured by the SOHIO process.

Exercise 10.1

Draw the structural formula of the monomers used to prepare the following polymers: (a) polyethylene, (b) poly(vinyl chloride), and (c) polystyrene.

Rubber: Natural and Synthetic

Natural rubber, a product of the *Hevea brasiliensis* tree, is a hydrocarbon with the empirical formula C_5H_8. When rubber is decomposed in the absence of oxygen, the monomer isoprene is obtained

$$CH_2{=}\overset{\displaystyle CH_3 \atop |}{C}{-}CH{=}CH_2 \quad \text{or} \quad H_2C{=}C(CH_3){-}C(H){=}CH_2$$

Isoprene (2-methyl-1,3-butadiene)

Natural rubber occurs as *latex* (an emulsion of rubber particles in water), which oozes from rubber trees when they are cut. Precipitation of the rubber particles yields a gummy mass that is not only elastic and water-repellent but also very sticky, especially when warm. In 1839, after five years of work on this material, Charles Goodyear (1800–1860) discovered that heating gum rubber with sulfur produces a material that is no longer sticky but is still elastic, water-repellent, and resilient.

Vulcanized rubber, as the type of rubber Goodyear discovered is now known, contains short chains of sulfur atoms that bond together the polymer chains of the natural rubber and reduce its unsaturation. The sulfur chains help to link the polymer chains, so the material does not undergo a permanent change when stretched but springs back to its original shape and size when the stress is removed (Figure 10.9). Substances that behave this way are called **elastomers**.

In later years chemists searched for ways to make a synthetic rubber so we would not be completely dependent on imported natural rubber during emergencies, such as during the first years of World War II. In the mid-1920s, German chemists polymerized butadiene (obtained from petroleum and structurally similar to isoprene, but without the methyl-group side chain).

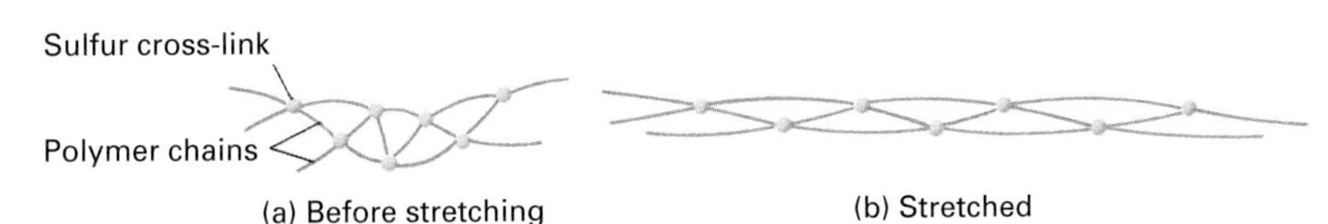

Figure 10.9 Stretching vulcanized rubber. After stretching, the cross-links pull it back into the original arrangement.

$$n\ \mathrm{CH_2{=}CH{-}CH{=}CH_2} \xrightarrow[\text{polymerization}]{\text{Addition}} \mathrm{{-}CH_2{-}CH{=}CH{-}CH_2{-}\!\left(CH_2{-}CH{=}CH{-}CH_2\right)_n\!{-}CH_2{-}CH{=}CH{-}CH_2{-}}$$

1,3-Butadiene — Polybutadiene

Polybutadiene is used in the production of tires, hoses, and belts.

The behavior of natural rubber (polyisoprene), it was learned later, is due to the specific molecular geometry within the polymer chain. We can write the formula for polyisoprene with the CH_2 groups on opposite sides of the double bond (the *trans* arrangement)

$$\mathrm{{-}CH_2{-}C(CH_3){=}CH{-}CH_2{-}\!\left(CH_2{-}C(CH_3){=}CH{-}CH_2{-}CH_2{-}C(CH_3){=}CH{-}CH_2\right)_n\!{-}}$$

Poly-*trans*-isoprene (the $-CH_2-CH_2-$ groups are *trans*)

or with the CH_2 groups on the same side of the double bond (the *cis* arrangement).

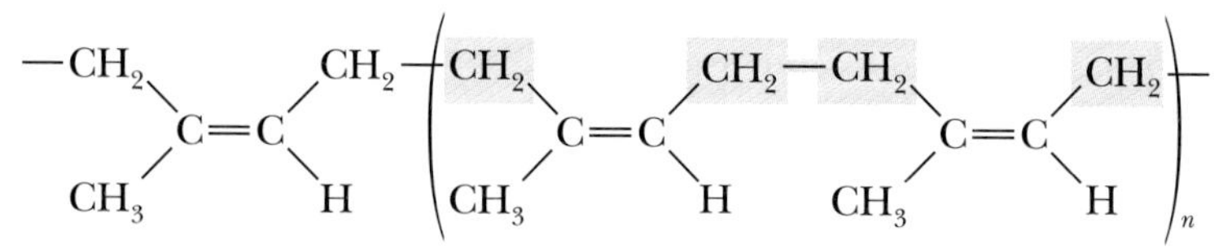

Poly-*cis*-isoprene (the $-CH_2-CH_2-$ groups are *cis*)

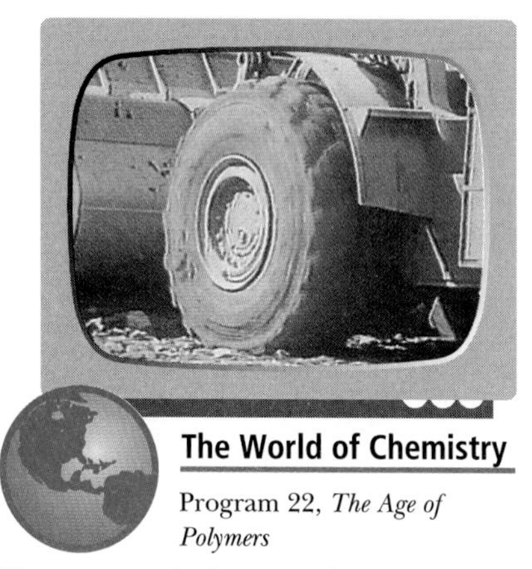

The World of Chemistry

Program 22, *The Age of Polymers*

Tires are made from synthetic rubber polymers.

Natural rubber is poly-*cis*-isoprene. However, the *trans* material also occurs in nature in the leaves and bark of the sapotacea tree and is known as *gutta-percha.* It is brittle and hard and is used for golf ball covers, electrical insulation, and other applications not requiring the stretching properties of rubber. Without an appropriate catalyst, polymerization of isoprene yields a solid that is like neither rubber nor gutta-percha because it is a random mixture of the *cis* and *trans* geometries. Neither the *trans* polymer nor the randomly arranged material is as good as natural rubber (*cis*) for making automobile tires.

■ Stereoregulation catalysts catalyze reactions that favor the formation of one geometric isomer.

In 1955, chemists at the Goodyear and Firestone companies almost simultaneously discovered how to use **stereoregulation** catalysts to prepare synthetic poly-*cis*-isoprene. This material is structurally identical to natural rubber. Today, synthetic poly-*cis*-isoprene can be manufactured cheaply and is used almost equally well (there is still an increased cost) when natural rubber is in short supply. More than 10 million tons of synthetic rubber are produced worldwide every year.

■ SELF-TEST 10C

1. The individual molecules from which polymers are made are called ____________.
2. The initiation step of an addition polymerization uses ____________ ____________ such as ____________ ____________.

3. Monomers must have a ______________ bond in their structure if they are to participate in an addition reaction.
4. The double bond in ethylene is converted to a ______________ bond during an addition reaction.
5. The monomer in teflon is CF_2CF_2, 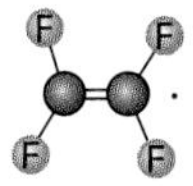 . (a) True, (b) False.
6. PVC is poly(vinyl chloride). It is built from the monomer ______________.
7. Natural rubber is poly- ______________ -isoprene.
8. Three different types of polyethylene are LDPE, HDPE, and CLPE.
   (a) LDPE stands for ______________.
   (b) HDPE stands for ______________.
   (c) CLPE stands for ______________.

Condensation Polymers

A **condensation reaction** is one in which two molecules react by forming a larger molecule and eliminating a small molecule. The reaction of alcohols with carboxylic acids to give esters and water (Section 10.3) is an example of condensation. This important type of chemical reaction does not depend on the presence of a double bond in the reacting molecules. Instead, it requires the presence of two different kinds of functional groups on two different molecules. If each reacting molecule has two functional groups, both of which can react, it is possible for condensation reactions to produce long-chain polymers.

Polyesters

A molecule with two carboxylic acid groups, such as terephthalic acid, and another molecule with two alcohol groups, such as ethylene glycol, can react with each other at both ends.

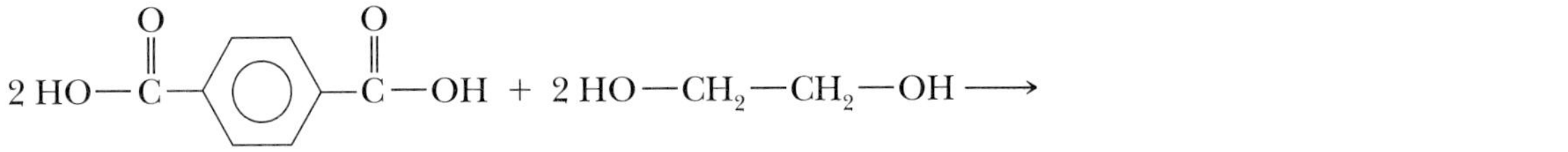

$$2\,HO-\overset{O}{\overset{\|}{C}}-C_6H_4-\overset{O}{\overset{\|}{C}}-OH + 2\,HO-CH_2-CH_2-OH \longrightarrow$$

Terephthalic acid — Ethylene glycol

$$HO-\overset{O}{\overset{\|}{C}}-C_6H_4-\overset{O}{\overset{\|}{C}}-O-CH_2-CH_2-O-\overset{O}{\overset{\|}{C}}-C_6H_4-\overset{O}{\overset{\|}{C}}-O-CH_2-CH_2-OH + 2\,H_2O$$

If n molecules of acid and alcohol can react in this manner, the process will continue until a large polymer molecule, known as a **polyester**, is produced.

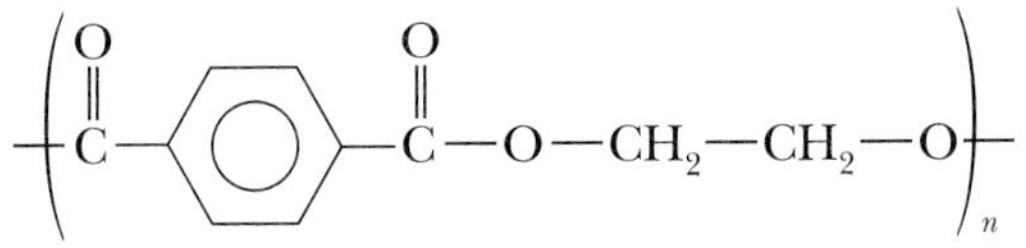

$$\left(\overset{O}{\overset{\|}{C}}-C_6H_4-\overset{O}{\overset{\|}{C}}-O-CH_2-CH_2-O\right)_n$$

Poly(ethylene terephthalate)

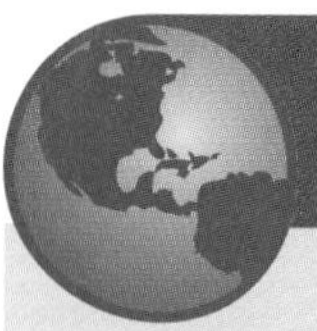

THE WORLD OF CHEMISTRY

Inventor of the Poly(ethylene terephthalate) Bottle

Program 22, *The Age of Polymers*

The inventor of the poly(ethylene terephthalate) soft drink bottle is Nathaniel Wyeth, who comes from the internationally famous family of artists. His brother, Andrew Wyeth, expresses his creativity on canvas, but Nat Wyeth expresses his through chemical engineering. He has an intriguing story:

I got to thinking about the work that Wallace Carothers did for DuPont way back in the days when nylon was born, where he found that, if you took a thread of nylon when it was cold—that is, below the melt point—and stretched it, it would orient itself. That is, the molecules of the polymer would align themselves. This is what you're doing to the molecules when you orient them—you're lining them up so they can give you the most strength. They're all pulling in the direction you want them to pull in.

Wyeth tried this approach to make a poly(ethylene terephthalate) bottle, but the bottles kept splitting. Wyeth estimates that he made 10,000 tries and had 10,000 failures before he made a simple observation:

Well, then I realized what we've got to do now is to align these molecules in the sidewall of the bottle, not only in one direction, but in two directions. So I thought I'd play a trick on this mold, on this problem. I took two pieces of poly(ethylene terephthalate) and turned one of them ninety degrees with the other. So then I had one that would split in this direction and one that would split in that direction. Well, one piece reinforced the other. As soon as I did that, I could blow bottles. That seems almost dirt simple. But as I've often said, quoting Einstein, the biggest part of a problem, and the easiest way to solving a problem, is to understand it, have the problem in a form you can understand what's going on. And what I was doing here was learning about what was going on. Once I knew, it was simple to solve.

Nathaniel Wyeth

More than 2 million tons of poly(ethylene terephthalate), commonly referred to as PET, are produced in the United States each year for use in beverage bottles, apparel, tire cord, film for photography, food packaging, coatings for microwave and conventional ovens, and home furnishings. Polyester textile fibers are marketed under such names as Dacron and Terylene. Films of the same polyester, when magnetically coated, are used to make audiotapes and videotapes. This film, Mylar, has unusual strength and can be rolled into sheets 1/30 the thickness of a human hair.

The inert, nontoxic, nonallergenic, noninflammatory, and non-blood-clotting properties of Dacron polymers make Dacron tubing an excellent substitute for human blood vessels (Figure 10.10) in heart bypass operations. Dacron sheets are used as a skin substitute for burn victims.

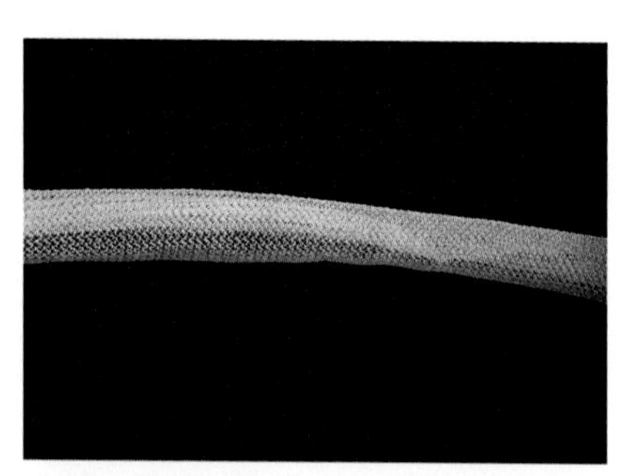

Figure 10.10 Dacron mesh used to replace damaged arteries. *(Dan McCoy/Rainbow)*

Polyamides (Nylons)

Another useful and important type of condensation reaction is that between a carboxylic acid and a primary **amine**, which is an organic compound containing an $—NH_2$ functional group. Amines can be considered derivatives of ammonia (NH_3), and most of them are weak bases, similar in strength to ammonia. An amine reacts with a carboxylic acid to split out a water molecule and form an **amide**, for example,

$$\underset{\text{Carboxylic acid}}{R-\overset{\overset{\displaystyle O}{\|}}{C}-OH} + \underset{\text{Amine}}{H-\underset{\underset{\displaystyle H}{|}}{N}-R} \xrightarrow{\text{Heat}} \underset{\text{Amide}}{R-\overset{\overset{\displaystyle O}{\|}}{C}-\underset{\underset{\displaystyle H}{|}}{N}-R} + H_2O$$

Polymers are produced when diamines react with dicarboxylic acids. Reactions of this type yield a group of polymers that perhaps have had a greater impact on society than any other type. These are the **polyamides**, or nylons.

In 1928, the DuPont Company embarked on a program of basic research headed by Dr. Wallace Carothers (1896–1937), who came to DuPont from the Harvard University faculty. His research interests were high-molecular-weight compounds, such as rubber, proteins, and resins, and the reaction mechanisms that produced these compounds. In February 1935, his research yielded a product known as nylon-66 (Figure 10.11), prepared from adipic acid (a diacid) and hexamethylenediamine (a diamine).

■ The name of nylon-66 is based on the number of carbon atoms in the diamine and diacid, respectively, that are used to make the polymer. Since both hexamethylenediamine and adipic acid have six carbon atoms, nylon-66 is the product.

$$n\,\underset{\text{Adipic acid}}{HO-\overset{\overset{\displaystyle O}{\|}}{C}-(CH_2)_4-\overset{\overset{\displaystyle O}{\|}}{C}-OH} + n\,\underset{\text{Hexamethylenediamine}}{H_2N-(CH_2)_6-NH_2} \longrightarrow$$

$$\underset{\text{Nylon-66}}{\left(\underset{\underset{\displaystyle H}{|}}{N}-(CH_2)_6-\underset{\underset{\displaystyle H}{|}}{N}-\overset{\overset{\displaystyle O}{\|}}{C}-(CH_2)_4-\overset{\overset{\displaystyle O}{\|}}{C}\right)_n} + n\,H_2O$$

This material could easily be extruded into fibers that were stronger than natural fibers and chemically more inert. The discovery of nylon jolted the American textile industry at almost precisely the right time. Natural fibers were not meeting the needs of 20th-century Americans. Silk was not durable and was very expensive; wool was scratchy; linen crushed easily; and cotton did not have a high-fashion image. All four had to be pressed after cleaning. As women's hemlines rose in the mid-1930s, silk stockings were in great de-

■ In extrusion, a pliable substance is given a shape by being pushed through an opening. Toothpaste is extruded from a toothpaste tube.

Figure 10.11 Making nylon. The diacid is dissolved in hexane and the diamine is dissolved in water. They do not mix with each other. The layers form because hexane and water are immiscible. The nylon polymer forms at the interface between the two reactants and continues to form there as the nylon "rope" is pulled away. *(C. D. Winters)*

mand, but they were very expensive and short-lived. Nylon changed all that almost overnight. Nylon could be knitted into the sheer hosiery women wanted, and it was much more durable than silk. The first public sale of nylon hose took place in Wilmington, Delaware (the location of DuPont's main office), on October 24, 1939. World War II caused all commercial use of nylon to be abandoned until 1945, as the industry turned to making parachutes and other war materials. Not until 1952 was the nylon industry able to meet the demands of the hosiery industry and to release nylon for other uses.

■ The amide linkage in nylon is the same linkage found in proteins, where it is called the peptide linkage (Section 11.7).

Figure 10.12 illustrates another facet of the structure of nylon—hydrogen bonding—which explains why nylons make such good fibers. To have good tensile strength, the chains of atoms in a polymer should be able to attract one another, but not so strongly that the plastic cannot be initially extended to form the fibers. Ordinary covalent chemical bonds linking the chains together would be too strong. Hydrogen bonds, with a strength about one tenth that of an ordinary covalent bond, link the chains in the desired manner. Kevlar, another polyamide, is used to make bulletproof vests (Figure 10.13) and fireproof garments. Kevlar is made from *p*-phenylenediamine and terephthalic acid.

■ Hair, wool, and silk are examples of nature's version of nylon. However, these natural polymers have only one carbon between each pair of $-\overset{\overset{\Large O}{\|}}{C}-\underset{\underset{\Large H}{|}}{N}-$ units instead of the half dozen or so found in synthetic nylons.

THE PERSONAL SIDE

Stephanie Louise Kwolek (1923–)

Stephanie Kwolek (DuPont)

Stephanie Kwolek received a Bachelor of Science degree from Carnegie–Mellon University in 1946. Although she wanted to study medicine, she couldn't afford it and decided to take a temporary job at DuPont. She liked her work so well that she stayed for 40 years, retiring in 1986. During her career at DuPont, she received 17 U.S. patents and 86 foreign patents for her work on a variety of polymeric fibers. However, she is best known for her work on the development of Kevlar fiber, which is five times stronger than steel on a weight basis. The patent for Kevlar fiber was issued in 1965, but 15 years passed before it was fully commercialized. Although Kevlar is best known for its use in bulletproof vests, it has a number of other important uses that include brake linings, underwater cables, and high-performance composite materials. In 1994, DuPont featured Stephanie Kwolek in a television commercial about the use of Kevlar in bulletproof vests, which resulted in name recognition for both Kevlar and Kwolek by the American public. However, what isn't widely known are the accomplishments of a woman during a time when women often didn't receive appropriate recognition for their work. She has received many awards, including an honorary Doctor of Science Degree from Worcester Polytechnic Institute in 1981 for her contributions to polymer and fiber chemistry; the American Chemical Society Award for Creative Invention in 1980; election to the Engineering and Science Hall of Fame in Dayton, Ohio; and the 1997 Perkin Medal from the Society of Chemical Industry for her outstanding achievements in applied chemistry.

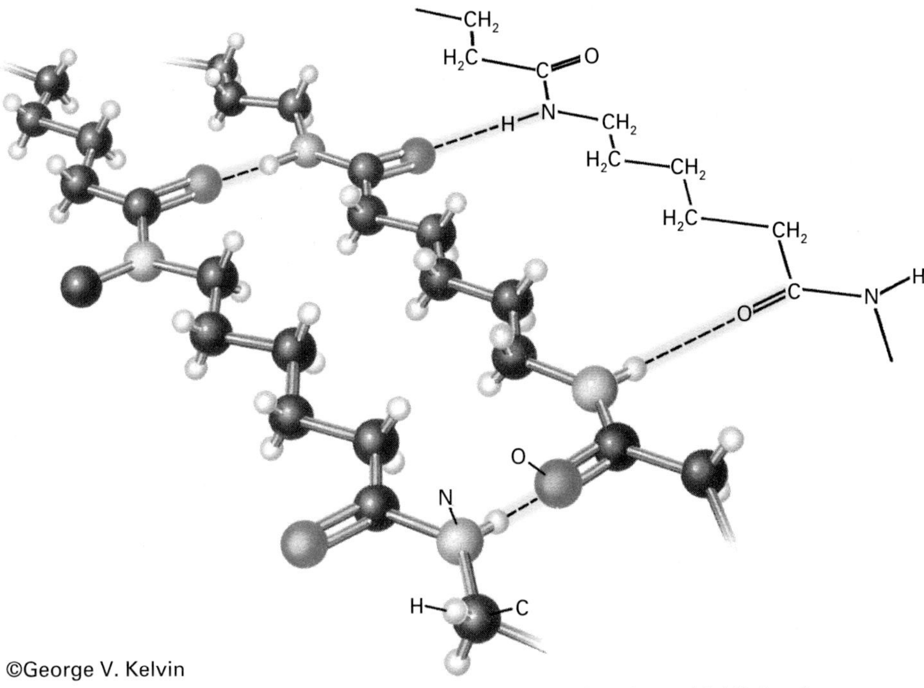

©George V. Kelvin

Figure 10.12 Hydrogen bonding in nylon-66. Hydrogen bonds are highlighted.

EXAMPLE 10.2 *Condensation Polymers*

Write the repeating unit of the condensation polymer obtained by combining $HOOCCH_2CH_2COOH$ and $H_2NCH_2CH_2NH_2$.

SOLUTION

The dicarboxylic acid reacts with the diamine to split out water molecules and form a polyamide. The repeating unit is

$$\left(\overset{O}{\overset{\|}{C}}CH_2CH_2\overset{O}{\overset{\|}{C}}\underset{\underset{H}{|}}{N}CH_2CH_2\underset{\underset{H}{|}}{N} \right)_n$$

Exercise 10.2

Draw the structure of the repeating unit of the condensation polymer obtained from reacting terephthalic acid with ethylene glycol.

Figure 10.13 Vests made of Kevlar have saved many policemen's lives. (Kevlar® is a registered trademark of DuPont for its aramid fiber.)

10.6 New Polymer Materials

Few plastics produced today find end uses without some kind of modification. For example, body panels for the GM Saturn and Corvette automobiles are made of **reinforced plastics**, which contain fibers embedded in a matrix of a polymer. These are often referred to as **composites**. The strongest geometry for a solid is a wire or a fiber, and the use of a polymer matrix prevents the fiber from bending or buckling. As a result, reinforced plastics are stronger than steel on a weight basis. In addition, the composites have a low density—from 1.5 g/cm^3 to 2.25 g/cm^3 compared with 2.7 g/cm^3 for aluminum, 7.9 g/cm^3 for steel, and 2.5 g/cm^3 for concrete. The only structural material with a lower density is wood, which has an average value of 0.5 g/cm^3. In addition, polymers do not corrode. The low density, high strength, and high chemical resistance of composites are the basis for their increased use in the automobile, airplane, construction, and sporting goods industries.

Glass fibers currently account for more than 90% of the fibrous material used in reinforced plastics because glass is inexpensive and glass fibers possess high strength, low density, good chemical resistance, and good insulating properties. In principle, any polymer can be used for the matrix material. Polyesters are the number-one polymer matrix at the present time. Glass-reinforced polyester composites have been used in structural applications such as boat hulls, airplanes, missile casings, and automobile body panels.

Other fibers and polymers have been used, and the trend is toward increased utilization of composites in automobiles and aircraft. For example, a composite of graphite fibers in a polymer matrix is used in the construction of the F-117 Stealth fighter and other military aircraft. Graphite–polymer composites are used in a number of sporting goods such as golf club shafts, tennis racquets, fishing rods, and skis. The F-16 military aircraft was the first to contain graphite–polymer composite material, and the technology has advanced to the point where many aircraft, such as the F-18, use graphite composites for up to 26% of the aircraft's structural weight. This percentage is projected to increase to 40% to 50% in future aircraft.

Although few automobiles contain exterior body panels made of plastics, a number of components are plastic. Examples include bumpers, trim, light lenses, grilles, dashboards, seat covers, and steering wheels—enough plastics

(a)

(b)

(c)

New polymer materials. (a) The flexible fender panel in a Saturn automobile. (b) An array of skis. (c) An F-16 fighter plane. *(a, Tom Pantages; b, Ann Kotowicz/Rainbow; c, Tom Hollyman/Photo Researchers)*

FRONTIERS IN THE WORLD OF CHEMISTRY

Elephants, Piano Keys, and Polymers

Until recently, no synthetic material has matched ivory in responding to the delicate touch of a concert pianist. Piano keys covered with plastic veneers have been rejected as too slippery and too cool. Eventually, however, a replacement for ivory must be found. In an effort to halt the slaughter of elephants, a global ban on trading in ivory was initiated in 1990 and is reported to be very effective.*

A team of scientists from Rensselaer Polytechnic Institute has patented a new material that may solve the problem. The essential step in their work was analysis of the surface of natural ivory at the microscopic level by a tribologist, an engineer who studies friction and materials that slide against each other. An ivory surface, they found, is covered with ridges, valleys, and tiny pores. When a sweaty finger slides over such a surface, it alternately sticks and slips, creating the feeling that pianists need for better control. The pores make an important contribution by absorbing sweat and oil from the finger.

To duplicate the ridges and valleys, the scientists worked with a finely made cast of a natural ivory surface. Duplicating the pores was challenging. Ultimately, they developed a synthetic ivory made from a mixture of a liquid polyester, a white titanium pigment, and finely powdered poly(ethylene glycol), a water-soluble polymer. When the material is soaked in hot water after it has hardened, the poly(ethylene glycol) is dissolved away, leaving behind pores like those in natural ivory. Their new material has met the ultimate test. Concert pianists have failed to detect a difference between Steinway pianos with keys covered by the new material, known as RPIvory, and those with keys of natural ivory.

Professor Henry Scarton of RPI with a piano key made of RPIvory. (Rensselaer Polytechnic Institute)

**The New York Times*, p. C3, May 25, 1993.

to account for an average of 250 lb per car. The increased emphasis on improving fuel efficiency will likely lead to the use of greater amounts of plastics in the construction of automobiles, both in interior components and exterior body panels.

10.7 Recycling Plastics

Recycling metals such as aluminum, iron, and lead has been occurring for years (Section 6.6). However, programs for recycling plastics developed much more slowly because of the costs associated with separating different types of plastics and producing usable recycled products from the used plastics.

Disposal of plastics has been the subject of considerable debate as municipalities face increasing problems in locating sufficient landfill space. The number-one waste is paper products, which make up about 40% of the volume in landfills. (Newspaper alone accounts for 16% of the volume.) Next are plastics, which make up about 20% of the volume in landfills. At the present time, about 3% of plastics waste is being recycled as compared to recycling of 65% of aluminum cans, 20% of paper, and 10% of glass.

A planter made of plastic lumber, which can be worked like wood, but does not rot, splinter, need paint, or get eaten by termites. A mixture of several kinds of recycled plastics is the raw material for this lumber. *(Courtesy of Obex, Inc., Stamford, CT)*

Four phases are needed for a successful recycling of any waste material: collection, sorting, reclamation, and end use. Public enthusiasm for recycling and state laws requiring recycling have resulted in a dramatic increase in the

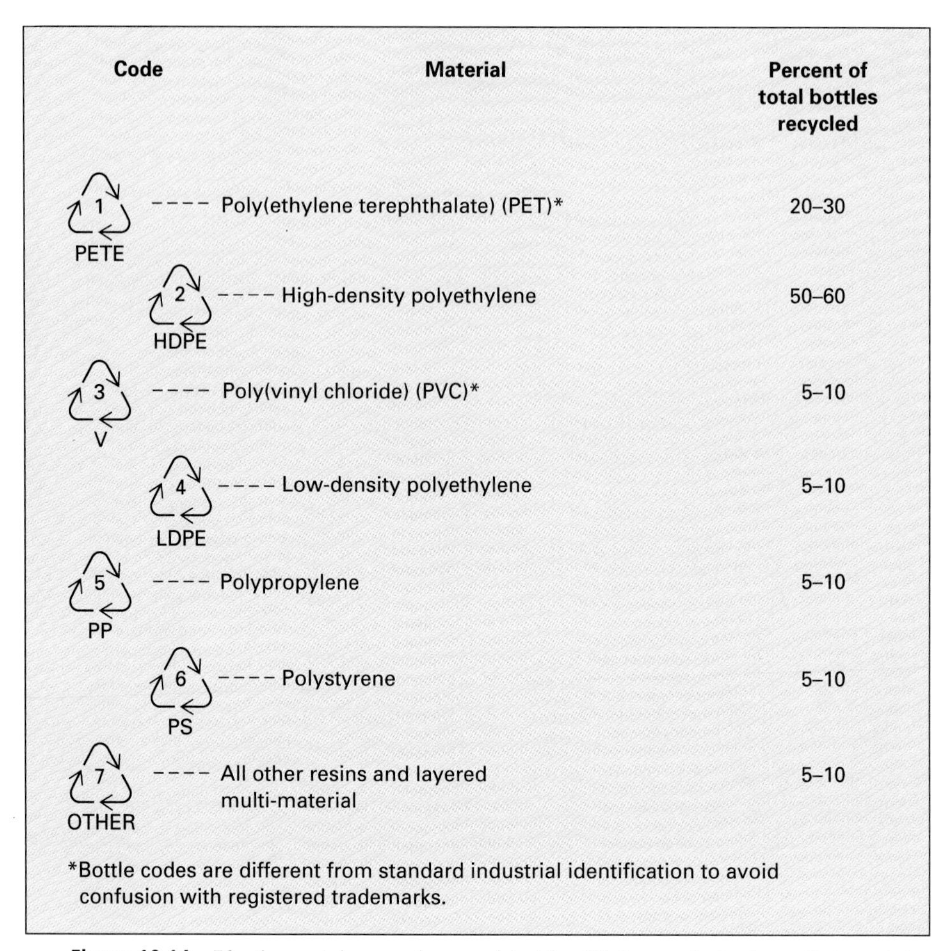

Code	Material	Percent of total bottles recycled
1 PETE	Poly(ethylene terephthalate) (PET)*	20–30
2 HDPE	High-density polyethylene	50–60
3 V	Poly(vinyl chloride) (PVC)*	5–10
4 LDPE	Low-density polyethylene	5–10
5 PP	Polypropylene	5–10
6 PS	Polystyrene	5–10
7 OTHER	All other resins and layered multi-material	5–10

*Bottle codes are different from standard industrial identification to avoid confusion with registered trademarks.

Figure 10.14 Plastic container codes, used to identify types of plastic so that containers can be sorted for recycling.

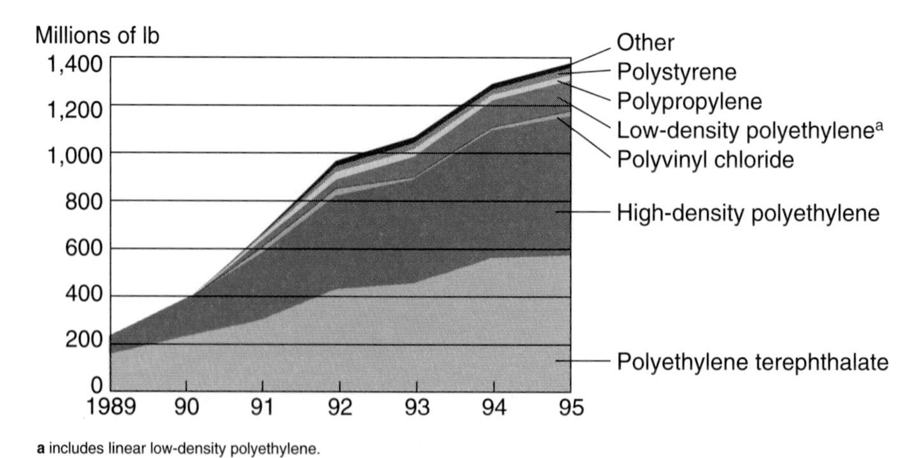

Figure 10.15 HDPE and PET are leaders in plastics recycling. *(Source: American Plastics Council and* Chemical and Engineering News, *p. 20, November 4, 1996.)*

collection of recycled items. This led to an annual average growth in recycling plastics of 32% per year for the 1988–1993 period, but it has slowed to 13% since then, primarily because of the continuing high cost of collecting and sorting recyclable plastics.

Codes are stamped on plastic containers to help consumers identify and sort their recyclable plastics (Figure 10.14). Poly(ethylene terephthalate) (PET), widely used as soft drink bottles, is the most commonly recycled plastic (Figure 10.15). Major end uses for recycled PET include fiberfill for ski jackets and sleeping bags, carpet fibers, and tennis balls. Coca Cola is using 2-L bottles made of 25% recycled PET. HDPE is the second most widely recycled resin. Milk, juice, and water jugs are the principal source of recycled HDPE. Some products made from recycled HDPE are trash containers, drainage pipe, garbage bags, and fencing.

SELF-TEST 10D

1. An example of the formation of a condensation polymer is the reaction of ______ with ______ to give ______ with the elimination of ______.
2. Nylon is an example of a ______ polymer.
3. Polyamides are formed when ______ is split out from the reaction of many organic groups and many amine groups.
4. Polyesters are formed by (a) addition reactions, (b) condensation reactions.
5. Successful recycling involves the four phases of collection, sorting, reclamation and end use. (a) True, (b) False.
6. ______ fibers currently account for over 90% of the fibrous material used in reinforced plastics.
7. Graphite fiber in a ______ matrix is a composite used to make tennis racquets, fishing rods, and aircraft components.

MATCHING SET

___ 1. Aldehyde
___ 2. Ketone
___ 3. Carboxylic acid
___ 4. Ester
___ 5. Monomer
___ 6. Present primary source of organic chemicals used as raw materials
___ 7. Likely future primary source of organic chemicals used as raw materials
___ 8. Poly-*cis*-isoprene
___ 9. Cross-linking via reaction with sulfur
___ 10. Polyester
___ 11. Nylon
___ 12. Alcohol

a. Contains —OH
b. Polyamide
c. Vulcanization
d. Contains $R-\overset{O}{\overset{\|}{C}}-R'$
e. Contains $R-\overset{O}{\overset{\|}{C}}-H$
f. Contains $R-\overset{O}{\overset{\|}{C}}-OH$
g. Contains $R-\overset{O}{\overset{\|}{C}}-O-R'$
h. Petroleum
i. Formed from a dialcohol and a diacid
j. Coal
k. Natural rubber
l. Building block for polymer

QUESTIONS FOR REVIEW AND THOUGHT

1. Why are there more than 11 million organic compounds?
2. Draw the structures of four alcohols that have the formula $C_4H_{10}O$.
3. Identify the class of each of the following compounds:
   (a) $CH_3CH_2CH_2CH_2COOH$
   (b) $CH_3CH{=}CHCH_2CH_3$
   (c) $C_2H_5OC_2H_5$
   (d) $CH_3CH_2CHOHCH_2CH_3$
   (e) $(CH_3)_3COH$
   (f) $CH_3\overset{O}{\overset{\|}{C}}H$
4. Name the compounds in Question 3.
5. Draw the condensed structure and give the name of
   (a) a carboxylic acid (b) an ether
   (c) an alcohol (d) a ketone
   (e) an aldehyde (f) an ester
6. What is meant by the following terms?
   (a) Proof rating of an alcohol
   (b) Denatured alcohol
7. What volume percent of ethanol does 90-proof gin contain?
8. What is the proof of pure ethanol?
9. What is the difference in the structures of ethanol, ethylene glycol, and glycerol? Give one use of each alcohol.
10. Many naturally occurring carboxylic acids have more than one acid group (Table 10.7). What other functional group is often present?
11. Identify each of the following molecules as an alcohol, an ether, an aldehyde, a ketone, a carboxylic acid, or an ester:
   (a) $H\overset{O}{\overset{\|}{C}}CH_2CH_3$
   (b) CH_3CH_2OH
   (c) $CH_3\overset{O}{\overset{\|}{C}}CH_2CH_3$
   (d) $CH_3CH_2CH_2\overset{O}{\overset{\|}{C}}{-}OH$
   (e) $CH_3\overset{O}{\overset{\|}{C}}{-}OCH_2CH_3$
   (f) CH_3OCH_3
   (g) $CH_3\overset{O}{\overset{\|}{C}}H$
12. Write the equation for the formation of an ester from acetic acid (CH_3COOH) and ethanol (C_2H_5OH), and give its name.
13. How do the structures of primary, secondary, and tertiary alcohols differ?
14. Give examples of
   (a) two naturally occurring esters and where they are found
   (b) two naturally occurring carboxylic acids and where they are found
15. Explain how the liver detoxifies ethanol.
16. Summarize the Eastman Kodak process that produces acetic anhydride from coal.
17. In what ways is a railroad train like polystyrene?
18. What is the origin of the word "polymer"?
19. What is meant by the term "macromolecule"?
20. What do the following terms mean?
   (a) Monomer (b) Polymer
21. What structural features must a molecule have to undergo addition polymerization?
22. What do the following terms mean?
   (a) Polyethylene (b) HDPE
   (c) LDPE (d) CLPE
23. What is the repeating unit of natural rubber?
24. What is the difference in the structures of poly-*trans*-isoprene and poly-*cis*-isoprene? Sketch the arrangements for the two polymers. How do the two differ in their physical properties?
25. What is the role of sulfur in the vulcanization process?
26. What is the effect of vulcanization on the physical properties of natural rubber?
27. What property does a polymer have when it is extensively cross-linked?
28. By using polyethylene as an example, draw a portion of
   (a) a linear polymer
   (b) a branched polymer
   (c) a cross-linked polymer
29. Indicate which of the following can undergo an addition reaction and which cannot. Explain your choices.

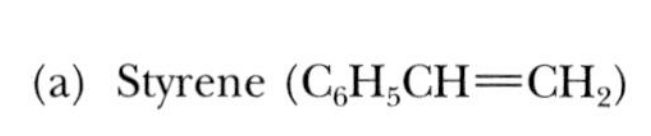
(a) Styrene ($C_6H_5CH{=}CH_2$)
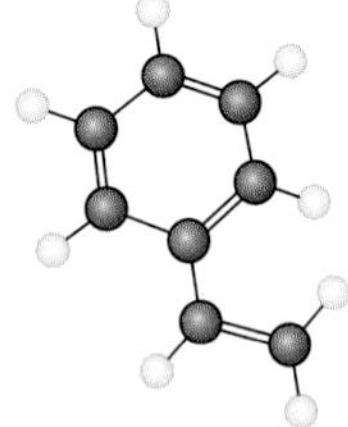

(b) Propene ($CH_2{=}CHCH_3$)
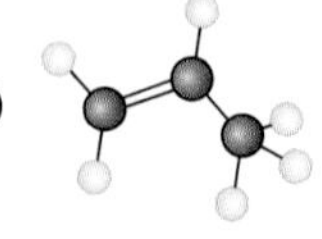

(c) Ethane (CH_3CH_3)
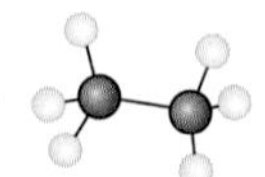

30. Explain how polymers could be prepared from each of the following compounds (other substances may be used):
   (a) $CH_3CH{=}CHCH_3$

(b) $HOOCCH_2CH_2COOH$
(c) $H_2NCH_2CH_2CH_2CH_2NH_2$

31. Draw the structural formula of the monomer used to prepare the following polymers:
    (a) Poly(vinyl chloride) (b) Polystyrene
    (c) Polypropylene
32. Orlon has the polymeric chain structure shown as follows:

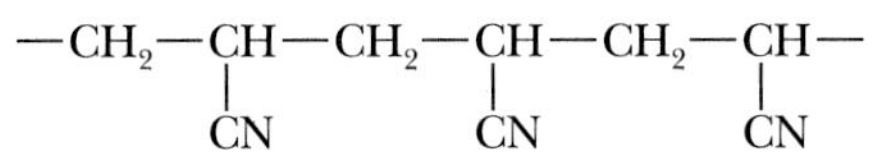

$$—CH_2—\underset{CN}{\underset{|}{CH}}—CH_2—\underset{CN}{\underset{|}{CH}}—CH_2—\underset{CN}{\underset{|}{CH}}—$$

    What is the monomer from which this structure can be made?
33. Give definitions for the following terms:
    (a) Polyester (b) Polyamide
    (c) Nylon-66 (d) Diacid
    (e) Diamine (f) Peptide linkage
34. What polymer is identified as PET?
35. What feature do all condensation polymerization reactions have in common?
36. What are the starting materials for nylon-66?
37. Which do you think is the source of most polymers used today, green plants or petroleum? Do you think this will ever change? Explain.
38. Do you think that plastic production will increase in the future? What advantages if any do you see when plastics are used in place of other materials?
39. What are the four phases involved in a successful recycling program? What happens if the public is cooperative in turning in recyclable materials but there isn't an adequate infrastructure to process the collected materials?
40. Composite materials are typically reinforced plastics. What properties do the embedded glass fibers or graphite fibers give to composites? Do you forsee any recycling problems with these materials?
41. Many organic compounds have more than one type of functional group. Identify the functional groups in the following naturally occurring molecules:

(a)

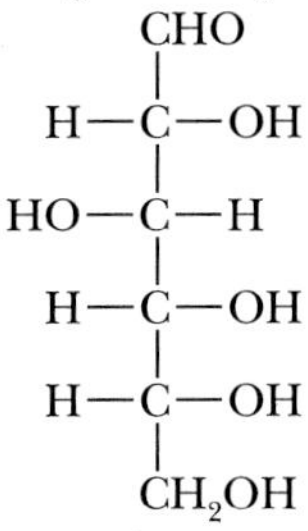

CHO
H—C—OH
HO—C—H
H—C—OH
H—C—OH
CH_2OH

D-Glucose

(b)

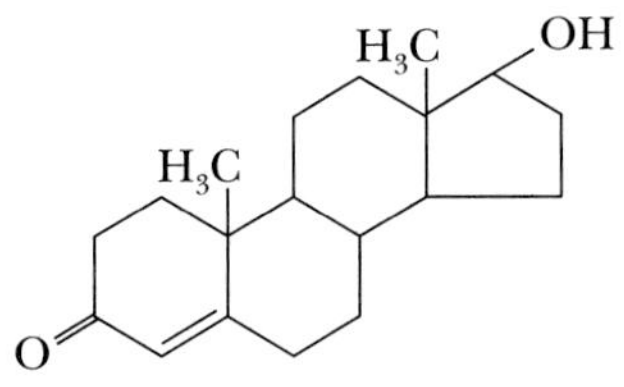

H_3C OH
H_3C
O

Testosterone

(c)

H O
C
OCH_3
OH

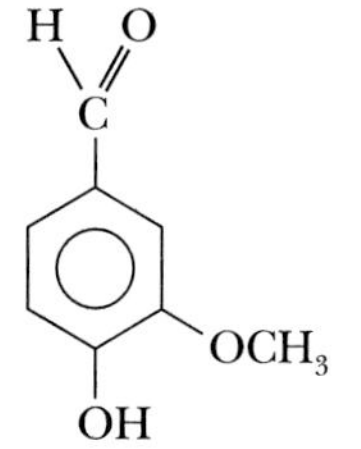

Vanillin
(vanilla bean)

(d)

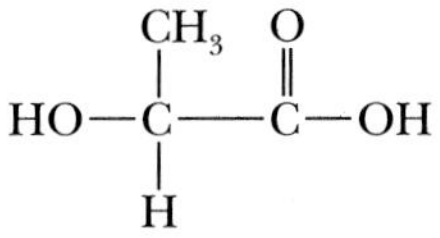

$$HO—\underset{H}{\underset{|}{\overset{CH_3}{\overset{|}{C}}}}—\overset{O}{\overset{\|}{C}}—OH$$

Lactic acid

42. Discuss what plastics are currently being recycled, and give examples of some products being made from these recycled plastics.
43. What are polymer composite materials? Give two examples that illustrate the importance of polymer composites in the manufacture of consumer products.
44. What is the polymer formed by the reaction of the following monomers?

(a) Ethylene glycol ($HOCH_2CH_2OH$)

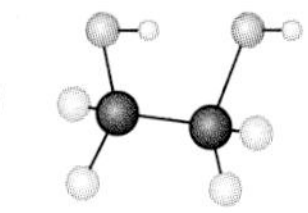

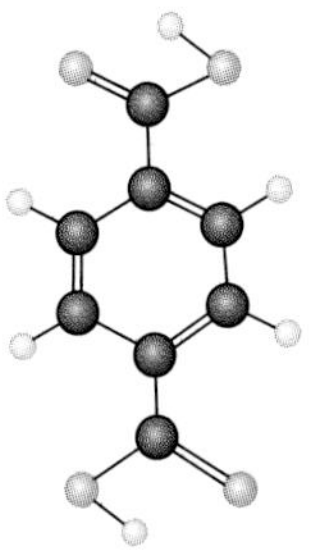

and terephthalic acid, HOOC—C_6H_4—COOH

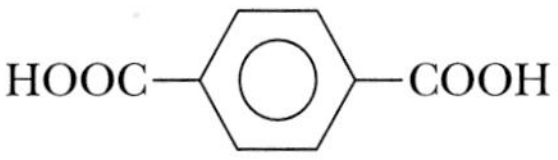

(b) Ethylene ($CH_2{=}CH_2$)

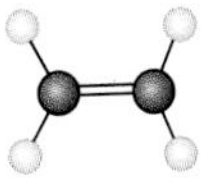

(c) Styrene ($C_6H_5CH{=}CH_2$)

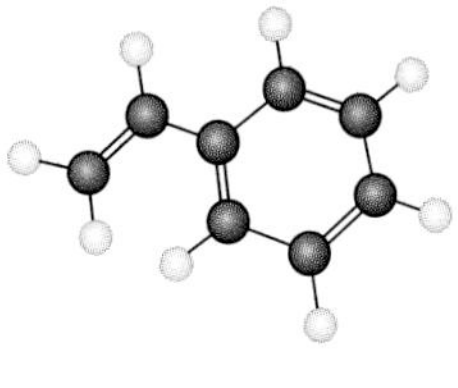

PROBLEMS

1. How many ethylene monomer units, $CH_2{=}CH_2$, are linked together in a polyethylene molecule with a molar mass of 280,000?

2. How many propylene monomer units—propene, ($CH_2{=}CHCH_3$) 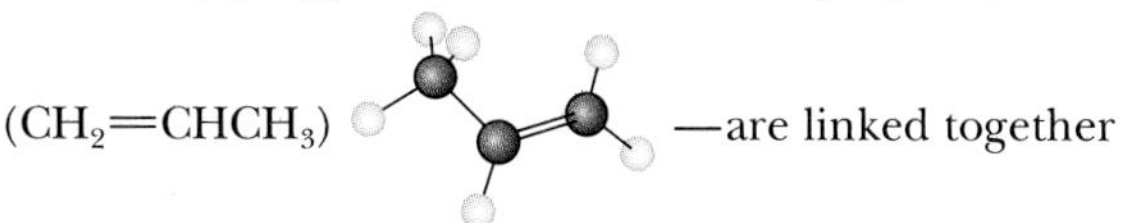 —are linked together in a polymer molecule with molar mass of 84,000?

Two people with identical DNA, the molecule that controls our biochemistry. *(Gunther/Explorer/Photo Researchers)*

CHAPTER 11

The Chemistry of Life

Cooking, nutrition, personal health, drugs, medicine and dentistry, agriculture, our natural environment—biochemistry is fundamental to them all. Sometimes biochemistry is applied in a practical fashion, for example, when a cook scrambles an egg or someone grabs for an aspirin tablet to quiet a headache. Sometimes it is applied by a practicing professional or research scientist, for example, when an agriculture expert recommends the appropriate pesticide to a farmer or a pharmaceutical chemist designs a molecule to combat disease. In this chapter you will be introduced to the major classes of biochemicals and their functions in the body. Some of the applied aspects of biochemistry are also included. Questions that will be addressed in this chapter include

- What are optical isomers, and why are they important in biochemical reactions?
- What are the different types of sugars?
- What is the difference between starch and cellulose?
- What are triglycerides?
- What are the differences among saturated, monounsaturated, and polyunsaturated fatty acids?
- What are steroids?
- What are soaps and synthetic detergents, and in what kinds of products do we find them?
- What are amino acids and proteins?
- What is the structure of hair?
- What do enzymes do?
- What are DNA and RNA, and how do they relate to the genetic code?

Figure 11.1 Mirror images. The mirror image of your left hand, as shown at the top, looks like your right hand, shown at the bottom. But if you place one hand over the other with palms up, they are not identical. This shows that they are nonsuperimposable mirror images. *(C. D. Winters)*

The chemistry of life is referred to as **biochemistry**, and the organic chemicals found in living things are called **biochemicals**.

A petrochemical manufacturing plant and the cells in your body are doing the same thing—taking in raw materials, using energy and carefully controlled conditions to perform chemical reactions, and putting out valuable products and waste materials. Many biochemicals are polymers. Starches are condensation polymers of simple sugars; proteins are condensation polymers of amino acids; and nucleic acids are condensation polymers of simple sugars, nitrogenous bases, and phosphoric acid groups. All these very large biomolecules and other essential, smaller ones must be assembled from the raw materials in food, our equivalent of petroleum for the industrial plant. At the same time, burning food must provide energy. To carry out these functions simultaneously, living things have evolved an exquisite system for breaking down food molecules and putting them back together again. The overview of biochemistry in this chapter focuses on the molecular structure of the most important kinds of biomolecules. Many of them exhibit "handedness," another type of isomerism, so that's where we begin.

11.1 Handedness and Optical Isomerism

Are you right-handed or left-handed? Regardless of our preference, we learn at a very early age that a right-handed glove doesn't fit the left hand, and vice versa. Our hands are not identical; they are mirror images of one another and are not superimposable (Figure 11.1). *An object that cannot be superimposed on its mirror image is called* ***chiral.*** Objects that are superimposable on their mirror images are **achiral**. Stop and think about the extent to which chirality is a part of our everyday life. We've already discussed the chirality of your hands (and feet). Helical seashells are chiral, and most spiral to the right like a right-handed screw. Many vines show a chirality when they wind around trees.

■ Chiral is pronounced "ki-ral" and is derived from the Greek *cheir*, meaning "hand."

■ Enantiomers have the same set of atoms connected by the same set of chemical bonds, but the atoms have a different three-dimensional arrangement in space—just like your left and right hands do.

What is not as well known is that a large number of the molecules in plants and animals are chiral, and usually only one form of the chiral molecule (left-handed or right-handed) is found in nature. For example, all but one of the 20 naturally occurring amino acids are chiral, and only the left-handed amino acids are found in nature. Most of the natural sugars are right-handed.

A chiral molecule and its non-super-imposable mirror image are called **enantiomers**. Enantiomers are two different molecules, just as your left hand and right hand are different. To have enantiomers a molecular structure must be **asymmetrical** (without symmetry). The simplest case is a tetrahedral carbon atom bonded to four *different* atoms or groups of atoms. Such a carbon atom is asymmetrical and is said to be chiral. A molecule that contains a chiral atom is likely to be a chiral molecule.

Some compounds are found in nature in both enantiomeric forms. For example, both forms of lactic acid are found in nature. During the contraction of muscles the body produces only one enantiomer of lactic acid; the other enantiomer is produced when milk sours. Let's look at the structure of lactic acid to see why two different isomers or chiral molecules might be possible. The central carbon atom of lactic acid has four different groups bonded to it: $—CH_3$, $—OH$, $—H$, and $—COOH$.

H
|
C
HO ⁄ COOH
CH_3
Lactic acid

A chiral seashell with a right-hand spiral. *(C. D. Winters)*

As a result of the tetrahedral arrangement around the central carbon atom, it is possible to have two different arrangements of the four groups. If a lactic acid molecule is placed so the C—H bond is vertical, as illustrated in Figure 11.2, one possible arrangement of the remaining groups would be that —OH, —CH_3, and —COOH are attached in a clockwise manner (isomer I). Alternatively, these groups can be attached in a counterclockwise fashion (isomer II). To see further that the arrangements are different, we place isomer I in front of a mirror (Figure 11.2*b*). Now you see that isomer II is the mirror image of isomer I. What is important, however, is that these mirror-image molecules *cannot be superimposed* on one another, no matter how the molecules are rotated. *These two nonsuperimposable, mirror-image chiral molecules are enantiomers.*

The "handedness" of enantiomers is sometimes represented by D for right-handed (D stands for "dextro," from the Latin *dexter* meaning "right") and L for left-handed (L stands for "levo," from the Latin *laevus* meaning "left"). In the case of lactic acid, the D-form is found in souring milk, and the L-form is found in muscle tissue, where it accumulates during vigorous exercise and can cause cramps.

Enantiomers of a chiral compound have the same melting point, the same boiling point, the same density, and many other identical physical and chemical properties. However, they always differ with respect to one physical property: They rotate a beam of **plane-polarized light** in opposite directions. For

■ Plane-polarized light consists of electromagnetic waves with their electric and magnetic fields oscillating in the same direction.

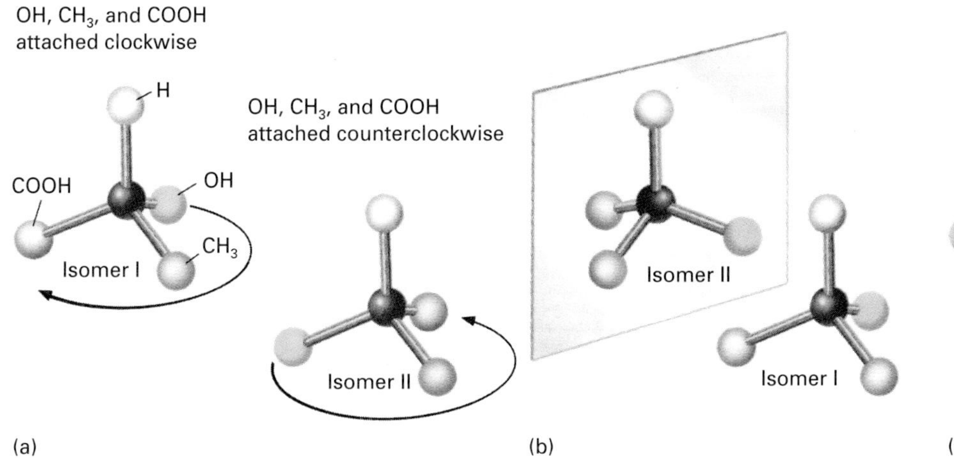

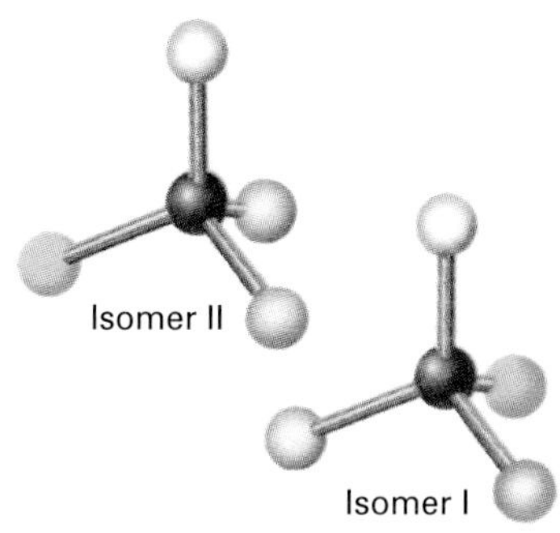

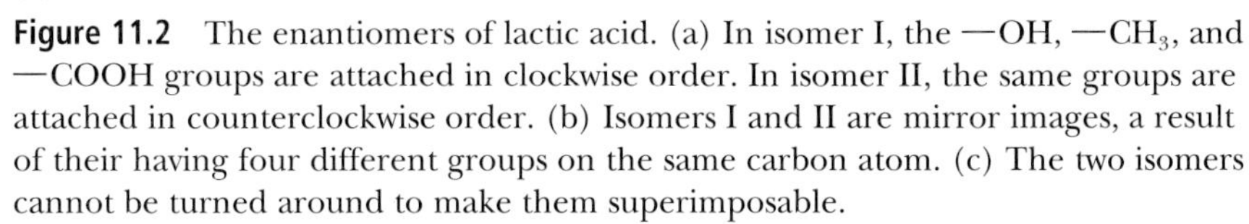

Figure 11.2 The enantiomers of lactic acid. (a) In isomer I, the —OH, —CH_3, and —COOH groups are attached in clockwise order. In isomer II, the same groups are attached in counterclockwise order. (b) Isomers I and II are mirror images, a result of their having four different groups on the same carbon atom. (c) The two isomers cannot be turned around to make them superimposable.

this reason, chiral molecules are sometimes referred to as **optical isomers** and are said to be ***optically active***.

Enantiomers also differ with respect to their biological properties. They react at different rates in a chiral environment, and many of the molecules in plants and animals are chiral. To understand why this difference in activity might exist, think of the hand-in-glove analogy. Although you can put a right-handed glove on your left hand, it takes longer to put it on and it doesn't fit very well. Since nature has a preference for L-isomers of amino acids and since enzymes, the catalysts for biochemical reactions, are proteins made from L-amino acids, enzymes are chiral molecules. The catalytic activity of enzymes is dependent on their three-dimensional structure, which in turn depends on their L-amino acid sequence. As a result, enzymes have a binding preference for one enantiomer of a reactant.

■ Since enantiomers rotate the plane of polarized light an equal amount but in opposite directions, a solution of a racemic mixture will not rotate the plane of polarized light.

Even though nature has a preference for one enantiomer, laboratory synthesis of a chiral compound gives a mixture of equal amounts of the enantiomers, which is called a **racemic mixture**. The separation and purification of enantiomers are difficult because of the similarity in their physical properties. The first separation of enantiomers by Louis Pasteur in 1848 was based on his observation of different kinds of crystals in a racemic mixture. This method is rarely an option, since racemic modifications seldom form mixtures of crystals recognizable as mirror images. The usual method of separating enantiomers is to react them with optically active reagents that have a greater affinity for one enantiomer than the other. Pasteur's discovery of a mold that would selectively destroy one enantiomer of tartaric acid was the first example of this approach.

*Chiral carbon

Thalidomide

The chirality of sugars, proteins, and DNA makes the human body highly sensitive to enantiomers. For example, one enantiomer of a drug is usually more active (or more toxic) than the other. A tragic example that called

THE WORLD OF CHEMISTRY

Molecular Architecture

Program 9, *Molecular Architecture*

There are various types of isomerism common to organic chemistry. Of these, optical isomers are among the most fascinating. This is because they play such important roles in life processes. In living things, chiral molecules exist in only one form or the other, but not both. How did the selection of one optical isomer over the other occur in nature? This question is addressed by Nobel laureate Christian Anfinsen.

How this selection began in nature is anybody's guess. One assumption is that some naturally occurring minerals, for example, might have been involved in binding one form and not the other. In the process a concentration of the form we have now was built up so that when life started, it was stuck with that form. In nature we're stuck pretty much with one isomer. The world has become so evolved that living things are in general composed of one of the two possible mirror images of the basic compounds.

Christian Anfinsen.

FRONTIERS IN THE WORLD OF CHEMISTRY

Chiral Drugs

One enantiomer of a chiral drug is usually more active than the other, but 80% of chiral drugs are still sold as racemic mixtures. Only in those cases where one enantiomer is toxic or has harmful side effects is the single-enantiomer drug used. The primary reason for this is the large increase in cost required to isolate an enantiomer from a racemic mixture. Before long, though, many more drugs may be brought to market as the pure optical isomers. You may soon be seeing advertisements for a drug that is *optically pure or twice as effective as . . . [the racemic mixture, half of which is inactive]* or a *new and improved* version of a familiar over-the-counter medication. Techniques for separating enantiomers have improved greatly. In 1992, the U.S. Food and Drug Administration (FDA) released long-awaited guidelines for the marketing of chiral drugs. The decision about whether to sell a chiral drug in the racemic mixture or the enantiomerically pure form has been left to the drug's manufacturer (although the decision is subject to FDA approval). With the regulations finally in place, drug companies will be looking for situations in which production of a single enantiomer can give them a competitive edge. The racemic mixture, however, may remain the best choice in most cases. For example, ibuprofen, the pain reliever contained in Advil, Motrin, and Nuprin, is now sold as a racemic mixture. The left-handed enantiomer of ibuprofen is the active pain reliever; the right-handed isomer is inactive. Since D-ibuprofen is converted to L-ibuprofen in the body, there is probably no therapeutic advantage to the patient to switch from the racemic mixture to the more costly L-ibuprofen.

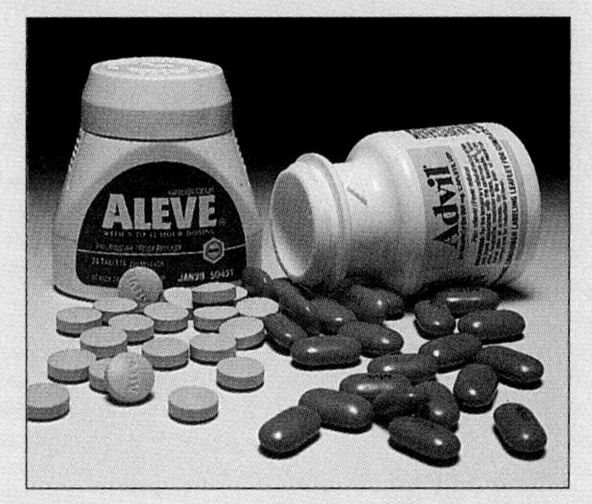

Ibuprofen (e.g., Advil) is sold as a racemic mixture and naproxen (e.g., Aleve) is sold as a single enantiomer. (C. D. Winters)

Naproxen is an example of a chiral drug sold as an enantiomer instead of the racemic mixture. In this case, one enantiomer is a pain reliever and the other causes liver damage. Originally available only by prescription, naproxen is now available over the counter as a pain killer and an anti-inflammatory (Aleve). The sale of enantiomeric drugs is already big business. In 1996 worldwide sales were $73 billion. (More information about classes of drugs is given in Chapter 13).

CH_3
H_3CO — (naphthalene ring) — *CH—COOH

*Chiral carbon

Naproxen

World Sales of Enantiomeric Drugs in 1996

Drugs	Sales (in Billions)*
Cardiovascular	$17.5
Antibiotics	19.5
Hormones	8.0
Central nervous system	4.0
Anti-inflammatory	5.2
Anticancer	7.5
Analgesics	0.9
Other†	10.4
TOTAL	$73.0

* Wholesale finished dosage forms.
† Includes vitamins, amino acids, immunosuppressants, peptide drugs, antiglaucoma drugs, and plant extracts.

Reference *Chemical and Engineering News,* October 20, 1997, p. 54.

attention to the need for testing both enantiomers occurred in 1963, when horrible birth defects were induced by thalidomide. After this was discovered, it was determined that one enantiomer cured morning sickness and the other enantiomer caused birth defects.

Often, the difference is one of activity or effectiveness with no difference in toxicity. Aspartame (NutraSweet), widely used as an artificial sweetener, has two enantiomers. However, one enantiomer has a sweet taste, while the other

■ "*Metabolism*" is a general term for the sum of all the chemical and physical processes in a living organism. When something is "metabolized," it is changed by these processes.

enantiomer is bitter. This indicates that the receptor sites on our taste buds must be chiral, since they respond differently to the "handedness" of aspartame enantiomers. This becomes clearer when looking at the properties of the simple sugars. D-Glucose is sweet and nutritious, whereas L-glucose is tasteless and cannot be metabolized by the body.

The World of Chemistry

Program 9, *Molecular Architecture*

Models of the optical isomers of aspartame.

EXAMPLE 11.1 *Chiral Molecules*

For each of the following molecules, decide whether the underlined carbon atom is or is not a chiral center: (a) $\underline{C}H_2Cl_2$ (b) $H_2N—\underline{C}H(CH_3)—COOH$, and (c) $Cl—\underline{C}H(OH)—CH_2Cl$.

SOLUTION

To be a chiral center, an atom must be bonded to four different groups. The underlined carbon atoms in molecules (b) and (c) meet this condition and are chiral centers. The underlined carbon in (a) is bonded to a pair of H atoms and a pair of Cl atoms and is therefore not chiral.

Exercise 11.1

Which of the following molecules is chiral? Draw the enantiomers for any chiral molecule.

(a) C bonded to OH, Cl, Cl, CH_3 (b) C bonded to OH, H, Cl, CH_3 (c) C bonded to H, H_2N, H, COOH

11.2 Carbohydrates

The word "carbohydrate" literally means "hydrate of carbon." Carbohydrates have the general formula $C_x(H_2O)_y$, in which x and y are whole numbers. However, even though the reaction of the carbohydrate sucrose with sulfuric acid produces carbon (Figure 11.3), this does not mean that sucrose is a simple combination of carbon and water. The carbon, hydrogen, and oxygen in carbohydrates are arranged primarily into three organic functional groups

—O—H (Alcohol hydroxy group) —C(=O)—H (Aldehyde group) —C(=O)— (Ketone group)

Figure 11.3 Reaction of sulfuric acid with sugar. As the sulfuric acid is poured onto the sugar, carbon can be seen forming. *(C. D. Winters)*

■ *Hydrolysis* is the term used for chemical reactions in which chemicals are decomposed through their reaction with water.

Carbohydrates are divided into three groups, depending on how many monomers are combined by condensation polymerization: **monosaccharides** (Latin *saccharum*, "sugar"), **disaccharides**, and **polysaccharides**. Monosaccharides are simple sugars that cannot be broken down into smaller carbohydrate units by acid hydrolysis. In contrast, hydrolysis of a disaccharide yields two monosaccharides (either the same or different), while complete hydrolysis of a polysaccharide produces many monosaccharides (sometimes thousands of them).

Carbohydrates, which make up about half the average human diet, form an essential part of the energy cycle for living things (Section 11.10). Besides energy storage in plants and energy production in animals, carbohydrates serve many other biological purposes. Cellulose is the main structural component of plants. The nucleic acids (Section 11.11) incorporate carbohydrate units in their repeating structure.

Monosaccharides

The most common simple sugar is D-glucose (also known as dextrose, grape sugar, and blood sugar), which is found in fruit, blood, and living cells. A solution of glucose is often given intravenously when a source of quick energy is needed to sustain life. D-Glucose, along with D-galactose and D-fructose, are the three common monosaccharides found in the body. They have different structures but the same molecular formula, $C_6H_{12}O_6$.

■ Simple sugars such as glucose are optical isomers, as indicated by the "D" prefix in D-glucose.

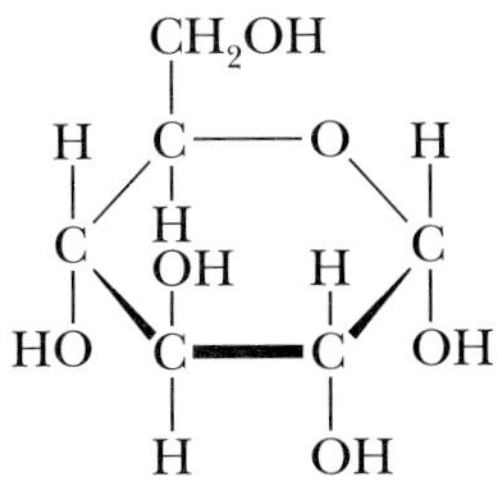

D-Glucose 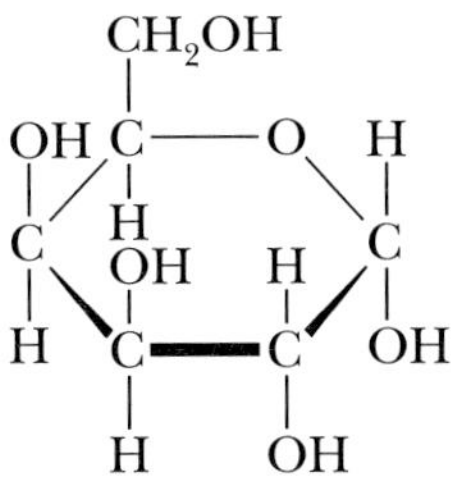 D-Galactose 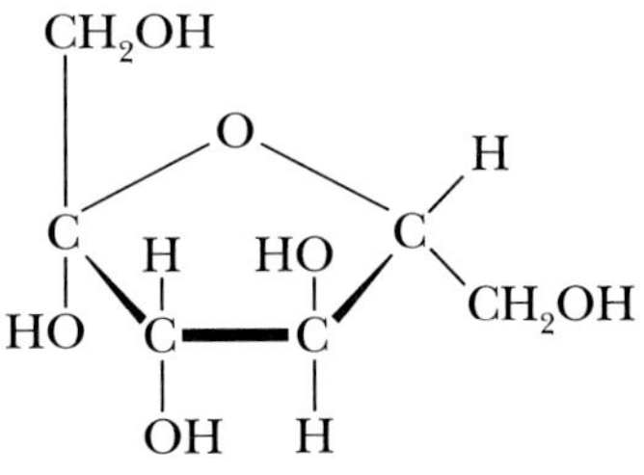 D-Fructose

Disaccharides

Disaccharides are two monosaccharides joined together with the elimination of water to give compounds with the general formula, $C_{12}H_{22}O_{11}$. The three most common disaccharides are

■ Disaccharides are condensation products formed from monosaccharides.

Sucrose (from sugarcane or sugar beets), composed of a D-glucose monomer and a D-fructose monomer—table sugar

Maltose (from starch), composed of two D-glucose monomers—used as a sweetener in prepared foods

Lactose (from milk), composed of a D-glucose monomer and a D-galactose monomer—used in formulating drugs and infant foods, in baking, and in making yeast.

The formula for these disaccharides ($C_{12}H_{22}O_{11}$) is not simply the sum of two monosaccharides, $C_6H_{12}O_6 + C_6H_{12}O_6$. A water molecule is eliminated as two monosaccharides are united to form the disaccharide. In the body, enzymes catalyze the breakdown (hydrolysis) of disaccharides to their monosaccharides.

Sucrose $+ H_2O \longrightarrow$ D-Glucose + D-Fructose

Sucrose is produced in a high state of purity on an enormous scale—more than 80 million tons per year. About 40% of the world sucrose production comes from sugar beets and 60% from sugarcane. A comparison of the sweetness of common sugars and artificial sweeteners relative to sucrose is given in Table 11.1. Honey, which is a mixture of the monosaccharides glucose and fructose, has been used for centuries as a natural sweetener for foods and is sweeter than sucrose, or cane sugar (Table 11.1). To convert cane sugar into glucose and fructose requires treatment with acid or with a natural enzyme called "invertase." The product, known as *invert sugar,* is often used as a sweetener in commercial food products.

Artificial Sweeteners

Saccharin was the first common artificial sweetener. Saccharin passes through the body undigested and consequently has no caloric value. It has a somewhat bitter aftertaste that is offset in commercial products by the addition of small amounts of naturally occurring sweeteners. Such products do have a small caloric value because of the natural sweeteners added.

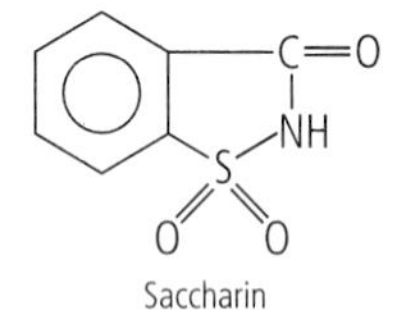

Saccharin

Aspartame (NutraSweet), which has replaced saccharin as the principal artificial sweetener, is used in more than 3000 products and accounts for 75% of the one-billion-dollar worldwide artificial sweetener market.

A pile of sugar beets awaiting collection. *(Nigel Cattlin/Photo Researchers)*

TABLE 11.1 ■ Sweetness of Common Sugars and Artificial Sweeteners Relative to Sucrose

Substance	Sweetness Relative to Sucrose as 1.0
Lactose	0.16
Galactose	0.32
Maltose	0.33
Glucose	0.74
Sucrose	1.00
Fructose	1.17
Aspartame*	180
Saccharin*	300

* Artificial sweeteners.

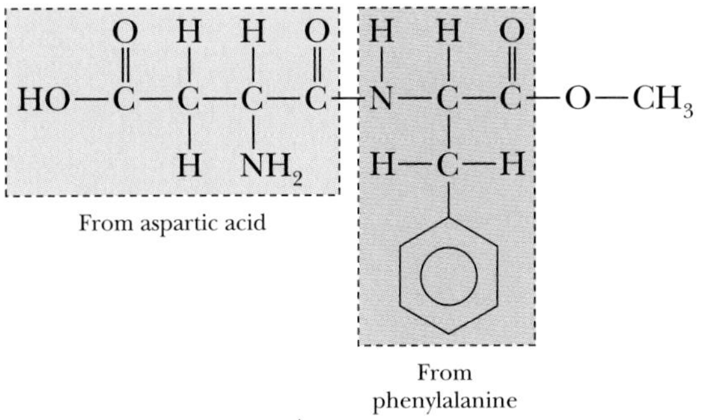

```
      O   H   H   O    H   H   O
      ||  |   |   ||   |   |   ||
HO — C — C — C — C — N — C — C — O — CH3
          |   |            |
          H   NH2        H—C—H
                           |
                       (benzene ring)
```

From aspartic acid

From phenylalanine

Aspartame

It is a dipeptide derivative made from aspartic acid and the methyl ester of phenylalanine. Aspartame can be digested, and its caloric value is approximately equal to that of proteins. However, since much smaller amounts of aspartame than of table sugar are needed for sweetness, many fewer calories are consumed in the sweetened food. Aspartame is not stable at cooking temperatures, which limits its use as a sugar substitute to cold foods and soft drinks.

Polysaccharides

Nature's most abundant polysaccharides are the **starches**, **glycogen**, and **cellulose**. Some polysaccharides have more than 5000 monosaccharide monomers and molar masses over 1 million. The monosaccharide most commonly used to build polysaccharides is D-glucose.

Starches and Glycogen

Plant starch is found in protein-covered granules. If these granules are ruptured by heat, they yield a starch, *amylose*, that is soluble in hot water. Amylose constitutes about 25% of most natural starches.

Structurally, amylose is a straight-chain condensation polymer with an average of about 200 glucose monomers per molecule. Each monomer is bonded to the next with the loss of a water molecule, just as the two units are bonded in maltose. A representative portion of the structure of amylose is shown in Figure 11.4*a* (p. 268).

Glycogen is an energy reservoir in animals, just as starch is in plants. Glycogen is stored in the liver and muscle tissues and is used for "instant" energy until the process of fat metabolism can take over and serve as the energy source.

Cellulose

Cellulose is the most abundant organic compound on the earth, and its purest natural form is cotton. The woody part of trees, the supporting material in plants and leaves, and the paper we make from them are also mainly cellulose. Like amylose, it is composed of D-glucose units. The difference between cellulose and amylose lies in the bonding between the D-glucose units (Figure

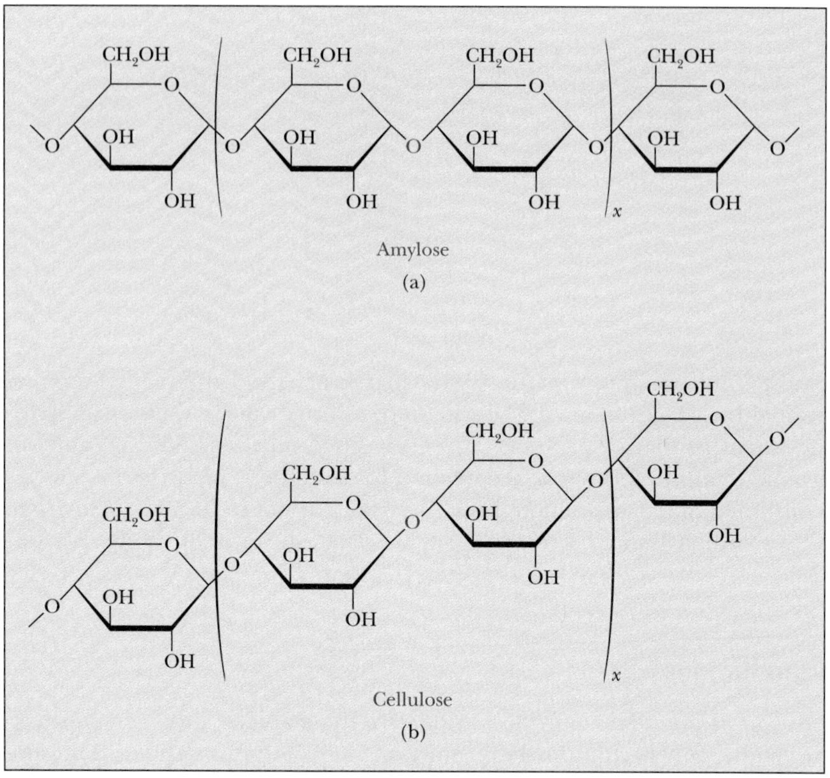

Figure 11.4 (a) Amylose structure. From 60 to 300 glucose units bond together in the manner shown. (b) Cellulose structure. From 900 to 6000 glucose units bond together in the pattern shown here. Source: *World of Chemistry: Essentials*, 1st edition.

■ Small groups of cellulose are held together by hydrogen bonding.

11.4*b*). The angle around the oxygen atoms connecting the glucose rings is 180° in cellulose and 120° in amylose. This subtle structural difference is the reason we cannot digest cellulose. Human beings do not have the necessary enzymes to hydrolyze cellulose. On the other hand, termites, a few species of cockroaches, and ruminant mammals (such as cows, sheep, goats, and camels) are able to digest cellulose because bacteria living in their guts provide the necessary enzyme. D-Glucose can be obtained from cellulose by heating a suspension of the polysaccharide in the presence of a strong acid.

11.3 Lipids

A lipid is an organic substance found in living systems that is insoluble in water but soluble in organic solvents. Because their classification is based on insolubility in water rather than on a structural feature such as a functional group, lipids vary widely in their structure and, unlike proteins and polysaccharides, are not polymers. Lipids include **fats** and **oils**, **steroids**, and **waxes**. The predominant lipids are fats and oils, which make up 95% of the lipids in our diet. The other 5% are steroids and several other lipids that are important to cell function.

Fats and Oils

Fats and oils are **triglycerides**—esters of glycerol (glycerin) and fatty acids. The general equation for the formation of a triester of glycerol is

$$
\begin{array}{ccccccc}
H_2C\!-\!OH & & HO\!-\!\overset{\displaystyle O}{\overset{\|}{C}}\!-\!R & & H_2C\!-\!O\!-\!\overset{\displaystyle O}{\overset{\|}{C}}\!-\!R & & \\
| & & & & | & & \\
HC\!-\!OH & + & HO\!-\!\overset{\displaystyle O}{\overset{\|}{C}}\!-\!R' & \rightleftharpoons & HC\!-\!O\!-\!\overset{\displaystyle O}{\overset{\|}{C}}\!-\!R' & + & 3\,H_2O \\
| & & & & | & & \\
H_2C\!-\!OH & & HO\!-\!\overset{\displaystyle O}{\overset{\|}{C}}\!-\!R'' & & H_2C\!-\!O\!-\!\overset{\displaystyle O}{\overset{\|}{C}}\!-\!R'' & & \\
\text{Glycerol} & & \text{Fatty acids} & & \text{Fat or oil} & & \text{Water} \\
\text{(one molecule)} & & \text{(three molecules that may} & & \text{(one molecule)} & & \text{(three molecules)} \\
 & & \text{or may not have the same R group)} & & & &
\end{array}
$$

An assortment of vegetable oils. *(C. D. Winters)*

The three R groups can be the same or different groups within the same fat or oil, and they can be saturated or unsaturated. The most common fatty acids in fats and oils are listed in Table 11.2.

Fatty acids such as oleic acid, which contain only one double bond, are referred to as **monounsaturated acids**. One of the unsaturated acids, linoleic acid, is an **essential fatty acid**. The human body cannot produce this acid, but it is required for the synthesis of the *prostaglandins*, an important group of more than a dozen related compounds. The prostaglandins have potent effects on physiological activities such as blood pressure, relaxation and contraction of smooth muscle, gastric acid secretion, body temperature, food intake, and blood platelet aggregation. Many prostaglandins cause inflammation and fever. The fever-reducing effect of aspirin results from the inhibition of cyclooxygenase, the enzyme that catalyzes the synthesis of prostaglandins.

The term "fat" is usually reserved for solid triglycerides (such as butter and lard), and "oil" is the term for liquid triglycerides (olive, soybean, and corn oils, for example). The R groups in the fatty acid portions of fats are

TABLE 11.2 ■ Common Fatty Acids in Fats and Oils

Acids		MP °C	Source
Saturated (All Solids at Room Temperature)			
Lauric	$CH_3(CH_2)_{10}COOH$	44	Coconut oil
Palmitic	$CH_3(CH_2)_{14}COOH$	63	Animal and vegetable fats
Stearic	$CH_3(CH_2)_{16}COOH$	69	Animal and vegetable fats
Unsaturated (all Liquids at Room Temperature)			
Oleic	$CH_3(CH_2)_7CH{=}CH(CH_2)_7COOH$	4	Animal and vegetable fats
Linoleic*	$CH_3(CH_2)_4CH{=}CHCH_2CH{=}CH(CH_2)_7COOH$	−5	Linseed oil, cottonseed oil
Linolenic	$CH_3CH_2CH{=}CHCH_2CH{=}CHCH_2CH{=}CH(CH_2)_7COOH$	−11	Linseed oil

* An essential fatty acid that must be part of the human diet.

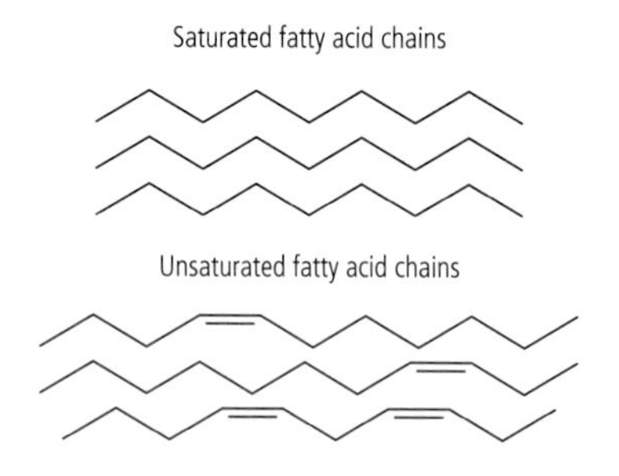

generally saturated, with only C—C single bonds. The R groups in oils are usually either monounsaturated (one C=C double bond) or polyunsaturated (two or more C=C double bonds). Since the C=C bonds interrupt the zigzag pattern of tetrahedral angles with 120° angles, the molecules are irregular in shape and do not pack together efficiently enough to form a solid. Table 11.3 illustrates the percentages of saturated and unsaturated fat found in common dietary oils and fats.

Hydrogen can be added catalytically to the double bonds of an oil to convert it into a semisolid fat. For example, liquid soybean and other vegetable oils are **hydrogenated** to produce cooking fats and margarine.

$$\begin{array}{l} H_2C{-}O{-}C(=O){-}(CH_2)_7CH{=}CH(CH_2)_7CH_3 \\ HC{-}O{-}C(=O){-}(CH_2)_7CH{=}CH(CH_2)_7CH_3 \\ H_2C{-}O{-}C(=O){-}(CH_2)_7CH{=}CH(CH_2)_7CH_3 \end{array} \xrightarrow[200^\circ C]{H_2,\ Ni} \begin{array}{l} H_2C{-}O{-}C(=O){-}(CH_2)_7CH_2CH_2(CH_2)_7CH_3 \\ HC{-}O{-}C(=O){-}(CH_2)_7CH_2CH_2(CH_2)_7CH_3 \\ H_2C{-}O{-}C(=O){-}(CH_2)_7CH_2CH_2(CH_2)_7CH_3 \end{array}$$

Triolein (a liquid oil) Tristearin (a solid fat)

■ The hydrogenation process also forms unnatural *trans* fatty acids from natural *cis* fatty acids (see *Science and Society, next page*)

If it is better to consume unsaturated fats instead of saturated ones, why do food companies hydrogenate oils to reduce their unsaturation? There are several answers to this question. First, the double bonds in the fatty acid are

TABLE 11.3 ■ Amounts of Saturated and Unsaturated Fatty Acids in Fats and Oils

Dietary Oil/Fat	Saturated Fat	Polyunsaturated Fat	Monounsaturated Fat
Canola oil	6%	36%	58%
Safflower oil	9%	78%	13%
Sunflower oil	11%	69%	20%
Corn oil	13%	62%	25%
Olive oil	14%	9%	77%
Soybean oil	15%	61%	24%
Peanut oil	18%	34%	48%
Lard	41%	12%	47%
Palm oil	51%	10%	39%
Butterfat	66%	4%	30%
Coconut oil	92%	2%	6%

SCIENCE AND SOCIETY

Cis *and* Trans *Fatty Acids and Your Health*

The direct link between diets high in saturated fats and heart disease is well known. Saturated fatty acids are usually found in solid or semisolid fats, while unsaturated fatty acids are usually found in oils. As a result, nutritionists recommend the use of liquid vegetable oils for cooking, and people are generally aware of this. However, what is less well known is that partially hydrogenated oils are a health hazard because of the formation of *trans* fatty acids during the process by which margarine and semisolid cooking fats are made. Hydrogen is added to the C=C double bonds in unsaturated fats (oils) to convert them to a solid or semisolid fat that has better consistency and less chance for spoilage. This hydrogenation process decreases the number of double bonds but also forms unnatural *trans* fatty acids from natural *cis* fatty acids. The *trans* fatty acids are not easily metabolized in the human system. Since the *trans* fatty acids are "straight" molecular structures, they pack together like the saturated fatty acids. By contrast, the *cis* fatty acids are bent and do not pack well (see (*b*) and (*c*) that follow).

You can reduce the health risks of *trans* fatty acids by not eating processed vegetable fats. How do you know which products were made with processed vegetable fats? Read the label. For example, a box of cookies or crackers may say on the label "made with 100% pure vegetable shortening . . . (partially hydrogenated soybean oil with hydrogenated cottonseed oil)." Although soybean oil and cottonseed oil are low in saturated fat, the hydrogenation process converts cottonseed oil to a saturated fat, and the partially hydrogenated soybean oil contains *trans* fatty acid.

Some of the many products that contain partially hydrogenated vegetable oils. (C. D. Winters)

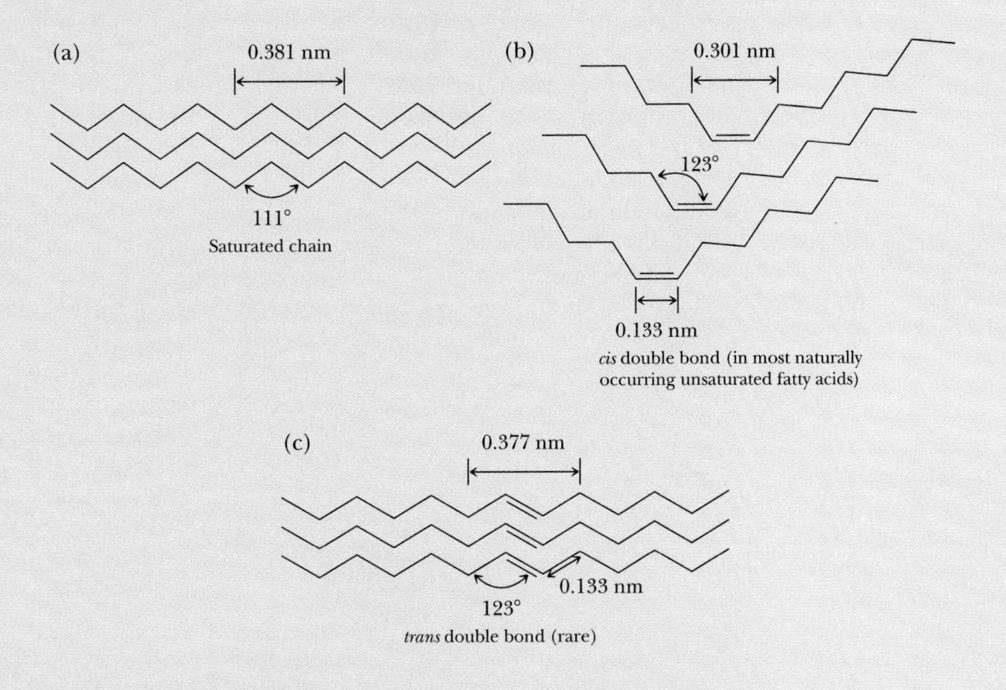

Saturated fatty acids (a) and *trans* fatty acids (c) pack more tightly than *cis* fatty acids (b).

a reactive functional group, and oxygen can attack the fat at these bonds. When the oil is oxidized, unpleasant odors and flavors develop. Hydrogenating an oil reduces the likelihood that the food will oxidize and become rancid. Second, hydrogenating an oil makes it less liquid. There are many times when a food processor needs a solid fat in a food to improve its texture and consistency. For example, if liquid vegetable oil were used in a cake icing, the icing would slide off the cake. Instead of using animal fat, which also contains cholesterol, the manufacturer turns to a hydrogenated or partially hydrogenated oil.

Steroids

Steroids are found in all plants and animals and are derived from the following four-ring structure:

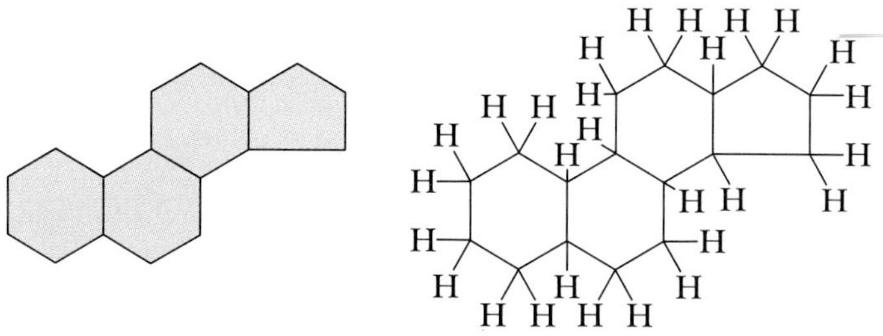

The skeletal four-ring structural drawing on the left is chemical shorthand similar to that described for cyclic hydrocarbons in Section 9.4. There is a carbon at each corner, and the lines represent C—C bonds. Since every carbon atom forms four bonds, additional bonds between carbon atoms and hydrogen atoms are understood to be present whenever the skeletal structure shows fewer than four bonds. The structure on the right shows the hydrogen atoms understood to be present in the four-ring structure shown on the left. Although all the rings in the skeletal drawing are shown as saturated rings, steroids often have one ring that is unsaturated or aromatic. For example, cholesterol has one double bond in the second ring. Note that its structure also includes alkyl groups and an alcohol group. The functional groups replace hydrogen atoms in the skeletal representation.

■ "Aromatic" means a benzene-like structure.

■ Heinrich Otto Wieland (1877–1957) and Adolf Wintaus (1876–1959) both received Nobel Prizes in chemistry in 1927 and 1928, respectively, for their work leading to the determination of the structure of cholesterol.

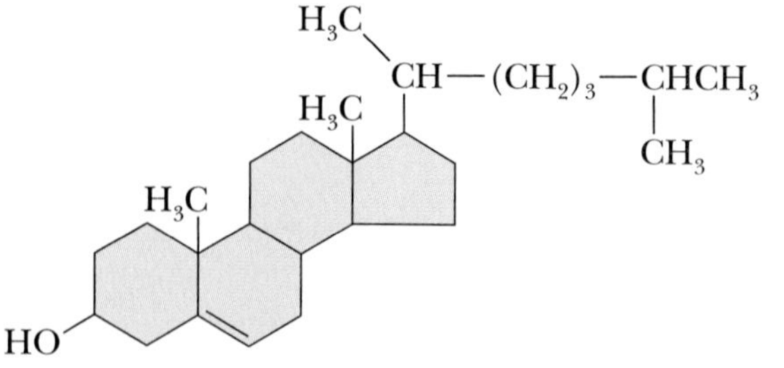

Cholesterol

Cholesterol is the most abundant animal steroid. The human body synthesizes cholesterol and readily absorbs dietary cholesterol through the intestinal wall. An adult human contains about 250 g of cholesterol. Cholesterol receives a lot of attention because high blood cholesterol levels are associated with heart disease (Section 13.11). Proper amounts of cholesterol are essential to our health, because cholesterol undergoes biochemical alteration to give milligram amounts of many important hormones, such as vitamin D, cortisone, and sex hormones.

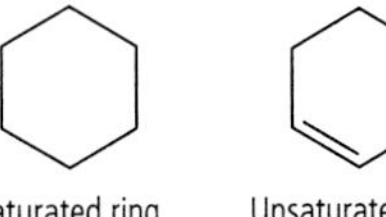
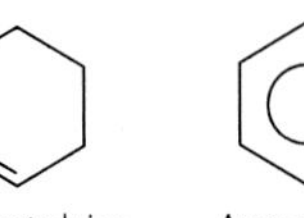

Sex Hormones

Cholesterol is the starting material for the synthesis of steroid sex hormones. One female sex hormone, **progesterone**, differs only slightly in structure from the male hormone, **testosterone**.

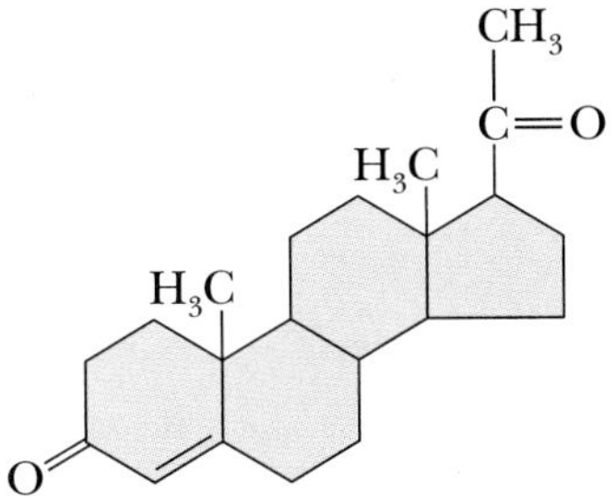

Progesterone

Testosterone

Other female hormones are estradiol and estrone, together called **estrogens**. The estrogens contain an aromatic ring, which differentiates them from the steroids progesterone and testosterone.

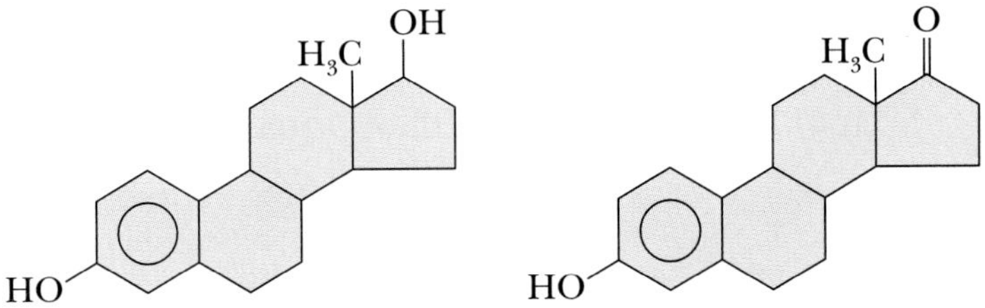

Estradiol

Estrone

The estrogens and progesterone are produced by the ovaries. Estrogens are important to the development of the egg in the ovary, whereas progesterone causes changes in the wall of the uterus and after fertilization prevents release of a new egg from the ovary (ovulation). Birth control drugs use derivatives of estrogens and progesterone to simulate the hormonal process resulting from pregnancy and thereby prevent ovulation (Section 13.4).

Waxes

Waxes are esters formed from long-chain (16 or more carbon atoms) fatty acids and long-chain alcohols. The general formula of a wax is the same as that of a simple ester, RCOOR′, with the qualification that R and R′ are limited to alkyl groups with a large number of carbon atoms. Natural waxes are usually mixtures of several esters. Wax coatings on leaves help to protect the leaves from disease and also help the plant to conserve water. The feathers of birds are also coated with wax. Our ears are protected by wax. Several natural waxes have been used in consumer products. These include carnauba wax (from a Brazilian palm tree), which is used in floor waxes, automobile waxes, and shoe polishes; and lanolin (from lamb's wool), which is used in cosmetics and ointments. Lanolin also contains cholesterol.

SELF-TEST 11A

1. To have optical isomers in carbon compounds, a carbon atom must have ______________ different groups attached.
2. The complete hydrolysis of a polysaccharide yields ______________.
3. When a molecule of sucrose is hydrolyzed, the products are one molecule each of the monosaccharides ______________ and ______________.
4. Starch and cellulose are condensation polymers built of ______________ monomers.
5. ______________ bonding holds polysaccharide chains together, side by side, in cellulose.
6. The sugar referred to as blood sugar, grape sugar, or dextrose is actually the compound ______________.
7. Fats and oils are esters of ______________ and ______________.
8. The structural difference between a saturated fat and an unsaturated fat is ______________.
9. Cholesterol is a (a) steroid, (b) protein, (c) sex hormone, (d) carbohydrate.
10. Cholesterol is essential to our health. (a) True, (b) False.
11. Waxes are esters of ______________ and ______________.

Making soap the old-fashioned way. *(North Wind Picture Archives)*

11.4 Soaps, Detergents, and Shampoos

In strongly basic solutions, fats and oils undergo hydrolysis to produce glycerol and salts of fatty acids. Such reactions are called **saponification** reactions, and the sodium or potassium salts of the fatty acids formed are **soaps**.

$$\begin{array}{l} CH_3{-}(CH_2)_{16}{-}COO{-}CH_2 \\ \qquad\qquad\qquad\qquad\quad | \\ CH_3{-}(CH_2)_{16}{-}COO{-}CH \\ \qquad\qquad\qquad\qquad\quad | \\ CH_3{-}(CH_2)_{16}{-}COO{-}CH_2 \end{array} + 3\,NaOH \longrightarrow 3\,CH_3{-}(CH_2)_{16}{-}COO^-Na^+ + \begin{array}{l} HO{-}CH_2 \\ \qquad\ \ | \\ HO{-}CH \\ \qquad\ \ | \\ HO{-}CH_2 \end{array}$$

Tristearin (an animal fat) — Sodium hydroxide — Sodium stearate (a soap) — Glycerol (glycerin)

■ Principal fats and oils for soap making are tallow from beef and mutton, coconut oil, palm oil, olive oil, bone grease, and cottonseed oil.

Pioneers prepared their soap by boiling animal fat with an alkaline solution obtained from the ashes of hardwood containing potassium carbonate (referred to as "lye"). The resulting lye soap could be "salted out" by adding sodium chloride, because soap is less soluble in a salt solution than in water. The crude soap made this way contained considerable caustic material in addition to the soap molecules, but it did its job of cleaning quite well.

Substances that are water soluble can readily be removed from the skin or a surface by simply washing with an excess of water. To remove a sticky sugar syrup from your hands, you can dissolve the sugar in water and rinse it away. Many times the material to be removed is oily, and water will merely run over the surface of the oil. Since the skin has natural oils, even substances such as ordinary dirt that are not oily themselves can adhere quite strongly to the skin and to clothing containing these oils. The hydrogen bonding holding water molecules together is too large to allow the oil and water to intermingle (Figure 11.5), so something like soap is needed to loosen the dirt and wash it away.

The cleaning action of soap is explained by its molecular structure. When present in an oil–water system, the stearate anion in the soap moves to the interface between the oil and the water

$$CH_3CH_2CH_2CH_2CH_2CH_2CH_2CH_2CH_2CH_2CH_2CH_2CH_2CH_2CH_2CH_2CH_2\ COO^-Na^+$$

Sodium stearate molecule

The hydrocarbon chain, which is a nonpolar organic structure, mixes readily with the nonpolar oil or grease molecules, whereas the highly polar $—COO^-$ group is attracted to water molecules (Figure 11.5).

■ Stearates include sodium—hard soap; potassium—soft soap; and ammonium—liquid soap.

One undesirable property of soaps is their tendency to form precipitates with Ca^{2+}, Mg^{2+}, and Fe^{2+} ions found in "hard" water. The resulting fatty-acid salts of these doubly positive ions are not as soluble in water as the Na^+ ion salts. These less-soluble molecules appear as a scum that sticks to laundry and bathtubs, often containing trapped dirt, which makes it appear even worse.

■ Floating soaps float because of trapped air.

Synthetic detergents are derived from organic molecules designed to have even better cleaning action than soaps but less reaction with the doubly positive ions found in hard water. As a consequence, synthetic detergents are often more economical to use and are more effective in hard water than soap. There are many different synthetic detergents on the market. An inventory of cleaning materials in a typical household might include half a dozen or more formulated products designed to be the most suitable for a specific job—cleaning skin, hair, clothes, floors, or the family car.

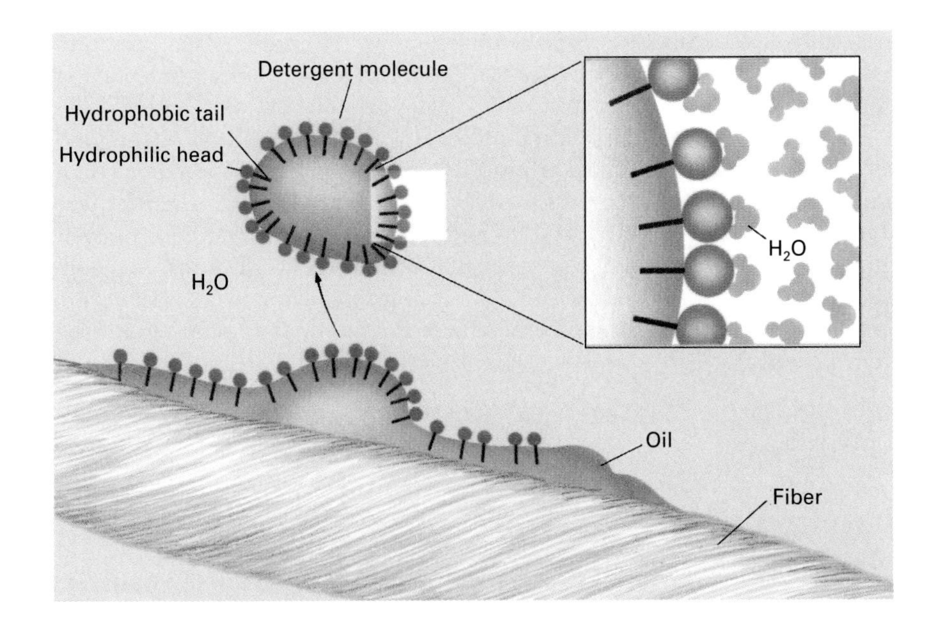

Figure 11.5 How soaps and detergents work. The hydrophilic, salt ends of the molecules interact strongly with water. The hydrophobic, hydrocarbon ends of the molecules avoid water and are drawn to the oily portion of the dirt. As greasy dirt is broken up by agitation, the particles are surrounded and isolated from one another. This prevents them from coming together again and allows them to be carried away in the wash water.

The molecular structure of a synthetic detergent molecule, like that of a soap, consists of a long oil-soluble (**hydrophobic**, meaning not liking, or fearing water) group and a water-soluble (**hydrophilic**, meaning liking or loving water) group.

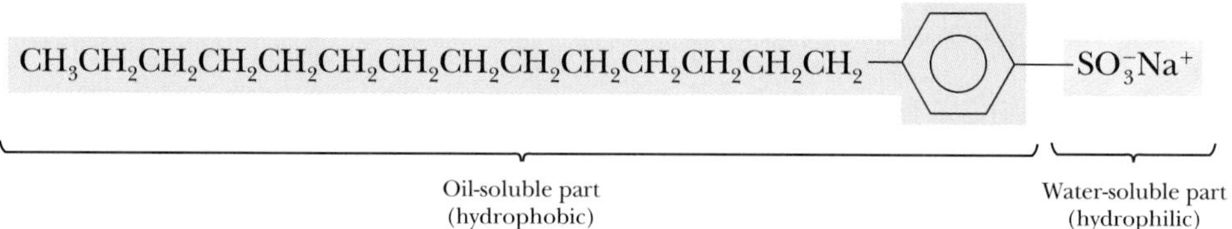

Oil-soluble part (hydrophobic) — Water-soluble part (hydrophilic)

A typical synthetic detergent molecule

Typical hydrophilic groups include negatively charged sulfate ($-OSO_3^-$), sulfonate ($-SO_3^-$), and phosphate ($-OPO_3^{2-}$) groups. Compounds with these groups are called **anionic surfactants**.

Cationic (positively charged) **surfactants** are almost all quaternary ammonium halides (four groups attached to the central nitrogen atom) with the general formula

$$R_1-\overset{\overset{R_2}{|}}{\underset{\underset{R_3}{|}}{N^+}}-R_4\ X^-$$

Most of us have a shelf like this. We rely on a wide variety of chemical surfactants to keep things clean. *(C. D. Winters)*

where one of the R groups is a long hydrocarbon chain and another frequently includes an $-OH$ group. The X^- in the formula represents a halide ion such as chloride (Cl^-) or bromide (Br^-) ion.

■ In spite of the major impact of various kinds of synthetics on the detergent industry, soap is still the number-one surfactant, holding approximately 39% of the market.

Shampoos are generally more complex formulations than simple soap solutions, with a number of ingredients to satisfy different requirements for maintaining clean and healthy-looking hair. If you use soap to wash your hair, incomplete rinsing will result in a "soap film" on the hair, which makes it appear dull. If soap is used to wash hair in hard water, a very noticeable film can usually be seen and is very difficult to remove. In addition, soaps tend to produce solutions with basic pH values. These harsh conditions are damaging to the hair.

Most shampoos contain anionic detergents. These give the product good foaming characteristics because anionic detergents generally foam more than cationic and nonionic detergents. Nonionic surfactants, such as the products obtained by reacting diethanolamine and lauric acid, are often also present in shampoos. While not as good surfactants as anionic surfactants, the nonionics are useful as thickeners and foam stabilizers. They make a shampoo pour from the bottle more slowly and cause the lather to remain thick for a longer period.

$$\underset{\text{Diethanolamine}}{HN(CH_2CH_2OH)_2} + \underset{\text{Lauric acid}}{CH_3(CH_2)_{10}COOH} \longrightarrow$$

$$\underset{\substack{\text{Lauric diethanolamide}\\ \text{(an amide detergent)}}}{CH_3(CH_2)_{10}-\overset{\overset{O}{\|}}{C}-N(CH_2CH_2OH)_2} + H_2O$$

■ Lather has nothing to do with cleaning efficiency, but consumers have been taught to expect a good lather whenever a shampoo is used.

The polar end of the shampoo molecule interacts with water molecules while the nonpolar end interacts with the oil and dirt in hair.

Hair is more manageable, has a better sheen, and has less tendency to attract static charges (causing "fly-away" hair) if all the shampoo is removed after washing. An anionic detergent can be removed from the hair by using a *rinse*, or conditioner, containing a dilute solution of a cationic detergent, usually a quaternary ammonium compound, which electrically attracts the anions and facilitates their removal. The positive charged end of the quaternary ammonium surfactant also neutralizes negative charges on damaged hair ($—COO^-$ groups from disrupted protein chains), while the alkyl chain attaches to the hair and gives it a smooth feel.

■ Caution should be taken with the cationic rinse because of its possible irritation to the eyes.

An after-shampoo conditioner also attempts to put back into the hair some of the oils that were removed by the detergent. A typical conditioner has a water–alcohol dispersing medium as well as skin softeners, oils, waxes, resins, and even short amino acid polymers that can adhere to the hair to produce a more pliable and elastic fiber that is not as likely to become dry or be affected by atmospheric conditions. Holding the correct amount of moisture is the key to hair control, because too much water causes the hair to be limp and too little causes the individual hairs to attract static charge. Although many hair preparations make direct or indirect claims that various proteins and other beneficial ingredients can penetrate the hair and "repair" and even strengthen it, there is no scientific evidence for these claims. Protein molecules, for example, are simply too large to pass through the surface of the hair. Only much smaller molecules can do this. In fact, *if* hair preparations did function in this way, they would have to be classed as drugs by the Food and Drug Administration (FDA).

Lanolin and mineral oil (or their substitutes) are often added to shampoos to replace the natural oils in the scalp, thus preventing it from drying out and scaling. The presence of oil additives and stabilizers sometimes gives the shampoo a pearlescent appearance. These ingredients also make the shampoo less foamy, which is popular in European countries.

11.5 Creams and Lotions

If dry skin is treated with an oily substance after washing, the skin will be protected until enough natural oils have been regenerated. The oily substance used can be derived from animal oils, vegetable oils, or even oils from petroleum (the mineral oils). It is not uncommon to see some kind of rare animal oil, such as mink oil, used in a skin preparation; yet the oil from a mink is not any more effective at holding the skin's moisture than is a less expensive vegetable or mineral oil. When choosing a skin product containing an oil, perhaps a more important factor than the kind of oil is whether it will be soothing or irritating to the skin.

Any substance that holds moisture in the skin can be called a **moisturizer**. Since these substances are all oily in nature, they can lubricate the skin (restore the feeling of normal oiliness) and have the noticeable effect of softening and soothing the skin. These substances are also called **emollients**. All the oils listed in the previous paragraph that help to hold moisture in the skin would be called emollients.

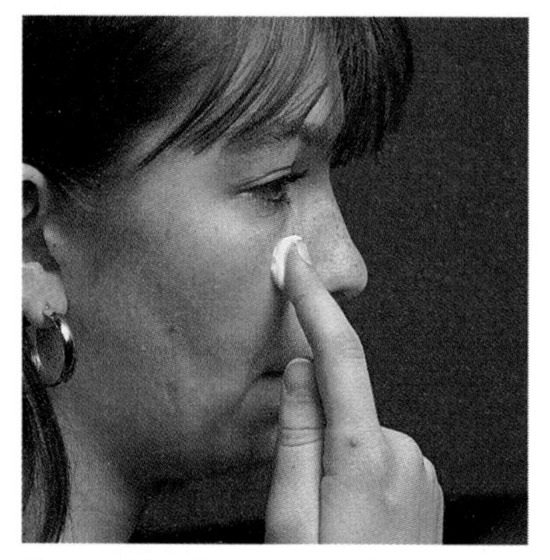

Figure 11.6 Cold cream, which has a thicker consistency than a lotion. *(C. D. Winters)*

Getting an emollient distributed evenly on your skin is not as easy as it may seem. Instead of just pouring oil over your body after a shower (this would leave your skin too oily), it would be better to use a mixture containing the emollient. Two kinds of mixtures are commonly used: *creams* and *lotions* are made by mixing an oily component with water and other ingredients in the right proportions to form a stable mixture that can be more like a solid (a cream) or more like a liquid (a lotion) (Figure 11.6). Mixtures like these are called **emulsions**. Emulsions consist of two substances that would normally not mix, such as an oil and water, but that are made to mix by an **emulsifying agent**, a compound whose molecules have a part that is soluble in water and a part that is soluble in oil. With its dual solubility, the emulsifying agent stabilizes the mixture. Thus, a cream can contain a rather high percentage of an oil and not feel oily or greasy because of the presence of the water. When the cream is applied to the skin, some of the water is absorbed by the skin, and the oil (emollient) remains on the surface to hold in the moisture.

■ The name "*colloid*" comes from the Greek word meaning glue. A close look at a typical glue will show that there are large particles dispersed in water. These particles are large, colloidal-sized aggregates of similar molecules.

Emulsions are examples of **colloids**, which are quite common. Fog, foams, foods such as milk, and aerosol sprays are all colloids (Table 11.4). Colloidal mixtures differ from solutions. The colloid particles (called the *dispersed phase*) distributed in the solventlike medium (called the *continuous phase*) are much larger than the molecules or ions that are the solutes in true solutions. Two kinds of emulsions can be formed between oil and water: oil droplets of colloid size dispersed in water and water droplets of colloid size dispersed in oil. As you might expect, oil-in-water emulsions have more of the properties of an aqueous solution, while the water-in-oil emulsions have more oil-like properties; for example, they tend to feel more greasy. An oil-in-water emulsion has tiny droplets of an oily or waxy substance dispersed throughout a water me-

TABLE 11.4 ■ Types of Colloids

Continuous Phase	Dispersed Phase	Type	Examples
Gas	Liquid	Aerosol	Fog, clouds, aerosol sprays
Gas	Solid	Aerosol	Smoke, airborne viruses, automotive exhaust
Liquid	Gas	Foam	Shaving cream, whipped cream
Liquid	Liquid	Emulsion	Mayonnaise, milk, face creams
Liquid	Solid	Sol	Milk of magnesia, mud
Solid	Liquid	Gel	Jelly, cheese, butter
Solid	Solid	Solid sol	Milk glass, some gemstones, many alloys such as steel

Note: Most colloids will separate unless stabilized. Food and cosmetic emulsions, foams, and aerosols are usually stabilized with emulsifying agents to give them a long shelf life and consistent properties such as color and texture throughout their life.

dium; homogenized milk is an example. A water-in-oil emulsion has tiny droplets of a water solution dispersed throughout an oil; examples are natural petroleum and butter. An oil-in-water emulsion can be washed off the skin surface with tap water, whereas a water-in-oil emulsion gives skin a greasy, water-repellent surface that resists being washed off by running water. With careful formulation of the emulsion, a *barrier cream* can be made that will effectively resist aqueous solutions that might contain harmful ingredients. Chemists in the laboratory often apply barrier creams to protect their hands from exposure to toxic chemicals.

The ingredient listed first on a cream or lotion label is an indication of the kind of emulsion present. If water is listed first, the emulsion is probably an oil-in-water type. If an oil is listed first, the emulsion is probably the water-in-oil type (Figure 11.7). A lotion might contain the same emollient as a cream

■ The molecules in the continuous phase of colloidal mixtures can have molar masses as high as several hundred thousand.

■ Spermaceti oil, once obtained from the head of the sperm whale, is chiefly cetyl palmitate, an ester of cetyl alcohol and palmitic acid. Identify the ester group in its formula: $CH_3(CH_2)_{14}COO(CH_2)_{15}CH_3$. Although some spermaceti oil is still derived from whales, it can be readily synthesized from other starting materials.

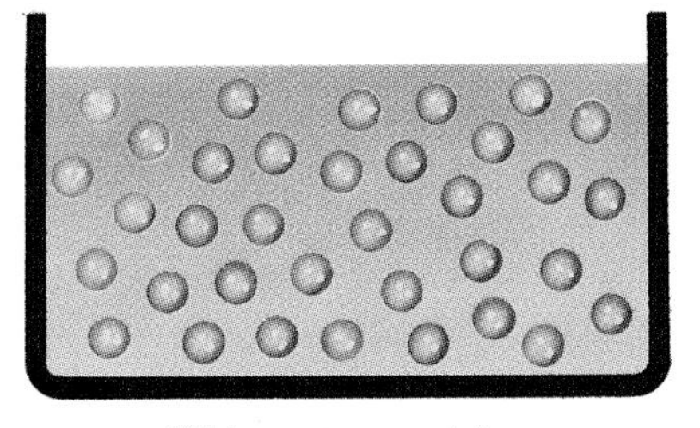

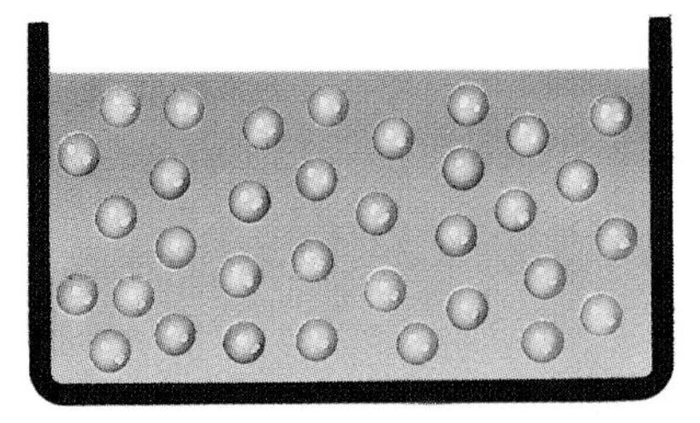

Figure 11.7 Two moisturizing lotions, one an oil-in-water emulsion and the other a water-in-oil emulsion. *(C. D. Winters)*

but contain a different emulsifier and a different ratio of water to oil. Whether an emollient is distributed on the skin in a cream or a lotion depends more on how the product is perceived by the user than on which kind of product is more effective. All creams and lotions have "shelf lives." Creams and lotions, like all colloids, can tend to "settle out" over a period of time, a property not observed in solutions.

SELF-TEST 11B

1. To make soap, a fat is treated with ____________.
2. The hydrocarbon end of a soap or detergent molecule is (hydrophilic, hydrophobic) while the polar end is (hydrophilic, hydrophobic).
3. Which is more likely to precipitate the hard-water ions (Ca^{2+}, Mg^{2+}, Fe^{3+}) as a sticky precipitate: (a) Traditional soaps or (b) Synthetic detergents?
4. Another name for a skin softener is a(an) ____________.
5. A skin cream is either an oil-in-water or water-in-oil ____________.
6. An example of a foam colloid is ____________.
7. An example of a solid aerosol colloid is ____________.
8. An example of a liquid aerosol colloid is ____________.
9. Milk is an example of a(an) ____________ colloid.

11.6 Amino Acids

All proteins are condensation polymers of **amino acids**: A large number of proteins exist in nature. For example, the human body is estimated to have 100,000 different proteins. What is amazing is that all these proteins are derived from only 20 different amino acids (Table 11.5, p. 282–283). Even more amazing is nature's preference for only the L-enantiomer of these amino acids. All but one of the 20 amino acids found in nature have the general formula

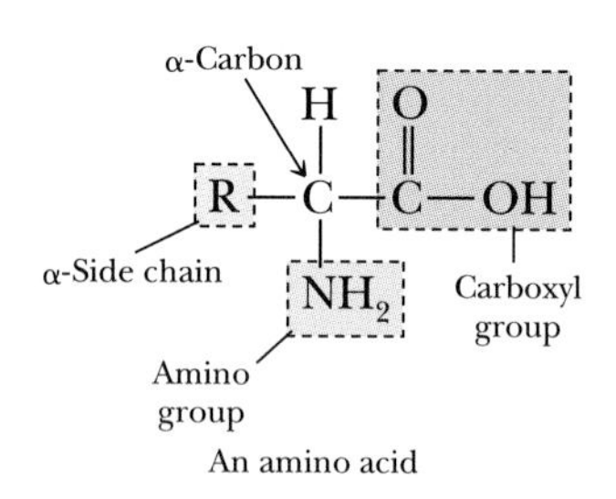

Crystalline phenylalanine, one of the essential amino acids, photographed under polarized light. *(Alfred Pasieka/SPL/Photo Researchers)*

with an amino ($—NH_2$) group, a carboxylic acid group (—COOH) and an R group attached to the *alpha*-carbon (the first carbon next to the —COOH). R is a characteristic group for each amino acid (see Table 11.5), and the α-carbon is a chiral carbon atom, with one exception—glycine. R is a hydrogen atom in glycine, the simplest amino acid. Therefore, glycine is achiral and is the only naturally occurring amino acid that does not have enantiomers. The polarity of the R groups in amino acids affects the structure and function of proteins. The amino acids are grouped in Table 11.5 according to whether the R group is nonpolar, polar, acidic, or basic.

■ For good nutrition we require all the essential amino acids in our daily diet, but the amount required does not exceed 1.5 g per day for any of them.

The **essential amino acids** must be ingested from food; they are indicated by asterisks in Table 11.5. A diet that includes meat, milk, eggs, or cheese provides all the essential amino acids. The other amino acids can be synthesized by the human body.

11.7 Peptides and Proteins

How are amino acids polymerized to give proteins? The formation of an amide from the reaction of an amine and a carboxylic acid is described in Section 10.5 in the discussion of polyamides such as nylons.

$$R-\overset{\displaystyle O}{\overset{\|}{C}}-OH + H_2NR' \longrightarrow R-\overset{\displaystyle O}{\overset{\|}{C}}-\overset{\displaystyle H}{\overset{|}{N}}-R' + H_2O$$

Since amino acids have both an amine group and a carboxylic acid group, the $-COOH$ of one amino acid can combine with the $-NH_2$ of a second amino acid.

$$H_2N-\underset{\underset{\displaystyle H}{|}}{\overset{\overset{\displaystyle R}{|}}{C}}-\overset{\overset{\displaystyle O}{\|}}{C}-\boxed{OH + H}-\underset{\underset{\displaystyle H}{|}}{N}-\underset{\underset{\displaystyle H}{|}}{\overset{\overset{\displaystyle R}{|}}{C}}-\overset{\overset{\displaystyle O}{\|}}{C}-OH \xrightarrow{-H_2O} H_2N-\underset{\underset{\displaystyle H}{|}}{\overset{\overset{\displaystyle R}{|}}{C}}-\boxed{\overset{\overset{\displaystyle O}{\|}}{C}-\underset{\underset{\displaystyle H}{|}}{N}}-\underset{\underset{\displaystyle H}{|}}{\overset{\overset{\displaystyle R}{|}}{C}}-\overset{\overset{\displaystyle O}{\|}}{C}-OH$$

A peptide bond

In the condensation reaction, one molecule of water is eliminated between the carboxylic acid of one amino acid and the amine group of another. The result is a **peptide** bond (called an amide group in simpler molecules), and the molecule is a **dipeptide**. When two different amino acids are bonded, two different combinations are possible, depending on which amine reacts with which acid group. For example, when glycine and alanine react, both glycylalanine and alanylglycine can be formed. Either end of the dipeptide can then react with another amino acid.

■ Names of peptides are written from left to right starting with the amino- or N-terminal end. The *-ine* ending of all amino acid residues (except the carboxyl or C-terminal end) is changed to *-yl*. For example, Gly-Ala-Ser is the tripeptide glycylalanylserine.

Peptide bonds

N-terminal end → $H_2N-\underset{\underset{\displaystyle H}{|}}{\overset{\overset{\displaystyle H}{|}}{C}}-\boxed{\overset{\overset{\displaystyle O}{\|}}{C}-\underset{\underset{\displaystyle H}{|}}{N}}-\underset{\underset{\displaystyle CH_3}{|}}{\overset{\overset{\displaystyle H}{|}}{C}}-COOH$ $\qquad$ $H_2N-\underset{\underset{\displaystyle CH_3}{|}}{\overset{\overset{\displaystyle H}{|}}{C}}-\boxed{\overset{\overset{\displaystyle O}{\|}}{C}-\underset{\underset{\displaystyle H}{|}}{N}}-\underset{\underset{\displaystyle H}{|}}{\overset{\overset{\displaystyle H}{|}}{C}}-COOH$ ← C-terminal end

Glycylalanine (Gly-Ala) $\qquad$ Alanylglycine (Ala-Gly)

The amino acids that have combined to form the peptide molecule are called **amino acid residues**. Since each dipeptide has a $-COOH$ and an $-NH_2$ group, a tripeptide can be formed from each dipeptide by reaction at either end, and the polymerization process can continue until a large **polypeptide** chain is formed (Figure 11.8, p. 284). Note that, as above, peptides are always written with the N-terminal end at the left.

Proteins are polypeptides containing from 50 to thousands of amino acid residues, and they vary greatly in structure and composition. They can be divided into two classes: simple and conjugated. **Simple proteins** consist only of amino acids, and two examples of simple proteins are insulin and chymotrypsin. Insulin, a hormone that is essential to controlling the concentration

TABLE 11.5 ■ Common L-Amino Acids Found in Proteins with the R group in Each Amino Acid Highlighted

Amino Acid	Abbre-viation	Structure
	Nonpolar R Groups	
Glycine	Gly	$H-CH-COOH$ NH_2 (on CH)
Alanine	Ala	$CH_3-CH-COOH$ NH_2 (on CH)
Valine*	Val	$CH_3-CH-CH-COOH$ $CH_3\quad NH_2$
Leucine*	Leu	$CH_3-CH-CH_2-CH-COOH$ $CH_3\qquad NH_2$
Isoleucine*	Ile	$CH_3-CH_2-CH-CH-COOH$ $CH_3\quad NH_2$
Proline	Pro	H_2C-CH_2 $H_2C\quad CHCOOH$ N H
Phenylalanine*	Phe	(benzene ring)$-CH_2-CH-COOH$ NH_2
Methionine*	Met	$CH_3-S-CH_2CH_2-CH-COOH$ NH_2
Tryptophan*	Trp	(indole ring)$CH_2-CH-COOH$ NH_2 N H
	Polar but Neutral R Groups	
Serine	Ser	$HO-CH_2-CH-COOH$ NH_2
Threonine*	Thr	$CH_3-CH-CH-COOH$ $OH\quad NH_2$
Cysteine	Cys	$HS-CH_2-CH-COOH$ NH_2

TABLE 11.5 ■ *Continued*

Amino Acid	Abbre-viation	Structure
	Polar but Neutral R Groups (cont)	
Asparagine	Asn	$H_2N-C(=O)-CH_2-CH(NH_2)-COOH$
Glutamine	Gln	$H_2N-C(=O)-CH_2CH_2-CH(NH_2)-COOH$
Tyrosine	Tyr	$HO-C_6H_4-CH_2-CH(NH_2)-COOH$
	Acidic R Groups	
Glutamic acid	Glu	$HO-C(=O)-CH_2CH_2-CH(NH_2)-COOH$
Aspartic acid	Asp	$HO-C(=O)-CH_2-CH(NH_2)-COOH$
	Basic R Groups	
Lysine*	Lys	$H_2N-CH_2CH_2CH_2CH_2-CH(NH_2)-COOH$
Arginine†	Arg	$H_2N-C(=NH)-NH-CH_2CH_2CH_2-CH(NH_2)-COOH$
Histidine	His	(imidazole ring, N, N–H)$-CH_2-CH(NH_2)-COOH$

* Essential amino acids that must be part of the human diet. The other amino acids can be synthesized by the body.

† Growing children also require arginine in their diet.

Source: *Chem World,* 2nd ed.

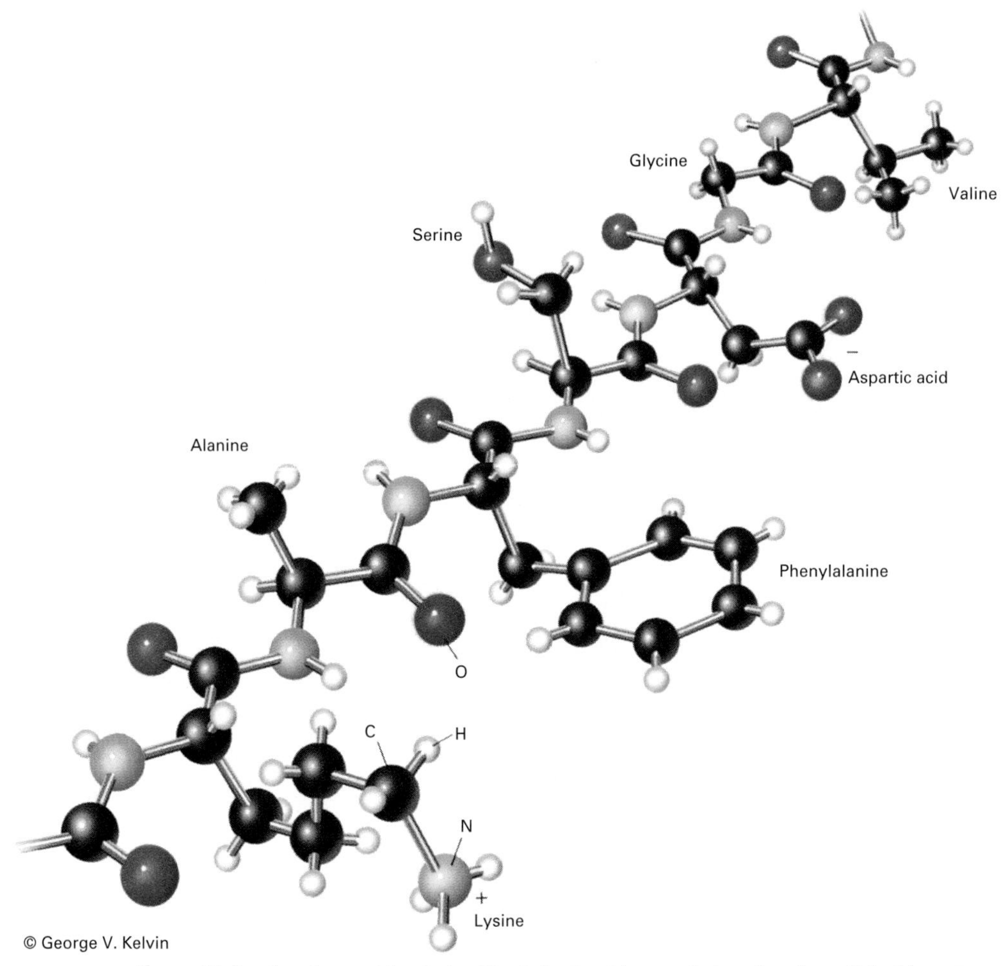

Figure 11.8 A polypeptide chain. Find the backbone chain of carbon (black) and nitrogen (blue) atoms, and then identify the planar peptide bonds and the R group of each amino acid.

of glucose in the blood, has 51 amino acid residues in two linked chains (Figure 11.9). Chymotrypsin, an enzyme that aids in the digestion of proteins in our diet, contains 245 amino acid residues.

Conjugated proteins contain other kinds of groups in addition to amino acids. Examples of conjugated proteins are hemoglobin and myoglobin, which bind oxygen in blood and muscles, respectively. Myoglobin has a single protein chain (Figure 11.10*a*). Human hemoglobin contains four protein chains, two identical ones having 141 amino acid residues and the other two, also identical, having 146 (Figure 11.10*b*). The site at which oxygen connects to myoglobin and hemoglobin is heme, an organic group that is the non–amino acid part of the molecules (Figure 11.10*c*). The oxygen binds with an iron ion (Fe^{2+}) in the center of the heme group.

To summarize, then, *proteins are polypeptides which are condensation polymers of amino acids.* They vary greatly in size, and some proteins include non–amino acid groups.

11.8 Protein Structure and Function

The order of the amino acid residues in a peptide or protein molecule is called the **amino acid sequence**. As the length of the chain increases, the number of variations in the sequence of amino acids quickly increases. Six tripeptides are possible if three amino acids (for example, glycine, Gly; alanine, Ala; and serine, Ser) are linked in all possible combinations:

Gly-Ala-Ser   Ser-Gly-Ala   Ala-Gly-Ser   Gly-Ser-Ala   Ser-Ala-Gly   Ala-Ser-Gly

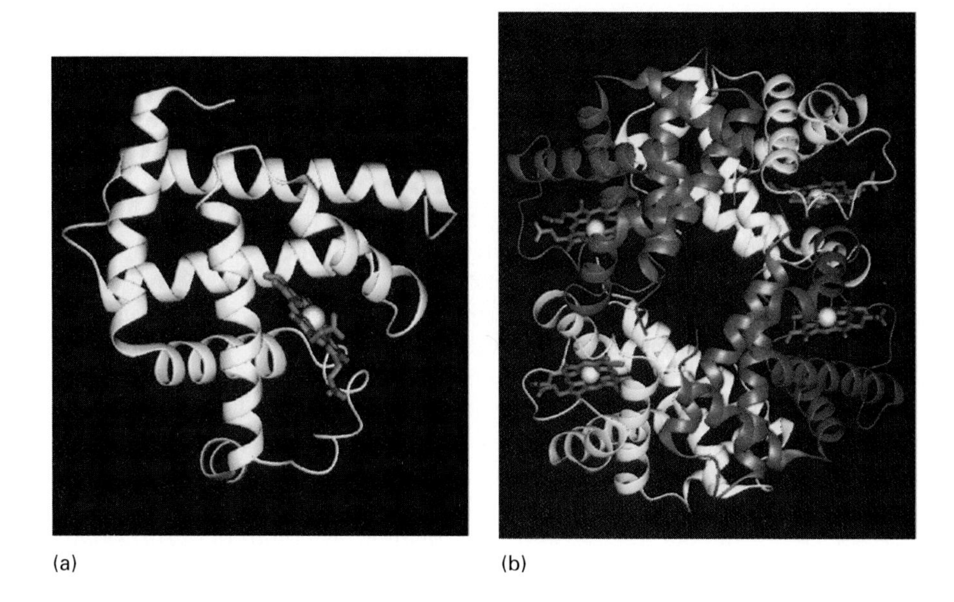

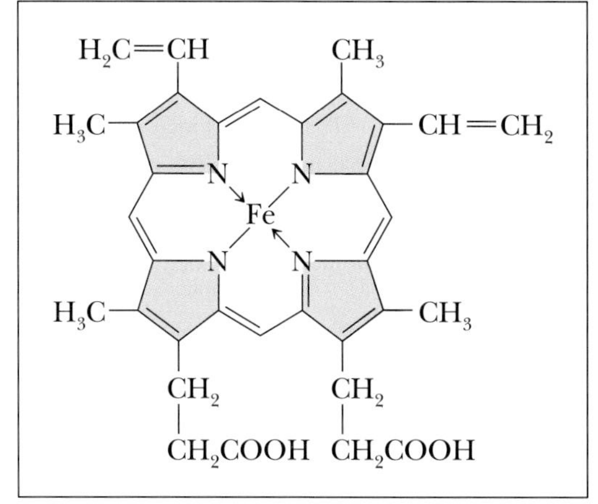

Figure 11.10 Myoglobin (a) and hemoglobin (b), which contain the nonprotein heme group (c). Myoglobin has a single protein chain (yellow) and one heme group (Fe ion shown in white). Hemoglobin has four protein chains of two different kinds (yellow and purple), with one heme group in each protein chain. In this type of representation of proteins, the alpha-helix parts of the protein chain are shown as ribbon-like spirals.

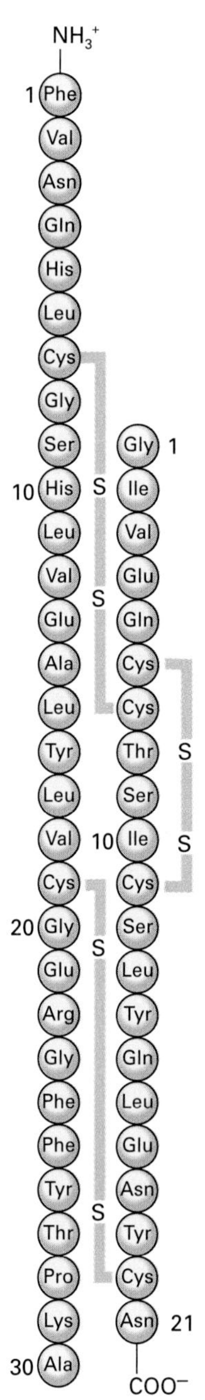

Figure 11.9 The amino acid sequence of insulin, which contains 51 amino acids in two chains linked by disulfide bonds between the side chains of cysteine residues.

Peptide	Amino acid sequence
Met-enkephalin	Tyr—Gly—Gly—Phe—Met
Leu-enkephalin	Tyr—Gly—Gly—Phe—Leu
β-endorphin	Lys—Arg—Tyr—Gly—Gly—Phe—Met—Thr—Ser—Glu—Lys—Ser—Glu—Thr—Pro—Leu—Val—Thr—Leu—Phe—Lys—Asn—Ala—Ile—Ile—Lys—Asp—Ala—Tyr—Lys—Lys—Gly—Glu

Figure 11.11 Enkephalins and endorphins, polypeptides that are natural opiates. Note that all three polypeptides have one identical amino acid sequence.

■ *n*!, called *n factorial*, is mathematical shorthand for the result of multiplying *n* by all the numbers between it and zero.

■ Neuropeptides are **neurotransmitters**, which are biomolecules that transmit chemical messages along nerve pathways. They act by connecting with other molecules (or parts of molecules) called receptors. The neurotransmitter–receptor interaction plays an important role in the effects on the body of both poisons and medicines.

If n amino acids are all different, the number of arrangements is $n!$. For four different amino acids, the number of different arrangements is 4!, or $4 \times 3 \times 2 \times 1 = 24$. For five different amino acids, the number of different arrangements is 5!, or 120. If all 20 different naturally occurring amino acids are bonded, the sequences alone make 2.43×10^{18} (2.43 quintillion) uniquely different 20-monomer molecules. *Since proteins can also include more than one molecule of a given amino acid, the possible combinations are essentially infinite.* However, of the many different proteins that could be made from a set of amino acids, a living cell will make only the relatively small, select number it needs.

Many short-chain peptides are important biochemicals. For example, enkephalins and endorphins (Figure 11.11) are referred to as " natural opiates" because they moderate pain in the same manner as opium derivatives. Our bodies synthesize enkephalins and endorphins to moderate pain, and our pain threshold is related to levels of these neuropeptides in our central nervous system. Individuals with a high tolerance for pain produce more of these neuropeptides and consequently tie up more receptor sites than normal; hence, they feel less pain. A dose of heroin temporarily bonds to a high percentage of the sites, resulting in little or no pain. Continued use of heroin causes the body to reduce or cease its production of enkephalins and endorphins. If use of the narcotic is stopped, the receptor sites become empty, and withdrawal symptoms occur.

EXAMPLE 11.2 *Peptides*

Use the structures of amino acids in Table 11.5 to draw the structure of the tripeptide represented by Ala-Ser-Gly, and give its name.

SOLUTION

The amino acid sequence in the abbreviated name shows that alanine should be written at the left with a free H_2N— group, glycine should be written at the right with a free —COOH group, and both should be connected to serine by peptide bonds:

$$
\begin{array}{ccccccccccccccc}
 & & H & & O & & & & H & & O & & H & & O \\
 & & | & & \| & & & & | & & \| & & | & & \| \\
H_2N & - & C & - & C & - & N & - & C & - & C & - N - & C & - & C - OH \\
 & & | & & & & | & & | & & & | & | & & \\
 & & CH_3 & & & & H & & CH_2OH & & & H & H & &
\end{array}
$$

The name is alanylserylglycine.

Exercise 11.2

Use the structures of amino acids in Table 11.5 to draw the structure of the tetrapeptide Cys-Phe-Ser-Ala.

Proteins are important in a wider variety of ways than other kinds of biomolecules. As *enzymes* they serve as catalysts in biological synthesis and degradation reactions. As *hormones* they serve a regulatory role, and as *antibodies* they protect us against disease. They make up the muscle fibers that contract so that we can move. And proteins are the major constituents of cellular and intracellular membranes, skin, hair, muscle, and tendons. Each individual protein has its own group of amino acids arranged in a definite molecular structure that is specific to the function of that protein.

The sequence of amino acids bonded to one another in a protein by peptide bonds is the protein's **primary structure**. Changing the sequence alters the properties of a protein, and just one change may produce a new protein unable to function like the original one. For example, *sickle cell anemia*, a reduction in the ability of hemoglobin to transfer oxygen, is caused by the replacement of only two amino acids out of 574 total amino acids in the four protein chains that make up the hemoglobin molecule (Figure 11.10*b*).

In some entire proteins and in parts of others, the shape of the backbone of the molecule (the chain containing peptide bonds) has a regular, repetitive pattern that is referred to as its **secondary structure**. The two most common secondary structures are the **α-helix** and the **β-pleated sheet**. The α-helix is held together by *intramolecular* (within the molecule) hydrogen bonding between backbone peptide bonds. An N—H group of one amino acid forms a hydrogen bond with the oxygen atom in the third amino acid down the chain (Figure 11.12).

The α-helix is the basic structural unit of the α-keratins in wool, hair, skin, beaks, nails, and claws.

Silk has the β-sheet structure (Figure 11.13, p. 288) in which several chains of amino acids are joined side-to-side by *intermolecular* (between protein chains) hydrogen bonds. The resulting structure is not elastic, because stretching the fibers would break either covalent bonds or the many hydrogen bonds holding the individual protein strands in the sheet. However, just as you can bend the stack of pages in this book, so too can the stack of protein sheets be bent.

Tertiary structure refers to how a protein molecule is folded. One kind of tertiary structure is found in collagen, a fibrous protein: three amino acid chains twisted into left-handed helices, which in turn are twisted into a right-handed superhelix to form an extremely strong fibril (Figure 11.14, p. 288). Bundles of fibrils make up the tough collagen.

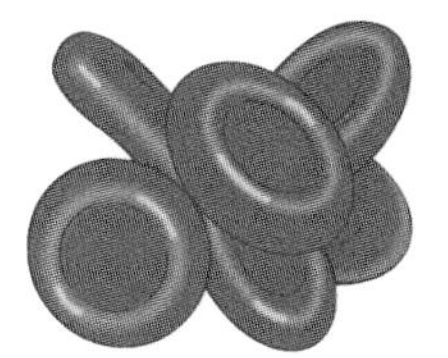

Disc-shaped

Sickle-shaped

Red blood cells (erythrocytes). Normally they are disc-shaped, but with the single amino acid defect of sickle cell anemia, they adopt the pointed, sickle shape. Sickle cell anemia is an inherited (genetic) disease caused by a flaw in DNA.

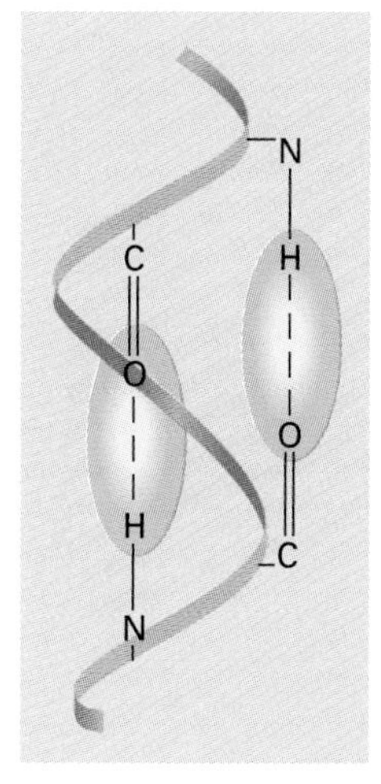

Figure 11.12 Helical protein structure. An illustration of how hydrogen bonding connects a peptide bond nitrogen atom to an oxygen atom in the third amino acid unit down the chain, resulting in a coiled structure.

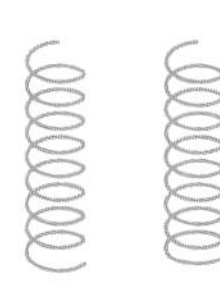

A coiled spring is helical in structure.

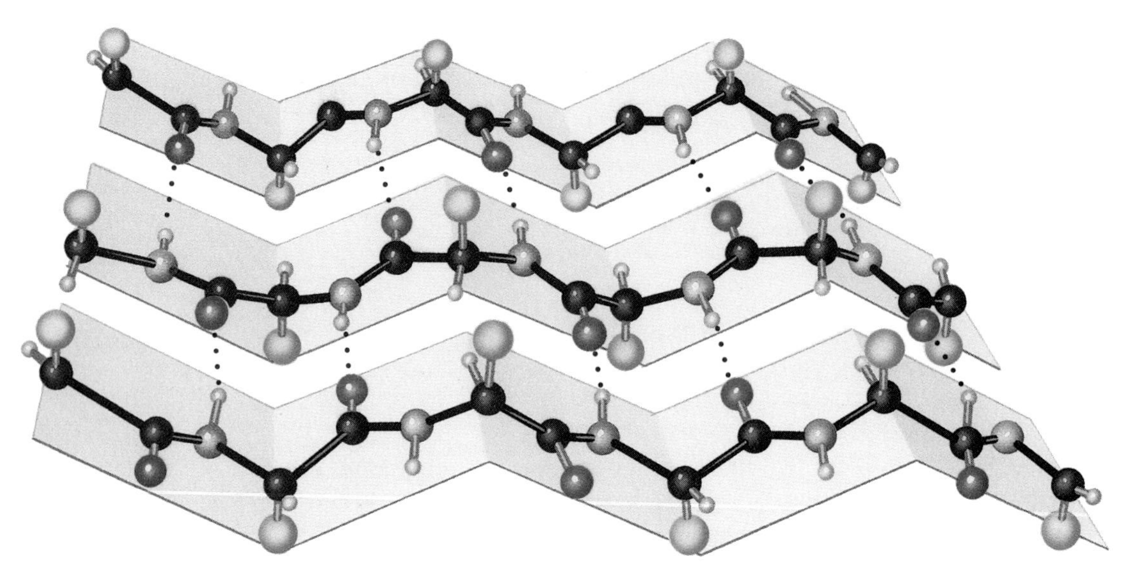

Figure 11.13 β-Pleated sheet protein structure. Hydrogen bonds are shown as dotted lines.

A second kind of tertiary structure is found in globular proteins, which contain regions where the polypeptide chain forms α-helices, other regions where parallel parts of the chain are organized into β-pleated sheets, and some parts that are just coiled randomly (e.g., Figure 11.10*a*).

Quaternary structure is the shape assumed by the entire group of chains in a protein composed of two or more chains. Hemoglobin (Figure 11.10*b*) is a prime example of quaternary structure—its four protein chains, together with the nonprotein heme molecule carried by each, can function only when combined in precisely the right shape.

All the structural features—primary, secondary, tertiary, and quaternary—are critical to the proper functioning of a protein and give a protein its "native" or "natural" structure (Figure 11.15). Any physical or chemical process that changes the native protein structure and makes it incapable of performing its normal function is called a **denaturation** process. For example,

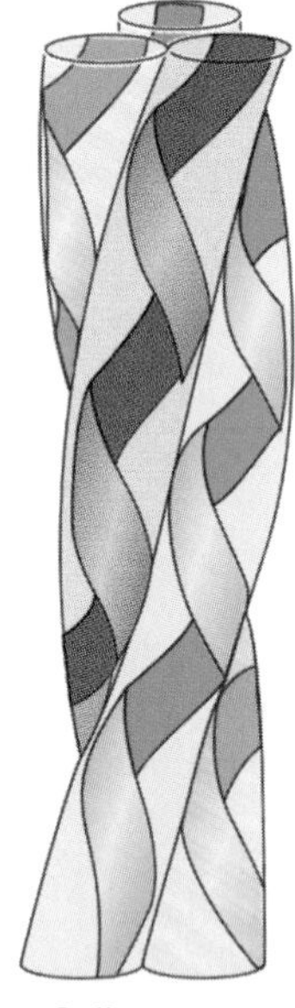

Figure 11.14 Collagen is composed of three polypeptide chains twisted together into a fiber like a rope.

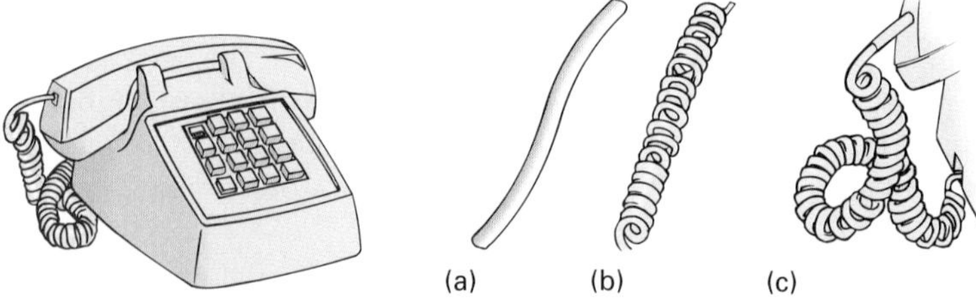

Figure 11.15 Structure of a telephone cord. (a) The straight cord is like the primary structure of a protein. (b) The curling of the cord is like the secondary structure of a protein. (c) When the curled cord is twisted up it is like the tertiary structure of a protein (which may include in its twisted form regions of α-helix and β-sheet).

heating an aqueous solution of a protein breaks hydrogen bonds in the secondary and tertiary structure and causes the protein molecule to unfold. Denaturing chemicals include reducing agents, which break disulfide linkages, and acids and bases, which affect the hydrogen bonds and ionic interactions between polypeptide chains. Whether denaturation is reversible depends on the protein and the extent of denaturation.

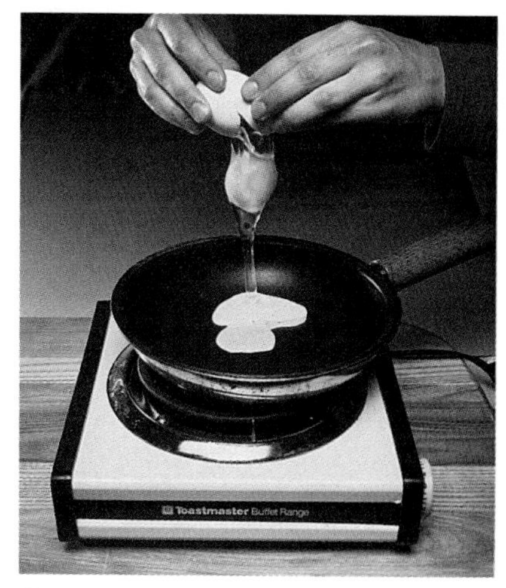

One way to denature a protein. *(C. D. Winters)*

Enzymes

Enzymes function as catalysts for chemical reactions in living systems. Many enzymes are globular proteins. One outcome of the highly organized, though seemingly random, structure of globular proteins is creation of the region that allows an enzyme to function as a catalyst (the active site). Like all catalysts, enzymes increase the rate of a reaction by lowering the energy of activation (Figure 6.3). This lowering occurs because the reaction path or process is changed. Enzymes are very effective catalysts and typically increase reaction rates by anywhere from 10^6 to 10^{16} times.

Many biomolecules are broken down during digestion by hydrolysis reactions, which are essentially the reverse of condensation reactions. In hydrolysis, a larger molecule is split into smaller molecules with the addition of H— and —OH of water where a bond was broken. The enzyme maltase catalyzes the hydrolysis of the sugar maltose into two molecules of D-glucose. This is the only function of maltase, and no other enzyme can substitute for it. Sucrase, another enzyme, hydrolyzes only sucrose. Some enzymes are less specific. The digestive enzyme trypsin, for example, primarily hydrolyzes peptide bonds in proteins. However, the structure and polarity of trypsin are such that it can also catalyze the hydrolysis of some esters.

■ The general equation for hydrolysis of an ester is

$$RCOOR + H_2O \longrightarrow RCOOH + ROH$$

Since the structure of the active site of an enzyme is important, the same factors that cause denaturation will also destroy the activity of the enzyme. For example, most enzymes are effective only over a narrow temperature range and a narrow pH range. Enzymes are denatured irreversibly at high temperatures or at pH values outside their effective range.

Many inherited, or **genetic**, diseases or defects affect how enzymes function. An example is found in **lactose intolerance**, which is an inability to digest lactose, the sugar present in milk from all mammals. Well over half the world's population has this problem to some degree. The pattern of inheritance is noticed largely in those with Asian or African ancestry and to a lesser extent in those of Northern European ancestry. When they are infants, people with lactose intolerance manufacture the enzyme lactase, which is necessary to digest lactose. As they grow older, their bodies stop producing this enzyme, and the ingestion of milk products containing lactose can lead to considerable discomfort in the form of stomachaches and diarrhea. The problem is avoided by eliminating milk products from the diet or by taking enzyme-containing tablets before eating any milk product.

■ Over-the-counter medications containing the lactase enzyme are available without prescription.

11.9 Hair Protein and Permanent Waves

Hair is composed principally of keratin. An important difference between hair keratin and other proteins is its high content of the amino acid cysteine, which contains the —SH group in its side chain.

■ Fingernails and toenails are composed of "hard" keratin, a very dense type of this protein. The epidermal cells of nails grow from epithelial cells lying under the white crescent at the growing end of the nail. Like hair, the nail tissue beyond the growing cells is dead.

THE WORLD OF CHEMISTRY

Unraveling the Protein Structure

Program 23, *Proteins: Structure and Function*

One of the key steps in unraveling the mystery of hydrogen bonds in protein structure involved Linus Pauling, a cold, and a Nobel Prize. Early in his career, Pauling worked on the structure of protein molecules. At that time there were several conflicting theories. Pauling and his colleagues thought that the first level of protein structure was a polypeptide chain. Then they asked themselves a fundamental question:

We asked: How is the polypeptide chain folded? We couldn't answer the question, but we said it's probably held together by hydrogen bonds. The conclusion we reached was that there are . . . polypeptide chains in the protein, which . . . are coiled back and forth, and that they are coiled into a very well-defined structure, configuration, with the different parts of the chain held together by hydrogen bonds. In 1937 I spent a good bit of the summer with models for—I assumed that I knew what a polypeptide chain looks like except for the way in which it's folded. And I wanted to fold it to form the hydrogen bonds. I didn't succeed. The fact is, I thought that there was something about proteins that perhaps I didn't know.

Pauling continued to work on this problem, but the solution eluded him. Then one day he had a crucial insight in a completely different and unexpected setting.

I had a cold. I was lying in bed for two or three days, and I read detective stories, light reading, for awhile, and then I got sort of bored with that. So I said to my wife, "Bring me a sheet of paper, and I'm going to—I think I'll work on that problem of how polypeptide chains are folded in proteins." So she brought me a sheet of paper and the slide rule and pencil, and I started working.

Using the knowledge gained from his years of model building, he drew the backbone of a polypeptide chain on a piece of paper. Then it occurred to him to try to fold the paper to see how hydrogen bonds could form along the polypeptide chain. The result was a structure that twisted around like a spring.

Well, I succeeded. It only took a couple of hours of work that day, March of 1948, for me to find the structure, called the alpha helix.

Linus Pauling. Along with R. B. Corey, Pauling proposed the α-helix and β-sheet structures for proteins. For his bonding theories and his work with proteins, Pauling was awarded the Nobel Prize in chemistry in 1954. For his fight against the dangers of nuclear weapons testing and fallout he received the Nobel Peace Prize in 1962.

The World of Chemistry

Program 23, *Proteins: Structure and Function*

Hair coloring in progress.

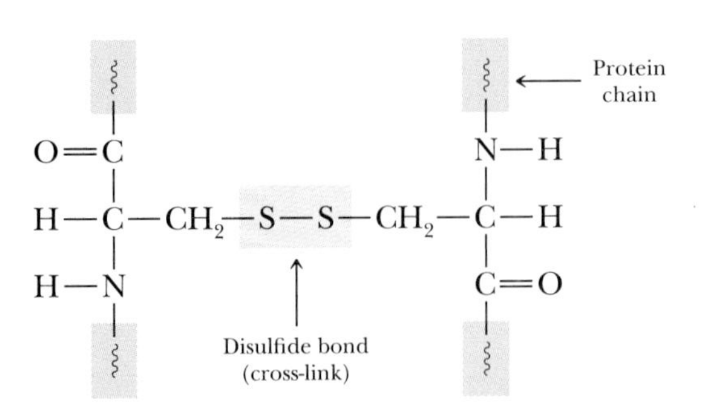

Cysteine plays an important role in the structure of hair by forming disulfide bonds (—S—S—) between protein chains. These protein chains are twisted together into spirals that group together to constitute individual strands of hair (Figure 11.16).

In addition to the disulfide bonds between hair protein structures, ionic bonds and hydrogen bonds between protein side chains affect the behavior of hair. In fact, these bonds explain bad hair days. Consider, for example, the interaction between a lysine $-NH_2$ group and a $-COOH$ of glutamic acid on a neighboring protein chain. The acidic $-COOH$ groups lose their protons, forming negatively charged $-COO^-$ groups, while the basic $-NH_2$ groups gain protons to form positively charged $-NH_3^+$ groups. When the $-NH_3^+$ and $-COO^-$ groups on adjacent chains approach each other, an ionic bond is formed that helps hold the two protein chains together.

$$\begin{array}{l} O{=}C \\ \quad | \\ H{-}C{-}CH_2{-}CH_2{-}CH_2{-}CH_2{-}NH_3^+ \;\; {}^-OOC{-}CH_2{-}CH_2{-}C{-}H \\ \quad | \\ H{-}N \end{array}$$

(Right chain: $N{-}H$ above, $C{=}O$ below the terminal carbon.)

Protein chain

Ionic bond

Ionic bond between two protein chains

Or consider how a hydrogen bond might form between side groups.

Protein chains

$$\begin{array}{l} O{=}C \qquad\qquad\qquad\qquad\qquad\quad N{-}H \\ H{-}C{-}CH_2{-}OH\cdots O{=}C{-}CH_2{-}C{-}H \\ H{-}N \qquad\qquad\qquad\quad OH \qquad\quad C{=}O \end{array}$$

Hydrogen bond

Hydrogen bond between two protein chains

Moisture affects hydrogen bonds between protein chains, and when their number and arrangement changes, hair changes. When hair is wet, it can be stretched to one and one half times its dry length. On a very moist day, enough bonds will break simultaneously for hair to lose its curl. On a very dry day, enough static electrical charge can build up for individual hairs to repel each other. Different people have bad hair on different days because the bonding patterns in their hair vary.

For more permanent curling, a more permanent chemical change in the hair structure is needed. The disulfide bonds between protein chains hold hair in its natural shape. In "permanent waving," these cross-links are broken by a reducing agent (step 1, Figure 11.16), which relaxes the tension. An oxidizing agent generates new cross-links, and the hair retains the shape of the roller, so it appears curly.

The most commonly used hair-waving reducing agent is the ammonium salt of thioglycolic acid ($HSCH_2COO^-NH_4^+$). A typical waving solution contains 5.7% thioglycolic acid, 2.0% ammonia, and 92.3% water. The usual oxidizing agent is hydrogen peroxide (H_2O_2) or a perborate compound that releases hydrogen peroxide in solution ($NaBO_3 \cdot 4\,H_2O$).

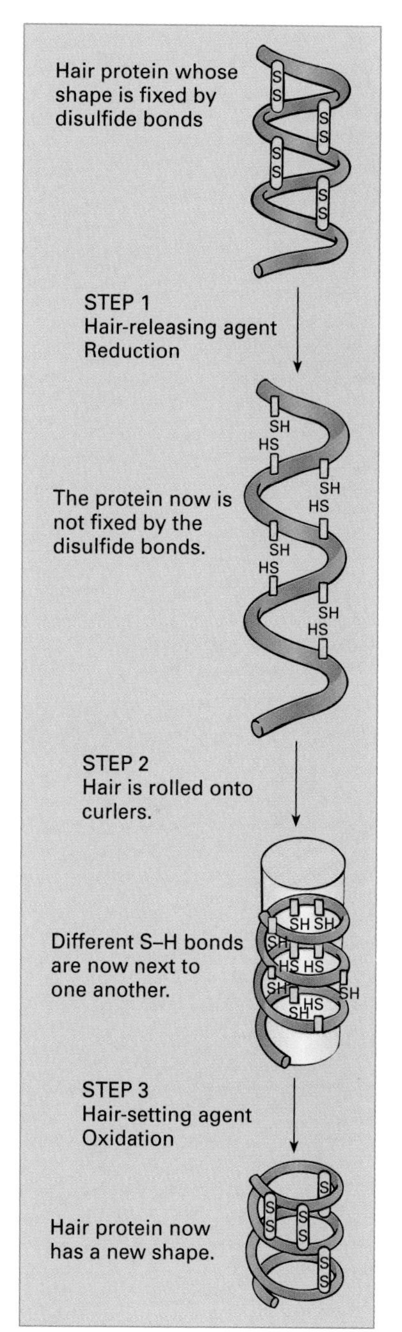

Figure 11.16 Permanent waving, a chemical oxidation–reduction process. The disulfide bonds in hair protein are broken by a reducing agent (step 1). The protein strands are twisted into a more curly shape (step 2). Then the new shape is fixed by an oxidizing agent that re-forms the disulfide bonds (step 3).

SELF-TEST 11C

1. All amino acids except ____________ have optical isomers.
2. Only left-handed optical isomers of amino acids are found in your body. (a) True, (b) False.
3. Amino acids that the body cannot synthesize from other molecules are called ____________ amino acids.
4. The peptide linkage that bonds amino acids together in protein chains has the structure ____________.
5. (a) If we have three different amino acids and can use each one three times in any given tripeptide, we can make a total of ________ different tripeptides.
   (b) If we can use each amino acid only once, there are still _______ possible different tripeptides.
6. (a) The primary structure of a protein refers to its ____________.
   (b) The secondary structure refers to its ____________.
   (c) The tertiary structure refers to ____________.
   (d) The quaternary structure refers to ____________.
7. The helical structure of proteins is caused by ____________ bonding.
8. That portion of the enzyme at which the reaction is catalyzed is called the ____________.
9. The best term to describe the general function of enzymes is (a) catalyst, (b) intermediate, (c) oxidant.
10. The activation energy of many biological reactions is decreased if a(an) ____________ is present.
11. Hair protein side chains are held together by ____________ and ____________ bonds.
12. Water easily breaks ____________ bonds found in hair structure, making wet hair more pliable.

11.10 Energy and Biochemical Systems

The energy to sustain all but a few forms of life comes from the sun. During photosynthesis, green plants absorb energy from the sun to make glucose and oxygen from carbon dioxide and water. Glucose is a major energy source for

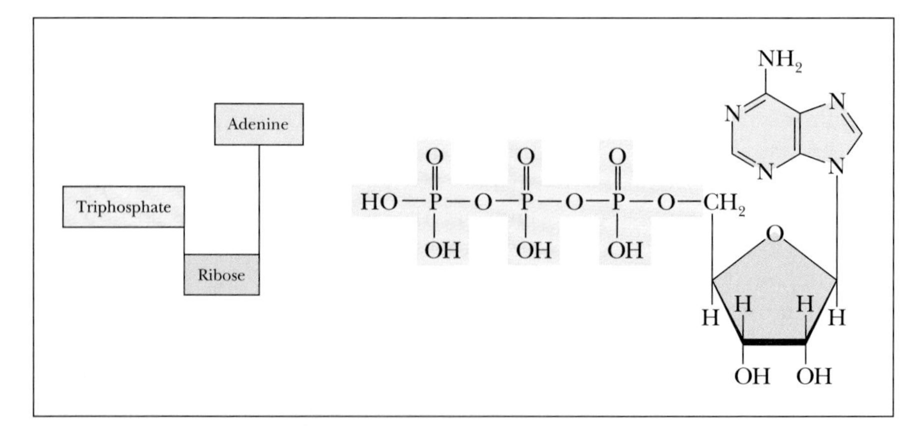

Figure 11.17 Adenosine triphosphate (ATP).

ATP + HOH $\xrightarrow{\text{catalyst}}$ HO—P(=O)(OH)—O—P(=O)(OH)—O—CH_2–(ribose–adenine) + H_3PO_4 + 7.3 kcal

Adenosine diphosphate (ADP)

Figure 11.18 Hydrolysis of ATP to give ADP (adenosine diphosphate).

all living organisms. During metabolism, much of the protein, fat, and carbohydrate in our diets is converted to glucose (Section 12.2). The energy stored in glucose is eventually transferred to the bonds in molecules such as adenosine triphosphate (ATP, Figure 11.17). When living organisms need energy, phosphate bonds in the ATP molecules are broken to give adenosine diphosphate (ADP, Figure 11.18) and energy for other biochemical reactions.

In the complex process of photosynthesis, carbon dioxide is reduced to make sugar, and water is oxidized to oxygen

$$\underset{\text{Carbon dioxide}}{6\ CO_2} + \underset{\text{Water}}{6\ H_2O} + \underset{\text{Energy (sunlight)}}{688\ \text{kcal}} \longrightarrow \underset{\text{Glucose}}{C_6H_{12}O_6} + \underset{\text{Oxygen}}{6\ O_2}$$

The oxygen produced in photosynthesis is the continuing source of all the oxygen in our atmosphere. We are dependent on the plant life of our planet, and we must live in balance with the oxygen output of that plant life, as well as with the food output of the same plant life. Photosynthesis is absolutely vital to life on Earth.

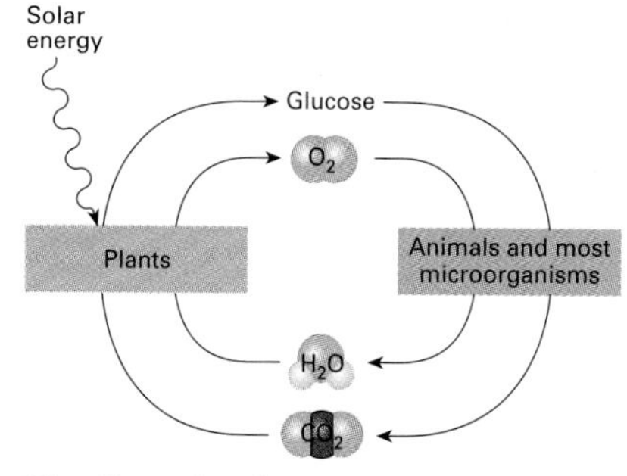

The flow of carbon atoms, oxygen atoms, and energy between plants and animals.

11.11 Nucleic Acids

The genetic information that makes each organism's offspring look and behave like its parents is encoded in molecules called **nucleic acids**. Together with a set of specialized enzymes that catalyze their synthesis and decomposition, the nucleic acids constitute a remarkable system that accurately copies millions of pieces of data with very few mistakes.

Like polysaccharides and polypeptides, nucleic acids are condensation polymers. Each monomer in these polymers includes one of two simple sugars, one phosphoric acid group, and one of a group of heterocyclic nitrogen compounds that behave chemically as bases. A particular nucleic acid is a **deoxyribonucleic acid (DNA)** if it contains the sugar 2-deoxy-D-ribose, and it is a **ribonucleic acid (RNA)** if it contains the sugar D-ribose.

D-Ribose

2-Deoxy-D-ribose

■ Notice the presence of the OH group on the 2 carbon in D-ribose. The 2-deoxy-D-ribose doesn't have this OH group.

The five organic bases that play a key role in the mechanism for information storage are ***adenine*** (A) and ***guanine*** (G), ***thymine*** (T), ***cytosine*** (C), and ***uracil*** (U). These bases are mentioned so often in any discussion of nucleic acid chemistry that to save space they are usually referred to only by the first letter of each name.

Adenine (A) Guanine (G) Thymine (T) Cytosine (C) Uracil (U)

Nucleic acids are found in all living cells, with the exception of the red blood cells of mammals. DNA occurs primarily in the nucleus of the cell, and RNA is found mainly in the cytoplasm, outside the nucleus. There are three major types of RNA, each with its own characteristic size, base composition, and function in protein synthesis (as described later in this section): messenger RNA (mRNA), transfer RNA (tRNA), and ribosomal RNA (rRNA).

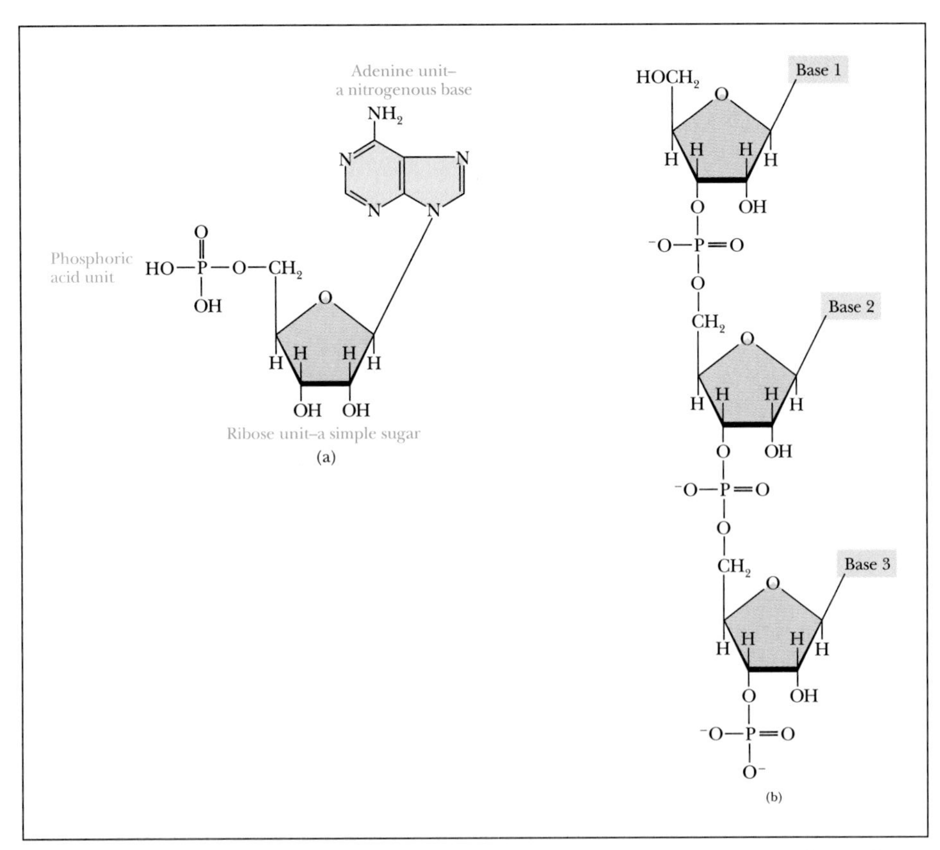

Figure 11.19 (a) A nucleotide. Because its sugar unit is ribose, this is a ribonucleotide. (b) Bonding in a trinucleotide. The sugars could also be deoxyribose units and the bases can be any of the five bases shown in the text. The primary structures of both DNA and RNA are extensions of this structure.

The monomers that polymerize to make both DNA and RNA are known as **nucleotides**. They have structures like the one shown in Figure 11.19*a*. Each nucleotide contains a phosphoric acid unit, a ribose sugar unit, and one of the five bases. The nucleotides of DNA and RNA have two structural differences: (1) they contain different sugars (deoxyribose and ribose); and (2) the base uracil occurs only in RNA, whereas the base thymine occurs only in DNA. The other bases—adenine, guanine, and cytosine—are found in both DNA and RNA.

When the phosphate group is absent, the remaining fragment (a ribose bonded to one of the five bases) is called a **nucleoside**.

Polynucleotides have molar masses ranging from about 25,000 for tRNA molecules to billions for human DNA. The sequence of nucleotides in the polymer chain (shown by the base sequence) is its primary structure. Polynucleotides are formed by the polymerization of nucleotides to make esters. As an example, Figure 11.19*b* shows three monomers condensed to a trinucleotide.

In 1953, James D. Watson and Francis H. C. Crick proposed a secondary structure for DNA that revolutionized our understanding of heredity and genetic diseases. They proposed that pairs of polynucleotides are arranged in a double helix stabilized by hydrogen bonding between the base groups lying opposite each other in the two chains. The critical point of the Watson–Crick model is that hydrogen bonding can best occur between specific bases. Adenine–thymine (A—T) and guanine–cytosine (G—C) pairs occur exclusively because they are very tightly hydrogen bonded. The hydrogen bonding between these specific bases, called **complementary hydrogen bonding**, is illustrated in Figure 11.20.

The World of Chemistry
Program 24, *The Genetic Code*

The DNA helix viewed from the top.

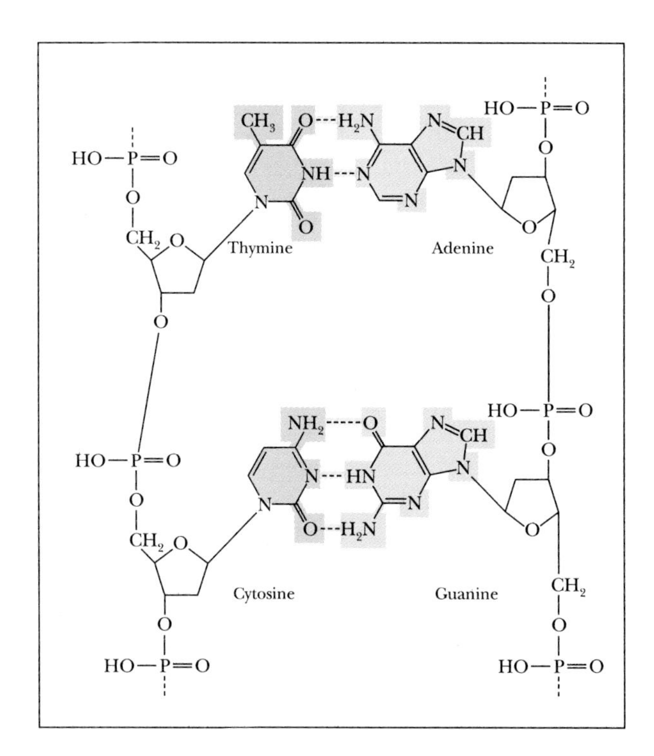

Figure 11.20 Complementary hydrogen bonding in the DNA double helix. Hydrogen bonds in the thymine–adenine (T—A) and cytosine–guanine (C—G) pairs stabilize the double helix.

■ RNA is generally a single strand of helical polynucleotide.

The function of polynucleotides is to transcribe hereditary information so that like begets like. The almost infinite variety of primary structures of polynucleotides allows an almost infinite variety of information to be recorded in the molecular structures of the strands of nucleic acids. The different arrangements of just a few different bases give the large variety of structures. In a somewhat similar fashion, the multiple arrangements of just a few language symbols convey the many ideas in this book. The coded information in the polynucleotide controls the inherited characteristics of the next generation as well as most of the continuous life processes of the organism.

■ The inherited traits of an organism are controlled by DNA molecules.

Double-stranded DNA forms the 46 human chromosomes, which have specialty heredity areas called **genes**. Genes are segments of DNA that have as few as 1000 or as many as 100,000 base pairs such as those shown in Figure 11.20. Each gene holds the information needed for the synthesis of a single protein. Human DNA (the human **genome**) is estimated to have up to 100,000 genes and about 3 billion pairs of bases. However, genes are estimated to make up only 3% of DNA, with each gene sandwiched between "junk" or noncoding DNA sequences. There are also short segments that act as switches to signal where the coding sequence begins.

■ The total sequence of base pairs in the cells of a plant or animal is called its *genome*.

Replication of DNA: Heredity

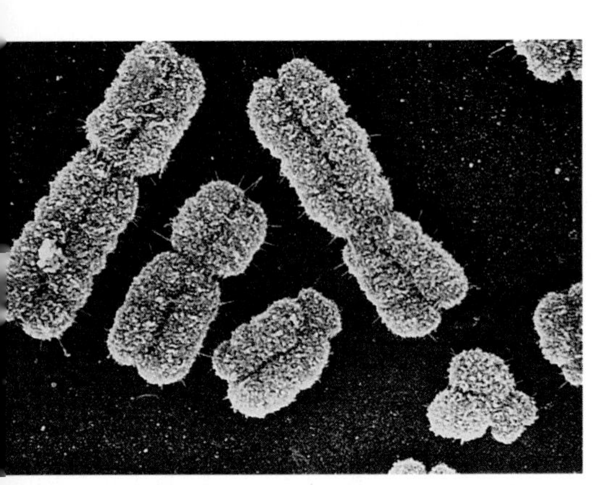

Human chromosomes (magnified about 8000 times). DNA assembles into chromosomes when a cell is about to divide. *(Biophoto Associates/Photo Researchers)*

The transfer of coded information begins with the replication of DNA and continues with natural protein synthesis. Almost all nuclei in an organism's cells contain the same chromosomal composition. This composition remains constant regardless of whether the cell is starving or has an ample supply of food materials. Each organism begins life as a single cell with this same chromosomal composition; in sexual reproduction, half of a chromosome comes from each parent. These well-known biological facts, along with recent discoveries concerning polynucleotide structures, have led scientists to the conclusion that the DNA structure is faithfully copied during normal cell division (mitosis—both strands) and that only half is copied in cell division that produces reproductive cells (meiosis—one strand).

In **replication** the double helix of the DNA structure unwinds, and each half of the structure serves as a template, or pattern, from which the other complementary half can be reproduced from the molecules in the cell environment (Figure 11.21). Replication of DNA occurs in the nucleus of the cell before the cell divides.

Natural Protein Synthesis

The proteins of the body are continually being replaced and resynthesized from the amino acids available to the body. The use of isotopically labeled amino acids (see Section 4.8) has made possible studies of the average lifetimes of amino acids as constituents in proteins—that is, the time it takes the body to replace a protein in a tissue. For a process that must be extremely complex, replacement is very rapid. Only minutes after radioactive amino acids are injected into animals, radioactive protein can be found. Although all the proteins in the body are continually being replaced, the rates of replacement vary. Half the proteins in the liver and plasma are replaced in six days; the time needed for replacement of muscle proteins is about 180 days; and replacement of protein in other tissues, such as bone collagen, takes even longer.

Recall that each organism has its own kinds of proteins. The number of possible unique arrangements of 20 amino acid units is 2.43×10^{18}, yet proteins characteristic of a given organism can be synthesized by the organism in a matter of a few minutes.

The DNA in the cell nucleus holds the code for protein synthesis, which is carried out in a series of steps summarized in Figure 11.22 (p. 298). First, messenger RNA, like all forms of RNA, is synthesized in the cell nucleus. The sequence of bases in one strand of the chromosomal DNA serves as the template from which a single strand of a messenger ribonucleotide (mRNA) is made in a process known as **transcription**. The bases of the mRNA strand complement those of the DNA strand. Two bases are complementary when each one fits the other and forms one or more hydrogen bonds. Messenger RNA contains only the four bases: adenine (A), guanine (G), cytosine (C), and uracil (U). DNA contains the four bases adenine (A), guanine (G), cytosine (C), and thymine (T). The bases in DNA are transcribed into bases in mRNA as follows:

DNA	mRNA
A	U
G	C
C	G
T	A

This means that, provided the necessary enzymes and energy are present, wherever a DNA has an adenine base (A), the mRNA will transcribe a uracil base (U), and so on.

After transcription, mRNA passes from the nucleus of the cell to a ribosome (which contains the rRNA), where mRNA serves as the template for the sequential ordering of amino acids during protein synthesis by the process known as **translation**. As its name implies, messenger RNA contains the sequence message, in the form of a three-base code (called a **codon**), for ordering amino acids into proteins (Table 11.6). Each of the thousands of different proteins synthesized by cells is coded by a specific mRNA or segment of an mRNA molecule.

Transfer RNAs carry the amino acids to the mRNA one by one. Each of the 20 amino acids found in proteins has at least one corresponding tRNA, and some have multiple tRNAs (Table 11.6, p. 299). Table 11.6 lists the RNA codes and shows in the first line that, for example, UUU codes for phenylalanine and UCU codes for serine.

In the schematic illustration in Figure 11.22, which summarizes DNA → RNA → protein, pick a three-base sequence on the bottom DNA strand of the two DNA strands at the top of the figure and follow it through to the bottom, where a tRNA attaches to a three-base codon on mRNA. Assume the DNA strand you have selected serves as the template for the synthesis of the single strand of mRNA shown in the figure. Do you agree with the changes that occur in the letters representing the three-base sequence?

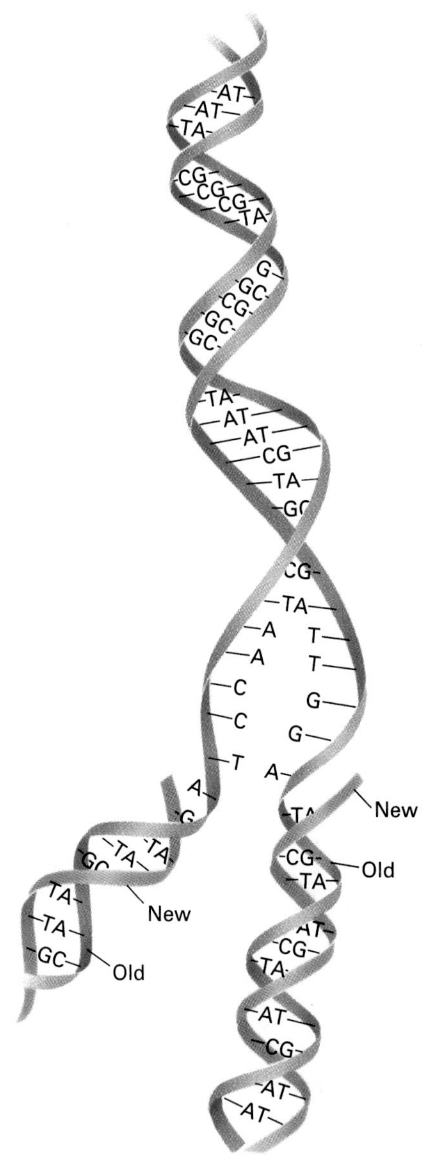

Figure 11.21 Replication of DNA. When the double helix of DNA (blue) unwinds, each half serves as a template on which to assemble nucleotides to form new DNA strands.

■ Protein synthesis is initiated by START codons and terminated by STOP codons (Table 11.6).

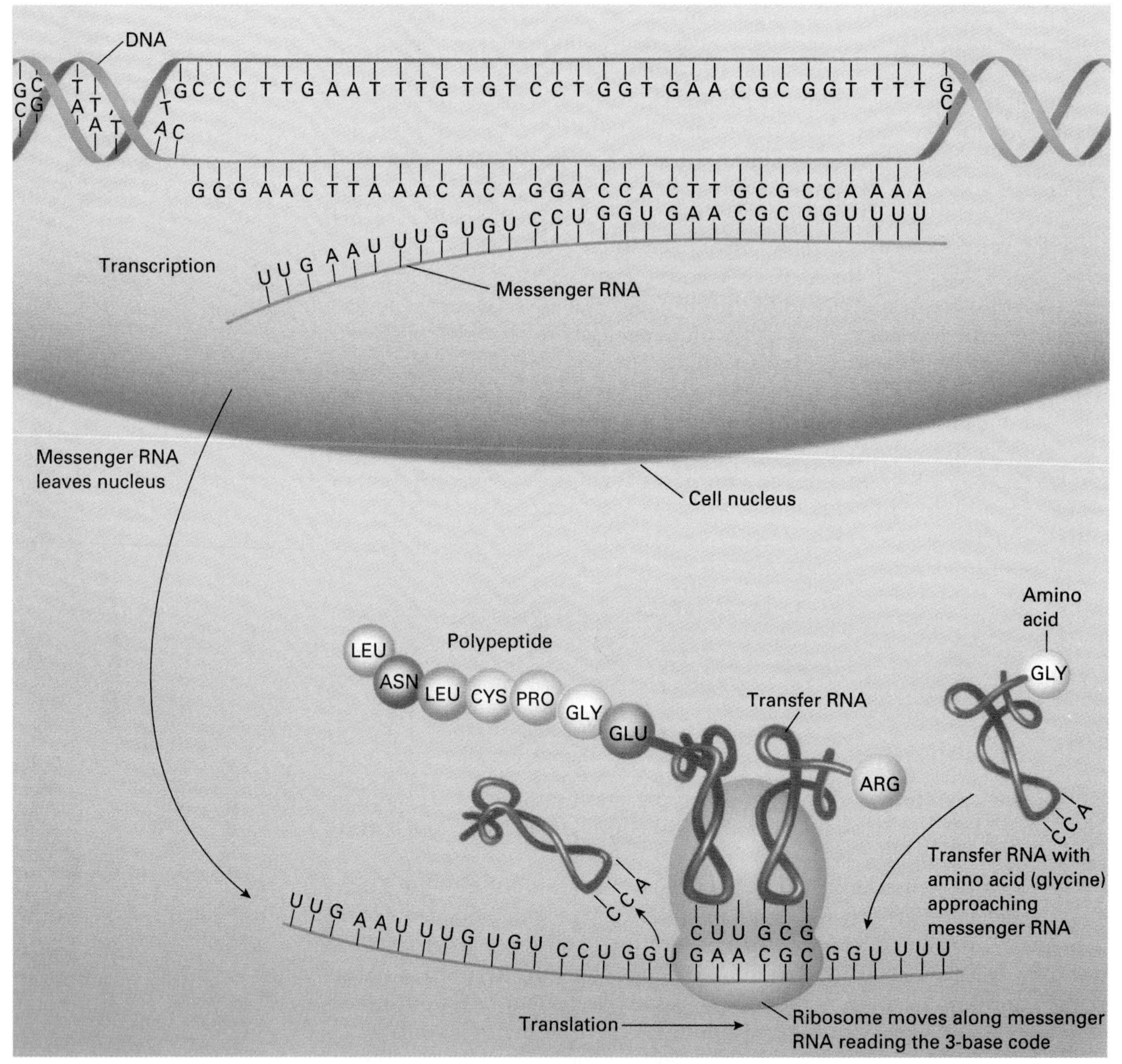

Figure 11.22 Protein synthesis. A section of DNA unwinds and transcription results in production of messenger RNA. Outside the cell nucleus, at a ribosome, the code carried by messenger RNA is translated into the amino acid sequence of a newly synthesized protein. Transfer RNA brings the amino acids into position one at a time.

EXAMPLE 11.3 *Nucleic Acids*

If the base sequence in a DNA segment is . . . GCTGTA . . . , what is the base sequence in the complementary mRNA? What is the order in tRNA?

SOLUTION

The base pairs between DNA and mRNA are G . . . C, A . . . U, and T . . . A. Therefore, the complementary mRNA segment for a DNA segment of GCTGTA is determined by using these allowed base pairs. Start with GCTGTA and place the correct base below it. The resulting mRNA segment is CGACAU. The order in tRNA is the complement of the mRNA segment. The allowed base pairs are G . . . C and A . . . U, so the order in tRNA is GCUGUA.

A tube full of DNA. A few dozen years ago, preparing this large a sample of DNA was almost impossible. *(Jean Claude Revy/Phototake)*

Exercise 11.3

If the sequence of bases along a mRNA strand is . . . UCCGAU . . . , what was the sequence along the DNA template?

TABLE 11.6 ■ Messenger RNA Codes for Amino Acids*

First Letter of Code	Second Letter of Code				Third Letter of Code
	U	C	A	G	
U	Phenylalanine	Serine	Tyrosine	Cysteine	U
	Phenylalanine	Serine	Tyrosine	Cysteine	C
	Leucine	Serine	STOP	STOP	A
	Leucine	Serine	STOP	Tryptophan	G
C	Leucine	Proline	Histidine	Arginine	U
	Leucine	Proline	Histidine	Arginine	C
	Leucine	Proline	Glutamine	Arginine	A
	Leucine	Proline	Glutamine	Arginine	G
A	Isoleucine	Threonine	Asparagine	Serine	U
	Isoleucine	Threonine	Asparagine	Serine	C
	Isoleucine	Threonine	Lysine	Arginine	A
	START or methionine	Threonine	Lysine	Arginine	G
G	Valine	Alanine	Aspartic acid	Glycine	U
	Valine	Alanine	Aspartic acid	Glycine	C
	Valine	Alanine	Glutamic acid	Glycine	A
	Valine	Alanine	Glutamic acid	Glycine	G

* In groups of three (called codons), bases of mRNA code the order of amino acids in a polypeptide chain. A, C, G, and U represent adenine, cytosine, guanine, and uracil, respectively. Some amino acids have more than one codon, and hence more than one tRNA can bring the amino acid to mRNA.

Human Genome Project

The Human Genome Project is an international collaborative project to determine the complete sequence of nucleic acids in the human genome, estimated to contain 3 billion base pairs. The 15-year project is expected to be completed in 2005. Many groups are focusing on the genes within the genome. Partial or complete sequences of about 30,000 of the estimated 100,000 genes have been completed. A number of human diseases have been traced to genetic defects, whose positions within the human genome have been identified. These include the cystic fibrosis gene; Huntington's disease gene; amyotrophic lateral sclerosis gene (Lou Gehrig's disease); and genes associated with the development of diabetes, brain cancer, breast cancer, and colon cancer. A complete sequencing of the human genome will improve our knowledge of the estimated 4000 hereditary diseases and lead to better diagnosis and treatment of them.

Recombinant DNA

The first successful gene-splicing and gene-cloning experiments produced recombinant DNA in the early 1970s. The basic idea is to use the rapidly dividing property of common bacteria, such as *Escherichia coli*, as a microbe factory for producing recombinant DNA molecules that contain the genetic information for the desired product. Bacteria have been produced that can synthesize specific proteins, such as human growth hormone and human insulin. To create such bacteria, a gene is removed from the bacterium, part of a gene from a human or other organism (the part that produces human insulin, for example) is spliced in, and the spliced gene is put back into the bacterium. The modified bacterium then serves as a microbe factory to make millions of copies of insulin-producing bacteria. The process of splicing and recombining genes is referred to as **recombinant DNA technology**, or **biogenetic engineering**.

One of the earliest benefits of recombinant DNA technology was the biosynthesis of ***human insulin*** in 1978. Millions of people with diabetes depend on the availability of insulin, but many are allergic to animal insulin, which was the only previous source. Biosynthesized human insulin is now being marketed. Biotechnology firms are also producing ***human growth hormone***, which is used in treating youth dwarfism.

A number of "transgenic animals" have been produced, including goats, rabbits, and mice; these animals are used as drug "pharms" since their new genes cause them to produce marketable quantities of desirable pharmaceuticals. For example, ***tissue plasminogen activator (TPA)***, which is used to dissolve blood clots in emergency treatment of heart attack victims, was originally obtained from a bioreactor of *E. coli*, then from the milk of transgenic mice, and now from the milk of transgenic goats.

Sometimes, transgenic animals are less than ideal. Transgenic pigs did produce leaner meat but they have arthritis, lethargy, and a low sex drive. A group of transgenic beef calves did better in general health, but it is as yet unclear if the quality of beef will be improved.

Biogenetic engineering has also led to developments in growing plants that produce their own insecticide, making plants resistant to viral and bac-

A pair of transgenic sheep. These sheep have the human gene for synthesis of α_1-antitrypsin. Persons with an inherited deficiency of this enzyme inhibitor are at high risk for developing emphysema, which affects lung function and causes difficulty in breathing. α_1-Antitrypsin isolated from the sheep's milk is being tested in the treatment of emphysema. *(Philippe Plailly/Eurelios/SPL/Photo Researchers)*

terial infections, and matching the chemistry of a plant with a protecting herbicide.

See Section 17.7 for discussion of genetically engineered plants.

SELF-TEST 11D

1. The reactants in the photosynthesis process are ____________ and ____________; ____________ must also be supplied.
2. Most of the energy obtained by the oxidation of food is used immediately to synthesize the molecule ____________.
3. The hydrolysis of ATP produces ____________ and phosphoric acid. ____________ is also released.
4. The basic code for the synthesis of protein is contained in the ____________ molecule.
5. The sugar in RNA is ____________, whereas the one in DNA is ____________.
6. A nucleotide contains ____________, ____________, and ____________.
7. The secondary structure of DNA is in the shape of a (an) ____________.
8. When DNA replicates itself, each base in the chain is matched to another one via ____________ bonds.
9. Base pairs in DNA are formed between A and ____________, G and ____________, T and ____________, and C and ____________.
10. Human DNA is estimated to have ____________ base pairs.

MATCHING SET

___ 1. Energy "cash" in the living cell
___ 2. Natural protein
___ 3. Product of ATP hydrolysis
___ 4. Saponification
___ 5. D-Glucose
___ 6. Enzymes
___ 7. Carbohydrate stored in animals
___ 8. Keratin
___ 9. Polypeptides
___ 10. DNA
___ 11. Fibrous protein
___ 12. Cysteine

a. ADP + energy
b. Amino acid common in hair
c. Structure determined by DNA and RNA
d. ATP
e. Hair protein
f. Proteins
g. Main sugar present in the blood
h. Polynucleotide
i. Biochemical catalysts
j. Glycogen
k. Collagen
l. Hydrolysis reaction in which a fat or oil produces glycerol and one or more fatty acids

QUESTIONS FOR REVIEW AND THOUGHT

1. What is meant by the term "chiral"?
2. What do the following terms mean?
   (a) L-Isomers (b) D-Isomers
   (c) Optically active (d) Polarimeter
3. What is a racemic mixture?
4. Give definitions for the following terms:
   (a) Carbohydrate (b) Monosaccharide
   (c) Disaccharide (d) Polysaccharide

5. What is an artificial sweetener?
6. What are the following substances?
   (a) Starches (b) Glycogen
   (c) Cellulose
7. What is a lipid?
8. What is an unsaturated fatty acid? What is a saturated fatty acid?
9. What is the process of hydrogenation of unsaturated molecules?
10. What are *cis* and *trans* fatty acids?
11. Describe how a surfactant such as a soap or detergent molecule works.
12. It is possible to mix oil and water and get a stable emulsion using a soap. It is also possible to make a stable oil-in-water emulsion using the yoke of an egg (mayonnaise). Explain how these two emulsifiers work.
13. Explain why synthetic detergents are more effective than soaps in hard water.
14. A certain cream imparts a somewhat greasy feel to the skin after its application. What kind of emulsion do you suspect this cream to be based on? (a) Water-in-oil (b) Oil-in-water
15. A skin cream has its ingredients listed on the back of the jar. Water is listed *after* several other ingredients, some of which have the word *oil* in their names. On what kind of emulsion do you think this cream is based?
16. What is an amino acid? Draw and label a generalized structure for an amino acid.
17. What is a peptide bond?
18. What is an essential amino acid?
19. What functional groups are always present in each molecule of an amino acid?
20. What is meant by the following terms?
    (a) Dipeptide (b) Polypeptide
    (c) Amino acid residue (d) Protein
21. What is meant by the term "amino acid sequence"?
22. Define "neuropeptides" and give two examples.
23. What are the meanings of the terms "primary," "secondary," "tertiary," and "quaternary" structures of proteins?
24. Which of the following biochemicals are polymers?
    (a) Starch, (b) Cellulose, (c) Glucose, (d) Fats, (e) Glycylalanylcysteine, (f) Proteins, (g) DNA, (h) RNA.
25. What is meant by the following terms?
    (a) α-Helix (b) β-Pleated sheet
26. What does it mean to "denature" a protein?
27. What is an active site on an enzyme?
28. Glutathione is an important tripeptide found in all living tissues. It is also named glutamylcystylglycine. Draw the structure of glutathione. Which enantiomeric forms of the amino acids would you predict are used to synthesize this peptide?
29. (a) How many tetrapeptides are possible if four amino acids are linked in different combinations that contain all four amino acids?
    (b) Write these combinations for tetrapeptides made from glycine, alanine, serine, and cystine. Use three-letter abbreviations for the amino acids. For example, one combination is Gly-Ala-Ser-Cys.
30. Use three-letter abbreviations to write the possible tripeptides that can be formed from phenylalanine, serine, and valine if each tripeptide contains all three amino acids.
31. Name the following tripeptide.

```
        H  O        H  O        H  O
        |  ‖        |  ‖        |  ‖
   H2N—C—C—N—C—C—N—C—C—OH
        |     |  |     |  |
        CH    H  CH2   H  CH2
       /  \      |        |
    CH3   CH3    SH     (C6H4)
                          |
                          OH
```

32. Draw the structure of alanylglycylphenylalanine.
33. The human body can metabolize D-glucose but not L-glucose. Explain.
34. What is lactose intolerance?
35. Describe the two kinds of bonding that hold protein strands together in hair.
36. What is the effect of water on the natural protein bonding in hair?
37. What is the disulfide linkage, and what role does it play in protein structure?
38. What three molecular units are found in nucleotides?
39. What are the differences in structure between DNA and RNA?
40. What is the purpose of ATP?
41. What are the fundamental components of nucleic acids?
42. Explain the term "codon."
43. What is meant by the term "complementary bases"?
44. How many hydrogen bonds are possible between the following complementary base pairs?
    (a) Cytosine and guanine (b) Thymine and adenine
45. What is the Human Genome Project?
46. A segment of a DNA strand has the base sequence . . . GCTGTAACCGAT. . . .
    (a) What is the base sequence in the complementary mRNA?
    (b) What is the order in tRNA?
    (c) Consult Table 11.6 and give the amino acid sequence in the portion of the peptide being synthesized.
47. A segment of a DNA strand has the base sequence . . . TGTCAGTGGGCCGCT. . . .
    (a) What is the base sequence in the complementary mRNA?

(b) What is the order in tRNA?
(c) Consult Table 11.6 and give the amino acid sequence in the portion of the peptide being synthesized.

48. Write the overall equation for photosynthesis.
49. What stabilizing forces hold the double helix together in the secondary structure of DNA proposed by Watson and Crick?
50. (a) What are the three major types of RNA?
    (b) What are their functions?
51. A base sequence in a DNA segment is . . . GTAGC . . . What is the base sequence in the complementary mRNA?
52. (a) What are the three bases present in both messenger RNA and DNA?
    (b) What is the fourth base found in mRNA?
    (c) What is the fourth base found in DNA but not in mRNA?
53. Give the mRNA complementary base sequence that matches the following DNA segment base sequence:

DNA sequence	**mRNA sequence**
G . . .	
A . . .	
C . . .	
A . . .	

CHAPTER 12

Nutrition: The Basis of Healthy Living

We are living in a health-conscious society. Newspapers, magazines, and television programs constantly offer us advice on what to eat. The advice ranges from responsible journalism to scare tactics and sensationalism. Fads about what to eat and what not to eat are rapidly translated into marketable products. And advertisements bombard us from every direction with information about food. The advertisements and food stories in the media are bound to attract our attention. After all, many times every day, we make choices about food.

How can we sort out the information so that these choices are good ones? The task is impossible without knowing something about the basic principles of nutrition, which requires starting with the fundamentals of biochemistry presented in Chapter 11. The questions addressed here are the following:

- How is our biochemical energy generated?
- What roles do carbohydrates, fats, and proteins play in our diets and health?
- What are current recommendations for a healthy diet?
- What is the relationship between diet and heart disease?
- What are the categories of vitamins and minerals?
- What are the major functions of vitamins and minerals in our diets?
- What are the types and functions of food additives?

Nutrition is the science that deals with diet and health. The old saying "we are what we eat" is true. The skin that covers us now is not the same skin that covered us seven years ago. The fat beneath our skin is not the same fat that was there just a year ago. Our oldest red blood cells are 120 days old. The entire lining of our digestive tract is renewed every three days. Many chemical reactions are required to replace these tissues, and the energy and raw materials for these reactions are supplied by what we eat.

Everyone should understand something about the chemistry of food and choosing what to eat. *(Jeff Greenberg/Photo Researchers)*

Nutrition, then, is concerned with the chemical requirements of the body—the nutrients and the chemical energy we get from them. **Nutrients** are chemical substances in foods that provide the energy and raw materials required by biochemical reactions. The five classes of nutrients are carbohydrates, fats, proteins, vitamins, and minerals. In addition, an average adult needs about 2.5 L of water each day.

12.1 Digestion: It's Just Chemical Decomposition

Even before you swallow your lunch, chemical reactions begin the process of **digestion**. Like all biochemical processes, digestion is under the control of enzymes (biochemical catalysts, Section 11.8). Amylase, an enzyme in saliva, begins the digestion of the carbohydrates in starch. Carbohydrates are polymers of simple sugars (Section 11.2), and their digestion is the reverse of their formation—bonds are broken to set free the simple sugars. The same is true for digestion of proteins (polymers of amino acids) and the triglycerides in fats and oils (composed of glycerol and fatty acids):

■ **Digestion** is the process of breaking food down into substances small enough to be absorbed into the body from the digestive tract.

$$\text{Carbohydrates} \xrightarrow{\text{Digestion}} \text{simple sugars (monosaccharides)}$$

$$\text{Proteins} \xrightarrow{\text{Digestion}} \text{amino acids}$$

$$\text{Fats (triglycerides)} \xrightarrow{\text{Digestion}} \text{glycerol + fatty acids}$$

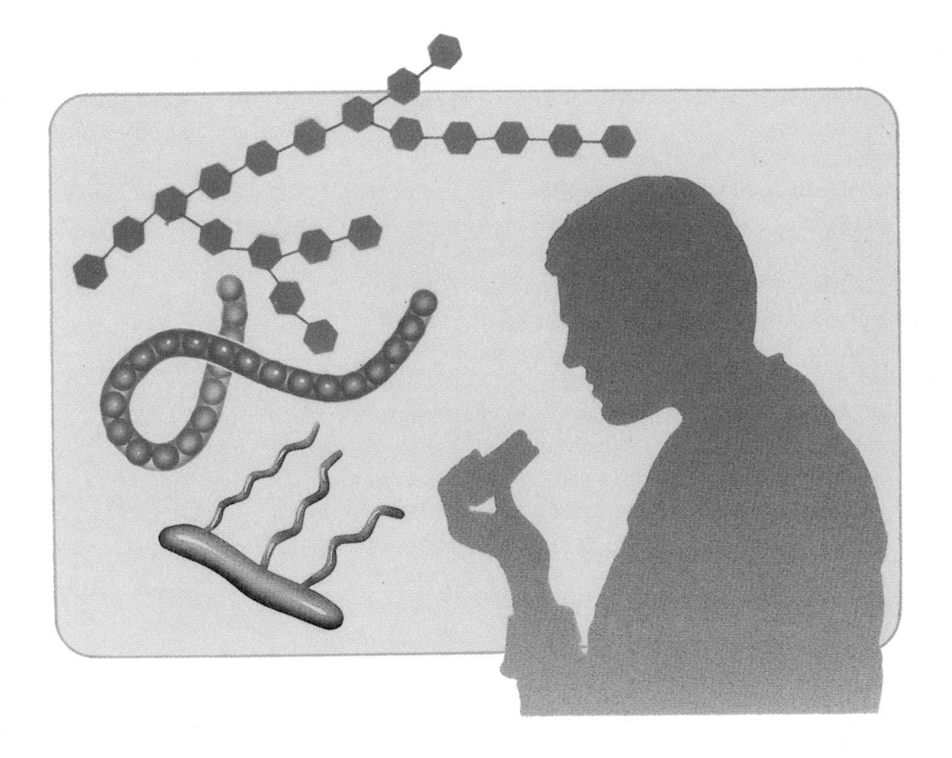

A sandwich and the structures of the carbohydrate, protein, and fat molecules it contains.

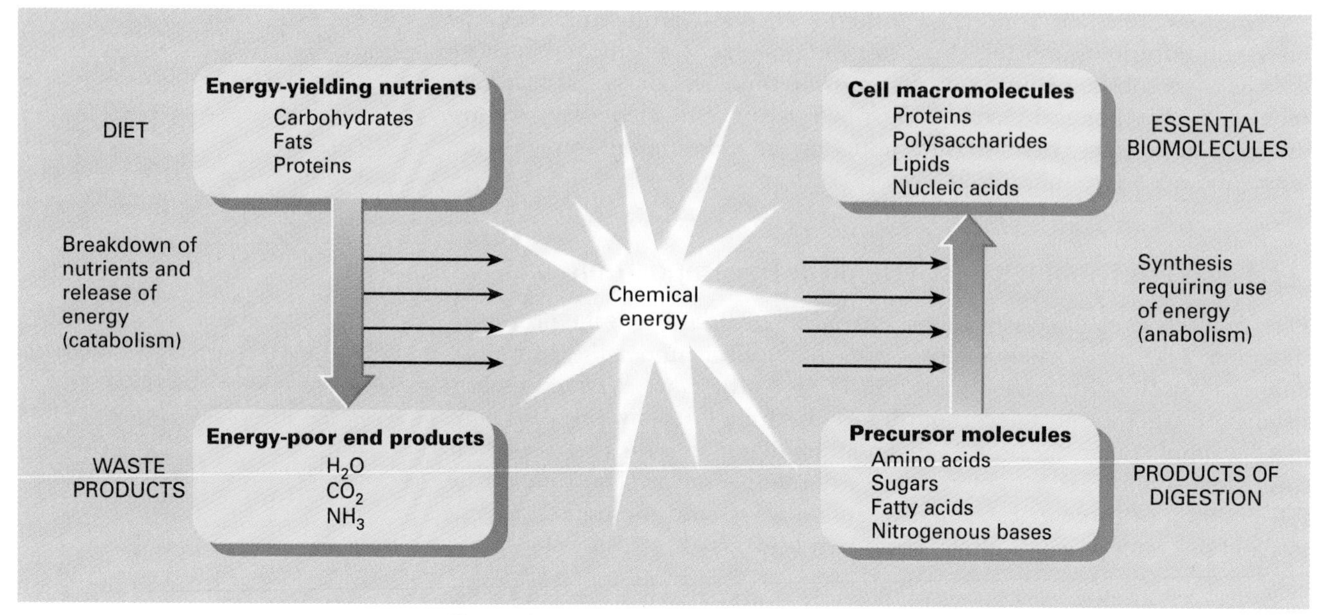

Figure 12.1 The release and use of chemical energy in the breakdown and synthesis of biomolecules.

Digestion begins in the mouth, continues in the stomach, and is completed in the small intestine. By the time a meal has been digested, all the nutrient molecules too large to cross cell membranes and enter the bloodstream have been converted into smaller molecules and absorbed.

Suppose you have a peanut butter and jelly sandwich for lunch. Digestion of starch from the bread begins as you chew and continues in your stomach. It is completed in the small intestine, where sugars from the jelly are also converted to monosaccharides. The indigestible carbohydrates (dietary fiber) from the bread and peanut butter pass unchanged through the small *and* large intestines.

■ A peanut butter and jelly sandwich on firm, toasted white bread contains approximately 20 g of protein, 86 g of carbohydrate, and 34 g of fat.

Digestion of protein from the bread and peanut butter starts in your stomach, where large polypeptides are broken down into smaller ones. The job is finished in the small intestine, where the individual amino acids enter the bloodstream. Lipids, mainly fats and oils from the peanut butter, are digested mainly in the small intestine.

Once digestion is finished, what happens to the small molecules that have been produced? They may undergo (1) complete breakdown to produce energy, carbon dioxide, and water; (2) recycling into new biomolecules; or (3) placement into storage as triglycerides (fat) for future use (Figure 12.1). The energy yield from foods is described in the next section. The following sections examine the roles of the nutrients in our diets.

12.2 Energy: Use It or Store It

How We Use Energy

A certain amount of energy is needed just to stay alive. While you are sitting completely still, your heart beats, your chest rises and falls as you breathe, your body temperature is maintained, chemical reactions proceed in cells, and messages that control these activities flow through your nervous system. The energy needed for these activities is the **basal metabolic rate (BMR)**. It is measured when a person is at rest at a comfortable temperature but not asleep, has not eaten for 12 hours, and has not engaged in vigorous activity for several hours.

The BMR is affected by many factors including body weight and activity level. An increased BMR can come from anxiety, stress, lack of sleep, low food intake, congestive heart failure, fever, increased heart activity, and the ingestion of drugs (including caffeine, amphetamine, and epinephrine). A decreased BMR can result from malnutrition, inactive tissue due to obesity, and low-functioning adrenal glands. Infants and children have higher BMRs than adults, and after early adulthood the BMR decreases about 2% to 3% per decade.

As soon as voluntary activity begins, the metabolic rate speeds up. Some examples are given in Table 12.1.

TABLE 12.1 ■ Energy Expended in a Variety of Activities

Activity	Energy (kcal/min)
Bicycling—5 mi/h	5
Chopping wood	7.5
Driving an automobile	2.8
Football, touch–tackle	8.8–12
Gardening, weeding	5.6
Listening to lecture	1.7
Pick and shovel work	6.7
Rollerskating, recreational–vigorous	5–15
Running 5 mi/h (12-min mi)	10
Skiing downhill	8–12
Squash and handball	10.0
Swimming, pleasure	6
Tennis, recreational–competitive	7–11
Volleyball, recreational–competitive	3.5–8.0
Walking 3.5 mi/h on road	5.6

Energy from Foods

Fats in the diet provide more than twice as many calories per gram as do carbohydrates and proteins.

Fats—9 kcal/g **Carbohydrates—4 kcal/g** **Proteins—4 kcal/g**

For example, if a steak is 49% water, 15% protein, 0% carbohydrate, 36% fat, and 0.7% minerals, then 3.5 ounces of steak (about 100 g) would provide about 384 kcal, or 384 food Cal. The caloric values of most foods are found as illustrated in the margin, and these are the values that are listed in diet books and on food labels. (Some representative values are given in Table 12.2.)

The human body, like everything else, is subject to the law of conservation of energy. In our case it translates into the following equation

$$\text{Energy taken in (food)} = \text{energy used} + \text{energy stored (fat)}$$

For most people, the secret to dieting is little more than applying this equation. When energy taken in exceeds energy used, the excess enters storage, mostly as triglyceride molecules in fat cells. When energy taken in remains the same, but more energy is used, less is stored as fat. When energy used is more than energy taken in, some must be removed from storage. To be genuinely successful at providing weight loss, a program must integrate management of both diet and exercise.

Glucose is our major fuel molecule. It is the end product of carbohydrate digestion, and if more is needed there are biochemical pathways for making it from stored fat and even protein. These pathways are essential because glucose is the only fuel that the brain and red blood cells can use. To provide energy, glucose molecules are "burned" throughout our bodies by a tightly integrated series of biochemical reactions whose final products are carbon dioxide and chemical energy stored in adenosine triphosphate (ATP), whose

■ 1 food Calorie (Cal) = 1 kilocalorie (kcal).

■ Fifteen percent (15%) of 100 g is 0.15 × 100 g.

■ *Energy from protein*

100 g of steak × 0.15 = 15 g protein

$$15 \cancel{\text{g protein}} \times \frac{4\text{ kcal}}{\cancel{\text{g protein}}} = 60\text{ kcal}$$

Energy from fat

100 g of steak × 0.36 = 36 g fat

$$36 \cancel{\text{g fat}} \times \frac{9\text{ kcal}}{\cancel{\text{g fat}}} = 324\text{ kcal}$$

Total energy 384 kcal

TABLE 12.2 ■ The Approximate Percentages of Carbohydrates, Fats, Proteins, and Water in Some Whole Foods as Normally Eaten

Food	Water	Protein	Fat	Carbo-hydrates	kcal/100 g
Vegetables					
Spinach, raw	90.7	3.2	0.3	4.3	26
Collard greens, cooked	89.6	3.6	0.7	5.1	33
Lettuce, Boston, raw	91.1	2.4	0.3	4.6	25
Cabbage, cooked	93.9	1.1	0.2	4.3	20
Potatoes, cooked	75.1	2.6	0.1	21.1	93
Turnips, cooked	93.6	0.8	0.2	4.9	23
Carrots, raw	88.2	1.1	0.2	19.7	42
Squash, summer, raw	94.0	1.1	0.1	4.2	19
Tomatoes, raw	93.5	1.1	0.2	4.7	22
Corn kernels, cooked on cob	74.1	3.3	1.0	21.0	91
Snap beans, cooked	92.4	1.6	0.2	5.4	25
Green peas, cooked	81.5	5.4	0.4	12.1	71
Lima beans, cooked	70.1	7.6	0.5	21.1	111
Red kidney beans, cooked	69.0	7.8	0.5	21.4	118
Soybeans, cooked	73.8	9.8	5.1	10.1	118
Meats and Fish					
Lean beef, broiled	61.6	31.7	5.3	0	183
Beef fat, raw	14.4	5.5	79.9	0	744
Lean lamb chops, broiled	61.3	28.0	8.6	0	197
Lean pork chops, broiled	69.3	17.8	10.5	0	171
Lard, rendered	0	0	100.0	0	902
Calf's liver, cooked	51.4	29.5	13.2	4.0	261
Beef heart, cooked	61.3	31.3	5.7	0.7	188
Brains	78.9	10.4	8.6	0.8	125
Chicken, whole, broiled	71.0	23.8	3.8	0	136
Cod, raw	81.2	17.6	0.3	0	78
Salmon, broiled	63.4	27.0	7.4	0	182
Meats and Fish (cont'd)					
Freshwater perch, raw	79.2	19.5	0.9	0	91
Oysters, raw	84.6	8.4	1.8	3.4	66
Grains and Grain Products					
Wheat grain, hard	13.0	14.0	2.2	69.1	330
Brown rice, dry	12.0	7.5	1.9	77.4	360
Brown rice, cooked	70.3	2.5	0.6	25.5	119
Whole wheat bread	36.4	10.5	3.0	47.7	243
White bread	35.8	8.7	3.2	50.4	269
Whole wheat flour	12.0	14.1	2.5	78.0	361
White cake flour	12.0	7.5	0.8	79.4	364
Dairy Products and Eggs					
Milk, whole	87.4	3.5	3.5	4.9	65
Yogurt, whole milk	89.0	3.4	1.7	5.2	50
Ice cream	62.1	4.0	12.5	20.6	207
Cottage cheese	79.0	17.0	0.3	2.7	86
Cheddar cheese	37.0	25.0	32.2	2.1	398
Eggs	73.7	12.9	11.5	0.9	163
Fruits, Berries, and Nuts					
Apples, raw	84.4	0.2	0.6	14.5	58
Pears, raw	83.2	0.7	0.4	15.3	61
Oranges, raw	86.0	1.0	0.2	12.2	49
Cherries, sweet	80.4	1.3	0.3	17.4	70
Bananas, raw	75.7	1.1	0.2	22.2	85
Blueberries, raw	83.2	0.7	0.5	15.3	62
Red raspberries, raw	84.2	1.2	0.5	13.6	57
Strawberries, raw	89.9	0.7	0.5	8.4	37
Almonds	4.7	18.6	54.2	19.5	598
Pecans	3.4	9.2	71.2	14.6	689
Walnuts	3.5	14.8	64.0	15.8	651

structure was given in Figure 11.17. ATP is often called the body's energy currency. It carries energy to be spent wherever in the body energy is needed.

To estimate your basal metabolic rate, multiply your weight in pounds by 10 kcal/lb. Then, to get a rough estimate of your daily caloric needs, choose your general level of activity according to Table 12.3 and multiply your basal metabolic rate by the appropriate factor given in the table.

TABLE 12.3 ■ Daily Energy Needs According to Physical Activity

Activity Level	Factor	
Very light		To estimate your daily energy needs, multiply your estimated BMR (10 kcal/lb × your weight in pounds) by the factor listed in the table that best represents your general level of daily activity:
Men	1.3	**Very light:** mostly sitting and standing activities
Women	1.3	**Light:** mostly walking activities
Light		**Moderate:** cycling, tennis, dancing
Men	1.6	**Heavy:** heavy manual digging, climbing, basketball, soccer
Women	1.5	
Moderate		
Men	1.7	
Women	1.6	
Heavy		
Men	2.1	
Women	1.9	

■ Section 12.6 explains how to interpret food labels. Some additional examples of food and energy arithmetic are given in Section 12.10.

EXAMPLE 12.1 *Daily Energy Needs*

Julie weighs 110 lb and is a college freshman—she plays tennis or skates several times a week and keeps in shape by running every day. By using Table 12.3, estimate her daily caloric needs.

SOLUTION

Based on her weight, Julie's estimated basal metabolism rate (BMR) is

$$110 \text{ lb} \times 10 \text{ kcal/lb} = 1100 \text{ kcal}$$

From what we know about Julie, her activity level is moderate. Therefore, using the factor of 1.6 from Table 12.3, her estimated daily caloric needs are

$$1100 \text{ kcal} \times 1.6 = 1800 \text{ kcal}$$

Exercise 12.1

Jack weighs 160 lb. As a graduate student he spends most of his time reading in the library or working at the computer. Estimate his daily caloric needs.

These two students have different levels of activity and different caloric needs. *(C. D. Winters)*

■ SELF-TEST 12A

1. Digestion converts carbohydrates, proteins, and triglycerides into __________, __________, and __________ plus __________, respectively.
2. The quantity of heat required to operate the body at rest is the __________, which in kilocalories is about ten times your __________.

Foods high in dietary fiber. *(C. D. Winters)*

■ Carbohydrate structures are given in Section 11.2.

■ In 1988–1989, with widespread information about the cholesterol-lowering properties of oat bran, sales of oat bran cereals increased 240%.

■ The results of an evaluation of health claims by the U.S. Food and Drug Administration (FDA) are given in *Frontiers in the World of Chemistry: Health Claims for Foods* on p. 312.

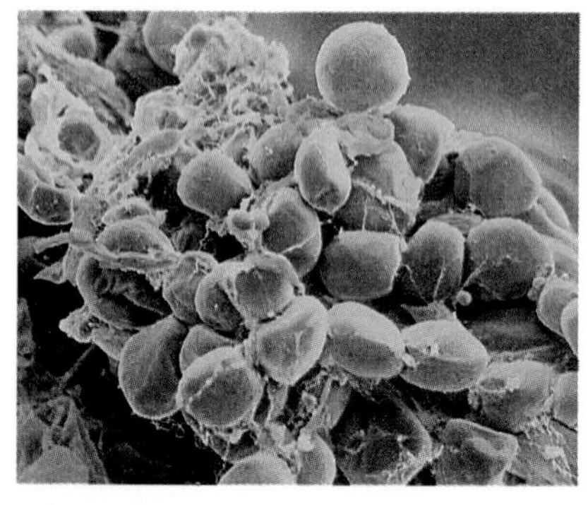

Where the fat goes. The yellow cells are adipocytes, the fat-storing cells of our bodies. Each adipocyte is almost entirely a single droplet of triglycerides. *(Prof. P. Motta/Dept. of Anatomy/University "La Sapienza," Rome/SPL/Photo Researchers)*

3. A food calorie has the same value as ____________ kilocalorie(s).
4. The kilocalories per gram are 4 kcal, 9 kcal, and 4 kcal for ____________, ____________, and ____________, respectively.
5. Our daily caloric needs are determined mainly by our weight and level of physical activity. (a) True, (b) False.
6. Which of the following is the body's major fuel molecule? (a) Triglyceride, (b) Glucose, (c) Maltose.
7. Energy is carried to where it is needed by the molecule abbreviated as ____________.

12.3 Sugar and Polysaccharides: Digestible and Indigestible

The major kinds of digestible carbohydrates in foods are the simple sugars (glucose and fructose), disaccharides (sucrose, maltose, and lactose), and polysaccharides (amylose and amylopectin in starch from plants, and glycogen from meat). The indigestible carbohydrates include cellulose and its derivatives, pectin (the substance that makes jam and jelly gel), and plant gums.

The indigestible polysaccharides are collectively referred to as **dietary fiber**. All dietary fiber comes from plants. There is insoluble fiber, mainly from the structural cellulose parts of plants, and soluble fiber—the gums and pectins. Barley, legumes, apples, and citrus fruits are foods with a high content of gums and pectins.

It has long been known that dietary fiber prevents constipation. Evidence has been accumulating that dietary fiber has significant other benefits. Adding soluble fiber to the diet decreases blood cholesterol levels, thereby decreasing the risk of heart disease. There is also evidence of a connection between *insufficient* dietary fiber and colorectal cancer. In our increasingly health-conscious society, such information provides a marketing advantage. We have all seen evidence of this in the bran muffins now found on every breakfast counter.

12.4 Lipids, Mostly Fats and Oils

Most of the lipids in the diet, which we usually refer to as "fats," are triglycerides. We get them from meats and fish, vegetables and vegetable oils, and dairy products. They may be solid fats or oils, and they may incorporate saturated or unsaturated fatty acids (Section 11.3). The fat content of some common foods is shown in Figure 12.2.

Fatty tissue is composed mainly of specialized cells, each featuring a large globule of triglycerides. When their energy is needed, the triglycerides in fat cells are hydrolyzed to give glycerol and free fatty acids, which leave the cells and are transported to the liver. There, the fatty acids are broken down to two-carbon molecular fragments that can enter the main energy-producing pathway.

Fat currently makes up about 38% of the average American diet (down from more than 40% in the 1980s). A fat content of 30% or less is strongly

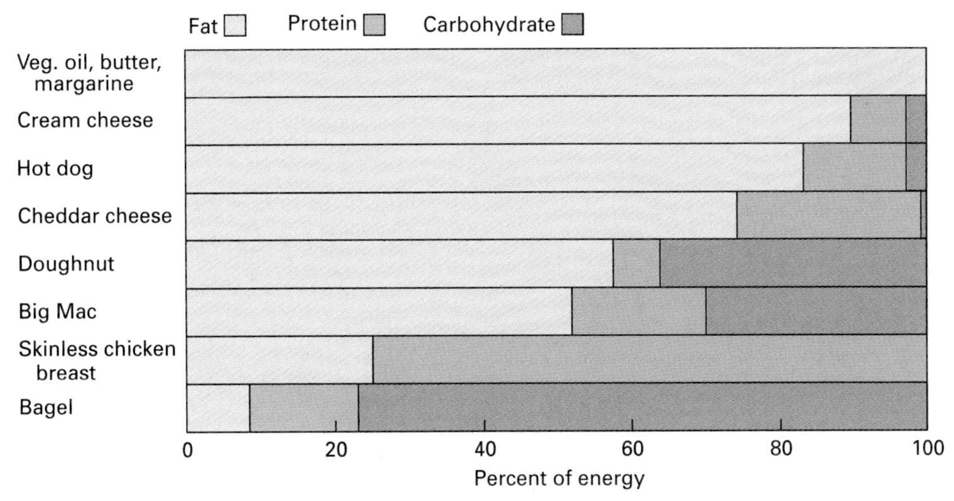

Figure 12.2 The percentage of the energy provided by the fats, proteins, and carbohydrates in some common foods.

recommended by most public health authorities. The fat in today's diet is about 40% saturated, 40% monounsaturated, and 20% polyunsaturated. Lowering the saturated and monounsaturated fat and raising the polyunsaturated fat content of the diet is also strongly recommended. What is the basis for these recommendations? Heart disease is the number-one cause of death in the United States (Section 13.1), and **atherosclerosis**, the buildup of fatty deposits called *plaque* on the inner walls of arteries, reduces the flow of blood to the heart. If a coronary artery is blocked by plaque, a heart attack occurs as a result of the reduced blood flow carrying oxygen to the heart. About 98% of all heart attack victims have atherosclerosis, and the major components of atherosclerotic plaque are saturated fatty acids and cholesterol.

■ Cholesterol occurs naturally *only* in dairy products, meat, and fish. There is *no cholesterol* in fresh vegetables, fruits, or vegetable oils.

The relation between blood levels of cholesterol and heart disease is well established. The more saturated fats and cholesterol in the diet, the higher the blood cholesterol is likely to be. Cholesterol, a lipid, has a waxy consistency, so to be transported in the bloodstream it must bond to a more water-soluble substance. Cholesterol combines with proteins to form **lipoproteins**, which are water soluble because of their many $—NH_3^+$ and $—COO^-$ ions. About 65% of the cholesterol in the blood is carried by low-density lipoproteins (**LDLs**), whereas about 25% is carried by high-density lipoproteins (**HDLs**). (The density difference is caused by the ratios of lipid to protein.)

LDLs are the "bad" cholesterol and HDLs are the "good" cholesterol referred to in discussions of heart disease. LDLs transport cholesterol away from the liver and throughout the body; they are therefore "bad" because they distribute cholesterol to arteries, where it can form the deposits of atherosclerosis. HDLs are "good" because they transport excess cholesterol from body tissues to the liver, where it is converted to bile acids that are needed in digestion.

■ The heart disease risk–blood cholesterol connection. The values given here are blood levels for persons more than 20 years of age. The values for children and adolescents are about 15% lower.

Risk Level	Total Cholesterol (mg/ 100 mL)	LDL Cholesterol (mg/ 100 mL)
Low	200 or lower	130 or lower
Border-line high	200–239	130–159
High	240 or higher	160 or higher

FRONTIERS IN THE WORLD OF CHEMISTRY

Health Claims for Foods

Major changes in food labels have been in place since 1994. At that time, the Nutrition Labeling and Education Act of 1990 became effective. To quote the U.S. Food and Drug Administration (the FDA):

The purpose of food label reform is simple: to clear up confusion that has prevailed on supermarket shelves for years, to help consumers choose more healthful diets, and to offer an incentive to food companies to improve the nutritive qualities of their products.

As part of the studies that preceded the new labeling act, the FDA, with the assistance of a large number of experts, evaluated the scientific evidence supporting a variety of health claims for foods. These original studies led to approval of seven general health claims sufficiently valid to be allowed on food labels. The claims must not be stated in absolute terms (e.g., the label must not indicate that eating a particular food will *cure* a particular disease) and the foods must meet certain criteria consistent with the claims. A sample acceptable label statement is

While many factors affect heart disease, diets low in saturated fat and cholesterol may reduce the risk of this disease.

Three of the original approved claims are for foods rich in fiber (the complex carbohydrates in grain products, fruits, and vegetables) that may contribute to a decreased risk of heart disease and cancer. One of these claims includes the roles of vitamins A and C along with fiber in decreased risk of cancer. In 1997, based on an ongoing review of new scientific research, a fourth claim for fiber was approved:

Foods with soluble fiber from whole oats may reduce heart disease risk when eaten as part of a diet low in saturated fat and cholesterol.

The new claim is based on evidence that the soluble fiber from oats is responsible for lowering the level of LDL (the low-density-lipoprotein, the bad one) in blood.

Two of the approved health claims relate to the *fat* content of foods: One is based on evidence that diets low in saturated fat and cholesterol (present only in animal products) decrease the risk of heart disease. The other is that, while recognizing that diet is only one of many proven risk factors for cancer (others are smoking, heredity, and environmental factors), a diet low in total fat may reduce the risk of some cancers.

The two remaining claims focus on minerals—the relation between increased dietary calcium and decreased osteoporosis risk (Section 12.8) and that between decreased sodium intake and lowered risk of the problems associated with high blood pressure.

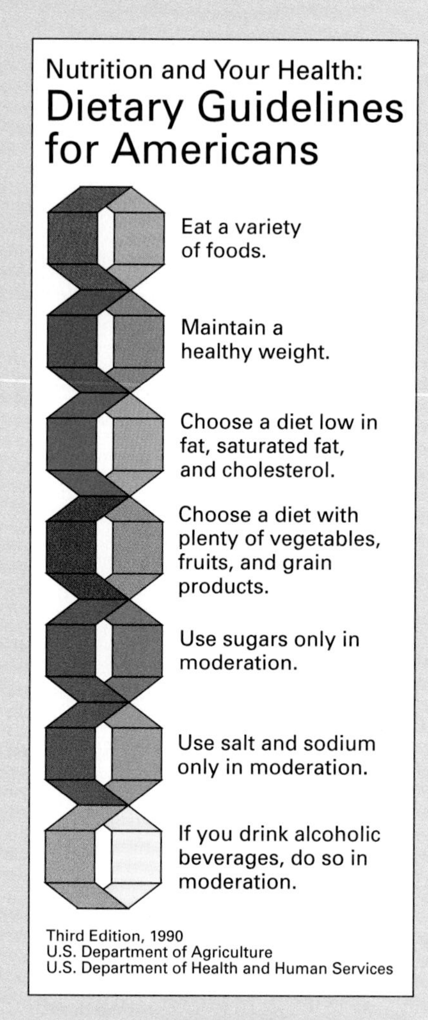

Nutrition and Your Health:
Dietary Guidelines for Americans
Eat a variety of foods.
Maintain a healthy weight.
Choose a diet low in fat, saturated fat, and cholesterol.
Choose a diet with plenty of vegetables, fruits, and grain products.
Use sugars only in moderation.
Use salt and sodium only in moderation.
If you drink alcoholic beverages, do so in moderation.
Third Edition, 1990
U.S. Department of Agriculture
U.S. Department of Health and Human Services

Reference The New Food Label Summaries, Food and Drug Administration, Department of Health and Human Services, Washington, D.C., January 6, 1993.

12.5 Proteins in the Diet

Meat, fish, eggs, cheese, and beans are high-protein foods. The major function of dietary protein is to provide amino acids for new protein synthesis. Proteins are also the major dietary source of nitrogen for the synthesis of other kinds of nitrogen-containing biomolecules.

Excess amino acids from dietary proteins are not stored in the body. The nitrogen is converted to ammonia and then excreted as urea, while the carbon atoms are cycled to glucose and energy generation or storage as fat. Therefore, it is necessary to eat some protein every day. The average amount of protein in the American diet is, however, well beyond what is necessary, and a protein-deficient diet is rare in the United States. *Protein–energy malnutrition,* a group of disorders due to various combinations of deficient protein and energy intake, is most common in children in underdeveloped countries.

■ The condition of extreme protein deficiency is called *kwashiorkor.* The name comes from an African dialect and translates as "the evil spirit that infects the first child when the second one is born." Kwashiorkor begins when the earlier child is no longer breastfed and is switched to a carbohydrate-based diet. If nutrition is improved before the condition has progressed too far, health can be restored.

12.6 Our Daily Diet

With implementation of the new food-labeling law in 1994, the most readily available information about the nutritive value of food is on food packages. An example of the Nutrition Facts label is given in Figure 12.3 (p. 314). The total Calories and Calories from fat are listed at the top. Next comes a section that lists the weight in grams or milligrams per serving, plus the *% Daily Value,* for the **macronutrients** judged most important to health. It is mandatory that this section includes total fat, saturated fat, cholesterol, sodium, total carbohydrate, dietary fiber, sugars, and protein, as illustrated. Except for dietary fiber and protein, these are nutrients that should be limited.

■ **Macronutrients** are those nutrients needed by the body in large amounts. **Micronutrients** are those needed by the body in small amounts—mostly the vitamins and minerals.

The *reference daily values* are established as listed at the bottom of the food label and as further explained in Table 12.4. The percentage of daily values on a label (the *DVs*) are calculated from these values, as represented in a 2000-Cal daily diet. This many calories is about right for moderately active women, teenage girls, and sedentary men. Teenage boys, active men, and very active or pregnant women have higher caloric requirements.

■ One reason the FDA settled on listings in terms of 2000 Cal per day is because it is a nice round number. Others must adjust their interpretation of the labels according to their own needs. Labels on large packages, where there is room, include the data for a 2500 Cal per day diet.

Beneath the heavy line on every nutrition facts label are mandatory listings for vitamin A, vitamin C, calcium, and iron. Here also, these required listings are for those vitamins and minerals deemed to be of greatest public health concern. The major deficiency diseases of today are anemia, associated with iron deficiency, and osteoporosis, associated with calcium deficiency. The two vitamins in the mandatory list are those that decrease cancer risk.

TABLE 12.4 ■ Basis for the Reference Daily Values on Nutrition Labels

Total Fat	30% of daily calories—an upper limit (65 g/2000 kcal daily diet)
Saturated Fat	10% of daily calories—an upper limit (20 g/2000 kcal)
Cholesterol	300 mg—a daily maximum for all diets
Sodium	2400 mg—a daily maximum for all diets
Total Carbohydrate	60% of daily calories (300 g/2000 kcal)
Dietary Fiber	23 g (based on 11.5 g/1000 kcal)
Sugars	No specified daily value
Protein	Listing the percentage daily value is not mandatory, but daily value is 10% of dietary calories from protein (50 g/2000 cal) for adults (excluding pregnant women and nursing mothers) and children over 4

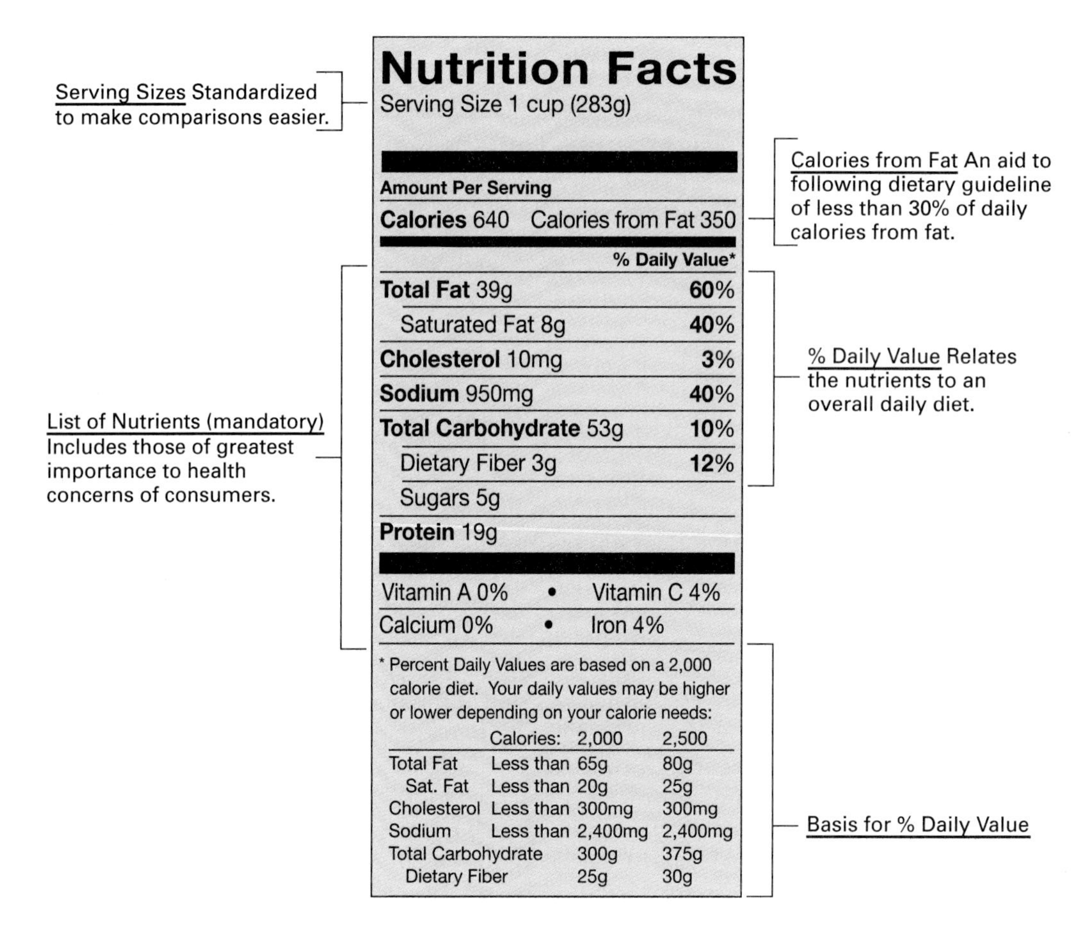

Serving Sizes Standardized to make comparisons easier.
List of Nutrients (mandatory) Includes those of greatest importance to health concerns of consumers.
Nutrition Facts
Serving Size 1 cup (283g)
Amount Per Serving
Calories 640 Calories from Fat 350
% Daily Value*
Total Fat 39g 60%
Saturated Fat 8g 40%
Cholesterol 10mg 3%
Sodium 950mg 40%
Total Carbohydrate 53g 10%
Dietary Fiber 3g 12%
Sugars 5g
Protein 19g
Vitamin A 0% • Vitamin C 4%
Calcium 0% • Iron 4%
* Percent Daily Values are based on a 2,000 calorie diet. Your daily values may be higher or lower depending on your calorie needs:
Calories: 2,000 2,500
Total Fat Less than 65g 80g
Sat. Fat Less than 20g 25g
Cholesterol Less than 300mg 300mg
Sodium Less than 2,400mg 2,400mg
Total Carbohydrate 300g 375g
Dietary Fiber 25g 30g
Calories from Fat An aid to following dietary guideline of less than 30% of daily calories from fat.
% Daily Value Relates the nutrients to an overall daily diet.
Basis for % Daily Value

Nutrients that may be listed at the manufacturers' option:

Calories from saturated fat	Soluble fiber
Stearic acid (meat and poultry products only)	Insoluble fiber
	Sugar alcohol (e.g., the sugar substitute sorbitol)
Polyunsaturated fat	Other carbohydrate
Monounsaturated fat	% Vitamin A as beta-carotene
Potassium	Other essential vitamins and minerals

Figure 12.3 The Nutrition Facts label. An example of the label required by law since 1994. Variations are allowed for small packages or foods that do not contain significant amounts of certain nutrients (e.g., vitamins and minerals need not be listed on canned soda).

■ SELF-TEST 12B

1. Indigestible polysaccharides are known as ______________ ______________.
2. Most of the lipids in our diet are ______________.
3. The ______________ are "good" cholesterol because they ______________.

4. The ______________ are "bad" cholesterol because they ______________.
5. Proteins are the major source of ______________ in the diet.
6. Which nutrient is recommended to be decreased in a healthy diet?
7. Which kinds of foods are recommended to be increased in a healthy diet?
8. A major goal in a healthy diet is to have no more than ______________ % calories from fat.

12.7 Vitamins in the Diet

A **vitamin** is an organic compound essential to health that must be supplied in small amounts in the diet. Vitamins provide no energy and are unchanged by digestion. The body does not synthesize vitamins, but needs them to function as **coenzymes**. There are two kinds of vitamins, fat soluble and water soluble.

A ***coenzyme*** is any nonprotein molecule that cooperates with an enzyme to make the enzyme function possible.

The fat-soluble vitamins—A, D, E, and K—can be stored in the fatty tissues of the body (especially the liver). They have nonpolar hydrocarbon chains and rings that are compatible with nonpolar oils and fats. It is important to store enough fat-soluble vitamins, but not too much because excesses are not excreted and can be toxic.

The water-soluble vitamins are the eight B vitamins and vitamin C. Because excesses are excreted rather than stored, it has been assumed that overdosing on water-soluble vitamins has no toxic effects. With the rise in popularity of megadoses of vitamins, however, some toxic reactions have been observed. It does, however, take much greater quantities of water-soluble vitamins to create harmful effects than for the fat-soluble vitamins.

We have our choice of a wide variety of vitamins. *(C. D. Winters)*

Fat-Soluble Vitamins

Vitamin A Vitamin A is essential to vision because it is a component of pigments in the eye. Carrots, long associated with good vision, provide beta-carotene, not vitamin A itself. As it passes through the intestinal wall, the beta-carotene is converted into vitamin A (also known as *retinol*). Vitamin A also aids in the prevention of infection by barring bacteria from entering and passing through cell membranes. A single dose of vitamin A greater than 200 mg (660,000 international units (IU), a unit for vitamin doses used by nutritionists) can cause acute *hypervitaminosis A* in an adult, with symptoms that include vomiting, fatigue, and headache.

Vitamin D Vitamin D is produced when ultraviolet light (UV) shines on the skin. Its major role is to help the body use calcium, and a deficiency causes rickets in children, the same condition caused by calcium deficiency. Vitamin D supplements are rarely needed, except by those who are almost never exposed to the sun. Both vitamin A and vitamin D are essential to normal growth and development. Overdosage of vitamin D can have serious consequences, however. Calcium deposits can form in the kidney, lungs, or tympanic membrane of the ear (leading to deafness). Infants and small children are especially susceptible to vitamin D toxicity.

Liver and carrots have a high vitamin A content. *(C. D. Winters)*

FRONTIERS IN THE WORLD OF CHEMISTRY

Olestra, A "Fat" for Calorie Counters

Take a look back at the structure of sucrose (a disaccharide of glucose and fructose, p. 266) and picture a long hydrocarbon chain attached to every —OH group in the molecule:

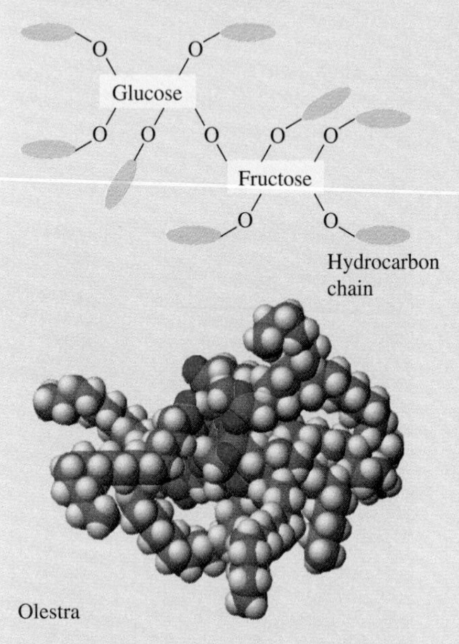

The result is the compound with the trade name Olestra, developed by Procter and Gamble at a cost of $200 million dollars. The long hydrocarbon chains give it the properties of a fat or oil that are desirable in foods, but it is sufficiently different from natural triglycerides that digestive enzymes cannot break it down. Olestra is the first synthetic nutrient to receive FDA approval for use in food products since aspartame (NutraSweet), which was allowed into the marketplace about 20 years ago. Unlike some artificial fats that have been studied, a major advantage of Olestra is that it does not break down at cooking temperatures.

Not only is the Olestra molecule immune to enzyme attack, but it is also so large that it is not absorbed into the bloodstream through the walls of the intestine. It passes through the body unchanged, and therefore adds zero calories to a food. The first approval for Olestra is limited to use in salty snack foods such as potato chips, crackers, and tortilla chips. A 1-ounce serving of potato chips made with Olestra adds no digestible fat and 75 Cal to the diet, as compared with 10 g of digestible fat and 150 Cal for regular potato chips. On further FDA review, Olestra may also be approved for other fat-laden foods such as pastries and doughnuts.

During its development, long-term studies with laboratory animals revealed no tendency for Olestra to cause cancer or other major health problems. In addition, tests showed that despite humans' inability to digest it, microorganisms common in soil and sewage can fully degrade it to carbon dioxide.

The dietary calorie reduction in Olestra-containing foods does not come without a couple of potential difficulties, however, and its use has been vigorously criticized by some watchdog groups for this reason. One possible negative result of consuming large quantities of Olestra is inhibition of absorption by the body of the fat-soluble vitamins, A, D, E, and K. To counter this effect, the FDA requires that these vitamins be added to Olestra-containing foods. Another possible problem for some people, and especially if they consume large quantities of Olestra, is abdominal cramping and diarrhea, difficulties that result from the nondigestible but oily nature of the compound. (Mineral oil, which contains no "minerals," is an old, traditional laxative that is an indigestible petroleum product.)

Package labels are required to carry a warning of the gastrointestinal problems and list a phone number for complaints to the manufacturer, which will be passed along to the FDA for use in further evaluation of Olestra. The FDA promises that after Olestra has been on the market for $2\frac{1}{2}$ years, the data will be formally reviewed.

Reference E. Doyle, *Journal of Chemical Education*, Vol. 74, pp. 370–372, April, 1997.

Vitamin E Vitamin E is now well established as an *antioxidant* (Section 8.3). It is particularly effective in preventing the oxidation of polyunsaturated fatty acids to form peroxides (which contain —O—O— groups). Perhaps this is why vitamin E is always found distributed among fats in nature. The fatty acid peroxides are particularly damaging because they can cause runaway oxidation in the cells. Vitamin E protects the integrity of cell membranes, which are composed mainly of triglycerides. No toxicity has been specifically associated with large doses of vitamin E, although caution in taking megadoses is advised.

■ Vitamin E is the only vitamin destroyed by the freezing of food. However, it survives cooking in water.

Vitamin K We need a steady supply of vitamin K because the body uses it up rapidly. Its major and essential role is in regulation of blood clotting, which goes on to repair not only cut skin but also small tears in blood vessels (on a daily basis). Because overdoses cause excessive blood clotting and danger of brain damage, vitamin K is the only vitamin for which a prescription is needed for supplements containing the vitamin alone.

Water-Soluble Vitamins

All the water-soluble vitamins are **coenzymes** that have polar —OH, $—NH_2$, and —COOH groups, as illustrated here by the structures of vitamin C and several of the B vitamins.

B Vitamins Vitamin B_6, considered the master vitamin, is known to be involved in 60 enzymatic reactions, many in the metabolism and synthesis of proteins. It is also needed for the synthesis of hemoglobin and the white blood cells of the immune system. Consuming more than 60 times the recommended daily quantity of vitamin B_6 causes nerve damage, and in huge doses (2–6 g per day) it can cause paralysis.

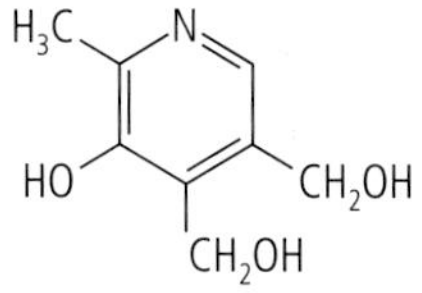

Large doses of niacin (50 mg per day or more) can also be toxic, with reactions ranging from a skin rash and nausea to abnormal liver function and changes in heart rhythm. Niacin deficiency is common in countries with a mainly corn diet.

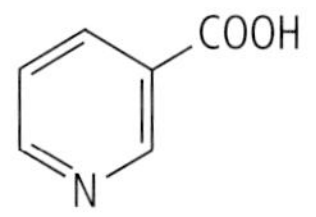

When the hulls of wheat, rice, and other grains are removed and the kernels ground to produce flour, a large proportion of riboflavin and certain other vitamins and minerals is lost, as is the soluble fiber. To counteract this loss, since the 1940s flour sold in the United States has been *enriched* by adding back some of the lost vitamins and minerals. Customarily, enriched wheat flour contains added riboflavin, thiamin, niacin, and iron. When enriched flour came into use in the United States, the incidence of pellagra, the niacin deficiency disease, almost disappeared.

H_3C N CH_2OH HO CH_2OH

Vitamin B_6

COOH N

Niacin

■ Once thought to be a single substance, the eight B vitamins are often found together and interact so that a deficiency of one can cause deficiencies of others. The B vitamins are B_6, B_{12}, folic acid, niacin, thiamin, riboflavin, biotin, and pantothenic acid.

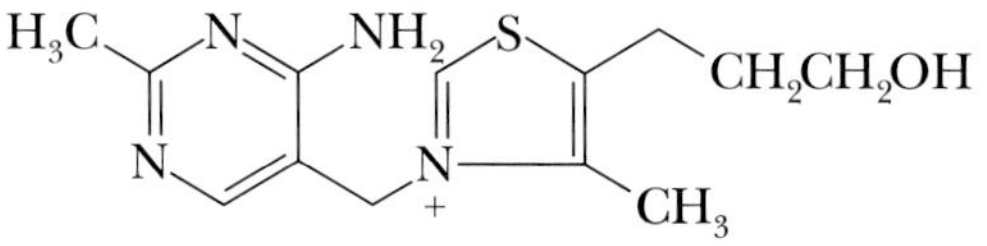

H_3C N NH_2 S CH_2CH_2OH N N + CH_3

Thiamine

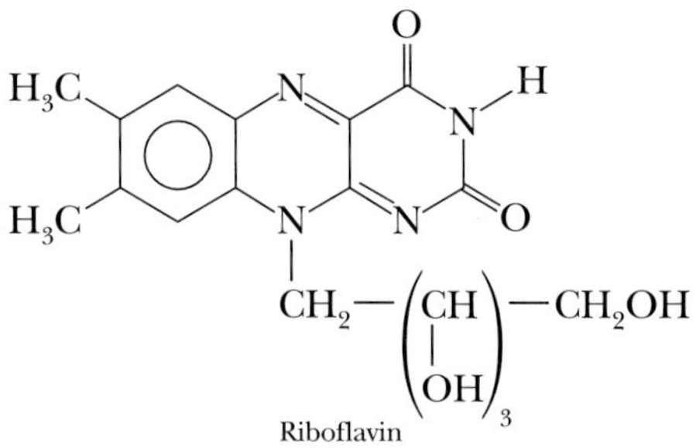

H_3C H_3C N O H N N N O CH_2—(CH(OH))$_3$—CH_2OH

Riboflavin

In 1996, based on growing scientific evidence, the FDA ruled that folic acid should also be added to enriched flour. Folic acid can reduce risks of certain birth defects, anemia, and heart disease.

Vitamin C Vitamin C, *ascorbic acid*, helps destroy invading bacteria; aids the synthesis and activity of interferon, which prevents the entry of viruses into cells; and combats the ill effects of toxic substances, including drugs and pollutants. It also aids in the synthesis of collagen, which is present in cartilage, bone, tendons, and connective tissue, and is therefore important in the healing of wounds and for infants and pregnant women. In addition, vitamin C is an antioxidant.

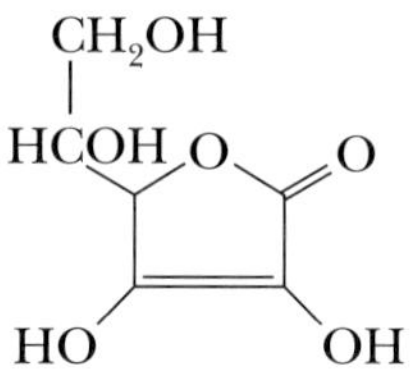

CH_2OH — HCOH — O — O — HO — OH

Vitamin C

■ Vitamin C is a strong enough acid to damage tooth enamel if vitamin capsules are chewed.

■ English sailors have been called "limeys" because the British admiralty in 1835 ordered a daily ration of lime juice to prevent scurvy, the vitamin C deficiency disease. The message did not hit home in America, however, where scurvy was common among troops during the Civil War.

The role of vitamin C in preventing the common cold has long been debated, and numerous studies have found it to be either effective or ineffective. Currently, there is more evidence in favor of its ability to decrease the severity of cold symptoms than for any ability to prevent colds. It is reported, however, that one third of the U.S. population takes vitamin C supplements to ward off colds. Daily doses of 1 g or more can cause diarrhea, nausea, and abdominal cramps in some people.

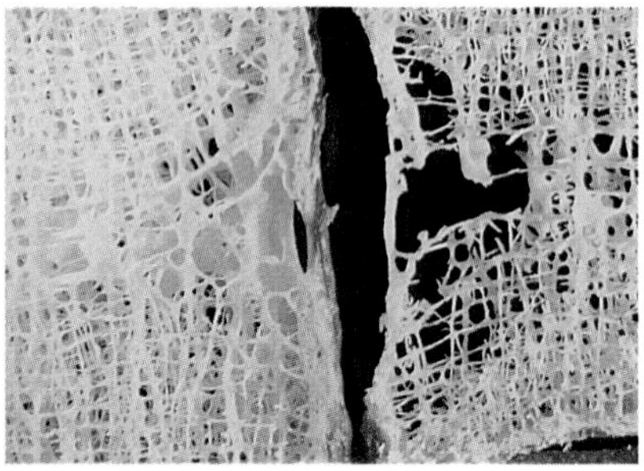

Osteoporosis. The effect of calcium loss on bone is shown by the comparison between a normal bone *(left)* and a bone afflicted with osteoporosis *(right)*. *(© Michael Klein/Peter Arnold, Inc.)*

12.8 Minerals in the Diet

As nutrients, **minerals** are elements other than carbon, hydrogen, nitrogen, and oxygen needed for good health. Although referred to on labels as "potassium" or "iodine" and so on, most minerals are present in foods, food supplements, and our bodies as ions. Not only does the human body need minerals, but the minerals must be maintained in balanced amounts, with no deficiencies and no excesses. Many of the body's minerals are excreted daily and must therefore be replenished each day.

The seven **macronutrient minerals** make up about 4% of body weight. They are calcium, phosphorus, magnesium, sodium, potassium, chlorine, and sulfur; and their sources and functions are listed in Table 12.5. Except for the role of long-term calcium deficiency in osteoporosis, deficiencies of these minerals are rare because of their abundance in a variety of foods.

Sodium, potassium, *and* ***chloride,*** as ions (Na^+, K^+, and Cl^-), are essential to *electrolyte balance* in body fluids. Electrolyte balance, in turn, is essential for fluid balance, acid–base balance, and transmission of nerve impulses. Table salt is the principal source of sodium and chloride ions, and dietary deficiencies are unlikely. When there is extreme fluid loss through vomiting, diarrhea, or traumatic injury, electrolytes must be supplied to restore their concentration in body fluids.

■ Electrolytes are substances that dissolve in water to produce ions that conduct electricity.

■ Potassium chloride (KCl) is sold as a salt (NaCl) substitute.

About 99% of the ***calcium ions*** in the body are in bones and teeth. Together with sodium and potassium ions, calcium ions also participate in transmission of nerve impulses and regulation of the heartbeat. To be absorbed, calcium must be present in solution as Ca^{2+}. Absorption is enhanced in the presence of lactose (milk sugar) and also by a fatty meal, which passes through the intestine more slowly, allowing more time for absorption. A deficiency of

TABLE 12.5 ■ The Macrominerals*

Name	Sources	Major Functions	Deficiency Symptoms	Groups at Risk of Deficiency
Sodium (Na^+)	Table salt, processed foods	Major extracellular ion, nerve transmission, regulates fluid balance	Muscle cramps	Those consuming a severely sodium-restricted diet
Potassium (K^+)	Fruits, vegetables, grains	Major intracellular ion, nerve transmission	Irregular heartbeat, fatigue, muscle cramps	Those consuming diets high in processed foods, those taking high blood pressure medication
Chloride (Cl^-)	Table salt	Major extracellular ion	Unlikely	No one
Calcium (Ca^{2+})	Milk, cheese, bony fish, leafy green vegetables	Bone and tooth structure, nerve transmission, muscle contraction, blood clotting	Increased risk of osteoporosis	Postmenopausal women, teenage girls, those with kidney disease
Phosphorus†	Meat, dairy, cereals, and baked goods	Bone and tooth structure, buffers, membranes, ATP, DNA	Bone loss, weakness, lack of appetite	Premature infants, elderly
Magnesium (Mg^{2+})	Nuts, greens, whole grains	Reactions involving ATP, nerve and muscle function	Nausea, vomiting, weakness	Alcoholics, those with kidney and gastrointestinal disease
Sulfur‡	Protein foods, preservatives	Part of amino acids and vitamins, glutathione, acid–base balance	None when protein needs are met	No one

* Adapted from L. A. Smolin and M. B. Grosvenor: *Nutrition: Science & Applications.* Philadelphia, Saunders College Publishing, 1997.

† Phosphorus is covalently bonded in organic compounds and is also present in phosphate ions (PO_4^{3-}).

‡ Sulfur is covalently bonded in organic compounds and is also present in sulfate ions (SO_4^{2-}).

vitamin D decreases calcium ion absorption and contributes to the bone deformities that accompany rickets.

A gradual loss of bone mass and density during adulthood is a normal process, and one of every three people older than 65 has some degree of **osteoporosis**. The role of a calcium-deficient diet over a lifetime in hastening osteoporosis is sufficiently well established that a food label health claim on this basis is allowed. Postmenopausal white or Asian women of slight body build with inactive lifestyles are at greatest risk for osteoporosis. After menopause, the body produces less estrogen, which has the ability to slow bone dissolution. A preventive measure against later osteoporosis is consumption of adequate calcium during the adolescent years (ages 12 to 18), when bone is forming. Estrogen replacement for postmenopausal women is effective in slowing bone loss (Section 13.4).

■ About ten years ago, an undergraduate chemistry student discovered that many calcium and vitamin pills dissolved hardly at all. Subsequent media reports claimed that undissolved vitamin pills could be observed in large-intestine X rays and septic tanks. The manufacturers of nutrition supplements have since adopted voluntary standards for disintegration and solubility of their products.

■ By definition, the micronutrient minerals, or trace elements, are those needed in quantities of 100 mg or less per day, or present in the body at less than 0.01% of body weight.

A number of minerals are essential in small amounts (**micronutrient minerals**). Of these, four—iron, copper, zinc, and iodine—are allowed to be listed on Nutrition Facts labels, iron being mandatory and the others optional. Other minerals known to be essential are selenium, manganese, fluorine, chromium, and molybdenum. Still other minerals (ten or more) are known to be essential in other mammals and are present and probably essential in humans. The list of trace minerals and our knowledge of their functions are constantly evolving.

Iodide ion has long been known to be essential to thyroid gland function. Hormones produced by the thyroid gland contain iodine and are responsible for growth, development, and maintenance of all body tissues. If there is a deficiency of iodine, the thyroid glands can become extremely enlarged, a condition called goiter. The routine use of iodized salt (0.1% potassium iodide) is the best way to ensure adequate iodine in the diet. More than half the table salt sold in the United States is iodized, and the use of such salt is mandatory in some countries.

■ Iron supplement pills are a frequent cause of poisoning due to excess intake by children. They should be stored out of reach.

■ A thorough medical evaluation of an individual with anemia is necessary because it can result from a variety of disease conditions as well as from nutritional deficiencies. Vitamin B_{12} and folic acid deficiencies can also cause anemia.

Iron deficiency is associated with *anemia*—a decrease in the oxygen-carrying capacity of the blood, as indicated by low red blood cell or heme concentrations. Iron-deficiency anemia is the most common nutritional deficiency disease in both developed and underdeveloped nations. Early symptoms include muscle fatigue and lethargy. The ability to fight off invading bacteria is also diminished. Continued anemia creates defects in the structure and function of skin, fingernails, mouth, and stomach.

Zinc deficiency is known to cause poor growth and development, decreased immune function, and poor wound healing. Because meat is a better source of zinc than plant-based foods, vegetarians are among those at risk of zinc deficiencies (others are the elderly and children with poor diets). Zinc supplements are often touted as able to improve immune function, enhance sexual performance, and increase fertility. It's a good idea to bring a degree of skepticism to such claims for zinc and other minerals. In the case of zinc, adding it to the diet of a person with a mild deficiency might improve wound healing, immune function, and appetite. Currently, however, there is little evidence for such changes in a healthy person with no existing zinc deficiency.

■ SELF-TEST 12C

1. Vitamins are synthesized by cells of the body. (a) True, (b) False.
2. Vitamins A, D, E, and K are ____________ soluble, while the B-complex vitamins and vitamin C are ____________ soluble.
3. In the body, beta-carotene is converted into vitamin ____________.
4. Vitamins A, C, and E, and beta-carotene are all ____________.
5. The B-complex vitamins function in the body as parts of ____________.
6. When vitamins are added to flour or other foods, the food is described as ____________.
7. The chemical form of minerals in the diet is mostly as ____________.
8. Sodium and potassium are essential to ____________ and ________ balance and the transmission of ____________ signals.
9. ____________ is caused by iron deficiency, and ____________ is caused by calcium loss.

12.9 Food Additives

Many chemicals with little or no nutritive value are added to commercially processed food (Figure 12.4). Some of these **food additives** serve to protect the food from being spoiled by oxidation, bacterial attack, or aging. Others add and enhance flavor or color. Still others control pH; prevent caking; or stabilize, thicken, emulsify, sweeten, leaven, or tenderize the food.

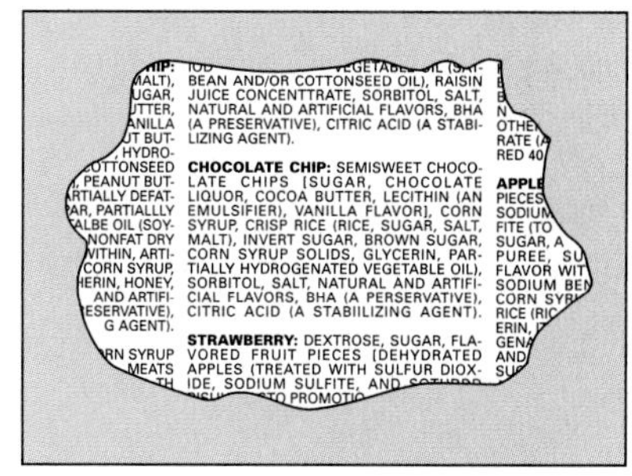

Note the variety of food additives in these cookies.

The GRAS List The Food and Drug Administration lists about 600 chemical substances "generally recognized as safe" (GRAS) for their intended use by the FDA. The GRAS list was published in several installments in 1959 and 1960. It was compiled from the results of a questionnaire asking experts in nutrition, toxicology, and related fields to give their opinions about the safety of various seasonings, artificial flavorings, and other substances customarily added to foods. Since its publication, few substances have been added to the GRAS list and some have been removed, mostly due to suspicion that they are cancer-causing agents. *New* food additives—substances not on the original GRAS list—must receive FDA approval for use in foods based on test results submitted by the manufacturers.

Food Preservation Oxidation and microorganisms (bacteria, fungi, and others) are the major enemies in the decomposition of food. Any process that prevents the growth of microorganisms or retards oxidation is generally an effective preservation process. Drying grains, fruits, and meat is one of the oldest preservation techniques. Drying is effective because water is necessary for both the growth of microorganisms and the chemical reactions of oxidation.

There are also chemical additives that can preserve food. Salted meat and fruit in a concentrated sugar solution are protected from microorganisms. The dissolved sodium chloride or sucrose creates a **hypertonic solution** in

■ A **hypertonic solution** is more concentrated than solutions in its immediate environment.

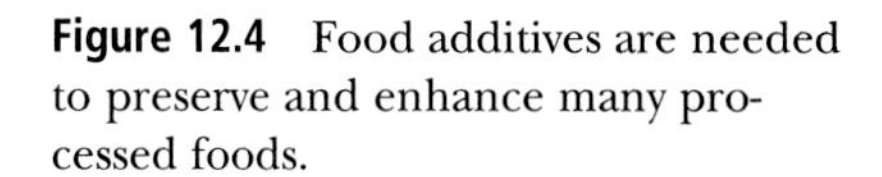

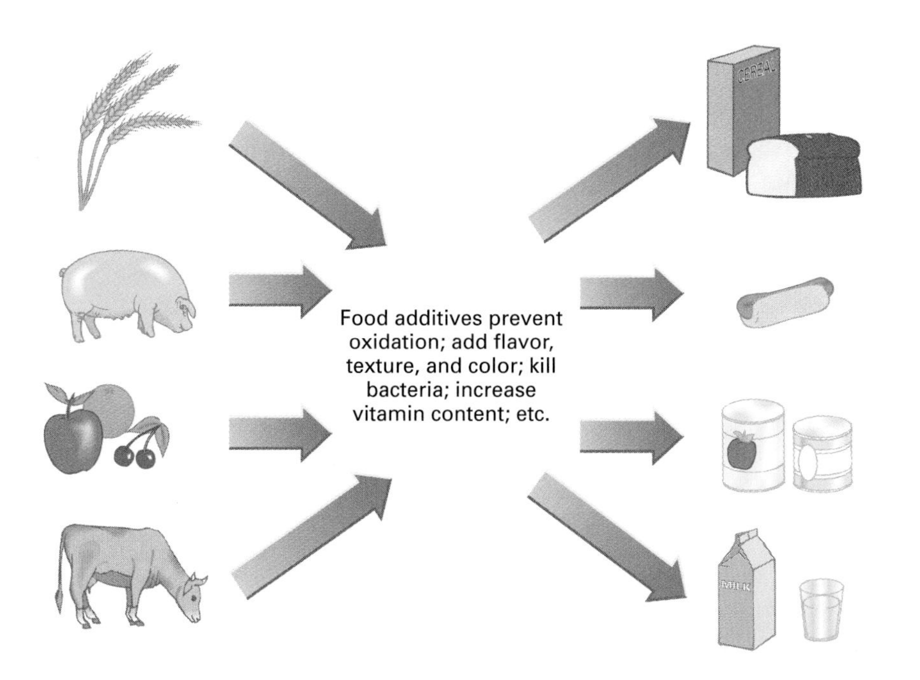

CEREAL
Food additives prevent oxidation; add flavor, texture, and color; kill bacteria; increase vitamin content; etc.
MILK

Figure 12.4 Food additives are needed to preserve and enhance many processed foods.

■ **Osmosis** is the flow of water from a more dilute solution through a membrane into a more concentrated solution (Section 15.12).

$CH_3CH_2C(=O)$—ONa

Sodium propionate

C_6H_5—C(=O)—ONa

Sodium benzoate

Because many individuals are allergic to sulfites, foods that contain more than 10 parts per million (ppm) of sulfites must have them listed on the label. The sulfites include chemicals such as sulfur dioxide (SO_2), sodium sulfide (Na_2S), and sodium bisulfite ($NaHSO_3$). *(C. D. Winters)*

■ A preservative must interfere with microbes but be harmless to the human system—a delicate balance.

■ Antioxidant additives in foods are serving the same purpose as antioxidant vitamins in the body (Section 8.3).

which water flows by **osmosis** from the microorganism to its environment. Thus, salt and sucrose have the same effect on microorganisms as dryness; both dehydrate them.

A preservative is effective if it prevents multiplication of microorganisms during the shelf life of the product. In general, food is spoiled by toxic substances secreted by the microorganisms. Sterilization by heat or radiation, or inactivation by freezing, is often undesirable, since it impairs the quality of the food. Chemical agents seldom achieve sterile conditions but can preserve foods for considerable lengths of time. Two common chemical preservatives in packaged foods are sodium benzoate (which is permitted in nonalcoholic beverages and in some fruit juices, fountain syrups, margarines, pickles, relishes, olives, salads, pie fillings, jams, jellies, and preserves) and sodium propionate (which can be used in bread, chocolate products, cheese, pie crust, and fillings).

Antioxidants The direct action of oxygen in the air is the chief cause of the destruction of the fats in food. Carbon–carbon double bonds in polyunsaturated fatty acids are particularly susceptible. Oxidation produces a complex mixture of volatile aldehydes, ketones, and acids that causes a rancid odor and taste. Foods kept wrapped, cold, and dry are relatively free of air oxidation. The most common antioxidant food additives are butylated hydroxyanisole (BHA) and butylated hydroxytoluene (BHT), which act by releasing a hydrogen atom from their —OH groups as a free radical (H·).

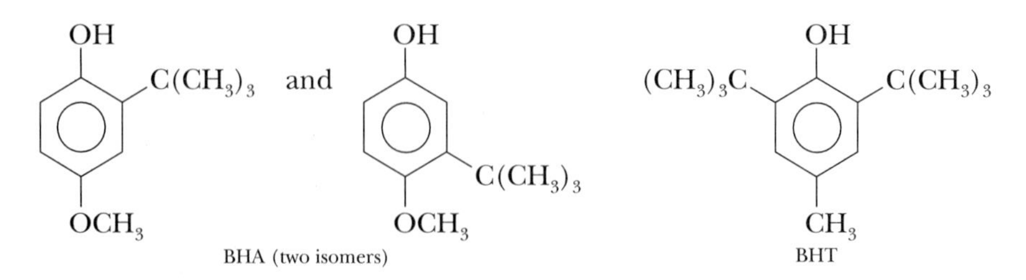

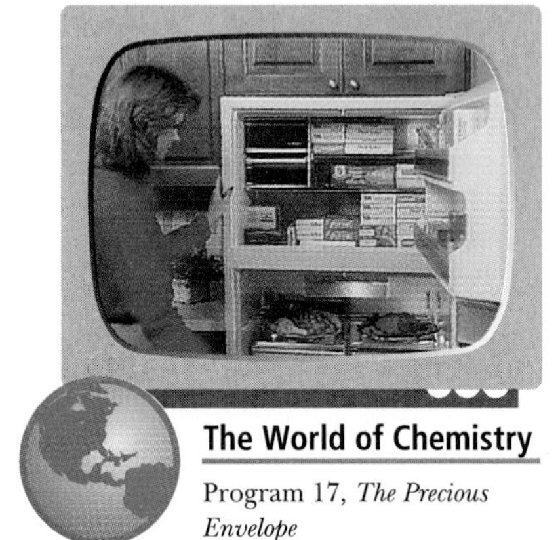

Food preservation by cooling, which slows down the rate of oxidation and enzyme-catalyzed chemical reactions.

Sequestrants Trace amounts of metals get into food from the soil and from machinery during harvesting and processing. Copper, iron, and nickel, as well

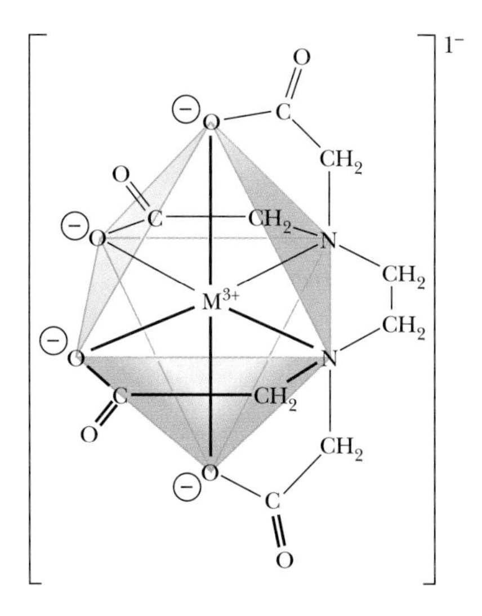

Figure 12.5 Ethylenediaminetetraacetic acid (EDTA). The EDTA molecule has the remarkable ability to "sequester," or isolate, a metal ion by forming six bonds to it—two from nitrogen atoms in amino groups and four from oxygen atoms in ionized carboxyl groups ($-COO^-$).

as their ions, catalyze the oxidation of fats. The class of food additives known as **sequestrants** react with trace metals in foods, tying them up in complexes so the metals will not catalyze decomposition of the food. With the competitor metal ions tied up (sequestered), antioxidants such as BHA and BHT can accomplish their task much more effectively.

The sodium and calcium salts of ethylenediaminetetraacetic acid (EDTA) are common sequestrants in many kinds of foods and beverages. The structural formula of EDTA bonded to a metal ion is shown in Figure 12.5.

■ To sequester means "to withdraw from use." The sequestering ability of EDTA accounts for its use in treating heavy-metal poisoning.

Food Flavors Much of the sensation of taste in food is from our sense of smell. For example, the flavor of coffee is determined largely by its aroma, which in turn is due to a mixture of more than 500 compounds, mostly volatile oils. Most flavor additives, like many perfume ingredients, originally came from plants. The plants were crushed, and the compounds were extracted with various solvents such as ethanol or carbon tetrachloride. Sometimes a single compound was extracted; more often, the residue contained a mixture of several compounds. By repeated efforts, relatively pure oils were obtained. Oils of wintergreen, peppermint, orange, lemon, and ginger, among others, are still obtained in this way. These oils, alone or in combination, are then added to foods to produce the desired flavor. Today synthetic preparations of the same flavors are common food additives.

■ Some 1700 natural and synthetic substances are used to flavor foods, making flavors the largest category of food additives.

Flavor Enhancers Flavor enhancers have little or no taste of their own but amplify the flavors of other substances. They exert synergistic and potentiation effects. "Synergism" is the cooperative action of two different substances (or people, or anything) such that the total effect is greater than the sum of the effects of each used alone. "Potentiators" do not have an effect themselves but exaggerate the effects of other chemicals. Some nucleotides (Section 11.11), for example, have no taste but enhance the flavor of meat and the effectiveness of salt. Flavor enhancers were first used in meat and fish but now are also

■ In some people MSG causes the so-called Chinese restaurant syndrome, an unpleasant reaction characterized by headaches and sweating that usually occurs after an MSG-rich Chinese meal. MSG is a natural constituent of many foods, such as tomatoes, strawberries, and mushrooms. These foods affect some individuals in the same way.

The World of Chemistry

Program 13, *The Driving Forces*

The kinds of reactions that cause food to spoil are not very different than what chemists study in pure chemical solutions. The way I always like to talk about food, it's the study of messy chemistry. And the reason I say that is that, in a food which has so many different organic compounds and inorganic compounds together, there are lots and lots of different reactions that could have caused spoilage. What we can do, however, is narrow them down to several classes. For example, when you bite into an apple, you see it start to brown. That's an enzyme reaction. Dr. Theodore Labuza, Food Chemist

■ The acid–base chemistry of buffers was discussed in Section 7.4. Buffers react as a base with an acid or as an acid with a base. The result is to maintain constant pH. (High pH is basic; low pH is acid.)

■ **Hygroscopic** substances absorb moisture from the air. You've probably had trouble getting salt out of the shaker during humid weather.

One use of FDA-approved food dyes. *(C. D. Winters)*

used to intensify flavors or cover unwanted flavors in vegetables, bread, cakes, fruits, and beverages. Three common flavor enhancers are monosodium glutamate (MSG), 5′-nucleotides, and maltol.

Food Colors Some 30 chemicals are in use as food colors. About half of them are laboratory-synthesized and half are extracted from natural materials. Most food colors are large organic molecules with several double bonds and aromatic rings. Such structures have electrons that can absorb certain wavelengths of light and pass the rest; the wavelengths passed give the substances their characteristic colors. Beta-carotene, the orange-red substance in carrots and a variety of plants (and also an antioxidant), is an example of a natural food color. Because one of the food colors, Yellow No. 5, causes allergic reactions (mainly rashes and sniffles) in susceptible individuals, the FDA requires manufacturers to list Yellow No. 5 on the labels of any food products containing it.

pH Control in Foods Weak organic acids are added to such foods as cheese, beverages, and dressings to give a mild acidic taste. They often mask undesirable aftertastes, but in some cases, such as in fruit-flavored sodas, the acidic taste is expected. Weak acids and acid salts react with bicarbonate to form CO_2 in the baking process. Buffers are also added to adjust and maintain a desired pH. Potassium acid tartrate, for example, is a buffer because it is a salt of an organic acid that can act as either an acid or a base.

The versatile acidulants also function as preservatives to prevent the growth of microorganisms, as antioxidants to prevent rancidity and browning, as viscosity modifiers in dough, and as melting point modifiers in such food products as cheese spreads and hard candy.

Anticaking Agents Anticaking agents are added to **hygroscopic** foods (in amounts of 1% or less) to prevent caking in humid weather. Table salt is particularly subject to caking unless an anticaking agent is present. The additive (magnesium silicate, for example) incorporates water into its structure as water of hydration and does not appear wet as sodium chloride does when it absorbs water physically on the surface of its crystals. As a result, the anticaking agent keeps the surface of sodium chloride crystals dry and prevents crystal surfaces from codissolving and joining together.

Stabilizers and Thickeners Stabilizers and thickeners improve the texture and blends of foods. Like carrageenan, most stabilizers and thickeners are polysaccharides, which have numerous hydroxyl groups as a part of their structure. The hydroxyl groups form hydrogen bonds with water and help to provide a more even blend of the water and oils throughout the food. Stabilizers and thickeners are particularly effective in icings, frozen desserts, salad dressings, whipped cream, confections, and cheeses. In reduced-fat foods, they replace some of the "fatty" texture that would otherwise be missing. The plant gum thickeners (e.g., gum arabic, guar gum, locust bean gum, gum tragacanth) are good sources of soluble dietary fiber (Section 12.3).

12.10 Some Daily Diet Arithmetic

Foods containing modified food starch, pectin, and carbohydrate gums as stabilizers and thickeners. *(C. D. Winters)*

Sometimes doing a little arithmetic helps to interpret food labels or identify your own food needs. A few examples of very practical kinds of diet and food calculations follow.

EXAMPLE 12.2 *Daily Grams of Fat*

A good way to keep track of Calories from fat is to monitor the total grams of fat in your diet every day. A moderately active 100-lb woman would get 30% of daily Calories by consuming 53 g of fat per day. She is deciding how many Yummie Bite cookies she can eat. Their nutrition label says that there are two cookies per serving and 45 Cal from fat per serving. How many cookies can she eat and leave 33 g of fat for other foods to be eaten that day? What percentage of her 53 g of fat will the cookies provide?

SOLUTION

To leave 33 g of fat for other foods means that she can eat enough cookies to contain

$$53 \text{ g fat} - 33 \text{ g fat} = 20 \text{ g fat}$$

With 9 Cal per gram of fat, the grams of fat per "serving" of cookies is

$$\frac{45 \text{ Cal}}{\text{serving}} \times \frac{1 \text{ g fat}}{9 \text{ Cal}} = 5 \text{ g fat per serving}$$

With two cookies and 5 g of fat per serving, she can enjoy eight Yummie Bite cookies (but other foods eaten that day will have to be pretty low in fat content). The 20 g of fat in the eight cookies provides 38% of her daily fat allowance.

$$\frac{20 \text{ g fat (cookies)}}{53 \text{ g fat (daily fat)}} \times 100 = 38\%$$

Exercise 12.2

Comparison of the labels gives the following information per serving of two kinds of soup

Healthy Chicken:	Calories	90	Calories from fat	15
Cream of Mushroom:	Calories	140	Calories from fat	80

How many grams of fat are there in one serving of each soup? For persons who wish to limit their daily fat consumption to 65 g, what percentage of this 65 g is in one serving of each?

How much of these popular foods should you eat? *(C. D. Winters)*

EXAMPLE 12.3 *Caloric Value of Food and Exercise*

A slice of pepperoni pizza contains 10 g of protein, 20 g of carbohydrate, and 7 g of fat. (a) How many kilocalories does the pizza slice provide? (b) What percentage of the total calories is from fat? (c) How long would a person have to run at 5 mi/h to burn off the calories in the pizza slice (consult Table 12.1)?

SOLUTION

(a) The total caloric value of the pizza is calculated from the quantities of protein, carbohydrate, and fat

$$10\ \cancel{\text{g protein}} \times \frac{4\text{ kcal}}{\cancel{\text{g protein}}} = 40\text{ kcal}$$

$$20\ \cancel{\text{g carbohydrate}} \times \frac{4\text{ kcal}}{\cancel{\text{g carbohydrate}}} = 80\text{ kcal}$$

$$7\ \cancel{\text{g fat}} \times \frac{9\text{ kcal}}{\cancel{\text{g fat}}} = 63\text{ kcal}$$

$$\text{Total} \quad 183\text{ kcal}$$

(b) Percent calories from fat

$$\frac{63\text{ kcal (fat)}}{183\text{ kcal (total)}} \times 100 = 34\%$$

(c) Running at 5 mi/h burns 10 kcal (total) min, therefore, to burn off this slice of pizza would require running for

$$183\text{ kcal} \times \frac{1\text{ min}}{10\text{ kcal burned}} = 18.3\text{ min}$$

Exercise 12.3

A slice of devil's food cake with chocolate frosting contains 3 g of protein, 8 g of fat, and 40 g of carbohydrate. (a) How many calories does the slice of cake provide? (b) What percentage of the calories is from fat? (c) How long could a person listen to a lecture fueled by the energy from this piece of cake?

■ SELF-TEST 12D

1. Two of the oldest means of preserving foods are ______________ and adding ______________.
2. GRAS is an acronym for ______________.
3. Antimicrobial preservatives make food sterile. (a) True, (b) False.
4. BHA and BHT are very common food additives because they function as ______________.

5. An important sequestrant goes by the initials ____________.
6. The flavor of a food can usually be traced to a single compound. (a) True, (b) False.
7. Most stabilizers and thickeners (e.g., carageenan) are from the class of nutrients known as ____________.
8. If a food yields 230 kcal per serving, of which 100 kcal are from fat, the percentage from fat is found from (____________ divided by ____________) × 100%.
9. If you require 2000 kcal per day, then to find how many calories from fat per day will give the maximum of 30%, you would do the following calculation: ____________ × 2000 kcal.

MATCHING SET

____	1. Chemical decomposition	a. Dietary saturated fat
____	2. Glucose	b. Goiter
____	3. Calorie value per gram of fat	c. Anemia
____	4. Calorie value per gram of protein	d. Digestion
____	5. Causes atherosclerosis	e. Sodium EDTA
____	6. Result of deficient dietary iodine	f. Monosodium glutamate
____	7. LDL	g. 4 kcal
____	8. HDL	h. Beta-carotene
____	9. Butylated hydroxyanisole	i. Major fuel molecule
____	10. Nutrient and food coloring	j. Affects daily caloric need
____	11. Sodium benzoate	k. Good cholesterol
____	12. Flavor enhancer	l. Osteoporosis
____	13. Loss of calcium from bone	m. Food preservative
____	14. Result of deficient dietary iron	n. 9 kcal
____	15. Physical activity level	o. Bad cholesterol
____	16. Sequesters metals	p. Antioxidant food additive

QUESTIONS FOR REVIEW AND THOUGHT

1. What is meant by the term "digestion"?
2. What is amylase? To what class of biochemical molecules does it belong?
3. What are the relationships between mass and energy for the following substances?
   (a) Fats (b) Proteins
   (c) Carbohydrates
4. What are the products formed by digestion of the following nutrients?
   (a) Carbohydrates (b) Fats
   (c) Proteins
5. What is meant by the term "BMR," or "basal metabolic rate"?
6. What is meant by the following terms?
   (a) Simple sugars
   (b) Digestible carbohydrates
   (c) Indigestible carbohydrates
   (d) Disaccharides
   (e) Polysaccharides
   (f) Dietary fiber
7. What are triglycerides?
8. Give definitions for the following terms as applied to fats and fatty acids:
   (a) Saturated (b) Monounsaturated
   (c) Polyunsaturated (d) Partially hydrogenated
9. What do the following terms mean?
   (a) Atherosclerosis (b) Cholesterol
   (c) Plaque (d) Lipoproteins
   (e) Low-density lipoproteins
   (f) High-density lipoproteins
10. What are the following?
    (a) FDA (b) reference daily values
11. Give definitions for the following terms:
    (a) Macronutrients (b) Micronutrients
12. What are the major categories of information required on a nutrition facts label?

13. Where do the following substances occur in nature?
    (a) Starches (b) Glycogen
    (c) Cellulose
14. What are micronutrient minerals? Which one must be listed on nutrition facts labels?
15. What is a vitamin? What is the function of a vitamin?
16. Overdoses of which of the following vitamins should definitely be avoided? Why?
    (a) Riboflavin (b) Vitamin C
    (c) Vitamin A (d) Vitamin B
    (e) Vitamin D
17. What is an electrolyte? What is meant when referring to the electrolyte balance of the body?
18. What is the GRAS list?
19. What are the seven most abundant macronutrient minerals in the body?
20. What is the function of each of the following food additives?
    (a) Sodium benzoate (b) Sodium propionate
    (c) BHA (butylated hydroxyanisole)
    (d) BHT (butylated hydroxytoluene)
    (e) Sodium EDTA
21. Why are sequestrants needed as food additives?
22. What is the function of each of the following food additives?
    (a) Monosodium glutamate (MSG)
    (b) Yellow No. 5
    (c) Acetic acid (d) Calcium silicate
    (e) Carrageenan (f) Lecithin
23. What percentage of fat is currently recommended for today's diet by most public health authorities?
24. Lactose is sometimes referred to as "milk sugar" because it is the principle carbohydrate in milk. The lactose structure follows. Is this a monosaccharide, disaccharide, or polysaccharide? Explain.

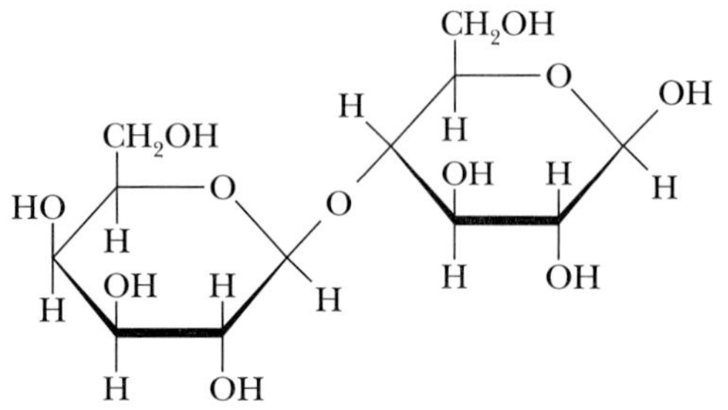

CH_2OH O CH_2OH O HO H OH H H OH H O H H OH OH H H H OH

25. Lactose intolerance occurs in people who are unable to hydrolyze milk sugar (lactose). This happens because they do not produce enough of the enzyme lactase. Normally infants and children have adequate amounts of lactase, but many adolescents and adults produce less. Should milk be promoted as a healthy food for the general population? Give one argument for and one against.
26. The body stores excess energy in small globules of fat in specialized cells of adipose tissue. The energy available from a gram of fat is 9 Cal. What would happen to the volume of these cells if the body stored energy by collecting carbohydrates instead of fats? Assume that the density of carbohydrate and fat are the same. (Recall that 1 g carbohydrate = 4 Cal.)
27. Cholesterol has the following structure. Why is cholesterol insoluble in water? Why is it recommended that people monitor their blood serum cholesterol levels? How many carbon atoms does a cholesterol molecule contain?

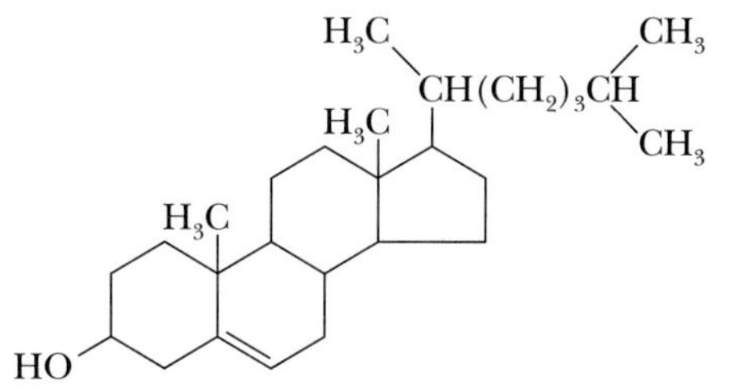

H_3C CH_3 $CH(CH_2)_3CH$ H_3C CH_3 H_3C HO

28. Folic acid is found in citrus fruits and leafy green vegetables. The FDA is now requiring that folic acid be added to cereals, breads, and pastas. The American Journal of Public Health says that such grain fortification would prevent 300 to 700 birth defects per year. At the level recommended for addition, approximately 3.25 million adults over 50 would receive too much folic acid. What question would you want answered before you would support such a nationwide program?
29. Why is osteoporosis a more significant problem for women after menopause? What preventive measures can be taken to reduce the chances of experiencing osteoporosis?
30. Should the government embark on a national program to add more calcium to our diets to reduce our chances of calcium deficiency and osteoporosis? Give one reason for such a program and one against.
31. What happens to the human thyroid gland if a person has an iodine deficiency? What is this condition called? What is the link between this problem and iodized salt?
32. When fats and fatty portions of food become rancid, what kind of reactions have probably occurred? What food additives are used to minimize these reactions?
33. Choose a label from a food item and try to identify the purpose of each food additive.
34. What foods have you eaten during the past week that did not contain any food additives?

PROBLEMS

1. What is the basal metabolic rate (BMR) for a 110-lb woman who has a moderate activity level?
2. What is the BMR for a 145-lb man who is sedentary, with very light activity? What would this person's BMR be if he started to exercise regularly so his activity increased to the moderate level?
3. How much time bicycling at 5 mi/h is needed to consume the 100 cal available from a serving of tomato soup? (*Note:* Energy expended bicycling at 5 mi/h is 5 Cal/min.)
4. A 100-g serving of ice cream contains 207 Cal. How long would you have to run at 5 mi/h to consume the energy available from the ice cream?
5. Walking on a road at 3.5 mi/h consumes 3.5 Cal/min. How long would you have to walk to consume the energy available from 100 g of white bread? (*Note:* 100 g white bread = 269 Cal.)
6. A blood serum cholesterol level of 200 mg/100 mL is supposed to give a low heart disease risk level. How much total cholesterol would be carried by the blood serum of an adult with a blood volume of 12 pints? (1 pint = 473 mL).
7. A blood serum cholesterol level of 240 mg/100 mL is supposed to give a high heart disease risk level. How much total cholesterol would be carried by the blood serum of an adult with a blood volume of 13 pints? (1 pint = 473 mL.)
8. The quantity of fat in a food is important to people who need to restrict fat intake. Which of the following would be the better low-fat choice, a piece of pecan pie weighing 113 g providing 459 Cal with 180 Cal from fat, or a piece of apple pie weighing 125 g providing 305 Cal with 108 Cal from fat?
9. How many Calories (kcal) are in a fast-food chicken sandwich that contains 27 g of protein, 46 g of carbohydrate, and 34 g of fat?
10. How many kcal are in a fast-food bacon cheeseburger supreme if it contains 34 g of protein, 44 g of carbohydrate, and 46 g of fat?
11. An exceptionally active female volleyball player who weighs 145 lb wants to maintain a diet of 30% calories from fat, 60% from carbohydrates, and 10% from protein. How many Calories does the player need per day? How many grams of fat, carbohydrate, and protein does she need?
12. By using the following nutrition facts label, determine what percentage of the calories in the product come from fat. What percentage of your daily calorie need is provided by a serving if your daily calorie need is 1800 Cal. Would this product be a good food item for a person on a low-sodium diet? Justify your answers.

Nutrition Facts	
Serving Size 1/2 cup (120mL) condensed soup	
Servings Per Container About 6	
Amount Per Serving	
Calories 100	Calories from Fat 20
	% Daily Value*
Total Fat 2g	**3%**
Saturated Fat 0g	**0%**
Cholesterol 0mg	**0%**
Sodium 730mg	**30%**
Total Carbohydrate 18g	**6%**
Dietary Fiber 2g	**8%**
Sugars 10g	
Protein 2g	
Vitamin A 10% •	Vitamin C 30%
Calcium 2% •	Iron 4%

CHAPTER 13

Chemistry and Medicine

Everyone is interested in developments that influence public health and, most particularly, their own personal health. For even a general understanding of the major public health issues the amount to be learned is large. Knowledge, however, is our personal first line of defense. With chemistry as the framework, this chapter takes a look at the most important classes of diseases and the medications available to treat them. The major questions addressed are the following:

- What are the major fatal diseases in the United States?
- How do antibiotics function?
- What is the cause of acquired immunodeficiency syndrome (AIDS)?
- What roles do hormones and neurotransmitters play in body chemistry?
- How do drugs enhance or counteract hormones and neurotransmitter actions?
- What are some important pain-killing and mood-altering drugs?
- What conditions can cold medications attack?
- What are the major chemical weapons against heart disease and cancer, and how do they work?

The average life expectancy for men in the United States rose from 53.6 years in 1920 to 72.6 years in 1995. During this same period, the life expectancy for women rose from 54.6 years to 79.0 years.

What roles do chemistry and chemical technology play in this ongoing increase in life expectancy and the accompanying improvements in the quality of life?

Most importantly, remarkable progress has been made in recent years in understanding the chemical reactions that regulate biological processes. As this understanding grows, the old trial-and-error method of screening chem-

Tools of the trade for an early pharmacist. Many medications were made on the spot from plants. *(C. D. Winters)*

icals for use as drugs is being replaced. Instead, drug molecules are being designed to have exactly the molecular shape and chemical reactivity needed to counteract disease. Another positive outcome is a growing understanding of what makes for a healthy lifestyle.

13.1 Medicines, Prescription Drugs, and Diseases: The Top Tens

Americans spend more than $34 billion a year on medicines. About $15 billion of this is for **over-the-counter drugs**—those that can be bought in any supermarket or drugstore. The rest is for **prescription drugs**, which cannot be purchased without instructions from a physician. A drug is classified as one or the other by the U.S. Food and Drug Administration (FDA). In general, a substance is available only by prescription if it has potentially dangerous side effects, if it should be used only by people with specific medical problems, or if it treats a condition so serious that a person with that condition should be under a doctor's care.

The top ten prescription drugs, based on the number of prescriptions written or refilled during 1996 (Table 13.1), show an interesting cross section

TABLE 13.1 ■ Ten Most Prescribed Drugs in the United States in 1996*

Trade Name	Generic Name (Drug Class)	Condition Treated
Premarin	Mixture of estrogens (hormones)	Menopausal symptoms
Trimox	Amoxicillin trihydrate (antibiotic, a penicillin)	Infectious disease
Synthroid	Levothyroxine sodium (hormone)	Thyroid hormone deficiency
Lanoxin	Digoxin (cardiac glycoside)	Heart disease
Zantac†	Ranitidine hydrochloride (acid secretion blocker)	Ulcers, heartburn
Vasotec	Enalapril (enzyme inhibitor)	Hypertension (high blood pressure)
Prozac	Fluoxetine (serotonin re-uptake blocker)	Major depression, obsessive compulsive disorders, bulimia, others
Procardia	Nifedipine (calcium channel blocker)	Heart disease
Hydrocodone and acetaminophen (generic mixture)	Hydrocodone and acetaminophen (narcotic analgesic and analgesic)	Pain relief, cough suppression
Coumadin sodium	Warfarin (oral anticoagulant)	Prevention of blood clot formation in, for example, heart disease, lung disease

* Source: *Pharmacy Times*, www.pharmacytimes.com

† As a prescription drug, in higher dosage than the over-the-counter product.

of medicinal uses. With increasing evidence that there are numerous benefits (such as decreased risk of heart disease and Alzheimer's disease) in addition to its initial purpose of counteracting the symptoms of menopause, Premarin, a hormone replacement, takes the lead. In second place is the most-prescribed of the antibiotics, Trimox (amoxicillin), a penicillin derivative. Of the other eight drugs, four (Lanoxin, Vasotec, Procardia, and Coumadin) play a role in treating heart disease and the related condition of **hypertension**—high blood pressure. Prozac, an antidepressant drug that has an ever-expanding variety of therapeutic uses, is a relative newcomer to the top-ten list.

As illustrated in Table 13.1, all drugs have a trade name and a generic name. The **trade name** (brand name) is the name used by the drug manufacturer. For example, the antibiotic amoxicillin is sold under trade names such as Amoxil, Amoxidall, Amoxibiotic, Infectomycin, Moxaline, Trimox, Utimox, and Wymox. *Amoxicillin* is a **generic name**, which is a drug's generally accepted common chemical name. Once the patent protection on a drug has expired, it can be manufactured and marketed competitively by many companies and prescribed by its generic name. Often, prescriptions written by the generic name are cheaper to fill.

■ A systematic chemical name for amoxicillin, from which a chemist should be able to write its complete structure, is 6-(*p*-hydroxy-α-aminophenylacetamido)penicillinic acid.

In 1900, five of the ten leading causes of death for individuals of all ages in the United States were infectious diseases (pneumonia and influenza, tuberculosis, gastrointestinal infections, kidney infections, diphtheria). Currently, pneumonia and influenza (tabulated together) are the only infectious diseases from this group remaining in this top ten, causing 3.7% of all deaths of individuals of all ages (Table 13.2). This dramatic decrease in deaths from infectious diseases can be attributed in large measure to the development of antibiotics.

Meanwhile, AIDS (acquired immunodeficiency syndrome), a new infectious disease that is caused by the human immunodeficiency virus (HIV) has come onto the scene as the eighth leading overall cause of death. In 1993, HIV infection hit the news by moving ahead of all other causes of death for

TABLE 13.2 ■ Leading Causes of Death in 1995*

	All ages†		25–44†	
	%	Rank	%	Rank
Diseases of the heart	31.9	1	11.4	4
Cancer	23.3	2	14.7	3
Stroke	6.8	3	2.3	8
Lung diseases	4.4	4	—	—
Accidents	4.0	5	18.2	2
Pneumonia and influenza	3.7	6	1.4	10
Diabetes mellitus	2.5	7	1.6	9
HIV infection	1.9	8	20.5	1
Suicide	1.4	9	8.5	5
Liver disease	1.1	10	2.9	7
Homicide	—		6.8	6

* Source: U.S. National Center for Health Statistics.

† Data are percentage of all deaths. Deaths from other causes are not included, so columns do not add up to 100%.

those in the 25- to 44-year-old age group, and it remains in this spot. Only three years earlier, in 1990, HIV infection placed third as the cause of death in this age group, with accidents and cancer coming first and second. Although vigorous study is underway and numerous leads are being pursued, HIV infection (discussed in Section 13.3) has not completely yielded to any drug. In 1994 and 1995, the big news in the death statistics was also in the social realm rather than in the scientific realm, as homicide dropped to 11th and then 12th place as a cause of death for those of all ages.

■ When the causes of death are broken down according to age groups, heart disease is number one for those 65 years old and older, cancer is number one for those 45 to 64 years old, HIV infection is number one for those 25 to 44 years old, and accidents are number one for those 15 to 24 years old. In the 15- to 24-year age group, over 75% of the accidents are motor vehicle accidents.

13.2 Drugs for Infectious Diseases

Modern **chemotherapy**—the treatment of disease with chemical agents—began with the work of Paul Ehrlich (see *The Personal Side*). In 1904, he concluded that **infectious diseases** could be conquered if chemicals could be found that attack disease-causing microorganisms without harming the host. After observing that dyes used to stain bacteria also killed the bacteria, he developed arsenic compounds similar to the dyes. One of these compounds *(arsphenamine)* was the first effective drug for an infectious disease. It revolutionized the treatment of syphilis at the time it was introduced. Syphilis is now treated with penicillin, the first of the antibiotics that are now our principal weapons against infectious diseases.

■ ***Infectious diseases*** are caused by microorganisms, which include viruses, bacteria, fungi, and other parasites. Ehrlich referred to his chemotherapeutic drugs for infectious diseases as "magic bullets"—they killed the microorganism but not the host.

Strictly speaking, an **antibiotic** is a substance produced by a microorganism that inhibits the growth of other microorganisms. Any compound, whether natural or synthetic, that acts in this manner is, however, usually referred to as an "antibiotic." A person falls victim to an infectious disease when invading microorganisms multiply faster than the body's immune system can destroy them. Antibiotics help the immune system by either destroying invaders or preventing their multiplication.

■ *Antimicrobial* is a frequently used, more accurate classification than *antibiotic.*

Penicillins The penicillins were discovered by Sir Alexander Fleming, a bacteriologist at the University of London. He was working in 1928 with cultures of *Staphylococcus aureus,* a bacterium that causes boils and some other infections. One day he noticed that one culture was contaminated by a blue-green mold. For some distance around the mold growth, the bacterial colonies had been destroyed. On further investigation, Fleming found that the broth in which this mold had grown had a similar lethal effect on many **pathogenic**

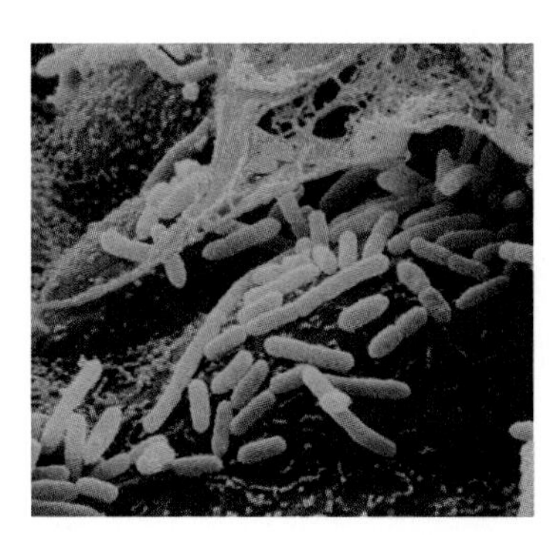

(a)

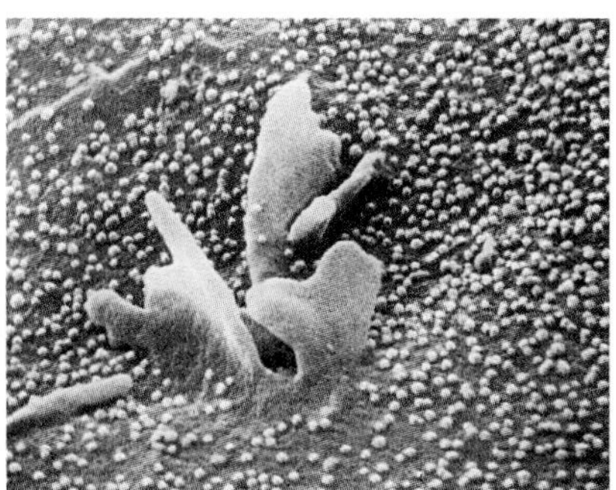

(b)

The enemies: bacteria and viruses. (a) The blue rods are *Haemophilus influenza* bacteria lying on human nasal tissue. This bacterium causes bronchitis, pneumonia, and a wide range of diseases in children including meningitis and blood poisoning. (b) The dots are *Rubella* viruses emerging from the surface of an infected cell, where the viruses are being reproduced. *Rubella* is the virus that causes German measles. *(a, © Dr. Tony Brain/SPL/Photo Researchers; b, © NIBSC/SPL/Photo Researchers)*

THE PERSONAL SIDE

Paul Ehrlich (1854–1915)

Paul Ehrlich. (SPL/Photo Researchers)

Paul Ehrlich, a German bacteriologist, was a pioneer in the application of chemistry to medicine. While he was a medical student, he became fascinated with chemistry and pursued learning about it on his own. He is an excellent example of a scientist driven by curiosity to do experiments that had incredibly successful practical applications. While investigating the biological properties of the many synthetic dyes available from the German chemical industry, he originated the microscopic study of blood cells. His work with the serum of immunized animals was fundamental to the then-young science of immunology (study of the body's response to and defense against disease-causing invaders). As noted in the text, he created the concept of chemotherapy.

Ehrlich was the first to explain the action of toxic substances and drugs. In what he called his "side-chain" theory, he explained that these substances affect only cells that have matching molecular fragments extending from their surfaces. Today we call these side chains receptors—the parts of cell surface molecules that extend outside the cell and interact with external messengers and drugs. They are still the object of intensive research.

■ Howard Florey and Ernest Chain, who isolated penicillin, and Fleming, who discovered it, shared the Nobel Prize for medicine and physiology in 1945.

■ Another *Penicillium* strain that proved to be an excellent source of a new antibiotic was discovered on a moldy cantaloupe in a Peoria, Illinois, market.

(disease-causing) bacteria. The mold was later identified as *Penicillium notatum* (the spores sprout and branch out in pencil shapes, hence the name). Although Fleming showed that the mold contained an antibacterial agent, which he called *penicillin,* he was not able to purify the active substance.

In 1940, the active ingredient, penicillin G, was identified, and by 1943 it was available for clinical use. As World War II drew to a close, penicillin G was saving many lives threatened by pneumonia, bone infections, gonorrhea, gangrene, and other infectious conditions. Since then, numerous penicillins have been developed. Amoxicillin, one of the most-prescribed drugs listed in Table 13.1, is a penicillin.

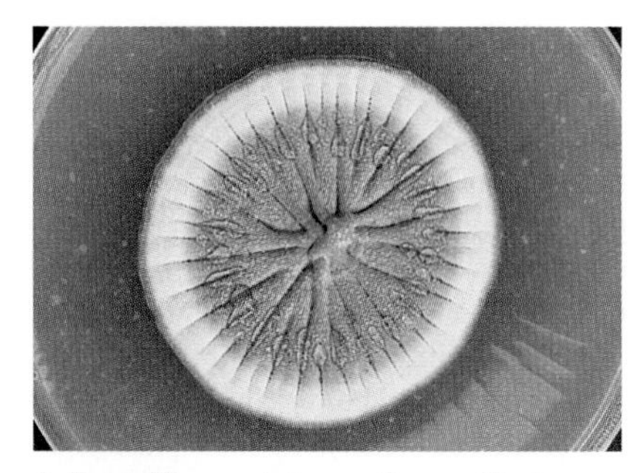

A *Penicillium notatum* culture. *(Andrew McClenaghan/SPL/Photo Researchers)*

$$\begin{array}{l} H_3C \\ \quad\;\; \rangle C - S - CH - CH - R \\ H_3C \quad | \qquad\quad\;\; | \qquad\; | \\ HOOC - CH - N - C{=}O \end{array}$$

General penicillin structure

Penicillin G, the first penicillin

$$R = -NH-\overset{\overset{\displaystyle O}{\|}}{C}-CH_2-C_6H_5$$

Amoxicillin, a broad-spectrum antibiotic—active against a variety of bacteria

$$R = -NH-\overset{\overset{\displaystyle O}{\|}}{C}-\underset{\underset{\displaystyle NH_2}{|}}{\overset{\overset{\displaystyle H}{|}}{C}}-C_6H_4-OH$$

All penicillins kill growing bacteria by preventing normal development of their cell walls. Unlike our cells and those of other mammals, bacteria rely on a rigid cell wall. The rigidity is maintained by cross-linking bonds between peptide chains. Penicillins inhibit the enzyme that forms these cross-links and prevent the reaction from occuring.

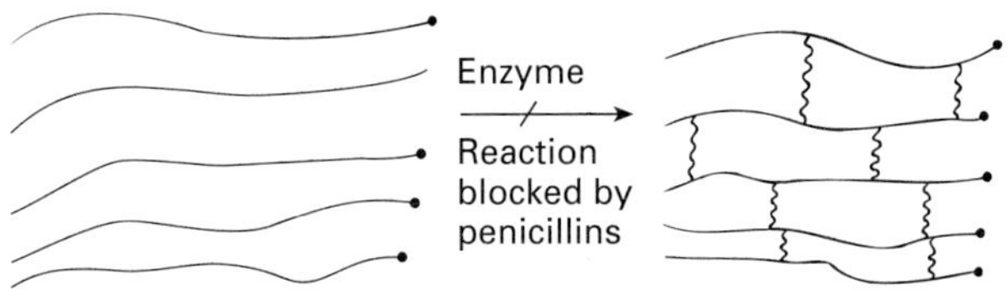

Other Classes of Antibiotics The *cephalosporins* are similar in structure to the penicillins and also act by disrupting cell wall synthesis. They have become the most widely used antibiotics in hospitals because of their low toxicity and broad range of antibacterial activity. All cephalosporins have the same general structure in which R_1 is often a hydrogen and R_2 and R_3 are substituents with several functional groups.

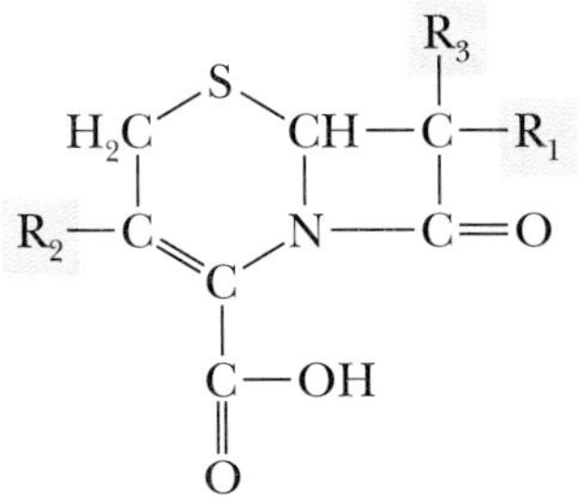

General cephalosporin structure

R groups in cephalexin, used to treat streptococcal infections

$R_1 = -H$

$R_2 = -CH_3$

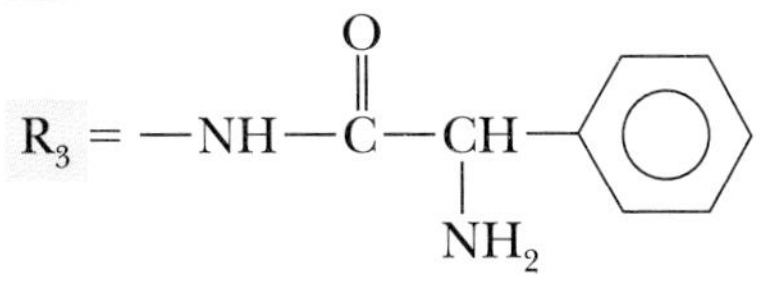

$R_3 = -NH-C(=O)-CH(NH_2)-C_6H_5$

Other kinds of antibiotics attack bacteria in different ways than the penicillins and cephalosporins. Many interfere with the synthesis or functioning of bacterial DNA. The *tetracylines* (e.g., acromycin and terramycin) and *erythromycin,* for example, inhibit bacterial protein synthesis, and *rifampin* (important in treating tuberculosis) inhibits RNA synthesis from DNA.

■ Some other antibiotics that act by inhibiting cell wall synthesis are bacitracin (used only on the skin, not internally) and vancomycin (important for treating bacteria resistant to other antibiotics).

The defeat of infectious diseases by antibiotics was once thought to be complete. That was a mistake. Strains of malaria, typhoid fever, gonorrhea, and tuberculosis have emerged that are resistant to the antibiotics that once defeated them. Researchers have begun to hunt for new antibiotics, in some cases resuming a search that had been abandoned because it was believed unnecessary. The urgency of the search was heightened with the discovery in 1997 of a strain of vancomycin-resistant staph *(Staphylococcus aureus),* the bacteria responsible for deadly infections among the seriously ill in hospitals. Vancomycin has been the only antibiotic able to cure such staph infections.

While the race for new antibiotics is on, the medical community endeavors to make everyone aware of some basic facts:

- Doctors should take care not to overprescribe antibiotics.
- People should not demand antibiotic treatment in cases where it is not necessary (e.g., antibiotics have no effect on viral infections).

- Antibiotics must always be taken for the full period of the prescription.
- Everyone should wash their hands frequently and thoroughly.

13.3 AIDS, A Viral Disease

Since its first appearance in the 1980s, HIV infection has spread rapidly and worldwide. In the United States, the first 100,000 cases of HIV infection were reported over the course of the 8 years from 1981 to 1989. The second 100,000 cases were detected in the following 2 years. By June 1996, over half a million cases of HIV infection and 343,000 deaths had been reported in the United States. The World Health Organization estimated in 1996 that 18 million adults and adolescents throughout the world were HIV-infected.

■ Fewer than a dozen drugs are available for attacking viral diseases (as opposed to treating the symptoms).

From a chemical perspective, it is important to understand that very few ways are known to combat viral diseases (other than vaccination). The reason for this is the method of attack used by viruses—they are essentially chemical parasites. They take over the DNA of human cells and put it to work for their own reproduction. It is difficult to attack the cells reproducing the virus without also attacking the host's own cells.

The HIV is a *retrovirus.* It consists of an outer double lipid layer surrounding a matrix containing proteins, an enzyme called *reverse transcriptase,* and RNA. The term "retrovirus" is used because the virus enzyme carries out RNA-directed synthesis of DNA rather than the usual DNA-directed synthesis of RNA (see Figure 11.22).

THE WORLD OF CHEMISTRY

Folk Medicine

Program 21, *Carbon*

There's a long history of folk medicine based on plants, especially in the Near and Far East. Now the question is, Which of these things really works, and how much of it is just a rumor? First, the chemist checks out the plant rumored to be good and analyzes its extracts to see if there is any activity. One such substance is called *fredericamycin.* Fredericamycin is an antibiotic of some interest as a possible antitumor compound. It comes from a soil organism, a bacterium, found in the soil in Frederick, Maryland.

Dr. Kathlyn Parker of Brown University, and other researchers like her, first analyze the naturally occurring substance in the lab. If they find an active component, they look further. As Dr. Parker says,

Then, if you are really interested in drug development, you have to isolate the active component, purify it, determine its structure, and then it gets handed over to the pharmaceutical people who decide how to package it. For some pharmaceuticals, what you really need is to be able to make a large amount of stuff really cheap. One solution is that you would develop methods so that it was so cheap to make something that you could distribute it to people in a way that they could afford it.

Kathlyn Parker.

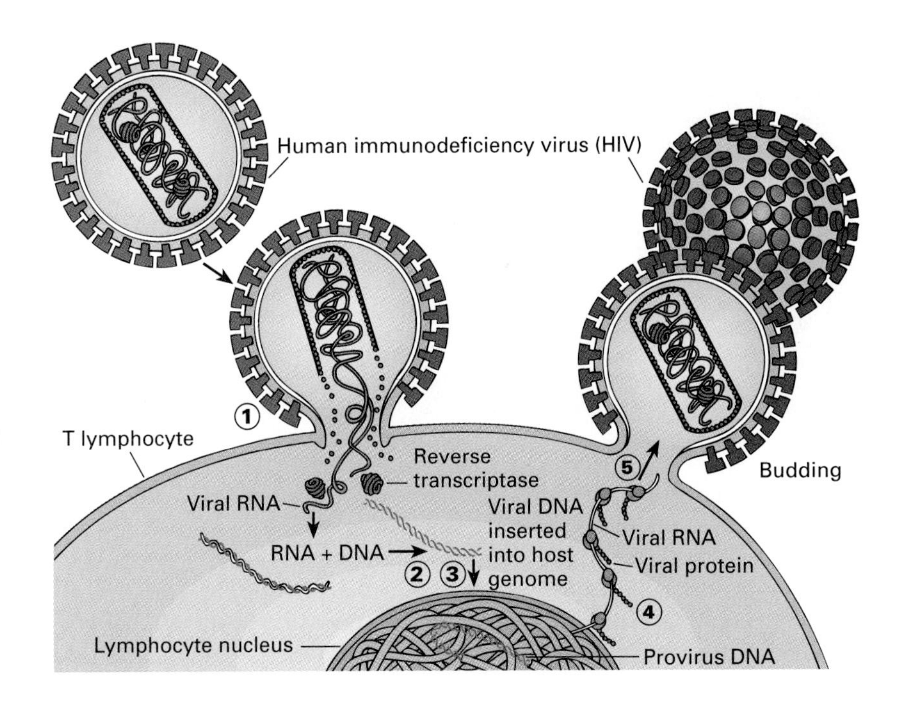

Figure 13.1 Attack of HIV virus on a T cell, a lymphocyte. The virus enters the lymphocyte (1) and produces its own DNA (2). The DNA then enters the lymphocyte nucleus (3), where it combines with the lymphocyte's DNA. Then, through the cell's normal processing, DNA is transcribed to messenger RNA and, back outside the nucleus, the viral RNA provides the template for the production of proteins (4) that assemble into a new virus particle (5).

The AIDS retrovirus penetrates the T cells of the immune system (1 in Figure 13.1). Once inside, the virus releases its contents, and the *reverse transcriptase* of the AIDS virus translates the RNA code of the virus into double-stranded DNA (2 in Figure 13.1). The virus DNA enters the T-cell nucleus and is incorporated into the cell's own DNA (3 in Figure 13.1) Then the T cell makes RNA from viral DNA, the proteins needed for a new virus are made from this RNA (4 in Figure 13.1), and the new virus is released (5 in Figure 13.1). Eventually the T cell swells and dies, releasing more AIDS viruses to attack other T cells. As their T cells are destroyed, individuals are attacked by diseases that are normally defeated by the body's immune system and thus are rare in healthy individuals.

A turning point was reached in the search for weapons against the HIV virus in 1996. Earlier efforts had yielded drugs that somewhat slow the progress of AIDS, but in no way approach a cure of the disease. In early 1996, three members of a new class of drugs directed at HIV were approved in rapid succession, the *protease inhibitors.* Convincing clinical trials had shown that a protease inhibitor taken in *combination* with older AIDS drugs dramatically lowers the concentration of the virus in patients' blood, often to below detectable levels. The drug combination diminishes the number of serious complications and decreases the death rate among seriously ill individuals. Apparently, a combination of drugs overcomes the major problem with single-drug therapy—if even a few virus particles are left alive by a drug treatment, the virus can very quickly adapt by converting to a drug-resistant form.

As their name implies, protease inhibitors act by inhibiting a protease enzyme. By fitting into the active site of the enzyme (Figure 13.2, p. 338), the inhibitors prevent the enzyme from functioning. Because this enzyme cuts a long protein chain into smaller proteins essential to survival of the HIV virus, halting its action kills the virus.

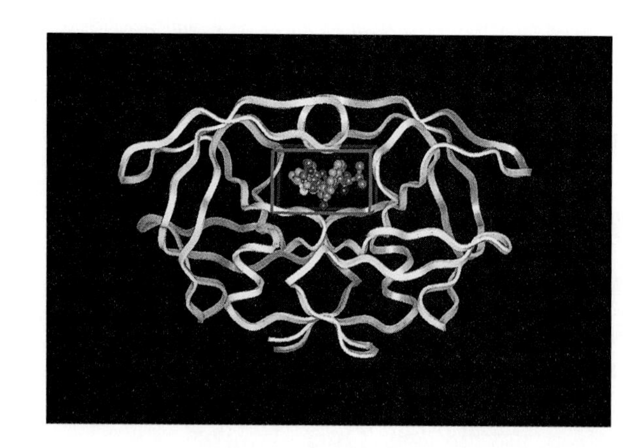

Figure 13.2 Ritonovir, a protease inhibitor, in the active site of the protease enzyme. The Ritonovir molecule is shown inside the red box. The ribbon-like structure represents the protein chain of the enzyme. *(Courtesy of Abbott Laboratories)*

■ AZT differs from the natural nucleoside by having an —N_3 group instead of an —OH group.

Azidothymidine (AZT)

The most successful older anti-AIDS drug, *azidothymidine* (AZT, also called *zidovudine*), is one of the drugs used in combination with a protease inhibitor. AZT (structure in the margin) is a derivative of deoxythymidine, a nucleoside. (Nucleosides are nucleotides without the phosphate group; Section 11.11). Because AZT resembles a natural nucleoside, it is accepted by the HIV reverse transcriptase enzyme and placed into the viral DNA. Once there, its structure prevents additional nucleosides being added to the chain, and cell division is halted.

Despite the new air of optimism, it is important to keep the situation in perspective relative to several facts. The new drug combination is not yet a cure and because it has been in use for a relatively short time, its effectiveness is not yet fully evaluated. Also, it is extremely expensive, it is unavailable to those in many third world countries, and there can be unpleasant side effects. An ominous further problem is provided by the need to take up to 20 pills a day on a precisely timed schedule and by the fact that wavering from the schedule may provide opportunity for new drug-resistant variations of the virus to develop.

■ SELF-TEST 13A

1. The ____________ name applies to the same drug no matter who makes it, but every manufacturer has its own ____________ name for the drug.
2. The infectious diseases now in the top ten leading causes of death for individuals of all ages in the United States are ____, ____, ____.
3. Infectious diseases are caused by ____________.
4. The general structures for penicillin and cephalosporin antibiotics have in common atoms of ____________ and ____________ in addition to the expected C, H, and O atoms, and also a ____________ -membered ring.
5. Penicillins act by blocking the enzyme that causes ____________ of peptides in bacterial cell walls.
6. Penicillin G, the first antibiotic, and amoxicillin, currently the most-prescribed antibiotic, differ in structure only by a(an) ____________ group and a(an) ____________ group.

7. Viruses are difficult to combat because they attack from ________ host cells rather than from ________ the cells.
8. Because the HIV virus carries out RNA-directed synthesis of DNA rather than DNA-directed synthesis of RNA, it is called a ________.
9. AZT, which is used to treat ________, fools the virus by becoming part of its ________.
10. The treatment of disease with chemical agents is referred to as ________.
11. The newest and most successful drugs to combat AIDS are known as ________ and the target of their action is a(n) ________ molecule.

13.4 Steroid Hormones

In considering hormones, we turn our attention from drugs that fight invading organisms to drugs that make up deficiencies in natural biochemicals or mimic their action.

Hormones are produced by glands and secreted directly into the blood. They serve as chemical messengers, regulating biological processes by interacting with **receptors** sometimes distant from where they are secreted. For example, the adrenal gland just above the kidney releases a group of hormones that act throughout the body to regulate the availability of glucose in the blood. Hormones are chemically diverse but are mostly proteins or steroids.

One of the most revolutionary medical developments of the 1950s was the worldwide introduction and use of "the pill." The development of synthetic analogs of the female sex hormones made reliable birth control available to women for the first time. The ongoing search for an equivalent to "the pill" for use by men has not yet been equally successful.

Most oral contraceptive pills used by women today contain a combination of two synthetic hormones. One, most commonly ethynyl estradiol, has estrogen-like activity in regulating the menstrual cycle. The other, most commonly norethindrone, has progesterone-like activity that establishes a state of false pregnancy that prevents ovulation. In a theme that is common in drug development, the successful contraceptives are very similar in molecular structure to the natural molecules with similar activity. (Compare the following structures with those in Section 11.3 p. 273.)

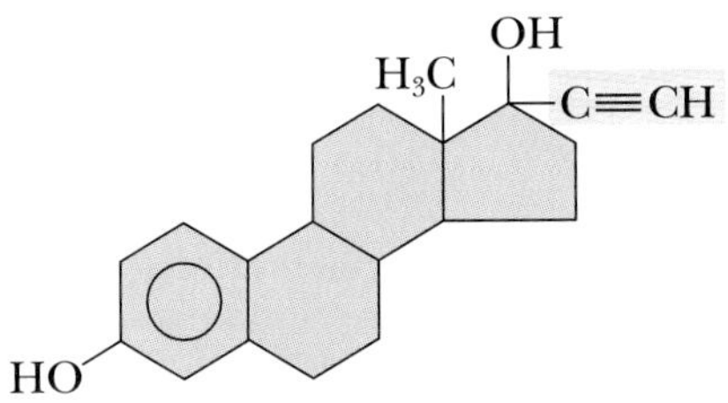

Ethynyl estradiol
(synthetic estrogen)

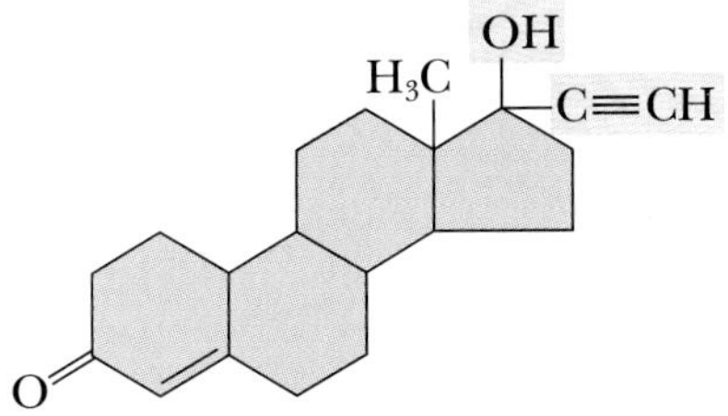

Norethindrone
(synthetic progesterone)

■ A **receptor** is a molecule or a portion of a molecule, often on the surface of a cell, that interacts with another molecule to cause some change in biochemical activity. Some drugs block receptors and prevent an undesirable change in activity. Other drugs activate a receptor to cause a desirable change in activity.

■ Cortisol, or hydrocortisone, is a steroid hormone. The ring structure common to all steroids is highlighted. Synthetic cortisol is used medically as an anti-inflammatory agent applied to the skin or injected into joints.

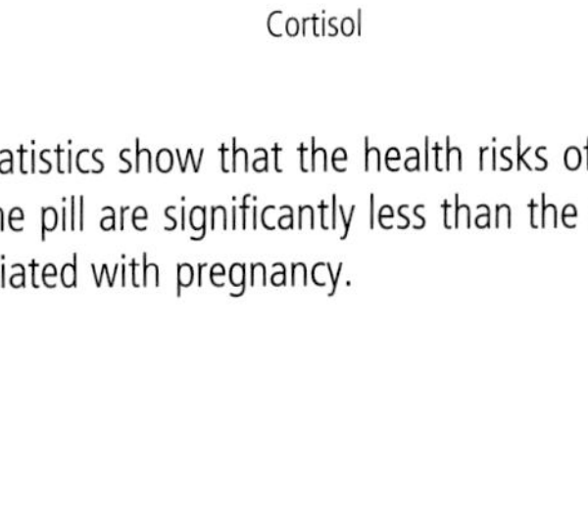

Cortisol

■ Statistics show that the health risks of using the pill are significantly less than the risks associated with pregnancy.

Synthetic steroids with estrogen-like activity are also used medically to replace natural hormones where there is a disease-related deficiency, after a hysterectomy, or after menopause. A major benefit is a slowing of the normal bone loss (osteoporosis) that occurs with age. Premarin, the number-one prescription drug, is a female hormone replacement medication.

13.5 Neurotransmitters

Hormones and neurotransmitters have in common their roles as chemical messengers within the body. Neurotransmitters, as illustrated in Figure 13.3, carry nerve impulses from one nerve to the next or to the location where a response to the message will occur.

The body has a variety of neurotransmitters, each with its own distinctive molecular receptors and functions. New information about neurotransmitters is being reported frequently and is of great interest because of its usefulness to medicine. Most drugs that affect the brain or the nervous system interact with neurotransmitters or their receptors.

Norepinephrine, Serotonin, and Antidepressive Drugs

Norepinephrine and serotonin are neurotransmitters with receptors throughout the brain. Norepinephrine helps to control the fine coordination of body movement and balance, alertness, and emotion; and also affects mood, dreaming, and the sense of satisfaction. Serotonin is involved in temperature

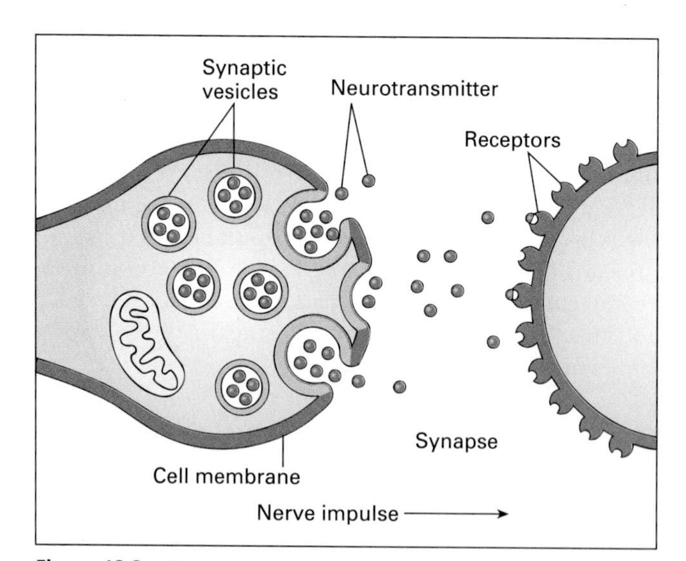

Figure 13.3 Transmission of a nerve impulse. The neurotransmitter is stored in the vesicles until it is needed. When a nerve impulse arrives, the vesicles move to the cell membrane and join with it so that the neurotransmitter is released. By crossing the synapse and binding to receptors on the surface of the adjacent nerve cell, the neurotransmitter transmits the impulse. The receptors in turn initiate chemical changes that allow the impulse to proceed.

A medication and its destination—a spatial fit with a receptor molecule that will activate a biochemical response.

and blood pressure regulation, pain perception, and mood. Serotonin and norepinephrine appear to work together to control the sleeping and waking cycle.

HO, HO—(benzene ring)—$CHCH_2NH_2$ with OH

Norepinephrine

HO—(indole ring, N)—$CH_2CH_2NH_2$

Serotonin

The normal cycle of neurotransmitter action at nerve *synapses,* the gaps between nerve endings (Figure 13.3) is as follows:

1. The neurotransmitter is released from the neuron.
2. The neurotransmitter crosses the synapse to interact with the receptor.
3. The neurotransmitter is inactivated, either by re-uptake by the neuron it came from or by conversion to an inactive form by an enzyme.

The biochemistry of mental depression is not fully understood, but a deficiency of norepinephrine and serotonin (possibly also dopamine; next section) almost certainly plays a role. Evidence is provided by the manner in which three classes of drugs, illustrated by the following three compounds, influence the action of these neurotransmitters.

$CHCH_2CH_2N(CH_3)_2$

Amitriptyline, a tricyclic antidepressant
(Elavil)

$-CH_2CH_2NHNH_2$

Phenelzine, an MAO inhibitor
(Nardil)

F_3C- $-O-CHCH_2CH_2NHCH_3$

Fluoxetine, an SSRI
(Prozac)

The tricyclic antidepressants such as Elavil (the trade name) prevent inactivation of neurotransmitters by preventing their re-uptake by neurons, which increases the concentration of the neurotransmitter in the synapse. The monoamine oxidase (MAO) inhibitors such as Nardil diminish the action of MAO, which is the enzyme that inactivates norepinephrine and serotonin. The third type of drug action is represented by Prozac, which prevents the recapture of serotonin by neurons that release it. Drugs of this type, known as selective *serotonin re-uptake inhibitors (SSRIs),* have become the drugs of choice for treating serious clinical depression. For each of these three classes of drugs, the major mechanism of action is to increase the concentration of neurotransmitters at synapses. In each case, there are multiple other modes of action not yet fully understood.

Dopamine

HO

HO— —$CH_2CH_2NH_2$

Dopamine
(L-dopa has a —COOH on the C atom next to —NH_2.)

Dopamine is produced in several areas of the brain, where it helps to integrate fine muscular movement as well as to control memory and emotion. An understanding of the brain chemistry of dopamine led to development of an effective treatment for Parkinson's disease. Patients with this disease experience trembling and muscular rigidity, among other symptoms, because of a deficiency of dopamine. Dopamine does not cross the blood–brain barrier (Figure 13.4) and thus cannot be administered as a drug. L-dopa, it was found, could cross the barrier and then be converted to dopamine in the brain. While

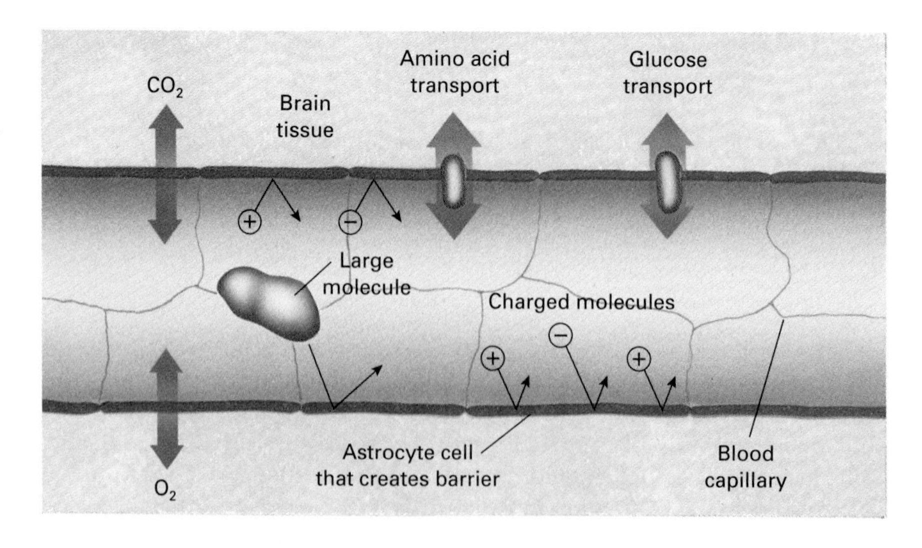

Figure 13.4 The blood–brain barrier. Openings in brain capillary membranes are small enough to keep out large molecules, and because the membrane contains mostly lipid molecules, ions cannot cross the membrane either. Only lipid (fat)-soluble molecules can cross unaided. These limitations provide the barrier that protects the brain from toxic compounds, but sometimes present a problem by also keeping out beneficial drugs.

it does not cure Parkinson's disease, L-dopa can completely alleviate symptoms for several years.

Dopamine has also been identified as the neurotransmitter that produces the feelings of well-being and reward associated with drug addiction (see *Frontiers in the World of Chemistry*, p. 349). Drugs that block dopamine receptors have been used to treat schizophrenia, which is, however, a complex condition not attributable solely to dopamine activity.

Epinephrine and the Fight or Flight Response

Epinephrine, or adrenalin, is both a neurotransmitter in the brain and a hormone released from the adrenal gland. Its sudden discharge when we are frightened produces the fight-or-flight response, which includes increased blood pressure, dilation of blood vessels, widening of the pupils, and erection of the hair. Because of its widespread and rapid effects, epinephrine has a number of medical uses, notably in crisis situations. It is administered to counteract cardiac arrest (by stimulating heart rate), to elevate dangerously low blood pressure (by constricting blood vessels), to halt acute asthma attacks (by dilating bronchial tubes), and to treat the extreme allergic reaction known as *anaphylactic shock*.

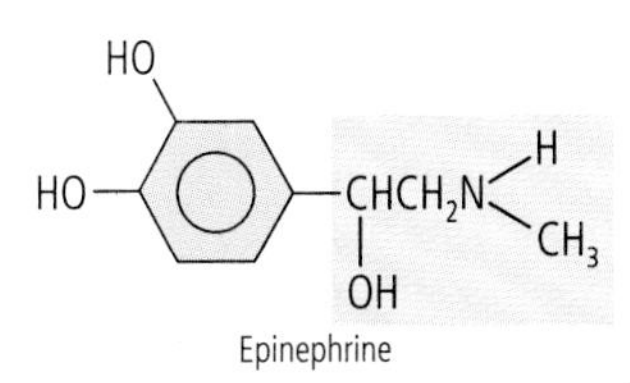

Epinephrine

■ The counteraction of the fight-or-flight reaction to the benefit of heart disease patients who take beta blockers is discussed in Section 13.11.

■ SELF-TEST 13B

1. Two main classes of chemical messengers within the body are __________ and __________.
2. Chemical messengers act only on cells that contain the appropriate __________.
3. Chemically, most hormones are __________ or __________.
4. Most oral contraceptives contain synthetic derivatives of which two of the following? (a) Progesterone, (b) An androgen, (c) An antihistamine, (d) An estrogen.
5. Patients with Parkinson's disease have a deficiency of the neurotransmitter __________.
6. Our bodies react to a sudden fright with a rapid production of __________.
7. Hormones are produced by __________ and secreted directly into the blood.
8. Several antidepressant drugs act by (increasing/decreasing) the concentration of neurotransmitters at synapses.

13.6 The Dose Makes the Poison

The amount of a chemical substance that enters the body is known as a *dose*. The substance might be a life-saving medication or it might be a poison. What accounts for the difference? *The dose makes the poison*. A German physician and chemist who referred to himself as Paracelsus recognized this in the 16th century and it is still true. Whether a dose of a given substance is poisonous or not depends on the size of the dose. Most substances can be poisonous in

TABLE 13.3 ■ LD_{50} Values for Several Chemicals

Chemical	LD_{50} (mg/kg administered orally to rat)
Aspirin	1750
Ethanol	1000
Morphine	500
Caffeine	200
Heroin	150
Lead	20
Cocaine	17.5
Sodium cyanide	10
Nicotine	2
Strychnine	0.8
Batrachotoxin	0.002*

* From a poisonous frog. LD_{50} in mice.

a sufficiently large dose. For a given individual, age, sex, weight, and general state of health also play a role in the effect of a given dose.

Doses of medications and poisons are customarily expressed as milligrams per kilogram of body weight (mg/kg). Aspirin, for example, is used to treat rheumatoid arthritis at a dosage of about 110 mg/kg per day. (A typical aspirin tablet contains just 325 mg; for an average 70-kg adult, two aspirin tablets amount to a dose of 9 mg/kg.)

A quantitative measure of toxicity is obtained by administering various doses of substances to be tested to laboratory animals (such as rats). The dose found to be lethal in 50% of a large number of test animals under controlled conditions is called the LD_{50} (lethal dose—50%) and is usually reported in milligrams of the substance per kilogram of body weight. Thus, if a statistical analysis of data on a large population of rats showed that a dose of 1 mg/kg was lethal to 50% of the population tested, the LD_{50} for this poison would be 1 mg/kg. Obviously, species differences can produce different LD_{50} values for a given poison. For this reason, defining risk to human beings based on animal data is difficult. It is, however, generally safe to assume that a chemical with a low LD_{50} value for several species will also be quite toxic to human beings. In Table 13.3, compare the LD_{50} values for substances you have probably ingested at one time or another with those for substances you would want to avoid.

13.7 Painkillers of All Kinds

Drugs that relieve pain are known as **analgesics**. They range from cocaine to morphine to aspirin. Some are illegal drugs, some are prescription drugs, and some are over-the-counter drugs.

Opioid structure

	R_1	R_2
Morphine	—O—H	—O—H
Codeine	$—O—CH_3$	—O—H
Heroin	$—O—\overset{\overset{O}{\parallel}}{C}—CH_3$	$—O—\overset{\overset{O}{\parallel}}{C}—CH_3$

Figure 13.5 Opioids. Morphine and codeine are natural alkaloids in the opium poppy, and heroin is a synthetic derivative with similar activity as a drug, but is dangerously addictive.

■ In an earlier time when few medications were available, Sir William Osler, a famous physician, called morphine "God's own medicine." It was named for the Greek god of dreams, Morpheus, by the German pharmacist who first isolated it from opium in 1803.

Opium and Its Relatives

Opium, obtained from the unripened seed pods of opium poppies, contains at least 20 different compounds. Chemically, they are **alkaloids**—organic compounds that contain nitrogen, are bases, and are produced by plants. About 10% of crude opium is ***morphine***, which is primarily responsible for the effects of opium. Morphine is medically valuable as a strong painkiller able to produce sedation and loss of consciousness. The term **opioid** is now applied to all compounds with morphine-like activity.

Heroin, the diacetate ester of morphine, does not occur in nature but can be synthesized from morphine. As shown in Figure 13.5, their structures differ in only one kind of functional group. Heroin is much more addictive than morphine and for that reason has no legal use in the United States. ***Codeine***, a methyl ether of morphine, is one of the alkaloids in opium and is used in cough syrup and for relief of moderate pain. Codeine is less addictive than morphine, but its analgesic activity is only about one fifth that of morphine. One of the most effective substitutes for morphine is *meperidine*, first reported structure in 1931 and now sold as Demerol. It is less addictive than morphine.

$CH_3CH_2OC(=O)$ — N—CH_3

Meperidine

Mild Analgesics

When milder general analgesics are required, few compounds work as well for many people as *aspirin*. Each year about 40 million pounds of aspirin are manufactured in the United States. Aspirin is also an **antipyretic** (fever reducer) and an **anti-inflammatory** agent. Aspirin inhibits cyclooxygenase, the enzyme that catalyzes the reaction of oxygen with polyunsaturated fatty acids to produce prostaglandins. Excessive prostaglandin production causes fever, pain, and inflammation—just the symptoms aspirin relieves.

Because aspirin is an acid, a dissolving aspirin tablet causes bleeding as it lies against the stomach lining. For most individuals, the blood loss from taking two 5-grain aspirin tablets is between 0.5 mL and 2 mL. Some persons, however, are more susceptible to blood loss. Early aspirin tablets were not particularly fast dissolving, which enhanced the bleeding problem. Today's aspirin tablets are formulated to disintegrate more quickly.

Seed capsule of an opium poppy. Crude opium is harvested from the sticky liquid that oozes through slits in the capsule. *(Dr. Jeremy Burgess/SPL/Photo Researchers)*

Willow trees, source of natural salicylates. Extracts from willow bark were known as pain killers in ancient times. Aspirin, first synthesized in the later 1800s, is a derivative of the natural salicylates, but with milder side effects. *(C. D. Winters)*

A bottle of aspirin tablets that has developed the vinegar-like odor of acetic acid should be discarded. Acetic acid is formed as aspirin ages and breaks down.

$$\text{C}_6\text{H}_4(\text{COOH})\text{–O–C(=O)–CH}_3 \xrightarrow{H_2O} \text{C}_6\text{H}_4(\text{COOH})\text{–OH} + \text{CH}_3\text{C(=O)–OH}$$

Acetylsalicylic acid (aspirin) — Salicylic acid — Acetic acid

Several over-the-counter aspirin alternatives are now available for pain sufferers. The three principal ones are ***acetaminophen (Tylenol), ibuprofen (Advil, Nuprin)***, and ***naproxen (Aleve)***.

$CH_3C(=O)–NH–C_6H_4–OH$
Acetaminophen (Tylenol)

$(CH_3)_2CHCH_2–C_6H_4–CH(CH_3)–COOH$
Ibuprofen (Advil, Motrin)

$CH_3O–C_{10}H_6–CH(CH_3)–COOH$
Naproxen (Aleve)

■ The reason that acetaminophen is the most used mild analgesic in hospitals is simple—it doesn't cause bleeding, which for hospital patients could be hazardous.

■ In some individuals, ibuprofen causes gastrointestinal problems, with cramping and diarrhea.

Because acetaminophen is not an acid, it is the only one that does not cause bleeding in the stomach. Acetaminophen is an effective analgesic and antipyretic, but it is not an anti-inflammatory agent. Aspirin, ibuprofen, and naproxen are *nonsteroidal anti-inflammatory drugs (NSAIDs)*, as distinguished from anti-inflammatory drugs that are steroids, such as cortisone. Ibuprofen, originally available only by prescription (Motrin), is similar to aspirin in its effectiveness but causes less bleeding. Naproxen, also originally a prescription drug (Anaprox or Naprosen), has as its principal advantage a long period of activity, making twice-a-day administration possible for round-the-clock pain relief.

13.8 Mood-Altering Drugs, Legal and Not

Everyone who drinks coffee or alcoholic beverages has personal experience with the mild effects of drugs classified as stimulants (the caffeine in coffee) or depressants (ethyl alcohol). Stimulants and depressants with stronger activity are among the drugs that are often *abused*—used in ways that are socially unacceptable and/or harmful. Although moderate alcohol consumption is an accepted social custom in the United States, alcohol too is an abused drug for many individuals. Other kinds of drugs that are subject to abuse are opioids and hallucinogens.

Although the list of drugs of abuse is wide ranging, a central theme is their relationship to brain chemistry and the effect of the drug on neurotransmitters or their receptor sites. Before discussing the mood-altering drugs, it's important to understand the legal control of abused drugs. Table 13.4 lists

TABLE 13.4 ■ Classification of Drugs

Designation	Description	Examples
Over-the-counter (OTC)	Available to anyone	Antacids, aspirin, cough medicines
Prescription drugs	Available only by prescription	Antibiotics
Unregulated nonmedical drugs	Available in beverages, foods, or tobacco	Ethanol, caffeine, nicotine†
Controlled substances*		
Schedule 1	Abused drugs with no medical use	Heroin, LSD, mescaline, marijuana
Schedule 2	Abused drugs that also have medical uses	Morphine, amphetamines, cocaine, codeine
Schedule 3	Prescription drugs that are often abused‡	Valium, phenobarbital

* The sale, distribution, and possession of drugs classified as controlled substances are controlled by the Drug Enforcement Administration of the U.S. Department of Justice.

† In the 1997 negotiations between tobacco companies and state's attorneys general, regulation of nicotine by the FDA was proposed.

‡ Prescription can be refilled only up to five times and within six months.

the various classifications of the U.S. Drug Enforcement Administration. Schedule 1 drugs are not legally available in any manner. Schedule 2 drugs can be dispensed only once from a written prescription. A refill requires a new written prescription.

Depressants

The effect of central nervous system depressants depends more on the dose than on the particular drug. The sequence proceeds from **sedation**, or relaxation, to sleep, to general anesthesia, to coma and death. Medically, depressants are used to treat anxiety and insomnia.

■ *Synergism* is the working together of two things to produce an effect greater than the sum of the individual effects.

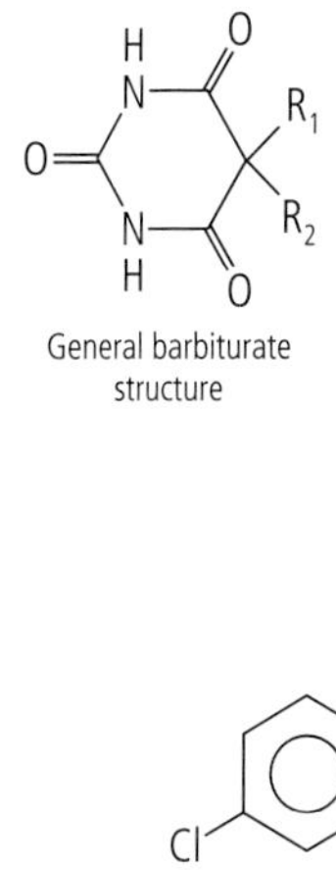

The best-known depressants are the *barbiturates.* Variations in chemical structure produce a range of barbiturates from very short acting to very long acting. Barbiturates are potential drugs of abuse, and overdoses cause many deaths. They are especially dangerous when ingested with ethyl alcohol, another depressant, because the two together give a synergistic effect.

$R_1 = -CH_2CH_3$

$R_2 = -C_6H_5$

General barbiturate structure

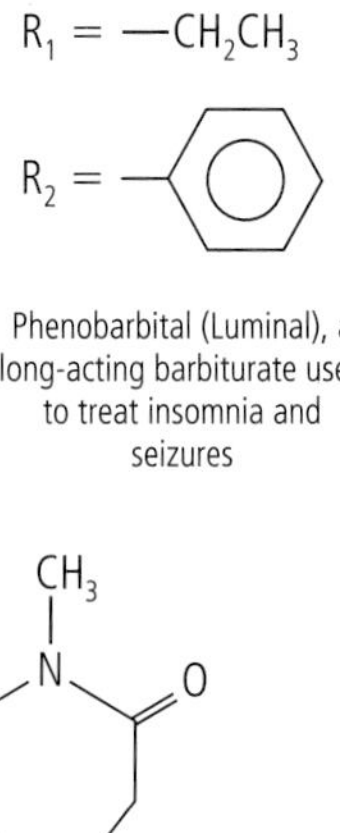

Phenobarbital (Luminal), a long-acting barbiturate used to treat insomnia and seizures

The second major class of depressants is the *benzodiazepines.* The familiar *tranquilizers* Librium (chlordiazepoxide) and Valium (diazepam) are members of this group. Both the barbiturates and the benzodiazepines act at receptors for the neurotransmitter gamma-aminobutyric acid (GABA, $HOOCCH_2CH_2CH_2NH_2$), which normally *inhibits* rather than excites transmission of nerve impulses. Because of differences in their molecular mechanism of action, benzodiazepines are safer and less subject to abuse than barbiturates.

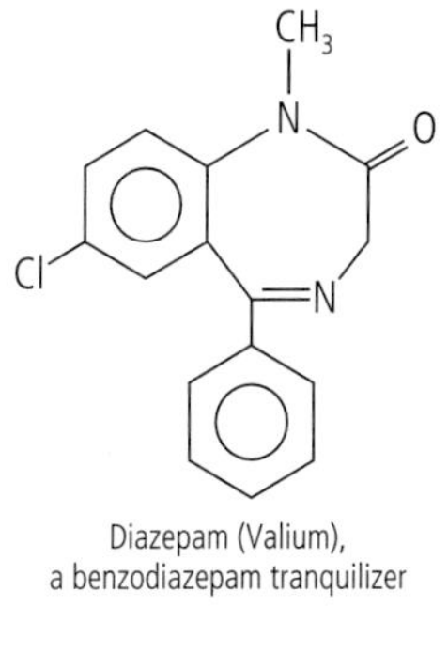

Diazepam (Valium), a benzodiazepam tranquilizer

Ethyl alcohol (CH_3CH_2OH), like barbiturates, enhances the action of the neurotransmitter GABA. Unlike barbiturates, however, the effects are dose dependent. At low doses, higher brain functions are affected, producing decreased inhibitions, altered judgment, and impaired control of motion. As dosage increases, reflexes diminish, and consciousness diminishes to the point of coma and death.

Stimulants

Stimulants are drugs that excite the central nervous system. They stimulate production of the neurotransmitters norepinephrine, serotonin, and dopamine in the brain, heart (which is stimulated to beat faster), and veins (which become constricted). Note the similarity in the structures of amphetamine and methamphetamine to epinephrine and norepinephrine (pp. 341, 343), which helps to explain their action.

$C_6H_5{-}CH_2{-}CH(CH_3){-}NH_2$

Amphetamine

$C_6H_5{-}CH_2{-}CH(CH_3){-}NH{-}CH_3$

Methamphetamine

Amphetamines were once available in over-the-counter preparations used to stay awake. Now, the only approved ingredient for "stay-awake" pills is caffeine. Amphetamines have become controlled substances because of their great potential for abuse. The only generally accepted medical use of amphetamines is to treat *narcolepsy*, a condition of uncontrollable attacks of sleep.

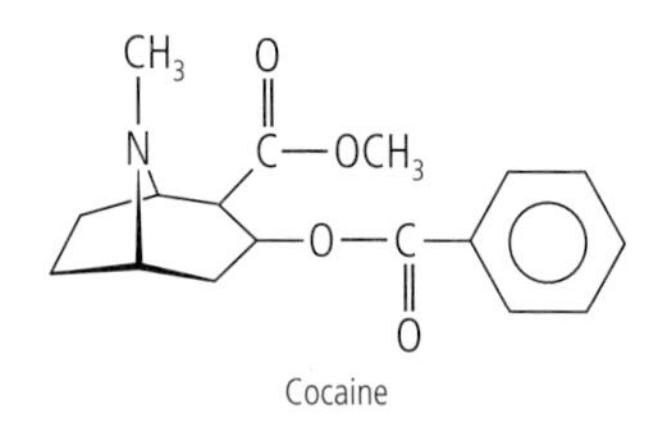

Cocaine

Cocaine, derived from the leaves of the coca plant of South America, is a stimulant and a Schedule 2 drug (used medically as a local anesthetic). By preventing the removal of norepinephrine from nerve endings, it causes uncontrolled firing of the nerves. *Crack* is a form of cocaine obtained by heating a mixture of cocaine and sodium bicarbonate. The reaction is an acid–base reaction since the base, sodium bicarbonate, is neutralizing cocaine hydrochloride, the usual form of cocaine. The term "crack" refers to the crackling sound of the heated mixture during the release of carbon dioxide. The appearance of crack cocaine on the illegal drug market has caused an increase in the number of cocaine addicts because crack is much more addictive than cocaine. The "high" lasts less than 10 min, creating the need to use crack repeatedly over a short period. Many users become addicted after only one try, and there is a high risk of taking a lethal dose.

Hallucinogens

Hallucinogens are chemicals that cause vivid illusions, fantasies, and hallucinations. Many have been found in plants, including *mescaline*, which comes from the fruit of the peyote cactus, and *lysergic acid diethylamide (LSD)*, which is made from lysergic acid derived from either the morning glory or ergot, a fungus that grows on grasses. These drugs are not addictive in the same manner as cocaine, but sometimes cause destructive behavior and lingering psychological problems. There are no therapeutic uses of the hallucinogens, which are believed to activate serotonin receptors.

Peyote cactus, source of mescaline. *(© R. Konig/Jacana/Photo Researchers)*

Marijuana is a mild hallucinogen and sedative made from the hemp plant, *Cannabis sativa*. Although the millions of marijuana users regard it as a "safe" drug, this is a controversial conclusion. At high doses, marijuana is moderately addictive and can create paranoia and intense anxiety. Long-term use at moderate doses causes a general state of disinterest in personal achievements. In addition, tetrahydrocannabinol (THC), the active ingredient in marijuana smoke, can cause damage to the lungs, impede brain function, and hamper the immune system. Animal studies also suggest it may produce birth defects in offspring.

Phencyclidine (PCP, known as "angel dust") is an especially dangerous drug with a unique pattern of effects. At low doses, its effects resemble those of alcohol. With higher doses, hallucinations set in and behavior can become hostile and self-destructive, promoting psychoses that can last for weeks. Physical effects include seizures, coma, and death from cardiac arrest.

Morning glory, whose seeds contain lysergic acid. *(© Carolina Biological Supply Company, Phototake NYC)*

FRONTIERS IN THE WORLD OF CHEMISTRY

Addictive Drugs

What role does brain chemistry play in the action of habit-forming drugs? Answering this question may open the door to combating the social, financial, and emotional costs of drug addiction.

Evidence is accumulating that all addictive drugs have an end result in common. In one way or another they increase the concentration of dopamine at nerve synapses in the brain. For example, cocaine blocks reuptake of dopamine from the nerve synapse and amphetamines accelerate release of dopamine. In numerous studies, it has been demonstrated that dopamine is the neurotransmitter responsible for feelings of reward and satisfaction. The higher the concentration of dopamine in the brain, the greater the reward—the "high" of drug addiction.

Not surprisingly, the first studies of the brain chemistry of addictive drugs were performed with laboratory animals. Thanks to modern noninvasive techniques for imaging the brain, the connection between dopamine and cocaine addiction in humans was verified early in 1997. Brain scans were performed as cocaine addicts received infusions of cocaine. The addicts receiving cocaine reported the expected pleasant feelings after at least half of their dopamine receptors were occupied by dopamine, as revealed in the scans.

A marijuana plant. (© 1988 Scott Camazine/Photo Researchers)

There is an ongoing debate about whether marijuana is an addictive drug. One argument that it is not addictive is based on the fact that traumatic withdrawal symptoms do not result from stopping its use. A probable explanation of this absence of withdrawal lies in the relatively long time that THC, the active ingredient of marijuana, persists in the bloodstream. Because its concentration decreases gradually rather than very suddenly, withdrawal is not experienced in the same manner as with drugs that are rapidly decomposed in the body. Research reported in 1997 demonstrated that in at least one respect—a surge in concentrations of brain dopamine—marijuana is no different than cocaine or heroin. The levels of dopamine in the brains of rats were observed to double when they were injected with THC, the active ingredient in marijuana.

In the hope of combating addiction of all kinds, the search is on for a new generation of drugs that will block the effects of increased dopamine in the brain.

CH_3, OH, CH_3, CH_3, O, $CH_2(CH_2)_3CH_3$

Tetrahydrocannabinol, THC (active ingredient in marijuana)

Reference N. D. Volkow et al.: *Nature,* Vol. 386, pp. 827, 830, April 24, 1997; G. Tanda et al.: *Science,* Vol. 276, pp. 2048–2050, July 4, 1997.

13.9 Colds, Allergies, and Other "Over-the-Counter" Conditions

What is the first thing we do when we feel sick? According to estimates, 75% of all illnesses are treated with products from the drugstore shelves. There are more than 300,000 over-the-counter (OTC) products on the market. But note that there are about 800 active ingredients packaged in all these different combinations. The top three categories of nonprescription medications in terms of sales dollars are used to treat allergies and colds, pain, and gastrointestinal problems.

About one person in ten suffers from some form of allergy. Allergic symptoms occur when special cells in the nose and breathing passages release histamine. Histamine is a neurotransmitter that accounts for most of the symptoms of hay fever, bronchial asthma, and other allergies. In a familiar theme in drug design, ***antihistamines*** are medications that block histamine receptors. Chlorpheniramine and compounds with very similar chemical structures (e.g., brompheniramine, diphenhydramine) are ingredients in many OTC preparations for treating hay fever and the "stuffy" noses associated with allergies. A frequent side effect of many antihistamines is drowsiness. In fact, many OTC sleep aids contain antihistamines.

$CH_2CH_2NH_2$

Histamine (neurotransmitter that causes allergic reaction)

$CHCH_2CH_2N(CH_3)_2$

Chlorpheniramine (antihistamine, e.g., in Chlortrimeton)

$CHOCH_2CH_2N(CH_3)_2$

Diphenhydramine (antihistamine used in sleep aids, e.g., Sleep-Eze)

The common cold is caused by a virus, and like other more serious viral diseases, it cannot truly be cured by any known method. The best we can do is treat the symptoms. Most OTC preparations for colds contain a variety of ingredients, usually two or more of those listed in Table 13.5. We have already discussed antihistamines and analgesics. There is some doubt about how useful an antihistamine is in treating a cold.

(a)

(b)

Two plants that send many people to the drugstore: (a) goldenrod, a major cause of hayfever; (b) poison ivy ("leaflets three, let it be"). *(a, Coco McCoy/Rainbow; b, C. D. Winters)*

TABLE 13.5 ■ Typical Ingredients in a Combination OTC Cold Medication

Type of Ingredient (Purpose)	Common Example *Name*	Common Example *Structure*
Decongestant (shrink nasal tissues)	Pseudoephedrine	C_6H_5—CH(OH)—CH(CH_3)NHCH_3
Antitussive (prevent cough)	Dextromethorphan	CH_3O-substituted ring system with N—CH_3
Expectorant (loosen fluids in cough)	Guaifenesin	C_6H_4(OCH_3)—OCH_2CH(OH)CH_2OH
Analgesic (diminish pain, fever)	Acetaminophen	HO—C_6H_4—NH—C(=O)—CH_3
Antihistamine (counteract allergic reaction)	Chlorpheniramine	(pyridyl)(Cl—C_6H_4)CHCH_2CH_2N(CH_3)$_2$

Decongestants shrink nasal passages to relieve the stuffiness that goes with colds and allergies. The decongestants activate receptors for epinephrine (an ingredient in some cold medications) and similar neurotransmitters. ***Antitussives*** suppress coughing. Opioids all have antitussive activity, and codeine is often used for this purpose. Dextromethorphan (see Table 13.5) is an opioid-related compound that has antitussive activity without the analgesic and other effects of opioids. ***Expectorants*** are meant to stimulate secretions in the respiratory tract so that mucus is dislodged in coughing. Most likely, guaifenesin is the only effective ingredient of this type.

A few guidelines for selecting and using OTC products are recommended by numerous groups concerned with public health:

- Choose single-ingredient products specific to the condition you have.
- Cut down on unnecessary expense by choosing generic products. (The chemical ingredients are the same.)

Over-the-counter drugs—we have a great many to choose from. *(C. D. Winters)*

- *Read* labels and *follow instructions* for dosage.
- *Pay attention* to cautions with respect to drowsiness, or interactions with alcohol or other medications.

13.10 Preventive Maintenance: Sunscreens and Toothpaste

Many drugstore products are directed toward preventing bodily harm rather than curing an existing condition. Among the most important are sunscreens, which protect against skin cancer, and toothpaste, which prevents not only cavity formation but also deterioration of the gums and eventual tooth loss.

Our world is bathed in ultraviolet (UV) radiation that is sufficiently energetic to harm living things exposed to it. The natural protective mechanism against UV radiation is an increase in the skin of the pigment known as melanin, producing what we call a "tan." The melanin molecules absorb some of the UV energy and convert it to heat, thus diminishing damage to the molecular structure of the skin. However, even with melanin's protection, exposure to sunlight causes trouble. Especially in fair-skinned people whose skin contains smaller amounts of melanin, the result is a visible reddening—*erythema* or in everyday language, sunburn. Although it is less visibly noticeable, dark-skinned people can also experience sunburn. Evidence has been mounting that the risk of developing skin cancer rises with the amount of lifetime exposure to the sun's rays.

In June 1994 the National Oceanic and Atmospheric Administration (NOAA) and the Environmental Protection Agency (EPA) began offering a UV index as part of the regular weather reports in 58 selected cities in the United States. The index uses satellite measurements as well as ground-based observations, which are fed into a computer model to forecast peak noontime UV levels. The computer model produces a 15-point index that corresponds to six exposure categories set by the EPA (Table 13.6).

If you are going to be exposed to direct sunlight, it is a good idea to protect yourself from as much UV exposure as you can. Besides physical barriers such as long-sleeve shirts and wide-brimmed hats, there are a variety of

TABLE 13.6 ■ UV Exposure Categories Used by EPA for the UV Index*

Exposure Categories	Index Values	Minutes to Burn for "Never Tans"—Most Susceptible	Minutes to Burn for "Rarely Burns"—Least Susceptible
Minimal	0–2	30	>120
Low	4	15	75
Moderate	6	10	50
High	8	7.5	35
Very high	10	6	30
Extreme	15	<4	20

* UV rays are only about half as intense 3 h before and after the peak. Physical surroundings such as snow, sand, and water reflect more UV and intensify exposure. Latitude and altitude also play a role; exposure increases with proximity to the equator and with altitude. Source: *Science News,* p. 61, July 23, 1994.

chemicals that will selectively absorb UV light. These *sunscreens* can be applied as oils, creams, or lotions. The sunscreen must function in a manner similar to melanin by absorbing UV light and converting it into heat energy. Many sunscreens contain *p*-aminobenzoic acid (PABA) and other chemicals with similar structures as the active ingredients. These compounds absorb an appreciable amount of the most harmful UV radiation.

Sunscreens have sun protection factors listed prominently on their labels. The *sun protection factor (SPF)* is defined as the ratio of time required to produce a perceptible erythema on a site protected by a specified dose of the sunscreen to the time required for minimal erythema development on unprotected skin. An SPF of 4, for example, would provide four times the skin's natural sunburn protection. By looking at Table 13.6 again, it is possible to *estimate* your safe exposure to the sun, given the exposure index and the SPF of the sunscreen you are using. An SPF of 4 would mean that someone who is most susceptible to the sun's rays could remain exposed for about 2 h (4×30 min) when the exposure index was between 0 and 2. Of course, other factors, such as how well you applied the sunscreen, should be considered as well.

The World of Chemistry
Program 10, *Signals from Within*

Sunscreen testing.

Keeping teeth clean requires ***toothpaste***, a mixture of detergents and ***abrasives***, which are hard substances that help remove unwanted materials on the tooth surface. The structure of tooth enamel is essentially that of a stone composed of calcium carbonate ($CaCO_3$) and calcium hydroxy phosphate (apatite—$Ca_{10}(PO_4)_6(OH)_2$). Despite being the hardest substance in the human body, tooth enamel is readily attacked by acids. Because the decay of some food particles produces acids, it is important to keep teeth clean.

Within moments after you clean your teeth, a transparent film composed of proteins from saliva begins to coat the teeth and gums. This coating offers a place for food debris to collect and for oral bacteria to multiply. These bacteria convert dextrins, from the breakdown of sugars, into acids. At the same time, a tenacious film composed of these bacteria, food particles, and their breakdown products begins to form. As the film hardens, it becomes dental ***plaque***. If this plaque is not removed regularly and completely from the surface of the teeth and beneath the gum line by brushing and flossing, the generation of acids and other harmful substances continues, destroying the tooth or the gum and eventually the bone that holds it in place.

Plaque that is not removed from the teeth becomes calcified from minerals in the saliva. The calcified plaque is known as ***tartar***. It is possible to control tartar buildup by using toothpastes containing sodium pyrophosphate ($Na_4P_2O_7$), which interferes with the mineral crystallization that causes tartar buildup. Beneath the gum line, tartar is a special problem because its presence makes it easier for plaque to grow, which irritates gum tissue and allows the gum to become diseased. Only a dentist or oral hygienist can remove tartar from beneath the gum line. By keeping teeth free from plaque and from prolonged contact with the acids produced by plaque bacteria, we can preserve the hard, stonelike enamel of the tooth.

You can't see them, but they're there. Bacterial plaque on the surface of a tooth. *(David Scharf/Peter Arnold, Inc.)*

The abrasive material in toothpaste serves to cut into the surface deposits, and the detergent assists in suspending the particles in the rinse water. Abrasives commonly used in toothpastes include hydrated silica (a form of sand, $SiO_2 \cdot nH_2O$), hydrated alumina ($Al_2O_3 \cdot n\ H_2O$), and calcium carbonate ($CaCO_3$). It is difficult to select an abrasive that is hard enough to cut the surface contamination yet not so hard as to cut the tooth enamel. The choice of detergent is easier; any good detergent will do quite well. Because the necessary ingredients in toothpaste are not very palatable, it is not surprising to see that various flavorings, sweeteners, thickeners, and colors are included to appeal to our senses.

Tooth decay occurs when bacteria eat food particles that remain in tiny fissures and crevices in the teeth and produce acids that attack the tooth enamel. Fortunately, it is possible to modify the crystalline structure of tooth enamel and make it more resistant to decay by the addition of fluoride ion (F^-) to toothpaste. When regularly applied, some of the fluoride ions actually replace the hydroxide ions in the hydroxyapatite structure to form fluoroapatite ($Ca_{10}(PO_4)_6F_2$). The fluoride ion forms a stronger ionic bond in the crystalline structure than the hydroxide ion because of its high concentration of negative charge; and as a result, the fluoroapatite is harder and less subject to acid attack than the hydroxyapatite. Hence, there is less tooth decay. Fluoride ion is introduced into essentially all the public water supplies in the United States for this purpose. Concentrations of fluoride ion of 1 part per million (ppm) have proved safe and efficient for reducing tooth decay. About 80% of all toothpaste sold in this country contains fluoride ions in some form. Compounds such as stannous fluoride (SnF_2) and sodium monofluorophosphate (Na_5FPO_3) provide a low level of fluoride ion concentration in toothpastes.

In the United States more teeth are now lost as a result of gum disease than from decay. Gum disease results from the lack of proper massage, from irritating deposits below the gum line, from bacterial infection, and from poor nutrition. More attention is being given to toothpastes containing disinfectants such as peroxides in addition to the other ingredients.

SELF-TEST 13C

1. ______________ is a derivative of morphine that is not found in nature.
2. Codeine is a derivative of morphine that is found in nature and that (does/does not) have a medical use.
3. Which of these three terms refers to each of the following products: (i) a pain killer, (ii) a medication that combats fever, (iii) a medication that combats the condition that causes muscle pain, swelling, and other symptoms?
   (a) Antipyretic (b) Analgesic
   (c) Anti-inflammatory
4. The chemical name for aspirin is ______________.
5. Acetaminophen differs from ibuprofen and aspirin in two important ways: it does not contain a (an) ______________ functional group, which means it does not cause intestinal bleeding, and it does not act as an anti- ______________.
6. A Schedule 1 controlled substance is a drug that is ______________ and has no ______________.
7. Barbiturates and benzodiazapines are two major classes of ______________.
8. Crack is a purified form of ______________ and is dangerously even more ______________.

9. Histamine is a neurotransmitter that causes which of the following? (a) Pain, (b) Upset stomach, (c) Allergic reaction.
10. A cold remedy can cure a cold. (a) True, (b) False.
11. Name the five classes of ingredients found in cold medications.
12. Sunscreens function by absorbing ____________ ____________.
13. The SPF, which stands for ____________ ____________ ____________, is a guide to choosing the correct sunscreen for yourself.
14. A toothpaste may contain (a) a detergent, (b) an abrasive, (c) a fluoride (d) sodium pyrophosphate. Match each of these ingredients with the condition it helps to prevent: (i) tooth decay (cavities), (ii) tartar buildup, (iii) accumulation of food debris and bacteria.
15. Whether a chemical is a drug or a poison is determined by the ____________.

13.11 Heart Disease

Many drugs and surgical techniques now in use are able to decrease the death rate and improve the quality of life for persons suffering from heart disease. However, heart disease remains the number-one killer of Americans. Known medically as **cardiovascular disease**, heart disease results from any condition that decreases the flow of blood, and consequently oxygen, to the heart or diminishes the ability of the heart to beat regularly and function in a normal manner.

The most common cause of heart disease is the plaque buildup on artery walls known as *atherosclerosis*, which was discussed in Section 12.4 in conjunction with the role of diet in plaque formation. If changes in lifestyle do not successfully combat this condition, the next step is cholesterol-lowering drugs, which include *lovastatin* and *cholestyramine*. Lovastatin acts by interfering with cholesterol synthesis in the liver. Cholestyramine acts by binding to bile acids in the intestines and accelerating their excretion. This causes the liver to convert more cholesterol into bile acids, leaving less to enter the circulatory system.

■ Bile acids are secreted into the small intestine to aid in the digestion of food. They are steroids (Section 11.3), and like all steroids, are synthesized from cholesterol.

A result of plaque buildup in blood vessels can be the chest pain during exertion known as *angina*, which occurs because of insufficient oxygen delivery to the heart muscle. The attacks are brought on when the heart must work harder and thus increase its oxygen demand. To treat angina, *vasodilators*, drugs that *dilate* veins (make them open wider), are used. By dilating the veins, the blood pressure against which the heart must work is reduced. The classic vasodilators are organic nitrogen compounds such as nitroglycerin or amyl nitrite.

■ Isn't it interesting that nitroglycerin is also a powerful explosive?

$$\begin{array}{l} CH_2—ONO_2 \\ | \\ CH—ONO_2 \\ | \\ CH_2—ONO_2 \end{array} \qquad C_5H_{11}NO_2$$

Nitroglycerin   Amyl nitrite

High blood pressure also contributes to heart disease. For this condition the next step after lifestyle changes is use of a **diuretic**, most commonly a *thiazide* (e.g., Diazide), which stimulates the production of urine and excretion of Na^+. With increased urine output, blood volume and consequently blood pressure are decreased.

The development of beta-blockers, one of the relatively new classes of drugs used to treat angina and other aspects of heart disease, illustrates how understanding the biochemistry of disease can lead to design of drugs.

In the 1960s two types of receptors that are part of the natural regulatory system for heart rate were discovered and named *beta receptors.* The beta-1 receptors are located primarily in the heart—stimulation of these sites speeds up the rate at which the heart beats. The beta-2 receptors are located in the peripheral blood vessels and the bronchial tubes. Stimulation of the beta-2 receptors relaxes muscle fibers, opening up the blood vessels and bronchial tubes so that blood flows more easily and it is easier to breathe deeply and quickly. These receptors are stimulated by the natural hormones epinephrine and norepinephrine during the fight-or-flight response (Section 13.5).

Armed with this information, chemists began to search for chemicals that would compete with epinephrine and norepinephrine at the beta receptor sites. If these sites could be blocked, stimulation of the heart muscle could be prevented. For a heart already overworked from the buildup of plaque in the arteries, this might produce enough relaxation to avoid an impending heart attack. In addition, these drugs might be able to relieve high blood pressure.

The first successful drug of this type, a *beta-blocker,* was *propranolol (Inderal),* now used to treat cardiac arrhythmias, angina, and hypertension. Look back at pages 341 and 343 to see its similarity in structure to the compounds whose action it blocks. Propranolol and the other beta-blockers have become widely prescribed drugs.

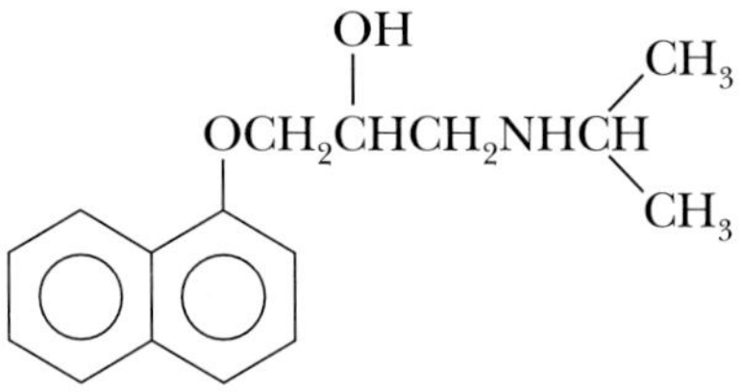

OH, CH_3, OCH_2CHCH_2NHCH, CH_3

Propranolol (Inderal)

■ It is important to realize that while death from heart attacks is most common among older people, the condition of atherosclerosis begins many years earlier.

A heart attack (a *myocardial infarction*) results from a reduction in blood flow to the heart muscle. About 98% of all heart attack victims have atherosclerosis, and the reduction in blood flow can usually be traced a clot in the plaque of a coronary artery. If prolonged, the blockage causes part of the heart muscle to die from lack of oxygen; and if the damage is sufficient, the heart attack can be fatal.

New clot-dissolving drugs given to heart attack victims in the emergency room or the ambulance show promise in reducing the death rate. These drugs are enzymes that act on *plasminogen,* a natural factor in the blood, by converting it to *plasmin.* Once this happens, plasmin proceeds to dissolve blood clots by the body's own natural mechanism. Three enzymes have been developed as drugs that catalyze the plasminogen → plasmin conversion: (1) *urokinase,* a natural enzyme isolated from human urine; (2) *streptokinase,* isolated from a *Streptococcus* bacterium, and (3) *tissue plasminogen activator (TPA),* one of the first drugs produced by recombinant DNA technology to receive government approval. TPA is made in genetically altered hamster ovary cells and is identical to the natural human enzyme that activates plasminogen.

13.12 Cancer, Carcinogens, and Anticancer Drugs

Cancer is not one but perhaps 100 different diseases. A cancer begins when a cell in the body starts to multiply without restraint and produces descendants that invade other tissues. It seems reasonable then that drugs might be able either to stop this undesirable spreading of cancer cells or to prevent cancer from happening at all. A major obstacle to successful drug treatment for cancer is that its biochemistry is not well understood. There is, however, general agreement that cancer is initiated by damage to DNA, which may be done by physical (e.g., ionizing radiation), biological (e.g., viruses), or chemical agents (e.g., compounds in cigarette smoke).

Every cancer comes from a single cell—one that is a modification of a normal cell. A normal cell functions according to directions stored in its genetic data bank, the DNA, and when a cell divides, each new cell gets its own exact copy of the parent DNA. If anything disrupts this DNA replication process, the genetic code in one of the descendent cells may cause that cell to grow and function differently from a normal cell. **Carcinogens** are chemicals that cause cancer, which manifests itself in at least three ways: (1) The *rate of cell growth* (that is, the rate of cellular multiplication) in cancerous tissue differs from the rate in normal tissue. Cancerous cells may divide more rapidly or more slowly than normal cells. (2) Cancerous cells *spread to other tissues;* they know no bounds. Normal liver cells divide and remain a part of the liver. Cancerous liver cells may leave the liver and be found, for example, in the lung. (3) Most cancer cells show *partial or complete loss of specialized functions.* Although located in the liver, cancerous cells no longer perform the functions of the liver.

Attempts to determine the chemical causes of cancer have evolved from early studies in which the disease was linked to a person's occupation. We now know that a person's lifestyle plays a role as well. In 1775, Dr. Percivall Pott, an English physician, first noticed that people employed as chimney sweeps had a higher rate of skin cancer than the general population. It was not until 1933 that benzo(α)pyrene ($C_{20}H_{12}$, an aromatic hydrocarbon containing five fused carbon rings, p. 209) was isolated from coal dust and shown to be metabolized in the body to produce one or more carcinogens.

Carcinogenesis is often a two-stage process. In the first stage, ***initiation***, a chemical, physical, or viral agent alters the cell's DNA. Sometimes a single exposure to some carcinogen causes a rapid onset of a tumor that is composed of rapidly growing, uncontrolled cells, but usually the abnormal cells continue to reproduce in about the same way as normal cells around them. Then a ***promotion*** occurs. This is the second stage and may occur days, months, or years after the initiation. This promotion may be a physical irritation or exposure to some toxic chemical that is itself not a carcinogen. In either case, the promotion results in the killing of a large number of cells. The destruction of cells is almost always compensated for by a sudden growth of new cells, and

■ When a cancer spreads from one site to another, the process is called *metastasis.*

Smoking is thought to play both an initiation and a promotion role in cancer causation.

the abnormal cells begin to grow in ways the original DNA coding never intended. The cancer has started.

To illustrate the initiation and promotion aspects of carcinogenesis, consider some experiments performed in 1947 at Oxford University in England. First, very small doses of dimethylbenzanthracene (DMBA), a known carcinogenic component of coal tar, were applied to the skin of a group of mice. These mice were then separated into two groups. In one group the exposed skin was daubed with croton oil, a strongly irritating natural oil. The other group of mice had their skin daubed with croton oil four months later. Almost every mouse in both groups developed a tumor where the DMBA had been applied. In other groups of mice tested, neither croton oil nor DMBA alone produced any tumors, and if croton oil was applied first to the skin, followed by the DMBA, tumors failed to appear. Apparently DMBA had an initiation effect, while croton oil had a promotion effect.

The mouse has come to be the classic animal for studies of carcinogenicity. Strains of inbred mice and rats have been developed that are genetically uniform and show a standard response to various chemicals.

Cancers are treated by (1) surgical removal of cancerous growths and surrounding tissue; (2) irradiation to kill the cancer cells; and (3) chemicals that kill the cancer cells, referred to as cancer chemotherapy. Cancer patients are considered cured if, after their treatment, they die at about the same rate as the general population. Another definition of success in cancer therapy is given by the number of patients who survive for five years after the treatment. In the 1930s fewer than 20 cancer patients in 100 were alive five years after treatment; in the 1940s, it was 25 in 100; in the 1960s, it was 33 in 100; and today it is close to 60 in 100.

The five-year cancer survival rate data are useful for tracking progress in detection and treatment of cancer. The American Cancer Society notes that the five-year rate is "less informative when used to predict individual progress."

Changes in the incidence and death rates from cancer between 1973 and 1994 (Figure 13.6) vary with the type of cancer. The overall rates continue to increase, principally because of the increase in lung cancer. Another contributer to the overall increase in rates is the increase in life span, which is accompanied by increases in the cancers of old age. If lung cancer is excluded from the data, however, the annual cancer death rate has begun to decline in recent years.

During World War I the toxic effects of the mustard gases were found to include damage to bone marrow and changes in DNA (mutations) that created abnormal offspring. In addition, the so-called *nitrogen mustards* caused cancers in some animals.

The name "mustard gas" comes from its mustard-like odor; mustard "gas," however, is not a gas but a high-boiling liquid that was dispersed as a mist of tiny droplets.

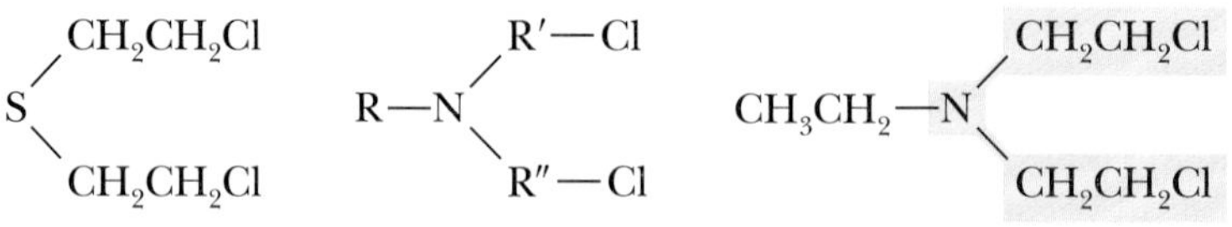

$S(CH_2CH_2Cl)_2$

Mustard gas

$R{-}N(R'{-}Cl)(R''{-}Cl)$

Nitrogen mustard (general formula)

$CH_3CH_2{-}N(CH_2CH_2Cl)_2$

A nitrogen mustard

When the wartime-imposed secrecy surrounding these chemicals ended, it occurred to some researchers that cancers might be treated with similar compounds. The result might be to alter the DNA in cancer cells to the extent that these cells could be destroyed selectively.

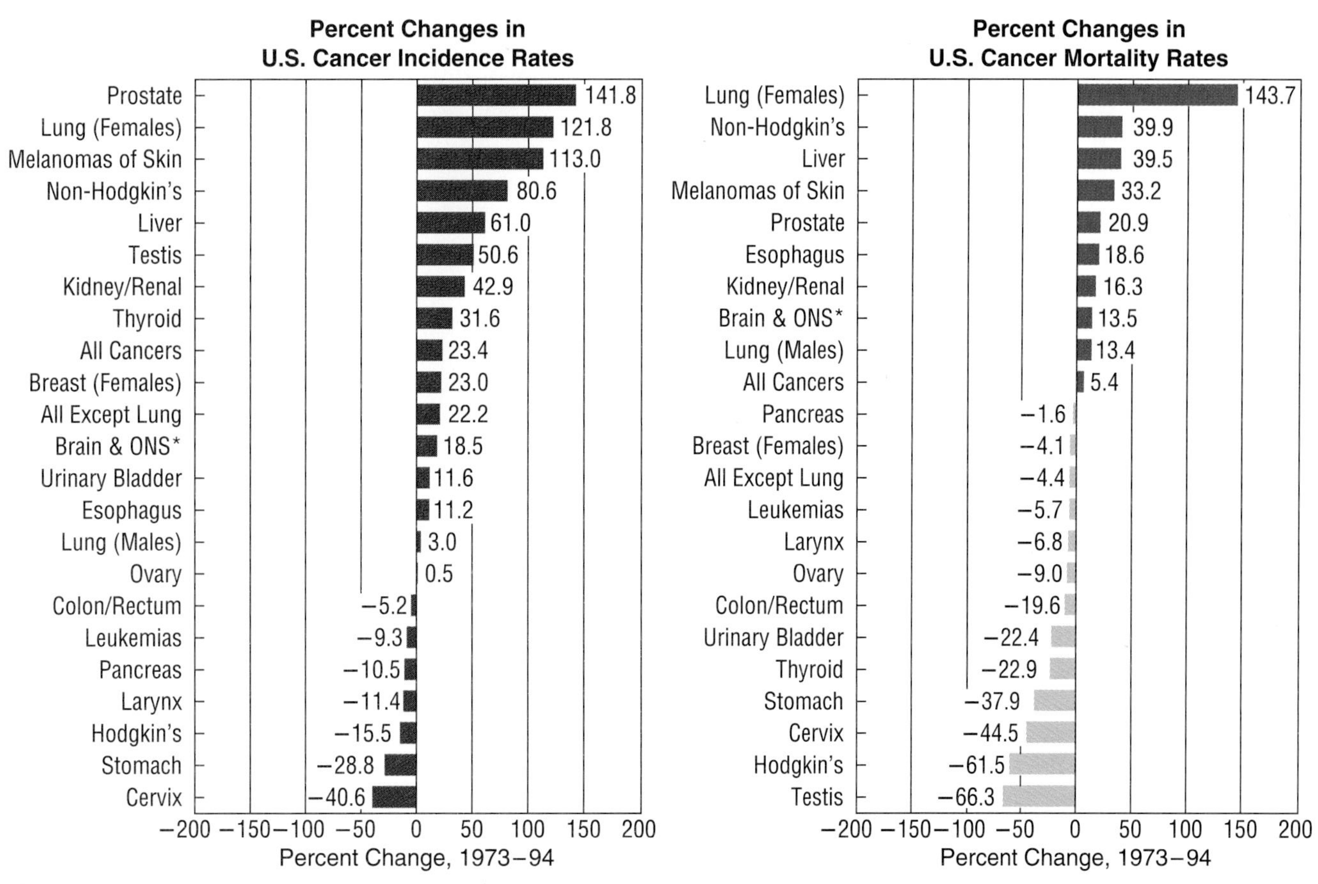

Figure 13.6 Percentage of change between 1973 and 1994 in the incidence and mortality rates for various cancers. The sites listed are the primary sites, that is, the sites at which the cancers began. (Non-Hodgkin's and Hodgkin's are lymphomas; ONS* = other nervous system sites.) Source: National Cancer Institute.

One of the most widely used anticancer drugs is now cyclophosphamide, a compound that contains the nitrogen mustard group.

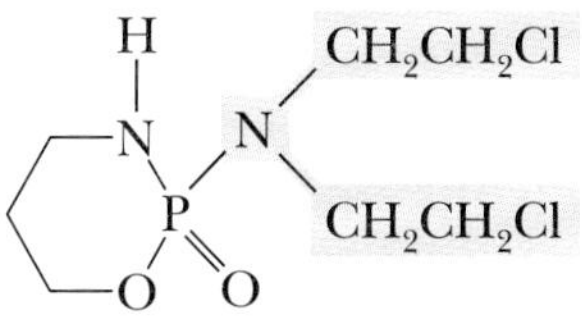

Cyclophosphamide

Cyclophosphamide, and other anticancer drugs that act in the same manner, are **alkylating agents**—reactive organic compounds that transfer alkyl groups in chemical reactions. Their anticancer activity results from the transfer of alkyl groups (e.g., CH_3CH_2—) to the nitrogen bases in DNA, often to guanine. The alkyl group physically gets in the way of base pairing and prevents DNA

THE WORLD OF CHEMISTRY

Toxic Substances in Perspective

Program 25, *Chemistry and the Environment*

How do scientists perceive the risk from chemicals and the environment? Interestingly, most chemicals, even vitamins, can be toxic at large doses. Most chemicals, in fact, have a threshold dose of toxicity. A dose above the threshold is dangerous; a dose below it is not. There is a great deal of debate about whether potential carcinogens have such thresholds, that is, concentrations below which they won't cause cancer. Thus, given conflicting and imprecise evidence, some scientists talk about carcinogens in terms of risk or probabilities. Risk is relative. For example, some think the cancer risk from hazardous waste is negligible compared with the naturally occurring carcinogens we successfully resist everyday.

The originator of a test for determining if a chemical has the potential to cause cancer is Dr. Bruce Ames, biochemist at the University of California, Berkeley. He claims,

People have been very worried about toxic waste dumps but, in fact, the evidence that they're really causing any harm is really minimal; there's not very much evidence. And the levels of chemicals are very tiny, so we don't really know whether there's no hazard or a little bit of hazard. The world is full of carcinogens, because half the natural chemicals they've tested have come out as carcinogens. Some plants have toxic chemicals to keep off insects, and we are eating those every time we eat a tomato or potato. Mushrooms have carcinogens, celery has carcinogens, and an apple has formaldehyde in it. So there are an incredible number of carcinogens in nature; we're getting much more of those than man-made chemicals.

Bruce Ames.

Dr. Halina Brown, Professor of Toxicology at Clark University, has stated,

If we accept those risks, why can't we accept small risks from chemical carcinogens in the environment? It's a valid argument. But then there is, of course, the counter argument. The counter argument is, we cannot do much about trace amounts of carcinogens that are present in food; why should we add to this burden that we already have by increasing the amount of exposure to carcinogens? But then it boils down to money. Unfortunately, it takes tremendous resources to reduce the levels of exposure in the environment to carcinogens, especially when you get to very low levels. Reducing it by another order of magnitude may take millions of dollars at one hazardous waste site. And the pie is not unlimited. Even those who don't consider hazardous waste dumps a health threat think the money should be spent to clean them up.

Dr. Bruce Ames also states,

I mean, if Congress has put $10 billion for cleaning up toxic waste dumps, you might as well find the worst ones and clean them up. Now, whether you're getting anything—whether you're gaining much in public health for cleaning them up—is something one could argue about. I think probably very little. But, in any case, you can—you might as well spend the money cleaning up the worst dumps.

replication, which stops cell division. Although alkylating agents attack both normal cells and cancer cells, the effect is greater for rapidly dividing cancer cells, a criterion that applies to all cancer chemotherapy agents.

From nitrogen mustard

$Cl-CH_2-CH_2-N(CH_3)-CH_2-CH_2-$ (bonded to N of guanine)

Guanine

Bonded to rest of DNA

An alkylated guanine in DNA. You can see how the group alters the size of the guanine residue.

Another class of chemotherapy drug, the **antimetabolites**, interferes with DNA synthesis because they are similar in molecular structure to compounds required for DNA synthesis (metabolites). Methotrexate, for example, is an antimetabolite similar in structure to folic acid (a B vitamin), used in the synthesis of nucleic acids. Methotrexate prevents reduction of folic acid in the first step of nucleic acid synthesis by strongly binding to the enzyme for that reaction. (Methotrexate and folic acid are large molecules. Methotrexate differs from folic acid by the addition of one $—CH_3$ group and replacement of a C=O by a $C—NH_2$.)

Yet another way to attack DNA is by physically disrupting its shape. A group of drugs accomplish this by fitting a planar ring portion of their molecular structure in between the base pairs that form the ladder structure of DNA (see Figure 11.21). Doxorubicin (structure in margin) is a most valuable chemotherapeutic agent of this type because it is active against an unusually large number of kinds of cancer.

All cancer chemotherapy is tedious and has risks and unpleasant side effects. The problem is that no agent has yet been found that kills only cancer cells, so there is always a balance to be struck between killing healthy cells and killing cancer cells. Because chemotherapy drugs kill actively dividing cells, the goal is to kill many more cancer cells than normal cells (ideally, of course, 100% of the cancer cells). In addition to being highly toxic, most of the cancer chemotherapy drugs are themselves carcinogenic and very high doses are usually necessary. As a result, single-agent chemotherapy has largely given way to combination chemotherapy because of positive additive, or even synergistic, effects when two or more drugs are used together. Because of their synergistic action, lower doses of each compound can be used when they are given together; and this reduces the harmful side effects. Chemotherapy alone is most successful in treating cancers such as leukemia or lymphoma, in which the cancer cells are widely dispersed in the body.

■ Because cells in hair follicles divide rapidly, they are often killed during cancer chemotherapy, resulting in hair loss. (When new cells are generated, hair returns.)

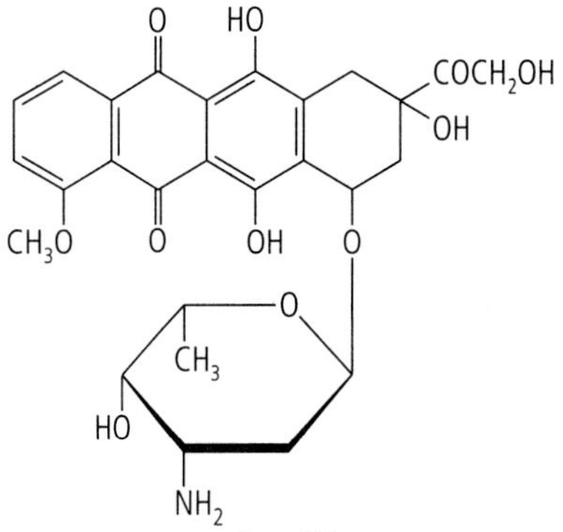

Doxorubicin

■ SELF-TEST 13D

1. Heart disease is known medically as __________.
2. During a heart attack, heart muscle can die from lack of __________.
3. When beta receptors in the heart are stimulated by epinephrine, the heart beats (a) faster, (b) slower.
4. Clot-dissolving drugs used in emergency treatment of heart attacks are classified biochemically as __________.
5. Which of the following can initiate cancer? (a) Smoking, (b) Viruses, (c) Ionizing radiation.
6. Most anticancer drugs can also cause cancer. (a) True, (b) False.
7. The nitrogen mustards act on cancer cells by blocking __________ replication.
8. An ideal cancer chemotherapy agent would be able to __________ cancer cells without doing the same to noncancerous cells.

MATCHING SET

___ 1. Can be bought without a prescription
___ 2. Interfere with DNA synthesis
___ 3. Beta-blockers
___ 4. Prevents serotonin reuptake
___ 5. Receptor
___ 6. Amoxicillin
___ 7. Action of a penicillin
___ 8. Neurotransmitter that causes allergic reaction
___ 9. Dopamine
___ 10. Common chemical name of a drug
___ 11. Morphine, codeine, and heroin
___ 12. Infectious disease
___ 13. Acetaminophen
___ 14. A very poisonous substance
___ 15. Protease inhibitor
___ 16. Acetylsalicylic acid

a. Treated with antibiotics
b. Frequently the site of action of a drug
c. Class of drugs for treating angina
d. Generic name
e. Antimetabolite for cancer chemotherapy
f. Over-the-counter drug
g. Opioids with very similar chemical structures
h. Most prescribed antibiotic in 1996
i. Newest type of drug for AIDS treatment
j. Has a very small LD_{50}
k. Histamine
l. Neurotransmitter associated with drug addiction and Parkinson's disease
m. Prozac
n. Aspirin
o. Inhibit formation of cross-linking bonds in bacterial cell wall
p. An analgesic

QUESTIONS FOR REVIEW AND THOUGHT

1. Give examples of a bacterial disease and a viral disease.
2. Give definitions of the following terms:
   (a) Over-the-counter drug
   (b) Prescription drug
   (c) Generic name for a drug
   (d) Trade name for a drug
3. Name the agency responsible for classifying drugs in the United States.
4. What are antibiotics?
5. Name three major classes of antibiotics.
6. What do the following three acronyms represent?
   (a) AIDS (b) HIV
   (c) AZT
7. What is chemotherapy?
8. Describe how penicillin kills bacteria.
9. What is a retrovirus?
10. For what disease or condition is each of the following classes of drugs used?
    (a) Analgesics (b) Antipyretics
    (c) Antibiotics (d) Antihistamines
11. For what disease or condition is each of the following classes of drugs used?
    (a) Vasodilators (b) Alkylating agents
    (c) Beta-blockers (d) Antimetabolites
12. Describe the role of a receptor in biochemistry.
13. What two classes of natural biomolecules require receptors for their action?
14. To what classes of drugs do the following compounds belong?
    (a) Barbiturates (b) Benzodiazepines
    (c) Ethanol (d) Amphetamines
15. To what classes of drugs do the following compounds belong?
    (a) Lovastatin
    (b) Methotrexate
    (c) Chlorphenirimine
    (d) Pseudoephedrine
16. Describe the following and give an example of each:
    (a) Schedule 1 drugs (b) Schedule 2 drugs
    (c) Schedule 3 drugs
17. Which of the following terms apply to codeine?
    (a) Analgesic
    (b) Antibiotic
    (c) Opioid

(d) A scheduled drug with a potential for abuse
(e) An antitussive

18. How do antihistamines work in the body? What, if any, side effects do antihistamines have?
19. Describe the symptoms experienced in
    (a) angina (b) arrhythmia
20. Name a drug that might be used to treat the symptoms described in question 19.
21. What is a barbiturate? What are the physiological effects of barbiturates?
22. Nitrogen mustards are alkylating agents, drugs that interfere with DNA replication. Explain what this means.
23. What happens when beta receptor sites in heart muscle are stimulated?
24. Name the class of biomolecule that includes dopamine, norepinephrine, and serotonin.
25. The estrogen estradiol has the following structure:

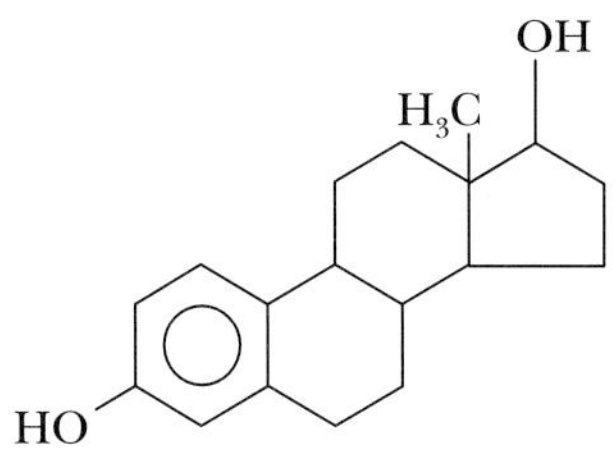

OH
H_3C
HO

What functional groups are different in ethynyl estradiol?

26. What is the function of Premarin? What relationship does it have to osteoporosis?
27. Give the functions for the following:
    (a) Dopamine (b) Epinephrine
28. What are the applications for the following drugs?
    (a) Tylenol (b) Ibuprofen
    (c) Naproxen
29. What are the four classifications of drugs in terms of Drug Enforcement Administration regulations?
30. How are cocaine and crack related?
31. Classify each of the following substances as either an hallucinogen, an anti-depressant, or a depressant:
    (a) Mescaline
    (b) Lysergic acid diethylamide (LSD)
    (c) *Cannabis sativa*
    (d) Phencyclidine (PCP)
    (e) Barbiturates
    (f) Amphetamines
32. What are the functions of the following over-the-counter drugs?
    (a) Antihistamines (b) Analgesics
    (c) Decongestants (d) Antitussives
    (e) Expectorants
33. What disease is treated with the following drugs? Tell the function of each drug.
    (a) Cholesterol lowering drugs
    (b) Vasodilators
    (c) Diuretics
34. What are nitrogen mustards? What was their original purpose? What is their current medical use?
35. What are the three modes of treatment for cancers?
36. Penicillins have the general formula shown as follows. What is the R group in penicillin G? Why is it necessary to have a number of different penicillins?

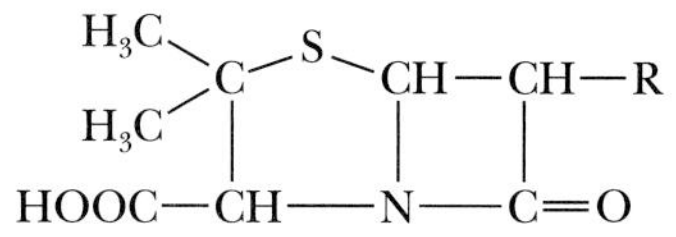

H_3C S
C CH—CH—R
H_3C
HOOC—CH——N——C=O

37. What is the effect of a hallucinogen? Name two examples of hallucinogens.
38. Describe the normal steps in the action of a neurotransmitter at a nerve synapse.
39. Of the three compounds heroin, morphine, and codeine,
    (a) which is the most effective pain killer?
    (b) which is not a natural alkaloid?
    (c) which is so addictive that its sale and use are illegal in the United States?
40. Identify by chemical name the oxygen-containing functional groups in morphine, codeine, and heroin (see Figure 13.5).
41. What is the physiological effect of nitroglycerine? What disease is it used to treat?
42. The FDA requires extensive testing of a prospective drug. The period of development of a new drug can take almost 12 years from initial discovery to final approval. Other countries have much shorter approval processes. Imagine that you have a close relative who is suffering from a serious illness and you have learned that an effective drug for the illness is available only outside the United States. What are the possible courses of action? What would you do?
43. Explain the role of the blood–brain barrier. What does it mean for the design of a drug that must be active in the brain?
44. During the time period represented in Figure 13.6, identify
    (a) the type of cancer whose incidence has decreased the most
    (b) the type of cancer for which mortality has increased the most
    (c) the type of cancer for which the incidence has increased the most

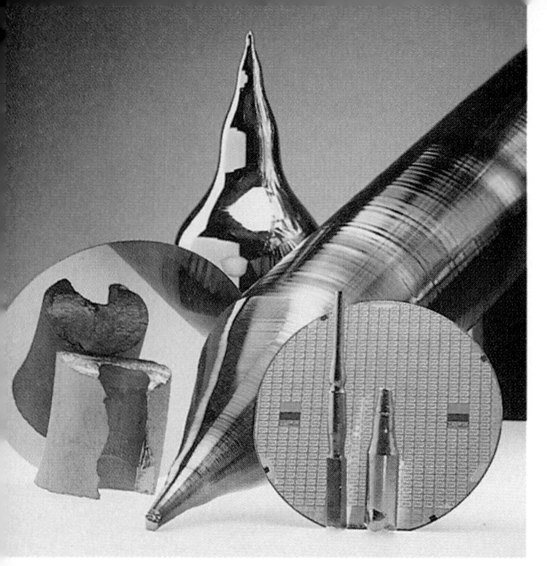

CHAPTER 14

The Chemistry of Useful Materials

The long view from space has dramatized what we already knew—the crust of the earth is a very unusual environment, uniquely suited, at least in this solar system, for the production and support of life. Our environment is also quite heterogeneous. Mixtures abound; everywhere we look, the elements and compounds are almost lost in the complicated array of mixtures produced by natural forces acting over very long periods.

Throughout most of history, we had not developed the power to alter our environment significantly. Early everyday objects, such as stone hammers or wooden plows, were only physically changed from the natural material. Then came the chemical reduction of copper from natural minerals, followed by iron, and now the flood of new materials produced each year. We have developed, beyond question, the power to change the earth's natural chemical mixtures in almost any way we choose.

In this chapter, our focus is on inorganic substances—the elements other than carbon and their compounds. We look at the origins of the raw materials for our pots and pans, homes and office buildings, automobiles and airplanes, and a multitude of other manufactured items.

- How are chlorine and sodium hydroxide made from salt?
- How are metals extracted from their minerals and ores?
- What causes metals to be conductors of electricity?
- What are semiconductors and superconductors, and what are some of their uses?
- What are the general compositions, properties, and methods of production of glass, ceramics, and cement?

Silicon in its various forms of purity leading to silicon "chips" containing integrated circuits, so important in modern computer applications.

14.1 The Whole Earth

This chapter illustrates how elements and compounds are separated from the hydrosphere and crust of the earth, and put to use. The **hydrosphere**, which includes salt waters and freshwaters above and below the earth's surface, must supply the water necessary to sustain life. The soluble salts in the oceans are a commercial source of magnesium; bromine; and sodium chloride, which is not only table salt but also an essential chemical raw material.

The portion of the solid earth available to us is a very small part of the whole, less than 1% by mass. The deepest mine extends only 3.8 km beneath the surface. Geologists define the earth's crust as a region between the surface and a depth of about 5 km to 35 km that lies over regions of greater density (Figure 14.1).

Three major types of rocks are found in the earth's crust: *igneous rocks*, formed by solidification of molten rock (e.g., basalt); *sedimentary rocks* (e.g., sandstone, which is cemented sand), formed by deposition of dissolved or suspended substances from oceans and rivers; and *metamorphic rocks* (e.g., marble), formed by the action of heat and pressure on existing rocks. Figure 14.2 (p. 366) gives the average composition of the earth's crust. The most abundant substances in rocks are silicates, which are composed of silicon, oxygen, and positive metal ions (Section 14.5). The more than 2000 kinds of known **minerals** fall into a few major classes (Table 14.1).

Fortunately for the mining industry, the composition of the crust is not uniform. Natural forces have concentrated different minerals in different places. For example, as molten rock gradually cools, the minerals that solidify first (those with the higher melting points) can sink in the remaining liquid and become concentrated. Or minerals can be redistributed according to variations in their solubility in natural waters.

The western United States was once covered by a large, land-locked sea. The water evaporated, leaving huge deposits of sodium carbonate, a soluble salt that is a valuable chemical raw material. While most nations must manufacture sodium carbonate from other chemicals, the United States meets a

■ *Ecology* is the study of the complex interrelationships among living things, the atmosphere, the hydrosphere, and the earth's solid surface.

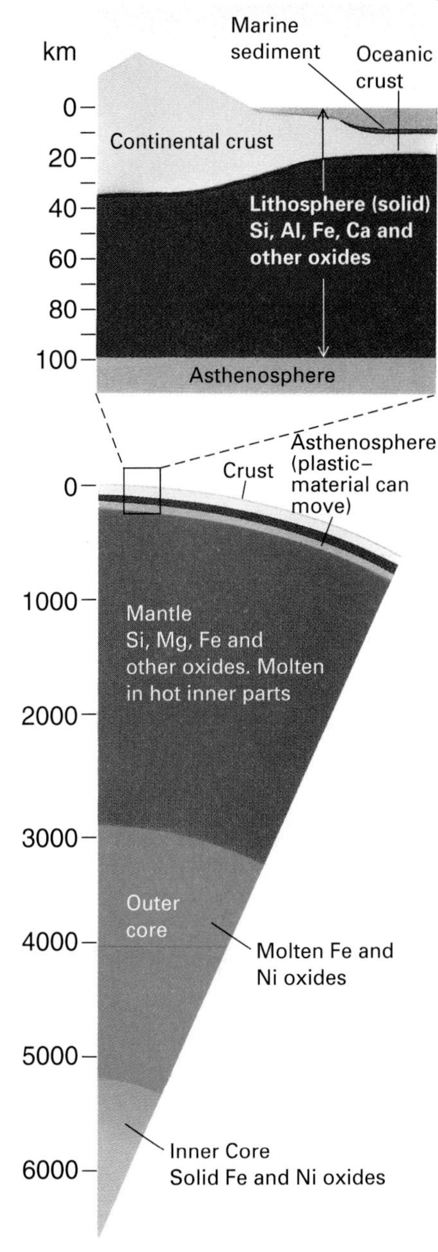

Figure 14.1 A cross section of the earth. Geologists customarily list the composition of the earth in terms of oxides, as shown here.

TABLE 14.1 ■ Major Mineral Groups in the Earth's Crust

Mineral Group	Example	Formula	Uses
Silicates	Quartz	SiO_2	Glass, ceramics, alloys
	Feldspar	$KAlSi_3O_8$	Ceramics
Oxides	Hematite	Fe_2O_3	Iron ore, paint pigment
Carbonates	Calcite	$CaCO_3$	Optical instruments (pure crystals), industrial chemicals
Sulfides	Galena	PbS	Lead ore, semiconductors
Sulfates	Gypsum	$CaSO_4 \cdot 2\,H_2O$	Cement, plaster of paris, wallboard, paper sizing
Halides	Fluorite	CaF_2	Lasers and electronics (pure crystals), source of fluorine (F_2)

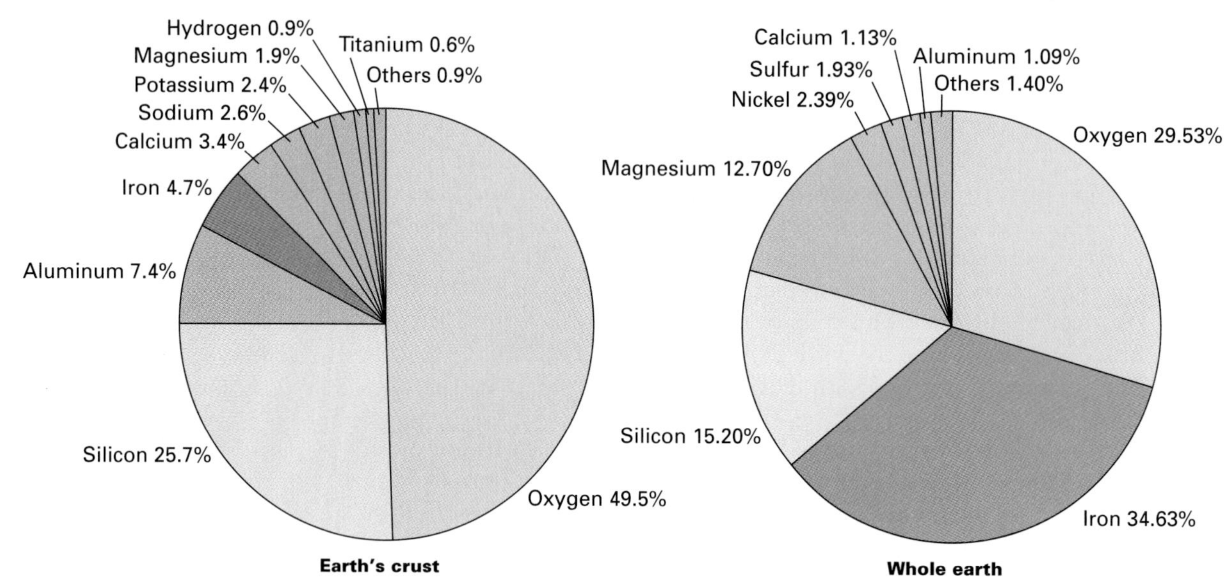

Figure 14.2 Relative abundance (by mass) of elements in the earth's crust compared to abundance in the whole earth.

■ **Minerals** are naturally occurring, solid inorganic compounds; they have definite internal structure and chemical composition. Most of the earth's crust consists of *rocks*, which may be single minerals or mixtures of minerals. A natural material that contains a sufficient concentration of an element or compound to economically justify mining it is an **ore**.

large proportion of its needs by mining. Other minerals are concentrated elsewhere, of course. Most of the known deposits of nickel are in New Caledonia and Zimbabwe. Most of the chromium is in Botswana and Turkey. Each country must rely on imports of one kind or another.

14.2 Chemicals from the Hydrosphere

■ Some other uses for sodium chloride include food processing, highway snow melting; animal feed; domestic table salt; and rubber, oil, paper, and textile manufacture.

Separation of salt from seawater by evaporation in San Francisco Bay. *(Christine L. Case/Visuals Unlimited)*

A single mouthful of seawater is enough to convince anyone that it is salty. Indeed, sodium chloride is the major mineral in seawater. But consider that other dissolved minerals are constantly being deposited into the oceans from rivers, undersea volcanoes, and thermal vents. In addition to sodium and chlorine, the major elements present in ions in solution are magnesium, sulfur, calcium, potassium, bromine, carbon, nitrogen, and strontium. Lower concentrations of other elements in seawater could also provide huge quantities of economically important metals such as uranium, copper, manganese, and gold.

The lower the concentration of a metal ion in seawater, of course, the higher the cost of isolating the metal is likely to be. Nevertheless, as high-quality mineral deposits on the land are depleted, the economics of mining the sea might become more attractive. Marine organisms may help to solve the problem. For example, one family of such organisms (the *tunicates*) accumulates vanadium to more than 280,000 times its concentration in seawater. Perhaps aquatic farming of such creatures could be put to work extracting metals.

Salt and the Chloralkali Industry

The majority of our salt is obtained from natural waters. In coastal regions salt is separated from seawater by evaporation of the water in large lagoons open to the sun. The natural **brines**, or salty waters, found in wells and lakes such as the Great Salt Lake in Utah are other sources of salt. After isolation, the salt is purified to the degree required for its use. Much of it is destined for the *chloralkali industry*, which produces chlorine and sodium hydroxide.

■ Salt has been important throughout history. The Romans gave a *salarium* to those who were "worth their salt"—the origin of the word "salary."

Chlorine is used to disinfect drinking water and sewage, and in the production of organic chemicals such as pesticides and vinyl chloride, the building block of plastics called poly(vinyl chlorides) (PVCs, Section 10.5). In 1996, chlorine was fifth on the list of chemicals produced in the United States. Almost all chlorine gas is made by electrolysis of aqueous sodium chloride. The other product of sodium chloride electrolysis, sodium hydroxide, is equally valuable because it is the most commonly used base in industrial processes. The reaction in electrolysis of aqueous NaCl is

■ For many years, cells with mercury cathodes were used in the chloralkali industry. Mercury is not very reactive or soluble and was thought to be harmless in the environment. Then some individuals who ate fish from mercury-contaminated waters became seriously ill. This event brought to light the fact that aquatic microorganisms convert metallic mercury to a toxic, water-soluble compound that enters the food chain (Section 15.6).

$$2\,NaCl(aq) + 2\,H_2O(\ell) \xrightarrow{\text{Electrical energy}} 2\,NaOH(aq) + H_2(g) + Cl_2(g)$$

A complicating factor in designing electrochemical cells for this reaction is that the chlorine and sodium hydroxide, if they remain in contact, react with each other.

A modern electrochemical cell for producing chlorine and sodium hydroxide is illustrated in Figure 14.3. The two electrode compartments are separated by a synthetic membrane that allows only sodium ions to pass through it. The reactions are

■ Electrochemical cells and electrolysis are described in Section 8.5.

$$\begin{array}{rll} 2\,Cl^-(aq) & \longrightarrow Cl_2(g) + 2\,e^- & \text{Oxidation} \\ 2\,H_2O(\ell) + 2e^- & \longrightarrow H_2(g) + 2\,OH^-(aq) & \text{Reduction} \\ \hline 2\,Cl^-(aq) + 2\,H_2O(\ell) & \longrightarrow H_2(g) + Cl_2(g) + 2\,OH^-(aq) & \text{Overall reaction} \end{array}$$

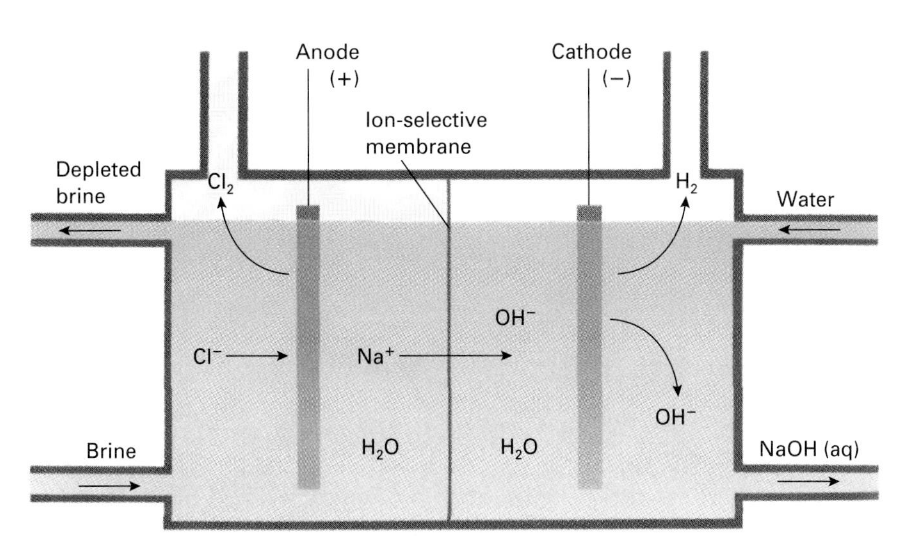

Figure 14.3 A chloralkali cell. Large banks of these cells produce gaseous chlorine and aqueous sodium hydroxide solution, both important industrial chemicals. Because of the need for cheap electricity, chloralkali plants are located near hydroelectric plants, for example, at Niagara Falls.

Brine is introduced into the anode compartment, and chloride ion oxidation occurs there. To maintain charge balance within the cell, as Cl^- ions are oxidized, Na^+ ions must pass from the anode to the cathode compartment. Since OH^- ions are produced in the cathode compartment, the product there is aqueous NaOH, with a concentration of 20% to 35% by weight. Most sodium hydroxide is used in industrial processes, although it is also present in some oven, drain, and sewer-pipe cleaners.

Magnesium from the Sea

■ An alloy is a mixture of two or more metals. Some alloys are homogeneous mixtures and others are heterogenous.

Magnesium, with a density of 1.74 g/cm^3, is the lightest structural metal in common use. Many **alloys** designed for light weight and great strength contain magnesium. Most manufactured aluminum objects, for example, contain about 5% magnesium, which is added to improve the mechanical properties and corrosion resistance of the aluminum under alkaline conditions. There are also alloys that have the reverse formulation, that is, more magnesium than aluminum. These alloys are used where a high strength-to-weight ratio is needed and where corrosion resistance is especially important. In the early 1990s, American automobiles contained about 4% magnesium by weight. This percentage is expected to increase as manufacturers produce lighter weight cars to meet new federal fuel–economy standards.

Because there are 6 million tons of magnesium present as Mg^{2+} in every cubic mile of seawater, the sea can furnish an almost limitless amount of this element. The recovery of magnesium from seawater (Figure 14.4) provides

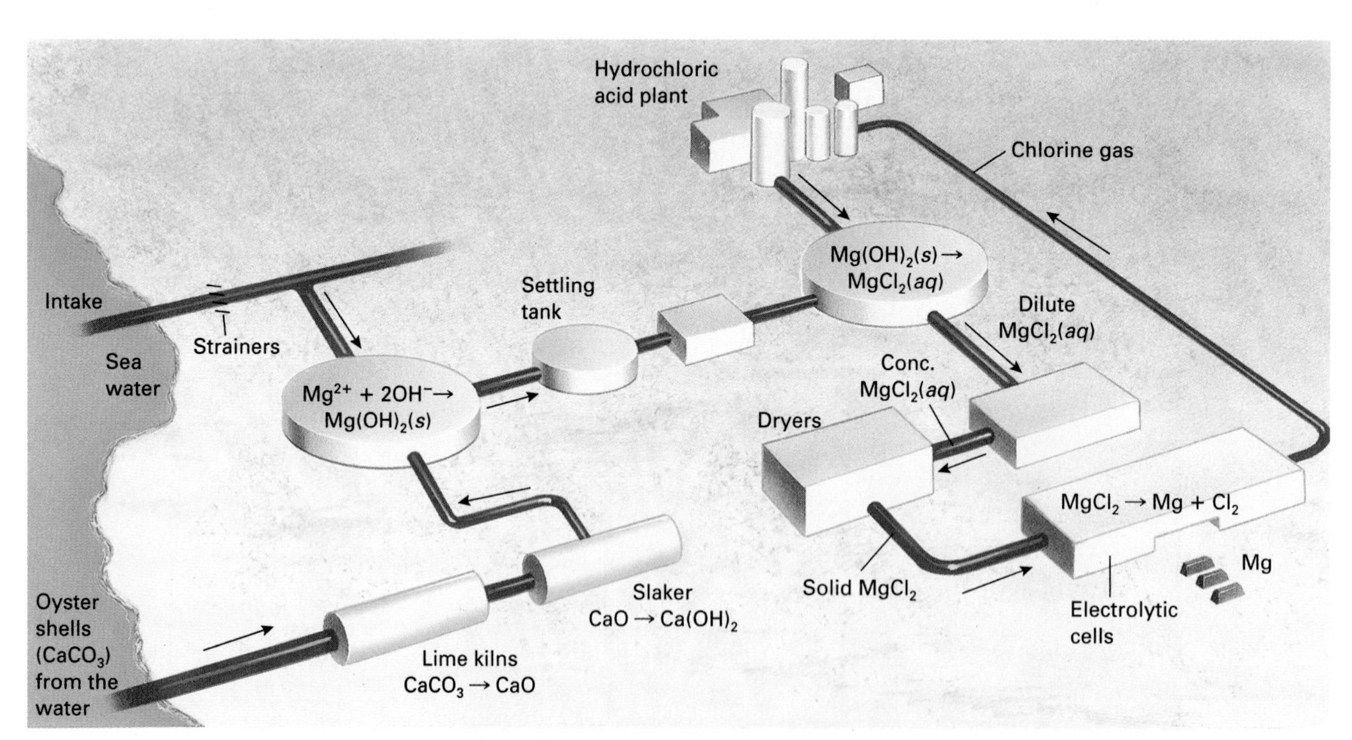

Figure 14.4 Recovery of magnesium from seawater.

an excellent illustration of the stepwise production of a chemical. The process begins with the **precipitation** of the insoluble magnesium hydroxide ($Mg(OH)_2$). The only thing needed for this step is a ready supply of an inexpensive base (a source of OH^- ions), a need fulfilled nicely by seashells, which contain calcium carbonate ($CaCO_3$). Heating the calcium carbonate converts it to lime, which then reacts with water to give calcium hydroxide, the base used in the precipitation.

■ When the product of a reaction between ions dissolved in water is not soluble, the reaction is called **precipitation**. The insoluble reaction product *precipitates* out of the solution.

$$\underset{\text{Seashells}}{CaCO_3(s)} \xrightarrow{\text{Heat}} \underset{\text{Lime}}{CaO(s)} + CO_2(g)$$

$$CaO(s) + H_2O(\ell) \longrightarrow Ca(OH)_2(aq)$$

$$Mg^{2+}(aq) + Ca(OH)_2(aq) \longrightarrow Mg(OH)_2(s) + Ca^{2+}(aq)$$

The solid magnesium hydroxide is isolated by filtration and then neutralized by another inexpensive chemical, hydrochloric acid.

$$Mg(OH)_2(s) + 2\ HCl(aq) \longrightarrow MgCl_2(aq) + 2\ H_2O(\ell)$$

When the water is evaporated, solid hydrated magnesium chloride is left. After drying, it is melted (at 708°C) and then electrolyzed in a huge steel pot that serves as the cathode. Graphite bars serve as the anodes. The electrode reactions are

■ Notice how many kinds of chemical reactions are organized into the production of magnesium (Figure 14.4)—decomposition by heat, neutralization, precipitation, and oxidation–reduction by using electrical current (electrolysis).

$$2\ Cl^- \longrightarrow Cl_2(g) + 2\ e^- \qquad \text{Oxidation}$$

$$Mg^{2+} + 2\ e^- \longrightarrow Mg(\ell) \qquad \text{Reduction}$$

$$Mg^{2+} + 2\ Cl^- \longrightarrow Mg(\ell) + Cl_2(g) \qquad \text{Overall reaction}$$

As the molten magnesium forms, it is removed. The chlorine produced in the electrolysis is recycled into the process by reacting it with hydrogen to produce hydrochloric acid.

■ The electrolysis of aluminum ions is discussed in Section 8.5.

■ SELF-TEST 14A

1. The most abundant element in the earth's crust is ____________.
2. The most abundant metal in the earth's crust is ____________.
3. Identify the source of each of the following substances as (i) the hydrosphere, or (ii) the earth's crust:
   (a) Chlorine (b) Sodium chloride
   (c) Magnesium (d) Sand
   (e) Marble
4. Which metal is sufficiently concentrated in the oceans so that it is currently extracted commercially: aluminum, copper, magnesium, or iron?
5. Magnesium ions are (reduced/oxidized) to produce magnesium metal.

14.3 Metals and Their Ores

According to the U.S. Bureau of Mines, the average American in a lifetime will require the quantities of minerals listed in Table 14.2 (p. 370). In this

TABLE 14.2 ■ Average Lifetime (73 Years) Supply of New Raw Materials from the Earth's Crust for an American

Element or Mineral	Quantity (pounds)	Major Uses
Stone, sand, gravel	1.3 million	Roads and buildings
Coal	500,000	Generating electricity; iron, steel, and chemicals manufacture
Iron and steel	91,000	Automobiles and ships, structural support
Clays	27,000	Bricks, paper, paint, glass, pottery
Salt	26,000	De-icing, detergents, cooking, chemical manufacturing
Aluminum	3,200	Food and beverage cans, household items, vehicles
Copper	1,500	Electrical motors, wiring
Zinc	840	Brass, galvanized iron, steel
Lead	800	Auto batteries, solder, electronics parts

section, we focus on the metals. Some of the chemistry of familiar silicon-based materials—glass, ceramics, cement—is discussed in Section 14.5.

Less active metals such as copper, silver, and gold can be found as free elements. The somewhat more reactive metals are present as sulfides formed early in the earth's existence (e.g., CuS, PbS, and ZnS). Because of their extremely low water solubility, sulfides resist oxidation and reactions with water and other ions. The still more reactive metals have been converted over millennia into oxides (e.g., MnO_2, Al_2O_3, and TiO_2) and are mined in that form.

The most reactive metals, such as sodium and potassium, are present in nature as soluble salts in the ocean and mineral springs; in solid deposits of these salts; or in insoluble, stable aluminosilicates, such as albite ($NaAlSi_3O_8$) and orthoclase ($KAlSi_3O_8$). Such silicates are found in all parts of the world, but because of their great stability they are not currently used as sources of the metals they contain. These minerals, like the ocean, represent a resource that may have to be tapped when richer and more easily processed ores are depleted.

A selection of beautiful minerals that are also ores is pictured in Figure 14.5. (A mineral is an ore if separating the metal from it is possible and economically practical.) The preparation of metals from minerals requires chemical reduction—the conversion of positive metal ions to free metals. To reduce a metal ion requires a source of electrons, which can be either an electrical current (as in the case of magnesium production mentioned earlier) or a chemical reducing agent.

Iron

Iron is the fourth most abundant element in the earth's crust and the second most abundant metal. Our economy depends on iron and its alloys, particularly steel. Most of the world's iron is located in large deposits of iron oxides

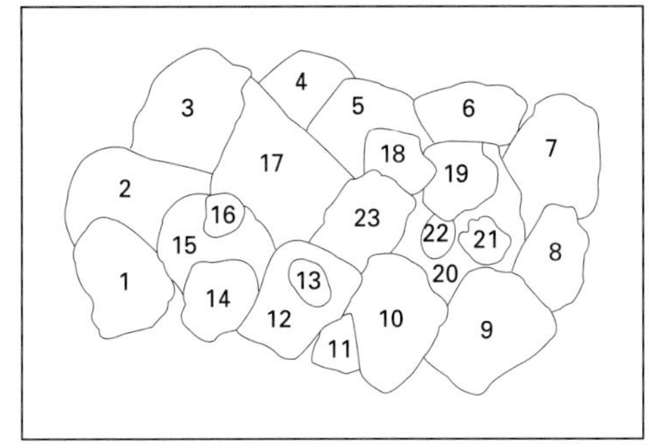

Figure 14.5 A collection of native metals and minerals. The minerals serve as ores for the metals indicated in the following key. *(© 1988 Paul Silverman/Fundamental Photographs)*

1. Bornite (iridescent)—COPPER
2. Dolomite (pink)—MAGNESIUM
3. Molybdenite (gray)—MOLYBDENUM
4. Skutterudite (gray)—COBALT, NICKEL
5. Zincite (mottled red)—ZINC
6. Chromite (gray)—CHROMIUM
7. Stibnite (top right, gray)—ANTIMONY
8. Gummite (yellow)—URANIUM
9. Cassiterite (rust, bottom right)—TIN
10. Vanadinite crystal on goethite (red crystal)—VANADIUM
11. Cinnabar (red)—MERCURY
12. Galena (gray)—LEAD
13. Monazite (white)—RARE EARTHS: cesium, lanthium, neodymium, thorium
14. Bauxite (gold)—ALUMINUM
15. Strontianite (white, spiny)—STRONTIUM
16. Cobaltite (gray cube)—COBALT
17. Pyrite (gold)—IRON
18. Columbinite (tan, gray stripe)—NIOBIUM, TANTALUM
19. Bismuth (shiny)
20. Rhodochrosite (pink)—MAGNESIUM
21. Rutile (shiny twin crystal)—TITANIUM
22. NATIVE SILVER filigree on quartz
23. Pyrolusite (black, powdery)—MANGANESE

in Minnesota, Sweden, France, Venezuela, Russia, Australia, and the United Kingdom.

Iron ore is reduced in a blast furnace (Figure 14.6*a*). The solid material fed into the top of the furnace is a mixture of iron oxide (Fe_2O_3), coke (C), and limestone ($CaCO_3$). A blast of heated air is forced into the furnace near the bottom. The major reactions that occur within the blast furnace result in reduction of iron oxide

$$2\,C(s) + O_2(g) \longrightarrow 2\,CO(g) + \text{heat}$$

$$Fe_2O_3(s) + 3\,CO(g) \longrightarrow 2\,Fe(s) + 3\,CO_2(g) + \text{heat}$$

and conversion of silica present in the ore to molten calcium silicate ($CaSiO_3$), known as **slag**.

$$CaCO_3(s) \xrightarrow{\text{Heat}} CaO(s) + CO_2(g)$$

$$CaO(s) + SiO_2(s) \xrightarrow{\text{Heat}} \underset{\text{Slag}}{CaSiO_3(\ell)}$$

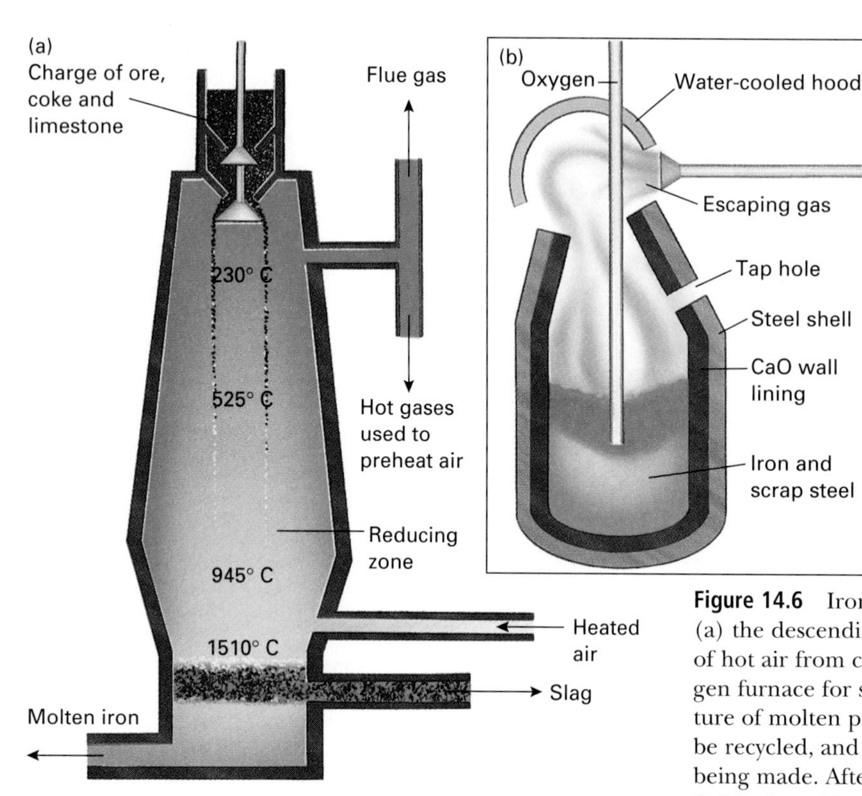

Figure 14.6 Iron and steel production. In the blast furnace (a) the descending materials in the charge are hit by a blast of hot air from coke burning in the heated air. The basic oxygen furnace for steel production (b) is charged with a mixture of molten pig iron from the blast furnace, steel scrap to be recycled, and other metals according to the type of steel being made. After oxygen is blown in for about 20 min, the finished steel is poured off through the tap hole, and the furnace is ready for another charge. The photo (c) shows molten iron being poured into a basic oxygen furnace. *(c, Bethlehem Steel)*

■ The mixture of nonmetallic waste and by-products from the refining of any metal is known as a **slag**.

■ About 75% of the weight of an automobile is iron and steel.

The World of Chemistry

Program 19, *Metals*

We make basic alloys, probably three or four dozen. But there are variations in which we put in a little pinch of this, a little pinch of that. We can make two—three—hundred varieties. I'm still accused of being a witch doctor, so to speak, with the little pinches of this and the little pinches of that. But we do know what we're doing. Gerald Gelazela, Senior Metallurgist, Armco Steel

Consequently, as the blast furnace operates, two molten layers collect in the bottom. The lower, denser layer is mostly liquid iron. The upper, less dense layer is the slag. From time to time the furnace is tapped at the bottom, and the molten iron is drawn off. Another outlet somewhat higher in the blast furnace can be opened to remove the liquid slag.

The iron that comes from the blast furnace, known as ***pig iron***, contains many impurities (up to 4.5% carbon, 1.7% manganese, 0.3% phosphorus, 0.04% sulfur, and 1% silicon). Iron reacts with the carbon impurity at the temperatures of the blast furnace to form cementite, an iron carbide (Fe_3C), which causes pig iron to be brittle.

$$3\,Fe(s) + C(s) \longrightarrow Fe_3C(s)$$

When molten pig iron is poured into molds of a desired shape (engine blocks, brake drums, transmission housings), it is called **cast iron**. However, pig iron and cast iron contain too much carbon and other impurities for most uses. The structurally stronger material known as **steel** is obtained by removing the phosphorus, sulfur, and silicon impurities and decreasing the carbon content.

THE WORLD OF CHEMISTRY

Materials from the Earth

Program 18, *The Chemistry of Earth*

In the South African bush veldt, an area the size of New England, interesting and valuable natural chemical separations have been made. Just beneath the surface lie some of the world's richest deposits of platinum, rhodium, and chromium. A mineral is a naturally occurring substance with a characteristic chemical composition. When minerals are concentrated and have economic value, they are called ores. Some places are rich with valuable ores, like South Africa, while others are not; and the fortunes of whole countries can rise and fall based on the wealth found in the natural treasuries.

Since prehistoric times, we have recovered and used a variety of minerals. How did our ancestors obtain the minerals and elements they needed from the earth? One of the techniques used in an ancient tin mine in Cornwall, England, was to build large fires at the base of the rock cliffs. The intense heat cracked the rocks, exposing the tin ore, which was then removed. Today we use dynamite and ammonium nitrate to blast rock away from ore-rich veins. But although we have greatly increased our power to remove useful ores from the earth, we are still limited by effective and economical mining techniques and, to an ever-increasing degree, the availability of exploitable ore deposits.

Ancient tin mines at Cornwall, England.

Steels

Many kinds of iron alloys are collectively known as *steels.* The most common is *carbon steel,* an alloy of iron with about 1.3% carbon. To convert pig iron into carbon steel, the excess carbon is burned out with oxygen.

In the *basic oxygen process* for steel production (Figure 14.6*b*), pure oxygen is blown into molten iron through a refractory tube, which is pushed below the surface of the iron. A *refractory* is a material that withstands high temperatures without melting. At elevated temperatures, the dissolved carbon reacts very rapidly with the oxygen to give gaseous carbon monoxide and carbon dioxide, which escape.

During steelmaking, silicon or transition metals such as chromium, manganese, and nickel can be added to give alloys with specific physical, chemical, and mechanical properties.

The properties of steel can also be adjusted by the temperature and rate of cooling used in its production. If the steel is cooled rapidly by quenching in water or oil, the carbon in the steel remains in the form of cementite (Fe_3C) and the steel will be hard, brittle, and light colored. Slow cooling favors the formation of crystals of carbon (graphite) instead of cementite. The resulting steel is more ductile (easily drawn into shape).

All the processes in steelmaking, from the blast furnace to the final heat treatment, use tremendous quantities of energy, mostly in the form of heat.

Steel structures in Chicago. Alexander Calder's steel sculpture, Flamingo, stands in a plaza surrounded by steel-frame office buildings. *(T. J. Florian/Rainbow)*

■ In carbon steel, the carbon atoms fit in the small spaces (interstices) between the larger iron atoms of the metal. The small structural change of even one part per thousand carbon makes slipping of the iron atoms more difficult. That's why steel is stronger than iron.

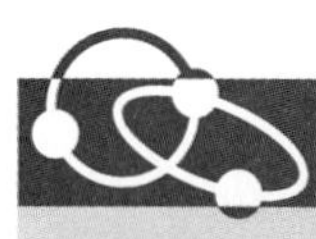

FRONTIERS IN THE WORLD OF CHEMISTRY

Swords of Damascus Leave Legacy of Superplastic Steel

About 3000 years ago, a type of steel called wootz steel was produced in India by heating a mixture of pure iron ore and wood in a sealed pot or crucible. Some of the carbon reduced the iron ore to metallic iron, which then absorbed some of the remaining carbon to form an excellent steel. The wootz steel later became famous as Damascus steel, used for making swords that retained their sharpness and strength after countless battles. The blacksmiths who forged this steel into swords used a process that they carefully kept secret—one that produced a steel that was much more pliable at high temperatures than normal steel. The knowledge of how to make this special steel was lost sometime in the 19th century.

Professor Oleg Sherby of Stanford University and specialists at the Lawrence Livermore National Laboratory have developed types of steels with properties similar to those of Damascus steel. The new steels have a higher percentage of carbon (about 1.8%) than normal carbon steels and are referred to as ultrahigh carbon steel alloys or superplastic steels.

Unlike most steels, which fail after being stretched to 2 times their original size, superplastic steel at elevated temperatures can be stretched to 11 times its size without cracking or pulling apart. As a result, heated superplastic steel can be formed into complex shapes. It can even flow like molasses and be poured into a mold. This property eliminates the need for machining, which typically results in about 50% scrap. Superplastic steel is also similar to stainless steel in its resistance to corrosion, but is made with less scarce and expensive materials than the nickel and chromium in stainless steel.

The superior properties of superplastic steel are attributed to a much finer grain structure than that in ordinary carbon steels, which are quite brittle. A group of industrial firms has banded together to develop the potential of using superplastic steel in manufactured objects. The first exploitation of the lower manufacturing cost and greater toughness of this material is expected to be in bulldozers, gears, and other complex parts subjected to high stress.

Stanford University Professor Oleg D. Sherby holds a 300-year-old sword of Damascus steel. (For further information, see *Chemecology*, p. 6 March 1992.) (Sam Forencich)

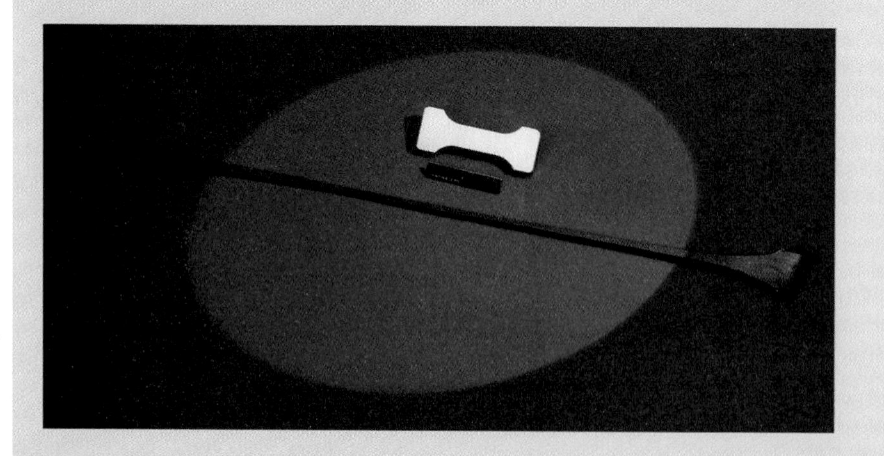

Superplastic, ultrahigh carbon steel can be stretched to 11 times its size at 900°C. The original size of the bottom steel piece was 1 inch, but it was pulled to a length of 14 inches. Conventional steels collapse when they are extended to two times their size. (James Stoots/Livermore Laboratory)

In the production of a ton of steel, approximately 1 ton of coal or its energy equivalent is consumed.

Copper

Although copper metal occurs in the free state in some parts of the world, the supply available from such sources is quite insufficient for the world's needs. The majority of the copper used today is obtained from various copper sulfide ores, such as chalcopyrite ($CuFeS_2$), chalcocite (Cu_2S), and covellite (CuS). Because the copper content of these ores is about 1% to 2%, the powdered ore is first concentrated by the flotation process.

In the flotation process, the powdered ore is mixed with water and a frothing agent such as pine oil. A stream of air is blown through the mixture to produce froth. The gangue in the ore, which is composed of sand, rock, and clay, is easily wetted by the water and sinks to the bottom of the container. In contrast, a copper sulfide particle is hydrophobic—it is not wetted by the water. The copper sulfide particle becomes coated with oil and is carried to the top of the container in the froth. The froth is removed continuously, and the floating copper sulfide minerals are recovered from it.

■ **Gangue** is the unwanted substances mixed with the desired mineral.

The preparation of copper metal from copper sulfide ore involves **roasting** the ore in air to convert some of the copper sulfide and any iron sulfide present to the oxides.

■ **Roasting** is a common step in recovery of metals from their ores. It consists of heating the ore, often a metal sulfide, in air to convert the sulfide to the oxide. If the metal oxide product from roasting is less stable than SO_2, the free metal is obtained by sulfide roasting. This is the case for Cu_2S, where further heating of the mixture of Cu_2S and Cu_2O produces Cu and SO_2. Unfortunately, if the SO_2 is vented to the atmosphere, it contributes to air pollution and is one source of "acid rain" (Section 16.9).

$$2\,Cu_2S(s) + 3\,O_2(g) \longrightarrow 2\,Cu_2O(s) + 2\,SO_2(g)$$

$$2\,FeS(s) + 3\,O_2(g) \longrightarrow 2\,FeO(s) + 2\,SO_2(g)$$

Subsequently the mixture is heated to a higher temperature, and some copper is produced by the reaction

$$Cu_2S(s) + 2\,Cu_2O(s) \xrightarrow{\text{Heat}} 6\,Cu(s) + SO_2(g)$$

The product of this operation is a mixture of copper metal, sulfides of copper, iron, and other ore constituents, and slag. The molten mixture is heated in a converter with silica materials. When air is blown through the molten material in the converter, two reaction sequences occur. In one, the iron is converted to a slag.

$$2\,FeS(s) + 3\,O_2(g) \longrightarrow 2\,FeO(s) + 2\,SO_2(g)$$

$$FeO(s) + SiO_2 \longrightarrow \underset{\text{Molten slag}}{FeSiO_3(s)}$$

In the other, the remaining copper sulfide is converted to copper metal by the preceding two reactions shown for Cu_2S. The copper produced in this manner is crude or "blister" copper, the blistered surface resulting from the escaping gas. The blister copper is later purified electrolytically.

Open-pit copper mining near Bagdad, Arizona. *(James Cowlin/Image Enterprises)*

In the electrolytic purification of copper, the anodes are crude copper and the cathodes are made of pure copper. As electrolysis proceeds, copper is oxidized at the anode, moves through the solution as Cu^{2+} ions, and is deposited on the cathode. The voltage of the cell is regulated so that more active impurities (such as iron) are left in the solution, and less active ones are not oxidized at all. The less active impurities include gold and silver, which collect as "anode slime," an insoluble residue beneath the anode. The anode slime is subsequently treated to recover the valuable metals.

The copper produced by the electrolytic cell is 99.95% pure and is suitable for use as an electrical conductor. Copper for this purpose must be pure because very small amounts of impurities, such as arsenic, considerably reduce the electrical conductivity of copper.

SELF-TEST 14B

1. The metal consumed in largest quantity during a person's lifetime is ____________.
2. The principal impurity in pig iron is ____________.
3. The reducing agent in the production of iron from iron ore is ____________.
4. Pig iron is brittle. (a) True, (b) False.
5. Copper usually occurs in its ores combined with the element ____________.
6. Copper can be purified using electrolysis. (a) True, (b) False.

14.4 Conductors, Semiconductors, and Superconductors

■ Section 6.6 discusses the recycling of metals.

Metals have some properties totally unlike those of other substances. Except for mercury, which is a liquid at room temperature, and gallium, which is a liquid at slightly above room temperature, all metals are solids. Some remain solids even at very high temperatures. Tungsten has a melting point of 3410°C. There are several properties common to the metals:

- ***High electrical conductivity.*** Metal wires easily carry electrical currents. The electrons in metals are quite mobile.
- ***High thermal conductivity.*** Some metals are much better conductors of heat than others. Try a sterling silver spoon in a cup of hot coffee or tea. Compare its thermal, or heat, conductivity with that of a stainless steel spoon.
- ***Ductility and malleability.*** Most metals can be drawn into wire (ductility) or hammered into thin sheets (malleability); gold is the most malleable metal. Extremely thin sheets of gold are used for decoration.
- ***Luster.*** Polished metal surfaces reflect light; most metals have a silvery white color because they reflect all wavelengths of light equally well.
- ***Insolubility in water and other common solvents.*** No metal dissolves in water, but a few, such as the Group IA and IIA metals, react with water to form hydrogen gas and solutions of metal hydroxides.

■ Electroplating, Section 8.5, is used to deposit protective and decorative coating of metals on various objects.

Any theory of the bonding of metal atoms must be consistent with these properties. Structural investigations of metals have led to the conclusion that

(a)

(b)

(c)

Putting the properties of metals to use. (a) Installation of electrically conducting wires. (b) A flask coated, like a mirror, with silver. (c) Very thin and stable gold leaf being applied to the cupola of a church. *(a,b: C. D. Winters; c: Paul Silverman/Fundamental Photographs)*

solid metals are composed of regular arrays, or *lattices*, of metal ions in which the bonding electrons are loosely held. Figure 14.7 illustrates one model for metallic bonding in which the regular array, or lattice, of positively charged metal ions is embedded in a "sea" of mobile electrons. These mobile valence electrons are delocalized over the entire metal crystal, and the freedom of these electrons to move throughout the solid is responsible for the properties associated with metals. In contrast to metals, the valence electrons in nonmetals are fixed in bonds between like atoms. This means that nonmetals are nonconductors of electricity.

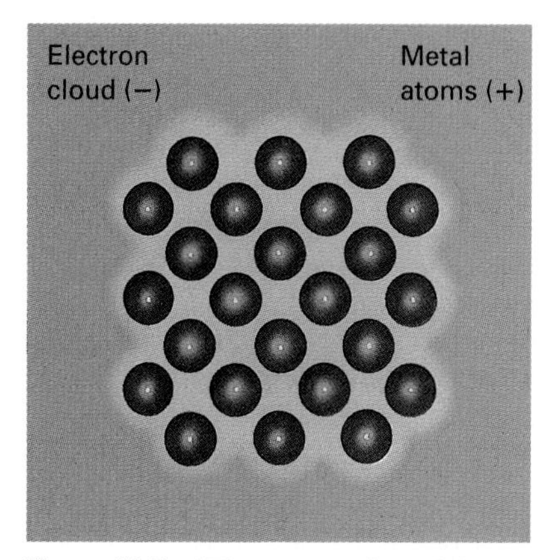

Figure 14.7 "Electron sea" model of bonding in metals. The positively charged metal atom nuclei are surrounded by a "sea" of negatively charged electrons.

Semiconductors: The Basis of Our Modern World

You should not be surprised that somewhere between the excellent electrical conductivity of most metals and the nonconductivity of nonmetals, there are some elements that are **semiconductors**. This means they will conduct electricity under certain conditions. Silicon, when it is in a highly purified state, is a semiconductor. Silicon acts like a nonmetal and fails to conduct a current until a certain voltage is applied; then it begins to conduct moderately well. This behavior interested electrical engineers, who recognized that silicon might act like a "gate" for electron flow in electrical circuits. This electron gatelike activity was not realized until a process known as **doping** was discovered. One common dopant is boron, a Group IIIA element just to the upper left of silicon in the periodic table. When boron is added to pure silicon, the boron atoms, with one fewer valence electron than silicon, introduce *positive holes* in the lattice arrangement of silicon atoms. The presence of these positive holes (which can move about in the solid just like electrons, but in the opposite direction) makes the doped silicon somewhat more conductive—in effect, it becomes a better electron gate. Silicon doped with an element that creates positive holes is called a *p-type* semiconductor. Another dopant is arsenic, from Group VA, which contains one *more* valence electron than silicon. Silicon doped with arsenic is called an *n-type* semiconductor because there are extra negative electrons present in the solid. These, of course, enhance

■ The elements shown in (blue) in the periodic table inside the front cover are referred to as the *metalloids* (Section 3.6).

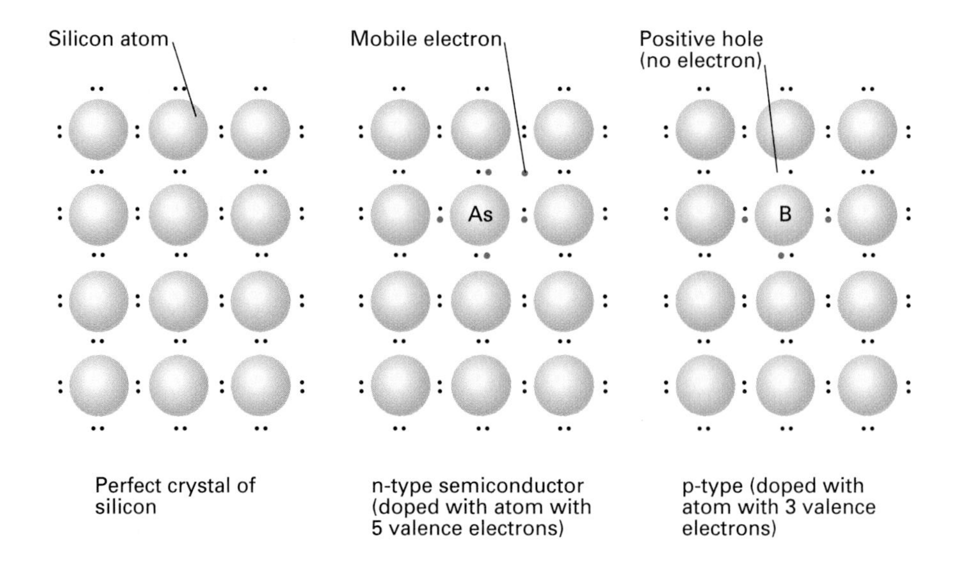

Figure 14.8 Doping of silicon in semiconducting devices. Adding atoms with five valence electrons (e.g., arsenic) introduces extra electrons that can move through the crystal. Adding atoms with three valence electrons introduces holes that can also move through the crystal.

the conductivity of the doped silicon. Through careful control of the amount and type of dopant, the conductivity of the silicon can be adjusted to a fine degree (Figure 14.8).

In 1947, a device consisting of a layer of p-type silicon sandwiched between two n-type layers was constructed by John Bardeen, Walter Brattain, and William Shockley at the Bell Laboratories. This device, called the **transistor**, has revolutionized our world (Figure 14.9). Because the transistor can control electron flow in circuits with such accuracy, yet is so small and requires so little power to operate, it is now possible to design electronic circuits to fit into extremely small volumes. Such objects as TV cameras as small as a pea, radios small enough to strap to the back of an ant, and other amazing devices can be made using transistors based on doped silicon. Of course, more mun-

■ Bardeen, Brattain, and Shockley shared the Nobel Prize in physics in 1956 for the discovery of the transistor. The importance of the transistor was recognized as soon as it was discovered. Although it was first demonstrated at the Bell Laboratories in December of 1947, it wasn't announced until July 1, 1948, after patent applications had been filed. John Bardeen was awarded a second Nobel Prize in physics in 1972, along with J. R. Schrieffer and Leon N. Cooper, for his work on the theory of superconductivity.

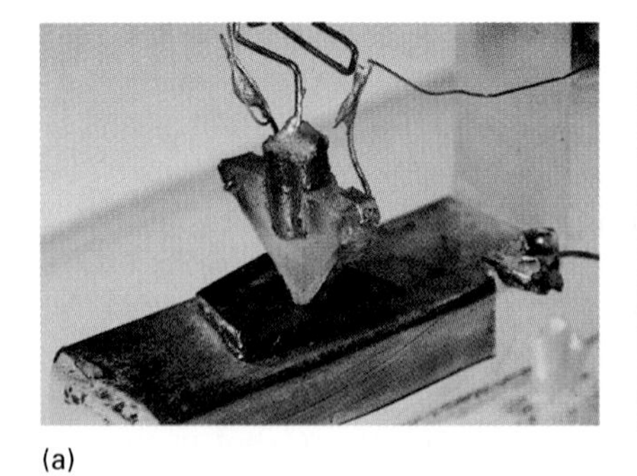

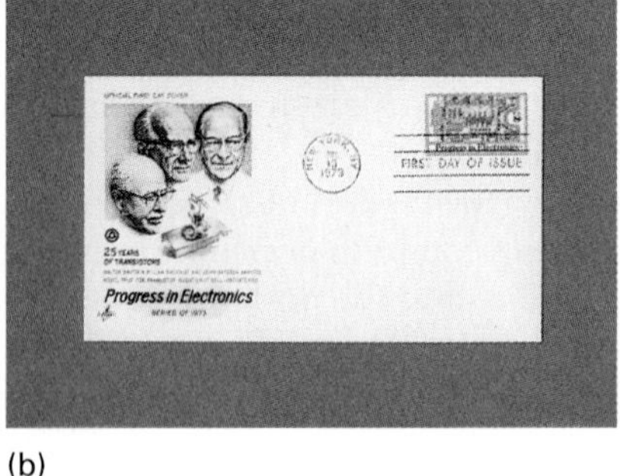

(a) (b)

Figure 14.9 The transistor and its inventors. (a) The first transistor, constructed in 1947 at Bell Laboratories. Electrical contact is made at a single point and the signal is amplified as it passes through a solid semiconductor; modern junction transistors amplify in a similar manner. (b) Envelope and stamp commemorating 25 years of the transistor, with portraits of its inventors Walter Brattain, William Shockley, and John Bardeen. *(Courtesy of AT&T Archives)*

dane items such as automatic cameras, microwave ovens, fax machines, and cellular phones also owe their existence to the transistor. The central processing unit (CPU) of computers consists of millions of transistors and other circuit elements fabricated on wafers of pure silicon (Figure 14.10). These devices are called integrated circuits.

Figure 14.10 A modern computer microprocessor—the Pentium Pro chip. *(Courtesy of Intel)*

Solar Energy to Electricity

The *solar cell* or photovoltaic cell converts energy from the sun into electron flow. Solar cells are now used in calculators, watches, spaceflight applications, communication satellites, and signals for automobiles and trains; and as the source of electricity in undeveloped regions throughout the world, where electrical power grids are virtually nonexistent. It has been estimated that all the electricity used in the United States could be made by solar cells having only 10% efficiency and covering about 13,000 km^2, which is 0.13% of the land area in the United States.

■ The United States land area is about 9×10^9 km^2.

One type of solar cell consists of two layers of almost pure silicon. The lower, thicker layer contains a trace of boron (B); and the upper, thinner layer has a trace of arsenic (As). As pointed out earlier, the As-enriched layer is an *n*-type semiconductor, with mobile electrons; and the B-enriched layer is a *p*-type semiconductor, with the electron deficiencies known as holes (see Figure 14.8). There is a strong tendency for the mobile electrons in the *n*-type layer to pair with the unpaired electrons in the holes in the *p*-type layer. If the two layers are connected by an external circuit (Figure 14.11) and light of sufficient energy strikes the surface and is absorbed, excited electrons can leave the *n*-type layer and flow through the external circuit to the *p*-type layer. As this layer becomes more negative because of added electrons, electrons

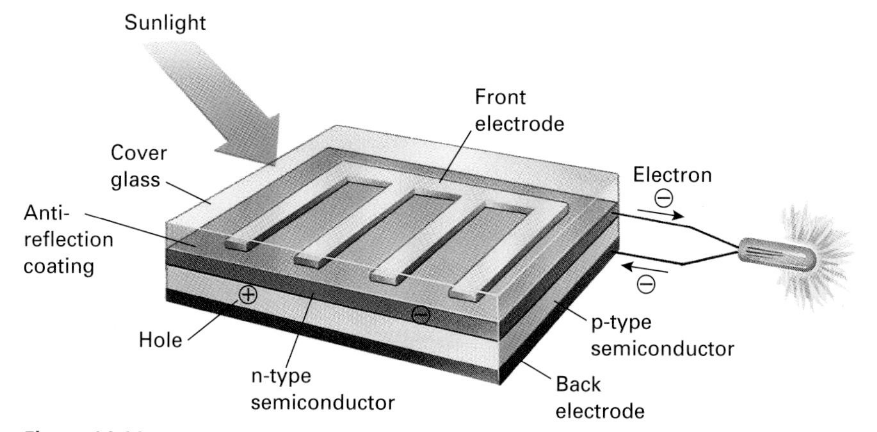

Figure 14.11 A solar cell. Beneath the outer glass is a metal grid that allows as much light as possible to strike the *n*-type semiconductor layer, while serving as the electrode at which electrons leave the cell. The *n*-type semiconductor layer is almost transparent. Beneath it is the *p*-type semiconductor layer and the electrode at which electrons re-enter the cell.

A refrigerator powered by the sun. Photovoltaic cells at the top of the refrigerator generate the electricity needed to keep vaccines cold as they are delivered to remote locations. *(Courtesy of Siemens Solar Industries, Camarillo, California 93011)*

are repelled internally back into the *n*-type layer, which is now positive and attracts the electrons. The process can continue indefinitely as long as the cell is exposed to sunlight.

Solar cells may be the next great technological breakthrough, perhaps comparable to the computer chip and genetic engineering. Experimental solar-powered automobiles are now available and many novel applications of solar cells already exist (such as powering an exhaust fan in an automobile to remove hot air accumulated in a car parked in the sun). The big breakthrough will be in the use of banks of solar cells at a utility power plant to produce huge amounts of electricity. One plant already in operation in California uses solar cells to produce 20 MW of electricity, which is enough to supply the electricity needs of a city of about 600,000 people.

Can Anything Conduct Better than a Metal?

When metals are heated, their electrical conductivity decreases. Lower conductivity at higher temperatures can be explained if the movement of the valence electrons is considered to be limited by rapidly vibrating atoms in the metal lattice. The kinetic molecular theory says that higher temperatures mean more motion, and this applies even when the atoms are in fixed positions, as they are in a metal. When the metal atoms are relatively stationary, as they are when the metal is cool, the electrons can move through the lattice much like a person moving through a room filled with a large number of other people quietly chatting with one another. When the temperature is elevated, the metal atoms begin to vibrate wildly, and the electrons have more trouble getting through the lattice, in much the same way that a person would have trouble moving through the room if all the other occupants suddenly became agitated.

From this picture of electrical conductivity of metals, you might assume that if a sufficiently low temperature were reached, conductivity might be quite high (almost zero resistance, in other words). In fact, the conductivity of a pure metal crystal does approach infinity (zero resistance) as absolute zero is approached. In many metals, however, a more interesting thing happens before absolute zero is reached. At a certain low temperature, the conductivity suddenly increases, as though absolute zero had already been reached. At this temperature, called the *superconducting transition temperature,* the metal becomes a **superconductor** of electricity. The superconductor offers no resistance whatever to electrical flow. This phenomenon means that electric motors made of superconducting wires would be 100% efficient, and electrical transmission lines could be made 100% efficient. Resistance to electron flow causes energy loss in motors, transmission lines, and other electrical devices. There is no good theory to explain the phenomenon of superconductivity, but the hindrance of electron flow by vibrating atoms in the metal lattice has been replaced by some kind of cooperative action that allows electron movement. Table 14.3 lists some of the metals that have superconducting transition temperatures. Not all metals display superconducting properties.

TABLE 14.3 ■ Superconducting Transition Temperatures of Some Metals

Metal	Transition Temperature (K)
Aluminum	1.183
Gallium	1.087
Lanthanum	4.8
Lead	7.23
Niobium	9.17

The relatively low transition temperatures of the metals shown in Table 14.3 mean that it would be impractical to make superconducting motors or transmission lines from them. Shortly after the superconductivity of metals was discovered, certain alloys (mixtures of metals) were prepared that had much higher transition temperatures than the metals themselves. Niobium alloys showed the most promise, but they still had to be cooled to below 23 K (−250°C) to exhibit superconductivity. To maintain such a low temperature would require liquid helium, which costs about $6 per liter—an expensive proposition for all but the most exotic applications.

■ The boiling point of helium is 4.35 K (−268.8°C).

In January 1986, K. Alex Müller and J. Georg Bednorz, scientists at an IBM laboratory in Switzerland, discovered that a barium–lanthanum–copper oxide became superconducting at 35 K. This discovery rocked the scientific world and provoked a flurry of activity that quickly resulted in a substance that became superconducting at 90 K. This compound, $LaBa_2Cu_3O_x$ (where x represents a varying number of oxygen atoms), can superconduct at temperatures above the boiling point of nitrogen (77 K). At less than 50 cents per liter, liquid nitrogen is a much cheaper refrigerant than liquid helium (Figure 14.12).

■ Müller and Bednorz shared the 1987 Nobel Prize in physics for their discovery of superconducting $LaBa_2Cu_3O_x$.

Why the great excitement over the potentialities of superconductivity? Superconducting materials are being used to build more powerful electromagnets, such as those used in nuclear particle accelerators (Section 4.6) and in magnetic resonance imaging (MRI) machines, which are used in medical diagnosis. One of the main factors affecting the efficiency of MRI machines is the heating of the electromagnet due to electrical resistance. Many scientists are saying that the discovery of high-temperature superconductors may prove to be more important than the discovery of the transistor because of its potential effect on electrical and electronic technology. For example, the use of superconducting materials for transmission of electric power could save as much as 30% of the energy now lost because of the resistance of the wire. Superchips for computers could be up to 1000 times faster than existing conventional silicon chips. Electromagnets could be both more powerful and smaller, which could hasten the development of a practical nuclear fusion reactor. Since there is a magnetic field around the superconducting material, it is conceivable that cars and trains could be moved along magnetic tracks so that no parts touch, thereby eliminating friction.

Figure 14.12 Superconductor. A pellet of a superconducting alloy suspended in air above magnets.

FRONTIERS IN THE WORLD OF CHEMISTRY

Solar Electricity for Developing Regions

For about 2 billion people—one third of the world's population—electricity and the comfort, safety, and entertainment it provides are unknown in their daily lives. They live in countries or regions that are too poor and underdeveloped to provide readily available electricity on a daily basis. The challenge is to provide electricity to these people at a cost they and their governments can afford. Large fossil fuel and nuclear power plants are out of the question because the huge investment would leave the countries with massive debt. Building dams to use the energy of rivers flowing through their land has had undesirable effects by displacing large numbers of people and by flooding fertile land. Transmission and distribution costs of electricity can also be staggering to economies that are hard pressed just to provide basic health care, pure water, and education.

The answer to providing electricity in developing countries seems to be small, affordable solar collection panels connected to storage batteries. Such small solar-powered systems (SSPS) can capture solar energy in the daytime, convert it to electricity, and store it for use at night when the family members have come home from work. Such systems, costing between $350 and $750, are already being made available to about 100,000 families, farms, and small businesses in countries such as Sri Lanka, Kenya, the Dominican Republic, and Zimbabwe. The challenges in putting this technology to use are numerous, and many discoveries will have to be made about how solar cells and batteries work and about how to make basic appliances that use electricity more efficiently. In addition to needing more efficient solar collectors, developers need batteries that can take deeper charge and discharge cycles without wearing out. Current systems typically consist of about 36 solar cells, each with a diameter of 10 cm, charging a lead–acid battery like those used in automobiles. When the sunlight is at full intensity, solar units like these can produce enough electricity to operate three fluorescent lamps for 3 h in addition to 3 h of combined radio listening and television viewing. Electricity could also be used to purify water. Additional research in lighting is helping to better utilize the stored solar electricity. For example, high-efficiency fluorescent lamps use polished aluminum reflectors to lower power requirements. One 9-W fluorescent bulb of this type can produce as much light as a 50-W incandescent bulb.

Although there are many opportunities for chemistry and applied chemical technology in improving the lives of people in developing countries, putting the applications into practice will probably depend on economic factors. It may cost $1 trillion to provide simple solar electrical systems to the developing world's population.

A community in Brazil where solar-powered electricity has been installed. Lead–acid batteries store electricity for use after sunset. (National Renewable Energy Laboratory (NREL))

Reference N. Williams, K. Jacobson, and H. Burris: Sunshine for light in the night, *Nature*, Vol. 362, p. 691, April 22, 1993.

■ SELF-TEST 14C

1. Many solar cells contain two types of semiconductors, sandwiched together. These types are called ____________ and ____________ types.
2. A substance whose resistance falls to almost zero at a certain low temperature is called a ____________.

3. As the temperature of a metal increases, its resistance to electrical flow (increases/decreases) ____________ .
4. Arsenic, a Group 5A element, can be used to dope silicon, a Group 4A element. This doping will produce (*n*-type/*p*-type) doped silicon.
5. Boron, a Group 3A element, will produce (*n*-type/*p*-type) doped silicon.
6. Doped silicon is used to make (a) magnets, (b) transistors, (c) electrical transmission wires.
7. Which would offer more promise, a compound whose resistance dropped to zero at or just above the boiling point of nitrogen, or one whose resistance dropped to zero at or just above the boiling point of helium?

14.5 From Rocks to Glass, Ceramics, and Cement

The earth's crust is largely held together by chemical bonds between silicon and oxygen, in either pure **silica** (SiO_2) or *silicate* minerals in which silicon–oxygen anions are combined with metal cations.

The most common of the many crystalline forms of pure silica is quartz. It is a major component of granite and sandstone and also occurs as pure crystals. The basic structural unit of quartz and most silicates is the tetrahedron. In quartz, every silicon atom is bonded to four oxygen atoms, and every oxygen atom is bonded to two silicon atoms. The result is an infinite array of tetrahedra sharing corners.

Most silicates consist of networks of silicon–oxygen tetrahedra linked together in ways that range from chains, rings, and sheets to three-dimensional networks.

The simplest network silicates are the *pyroxenes*, which contain extended chains of linked SiO_4 tetrahedra (Figure 14.13). If two such chains are laid

■ The serpentine form of asbestos is composed of two-dimensional sheets curled over into fibrous tubes.

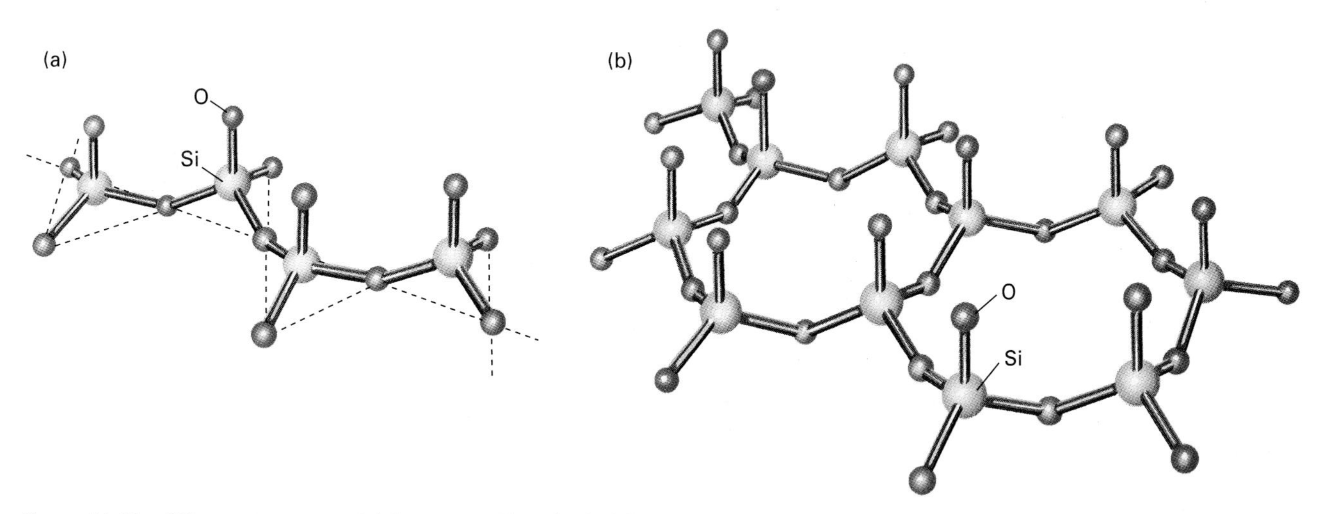

Figure 14.13 Silicate structures. (a) Pyroxene. Tetrahedral SiO_4 units are joined in chains by silicon–oxygen–silicon bonds. (b) Amphibole, which is asbestos. Chains of SiO_4 units are joined side-by-side by silicon–oxygen–silicon bonds.

TABLE 14.4 ■ Silicates

Si—O Tetrahedra Present as	Class Name	Structure
Individual anions	Orthosilicates	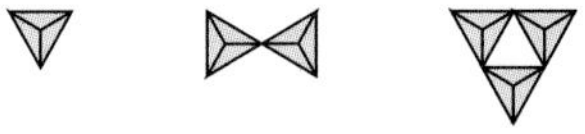
Chains	Pyroxenes (linear chains)	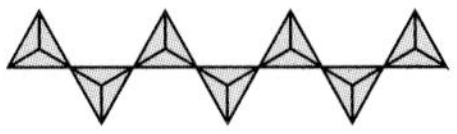
	Amphiboles (double chains)	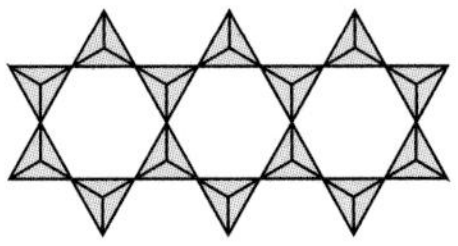
Sheets	Mica, talc, clays	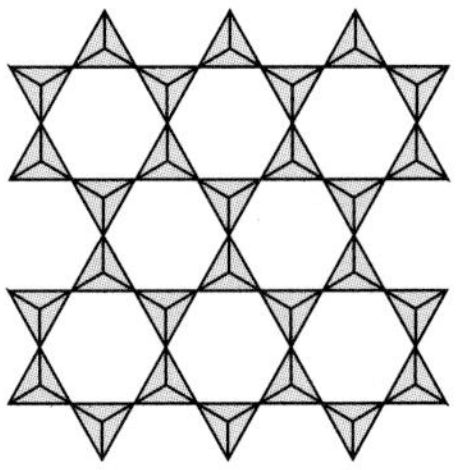
Three-dimensional networks	Silica	—
	Feldspars and zeolites	

side by side, they may link up by sharing oxygen atoms in adjoining chains. The result is an *amphibole.* Because of their double-stranded chain structure, the amphiboles are fibrous materials.

If the linking of silicate chains continues in two dimensions, sheets of SiO_4 tetrahedral units result (Table 14.4). Various clays and mica have this sheetlike structure. Clays, which are essential components of soils, are *aluminosilicates*—some Si^{4+} ions are replaced by Al^{3+} ions plus other cations that take up the additional positive charge. Feldspar, a component of many rocks and a network silicate, is weathered in the following reaction to form clay.

■ Feldspars make up 60% of the earth's crust.

$$\underset{\text{Feldspar}}{2\ KAlSi_3O_8(s)} + CO_2(g) + 2\ H_2O(\ell) \longrightarrow \underset{\text{Kaolinite (a clay)}}{Al_2(OH)_4Si_2O_5(s)} + 4\ SiO_2(s) + K_2CO_3(aq)$$

Some medications sold in the United States (e.g., Kaopectate) contain highly purified clay that absorbs excess stomach acid and possibly harmful bacteria that cause stomach upset.

Glass

When silica is melted, some of the bonds are broken and the tetrahedral SiO_4 units move with respect to each other. On cooling, reorganization into the same orderly arrangement in crystalline silica is hard to achieve because of the difficulty the groups experience in moving past one another. Instead, cooling produces a **glass**—a hard, noncrystalline transparent substance with an internal structure like that of a liquid. The random liquid-like molecular arrangement accounts for one of the typical properties of a glass: it breaks irregularly instead of splitting along a plane like a crystal.

■ Glass and other solid substances (e.g., tar) that lack the internal order and properties of crystals are known as **amorphous** substances.

Common window glass is made by melting a mixture of silica with sodium and calcium carbonates. Bubbles of carbon dioxide are evolved, and the cooled mixture is a glass composed of sodium and calcium silicates

■ The *n* in these formulas represents a large number and is necessary to show that glass contains SiO_3 groups linked together rather than individual SiO_3^{2-} ions.

$$n\ Na_2CO_3(\ell) + n\ SiO_2(s) \xrightarrow{\text{Heat}} Na_{2n}(SiO_3)_n + n\ CO_2(g)$$

$$n\ CaCO_3(\ell) + n\ SiO_2(s) \xrightarrow{\text{Heat}} Ca_n(SiO_3)_n + n\ CO_2(g)$$

White sand is the source of the silica for glassmaking. Even the best grade of sand contains a small proportion of iron(III) compounds that gives it a brown or yellow color. When this sand is made into glass, the iron is converted to a mixture of light green iron(II) silicates, explaining the green tint of some old bottles. Adding a manganese compound to the melt produces pink manganese silicates, which offset the green of the iron silicates, making the glass appear colorless.

Countless variations in glass composition and properties are possible. If part of the silica is replaced by boron oxide, the glass has less tendency to crack with changes in temperature. Pyrex, the trademarked glass common

(a)

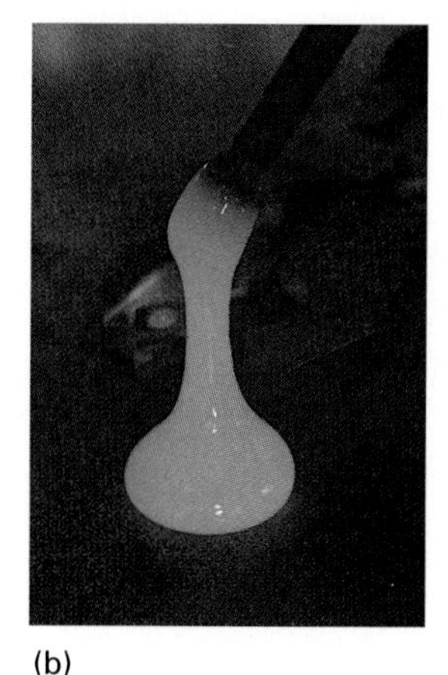

(b)

(c)

Glass. (a) Natural ingredients for making glass: sand (SiO_2), seashells for lime ($CaCO_3$), and seaweed for soda ash (Na_2CO_3). (b) Molten glass ready to be shaped. (c) A hand-blown goblet being turned in the annealing oven. *(a, © James L. Amos/Peter Arnold, Inc.; b and c, Tom Pantages)*

SCIENCE AND SOCIETY

Asbestos

The term "asbestos" is generally applied to two forms of fibrous natural silicates: the amphiboles and serpentine. The asbestos minerals do not burn, do not rot, and have low thermal and electrical conductivity. They can be woven into fabrics, compressed into mats, and mingled with such binders as rubber and asphalt to produce strong and dimensionally stable composites. These properties have led to widespread use of asbestos in fire-proofing materials, brake linings, floor tiles, pipes, and roofing materials.

The serpentine form of asbestos, known as *chrysotile*, is mined chiefly in Canada and the former Soviet Union; more than 90% of the asbestos used in the United States is in this form. The amphibole *crocidolite* is mined in small quantities, mainly in South Africa. The two minerals differ greatly in composition, color, shape, solubility, and persistence in human tissue. Crocidolite is blue, relatively insoluble, and persists in tissue. Its fibers are long, thin, and straight; and they penetrate narrow lung passages. In contrast, chrysotile is white, and it tends to be soluble and disappear in tissue. Its fibers are curly; they ball up like yarn and are more easily rejected by the body. Scientific studies of many types and by groups in many countries have shown that chrysotile asbestos is significantly less of a health hazard than other types. It is important to note that almost all manufactured materials in the United States contain only this form of asbestos.

Long-term occupational exposure to airborne asbestos fibers can, however, lead to a greater-than-average risk of lung cancer and certain other health problems, risks greatly enhanced by cigarette smoking. An understanding of this risk has fostered tight controls during the mining of asbestos and fabrication of asbestos-containing products.

The perceived risk also fostered an assortment of drives in the United States to ban all uses of asbestos and to remove all asbestos-containing products in existing public buildings and homes. One outcome of this activity came in 1989 when the U.S. Court of Appeals struck down the proposed U.S. Environmental Protection Agency (EPA) ban on asbestos, citing failure to provide sufficient evidence and failure to adequately consider alternative measures. EPA bypassed the opportunity to appeal this ruling. Another outcome was a congressionally mandated study of asbestos in buildings, which concluded in 1991 (after seven years of study) that asbestos-containing material in good repair inside buildings creates no higher asbestos fiber concentration in the air inside the buildings than in the air outside the buildings. Subsequently, an EPA advisory, while stressing assessment of asbestos risk in public buildings, stated:

> *Removal is often* not *a school district's or other buildings owner's best course of action to reduce asbestos exposure. In fact, an improper removal can create a dangerous situation where none previously existed.*

"An Advisory to the Public on Asbestos in Buildings," U.S. Environmental Protection Agency, 1991.

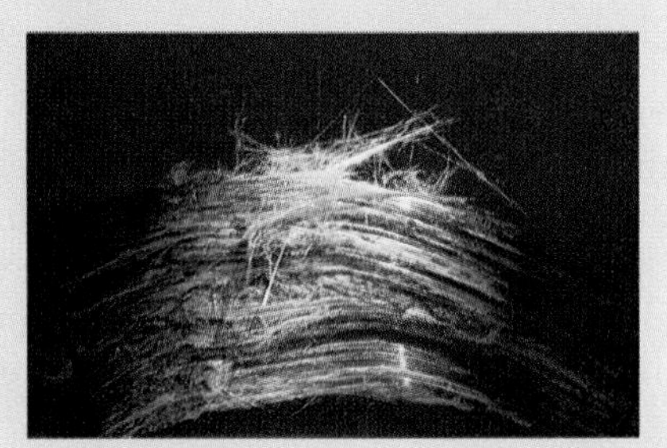

Chrysotile asbestos. (C. D. Winters)

Public fears about asbestos exposure were still in the news in 1993, when a school opening in New York City was delayed because of delays in removal. At that time, a group of 17 scientists and physicians from throughout the United States and the United Kingdom, all of them actively involved in asbestos study, stated:

> *Except under unusual conditions, such as demolition or extensive renovation, we strongly advise against asbestos removal in schools. It is scientifically unsound, economically wasteful, and medically imprudent . . . since it carries potentially unnecessary future risks.*

Letter to the Editor, *The New York Times*, December 23, 1993.

Armed with information of this kind, what would you do if your town proposed to spend $10 million of taxpayers money on asbestos removal in your local high school? What questions should be asked of the decision makers?

TABLE 14.5 ■ Substances Used to Color Glass

Substance	Color
Copper(I) oxide	Red, green, blue
Tin(IV) oxide	Opaque
Calcium fluoride	Milky white
Manganese(IV) oxide	Violet
Cobalt(II) oxide	Blue
Finely divided gold	Red, purple, blue
Uranium compounds	Yellow, green
Iron(II) compounds	Green
Iron(III) compounds	Yellow

in kitchens and laboratories, is a borosilicate glass. Many beautiful colors can be produced by adding the substances listed in Table 14.5. The composition and properties of some other types of glasses are listed in Table 14.6.

In the manufacture of glass, proper **annealing** is important. Annealing is the cooling schedule that a glass is put through on its way from a viscous, liquid state to a solid at room temperature. If a glass is cooled too quickly, bonding forces become uneven in local regions as small areas of crystallinity develop. This results in strain that will cause the glass to crack or shatter when subjected to mechanical shocks or sudden temperature changes. High-quality glass, such as that used in optics, must be annealed very carefully. The huge Mt. Palomar, California, observatory mirror was annealed from 500°C to 300°C over a period of nine months.

Ceramics

What do you know about "ceramics"? Perhaps you associate the term with pottery vases that you see at a crafts show, bathroom tile, or maybe components of electrical equipment. **Ceramics** are a large and diverse class of materials with the properties of nonmetals. What they have in common is that

TABLE 14.6 ■ Some Special Glasses

Special Addition or Composition	Desired Property
Large amounts of PbO with SiO_2 and Na_2CO_3	Brilliance, clarity, suitable for optical structures: crystal or flint glass
SiO_2, B_2O_3, and small amounts of Al_2O_3	Small coefficient of thermal expansion; borosilicate glass: Pyrex, Kimax, and others
One part SiO_2 and four parts PbO	Ability to stop (absorb) large amounts of X rays and gamma rays: lead glass
Large concentrations of CdO	Ability to absorb neutrons
Large concentrations of As_2O_3	Transparency to infrared radiation

A potter at his wheel, with already-fired pieces in the background. *(Blair Seitz/ Photo Researchers)*

all are made by baking or firing minerals or other substances, often including silicates and metal oxides.

Ceramic materials have been made since well before the dawn of recorded history. They are generally fashioned from clay or other natural earths at room temperature and then permanently hardened by heat. Silicate ceramics include objects made from clays, such as pottery, bricks, and table china. The three major ingredients of common pottery are clay (from weathering of feldspar as described previously), sand (silica), and feldspar (aluminosilicates). Clays mixed with water form a moldable paste because they consist of many tiny silicate sheets that can easily slide past one another. When the clay–water mixture is heated, the water is driven off, and new Si—O—Si bonds are formed so that the mass of platelets becomes permanently rigid.

A new type of glasslike ceramics with unusual properties has become widely available for home and industrial use. Ordinary glass breaks because once a crack starts, there is nothing to stop the crack from spreading. It was discovered that if glass objects produced in the usual manner are heated until many tiny crystals develop, the resulting material, when cooled, is much more resistant to breaking than normal glass. In molecular terms, the randomness of the glass structure has been partially replaced by the order of a crystalline silicate. The materials produced in this way (an example is Pyroceram) are generally opaque and are used for kitchenware and in other applications in which the material is subjected to stress or high temperatures.

Ceramic materials are attractive for several reasons. The starting materials for making them are readily available and cheap. Ceramics are lightweight in comparison with metals and retain their strength at temperatures above 1000°C, where metal parts tend to fail. They also have electrical, optical, and magnetic properties of value in the computer and electronic industries.

The one severely limiting problem in utilizing ceramics is their brittle nature. Ceramics deform very little before they fail catastrophically, the failure resulting from a weak point in the bonding within the ceramic matrix. However, such weak points are not consistent from object to object, so that the predictability of failure is poor. Since the stress failure of ceramic materials is due to molecular abnormalities resulting from impurities or disorder in the basic atomic arrangements, much attention is now being given to purer starting materials and the control of the processing steps. In addition, ceramic **composite materials**, mixtures of ceramic materials or of ceramic fibers with plastics, can overcome the tendency of plain ceramics to crack.

Cement and Concrete

A **cement** is a material that can bond mineral fragments into a solid mass. The most common cement, known as Portland cement, is made by roasting a powdered mixture of calcium carbonate (limestone or chalk), silica (sand), aluminosilicate mineral (kaolin, clay, or shale), and iron oxide at a high temperature in a rotating kiln. As the materials pass through the kiln, they lose water and carbon dioxide and ultimately form "clinker," in which the materials are partially melted together. The clinker is ground to a very fine powder after the addition of a small amount of calcium sulfate (gypsum). A typical composition of a Portland cement, expressed in terms of oxides, is 60% to 67% CaO; 17% to 25% SiO_2; 3% to 8% Al_2O_3; up to 6% Fe_2O_3; and small

■ Portland cement, patented in England in 1824, is named for its similarity to a rock native to the Isle of Portland in England. It is also called *hydraulic cement* because it sets even under water.

amounts of magnesium oxide, magnesium sulfate, and potassium and sodium oxides. As in glass, the oxides are not isolated into molecules or ionic crystals, and the submicroscopic structure is quite complex.

Many different reactions occur during the setting of cement. Initially the calcium silicates react with water to give a sticky gel. The gel has very large surface area and is responsible for the strength of concrete. Reactions with carbon dioxide in the air also occur at the surface. After the initial solidification, small, densely interlocked crystals begin to form, a process that continues for a long time and increases the compressive strength of the cement.

■ **Mortar** is a mixture of cement, sand, water, and lime. **Concrete** is a mixture of cement, sand, and aggregate (crushed stone or pebbles).

More than 800 million tons of cement are manufactured each year, most of which is used to make concrete. Concrete, like many other materials containing Si—O bonds, is virtually noncompressible but lacks tensile strength. If concrete is to be used where it will be subject to tension, it must be reinforced with steel.

■ *Tensile strength* refers to the resistance of a material to being stretched.

SELF-TEST 14D

1. The two principal nonmetals in glass are ____________ and ____________.
2. Which of the following apply to (i) silica and (ii) silicates?
   (a) SiO_2
   (b) Quartz
   (c) Contain metal ions
   (d) Clay
   (e) Built up from tetrahedral units
3. How does glass differ from silica?
4. What three silicate minerals are used to make common pottery?
5. A distinctive property of ceramics is their ability to withstand high ____________. Another is their tendency to fail in response to ____________.
6. The three principal ingredients of concrete are ____________, ____________, and ____________.

MATCHING SET

____	1. Glass	a. Reduction
____	2. SiO_4	b. Chlorine
____	3. Steel	c. Magnesium
____	4. Sodium hydroxide	d. Noncrystalline
____	5. Product of the blast furnace	e. Ore
____	6. $Fe^{3+} \rightarrow Fe$	f. In mortar and concrete
____	7. Valuable mixture of minerals	g. Pig iron
____	8. Ceramics and glass	h. Product of chloralkali industry
____	9. Superconductor	i. No resistance to electrical flow
____	10. Seashells	j. Building block of the earth's crust
____	11. Produced from sea water	k. B-enriched Si
____	12. Metal from the ocean	l. Oxidation
____	13. *p*-type semiconductor	m. Source of calcium carbonate
____	14. Portland cement	n. Alloy
____	15. $Fe \rightarrow Fe^{3+}$	o. Made of silicates

QUESTIONS FOR REVIEW AND THOUGHT

1. Define the following terms:
   (a) Hydrosphere (b) Igneous rock
   (c) Sedimentary rock
2. Define the following terms:
   (a) Slag (b) Ductile
   (c) Brine (d) Alloy
3. Define the following terms:
   (a) Annealing (b) Amorphous
   (c) Ceramic (d) Cement
   (e) Glass
4. What are the two products of the "chloralkali" process? Name two common uses for each of the products.
5. Name a common metal taken from seawater. Give several uses for this metal.
6. Complete and balance (if needed) the following equations:
   (a) $2\ H_2O(\ell) + 2\ e^- \rightarrow$ ________ $+ OH^-(aq)$
   (b) $Mg^{2+}(aq) + 2\ OH^-(aq) \rightarrow$ ________
   (c) $Cl^-(aq) \rightarrow Cl_2(g) +$ ________
   (d) $CaCO_3(s) \rightarrow$ ________ $+ CO_2(g)$
   (e) $Mg^{2+}(aq) +$ ________ $\rightarrow Mg(s)$
7. Name two elements that occur in the free metallic state in nature.
8. What are the three ingredients used in a blast furnace to make iron? Which one of these is the reducing agent?
9. Explain why the molten mixture in a blast furnace separates naturally into a layer of slag and a layer of molten iron.
10. What are the differences between pig iron, cast iron, and steel? Mention composition as well as properties and uses.
11. What is the principal impurity in pig iron?
12. What element is used to convert pig iron into steel?
13. Complete and balance (if needed) the following equations:
    (a) $Cu_2S(s) + O_2(g) \rightarrow$ ________
    (b) $FeS(s) + O_2(g) \rightarrow$ ________
14. How is pure copper produced?
15. Name four properties of metals and give an example of the use of a metal taking advantage of each property.
16. Describe the electron structure of metals in terms of their conductivity of electricity.
17. (a) Describe a p-type semiconductor made from silicon. Name two elements that might be used as dopants.
    (b) Do the same for an n-type semiconductor made from silicon.
18. Name an important application of p- and n-type semiconductors in addition to their application in the manufacture of transistors and integrated circuits.
19. Draw the structural unit SiO_4, found in many silicate minerals. What is the name for the shape of this structural unit?
20. Draw a structure consisting of SiO_4 units arranged in a chain with each unit sharing two of its oxygen atoms with neighboring SiO_4 units. What is this silicate structure named?
21. What is a superconductor and why is the discovery of a "room-temperature" superconductor an important goal?
22. The element silicon occurs in all clays. Name two other elements common in all clays.
23. What element causes glass to have a brown to yellow color? (*Hint:* This is the same element that causes some clays to be red in color.)
24. What is the purpose of annealing glass?
25. Compare ceramics with metals by filling in the following table with a yes or a no depending on whether the material has that property.

	Metal	**Ceramic**
Hardness	________	________
Strength	________	________
Ductility	________	________
Electrical conductivity	________	________
Brittleness	________	________

26. Write the formula for limestone.
27. Cements contain calcium oxide (lime), SiO_2 (sand), and various other metal oxides. Which of the oxides present reacts with CO_2 in the air, and which helps bond everything together?
28. How is the structure of a common glass different from the structure of a solid like sodium chloride?
29. When a molten mixture of different materials cools, will all the substances crystallize at the same time? Silicates and quartz (SiO_2) have the highest melting points. Which will crystallize first, the metallic elements or the silicates?
30. The extraction of iron from its ore is an oxidation–reduction reaction. The first reaction between iron oxide—$Fe_2O_3(s)$—and carbon monoxide—$CO(g)$—is

$$Fe_2O_3(s) + 3\ CO(g) \longrightarrow 2\ Fe(s) + 3\ CO_2(g) + \text{heat}$$

    What is oxidized and what is reduced in the reaction?
31. Would it make any difference in our ability to extract pure elements if they were uniformly distributed on the earth instead of being concentrated as many of them are?
32. Discuss the pros and cons of asbestos removal in schools and other public buildings.

Everyone wants this water to be clean.
(C. D. Winters)

CHAPTER 15

Water: Plenty of It, But of What Quality?

For a molecule so simple, water, made of two hydrogen atoms covalently bound to an oxygen atom, is extremely important in our world. Our lives and the lives of all the other creatures of Earth depend on the liquid we call water—a collection of these simple H_2O molecules. Too little water and we can die of thirst; too much and we can drown. If an area on Earth has too little water, it becomes an arid desert. If too much water is present, normal life for land dwellers becomes impossible, and *aquatic* animal life takes over. There is a lot of water on this planet, but much of it is either not in the liquid state—the physical state we find most useful—or contains dissolved substances that make it unfit for most uses. Our largest reserves of water are in the oceans. Most water we come in contact with daily is not *pure;* that is, it is not 100% water molecules and nothing else. Water dissolves all kinds of substances; it is a "universal solvent." To ensure a safe water supply, we must limit the dissolved substances in type and quantity to what is safe. In this chapter we shall look at the answers to these questions:

- How can there be a water shortage if water is such an abundant compound?
- How is water used and reused?
- What are the differences between clean water and polluted water?
- How is water pollution measured?
- What are some of the causes of water pollution?
- How is water made pure?
- How can we ensure that our water supplies remain pure?
- Will we be able to use the vast quantities of water found in the oceans?

Ensuring the quality of the water supply is a shared responsibility. The federal government enacts laws governing the quality of wastewater that can

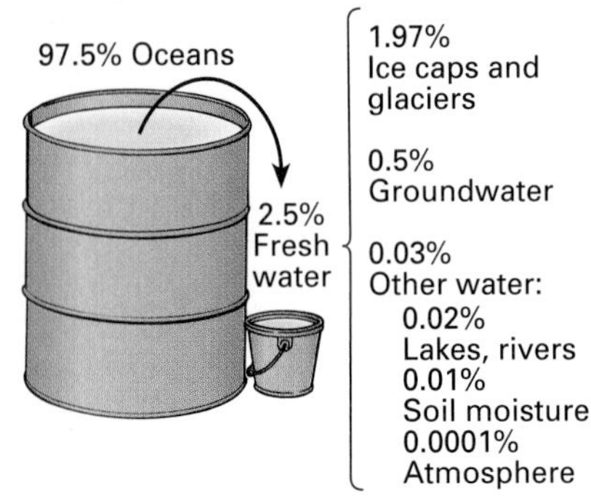

The water supply. Of the 2.5% freshwater, less than 1% is available as groundwater or surface water for human use.

be returned to the environment and also the quality of water that can be supplied for drinking. The U.S. Environmental Protection Agency (EPA) enforces these laws. State and local governments have their own laws and shoulder the responsibility for enforcing them as well as seeing to it that local industries, sewage treatment plants, and municipal water treatment plants meet federal standards.

The goal of these regulations is to keep water relatively pure so the next user of that water, whether that is a city wanting a water supply or a trout looking for a nice stream to live in, will find it suitable for its intended use.

Individual citizens also have a role to play in keeping the water clean. Voluntary action by an informed public is needed to halt water pollution that originates in our households. Laws alone cannot halt disposal of hazardous waste in the municipal garbage or pouring of potential water pollutants into toilets, sinks, or storm drains.

In this chapter we will look at how water is used and reused, how it becomes polluted, and how it is purified. This chapter will build on what you learned about the properties of water in Section 5.10. As you will see, understanding how water is used, how it becomes polluted, and how it can be purified depends on understanding its properties.

15.1 How Can There Be a Shortage of Something as Abundant as Water?

Water is the most abundant substance on the earth's surface. Oceans (with an average depth of 2.5 mi) cover about 72% of the earth. They are the reservoir of 97.5% of the earth's water. Only 2.5% is freshwater. Water is also the major component of all living things (Table 15.1). For example, the water content of human adults is 70%—the same proportion as for the earth's surface.

■ The Mississippi River discharges an estimated 400 billion gallons of water daily into the Gulf of Mexico.

An average of 4350 billion gallons of rain and snow fall on the contiguous United States each day (Figure 15.1). Of this amount, 3100 billion gallons return to the atmosphere by evaporation and transpiration. The discharge to the sea and to underground reserves amounts to 800 billion gallons daily, leaving 450 billion gallons of surface water each day for domestic and commercial use. The 48 contiguous states withdrew 40 billion gallons per day from natural sources in 1900, but that rose to over 400 billion gallons by 1993. It is estimated that the demand may be 900 billion gallons per day by the year 2000. You can see from these numbers that the demand for water by our growing population is already as great as the resupply by natural resources. This means that water must be reused on a daily basis. The water molecules you drink from a water fountain may have been in the municipal wastewater of another city just a few days earlier.

■ A shallow well generally goes no deeper than 50 ft. A deep well is generally drilled to between 100 and 200 ft below the surface. The deepest water well in the world is in Montana—it goes 7320 ft into the ground.

The two sources of usable water are **surface water**, such as rivers, lakes, and wetland waters, and **groundwater**, which is beneath the earth's surface. Figure 15.1 shows our water resources and the flow of groundwater. About 90 billion gallons of the total water withdrawn every day is groundwater drawn from wells drilled into **aquifers**, layers of water-bearing porous rock or sediment held in place by impermeable rock. These kinds of wells, which supply water to many cities in the Great Plains and along the East Coast, are called artesian wells.

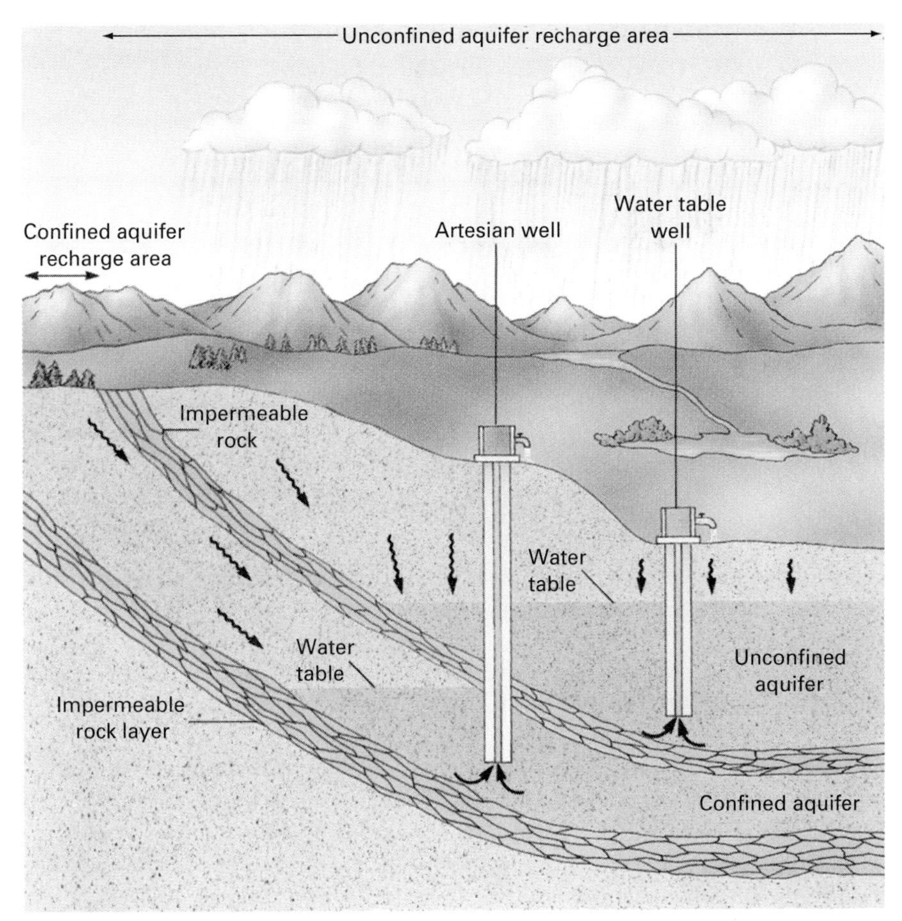

Figure 15.1 Water resources. Surface water includes water that collects in rivers, lakes, and wetlands. Groundwater may be in unconfined aquifers, which are recharged by surface water from directly above, or in confined aquifers, which lie between impermeable rock layers. Wells may tap into either kind of aquifer. Where water in a confined aquifer is under pressure, pumping of the water may not be needed.

TABLE 15.1 ■ Water Content

	Percentage (%)
Marine invertebrates	97
Human fetus (1 month)	93
Adult human	70
Body fluids	95
Nerve tissue	84
Muscle	77
Skin	71
Connective tissue	60
Vegetables	89
Milk	88
Fish	82
Fruit	80
Lean meat	76
Potatoes	75
Cheese	35

In the arid West, wells used to pump water for irrigation either are going dry or require drilling so deep that irrigation is no longer economically feasible. The huge Ogallala aquifer that stretches through eight states from South Dakota to Texas has 150,000 wells tapping it for irrigation of 10 million acres. As a result, the Ogallala aquifer is being drawn down at a rate that has reduced the average thickness of the aquifer from 58 ft in 1930 to 8 ft today.

The depletion of a major aquifer along the Eastern seaboard has caused large sinkholes in Georgia and Florida when the limestone rock strata of the aquifer collapse as the water is withdrawn (Figure 15.2). Many coastal cities are also experiencing problems with brackish water that comes from aquifers where the freshwater has been drawn off, causing seawater to flow into the depleted aquifer. Depletion of underground sources has also caused sinkholes in Texas. The entire city of Houston has sunk several feet over the years as the result of extensive use of the underground water sources in that area.

Figure 15.2 A sinkhole in Winter Park, Florida. As the result of aquifer depletion, the top of an underground cavern has collapsed, an event that can happen in the course of just a few minutes. *(M. Timothy O'Keefe/Tom Stack & Associates)*

TABLE 15.2 ■ Water Consumption in the United States (Typical) per Day

	Billion Gallons
Households	8.9
Industry	6.7
Steam–electric utilities	4.0
Agriculture	76
	95.6

Source: Statistical Abstracts of the United States, 114th ed., U.S. Department of Commerce, 1994.

TABLE 15.3 ■ Average Water Usage per Person per Day

Use	Gallons
Flushing toilets	30
Bathing	23
Laundering	11
Drinking and cooking	2
Miscellaneous	10
Dishwashing	6
Other*	9
Total	91†

* Includes car washing, swimming pools, lawn watering, public fountains, losses due to leaks in water mains, etc.

† Based on taking all treated water used per day and dividing by the total population.

Source: American Water Works Association.

■ Water that is returned to a waterway at a higher temperature than it was withdrawn contributes to thermal pollution of the waterway. Because the solubility of oxygen in water decreases with increased temperature, thermally polluted water cannot support aquatic life as well as cooler water can.

■ There are newer design toilets that use about 1.5 gal per flush.

15.2 Water Use and Reuse

Who's using the water? Table 15.2 shows the breakdown for water use in the United States. Agriculture is the major user of water, with all other uses being a distant second.

Industrial water usage can be directly related to production of finished products. One gallon of water, for example, is used to produce eight sheets of ordinary typing paper, while 80 gal of water are needed to produce a gallon of gasoline, and about 25 gal of water are needed to make a single box of nails. Of course, because none of these products *contain* water, the water used is either recycled into the environment or recycled within the facility. Industrial recycling has not always been the case. For example, the paper industry used more than 400,000 gal of water to produce a ton of paper in 1900, but modern processes have dropped the water usage to under 40,000 gal per ton of paper produced.

In many industrial locations the largest single use of water is in plant cooling systems. Water can absorb more heat than any other readily available liquid—it has a heat capacity of 1 cal/g, or 4.18 J/g. Recirculating cooling water is an important means of water conservation. A cooling tower allows the warmed water to lose its heat to the surrounding air and helps to reduce thermal pollution of the river or lake where the used water would be discharged. In addition, the high heat capacity of water enables the industrial user to recycle heat energy captured by the cooling water.

Groundwater and surface water sources each provide half of the more than 44 billion gallons of **potable** (drinkable) water that is used each day in the United States. Table 15.3 gives the average amounts for various personal uses. Only a small fraction of municipal water really needs to be of drinking water quality. The largest portion of a water supply could be disinfected and made bacteriologically safe, while avoiding the more costly treatment needed to meet drinking quality standards. Water treated in this way would be suitable for irrigation of parks and golf courses, air-conditioning, industrial cooling, and toilet flushing. Consider the data in Table 15.3, which illustrate the inefficiency of residential water systems. We use 33% of residential water for flushing toilets and 25% for bathing. Conventional showers use up to 10 gal/min, which can be reduced by as much as 70% by the installation of inexpensive water-saving showerheads. It is obvious that strong arguments could be made for dual water systems—one system that treats water for drinking and a different system for other uses.

Residential water conservation is a way to cut demand for freshwater supplies. Although Table 15.2 shows that residential use is a small part of the total, home water conservation can be an important step in cutting demand for freshwater supplies in large urban areas.

Almost all water molecules have been recycling since they were formed billions of years ago. Have you ever considered that your next glass of water may include some molecules that Aristotle or Abraham Lincoln drank, or some molecules that flooded into the Titanic when it sank? In spite of these possibilities, the idea of obtaining potable water from wastewater or even sewage, especially wastewater that was *recently* discharged, is psychologically difficult for many people to accept. Nevertheless, the technology has been developed and is currently being used in NASA's space shuttle flights. What this means is if water (pure or not) is available, it should be considered for reuse. In the southwestern United States the rate of depletion of aquifers has led to direct recycling of water from sewage, a process called **groundwater recharge**. For example, in El Paso, Texas, 10 million gallons of pure water per day from sewage effluent is pumped into the underground aquifer that is the main source of water for El Paso.

Surface level drop in San Joaquin, California. The markers on the utility pole show how the surface level has dropped over the years due to withdrawal of groundwater for irrigation. *(Courtesy of U.S. Geological Survey)*

15.3 What Is the Difference Between Clean Water and Polluted Water?

The term **pollution** is used to describe any condition that causes the natural usefulness of air, water, and soil to be diminished. Water that is judged unsuitable for drinking, washing, irrigation, or industrial uses is polluted water. The pollution (Table 15.4) may be heat, radioisotopes, toxic metal cations and anions, organic molecules, acids, alkalis, or organisms that cause disease (**pathogens**). Water suitable for some uses might be considered polluted and therefore unsuitable for other uses—you might go swimming in water you would not consider drinking. Water that is unsuitable for use has often been polluted by human activity, but natural processes can also pollute water. For example, water that contacts organic substances such as decaying leaves and animal wastes will pick up numerous organic compounds, many of which im-

■ Creeks often have names that indicate a content other than pure water. Here are a few creeks along with the metal compounds they contain. Creeks with these names can be found in many different states. The chemicals are usually the same.

Red Creek	Iron
Sulfur Creek	Sulfur of hydrogen sulfide
Copper Creek	Copper
Red Mountain Creek	Iron
Alum Creek	Aluminum compounds
Bitter Creek	Aluminum
Buttermilk Creek	Colloidal zinc carbonate

TABLE 15.4 ■ U.S. Public Health Service Classes of Pollutants

Pollutant	Example
Oxygen-demanding wastes	Plant and animal material
Infectious agents	Bacteria, viruses
Plant nutrients	Fertilizers such as nitrates and phosphates
Organic chemicals	Solvents, pesticides, detergent molecules
Other minerals and chemicals	Acids from mining operations, inorganic chemicals from metal-working operations
Sediment from land erosion	Clay silt from stream beds
Radioactive substances	Mining wastes, used radioisotopes
Heat from industry	Cooling water from electric generating plants

TABLE 15.5 ■ Dissolved Substances Found in "Pure" Water

Name	Formula	Comment
The following come from contact of water with the atmosphere:		
Carbon dioxide	CO_2	Makes water slightly acidic
Dust particles	—	Can be large amounts at times
Nitrogen	N_2	Along with oxygen, causes visible bubbles in hot water
Nitrogen dioxide	NO_2	Formed by lightning
Oxygen	O_2	Supports aquatic life
The following vary considerably, depending on the kinds of rock formations the water has contacted:		
Bicarbonate ions	HCO_3^-	Soils and rocks
Calcium ions	Ca^{2+}	From limestone
Chloride ions	Cl^-	Soils, clays, and rocks
Iron(II) ions	Fe^{2+}	Soils, clays, and rocks
Magnesium ions	Mg^{2+}	Soils, clays, and rocks
Potassium ions	K^+	Soils, clays, and rocks
Sodium ions	Na^+	Soils, clays, and rocks
Sulfate ions	SO_4^{2-}	Soils and rocks

part odors and color to the water and some of which might be pathogenic. Silt, consisting of colloidal-sized particles of dirt and sand, can also pollute water. Table 15.5 lists some of the substances that can be found in "pure" natural water. By looking at this list, it is clear that absolutely pure (100%) water is not a common commodity.

■ The Clean Water Act is also known as the Federal Water Pollution Control Act. It was revised in 1978, 1980, and 1988.

As human activities have continued to pollute water, governments have passed laws designed to keep our waters clean. The Clean Water Act of 1977 represented a major change in the thinking of Congress concerning who is responsible for keeping water clean. This act shifted the burden of producing water suitable for reuse from the user (a municipality, for example) to the wastewater discharger. Because it is easier to clean wastewater prior to dumping than to clean the water after the untreated waste has been discharged (Figure 15.3), the Clean Water Act was a major step in improving the quality of our natural waters. The Act requires the EPA to establish and monitor emission standards—the maximum amounts of water pollutants that can be discharged into natural bodies of water from factories, municipal sewage treat-

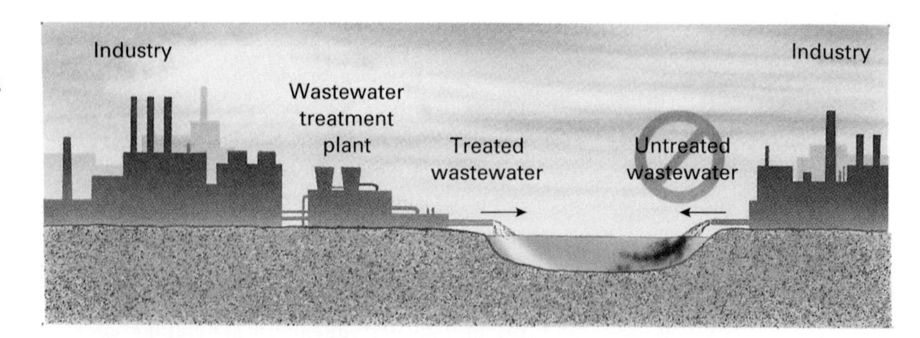

Figure 15.3 The EPA requires that virtually all industrial wastewaters be treated prior to discharge. This water is reasonably pure although it is not as pure as drinking water.

ment plants, and other facilities. As it turns out, the wastewater effluent from an industry can now often be clean enough to be used for such purposes as irrigation or industrial cooling.

SELF-TEST 15A

1. Approximately what percentage of the human body is water?
2. What is the major reservoir of water on the earth?
3. Water beneath the earth's surface is called ____________.
4. Water on the surface of the earth is called ____________.
5. The average person in the United States uses about ____________ gallons of water per day for all purposes.
6. The actual amount of potable (drinkable) water a person needs is ____________ per day.
7. Three common water pollutants are ____________, ____________, and ____________.
8. A water-bearing stratum of porous rock, sand, or gravel is called a(n) ____________.
9. What happens to most of the water that falls on the United States each day?

15.4 The Impact of Hazardous Industrial Wastes on Water Quality

Industrial processes, whether making paper, automobiles, or TV sets, produce waste materials. Table 15.6 lists some of the industrial pollutants that result from the manufacture of products important to us. For many years the disposal of solid wastes in *landfills* was considered good engineering practice. Many of the substances present in those wastes were partially dissolved by rainwater and became part of the groundwater, causing serious pollution of water supplies. Many of these older landfills also allowed surface runoff that contained dissolved substances from the wastes to be carried into natural bodies of water.

Corroding waste barrels. *(USDA/Soil Conservation Service)*

TABLE 15.6 ■ Important Industrial Products and Pollutants Associated with Their Manufacture

Products	Pollutants
Plastics	Solvents, organic chlorine compounds
Pesticides	Organic chlorine compounds, organic phosphate compounds
Medicines	Solvents, metals such as mercury and zinc
Paints	Metals, pigments, solvents, organic residues
Petroleum products	Oils, organic solvents, acids, alkalies
Metals	Metals, fluorides, cyanide, acids, oils
Leather	Chromium, zinc
Textiles	Metals, pigments, organic chlorine compounds, solvents

It was also common practice in the past to place waste and discarded chemicals into metal drums and bury them directly in the ground. After a few years the drums developed leaks due to corrosion, allowing the drum contents to leak into water that would ultimately become groundwater or surface water. In recognition of this common practice and what it was doing to water quality, the U.S. Congress in 1976 passed the Resource Conservation and Recovery Act (RCRA). In 1980, Congress established the "Superfund," a $1.6 billion program designed to clean up hazardous waste sites that were threatening to contaminate the nation's water supplies. Since 1980, the U.S. EPA has targeted many thousands of waste sites that have the potential of harming our water supplies, and some of these have been cleaned up. In general, the waste site cleanups have seen a lot of tax dollars spent and a lot of litigation involving "responsible parties" concerning who will pay for the cleanup. Currently, the EPA has identified more than 1200 sites in the United States where toxic wastes have been stored that should be cleaned up. The cleanup of these remaining waste sites will cost hundreds of billions of dollars.

■ When the Superfund was first established, over 40,000 sites were identified for cleanup.

The World of Chemistry

Program 25, *Chemistry and the Environment*

Workers in protective suits at a waste site.

While the Superfund is dealing with existing hazardous waste sites, the RCRA law and its regulations govern the disposal of newly generated wastes that have the potential to harm the environment. The EPA defines certain industrial wastes as **hazardous wastes** (Table 15.7) and closely regulates how they are generated, stored, transported, and disposed of. The RCRA law was designed to give "cradle-to-grave" (origin to disposal) responsibility to *generators* of hazardous wastes. Before RCRA, an industry could hire almost anyone to haul away its waste without regard to where it was taken or how it would be disposed of. Today, a generator of a hazardous waste must know who is transporting it, where it is going, and how it will be disposed of. Each shipment of

TABLE 15.7 ■ Hazardous Wastes as Defined by EPA (Causing Water Pollution)

Wastes containing the following metals and pesticides:
Arsenic, barium, cadmium, chromium, lead, mercury, selenium, silver, endrin, lindane, methoxychlor, toxaphene, 2,4-D, 2,4,5-TP (Silvex)
Wastes that have the following characteristic properties*:
Ignitible, corrosive, reactive, acutely toxic
Twenty-one wastes from nonspecific sources (such as):
Wastes containing the cyanide ion, distillation residues, used halogenated solvents such as carbon tetrachloride
Eighty-nine wastes from specific sources (such as):
Wastewater sludges from chloride production, wastewater from pesticide manufacture, sludges from the production of petroleum products
A large number of various discarded and off-specification chemicals, many of which are used in the chemical industry to manufacture pharmaceuticals, polymers, paints, dyes, automotive products, cosmetics, and so on

Note: Shipments of hazardous wastes are carefully monitored by EPA and state governments. In addition, all facilities receiving these wastes must have permits and licenses.

* Detailed definitions apply to these waste characteristics.

Source: EPA

FRONTIERS IN THE WORLD OF CHEMISTRY

Arsenic-Eating Microbes

Ordinarily, arsenic is toxic to most organisms. But a microbe has been discovered that actually thrives on arsenic, using its most highly oxidized form as a source of energy. Dianne Ahmann at the Massachusetts Institute of Technology in Boston and her co-workers isolated a microbe from arsenic-contaminated sediments of the Aberjona watershed in eastern Massachusetts. Up to the 1930s, arsenic-containing industrial waste had been dumped into the watershed. Normally, when exposed to the environment, arsenic is oxidized to arsenic(V) and, in the form of insoluble arsenate ions, binds to sediment particles by strong electrostatic forces. The arsenic would be expected to remain in the sediment for a long time, all the while having a toxic effect on any organisms that might come in contact with it.

Ahmann and her colleagues found that samples of the Aberjona sediment contained unusually high concentrations of arsenic(III), a less oxidized form of arsenic. After carefully screening samples of the sediment to determine what was producing the arsenic(III), they isolated a microbe, which they named MIT-13. Experiments with the microbe showed that the concentration of arsenic(V) in its environment decreased as the concentration of arsenic(III) increased; and that at the same time, cell growth increased. The microbe uses arsenic(V) as a source of energy, in other words, as a food. To show just how dependent MIT-13 is on arsenic(V), experiments were done in which everything was kept the same but with no arsenic(V) added. In these studies, minimum cell growth resulted.

What promise does a discovery like this hold for dealing with arsenic-containing wastes? If this microbe can be grown in sufficient quantity, perhaps it could be used routinely to help get rid of persistent arsenic(V) deposits in sediments and perhaps even in soil samples. Of course, there is the arsenic(III) to be dealt with. It is water soluble, but that property can be used to advantage. If a solution containing arsenic(III) is passed through an organic polymer containing numerous ionically charged sites (an ion-exchange resin), the arsenic(III) can be concentrated and eventually purified and reused. Scientists will continually be on the lookout for microbes like MIT-13 that can thrive on metals and organic compounds that are deadly to most organisms. To turn an old saying around, "one microbe's poison is another microbe's meat."

Dianne Ahmann holding a microbe-filled test tube. (Tom Pantages)

Reference D. Ahmann, et al., Microbe grows by reducing arsenic, *Nature*, Vol. 371, p. 750, October 27, 1994.

hazardous waste is accompanied by a **manifest**, a document listing the hazardous waste by name, how much is present, and how it will be disposed of. As the load of hazardous waste is transported, temporarily stored, and finally disposed of, parts of the manifests are signed by each responsible party and returned to the generator at each step. The states also receive copies of the manifests, and annual reports of hazardous waste activities are filed with EPA by everyone who handles hazardous wastes routinely.

Today, industrial hazardous wastes must be placed into *secure* landfills, incinerated, or treated in some way to render them nonhazardous. No hazardous waste is allowed to be disposed of in a way that could pollute the environment. Some hazardous wastes go to secure landfills with plastic linings

Figure 15.4 A hazardous waste landfill. Underneath, this landfill has several feet of clay covered by three plastic liners. Barrels of hazardous waste are placed above the liners and buried in soil. *(Dennis Barnes)*

(Figure 15.4) that prevent their contents from easily reaching surrounding water supplies. These landfills also have carefully spaced monitoring wells so any substances escaping from the landfill's contents may be detected, allowing the problem to be corrected. Other hazardous wastes may no longer be placed in secure landfills, but must be incinerated or destroyed in some other way. While incineration seems like a logical best choice to dispose of hazardous wastes, current incinerators are operating at near capacity, and it is difficult to get proper permits for new ones due to community opposition. (Incineration of *some* hazardous wastes does produce small quantities of other, even more harmful combustion products.)

■ Proposition 65 states, "No person in the course of doing business shall knowingly discharge or release a chemical known to cause cancer or reproductive toxicity into water or onto or into land where such chemical passes or probably will pass into any source of drinking water."

States like California that have severe water shortage problems have taken drastic steps to protect their ground and surface water. In California, Proposition 65, the Safe Drinking Water and Toxic Enforcement Act of 1986, lists approximately 200 chemicals or classes of chemicals known to cause cancer or reproductive toxicity and prohibits their discharge into any water that might become a drinking water supply. Other states have shown interest in creating statutes similar to California's Proposition 65.

15.5 Household Wastes that Affect Water Quality

■ The EPA estimates that each year 350 million gallons of waste motor oil are poured on the ground or flushed down the drain by individual citizens. That's 35 times more oil than the *Exxon Valdez* spilled in Alaska.

Often we do not think about the things we discard in our garbage, but what we throw away and how we do it can affect the quality of natural waters as much as what industry does. Household wastes that are incinerated can contribute to air pollution (see Chapter 16), but because the bulk of our household waste goes to landfills, we too can be responsible for causing pollution of groundwater as well as of rivers, streams, and lakes. Table 15.8 lists some common household products and the kinds of chemicals they contain. Because we are the consumers of industrial products, we can put the very same chemicals into the water as industry can. Although the individual amounts of harmful chemicals used in a household are less than those used in a large industry, the total amounts disposed of daily by all households can be very large, even for a medium-sized city.

TABLE 15.8 ■ Some Common Household Hazardous Wastes and Recommended Disposal

Type of Product	Harmful Ingredients	Disposal*
Bug sprays	Pesticides, organic solvents	Special
Oven and bathroom cleaners	Bases and acids	Drain
Furniture polish	Organic solvents	Special
Aerosol cans (empty)	Solvents, propellants	Trash
Nail polish and remover	Organic solvents	Special
Antifreeze	Organic solvents, metals	Special
Insecticides	Pesticides, solvents	Special
Auto battery	Sulfuric acid, lead	Special
Medicine (expired)	Organic compounds	Drain
Paint (latex)	Organic polymers	Special
Gasoline	Organic solvents	Special
Motor oil	Organic compounds, metals	Special
Drain cleaners	Bases	Drain
Paints (oil-based)	Organic solvents	Special
Household batteries	Heavy metals such as mercury, nickel, and cadmium	Special

* Special: Professional disposal as a hazardous waste. Drain: disposal down the kitchen or bathroom drain. Trash: Treat as normal trash—no harm to the groundwater. In most households, the items marked special are disposed of as normal trash, which results in groundwater pollution.

Source: "Household Hazardous Waste: What You Should and Shouldn't Do," Water Pollution Control Federation, 1986.

Municipal hazardous waste day. On a designated day, citizens bring their hazardous waste to a site where trained crews sort the materials for disposal. Every community should have such days or have access to a permanent hazardous waste disposal facility. *(Alan Pitcairn/Grant Heilman Photography)*

Households have a greater problem disposing of hazardous wastes than industry does. Even where there is an active recycling project for glass, paper, metals, and plastics, there is often no pickup of chemicals that should be separated from the ordinary trash destined for the landfill. If these chemicals are mixed with ordinary garbage, they go to the city landfill or incinerator. If they are poured out in the sink, driveway, or backyard, they will eventually reach natural waters.

■ Ordinary garbage costs about $27 per ton for disposal, whereas hazardous waste costs about $1000 per ton for proper disposal.

How can we dispose of hazardous household wastes without danger to the groundwater supply? We can ask our city's municipal waste authorities to provide disposal sites for these wastes or to sponsor periodic household hazardous waste days when these materials can be brought to a central site. In some U.S. cities and some European countries (such as the Netherlands), special trucks routinely pick up hazardous household wastes.

■ Suspended particles in water are colloidal in size and might include bacteria and viruses, as well as harmless soil particles.

15.6 Toxic Elements Often Found In Water

Heavy metal ions are perhaps the most common of all water pollutants. The heavy metals include such frequently encountered elements as lead and mercury, as well as many less common ones like cadmium, chromium, nickel, and copper. These metals can, at times, be acutely toxic, causing immediate symptoms, but often they are chronically toxic in very small quantities. Chronic toxicity is characterized by nagging symptoms that lessen normal body functions. Inadequate disposal of wastes from mining or industrial activities causes these metals to find their way into water supplies. In addition, some farming activities and the disposal of household wastes can contribute to the presence of heavy metals in our water supplies.

■ The heavy metals are considered "heavy" because they have higher atomic weights.

Mercury

Mercury is a fairly common heavy metal. Its elemental form at room temperature is that of a liquid that is volatile with a characteristic shininess of a metal. Mercury atoms are readily oxidized to Hg_2^{2+} (mercury(I) ion) and Hg^{2+} (mercury(II) ion). Both of these ionic forms are toxic; and their effects can be cumulative (i.e., repeated exposures will increase the toxic effects because the body does not easily rid itself of the element).

Mercury is widely used in industry and can still be found in homes in mercurial thermometers. The disposal of used fluorescent lamps represents a major source of mercury in water. Fluorescent lamps contain small amounts (about 60 mg per lamp) of mercury, which vaporizes inside the lamp and helps carry the electric current between the electrodes. Discarded lamps of this type are ultimately crushed when they are disposed of in landfills. This releases the mercury to become oxidized and then to dissolve in any water that comes in contact with the waste.

■ Mercury thermometers are being replaced by digital thermometers.

Children's sandal with step-sensitive light switch. Originally such light-up sandals and sneakers contained mercury switches and ran into trouble in states with strict laws against disposal of mercury in landfills. At a high cost to itself, the manufacturer withdrew the sandals and sneakers from the market in some states, offered prepaid envelopes to send discarded shoes to a mercury recycler, and eventually ceased using the mercury switches. *(C. D. Winters)*

Recently, several designs of shoes for small children contained small amounts of mercury (less than 50 mg) in "step-sensitive" switches that would turn on and off tiny lights in the soles of the shoes. On realizing the potential harmful environmental consequences of large numbers of these shoes being eventually discarded into landfills, the manufacturers made offers to recall the shoes.

Lead

Lead is another widely encountered heavy metal. Like mercury, lead tends to accumulate in the body on repeated exposures. Lead has been used for centuries in plumbing (the Latin name for lead is *plumbum*, the word from which plumber is derived). Until recently, lead-based solders were routinely used in almost all plumbing fixtures, including the valves in drinking fountains as well as shower fixtures and faucets found in both the bathroom and kitchen. The current lead level for drinking water allowed by the U.S. Environmental Protection Agency is 15 parts per billion (ppb).

■ Lead has historically been used in plumbing. The Romans used lead pipes to carry water into their homes.

In addition, lead has been widely used in various paint formulations. For the past 30 years, paints containing lead have been banned for interior use, but lead-containing paints are still available for outdoor and industrial use. When these paints are discarded, or objects coated with lead-containing paints are discarded, lead can find its way into water supplies.

TABLE 15.9 ■ Some Arsenic-Containing Insecticides

Name	Formula
Lead arsenate	$Pb_3(AsO_4)_2$
Monosodium methanearsenate	$CH_3—As(=O)(OH)—O^-Na^+$
Copper acetoarsenite (Paris green)	$3\ CuO \cdot 3As_2O_3 \cdot Cu(C_2H_3O_2)_2$

The World of Chemistry

Program 19, *Metals*

Soldering copper water pipes with lead-based solder, a common practice that has been the source of lead in drinking water.

Arsenic

Arsenic occurs naturally in small amounts in many foods. Shrimp, for example, contain about 19 parts per million (ppm) arsenic, and corn may contain 0.4 ppm arsenic. The amount of naturally occurring arsenic in foods depends on the surroundings where they are grown and the metabolism of the plant or animal. While many soils contain arsenic, which causes an accumulation of the element as a plant grows, some insecticides also contain arsenic (Table 15.9), which causes an arsenic residue when the insecticide is applied. The U.S. Food and Drug Administration (FDA) has set a limit of 0.15 mg of arsenic per pound of food (330 ppm), and this amount apparently causes no harm. In its ionic forms, arsenic is much more toxic than in its covalently bound compounds. The typical toxic arsenic compounds contain ions such as arsenate (AsO_4^{3-}) or arsenite (AsO_2^-).

Arsenic and the heavy metal ions are toxic primarily due to their ability to react with sulfhydryl groups (—SH) in enzymes (Section 11.8).

15.7 Measuring Water Pollution

Biochemical Oxygen Demand

Many organic compounds that find their way into water can easily be oxidized by microorganisms that are also there. This is a natural process that prevents a buildup of organic waste in natural waters. To change this organic material into simple substances (such as CO_2 and H_2O) requires oxygen. The amount of dissolved oxygen required is called the **biochemical oxygen demand (BOD)**, and it is a measure of the quantity of dissolved organic matter. The oxygen is necessary so that the bacteria and other microorganisms can metabolize the organic matter that constitutes their food. Ultimately, given near-normal conditions and enough time, the microorganisms will convert huge quantities of organic matter into the following end products.

$$\text{Organic carbon} \longrightarrow CO_2$$

$$\text{Organic hydrogen} \longrightarrow H_2O$$

$$\text{Organic oxygen} \longrightarrow H_2O$$

$$\text{Organic nitrogen} \longrightarrow NO_3^- \text{ or } N_2$$

TABLE 15.10 ■ Solubility of Oxygen in Water at Various Temperatures

Temperature (°C)	Solubility of O_2 (g O_2/L H_2O)
0	0.0141
10	0.0109
20	0.0092
25	0.0083
30	0.0077
35	0.0070
40	0.0065

These data are for water in contact with air at 760 mm Hg pressure.

■ Characteristic BOD levels (g O_2/L):

Untreated municipal sewage	0.1–0.4
Runoff from barnyards and feed lots	0.1–10
Food-processing wastes	0.1–10

■ Fish cannot live in water that has less than 0.004 g O_2/L (4 ppm).

At 68°F (20°C) the solubility of O_2 in water under normal pressure of 1 atmosphere (atm) is only 0.0092 g O_2/L (Table 15.10). But, a stream containing 10 ppm by weight (just 0.001%) of an organic material with the formula $C_6H_{10}O_5$ has a BOD of 0.012 g O_2/L of water. Clearly, this BOD value exceeds the equilibrium concentration of dissolved O_2 at this temperature. As the bacteria utilize the dissolved oxygen in a stream or lake with this BOD, the oxygen concentration of the water may soon drop too low to sustain any form of fish life. Whether this happens depends on the opportunities for new oxygen to become dissolved in the water. Life forms can survive in water where the BOD exceeds the dissolved oxygen if the water is flowing vigorously in a shallow stream (this facilitates the absorption of more oxygen from the air via aeration).

BOD values can be greatly reduced by treating industrial wastes and sewage with oxygen or ozone. Numerous commercial cleanup operations now being developed and used employ this type of "burning" of the organic wastes. Another benefit of treating wastewater with oxygen is that some of the nonbiodegradable material becomes biodegradable as a result of partial oxidation.

Highly polluted water often has a high concentration of organic material, with resultant large biochemical oxygen demand (Figure 15.5). In extreme cases, more oxygen is required than is available from the environment. The result is that fish and other aquatic life can no longer survive. The aerobic bacteria (those that require oxygen for the decomposition process) die. As a result, even more lifeless organic matter results, and the BOD soars. Nature, however, has a backup system for such conditions. A whole new set of microorganisms (anaerobic bacteria) takes over; these organisms take oxygen from oxygen-containing compounds to convert organic matter to CO_2 and water.

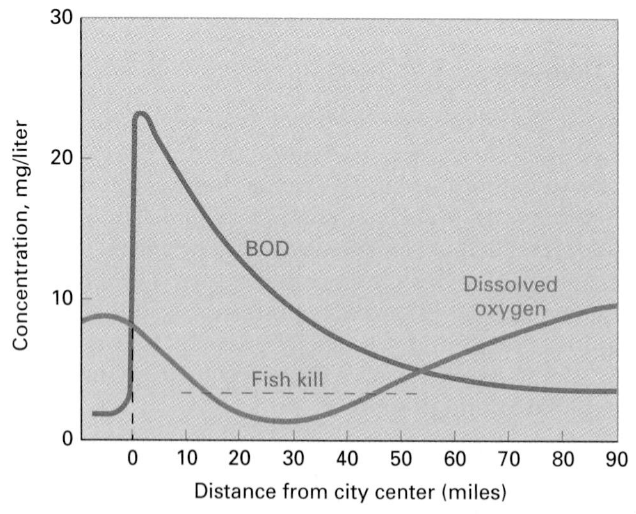

Figure 15.5 Oxygen depletion and rising BOD due to a sewage spill. The plots represent the result of a sewage spill at the center of a city (0 mi) into a river that flows at the rate of 750 gal/s. Near the spill the oxygen content drops and the BOD soars to 23 mg/L (0.023 g/L). Note that it takes 70 mi for the stream to recover, and that 15 mi to 45 mi downstream the oxygen content drops so low that fish will die.

Organic nitrogen is converted to elemental nitrogen by these bacteria. Given enough time, enough oxygen may become available, and aerobic oxidation will then return.

■ High concentrations of organic pollutants
↓
Low oxygen concentration
↓
Dead organisms
↓
Higher concentrations of organic pollutants
↓
Lower oxygen concentrations
↓
Anaerobic conditions
↓
Oxygen-requiring fish, shellfish, and other aquatic organisms leave or die

■ SELF-TEST 15B

1. The federal law requiring cleanup of hazardous waste sites that can pollute water is called ____________.
2. Hazardous wastes can be indiscriminately placed in landfills. (a) True, (b) False.
3. Name three household wastes that can contaminate groundwater with the same harmful chemicals as industrial wastes. Beside each, list the harmful chemical.

Household Waste	**Harmful Chemical**
____________	____________
____________	____________
____________	____________

4. List four household waste types that lend themselves to recycling.
5. The amount of oxygen required to oxidize a given amount of organic material is called the ____________ ____________ ____________, which is abbreviated ____________.
6. Name two industrial products whose manufacture introduces heavy metals into groundwater.
7. Name two industrial products whose manufacture introduces chlorinated organic compounds into groundwater.
8. Two heavy metals found in water are ____________, and ____________.

15.8 How Water Is Purified Naturally

Water is a natural resource that, within limitations, is continuously renewed. The water cycle offers a number of opportunities for nature to purify its water. The worldwide ***distillation*** process results in rainwater containing only traces of nonvolatile impurities, along with gases dissolved from the air. ***Crystallization*** of ice from ocean salt water results in relatively pure water in the form of icebergs. ***Aeration*** of groundwater as it trickles over rock surfaces, as in a rapidly running brook, allows volatile impurities to be released to the air and allows oxygen to be dissolved. ***Sedimentation*** (or ***settling***) of solid particles occurs in slow-moving streams and lakes. ***Filtration*** of water through sand rids the water of suspended matter such as silt and algae. Of very great importance are the ***oxidation*** processes carried out by bacteria and other microorganisms. Practically all naturally occurring organic materials—plant and animal tissue, as well as their waste materials—can be oxidized in surface waters as long as oxygen is available and their concentration is not too high. Finally, another natural process is ***dilution***. Most, if not all, pollutants found in nature are rendered harmless if reduced below certain levels of concentration by dilution.

■ Rainwater in clean air is very pure, containing only small amounts of dissolved gases such as N_2, O_2, and CO_2.

Pumping mud from the bottom to make room for deeper boats in the Mississippi River. When the bottom of a natural body of water is disturbed, the possibility exists of redistributing into the water pollutants that had been trapped in the mud. *(Grant Heilman Photography)*

Before the explosion of the human population and the advent of the Industrial Revolution, natural purification processes were quite adequate to provide ample water of very high purity in all but desert regions. Nature's purification processes can be thought of as massive but somewhat delicate.

Today, the activities of humans often push the natural purification processes beyond their limit, and polluted water accumulates. A simple example comes from dragging gravel from streambeds. The excavation leaves large amounts of suspended matter in the water, and for miles downstream, aquatic life is destroyed. Eventually, the solid matter settles, and normal life can be found again in the stream.

A more complex example—one that perhaps cannot be solved by relying on natural purification processes—is pollution by organic molecules that cannot be easily oxidized by microorganisms. A **biodegradable** substance is composed of molecules that are broken down to simpler ones by microorganisms. For example, cellulose suspended in water will be converted to carbon dioxide and water. A **nonbiodegradable** substance, on the other hand, cannot be easily converted to simpler molecules by microorganisms. If the conversion process is extremely slow, or if it cannot be done at all by natural microorganisms, nonbiodegradable substances tend to accumulate in the environment.

Some organic compounds, notably some of those produced synthetically, are nonbiodegradable. When these substances are introduced into the environment, they simply stay in the natural waters or are absorbed by life forms and remain intact for a long time. Branched-chain detergent molecules, for example, cannot easily be decomposed by microorganisms.

$$CH_3-\underset{\substack{|\\CH_3}}{CH}-CH_2-\underset{\substack{|\\CH_3}}{CH}-CH_2-\underset{\substack{|\\CH_3}}{CH}-CH_2-\underset{\substack{|\\CH_3}}{CH}-C_6H_4-SO_3^-\ Na^+$$

A branched-chain sodium alkylbenzenesulfonate detergent molecule

The first detergents that contained such molecules accumulated and caused noticeable foaming in rivers and streams. The branched-chain detergents were soon replaced by linear-chain alkylbenzenesulfonate detergents (Section 11.4), which are easily decomposed by microorganisms—they are biodegradable.

$$CH_3CH_2CH_2CH_2CH_2CH_2CH_2CH_2CH_2CH_2CH_2CH_2-C_6H_4-SO_3^-\ Na^+$$

A linear sodium alkylbenzenesulfonate detergent molecule

■ DDT is fat soluble.

Other examples of nonbiodegradable organic pollutants are the chlorinated and polychlorinated hydrocarbons. Many of these compounds are used as insecticides (Section 17.5). The insect-killing ability of dichlorodiphenyltrichloroethane (DDT) was first recognized in 1939. By the end of World War II its insecticidal properties were legendary due to its ability to kill everything from malaria-causing mosquitoes to lice. This success prompted the introduction of numerous other chlorinated hydrocarbons as insecticides, and by the early 1960s their use was widespread throughout the world. These compounds are broad-spectrum insecticides, killing most insects rather effectively; however, they also biodegrade very slowly, so they tend to accumulate in the environment. This persistence is especially troublesome since their slow biodegradation allows such compounds to accumulate in the food chain. Fish-eating birds, for example, can accumulate large quantities of these insecticides that have accumulated in fish that ate smaller organisms containing these compounds. The populations of falcons, pelicans, bald eagles, ospreys, and other birds have been endangered by persistent pesticides. DDT causes reproductive failure in birds by interfering with the mechanisms that produce strong eggshells. After the use of DDT was banned in the United States in 1972, the numbers of surviving hatchlings increased rather dramatically.

■ A hypothetical food chain might be

Plant ⟶ insect ⟶ fish ⟶ hawk

15.9 Water Purification Processes: Classical and Modern

The outhouses of some rural dwellers had their counterparts in city cesspools, which were basically holes in the ground into which sewage flowed. In cesspools, organic matter is decomposed by anaerobic bacteria, producing some pretty bad-smelling chemicals such as hydrogen sulfide, which has a characteristic rotten-egg odor. The terrible job of cleaning cesspools inspired the development of cesspools that could be flush-cleaned with water, followed by a connecting series of such pools that could be flushed from time to time. City sewage systems with no holding of the wastes were the next step.

■ Cesspools were an early and crude form of the modern activated sludge process.

■ Sewage includes everything that flows from the sinks, tubs, washing machines, and toilets in our homes, factories, and public buildings. It excludes wastewater treated separately by industrial facilities.

At first, city sewage systems did little but channel sewage water to rivers and streams, where natural purification processes were expected to clean the water for the next users downstream. Today, however, sewage is treated using a combination of methods that can render the treated municipal wastewater almost as clean as the natural waters into which it is being discharged. The simplest treatment method is ***primary wastewater treatment***, which copies two of nature's purification methods, settling and filtration. In primary treatment, sewage goes from the primary sedimentation tank shown in Figure 15.6 to the chlorinator.

■ Sewage is still 99.9% water.

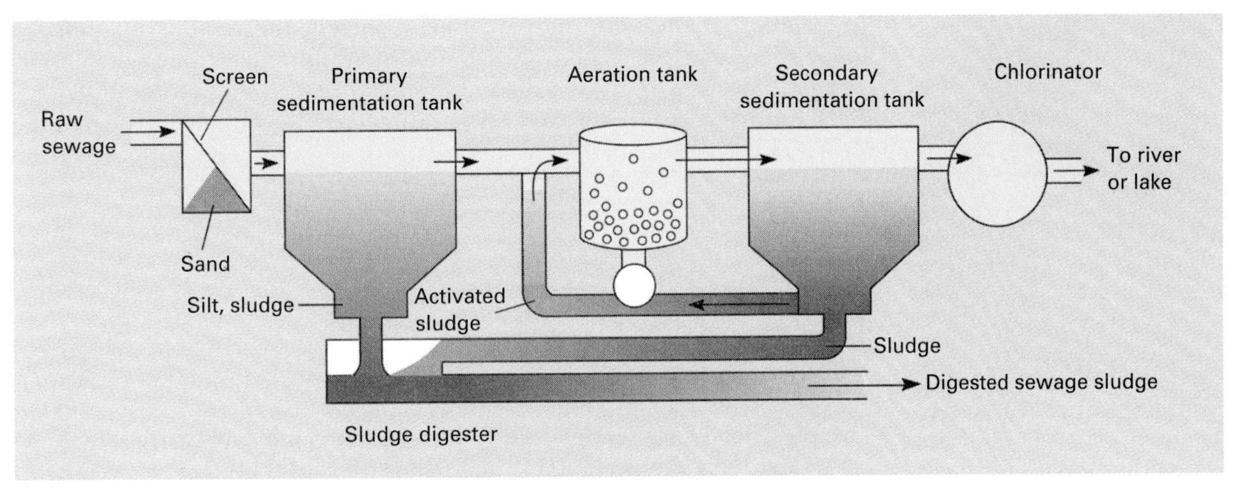

Figure 15.6 The steps in primary and secondary sewage treatment.

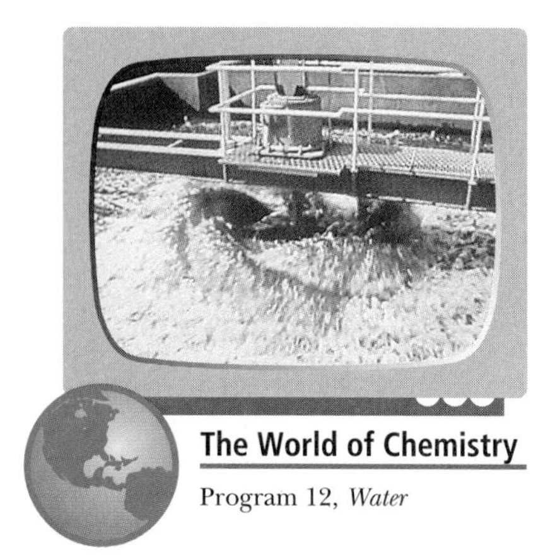

The World of Chemistry

Program 12, *Water*

Figure 15.7 The aeration tank in a sewage treatment plant.

■ Definitions of "pure water":
Chemist: "Pure H_2O—no other substance."
Parent: "Nothing harmful to my child."
Game and Fish Commission: "Nothing harmful to animals."
Sunday boater: "Pleasing to the eye and nose, no debris."
Ecologist: "Natural mixture containing necessary nutrients."

■ Carbon black is used in many processes for manufacturing extremely pure organic compounds such as pharmaceuticals or food additives. **Adsorption** is the process by which molecules are attracted and held onto a surface.

A sewage outfall, where treated water is returned to the environment. This water should be as clean as possible. *(Doug Wechsler)*

Primary treatment removes 40% to 60% of the solids present in sewage and about 30% of the organic matter present. Calcium hydroxide and aluminum sulfate are added to produce aluminum hydroxide, which is a sticky, gelatinous precipitate that settles out slowly, carrying suspended dirt particles and bacteria with it.

$$3\ Ca(OH)_2(aq) + Al_2(SO_4)_3(aq) \longrightarrow 2\ Al(OH)_3(s) + 3\ CaSO_4(s)$$

For many years municipal sewage treatment plants had only primary treatment, followed by chlorination of the treated wastewater (see Section 15.11) before it was discharged into a suitable river or stream. Chlorination kills any remaining harmful pathogens. Presumably, natural processes would get rid of the remaining solids and dissolved organic matter.

In realizing that this treatment was not sufficient to protect the public from contaminated water, the writers of the 1972 Clean Water Act required that sewage treatment plants also provide ***secondary wastewater treatment***, which revives the old cesspool idea but under more controlled conditions. Modern secondary treatment operates in an oxygen-rich environment (aerobic; Figure 15.7), whereas the cesspool operates in an oxygen-poor environment (anaerobic). The results are the same: The organic molecules that will not settle are consumed by microorganisms and the resulting sludge will settle.

Even a combination of primary and secondary wastewater treatment systems will not remove dissolved inorganic materials such as toxic metal ions, nutrients such as nitrate ions (NO_3^-) or ammonium ions (NH_4^+), or nonbiodegradable organic compounds such as chlorinated hydrocarbons. These materials can be removed by a variety of ***tertiary wastewater treatment*** methods that are selectively introduced where the nature of the wastewater requires them. One obstacle to tertiary treatment is the initial expense of modifying sewage treatment plants and the ongoing expense of additional treatment.

Filtration of the water through ***carbon black*** is a type of tertiary treatment effective for removing soluble organic compounds that are nonbiodegradable and thus remain in the water after secondary treatment. Carbon black consists of finely divided carbon particles with a large surface area on which solutes, including certain potentially toxic substances, can be *adsorbed.*

A different kind of tertiary treatment is needed to remove ammonia or ammonium ion. Because nitrogen is a nutrient for aquatic microorganisms, excessive nitrogen released to natural waters can cause a soaring BOD with the accompanying fish kills and other problems. The water is exposed to **denitrifying bacteria** that convert ammonium ion or ammonia to harmless nitrogen gas.

$$NH_4^+(aq) \text{ or } NH_3(aq) \xrightarrow{\text{Denitrifying bacteria}} N_2(g)$$

15.10 Softening Hard Water

The presence of Ca^{2+}, Mg^{2+}, Fe^{3+}, or Mn^{2+} ions will impart "hardness" to water. Hardness is objectionable because (1) it causes precipitates (scale) to form in boilers and hot-water systems, (2) it causes soaps to form insoluble

THE WORLD OF CHEMISTRY

Bioremediation of Hazardous Waste Sites

Program 25, *The Environment*

Bioremediation is the microbial detoxification or degradation of wastes. Companies that are designated by the EPA to clean up a hazardous waste site, such as a Superfund site, are faced with high costs to do the cleanup. In the early days of waste site cleanup, contaminated soil was often dug up and transported to a secure landfill. But as transportation costs and landfills began to increase, this method fell out of favor. (When you think about it, it makes more sense to clean up the site and restore it to its former state.) Bioremediation offers the possibility of getting rid of the (at times small amounts of) contaminants or converting them to less toxic substances, and the site is left very close to its original state. It can even possibly be sold to a new owner. (If all the soil is dug up and hauled away, the property might not be as desirable.)

In bioremediation, a cue is taken from the microbes that eat ordinary wastes—if bacteria can be found to eat that, then maybe some exist that can eat other chemicals, including those that are considered hazardous wastes. Bioremediation methods fall into three categories—land treatment, in which the bacteria are spread across the surface of the land; bioreactors, in which soil and liquids are removed from the surface and placed in a reaction vessel for a period of time; and *in situ* treatment, in which bacteria are intimately mixed with the soil or water.

Cleanup from leakage of underground oil storage tanks is one of the most direct applications of bioremediation. The leaked oil, a hydrocarbon, is food for many kinds of bacteria, and finding bacteria that are compatible with the oil at a site is a fairly straightforward task. One of the largest oil bioremediation projects to date has been the cleanup of the March 1989 spill of 11 million gallons of crude oil by the *Exxon Valdez*. Immediately after that spill, it was proposed to use the bacteria already on the beaches there, but one problem arose. There was plenty of food for the bacteria (the oil) but a limited amount of available nutrients such as nitrogen, phosphorus, and trace elements, which they required to thrive. This problem was solved when the EPA allowed the use of an oil-soluble slow-release fertilizer that had been developed a few years earlier by a French company when the tanker *Amoco Cadiz* went aground off the coast of Brittany. The combination of the bacteria and the fertilizer worked, and the shores of Prince William Sound became cleaner much more quickly and probably at much less cost.

According to the EPA, about 135 sites are being considered or planned, or are in operation for bioremediation.

curds (this reaction does not occur with some synthetic detergents—see Section 11.4), and (3) it can impart a disagreeable taste to the water.

Hardness due to calcium or magnesium, present as bicarbonates, is produced when water containing carbon dioxide trickles through limestone or dolomite.

$$\underset{\text{Limestone}}{CaCO_3(s)} + CO_2(g) + H_2O(\ell) \longrightarrow Ca^{2+}(aq) + 2\,HCO_3^-(aq)$$

$$\underset{\text{Dolomite}}{CaCO_3 \cdot MgCO_3(s)} + 2\,CO_2(g) + 2\,H_2O(\ell) \longrightarrow Ca^{2+}(aq) + Mg^{2+}(aq) + 4\,HCO_3^-(aq)$$

Such "hard water" can be softened by removing these ions. One of the methods for softening water is the lime–soda process. The lime–soda process takes advantage of the facts that calcium carbonate ($CaCO_3$) is much less soluble than calcium bicarbonate ($Ca(HCO_3)_2$) and that magnesium hydroxide is much less soluble than magnesium bicarbonate. The raw materials added to

The World of Chemistry

Program 12, *Water*

Over time, copper water pipes like this one become coated with deposits of the minerals dissolved in the water.

Chlorine contact tank in a sewage treatment plant. Whatever the level of treatment, the water is chlorinated to kill disease-causing organisms before it is returned to natural waterways. *(Runk/Schoenberger/Grant Heilman Photography)*

■ Hard water contains metal ions that react with soaps and give precipitates.

■ Soft water: <65 mg of metal ion per gallon. Slightly hard water: 65 mg to 228 mg. Moderately hard water: 228 mg to 455 mg. Hard water: 455 mg to 682 mg. Very hard water: >682 mg.

■ For people on low-sodium diets, lime–soda treated water might represent too high a daily dose of sodium.

the water in this process are hydrated lime ($Ca(OH)_2$) and soda (Na_2CO_3). In the system, several reactions then take place, which can be summarized as follows:

$$HCO_3^-(aq) + OH^-(aq) \longrightarrow CO_3^{2-}(aq) + H_2O(\ell)$$

$$Ca^{2+}(aq) + CO_3^{2-}(aq) \longrightarrow CaCO_3(s)$$

$$Mg^{2+} + 2\ OH^-(aq) \longrightarrow Mg(OH)_2(s)$$

The overall result of the lime–soda process is to precipitate almost all the calcium and magnesium ions and to leave sodium ions as replacements.

Iron present as Fe^{2+} and manganese present as Mn^{2+} can be removed from water by oxidation with air (aeration) to higher oxidation states. If the pH of the water is 7 or above (either naturally or through the addition of lime), the insoluble compounds $Fe(OH)_3$ and $MnO_2(H_2O)_x$ are produced and precipitate from solution.

The desire for and achievement of soft water for domestic use has sparked a rather heated health debate during the past two decades. Soft water is usually acidic and contains Na^+ ions in the place of di- and trivalent metal ions. An increased intake of Na^+ is known to be related to heart disease. The acidic soft water is also more likely to attack metallic pipes, joints, and fixtures, resulting in the dissolution of toxic ions such as Pb^{2+}. One way to avoid sodium ions in drinking water and to use less soap when washing would be to drink only naturally hard water and to do your washing in soft water.

15.11 Chlorination and Ozone Treatment of Water

With the advent of chlorination of water supplies in the early 1900s, the number of deaths in the United States caused by typhoid and other waterborne diseases dropped from 35 per 100,000 population in 1900 to 3 per 100,000 population in 1930.

Chlorine is introduced into water as the gaseous element (Cl_2), and it acts as a powerful oxidizing agent for the purpose of killing bacteria in water. This process is used in treating both water that will become tap water and wastewater before it is released. Chlorination largely prevents the principal waterborne diseases spread by bacteria, which include cholera, typhoid, paratyphoid, and dysentery.

In spite of chlorination, most city water supplies are not bacteria-free, but only rarely do these surviving bacteria cause disease. Today the most common waterborne bacterial disease is *giardiasis*, a gastrointestinal disorder. Most often this disease comes from surface water, but on occasion it can be traced to city water systems.

Chlorination of industrial wastewater and city water supplies presents a potential threat because of the reaction of chlorine with residual concentrations of organic compounds to produce **disinfection by-products**.

$$\text{Water containing organic compounds} \xrightarrow{\text{Chlorine}} \text{chlorinated organic disinfection by-products}$$

These disinfection by-products, which may be present at levels of a few parts per million or less, include dichloromethane, chloroform, trichloroethylene, and chlorobenzene—all suspected carcinogens. According to the EPA, mutagenic or carcinogenic chemicals have been found in 14 major river basins in the United States. It is estimated that more than 500 water systems in the United States exceed EPA's maximum of 0.1 ppm for chlorinated hydrocarbons. The presence of these chlorinated hydrocarbons can be prevented by more efficient removal of the organic matter that becomes chlorinated, but unfortunately even the best-designed purification systems (including carbon filtration) allow some organic compounds to pass through, only to become chlorinated.

One way to eliminate chlorinated hydrocarbons as disinfection by-products is to use ozone (O_3) as the disinfectant. Ozone is used in more than 1000 water treatment plants, mostly in Europe. The ozone is produced on site by passing oxygen or air through an electric discharge. This process normally gives about a 20% ozone–oxygen mixture that is a very strong oxidizer. Although the use of ozone in the United States has been minimal, 20 of the 25 ozone plants in the United States have been built in the last decade.

Ozonation, like chlorination, is also not without potentially harmful disinfection by-products. Bromide ion (Br^-), which is found in most natural waters, is oxidized by ozone to bromate ion (BrO_3^-), which is a suspected carcinogen. Generally, this single known harmful disinfection by-product is considered less of a risk factor than the numerous chlorinated hydrocarbons produced by chlorine disinfection.

The World of Chemistry
Program 12, *Water*

Ozone gas, a disinfectant, bubbling through water.

15.12 Freshwater from the Sea

Because seawater covers 72% of the earth, it is not surprising that it is considered a water source for areas where freshwater supplies aren't sufficient to meet the demand. The oceans contain an average 3.5% (35,000 ppm) dissolved salts by weight, a concentration too high to make ocean water useful for drinking, washing, or agricultural use (Table 15.11). The total of dissolved ions must be reduced to below 500 ppm before the water is suitable for human consumption.

The technology has been developed for the conversion of seawater to freshwater. The extent to which this technology is actually put to use depends on the availability of freshwater and the cost of the energy for the conversion. More than 3000 desalination plants were in operation throughout the world in the early 1990s. Two methods used to purify seawater are reverse osmosis and solar distillation.

TABLE 15.11 ■ Ions Present in Seawater at Concentrations Greater than 0.001 g/kg

Ion	g/kg Seawater
Cl^-	19.35
Na^+	10.76
SO_4^{2-}	2.71
Mg^{2+}	1.29
Ca^{2+}	0.41
K^+	0.40
HCO_3^-, CO_3^{2-}	0.106
Br^-	0.067
$H_2BO_3^-$	0.027
Sr^{2+}	0.008
F^-	0.001
Total	35.129

Reverse Osmosis

An extremely thin piece of material such as a sheet of synthetic polymer or animal tissue can allow molecules to pass through it. Such a material is called a membrane and is said to be **permeable** to those molecules and ions that can pass through. Permeability is dependent on the presence of tiny passages within the membrane. A membrane permeable to water molecules but not to ions or molecules larger than water molecules is called a **semipermeable** mem-

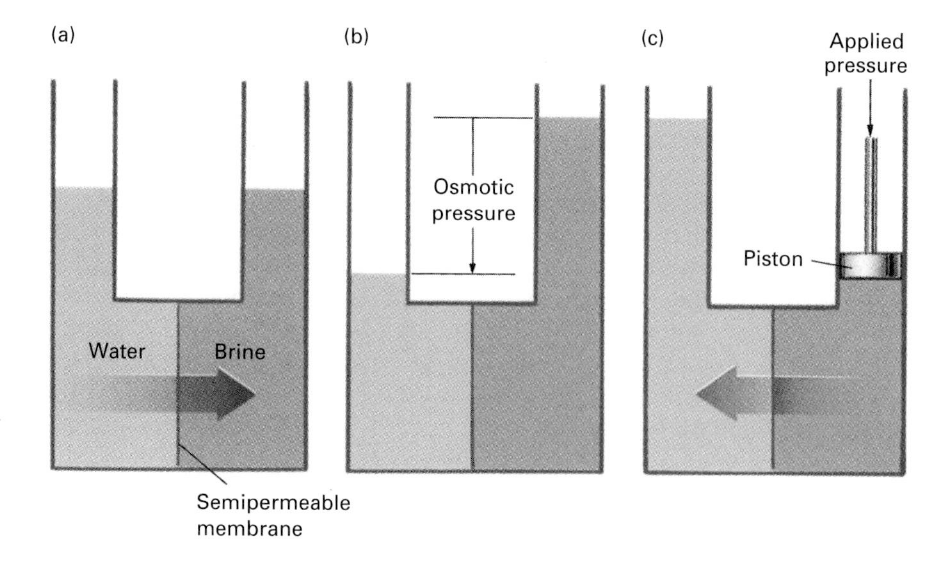

Figure 15.8 Osmosis (a, b) and reverse osmosis (c). (a) In normal osmosis, water molecules pass through the semipermeable membrane from the less concentrated into the more concentrated solution, in this case from pure water into the brine. (b) At a certain height of the brine solution in this apparatus, the pressure of the column of water is equal to the osmotic pressure and the flow stops. (c) In reverse osmosis, the application of external pressure greater than the osmotic pressure forces water molecules to the pure water side.

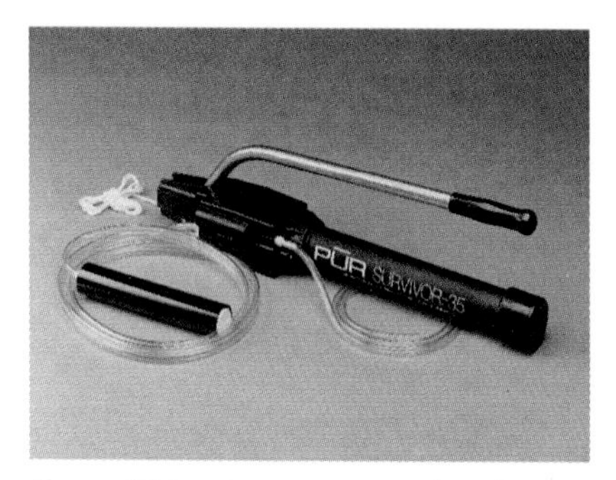

Figure 15.9 An emergency hand-operated water desalinator that works by reverse osmosis. It can produce 4.5 L of pure water per hour from seawater. Such devices can be very useful for persons adrift at sea. *(Courtesy of Recovery Engineering, Inc.)*

■ Irrigation water of desert fields dissolves about 2 tons of salt per acre per year. Irrigation wastewater carries the salt back to the Colorado River.

brane. Many membranes made from synthetic polymers have this characteristic. One such polymer is cellulose acetate. If a semipermeable membrane is placed between seawater and pure water, the pure water will pass through the membrane to dilute the seawater. This is a process called **osmosis**. The liquid level on the seawater side rises as more water molecules enter than leave, and pressure is exerted on the membrane until the rates of diffusion of water molecules in both directions are equal. **Osmotic pressure** is defined as the external pressure required to *prevent* osmosis. Figure 15.8 illustrates the concepts of osmosis and osmotic pressure.

Reverse osmosis is the application of pressure to cause water to pass through the membrane from the aqueous-solution side to the pure-water side (Figure 15.9). The osmotic pressure of normal seawater is 24.8 atmospheres (atm). As a result, pressures greater than 24.8 atm must be applied to cause reverse osmosis. Pressures up to 100 atm are used to provide a reasonable rate of reverse osmosis and to account for the increase in salt concentration that occurs as the process proceeds.

The largest reverse osmosis plant in operation today is the Yuma Desalting Plant in Arizona. This plant, which began operation in the 1980s, can produce 100 million gallons of water per day. The plant was built to reduce the salt concentration of irrigation wastewater in the Colorado River from 3200 ppm to 283 ppm. The project is part of a U.S. commitment to supply Mexico with a sufficient quantity of water suitable for irrigation. The Mediterranean island of Malta now uses four reverse osmosis plants that produce a total of 12 million gallons of freshwater per day from the sea. On Florida's Sanibel Island, increasing salinity in the well water led to the installation of a reverse osmosis system. This facility has a design capacity of 3.6 million gallons per day and has one of the lowest energy-consumption rates per 1000 gal of potable water of any comparably sized system in commercial use.

15.13 Pure Drinking Water for the Home

In spite of all the efforts taken to purify public water supplies, many consumers are concerned about the quality of the water that comes out of the taps in their homes, schools, and places of business. Parents of small children are especially worried about chemicals such as lead and carcinogenic organic compounds that are chlorine disinfection by-products. Many have turned to bottled water or home water treatment devices that offer some protection from these harmful trace pollutants.

■ Americans spend about $350 million a year on bottled water. Buyer beware: A very wide variety of standards exists for bottled water.

While some bottled water is untreated groundwater (sometimes called "spring" water), most bottled water has passed through one or more purification steps (Figure 15.10). The three purification methods (which can also be done at home)—distillation, carbon filtration, and reverse osmosis—have already been discussed as methods used for treating municipal water. The maximum levels of contaminants allowed by the EPA after these treatments are listed in Table 15.12 (p. 414). Each of these methods is expensive and results in a high cost per gallon of treated water. In the case of bottled water or home water treatment, the cost of treatment is not the major factor, since only the small amount of water needed for human consumption needs to be specially purified. Figure 15.10 shows which trace pollutants are removed or allowed to remain in the water by each treatment method, color-coded to

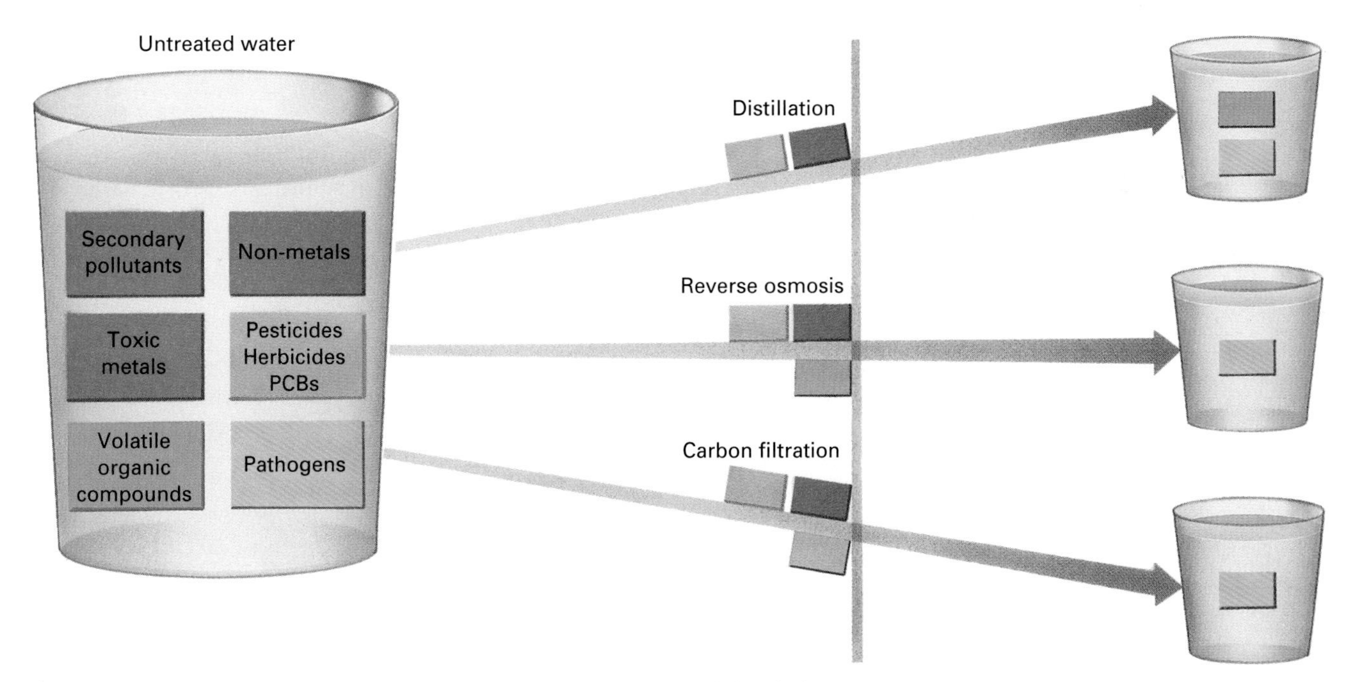

Figure 15.10 Final steps that can be used in water purification. Both bottled water and municipal tap water are purified in these ways. The color code shows which pollutants are removed by each method. Pathogenic bacteria can and do pass through all these methods. This is why municipal tap water must be treated with chlorine or some other disinfectant before release into the system.

TABLE 15.12 ■ A Partial List of Maximum Contaminant Levels (MCL) for Drinking Water Allowed by the EPA*

Metals	
Beryllium	0.004
Cadmium	0.005
Chromium	0.100
Copper	1.300
Lead	0.015
Metalloids	
Arsenic	0.050
Antimony	0.006
Nonmetals	
Fluoride	4.00
Nitrate	10.00
Volatile Organic Compounds	
Benzene	0.005
Carbon tetrachloride	0.005
Vinyl chloride	0.002
Trichloroethylene	0.005
Hexachlorobenzene	0.001
Styrene	0.100
Herbicides, Pesticides, PCBs	
Chlordane	0.002
Endrin	0.0002
Heptachlor	0.0004
Lindane	0.0002
Methoxychlor	0.040
Toxaphene	0.003
PCBs	0.0005
Secondary Contaminants	
Iron	0.30
Manganese	0.05
Zinc	5.00
Chloride	250.00
Sulfate	250.00
Total dissolved solids	500.00

* Values are in milligrams per liter.

examples of the types of contaminants in Table 15.12. Thus, the analysis on the label of the bottled water is highly important and should be read with care before purchase.

15.14 What About the Future?

Our water quality in the United States has improved substantially since the passing of the Clean Water Act of 1977 and its subsequent amendments. Municipal water is now monitored more carefully for heavy metals, pesticides, and chlorinated organic molecules than was even thought possible just 30 or 40 years ago. As a result of this continual monitoring, consumers can feel better about the water they drink. Industry is complying more with hazardous waste-reduction and pollution-discharge regulations, and much research is focused on finding ways to totally eliminate releases of harmful chemicals that might find their way into our water supplies.

In spite of this progress, some problems still exist. Household wastes continue to be a major potential source of water pollution. As long as households are not offered inexpensive alternatives to mixing hazardous chemicals such as paints, solvents, and waste oil with ordinary household garbage, groundwater and surface water will remain at risk. **Recycling** of wastes is growing, but the recycling of some items such as automobile batteries, mercury batteries, unused pesticides, solvents, and used lubricating oil is still severely limited by economic factors that overshadow a strong desire by citizens for purer water supplies.

The politics of water protection must improve in the future. Today, many communities are taking rather short-sighted approaches to hazardous waste disposal. In effect, these communities are saying "not in my backyard!" States that allow communities veto power over the location of hazardous waste sites have been singled out for retaliation by other states with active hazardous waste disposal sites located within their borders. These states are passing laws effectively banning another state's hazardous wastes if that state doesn't allow

A paper mill without pollution controls. Prior to the Clean Air Act and the Clean Water Act, paper mills in the United States were among the most notorious of polluters. *(Dan Guravich/Photo Researchers)*

hazardous waste disposal in its own borders or if it allows communities to have veto power over the location of disposal sites within their city or community limits. Political problems like these can only be solved when everyone recognizes the importance of the proper disposal of hazardous industrial and household wastes.

The future will see increased water conservation measures by everyone who uses water—industry, agriculture, and households. Recently, the city of Boston has encouraged water-efficient fixtures in homes as well as industry water audits and system-wide leak detection. These steps, coupled with public education about how to reduce annual water usage, have resulted in a 16% decrease in water demand. In Texas, farmers drawing water from the Ogallala aquifer (Section 15.1) have begun to use an old-fashioned furrow technique to introduce water more directly to the plant's roots, thus causing less water loss into the ground. We will probably see an expansion of the capacity for desalination of seawater, particularly for industrial and agricultural uses. Most large cities will have to replace leaky water mains and plumbing, which account for up to one third of their water use. This will mean higher water bills in the future.

The discussion in this chapter has focused on water quality in the United States. A combined program of water conservation, protection of water quality, and water recycling will help to alleviate the water crisis in the United States and other industrialized nations. However, contaminated water is still a serious problem for 75% of the world's population. It has been estimated that 80% of the sickness in the world is caused by contaminated water. For years, many countries and international organizations have provided financial and technical aid to help improve the water quality in developing countries. However, much work remains to be done to reduce sickness caused by contaminated water.

The NIMBY response (Not in My Backyard). This protest against a solid waste incinerator in East Liverpool, Ohio, in 1992 is typical of the reaction to plans for incinerators, landfills, and other repositories for waste. Sometimes the protests result from anger at being left out of the decision-making process, rather than fact-based conclusions that the site would be harmful to those in the area. *(Visuals Unlimited/Bill Beatty)*

SELF-TEST 16C

1. Which water purification process is not a natural process? (a) Distillation, (b) Aeration, (c) Filtration, (d) Reverse osmosis, (e) Settling.
2. Which word can be used to describe organic pesticides that are not readily biodegradable? (a) Permanent, (b) Persistent, (c) Nonvolatile.
3. Name two water purification methods that are part of primary wastewater treatment.
4. Secondary wastewater treatment operates under (aerobic/anaerobic) conditions.
5. A cesspool operates under (aerobic/anaerobic) conditions.
6. Select the ions that may cause water to be hard: (a) Sodium, (b) Calcium, (c) Magnesium, (d) Potassium.
7. Two methods used to kill harmful microorganisms in water are ______________ and ______________.
8. What are the four metal ions present in seawater at concentrations of 400 ppm or higher?
9. Water flows through a semipermeable membrane from a solution of low concentration to a solution of higher concentration. This process is called ______________.

MATCHING SET

____ 1. Sedimentation
____ 2. Biodegradable
____ 3. Clean Water Act of 1977
____ 4. BOD
____ 5. Aquifer
____ 6. Ammonium ion
____ 7. Unused paint
____ 8. Reverse osmosis
____ 9. Water hardness
____ 10. Ozone treatment
____ 11. Recycling
____ 12. Superfund
____ 13. Incineration
____ 14. Aeration
____ 15. Carbon adsorption
____ 16. Chlorination
____ 17. Auto battery

a. A measure of dissolved organic material in water
b. Can impart lead to water supplies when improperly disposed of
c. Source of groundwater
d. Caused by metal ions such as Ca^{2+} and Mg^{2+} in solution
e. A nutrient for microorganisms living in water
f. Provides for hazardous waste site cleanup
g. Alternative to landfills but can contribute to air pollution
h. Disinfection method commonly used for drinking water and wastewater
i. Removes organic compounds from water
j. Relieves dependence on landfills
k. Common household hazardous waste
l. Primary sewage treatment process
m. Disinfectant method that can produce bromate ion from bromide ion
n. Secondary sewage treatment process
o. Naturally decomposed to simpler compounds
p. Shifted responsibility for water purity to wastewater discharger
q. Uses pressure to purify water

QUESTIONS FOR REVIEW AND THOUGHT

1. Define the following terms:
   (a) Surface water
   (b) Groundwater
   (c) Aquifer
   (d) Brackish water
   (e) Pollution
   (f) Groundwater recharge
   (g) Potable
   (h) Dilution
   (i) Hazardous waste
2. Define the following terms:
   (a) BOD
   (b) Distillation
   (c) Aeration
   (d) Sedimentation
   (e) Biodegradable
   (f) Hard water
   (g) Reverse osmosis
   (h) Nonbiodegradable
   (i) Disinfection by-products
   (j) Semipermeable
3. Where does most of the water go that falls on the continental United States every day?
4. Explain how rainwater becomes groundwater.
5. Explain how groundwater can become contaminated with pollutants.
6. What is the origin of brackish water?
7. What activity is the largest single user of water?
8. What causes the level of an aquifer to drop? Cite some examples of the effects of aquifers dropping in level.
9. How much water do you think you use per day? List your uses and include water that might be used "for you," such as in food preparation in a restaurant.
10. What would you expect to find dissolved in "clean" water?
11. What does the term "groundwater recharge" mean? What is the source of water that is used for this purpose?
12. Name five kinds of pollution often found in water. Give a source for each.
13. Prior to the enactment of the Clean Water Act, who was responsible for ensuring that water being used was pure? After passage of this act, who is now responsible?
14. Both surface water and groundwater (natural waters) often contain dissolved ions. Name three positive and three negative ions that are often found in natural waters.
15. Name two methods of disposal for solid wastes from industry and households. Which one of these has the greater possibility to adversely impact water quality?
16. What is the "Superfund"? Explain how it is used to improve water quality.
17. Describe a way by which a landfill can be made more "secure" in terms of water quality protection.
18. Go to the hardware department in a large department store and choose five products. Then list the kinds of hazardous wastes the manufacture of these products might produce. Use Tables 15.6 and 15.7 as guides.
19. Name three common household wastes and the kinds of chemicals they contain that might be harmful to water quality.
20. Describe how measuring the biochemical oxygen demand (BOD) of a sample of water indicates something about its purity.
21. Describe how pure water can become contaminated with lead in the home.
22. Describe how the BOD of wastewater can be lowered.
23. Explain how distillation purifies a sample of water.
24. Explain how aeration purifies a sample of water containing dissolved organic compounds.
25. Explain how settling and filtration purify water samples.

26. What is the difference between "biodegradable" and "nonbiodegradable"? If you had a choice between using a biodegradable and a nonbiodegradable detergent to clean your clothes, which would you choose?
27. Chlorinated hydrocarbons and branched-chain hydrocarbons are nonbiodegradable. What happens to them when they are released into the environment?
28. Name the two methods of primary sewage treatment.
29. Too-high concentrations of nitrogen compounds like ammonia (NH_3) adversely affect water quality. What tertiary sewage treatment method gets rid of these compounds?
30. Name the ions commonly present in hard water. What kinds of problems do they cause?
31. How is chlorination of water similar to aeration of water? How are these different?
32. What are "disinfection by-products," and how are these potentially harmful?
33. Explain how reverse osmosis can be used to purify seawater.
34. Which method of purification of drinking water for the home would most likely get rid of dissolved organic compounds?

CHAPTER 16

Air: The Precious Canopy

Planet Earth is enveloped by a few vertical miles of chemicals that compose the gaseous medium in which we exist—the atmosphere. Close to Earth's surface and near sea level, the atmosphere is mostly nitrogen and life-sustaining oxygen. It is the few little fractions of a percentage point of other chemicals that make a difference in the quality of life in various places on Earth. Urbanization is the main culprit in causing problems in our atmosphere. With its vast number of vehicles and increases in industrialization, urbanization has produced an abnormal increase in some of the naturally occurring "minor" chemicals in the atmosphere—compounds such as nitrogen oxides, sulfur dioxide, carbon monoxide, carbon dioxide, and ozone. Increased amounts of these compounds in the atmosphere can create an unhealthful, unpleasant medium. An atmosphere containing these unwanted and harmful ingredients is called *polluted*. Some questions that will be answered in this chapter are the following:

- What is clean air, and what role does government play in helping to maintain clean air?
- What are the different kinds of air pollution, and what are their sources?
- What kinds of chemical reactions take place in the atmosphere that contribute to air pollution?
- What role does sunlight play in causing air pollution?
- What two kinds of industry-related air pollution have the potential to change our planet?
- What are some of the sources of indoor air pollution?

Viewed over the ocean in a photo taken from a space shuttle, the atmosphere appears as a thin dark blue band in the distance. *(Courtesy of NASA)*

The atmosphere of Earth is a fantastically large source of the elements nitrogen (N_2) and oxygen (O_2), with much smaller amounts of certain of the noble gases, including argon (Ar), neon (Ne), and xenon (Xe) (Table 16.1).

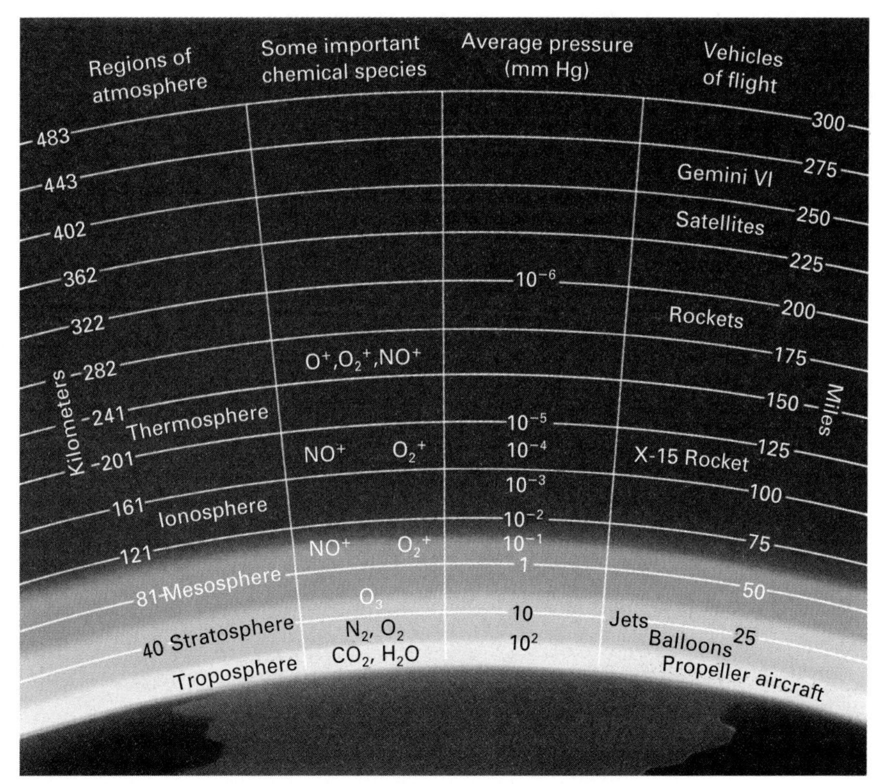

Figure 16.1 Some facts about our planet Earth's atmosphere.

TABLE 16.1 ■ Composition of Dry Air at Sea Level

Gas	Percentage by Volume
Nitrogen	78.084
Oxygen	20.948
Argon	0.934
Carbon dioxide	0.033*
Neon	0.00182
Hydrogen	0.0010
Helium	0.00052
Methane	0.0002*
Krypton	0.0001
Carbon monoxide	0.00001*
Xenon	0.000008
Ozone	0.000002*
Ammonia	0.000001
Nitrogen dioxide	0.0000001*
Sulfur dioxide	0.00000002*

* Trace gases of environmental importance discussed in this chapter

Figure 16.1 presents some of the basic facts about our atmosphere, including the naming of the stratified layers that compose the atmosphere and the chemical species present in those layers. Our main concerns are with the layers called the **troposphere** (the air we breathe and where our weather takes place) and the **stratosphere** (where the ultraviolet (UV)-protective ozone layer is found).

16.1 Air: A Source of Pure Gases

Before air can be separated into pure nitrogen, oxygen, and other gases by the process known as ***fractionation of air***, water vapor and carbon dioxide must be removed. This is usually done by precooling the air to separate ice and frozen carbon dioxide. Afterward, the air is compressed to more than 100 times normal atmospheric pressure, cooled to room temperature, and allowed to expand into a chamber. Overcoming intermolecular forces requires energy (Section 5.7), so as the gas expands, the molecules lose energy, they slow down, and the gas gets cooler. If the compression and expansion are repeated many times and controlled properly, the expanding air cools to the point of liquefaction. Once air has been liquefied, its pure gases can be separated by taking advantage of their different boiling points.

■ Air, like most gases, heats up when it is compressed and cools down when it expands.

When liquid air is allowed to warm up, nitrogen (bp −196°C) vaporizes first, and the liquid becomes more concentrated in oxygen and argon. Further processing allows separation of high-purity oxygen (bp −183°C) and argon (bp −189°C). The less-abundant noble gases (neon, bp −246°C; xenon, bp −108°C; and krypton, bp −153°C) are also separated from liquid air.

■ Because helium atoms are very light (4 amu), once released they can achieve velocities great enough to escape from the earth's gravity. The decay of radioactive elements resupplies helium, creating a roughly constant amount in the atmosphere.

Helium (bp −271°C) is not commercially recovered from air because it is cheaper to isolate it from natural gas, where it is sometimes present in as much as 7% by volume.

Oxygen

Most oxygen produced by the fractionation of liquid air is used in steel-making. Some is also used in rocket propulsion (to oxidize hydrogen) and in controlled oxidation reactions of other types.

■ A small but vital use of oxygen is in breathing therapy.

Liquid oxygen (LOX) can be stored and shipped at its boiling temperature of −183°C under atmospheric pressure. Substances this cold are called **cryogens** (from the Greek *kryos*, meaning "icy cold"). Cryogens represent special hazards, since contact produces instantaneous frostbite and structural materials such as plastics, rubber gaskets, and some metals become brittle and fracture easily at these low temperatures. Because liquid oxygen can accelerate oxidation reactions to the point of explosion, contact between it and substances that can ignite and burn in air must be prevented.

Nitrogen

Since nitrogen gas is so chemically unreactive, it is used as an inert atmosphere for applications such as welding and other high-temperature metallurgical processes. If air is not excluded from these processes, unwanted oxides of the hot metals would form.

■ Nitrogen ions and molecules ranked from the most oxidized (bonded to oxygen) to the most reduced (bonded to hydrogen) are NO_3^-, NO_2, NO_2^-, NO, N_2O, N_2, NH_3, and NH_4^+.

Liquid nitrogen is used in medicine *(cryosurgery)*, for example, in cooling a localized area of skin prior to removal of a wart or other unsightly or pathogenic tissue. Because of its low temperature and inertness, liquid nitrogen is also widely used in frozen food preparation and preservation during transit. Trucks or railroad boxcars with nitrogen atmospheres present health hazards, since they contain little (if any) oxygen to support life. Workers must either enter such areas with breathing apparatus or first allow fresh air to enter.

■ Nitrogen fixation requires breaking the strong triple bond connecting the two atoms in the nitrogen molecule N_2 (:N≡N:), as described in Section 5.3.

Nitrogen is an essential element for plants, but they cannot derive it from gaseous nitrogen. It must first be "fixed"—**nitrogen fixation** is the process of changing atmospheric nitrogen into compounds that dissolve in water and can be absorbed through plant roots (see Section 17.3).

Noble Gases

Approximately 250,000 tons of argon, the most abundant noble gas in the air, are recovered each year in the United States. Most of the argon is used to provide inert atmospheres for metallurgical processes. It is also used as a filler gas in incandescent light bulbs to prolong the life of the hot filament. Neon is used in many "neon" signs, but argon, krypton, and xenon are also used for this purpose.

16.2 Doing Something About Polluted Air

Nature pollutes the air on a massive scale with ash, mercury vapor, hydrogen chloride, and hydrogen sulfide from volcanoes; carbon dioxide and chlorinated organic compounds from forest and grassland fires; and reactive organic compounds from coniferous and deciduous plants. Human activity, however, especially in heavily populated (urban) areas, seems to have the most noticeable effects on the quality of the air we breathe. Automobiles, fossil fuel-burning power plants, smelting plants, other metallurgical plants, and petroleum refineries add significant quantities of polluting chemicals to the atmosphere. These atmospheric pollutants, especially in the concentrations found in urban areas, cause people to have burning eyes, coughing, and breathing difficulties. Air pollution is nothing new; Shakespeare wrote about it in the 17th century.

(a)

(b)

Air pollution by nature and industry. (a) Volcanoes, like the one pictured in Mauna Loa, Hawaii, emit ash and a variety of inorganic chemicals. (b) Industrial smokestacks, like these at a steel plant in Pennsylvania, also emit pollutants. Enforcement of the Clean Air Act has, however, been effective in greatly improving air quality near industrial sites. *(a, Dan McCoy/Rainbow; b, Jack Rosen/Photo Researchers)*

Prior to 1960, there was little concern about air pollution and little effort toward its control in the United States, in spite of some dramatic episodes in which many people suffered as a direct result of polluted air. For example, in October of 1948, the city of Donora, Pennsylvania, was overcome by five days of air pollution that caused almost 6000 residents to become ill and 18 to die. In the past, smoke, carbon monoxide, sulfur dioxide, nitrogen oxides, and organic vapors were emitted into the air from industrial facilities with little apparent thought about their harmful nature as long as they were scattered into the atmosphere and away from human smell and sight.

Early in the 1960s, air pollution became generally recognized as a problem in the United States, and this resulted in laws governing emissions of air pollutants by industry. In 1970, the first Clean Air Act was passed. This law helped in controlling air pollution from sources such as industry and automobiles, but it was not very comprehensive. The Clean Air Act was amended in 1977 to add stricter requirements, for example, on emissions from automobiles. In November of 1990, the President signed into law the 1990 Clean Air Act (CAA) amendments, a major overhaul of the earlier Clean Air Act. The 1990 CAA affects almost everything that is manufactured and consumed in this country, all in the name of cleaner, safer air. The substances regulated by the 1990 CAA include those discussed in this chapter—particulates, ozone, carbon monoxide, oxides of nitrogen and sulfur, hydrocarbons, volatile toxic substances, carbon dioxide, and stratospheric ozone-depleting chemicals. Let's begin by looking at the particles that obscure our vision, aggravate respiratory illnesses, and cause regional and global cooling by scattering sunlight.

■ A few decades ago, we operated on the principle that "Dilution is the solution to pollution."

16.3 Air Pollutants: Particle Size Makes a Difference

One of the most common forms of air pollution occurs as particles. Pollutant particles range in size from fly ash particles, which are big enough to see, down to individual molecules, ions, or atoms. Because of their polar nature, many pollutants are attracted into water droplets and form **aerosols**. Fogs and smoke are common examples of aerosols. Larger solid particles in the atmosphere are called **particulates**. The solids in an aerosol or particulate may be metal oxides, soil particles, sea salt, fly ash from electric generating plants and incinerators, elemental carbon, or even small metal particles. Aerosol particles range upward from a diameter of 1 nanometer (nm) to about 10,000 nm and may contain as many as a trillion (10^{12}) atoms, ions, or small molecules. Particles in the 2000-nm range are largely responsible for the deterioration of visibility.

■ 1 μm = 10^{-6} m, or 1000 nm.

■ Major contributors to the amount of atmospheric particulates were volcanic eruptions by Krakatoa, Indonesia, 1883; Mt. Katmai, Alaska, 1912; Hekla, Iceland, 1947; Mt. Spurr, Alaska, 1953; Bezymyannaya, U.S.S.R., 1956; Mt. St. Helens, Washington, 1980; and Mt. Pinatubo, 1991.

Aerosol particles are small enough to remain suspended in the atmosphere for long periods. Such particles are easily breathable and can cause lung diseases. They may also contain mutagenic or carcinogenic compounds. Because of their relatively large surface area, aerosol particles have great capacities to *adsorb* and concentrate chemicals on their surfaces. Liquid aerosols or particles covered with a thin coating of water may *absorb* air pollutants, thereby concentrating them and providing a medium in which reactions may occur. A typical urban aerosol and some of the reactions that can take place there are shown schematically in Figure 16.2.

Millions of tons of soot, dust, and smoke particles are emitted into the atmosphere of the United States each year. The average suspended particulate concentrations in the United States vary from about 0.00001 g/m^3 of air in rural areas to about six times as much in urban locations. In heavily polluted areas, concentrations of particulates may increase to 0.002 g/m^3.

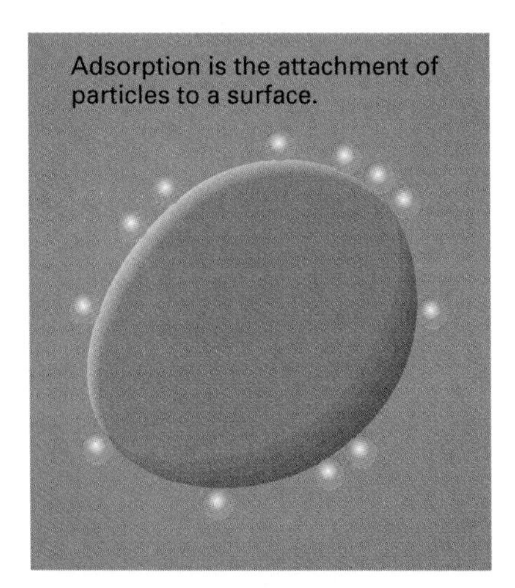

Adsorption is the attachment of particles to a surface.

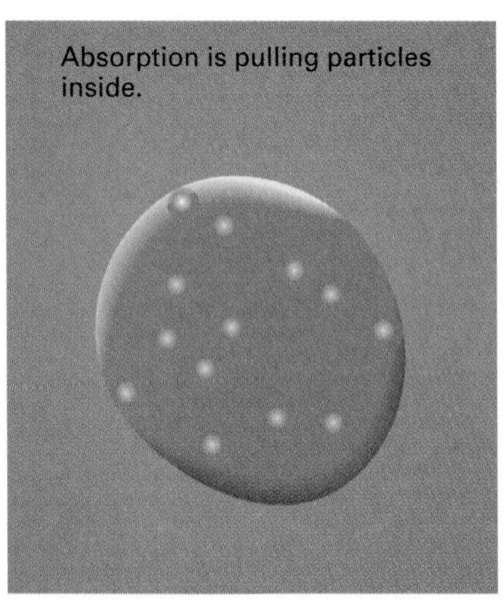

Absorption is pulling particles inside.

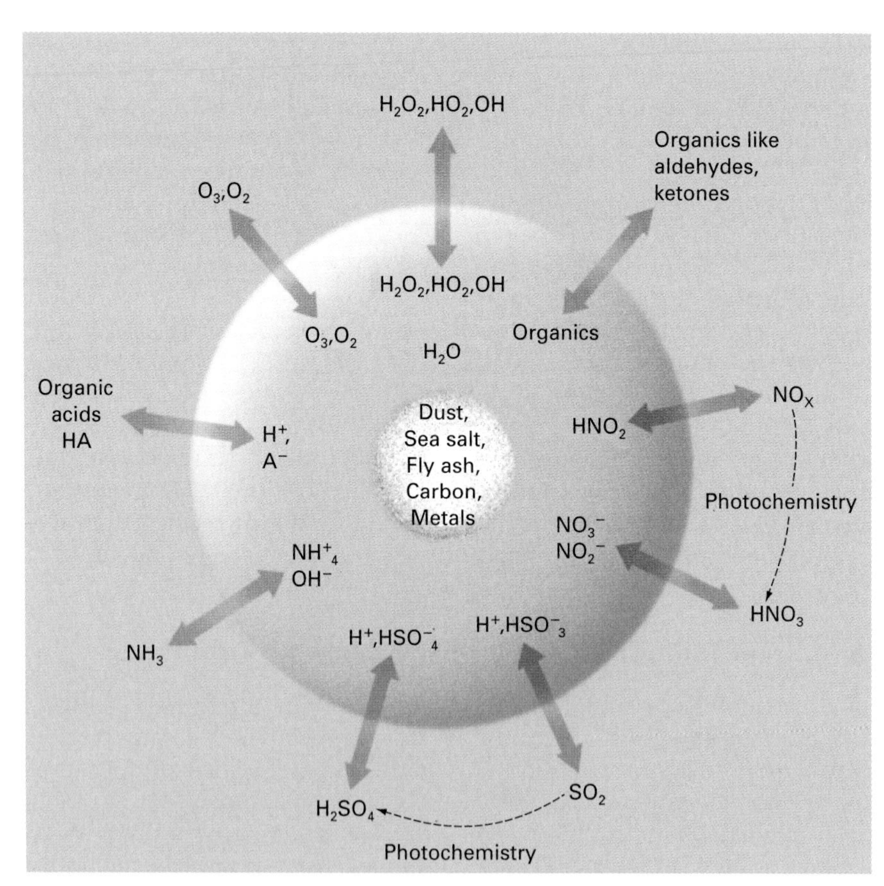

Figure 16.2 A typical urban aerosol particle, showing its composition and some of the chemical reactions of urban air pollutants.

Particulates in the atmosphere can cool Earth by scattering and partially reflecting light from the Sun. Large volcanic eruptions such as those from Mt. St. Helens in 1980 and Mt. Pinatubo in 1991 had measurable cooling effects on Earth.

Particulates and aerosols are removed naturally from the atmosphere by gravitational settling and by rain and snow. Industrial emissions of particulates can be prevented by treating the emissions with one or more of a variety of physical methods such as filtration, centrifugal separation, and scrubbing. Another method often used is electrostatic precipitation, which is more than 98% effective in removing aerosols and dust particulates even smaller than 1 μm from exhaust gases. The effects of an efficient electrostatic precipitator can be quite dramatic, as Figure 16.3 shows.

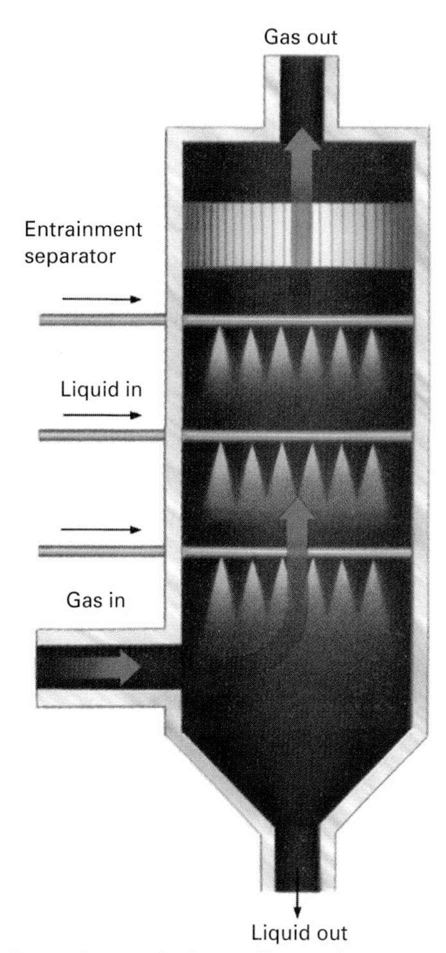

Removing particulate pollutants by scrubbing. The fine mist of water droplets traps particulates entering with the gas stream.

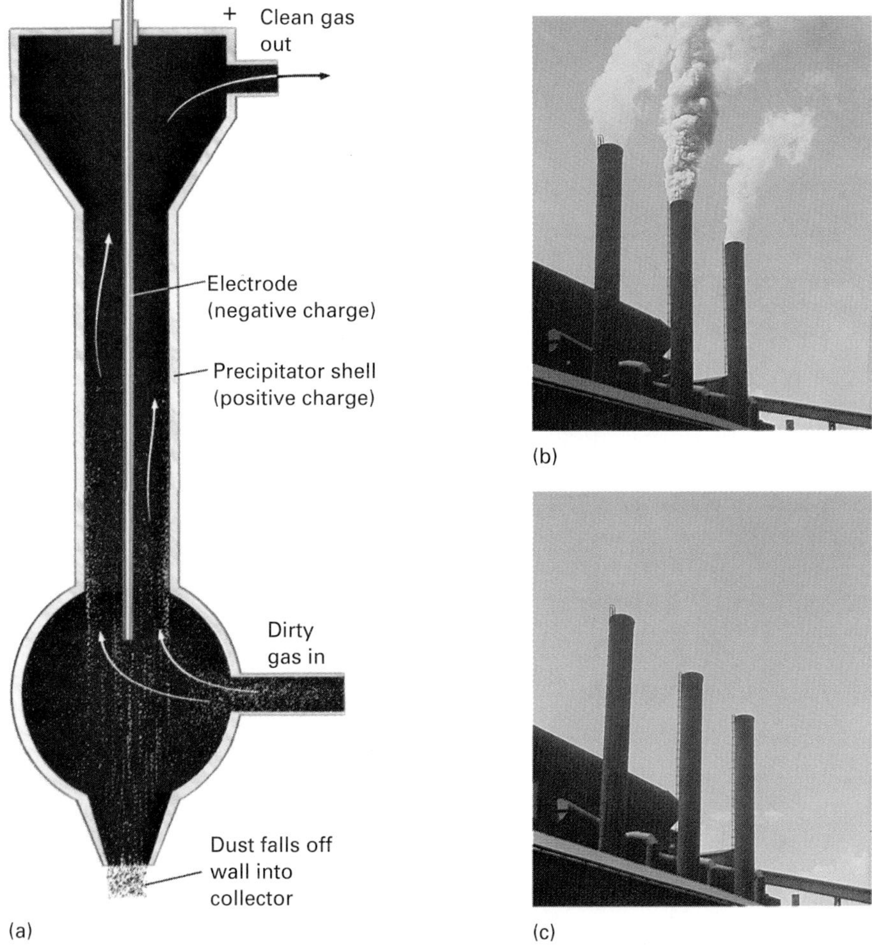

Figure 16.3 Electrostatic precipitation and its effectiveness. (a) An electrostatic precipitator. The central electrode is negatively charged and imparts a negative charge to particles in smoke that pass over it. These charged particles are then attracted to the positively charged walls and fall into the collector. (b) Smokestacks at a steel mill with the electrostatic precipitator turned off. (c) The same smokestacks with the electrostatic precipitator turned on. *(b and c, Visuals Unlimited/John D. Cunningham)*

16.4 Smog

The poisonous mixture of smoke, fog, air, and other chemicals was first called **smog** in 1911 by Dr. Harold de Voeux in his report on a London air pollution disaster that caused the deaths of 1150 people. Through the years, smog has been a technological plague in many communities and industrial regions.

What general conditions are necessary to produce smog? Although the chemical ingredients of smogs vary depending on the unique sources of the pollutants, certain geographical and meteorological conditions exist in nearly every instance of smog.

■ Thermal inversion is a mass of warmer air over a mass of cooler air.

First, there must be a period of windlessness so that pollutants can collect without being dispersed vertically or horizontally. This sets the conditions for a **thermal inversion**, which is an abnormal temperature arrangement for air masses (Figure 16.4). Normally, warmer air is on the bottom, nearer the warm earth and this warmer, less dense air rises and transports most of the pollutants to the upper troposphere, where they are dispersed. In a thermal inversion the warmer air is on top, and the cooler, denser air retains its position nearer Earth. The air becomes stagnated. If the land is bowl shaped (surrounded by mountains, cliffs, or the like), a stagnant air mass can remain in place for quite some time. When these atmospheric conditions exist, the pollutants supplied by combustion and evaporation in automobiles, electric power plants, and industrial plants accumulate to form smog.

Two general kinds of smog have been identified. One is the *chemically reducing type*, which is derived largely from the combustion of coal and oil; and contains sulfur dioxide mixed with soot, fly ash, smoke, and partially oxidized organic compounds. This type of smog is usually seen near industrial centers. Because it was first characterized in and around the city of London, it is sometimes called "London" smog. Smogs of this kind, also called industrial smogs because of their association with industrial activity, generally diminish in intensity and frequency as less coal is burned and more controls are installed on industrial emissions.

■ Industrial smog is fog + SO_2.

The main ingredient in industrial smog is sulfur dioxide. Laboratory experiments have shown that sulfur dioxide increases aerosol formation, partic-

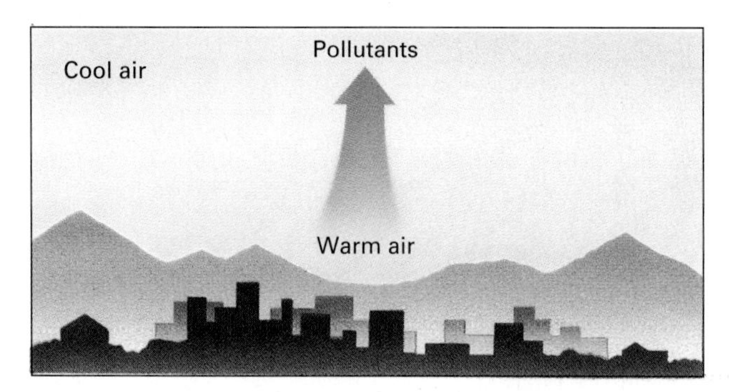

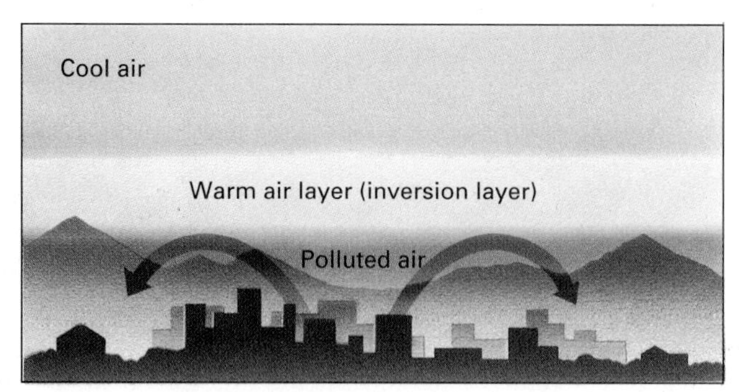

Figure 16.4 A thermal inversion. Normally, air that is warmed near the surface rises, carrying pollutants with it. During a thermal inversion, a blanket of warm air becomes stationary over a layer of cooler, denser air. The result is that pollutants are trapped near the surface.

ularly in the presence of mixtures of hydrocarbons and nitrogen oxides. For example, mixtures of 3 parts per million (ppm) hydrocarbons, 1 ppm NO_2, and 0.5 ppm SO_2 at 50% relative humidity form aerosols that have sulfuric acid as a major product. Breathing a sulfuric acid aerosol is very harmful, especially to people suffering from respiratory diseases such as asthma or emphysema. At a concentration of 5 ppm for 1 h, this kind of aerosol can cause constriction of bronchial tubes. A level of 10 ppm for 1 h can cause severe breathing distress.

A second type of smog is the *chemically oxidizing type*, typical of Los Angeles and other urban centers where exhaust fumes from internal combustion engines are highly concentrated in the atmosphere. This predominantly urban smog is called **photochemical smog** because light—in this instance sunlight—is important in initiating several chemical reactions that together make the smog harmful. Photochemical smog is practically free of sulfur dioxide but contains substantial amounts of nitrogen oxides, ozone, oxygenated and ozonated hydrocarbons, and organic peroxide compounds, together with unreacted hydrocarbons of varying complexity. The automobile is a direct or indirect source of many of the components of photochemical smog. Consider Figure 16.5. It shows how several components of photochemical smog increase during the rush hour in Los Angeles.

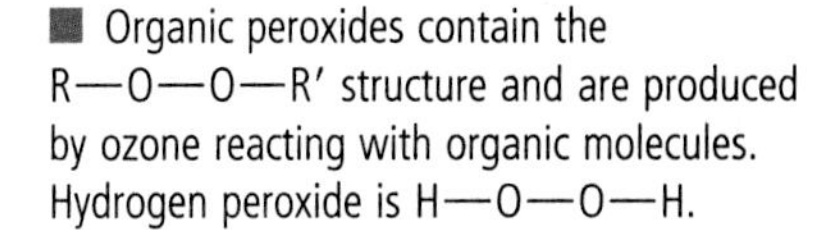

■ Organic peroxides contain the R—O—O—R′ structure and are produced by ozone reacting with organic molecules. Hydrogen peroxide is H—O—O—H.

Many of the chemical reactions that create photochemical smog take place in aerosol particles. These reactions produce **secondary pollutants**—pollutants that are not directly released from some source but are formed by reactions with other components in the air.

■ **Secondary pollutants** are formed in the air by chemical reactions.

The exact reaction scheme by which **primary pollutants** form the secondary pollutants of photochemical smog is still not completely understood. One process that is known begins with the absorption of light energy by a molecule of nitrogen dioxide. Nitrogen dioxide reacts with light, which can be written as $h\nu$, with a wavelength between 280 nm and 430 nm. This **photodissociation** reaction (*photo*, light; *dissociation*, breaking apart) produces nitric oxide and

■ **Primary pollutants** are emitted directly into the air from a source.

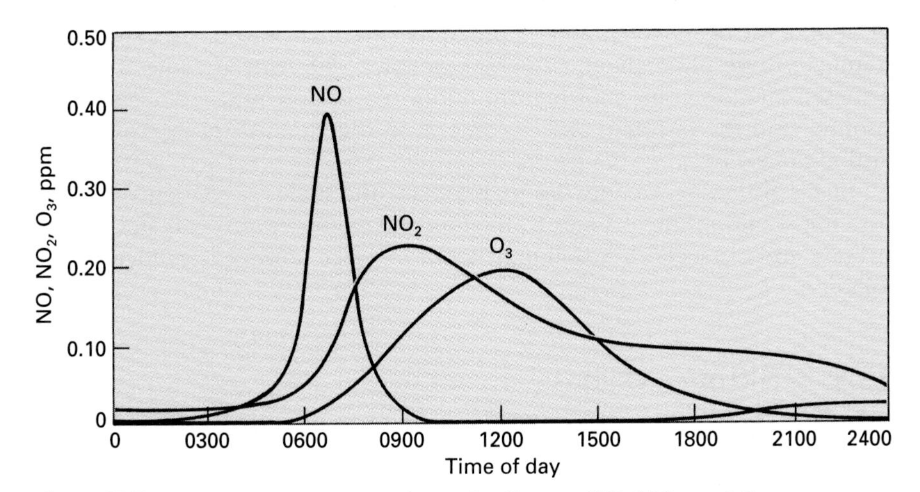

Figure 16.5 The average concentrations of pollutants NO, NO_2, and O_3 on a smoggy day in Los Angeles. The NO concentration builds up during the morning rush hour. Later in the day, the concentrations of NO_2 and O_3 build up.

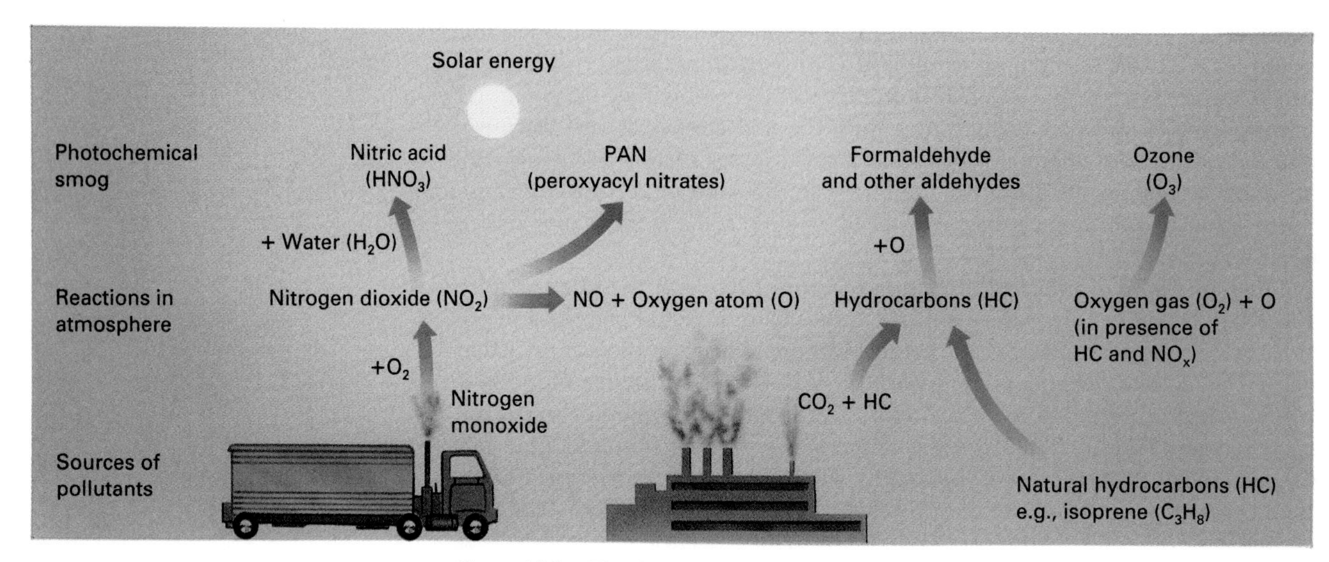

Figure 16.6 The formation of photochemical smog.

■ Both NO and oxygen atoms are free radicals. They are usually very reactive (see Section 8.3).

free oxygen atoms (O, oxygen free radicals) that can react with a molecule of oxygen to produce a molecule of ozone (O_3), which is an important secondary pollutant (Figure 16.6).

$$NO_2(g) + h\nu \longrightarrow NO(g) + O(g)$$
$$O_2(g) + O(g) \longrightarrow O_3(g)$$

Photochemical smog over San Diego. The layer of smog held in place by a thermal inversion layer is clearly visible here. The red-brown color of the smog shows the presence of nitrogen dioxide. *(Alan Pitcairn/Grant Heilman Photography)*

Atomic oxygen that doesn't react with oxygen molecules to form ozone can also react with hydrocarbons, such as olefins (molecules with double bonds) and aromatics, to form other chemicals such as aldehydes and ketones, which are toxic and also impart an odor to the air. On a sunny day only about 0.2 ppm of nitrogen oxides and 1 ppm of reactive hydrocarbons are sufficient to initiate these photochemical smog reactions. The hydrocarbons involved in these reactions come mostly from unburned petroleum products such as gasoline, and the nitrogen oxides come from the exhausts of internal combustion engines.

In the following sections we shall look at the major ingredients of photochemical smog—the primary pollutants, the oxides of nitrogen and hydrocarbons, and the secondary pollutant ozone—to see how they produce urban pollution.

16.5 Nitrogen Oxides

■ Photochemical smog is fog + NO_x + hydrocarbons.

There are eight known oxides of nitrogen, three of which are recognized as important components of the atmosphere: dinitrogen oxide (N_2O), nitric oxide (NO), and nitrogen dioxide (NO_2). These oxides of nitrogen are collectively known as "NO_x." About 97% of the nitrogen oxides in the atmosphere are naturally produced, and only 3% result from human activity. Certain bacteria can produce N_2O, so this oxide of nitrogen is also commonly found in the atmosphere in trace amounts.

Almost all the nitrogen chemically bound to oxygen begins as NO (nitric oxide), a colorless reactive gas. Nitrogen, normally a relatively inert gas, readily reacts with oxygen when there is a source of energy to produce a high temperature. The product is NO.

■ Another name for nitric oxide is "nitrogen monoxide."

$$N_2(g) + O_2(g) \xrightarrow{\text{Energy}} \underset{\text{Nitric oxide}}{2\,NO(g)}$$

Nitric oxide is formed in this manner during electrical storms, where lightning can supply the needed energy. Nitric oxide is also formed whenever combustion of fuels in air produces high temperatures. Because the formation of nitric oxide requires heat, it follows that a higher combustion temperature would produce relatively more NO. The combustion of gasoline vapors in an automobile engine is always accompanied by the production of NO because both nitrogen and oxygen are present in the combustion chamber.

The NO molecule is short-lived in the atmosphere because it reacts rapidly with atmospheric oxygen to produce NO_2, a brown gas.

$$2\,NO(g) + O_2(g) \longrightarrow \underset{\text{Nitrogen dioxide}}{2\,NO_2(g)}$$

Normally the atmospheric concentration of NO_2 is a few parts per billion (ppb) or less; most of the nitrogen oxides formed during lightning storms are washed out by rain. This is one of the ways nitrogen is made available to plants. Looking at all the sources of oxides of nitrogen (Table 16.2), it is apparent that combustion processes are their primary sources. In the United States, most oxides of nitrogen from sources other than nature are produced from fossil fuel combustion.

Nitrogen dioxide is a powerful corrosive agent. It will cause severe skin burns in high concentrations. Breathing a concentration of only 3 ppm NO_2 for 1 h causes bronchioconstriction in humans, and short exposures at high

TABLE 16.2 ■ Emissions of NO_x

	Emissions (millions of tons)	
Source	*United States*	*Global*
Fossil fuel combustion	66	231
Biomass burning	1.1	132
Lightning	3.3	88
Microbial activity in soil	3.3	88
Input from the stratosphere	0.3	5.5
Total (uncertainty in estimates)	74 (±1)	544.5 (±275)

Note: The large uncertainty for global emissions is due to incomplete data for much of the world.

Source: Stanford Research Institute.

levels (150–220 ppm) result in corrosive reaction with lung tissue that can be fatal. A seemingly harmless exposure to high concentrations of NO_2 one day can even cause death a few days later.

■ See pp. 425–426 for a discussion of the role of NO_2 in the formation of the pollutant ozone.

One of the primary roles of nitrogen dioxide as a pollutant is in the formation of the secondary pollutant ozone. Nitrogen dioxide can also react with water to form nitric acid and nitrous acid. This reaction takes place readily in aqueous aerosols, producing acids that help to stabilize the droplet.

$$2\,NO_2(g) + H_2O(\ell) \longrightarrow \underset{\text{Nitric acid}}{HNO_3(aq)} + \underset{\text{Nitrous acid}}{HNO_2(aq)}$$

■ Nitrates are important components of fertilizers.

Of course, breathing air containing these aerosol droplets is harmful because of the corrosive nature of the acids. The acids in turn can react with ammonia or metallic particles in the atmosphere to produce nitrate or nitrite salts. For example,

$$\underset{\text{Ammonia}}{NH_3(g)} + HNO_3(aq) \longrightarrow \underset{\text{Ammonium nitrate (a salt)}}{NH_4NO_3(aq)}$$

Both the acids and the salts stabilize the aerosol particles, which eventually settle from the air or dissolve in larger raindrops. Nitrogen dioxide, besides causing the formation of ozone, is a primary cause of haze in urban or industrial atmospheres because of its participation in the process of aerosol formation.

16.6 Ozone and Its Role in Air Pollution

■ The odor of ozone can be detected by most people at concentrations as low as 0.02 ppm.

Ozone consists of three oxygen atoms bound together in a molecule with the formula O_3. It has a pungent odor that we often smell near sparking electrical appliances or after a thunderstorm when lightning-caused ozone washes out with the rainfall.

As you will see in this chapter, there is "good" ozone and "bad" ozone. The bad ozone is that found in the air we breathe, whereas the good ozone is found in the stratosphere, where it forms a protective blanket, absorbing harmful UV radiation (see Section 16.11).

Being a secondary pollutant, ozone is one of the most difficult pollutants to control. According to the Environmental Protection Agency (EPA), the upper limit for ozone of 0.12 ppm was exceeded in many of the urban areas of the United States (Table 16.3) during the 1980s and early 1990s. These high ozone concentrations were primarily the result of the excess nitrogen oxide emissions from automobiles, buses, and trucks. Most major urban areas have vehicle inspection centers for passenger automobiles in an effort to control nitrogen oxide emissions as well as emissions of carbon monoxide and unburned hydrocarbons. In spite of these efforts, large urban centers still have high NO_2 concentrations, which result in too much ozone being formed.

■ Lower FEV_1 accelerates the aging of the lungs.

In July 1997, the EPA announced a new, lower ozone standard. This standard was based on the fact that exposures to concentrations of ozone at or near 0.12 ppm lower the volume of air a person breathes out in 1 s (the forced expiratory volume, FEV_1). Children who were exposed to ozone concentra-

TABLE 16.3 ■ Urban Areas with the Worst Ozone Air Quality in the 1980s and Early 1990s

Extreme (0.28 ppm O_3)	***Severe (0.18–0.19 ppm O_3)***
Los Angeles and south coast air basin, California	Baltimore and the state of Maryland
	Philadelphia, Pennsylvania
Very Severe (0.19–0.28 ppm O_3)	Wilmington, Delaware
Chicago and Gary and Lake Counties, Indiana	Trenton, New Jersey
Houston, Galveston, and Brazoria, Texas	San Diego, California
Milwaukee and Racine, Wisconsin	Ventura County (between Santa Barbara and Los Angeles), California
New York City and Long Island, New York	
Northern New Jersey and Connecticut	
Southeast Desert, California	

tions close to the previous EPA standard, but not exceeding it, showed a 16% decrease in the FEV_1. The new standard is based on 8 h at 0.08 ppm.

No matter what the standard is, present ozone concentrations in many urban areas represent health hazards to children at play, joggers, others doing outdoor exercise, and older people who may have diminished respiratory capabilities. The only effective way to limit ozone is to limit NO_x emissions. In some areas this means possibly limiting the numbers of automobiles in use on any given day or limiting the number of automobiles with internal combustion engines. This idea has given rise to a state rule in California setting a number of electric-powered automobiles that must be sold (see Chapter 8, *Science and Society*—Electric Automobiles Moving Off the Drawing Boards).

Rush hour during a transit strike in New York City. These bicycle commuters are not contributing to air pollution, but they are probably breathing in more of it than is healthy. *(© James H. Kerales/Peter Arnold, Inc.)*

16.7 Hydrocarbons and Air Pollution

Hydrocarbons enter the atmosphere from both natural sources and human activities. Certain natural hydrocarbons are produced in large quantities by both coniferous and deciduous trees. Methane gas (CH_4) is produced by such diverse sources as rice growing, ruminant animals, termites, ants, and decay-causing bacteria acting on dead plants and animals. Human activities such as the use of industrial solvents, petroleum refining and its distribution, and the release of unburned gasoline and diesel fuel components account for a large amount of hydrocarbons in the atmosphere.

■ In 1988, William Chameides, of Georgia Tech in Atlanta, published a report in *Science* magazine, in which he stated that in some cities trees may account for more hydrocarbons in the atmosphere than those produced from human activities. The EPA has since found this to be true. This fact is causing a rethinking about how to control urban pollution.

In addition to simpler hydrocarbons like alkanes, alkenes, and alkynes, a large number of polynuclear aromatic hydrocarbons (PAH) are released into the atmosphere, primarily from motor vehicle exhaust. The greatest danger of these pollutants is their toxic properties. One PAH, benzo(α)pyrene (BAP), is a known carcinogen (see Section 9.4). Concentrations of BAP as high as 60 $\mu g/m^3$ of air have been found in urban air.

■ For every million tons of coal burned, about 750 tons of benzo(α)pyrene can be produced. Coal smoke contains about 300 ppm benzo(α)pyrene.

Hydrocarbons also contribute to ozone formation. Some of the oxygen atoms formed during the photodecomposition of NO_2 can react with water, forming hydroxyl free radicals (OH).

$$O(g) + H_2O(g) \longrightarrow \underset{\text{Hydroxyl free radical}}{2\,OH(g)}$$

FRONTIERS IN THE WORLD OF CHEMISTRY

Smog-Eating Radiators

What could be better? Driving your car *and* destroying more pollutants than you produced. That just might be possible someday if a new catalytic *radiator* being tested by Ford Motor Company proves successful. Radiators found under the hoods of cars allow engine coolant to circulate and be cooled by outside air. Even on hot days, fast-moving air passing through openings in the radiator provide enough cooling to keep a typical engine from overheating. (When the car isn't moving, as in heavy traffic, an electric fan turns on and blows air through the radiator and removes heat from the engine coolant.) The fact that the very air passing through the radiator is also the polluted air we breathe caused some scientists at Engelhard Corporation to see an interesting solution to the air pollution problem. (Engelhard is one of the major world suppliers of platinum-based catalysts for automotive catalytic converters.)

By using their knowledge about how catalysts work, they developed a special platinum-based coating for the automotive radiator. This coating converts ozone, one of the main ingredients of smog, into oxygen. Carbon monoxide is also converted into carbon dioxide as the outside air passes through the radiator. The coating worked so well that about 90% of all the ozone and carbon monoxide passing over it were converted.

Recognizing a potentially good thing, Ford engineers have placed these PremAir catalytic-coated radiators on a test fleet of vehicles with the hopes that everyday driving will show them to be net pollution eliminators rather than pollution generators. After determining how well the catalytic radiators hold up under salt, dirt, insects, high altitudes, and ice and snow, Ford will determine how much the radiators will cost car buyers. It is expected that the radiators will be "significantly under $1000 per vehicle."

As the new standards of the federal Clean Air Act requiring significant numbers of nonpolluting cars by 1998 go into effect, cars equipped with pollution-reducing radiators might offer buyers a chance to continue to buy gasoline engine cars instead of models powered by batteries.

Reference Associated Press, June 1995.

These hydroxyl radicals in turn react with hydrocarbon molecules, producing a number of compounds including aldehydes and ketones, and NO_2. Of course, NO_2 is easily photodecomposed, producing oxygen atoms, which go on to form ozone.

Although it is practically impossible to control hydrocarbon emissions from living plants and other natural sources, hydrocarbon emissions from automobiles can be controlled. Two means of control are being used at present. First, the spouts and hoses on gasoline pumps have been redesigned to prevent gasoline from entering the air. Second, catalytic converters that reduce emissions of hydrocarbons, CO and NO_x, are now part of every automobile's exhaust system. Careful control of the engine fuel–air ratio is required for these catalysts to perform well. This is accomplished by means of an oxygen sensor in the engine. When it operates properly, the fuel–air ratio is correct, but if it malfunctions, some automobiles will not run until the sensor is replaced.

Trees in urban environments may emit as many reactive hydrocarbons as do automobiles.

The effectiveness of these catalytic converters, which have been on automobiles sold in the United States since the mid-1970s, can be seen by comparing the emissions of hydrocarbons, CO and NO_x, in grams per vehicle-mile in 1960, before there were controls, to the same values in the 1990s (Table 16.4).

TABLE 16.4 ■ Emission Rates for Hydrocarbons (HC), Carbon Monoxide (CO), and NO_x

1960 (Precontrol—no catalytic mufflers installed)		*1993 (Catalytic mufflers required on all automobiles)*		*1996 Standards*	
HC	10.6 g/mi	HC	0.41 g/mi	HC	0.25 g/mi
CO	84.0 g/mi	CO	3.4 g/mi	CO	3.4 g/mi
NO_x	4.1 g/mi	NO_x	1.0 g/mi	NO_x	0.4 g/mi

Note: As of December 31, 1995, there were 201.5 million motor vehicles registered in the United States. Of these, about 76% were cars; about 24% were trucks, buses, motorcycles, and other vehicles.

■ SELF-TEST 16A

1. What has been the general trend in air pollutants for approximately the past decade? (a) Increase, (b) Decrease.
2. Name a chemical that is considered both an air pollutant and a beneficial chemical.
3. Because of their large surface areas, aerosol particles can (absorb/adsorb) chemicals onto their surfaces.
4. A liquid aerosol particle will probably (adsorb/absorb) a chemical.
5. A thermal inversion occurs when (warm/cool) air is above (warm/cool) air below.
6. Industrial-type smog is often associated with (a) coal burning, (b) sunlight.
7. In all combustion processes in air, some nitrogen ____________ are formed.
8. What are the products of the photodissociation of nitrogen dioxide?
9. What species reacts with molecular oxygen to form ozone?

■ Hydrocarbon emissions from vehicles in California in 2003 will have to be much lower than the national standards. By 2003, hydrocarbon emissions for 75% of all vehicles must be no greater than 0.075 g/mi, and no greater than 0.04 g/mi for 15% of all vehicles; and 10% of all vehicles must have zero hydrocarbon emissions. That means they will probably be electric vehicles.

16.8 Sulfur Dioxide: A Major Primary Pollutant

Sulfur dioxide is produced when sulfur or sulfur-containing compounds are burned in air.

$$S(s) + O_2(g) \longrightarrow SO_2(g)$$

While volcanoes put large amounts of SO_2 into the atmosphere annually, human activities probably account for up to 70% of all emissions on a global basis. In the United States, about 21 million tons of sulfur are released annually. Once formed, SO_2 generally becomes distributed in aerosol droplets, which are numerous enough to contribute to significantly reduced visibility and can affect both global and regional climate by causing the scattering of sunlight that would otherwise warm the earth. Emissions of SO_2 cause the mean temperature in the United States to be about 1°C cooler than it would be otherwise.

Most of the coal burned in the United States contains sulfur in the form of the mineral pyrite (FeS_2). The weight percent of sulfur in this coal ranges from 1% to 4%. The pyrite is oxidized as the coal is burned, forming SO_2.

$$4\ FeS_2(s) + 11\ O_2(g) \longrightarrow 2\ Fe_2O_3(s) + 8\ SO_2(g)$$

Large amounts of coal are burned in this country to generate electricity. A 1000-megawatt (MW) coal-fired generating plant can burn about 700 tons of coal an hour. If the coal contains 4% sulfur, that equals 56 tons of SO_2 an hour, or 490,560 tons of SO_2 every year. About 800 million tons of coal are burned each year to produce electricity.

Oil-burning electric generating plants can also produce comparable amounts of SO_2 because some fuel oils can contain up to 4% sulfur. The sulfur in the oil is in the form of compounds in which sulfur atoms are bound to carbon and hydrogen atoms.

States that rely mainly on coal for their electricity production and industrial furnaces have the highest SO_2 emissions in the United States. Operators of all coal-fired burners are under EPA orders to eliminate most of the SO_2 before it reaches the stack. The 1990 Clean Air Act requires that by the year 2000, SO_2 emissions from *all* power-generating sources will be no greater than 8.9 million tons per year. That's a 10 million ton per year reduction from 1980 levels.

The removal of sulfur from high-sulfur coal is costly and incomplete. One method is to pulverize the coal to the consistency of talcum powder and remove the pyrite (FeS_2) by magnetic separation. Reducing the sulfur content of fuel oil is also costly. It involves the formation of hydrogen sulfide (H_2S) by bubbling hydrogen through the oil in the presence of metal catalysts.

At present, most sulfur-containing coal is burned without prior treatment, and SO_2 is removed from the exhaust gases. In one method, lime reacts with SO_2 to form calcium sulfite, a solid particulate, which can be removed from an exhaust stack by an electrostatic precipitator.

$$\underset{\text{Limestone}}{CaCO_3(s)} \xrightarrow{\text{Heat}} \underset{\text{Lime}}{CaO(s)} + CO_2(g)$$

$$CaO(s) + SO_2(g) \longrightarrow \underset{\text{Calcium sulfite}}{CaSO_3(s)}$$

In another method, the exhaust gases containing SO_2 are passed through molten sodium carbonate, and solid sodium sulfite is formed (Figure 16.7).

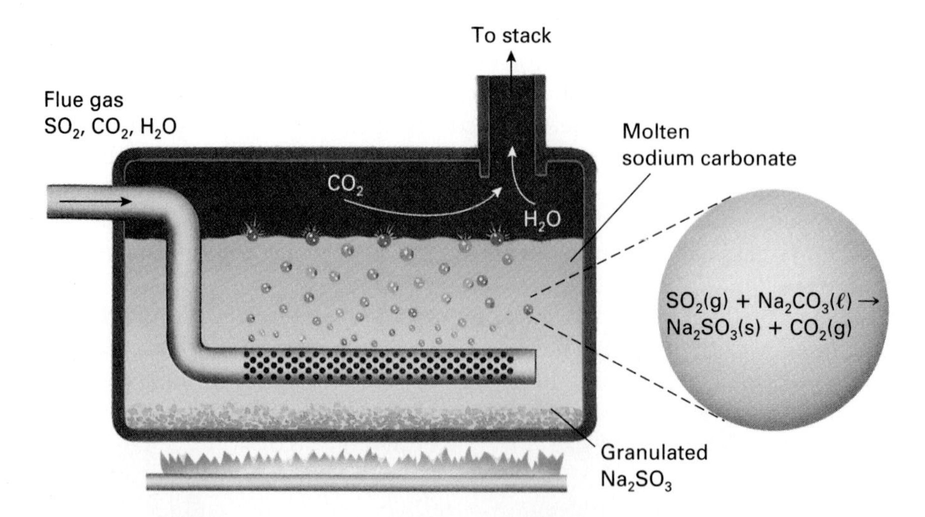

Figure 16.7 Removal of SO_2 from flue gas by reaction with molten sodium carbonate.

Notice how both of these methods of SO_2 removal emit additional CO_2, also a pollutant (Section 16.12), into the atmosphere. Newer technology to address this problem is being developed.

A less desirable method of lowering the effects of SO_2 emissions, but one still being used, is sending the smoke up very tall exhaust stacks. Tall stacks emit SO_2 at a high elevation and away from the immediate vicinity, which allows the SO_2 to be diluted before forming aerosol particles. The fact remains, however, that the longer it stays in the air, the greater chance SO_2 has to become sulfuric acid. A ten-year study in Great Britain showed that although SO_2 emissions from power plants increased by 35%, the construction of tall stacks decreased the ground-level concentrations of SO_2 by as much as 30%. The question is, who got the SO_2? In this case, Britain's solution was others' pollution. In the United States, the EPA may have added to a pollution problem unwittingly with rules in 1970 that caused plants to increase the height of smokestacks and caused pollutants to be carried longer distances by winds. Currently, about 179 stacks in the United States that are 500 ft tall or higher and 20 stacks that are 1000 ft or more tall are in use.

Most of the SO_2 that does get into the atmosphere reacts with oxygen to form sulfur trioxide (SO_3). The SO_3 has a strong affinity for water and dissolves in aqueous aerosol particles, forming sulfuric acid, a strong acid.

$$SO_3(g) + H_2O(\ell) \longrightarrow H_2SO_4(aq)$$

16.9 Acid Rain

The term **acid rain** was first used in 1872 by Robert Angus Smith, an English chemist and climatologist. He used the term to describe the acidic precipitation that fell on Manchester, England, just at the start of the Industrial Revolution. Although neutral water has a pH of 7, rainwater becomes naturally acidified from dissolved carbon dioxide, a normal component of the atmosphere. The carbon dioxide reacts reversibly with water to form a solution of the weak acid carbonic acid.

■ Reversibility and equilibria are discussed in Section 6.4. Weak acids are discussed in Section 7.2. See Figure 7.2 for a review of pH values.

$$2\ H_2O(\ell) + CO_2(g) \rightleftharpoons H_3O^+(aq) + HCO_3^-(aq)$$

At equilibrium, the pH of a solution of CO_2 from the air is 5.6. Any precipitation with a pH below 5.6 is considered to be acid rain.

As you have seen, NO_2 and SO_2 can both react with water in the atmosphere to produce acids: NO_2 produces nitric acid (HNO_3) and nitrous acid (HNO_2); SO_2 produces sulfuric acid (H_2SO_4) and sulfurous acid (H_2SO_3). When conditions are favorable, these acidic water droplets precipitate as rain or snow with a low pH. Ice core samples taken in Greenland and dating back to 1900 contain sulfate (SO_4^{2-}) and nitrate (NO_3^-) ions. This indicates that at least from 1900 onward, acid rain has been commonplace.

■ A more adequate term for acid rain might be "acid precipitation." Some scientists use "acid deposition."

Acid rain is a problem today due to the large amounts of these acidic oxides being produced by human activities and put into the atmosphere annually. When this precipitation falls on natural areas that cannot easily tolerate such acidity, serious environmental problems occur. The average annual pH of precipitation falling on much of the northeastern United States and northeastern Europe is between 4 and 4.5. Specific rainstorms in

■ The March 1991 eruption of Mt. Pinatubo in the Philippines injected more than 10^8 kg of SO_2 into the stratosphere. The SO_2 eventually came down as acid rain. During the period from 1991 to 1994, aerosol particles from these eruptions enhanced the beauty of sunrises and sunsets by scattering sunlight more than normal.

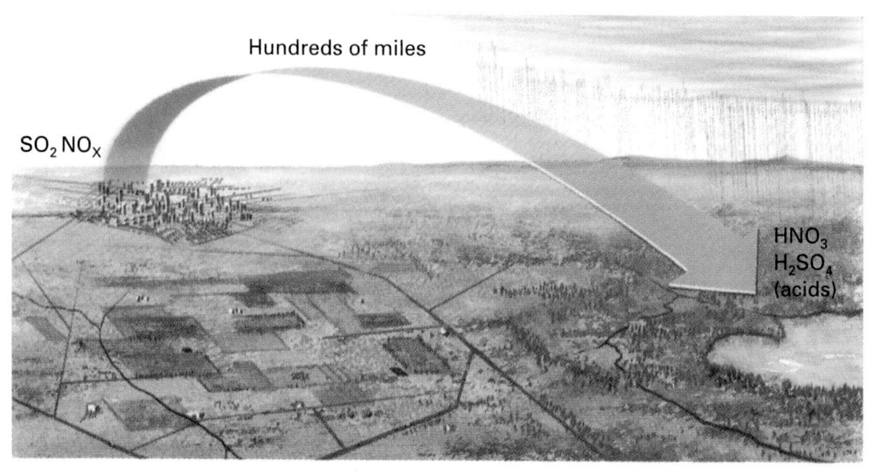

Figure 16.8 Pollutants can cause acid rain to fall far from where the pollutants are generated.

some areas where there are numerous sources of SO_2 and NO_x have had pH values as low as 1.5. Further complicating matters is the fact that acid rain is an international problem—rain and snow don't observe borders (Figure 16.8). Many Canadian residents are offended by the government of the United States because some of the acid rain produced in the United States falls on Canadian cities and forests (Figure 16.9).

The extent of the problems with acid rain can be seen in "dead" (fishless)

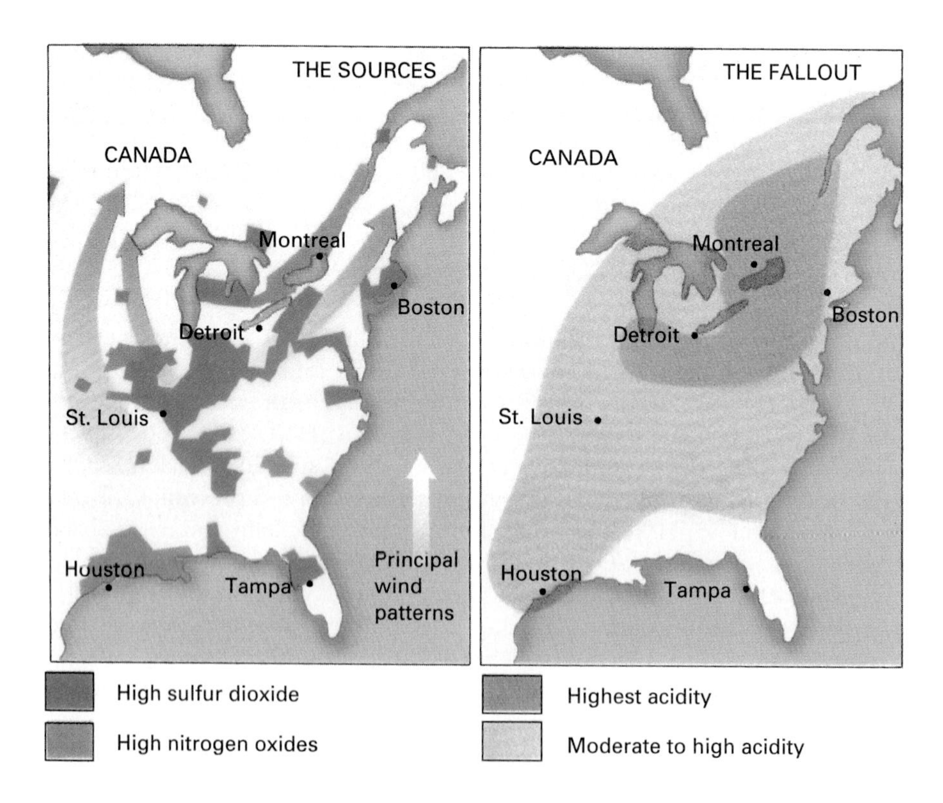

Figure 16.9 Distribution of sulfur dioxide, nitrogen oxides, and acid rain. Prevailing winds carry the acid droplets over the Northeast and into Canada.

ponds and lakes, dying or dead forests, and crumbling buildings. Because of wind patterns, Norway and Sweden have received the brunt of western Europe's emission of sulfur oxides and nitrogen oxides as acid rain. As a result, of the 100,000 lakes in Sweden, 4000 have become fishless, and 14,000 other lakes have been acidified to some degree. In the United States, 6% of all ponds and lakes in the Adirondack Mountains of New York are now fishless, and 200 lakes in Michigan are dead. For the most part, these dead lakes are still picturesque, but no fish can live in the acidified water. Lake trout and yellow perch die at pH values below 5.0, and smallmouth bass die at pH values below 6.0. Mussels die when the pH is below 6.5.

Effects of acid rain on a forest in one of the most polluted parts of Europe. The devastation has been caused by emission of sulfur dioxide and nitrogen oxides from factories in the former East Germany and Czechoslovakia. *(Simon Fraser/SPL/Photo Researchers)*

Acid rain damages trees in several ways. It disturbs the stomata (openings) in tree leaves and causes increased transpiration and a water deficit in the tree. The surface structures of the bark and the leaves can also be destroyed by the acid. Acid rainfall can acidify the soil, damaging fine root hairs and thus diminishing nutrient and water uptake. In addition, acid rain dissolves minerals that are insoluble in groundwater and surface waters of normal pH, and many of these minerals contain metal ions toxic to plant life. For example, acid rain dissolves aluminum hydroxide in the soil, allowing aluminum ions (Al^{3+}) to be taken up by the roots of plants, where they have toxic effects.

$$Al(OH)_3(s) + 3\ H^+(aq) \longrightarrow Al^{3+}(aq) + 3\ H_2O(\ell)$$

■ The leaching of toxic metal ions into groundwater by acid rain may also increase groundwater pollution.

The effects of acid rain and other pollution on stone and metal structures are especially devastating because of their irreversibility. By damaging stone buildings in Europe, acid rain is slowly but surely dissolving the continent's historical heritage. The bas-reliefs on the Cologne Cathedral in Germany are barely recognizable today. The Tower of London, St. Paul's Cathedral, and Lincoln Cathedral in Great Britain have suffered the same fate. Other beautifully carved statues and bas-reliefs on buildings throughout Europe and the eastern part of the United States and Canada are slowly passing into oblivion by the action of pollutants, in particular, acid rain.

What can be done about acid rain? Obviously, eliminating the emissions of the oxides of nitrogen and sulfur would be the answer. This is not easy, however. Some stopgap measures are being taken, such as spraying hydrated lime, $Ca(OH)_2$, into acidified lakes to neutralize at least some of the acid and raise the pH toward 7.

$$Ca(OH)_2(s) + 2\ H^+(aq) \longrightarrow Ca^{2+}(aq) + 2\ H_2O(\ell)$$

Some lakes in the problem areas have their own safeguard against acid rain by having limestone-lined bottoms, which supply calcium carbonate ($CaCO_3$) for neutralizing the acid from acid rain (just as an antacid tablet relieves indigestion).

Effects of acid rain on a tombstone in England that was carved in 1817. *(Bruce F. Molnia/Terraphotographics)*

The ultimate answers to acid rain problems lie with those industries that produce the oxides of sulfur and nitrogen and with the regulatory agencies that govern them. As you read in the earlier section, methods exist for the control of SO_2 emissions, although some of these are costly. In the final analysis, the consumer will bear those costs. The control of oxides of nitrogen is more difficult because there are so many sources of combustion exhaust gases. Catalytic mufflers help control NO_x emissions from automobiles, but most

■ The government of Sweden is spending $40 million a year to neutralize the acid in some of its lakes.

home furnaces, industrial boilers, and even electrical generating plants do not have adequate controls. Fortunately for acid rain production, NO_2 is so reactive in the troposphere that it is not the major contributor that SO_2 is to acid rain.

16.10 Carbon Monoxide

At least ten times more carbon monoxide enters the atmosphere from natural sources than from all industrial and automotive sources combined. Of the 3.8 billion tons of carbon monoxide emitted every year, about 3 billion tons are emitted by the oxidation of decaying organic matter in the topsoil. In spite of this fact, carbon monoxide is considered an air pollutant, primarily because so much of it is produced by human activities in the urban environment.

Like ozone, carbon monoxide is one of the most difficult pollutants to control. Cities such as Los Angeles and other highly populated urban centers with their high densities of automobiles tend to be repeatedly cited by the EPA for not attaining the required ambient air quality for carbon monoxide.

Carbon monoxide is always produced when carbon or carbon-containing compounds are oxidized using an insufficient quantity of oxygen.

$$2\,C(s) + O_2(g) \longrightarrow 2\,CO(g)$$

■ The abbreviation "ppm" means parts per million—a measure expressing concentration; 30 ppm CO means 30 mL of CO for every million milliliters of air. To convert ppm to percent, divide by 10,000. To convert percent to ppm, multiply by 10,000.

Gasoline engines are notorious sources of CO. This happens because the rapid combustion inside the combustion chamber does not burn all the carbon to CO_2 before the exhaust gases are swept out. Modern catalytic converters convert much of this carbon monoxide to carbon dioxide, but the amounts that are not converted make being near a heavily traveled street dangerous because of the carbon monoxide concentrations. At peak traffic times, concentrations as high as 50 ppm are common. In the countryside, carbon monoxide levels are closer to the global average of 0.1 ppm. The only effective means of controlling carbon monoxide concentrations in urban air is to control the major emitters—automobiles. Of course, transportation that does not depend on burning hydrocarbon fuels would not emit *any* CO.

In some occupations, it is difficult to avoid long-term exposure to carbon monoxide. *(© Yoav Levy/Phototake NYC)*

Carbon monoxide is a colorless, odorless gas at room temperature. This means carbon monoxide has no warning properties. Because it is a gas, it mixes with air, is inhaled, and comes in contact with the blood while in the lungs. The interference of carbon monoxide with oxygen transport in the blood is one of the best understood kinds of metabolic poisoning. Carbon monoxide, like oxygen, combines with the hemoglobin in red blood cells.

$$O_2(g) + \text{hemoglobin}(aq) \rightleftharpoons \text{oxyhemoglobin}(aq)$$
$$CO(g) + \text{hemoglobin}(aq) \rightleftharpoons \text{carboxyhemoglobin}(aq)$$

Both of these reactions are reversible, as indicated by the double arrow ($\rightleftharpoons$) (see Section 6.4). That is, oxyhemoglobin can give up its oxygen (that's what it does when it transports oxygen to a cell in the body), and carboxyhemoglobin can give up its carbon monoxide molecule, although not as easily. Laboratory tests show that carboxyhemoglobin is 140 times more stable than oxyhemoglobin.

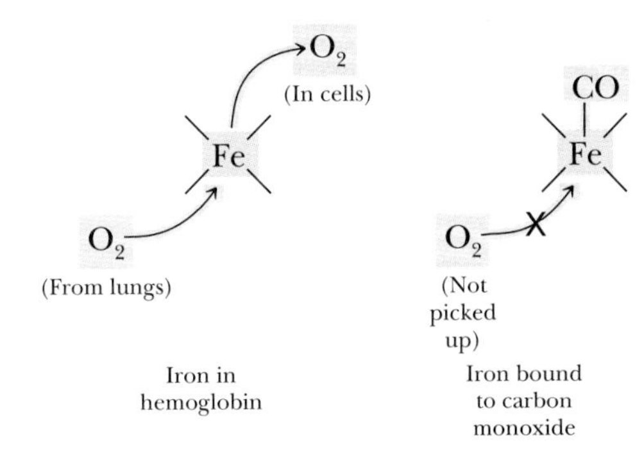

Both O_2 and CO bond to the iron atom in the hemoglobin molecule. The greater stability of carboxyhemoglobin occurs because the Fe—CO bond is stronger than the Fe—O_2 bond. Since hemoglobin is so effectively tied up by carbon monoxide, those hemoglobin molecules that contain a CO molecule cannot perform their vital function of transporting oxygen. Breathing air with a concentration of 30 ppm of CO for 8 h is sufficient to cause headache and nausea for most people. Breathing air that is 0.1% (1000 ppm) carbon monoxide for 4 h converts approximately 60% of the hemoglobin of an average adult to carboxyhemoglobin, and death is likely to result unless the carboxyhemoglobin molecules can be freed of the attached CO molecules.

■ Air is 21% O_2 by volume; in 1 million "air molecules" there would be 210,000 O_2 molecules.

Carbon monoxide exposures are quite common. In fact, low exposures are almost impossible to avoid. This means that some of the hemoglobin molecules in your blood are always bound to carbon monoxide. Any organic material that undergoes incomplete combustion will always liberate carbon monoxide. Sources include auto exhausts, smoldering leaves, lighted cigars or cigarettes, and charcoal burners. In the United States alone, combustion sources of all types dump about 200 million tons of carbon monoxide per year into the atmosphere, where it mixes with the other molecules in the air.

■ Carbon monoxide poisoning can occur when kerosene heaters or charcoal burners are used indoors without proper ventilation.

■ For incomplete combustion of carbon, the ratio is 2 C to 1 O_2: $2\,C(s) + O_2(g) \rightarrow 2\,CO(g)$. For complete combustion of carbon, the ratio is 1 C to 1 O_2: $C(s) + O_2(g) \rightarrow CO_2(g)$.

Since the reactions of both carbon monoxide and oxygen with hemoglobin are reversible, the concentrations of the two gases, in addition to the relative strengths of bonds, affect how much hemoglobin will be combined with either molecule.

■ High O_2 concentrations favor oxyhemoglobin, while high CO concentrations favor carboxyhemoglobin.

$$\text{Carboxyhemoglobin}(aq) + O_2(g) \rightleftharpoons \text{oxyhemoglobin}(aq) + CO(g)$$

As with all chemical equilibria, if the concentration of one of the reactants or one of the products is increased, the equilibrium will shift so that the increased reactant or product will be used and equilibrium is re-established (Le Chatelier's principle, Section 6.4). In air that contains 0.1% or less CO, oxygen molecules outnumber CO by at least 200 to 1. This higher concentration of O_2 molecules helps to counteract the stronger binding between CO and hemoglobin, so the equilibrium favors the formation of oxyhemoglobin. If a person breathes air that has a CO concentration higher than about 0.1%, the equilibrium begins to favor the formation of carboxyhemoglobin. When a victim of carbon monoxide poisoning is exposed to fresh air or, still better, pure oxygen, the equilibrium shifts back in the favor of oxyhemoglobin.

A bit of a mystery concerning carbon monoxide is that its global level does not seem to be changing, as is the case with some pollutants. Although polar carbon monoxide molecules dissolve readily in water, they react very slowly with oxygen to form carbon dioxide. The fate of atmospheric carbon monoxide is the subject of ongoing research in atmospheric chemistry.

16.11 Chlorofluorocarbons and the Ozone Layer

■ The low toxicity of CFCs allowed them to be used as propellants for aerosols, dispensing such things as hair sprays, deodorants, and even medicines. Virtually all these uses, except for some medical applications, have been banned.

Most pollutants are adsorbed onto surfaces, are absorbed into water droplets and react, or react in the gas phase with other pollutants in the lower atmosphere (troposphere) and eventually wash out in precipitation. There is one class of industrial pollutants, the halogenated hydrocarbons collectively called **chlorofluorocarbons (CFCs)** that are relatively unreactive and are not easily or quickly eliminated in the troposphere. Being unreactive, CFCs are also virtually nontoxic, so their presence in the troposphere causes none of the problems associated with carbon monoxide or the oxides of sulfur and nitrogen. Instead, as a direct result of their nonreactivity, these compounds have a chance to mix with air in the stratosphere, where they can reside for many years.

■ Approximately two thirds of U.S. households rely on air-conditioning systems that use CFC refrigerants.

After the discovery of the refrigerant gas properties of the CFCs (see Section 1.3), their use became widespread in applications such as automotive and home air-conditioning as well as in the manufacture of formed plastics for insulation. During the time CFCs were in common use, little effort was made to prevent them from escaping into the atmosphere. For example, if your automobile air conditioner needed repair, it was common practice to vent all of the CFC refrigerant gas to the atmosphere before any work was done. In fact, probably most of the CFCs ever manufactured have been released into the atmosphere. The industrialized countries were the first to use these compounds, but by the late 1970s they were also being extensively used in developing countries as well (Figure 16.10). Because of their excellent solvent properties for greases and oils, many of the CFCs also have been used as degreasers during the manufacture of printed circuit boards used in computers, television sets, and other kinds of appliances. In effect, the increased use of CFCs

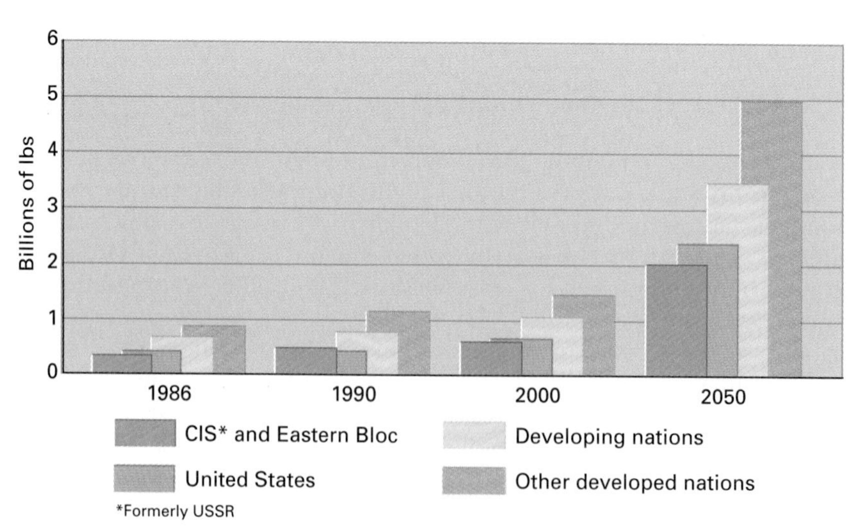

Figure 16.10 Projection of growth in CFC use by geographical region for 1986 to 2050 if no international controls were to be exercised. *(Environmental Protection Agency)*

paralleled the growth of modern, urbanized, industrial society with its indoor climate control, electronics, and electrical appliances.

$$\begin{array}{c} Cl \\ | \\ Cl-C-F \\ | \\ Cl \end{array} \qquad \begin{array}{c} F \\ | \\ Cl-C-F \\ | \\ Cl \end{array}$$

CFC-11 Trichlorofluoromethane   CFC-12 Dichlorofluoromethane

The dangers of CFCs were first announced in 1974, when M. J. Molina and F. S. Rowland of the University of California, Irvine, published a scientific paper in which they predicted that continued use of CFCs would lead to a serious depletion of Earth's protective stratospheric ozone layer. Depletion of the ozone layer is important because for every 1% decrease in stratospheric ozone, an additional 2% of the Sun's most damaging UV radiation reaches Earth's surface. The result is increased skin cancer, damage to plants, and possibly other effects we know little about now. Let's examine how these CFCs destroy the ozone layer.

■ The numbers in the names CFC-11 and CFC-12 are industrial code numbers that identify the compounds without using complicated chemical names.

■ In the United States, California leads all other states in releases of ozone-depleting substances, with more than 10,000,000 lb released in 1992.

In the stratosphere an abundance of ozone is produced because UV light with wavelengths below 280 nm readily breaks down oxygen molecules to produce oxygen atoms. These oxygen atoms, in turn, can react with oxygen molecules to produce ozone.

$$O_2 + h\nu \longrightarrow 2\ O$$

$$O + O_2 \longrightarrow O_3$$

Stratospheric ozone formed in this manner is so abundant (about 10 ppm) that it *absorbs between 95% and 99%* of the sunlight in the 200 nm to 300 nm wavelength range (the UV range). Light in this wavelength range is especially damaging to living organisms, so the stratospheric ozone layer is highly beneficial.

While the carbon–chlorine bond in a CFC molecule is not easily broken by reactions with acids, bases, or water, it *is* easily broken by UV light found in the stratosphere. This photodissociation produces a chlorine atom (Cl), for example,

■ Until the 1994 model year, CFC-11 was the refrigerant gas commonly used in automotive air conditioners. It is still used in most older (prior to 1993) automobiles.

$$\begin{array}{c} Cl \\ | \\ F-C-Cl \\ | \\ Cl \end{array} + h\nu \longrightarrow \begin{array}{c} Cl \\ | \\ F-C \\ | \\ Cl \end{array} + Cl$$

The chlorine atom is quite reactive. If it happens to collide with an ozone molecule, it forms a chlorine oxide (ClO) free radical and an oxygen molecule.

$$Cl + O_3 \longrightarrow ClO + O_2$$

SCIENCE AND SOCIETY

The Montreal Protocol on Substances that Deplete the Ozone Layer

It is rare for many different countries to get together and ban the use of a certain class of chemicals. With the exceptions of bans on chemicals used for warfare and bans on addictive drugs, this kind of activity almost never happens. Probably the reason it doesn't happen often is that most substances are available in such small quantities that the consequences of their use on a global scale would be insignificant. This means that local laws and regulations could control their use. Not so with the CFCs and other chemicals that can deplete the stratospheric ozone layer. These chemicals, as explained in the text, get involved in *cyclic* reactions that actually magnify their effect. That fact alone probably wasn't enough evidence to convince skeptics until observations of a decreasing ozone layer began to become common. By then, 11 years had passed since Roland and Molina first warned of a potential environmental disaster. In 1985, a conference called the Convention for the Protection of the Ozone Layer convened in Vienna. The United States and several Scandinavian countries wanted to freeze the use of ozone depleters and follow that with a ban on production. Of the attendees, only 27 countries, including the United States, the then Soviet Union, Japan, and the nations of the European community, signed on. Larger developing countries such as India, China, and Brazil were afraid that this action would harm their economic

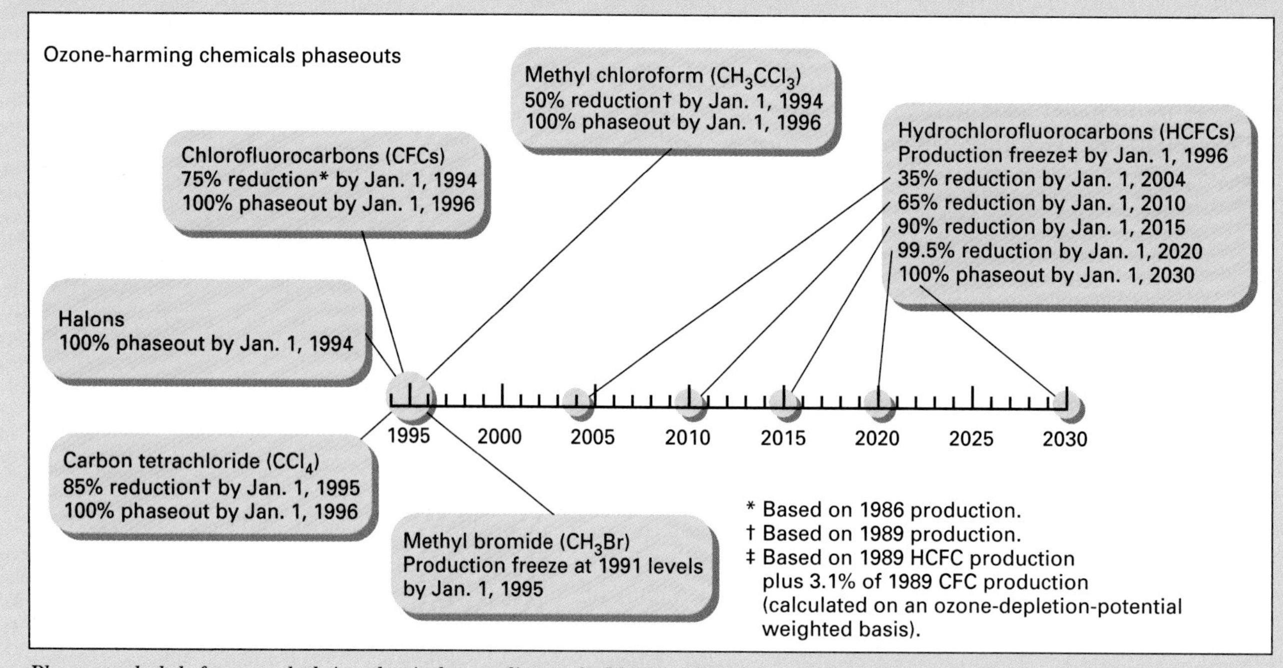

Phaseout schedule for ozone-depleting chemicals according to the Montreal Protocol.

Since there are plenty of ozone molecules present, this reaction is very likely to occur—and each time an ozone molecule is destroyed. If this were the *only* ozone molecule destroyed by the photodissociation of a single CFC, there would be little danger to the ozone layer. However, the ClO free radical can react with oxygen atoms and produce the chlorine atom again, which is ready to react with yet another ozone molecule.

growth. In May of 1985, however, the British Antarctic Survey announced that a 40% loss in ozone was occurring every fall over Halley Bay, Antarctica, and had been since the 1960s. It was shortly after this discovery that the public's attention was drawn to the fact that stratospheric ozone acted as a protective layer, filtering out harmful UV-B radiation. With a growing sense of urgency, nations began signing on to the Vienna Conference pact; and by 1987 another, larger conference was held in Montreal. The agreement, which became effective in January 1989, was called the Montreal Protocol. One of its strongest inducements to sign was trade restrictions that were to be imposed on nations who did not sign the protocol. By 1994, 134 nations had signed on. There have been amendments to the Protocol, but the real enforcement of restrictions on the manufacture, use, and distribution of substances that can deplete the ozone layer is contained in the 1989 agreement.

This agreement on the worldwide control of a class of chemicals that are capable of harming the entire global environment represents an interesting history in the handling of scientific data, its interpretation, drawing conclusions about the implications of those interpretations, and taking drastic steps to correct a problem.

Key Dates in the History of Ozone-Depleting Substances

Dec. 1973:	Rowland and Molina made their discovery.
Oct. 1978:	The use of CFCs in aerosols was banned in United States.
Oct. 1984:	British team reports 40% loss of ozone over Antarctica during austral (Southern hemisphere) spring.
Sept. 1987:	Montreal Protocol—representatives from 43 nations agree to CFC reductions of 50% by 2000.
Oct. 1987:	Antarctic expedition verifies huge losses of ozone over Antarctica during austral spring.
Mar. 1988:	United States ratifies Montreal Protocol; large U.S. manufacturers announce they will cease production of CFCs.
Apr. 1988:	Plastic foam manufacturers announce they will stop using CFCs.
Mar. 1989:	Seven hundred representatives from 124 countries attend London conference on saving the ozone layer.
June 1990:	Environment ministers from 93 nations agree to strengthen Montreal Protocol with complete phaseout of CFCs by 2000 (and HCFCs by 2040).
Oct. 1990:	U.S. Congress passes revised Clean Air Act that includes phaseout of CFCs by 2000.
Jan. 1991:	Environment ministers of European Community agree to complete CFC ban by 1997.
Jan. 1992:	Increased concentrations of ozone-depleting chemicals are found over populated areas in the Northern Hemisphere.
Feb. 1992:	U.S. president moves target date for phaseout of CFCs from the year 2000 to 1995.
Jan. 1, 1994:	Halon production stops. EPA formally asks some companies to continue production of CFCs through 1995 to meet "consumer needs" in automotive air conditioners.
Oct. 1994:	Antarctic ozone hole appears earlier than normal—covers 23 to 24 million square kilometers.
Jan. 1995:	NASA satellite data confirm CFC link to ozone hole.

$$ClO + O \longrightarrow O_2 + Cl$$

$$Cl + O_3 \longrightarrow O_2 + ClO$$

These reactions are summarized as follows, with the overall, or *net* reaction showing the destruction of an ozone molecule for every reaction cycle in which the chlorine atom participates

THE PERSONAL SIDE

Susan Solomon (1956–)

In 1985, a British team at Halley Bay Station, Antarctica, discovered the existence of a hole in the ozone layer above that continent. This totally unexpected phenomenon needed an explanation, and Susan Solomon—a young National Oceanic and Atmospheric Administration (NOAA) scientist—first proposed a good theory for it. While attending a lecture on polar stratospheric clouds, she realized that ice crystals in the clouds might do more than just scatter light over the Antarctic. Her chemist's intuition told her that the ice crystals could provide a surface on which chemical reactions of CFC compounds could take place.

Susan Solomon in the Dry Valleys. (Courtesy of S. Solomon)

In 1986, the National Aeronautics and Space Administration (NASA) chose Solomon (then 30 years old) to lead a team to Antarctica to sort out the right explanation for the ozone hole. Experiments during that visit to Antarctica showed that her cloud theory was correct, and a second expedition the next year added further evidence of its validity. Solomon's team and their experiments led to the first solid proof that there is a connection between CFCs and ozone depletion.

Susan Solomon is one of the youngest members of the National Academy of Sciences. She decided to become a scientist at age 10, having been influenced by watching Jacques Cousteau on TV. At age 16 she won first place in the Chicago Science Fair for a project called "Using Light to Determine Percentage of Oxygen," and went on to place third in the national science fair that year. She said that her winters as a young girl in Chicago prepared her for her visits to Antarctica.

$$
\begin{array}{r}
Cl(g) + O_3(g) \longrightarrow ClO(g) + O_2(g) \\
ClO(g) + O(g) \longrightarrow O_2(g) + Cl(g) \\
\hline
\text{Net: } O_3(g) + O(g) \longrightarrow 2\ O_2(g)
\end{array}
$$

The result of this reaction cycle is that a single chlorine atom may react up to 100,000 times before it eventually reacts with a water molecule to form HCl, which then mixes into the troposphere and washes out in acidic rainfall. This chlorine atom chain is thought to account for about 80% of the known losses of stratospheric ozone.

Molina and Rowland's warning more than 20 years ago concerning CFCs and their potential for depleting the stratospheric ozone layer has proved to be correct. Satellite and ground-based measurements since 1978 indicate that global concentrations of ozone in the stratosphere have been decreasing. The early stratospheric ozone concentration data showed an average decrease of about 2.5% annually in the decade from 1978 to 1988. Since then, studies over North America in 1993 have shown decreases of 12% to 15% below normal levels. These levels were lower than any measured over the past 35 years. Even larger decreases have been recorded over Europe. In addition, near the North and South Poles ozone losses have been so massive that "holes" in the ozone layer have been observed (Figure 16.11). These ozone holes are of special concern because many scientists believe that they may happen at the mid-latitudes in the future.

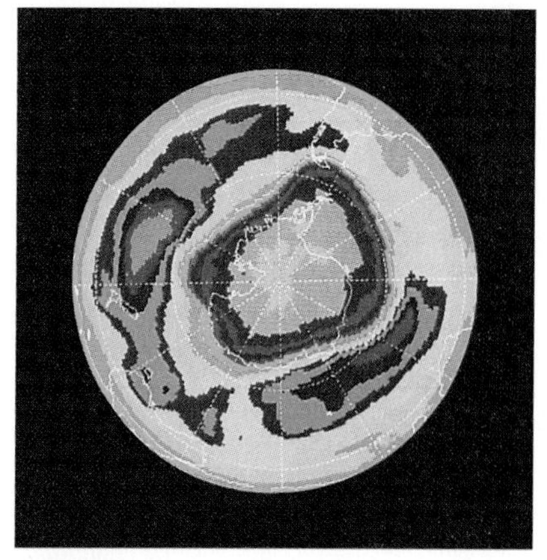

Figure 16.11 Ozone hole over Antarctica in October 1994. The land is outlined in white, and the "hole" is shown by concentrations of ozone represented by gray for the lowest, followed by pink, and purple. The hole covers 24 million square kilometers. Higher concentrations of ozone are represented by yellow, green, and brown, in that order. *(NASA/SPL/Photo Researchers)*

Alternatives to Ozone-Depleting Chemicals

The alarming drops in ozone concentration along with the regulatory actions taken by countries who signed the Montreal Protocol have led scientists to look for alternatives to compounds that deplete the ozone layer. In the early 1990s, it was believed that finding suitable alternatives for the CFCs most commonly used as refrigerant gases would be almost impossible, yet by 1993 substitutes were found. Today CFC-12, once the most common automotive refrigerant, has been almost totally replaced by HCFC-134a, while CFC-11 has been replaced by HCFC-141b.

```
     F   F              H   Cl
     |   |              |   |
 H — C — C — F      H — C — C — F
     |   |              |   |
     H   F              H   Cl
   HCFC-134a          HCFC-141b
```

These compounds, *hydrochlorofluorocarbons* (HCFCs), still have some ozone-depleting capabilities, but they are much more reactive in the lower atmosphere, which lessens their chances of getting into the stratosphere. Under current regulations, the EPA will allow the use of HCFCs as refrigerants until the year 2030. By that time, it is believed that even better alternatives will be found. When it became apparent that existing automotive air conditioners would have to be "retrofitted" with HCFCs as they were repaired or replaced, fears of high consumer costs sent engineers to the labs searching for solutions. Now, it is known that HCFCs are more compatible with existing air-conditioning systems on older cars and that repairs and retrofits should not cost as much as once feared. Still, retrofitting cars to use HCFC-134a can cost from $200 to $800. (Prices like these will undoubtedly cause some car owners to choose to drive with their windows open during the hot months.) In spite of the availability of substitutes, the EPA will allow older cars to be repaired using CFCs. The American Automobile Manufacturers Association has estimated a need for 200,000 metric tons of CFC-12 from 1996 to 2005 to service older vehicles.

■ Almost all new cars sold since 1994 have air conditioners using HCFC-134a (CH_2FCF_3).

The electronics industry has also made significant advances in eliminating CFCs and similar compounds in parts-cleaning operations. One large electronics plant in California had been releasing more than 1.5 million pounds of CFCs annually as part of its normal operations. As a result of the bans on such chemicals, it found a way to get its parts just as clean using soapy water, rinsing, and blow-drying with hot air.

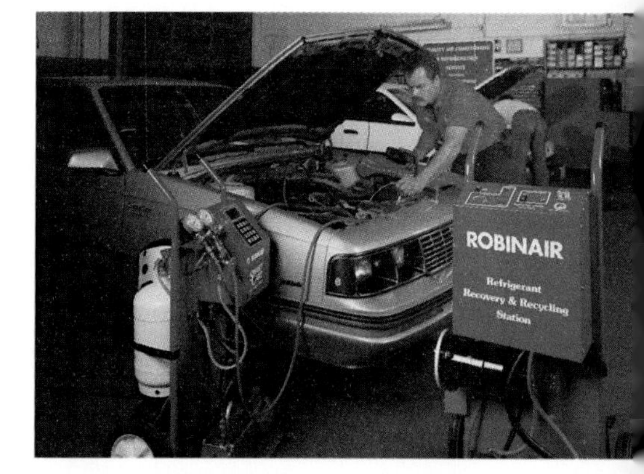

Recycling CFCs during repair of an automotive air conditioner. At many service centers, CFCs have simply been vented into the atmosphere. *(Courtesy of Robinair)*

A class of compounds known as **halons** must also be replaced to protect the ozone layer. The halons are structurally similar to CFCs, but contain a

Halons

Name	Formula	Uses
Halon-1211	CF_2BrCl	Portable fire extinguishers
Halon-1301	CF_3Br	Fixed fire extinguishers, aircraft fire extinguishers

carbon–bromine bond. Like the CFCs, the halons photodissociate in the stratosphere, where they produce bromine oxide radicals (BrO) that are destructive to ozone. Finding halon substitutes has presented quite a challenge. The halons are superior fire-fighting agents. They are used on aircraft, for example, and have been credited with saving many lives. Existing halon supplies are being allowed to remain in place, but halon substitutes such as perfluorobutane, bromodifluoromethane, and chlorotetrafluoroethane are being employed where halons were once used and in new systems.

■ The "per-" in perfluorobutane indicates that all the hydrogens in the butane molecule have been replaced, in this case by fluorine atoms. Another common "per-" compound is perchloroethylene ($CCl_2{=}CCl_2$), also known as "perc," which has been widely used as a dry-cleaning solvent.

$$\begin{array}{ccccccccc} & & F & & F & & F & & F & & \\ & & | & & | & & | & & | & & \\ F & - & C & - & C & - & C & - & C & - & F \\ & & | & & | & & | & & | & & \\ & & F & & F & & F & & F & & \end{array}$$

Perfluorobutane

$$\begin{array}{ccccc} & & F & & \\ & & | & & \\ H & - & C & - & Br \\ & & | & & \\ & & F & & \end{array}$$

Bromodifluoromethane

$$\begin{array}{ccccccc} & & Cl & & F & & \\ & & | & & | & & \\ H & - & C & - & C & - & F \\ & & | & & | & & \\ & & F & & F & & \end{array}$$

Chlorotetrafluoroethane

These compounds do not have the ozone-depleting potential that the other halons have. Their main disadvantage, and the prime reason they were not used earlier in the place of the other halons, is their higher cost and somewhat lower ability to stop the reactions going on during a fire.

The Future of the Stratospheric Ozone Layer

■ Unfortunately, a black market for CFCs has developed with unscrupulous importers smuggling loads of CFCs into the United States and other countries.

It is generally agreed that controls on emissions of CFCs and other compounds that can deplete the stratospheric layer are going to have the desired effect. Indeed, data now indicate that smaller decreases in ozone concentrations are occurring than just a few years ago. Natural processes that produce ozone-depleting compounds will continue, but it appears the stratospheric concentrations of all other ozone depleters will begin to decrease before 1999 and will probably be back to 1978 levels (about 50% over natural levels) by the year 2050. As markets for CFCs disappear, even reserves of those compounds (and hoarded quantities) will become less valuable. As that happens, they will eventually disappear. The present regulatory climate in the United States (Clean Air Act of 1990) as well as in other countries makes it unappealing to illegally manufacture or import CFCs and similar compounds. Industry has certainly indicated that it can "change with the times" and find substitutes even when no substitutes come to mind.

■ SELF-TEST 16B

1. When coal and fuel oil are burned, what two primary pollutants are formed?
2. When sulfur dioxide reacts with oxygen, what oxide of sulfur is formed? When this oxide reacts with water, what acid is formed?
3. Which chemical would be most likely to react with sulfur dioxide and remove it from combustion gases? (a) Sodium chloride (NaCl), (b) Lime (CaO), (c) Nitric oxide (NO).
4. Name two acids found in acid rain. One must contain sulfur, and the other, nitrogen.

5. What is the pH of normal rainfall? What dissolved chemical causes this pH to be below pH 7?
6. Approximately when was acid rainfall first observed?
7. Which chemical bond is broken by a photon of UV light in a typical CFC molecule?
8. Write the reaction producing ozone from oxygen.
9. What is the chemical species containing chlorine that destroys ozone molecules?
10. Over what continent have scientists found an ozone "hole"?
11. If ozone in the stratosphere is destroyed, what form of radiation will pass through to Earth below?

16.12 Carbon Dioxide and the Greenhouse Effect

How can carbon dioxide (CO_2) be considered a pollutant when it is a natural product of respiration and fossil fuel burning and is a required reactant for photosynthesis? Actually, CO_2 is not a pollutant in the strictest meaning of the term, but the fact that it is increasing in Earth's atmosphere is cause for deep concern. Consequently, it is treated as a pollutant. Without human influences, the flow of carbon between the air, plants, animals, and oceans would be roughly balanced. However, between 1900 and 1970, the global concentration of CO_2 increased from 296 ppm to 318 ppm, an increase of 7.4%. By 1994, the concentration was 360 ppm, and expectations are that the CO_2 concentration will continue to increase (Figure 16.12). For example, since the end of World War II, a world energy growth rate of about 5.3% per year took place until the OPEC oil embargo in the mid-1970s. For a while after that rates of energy use decreased, but that trend is now being reversed.

■ OPEC stands for Oil Producing and Exporting Countries.

Population pressures are contributing heavily to increased CO_2 concentrations. In the Amazon region of Brazil, for example, extensive cut-and-burn practices are being used to create cropland. This is causing a tremendous burden on the natural CO_2 cycle, because CO_2 is being added to the atmosphere during burning while there are fewer trees present to photosynthesize this additional CO_2 into plant nutrients.

Deforestation in Brazil. This satellite photo taken over the Amazon Basin shows the dramatic extent of the cut-and-burn creation of cropland. The remaining rain forest is dark green, and the leveled forest shows in pale greens and browns. The ladder-like pattern at the upper left is a typical result of the cut-and-burn practices in this region. *(Geospace/SPL/Photo Researchers)*

Counting all forms of fossil fuel combustion worldwide, the amount of CO_2 added to the atmosphere is about 50 billion tons a year. About half of this remains in the atmosphere to increase the global concentration of CO_2. The other half is taken up by plants during photosynthesis and by the oceans, where CO_2 dissolves to form carbonic acid, which then can form bicarbonates and carbonates.

$$CO_2(g) + H_2O(\ell) \rightleftharpoons H^+(aq) + \underset{\text{Bicarbonate ion}}{HCO_3^-(aq)}$$

$$HCO_3^-(aq) \rightleftharpoons H^+(aq) + \underset{\text{Carbonate ion}}{CO_3^{2-}(aq)}$$

To see how easily our everyday activities affect the amount of CO_2 being put into the atmosphere, consider a round-trip flight from New York to Los Angeles. Each passenger pays for about 200 gal of jet fuel, which weighs

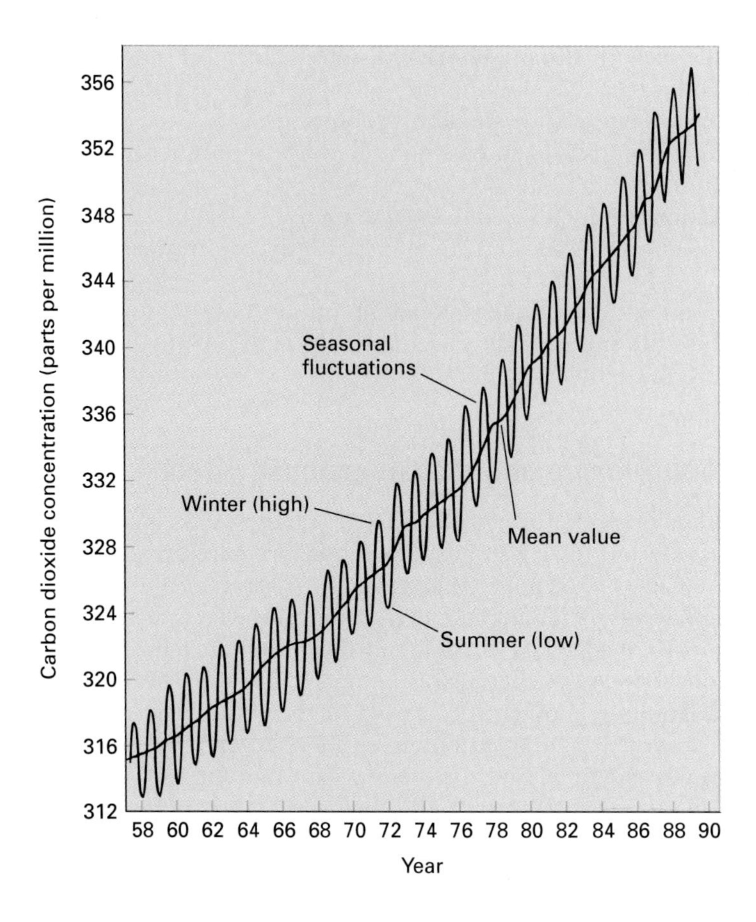

Figure 16.12 Carbon dioxide concentration in the atmosphere, showing the steady increase over the years. The seasonal variation results from high CO_2 in the summer due to greater photosynthesis and low CO_2 in the winter when photosynthesis diminishes. These measurements are taken at the Mauna Loa Observatory in Hawaii, which is located far from urban areas with their industrial CO_2 emissions.

1400 lb. When burned, each pound of jet fuel produces about 3.14 lb of carbon dioxide. So 4400 lb, or 2.2 tons, of carbon dioxide are produced per passenger during that trip. It seems reasonable that if we are rapidly burning fossil fuels that took millions of years to form, we are then going to be adding CO_2 back into the atmosphere at a more rapid pace than it can be used up in natural processes.

What is the problem with increasing atmospheric CO_2? When solar radiation arrives at Earth's atmosphere, about half of the visible light (400–700 nm) is reflected back into space. (That's a good thing; otherwise, the temperature of Earth would be far too hot to support life as we know it.) The remainder reaches Earth's surface and causes warming (Figure 16.13). The warmed surfaces (average temperature about 27°C) then reradiate this energy as heat energy in the infrared portion of the spectrum. Water vapor, CO_2, ozone, and methane (CH_4) readily absorb some of this reradiated energy and in turn *warm* the atmosphere, creating what is called the **greenhouse effect**. A botanical greenhouse works on the same principle. Glass transmits visible light but blocks infrared radiation trying to leave. The effect is a warming of the air inside the greenhouse. In warm weather, the windows of a greenhouse must be opened or the plants inside will overheat and die.

All four of the "greenhouse gases" act as an absorbing blanket that prevents radiation losses and keeps Earth's atmospheric temperature comfortable

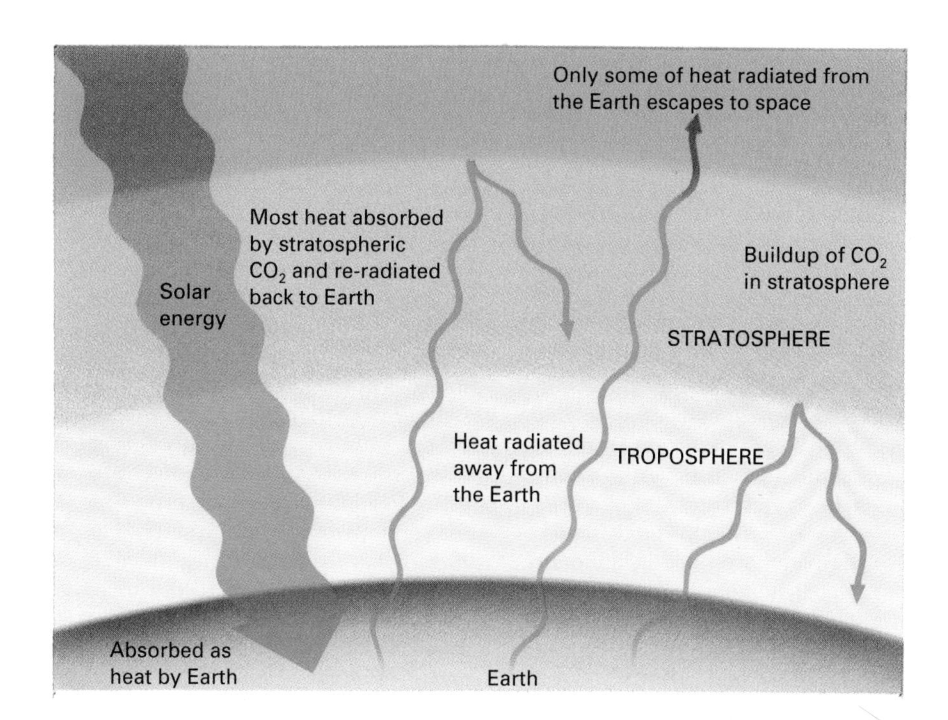

Figure 16.13 The greenhouse effect. Greenhouse gases effectively form a barrier that prevents heat from escaping from planet Earth's surface.

(although not in all locations at the same time). Water vapor in the atmosphere is subject to such vast cycles that human activity doesn't seem to bother it. Because ozone is present in relatively low concentrations, and because methane is produced naturally in vast quantities, our attention is focused on CO_2, the greenhouse gas whose atmospheric concentration is most closely related to human activity.

Recently, Russian scientists took ice core samples dating back 160,000 years. In these ice samples were tiny pockets of air that could be analyzed for CO_2 content. They found a direct correlation between CO_2 and geological temperatures known by other means. As the CO_2 increased, the global temperature increased; and as the CO_2 decreased, the global temperature decreased. It is generally agreed that rising CO_2 concentrations will probably lead to increasing global temperatures and corresponding changes in climates. If predictions by the National Academy of Sciences prove correct, when and if the global concentration of CO_2 reaches 600 ppm, the average global temperature will have risen by 1.5°C to 4.5°C (2.7° to 8.1°F). Even a 1.5°C warming would produce the warmest climate seen on Earth in the past 6000 years, and a 4.5°C warming would produce world temperatures higher than any since the Mesozoic era—the time of dinosaurs.

Clearly, **global warming** is a major potential problem, and its effects appear to be measurable on a human time scale. It has been calculated that stabilizing the CO_2 concentration in the atmosphere at 360 ppm will require limiting global industrial emissions to something less than 2×10^9 tons CO_2 per year, which is well below the current emissions of 6×10^9 tons CO_2 per year. Scientists who study the atmosphere fear that at the current rate of emissions, CO_2 concentrations will rise to about 550 ppm before they begin to

SCIENCE AND SOCIETY

Global Warming Controls Coming

In December 1997, representatives from most of the world's countries met to hammer out an agreement on reducing the emissions of greenhouse gases that cause global warming. The negotiations centered mainly on how developed nations would respond to emission controls compared to the responses of underdeveloped and developing nations. The treaty, agreed to by all the participants, calls for the United States to cut its emissions of six greenhouse gases (see the following table) to a level of 7% below 1990 levels by the years 2008 to 2012. European Union countries would have to cut their emissions by 8% of the 1990 levels and Japan by 6%. Other countries would face smaller reductions. Overall, emissions will have to be reduced by 5% when all the countries signing the treaty are considered. For highly developed countries like the United States, these reductions will be equivalent to a 30% reduction from 1990 levels when predicted growth is factored in.

Not surprisingly, predictions of dire consequences met the announcements of the signing of the treaty. Unlike similar treaties involving ozone-depleting gases, most of the greenhouse gases and their emissions are directly connected to large-scale economic activity and growth. This includes groups as diverse as utilities, auto manufacturers, farm organizations, and unions. To make the treaty more acceptable, it contains provisions for emissions trading, where one country can purchase excess "quota" from other countries that have met their goals. In addition, there are provisions for creating more "carbon sinks," like forests, that can utilize carbon dioxide. Enforcement provisions in the treaty were not agreed on at the December 1997 signing.

Treaty Greenhouse Gases	Comment
Carbon dioxide	Linked to power generation
Methane	Linked to fuel production
Nitrous oxide	A product of combustion
Hydrofluorocarbons	Also an ozone-depleting gas with a long life in the atmosphere
Sulfur hexafluoride	Long atmospheric life, used in some electrical transformers
Perfluorocarbons	Long atmospheric life, used mainly in industry

Source: *The Wall Street Journal*, December 9 and 12, 1997.

level off. Controlling CO_2 emissions worldwide will undoubtedly prove to be more difficult than controlling CFCs or the precursors to acid rain. At the 1992 Earth Summit in Rio de Janeiro, Brazil, a number of countries signed a treaty to limit CO_2 emissions by the year 2000 to 1990 levels in their countries. This goal is proving very difficult to achieve. Industrialized countries emit so much CO_2 from so many diverse sources that reducing emissions may be almost impossible unless radical societal changes are made. One of these would be in the ways energy is produced and distributed. Strong arguments can be and are being made for rapid conversion to solar energy as well as to nuclear energy, neither of which contributes significantly to global warming, but each has its own unique negative environmental consequences. Unfortunately, developing countries appear to have no other choice but to continue to emit CO_2 as they try to raise their standards of living by producing food, energy, clothing, and shelter. Most planners agree that global CO_2 emissions will depend on both population growth and world economic growth. By using projections of large population growth (to just over 11.5 billion by the year 2050) and moderate world economic growth (3% per year until 2050), a United Nations panel on climate control has estimated the mean global temperature could increase by almost 3°C.

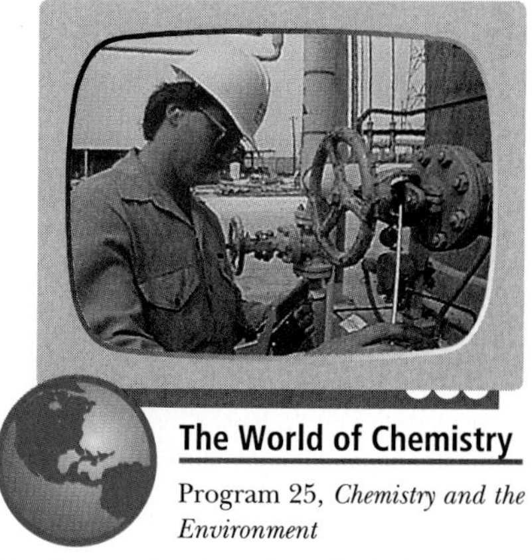

The World of Chemistry

Program 25, *Chemistry and the Environment*

Testing a valve in a chemical plant for leaks.

16.13 Industrial Air Pollution

Industrial activity pollutes the atmosphere by emitting a wide variety of solvents, metal particulates, acid vapors, and unreacted monomers in addition to CO_2, CFCs, NO_x, and SO_2. The extent to which this takes place became evident in 1989 when the first summary of annual releases was published from data received by the EPA. This report was a part of the Superfund Reauthorization and Amendments Act of 1986, regulations that resulted in part from a tragic release of a toxic chemical in Bhopal, India, where more than 4000 people were killed and many thousands more were injured. These release-reporting regulations were placed on manufacturers who use any of a group of about 320 chemicals and classes of compounds representing special health hazards. The reporting was divided into releases to air, water, and land. Recently, the list of reportable chemicals was expanded by the EPA to include a total of 654 compounds in an effort to give citizens an even clearer picture of chemicals that affect their community.

■ Water and land releases directly affect surface and groundwater purity. (See Chapter 15.)

■ These regulations have been called "community right-to-know" regulations because they inform communities about releases of harmful chemicals in their areas.

In a release report, an industrial facility must list all releases of a reportable chemical to the atmosphere, regardless of the type of release. This means that leaky valves and fittings, accidental spills, vapor losses while filling tank trucks and rail tank cars, emissions at stacks, and so forth are all added together on the report. As expected, heavily industrialized states and states with a lot of chemical industry have high releases, but the amounts of some chemicals released have also been surprisingly large (Table 16.5). These data are now summarized annually and are available by means of publicly accessible computerized databases. Interested persons may call the EPA at 1-800-535-0202 for more information about chemical releases in their community.

■ Since toxic chemical releases to the atmosphere have been measured, the top two states have been Louisiana and Texas.

What do these releases of harmful chemicals into the atmosphere mean? Compared with the vast quantities of matter comprising the atmosphere, industrial releases of chemicals seem low. Even compared with the releases of certain classes of chemicals—such as halogenated compounds by marine organisms and methane by ruminant animals, ants, and termites—industrial releases seem small. The problem is that industrial releases are usually concentrated close to home, that is, near population centers where the quality of the atmosphere becomes lowered before the released chemicals ever have a chance to be diluted by all the surrounding atmosphere. (Compare this with the release of some 10^9 tons annually of methyl iodide (CH_3I) by marine organisms over the entire surface of the oceans.) In addition, all those released compounds represent *financial losses* for the companies involved. If anything, the EPA release rules helped focus industry's attention on the need to reduce emissions.

TABLE 16.5 ■ Top Five Chemicals Released into the Environment—1995

Chemical	Total 10^6 lb
1. Methanol	245.0
2. Ammonia	195.1
3. Toluene	145.9
4. Nitrate compounds	137.7
5. Xylenes	95.7

Source: U.S. EPA. Toxic Release Report, 1995.

16.14 Indoor Air Pollution

As if the data about pollutants in the outside air were not enough to concern us, the air inside our homes and workplaces is also contaminated, usually by the same chemicals emitted by industry. Some scientists have concluded that air in our homes may be *more* harmful than the air outdoors, even in heavily industrialized areas. A study by the EPA indicated that indoor pollution levels in rural homes were about the same as in homes in industrialized areas. One cause for this is the emphasis on tighter, more energy-efficient homes, which tend to trap air inside for long periods.

■ We shouldn't be surprised that air in our homes is contaminated by industrial chemicals—after all, we bring industry's products into our homes.

What are the sources of home air pollution? (See Figure 16.14) Tobacco smoke, if present, is an obvious source. Benzene, a known carcinogen, occurs at 30% to 50% higher levels in homes of smokers than in homes of nonsmokers. Building materials and other consumer products are also sources of

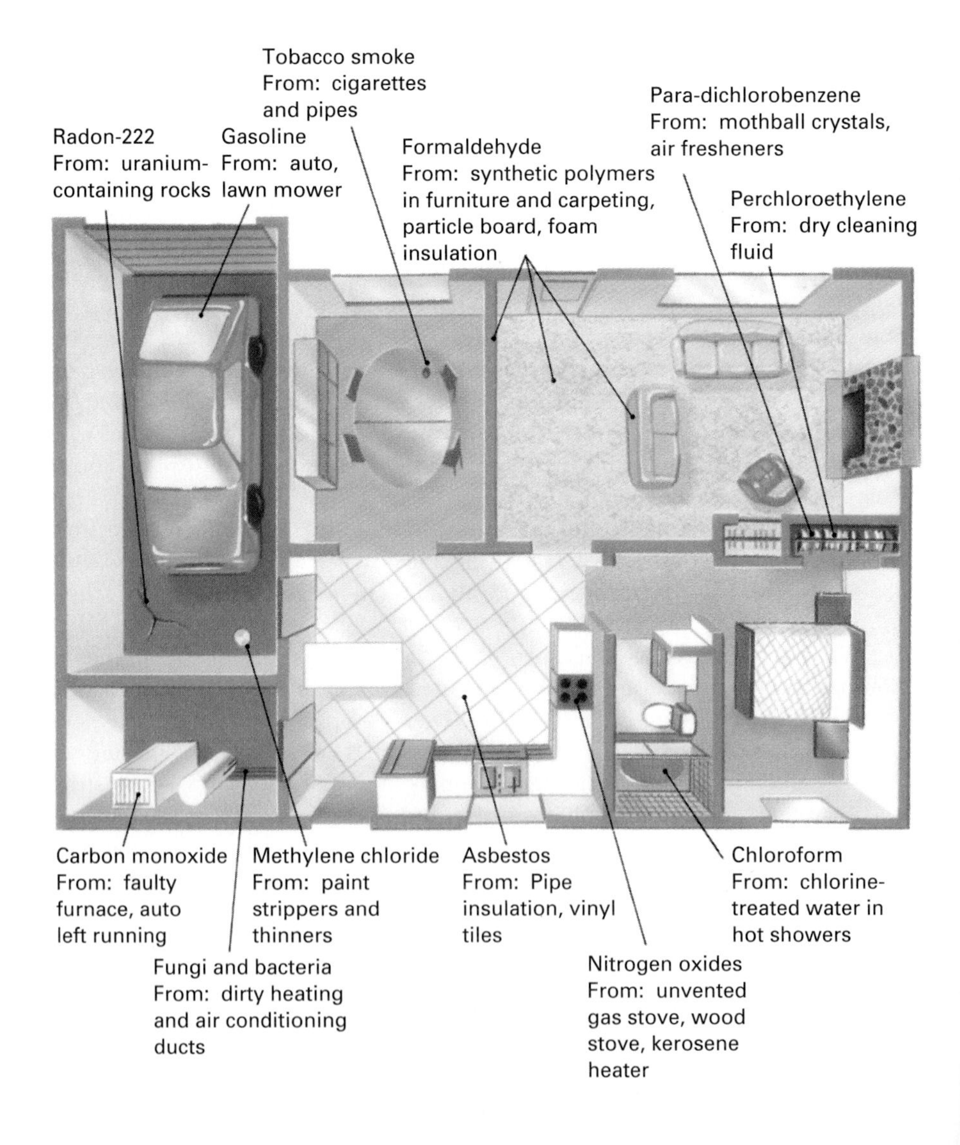

Figure 16.14 Sources of some indoor air pollutants.

pollutants. Entire buildings can acquire a "sick building syndrome" when a particular chemical or group of chemicals is found in sufficiently high concentration to cause headaches, nausea, stinging eyes, itching nose, or some combination of these symptoms. Usually, the best cure for all forms of indoor air pollution is to limit the introduction of the offending chemicals and to have better exchange with the outside air. Of course, this solution comes at a cost. If you exchange indoor air with outside air when the temperature differences between inside and outside are great, you will be paying a larger heating or cooling bill.

SELF-TEST 16C

1. Name three different human activities that produce large amounts (millions of tons per year) of CO_2.
2. What are two principal processes whereby CO_2 is consumed?
3. Name three major greenhouse gases found in the atmosphere.
4. Which greenhouse gas is most closely associated with human activities?
5. Global temperature seems to follow carbon dioxide concentrations. (a) True, (b) False.
6. Approximately what is the current global CO_2 concentration?
7. What state had the highest air releases of toxic chemicals in 1995, the latest year for which such data are available?
8. What chemical was released in greatest amount nationwide in 1995, according to the EPA release report?
9. What single activity inside the home can account for increased concentrations of benzene, a known human carcinogen?

MATCHING SET

___ 1. Gas with the highest percentage by volume in the atmosphere
___ 2. Consisting mostly of very small water droplets
___ 3. Abnormal temperature arrangement for air masses
___ 4. A mixture of smoke, fog, air, and various other chemicals
___ 5. A pollutant caused by reactions of other chemicals
___ 6. Second most abundant gas in the atmosphere
___ 7. A pollutant that is directly discharged into the atmosphere
___ 8. A chemical species with an unpaired valence electron
___ 9. Main ingredient in industrial smog
___ 10. The product of a reaction between an oxygen atom and an oxygen molecule
___ 11. An oxide of nitrogen produced by certain bacteria
___ 12. The oxide of nitrogen produced by lightning, forest fires, and internal combustion engines
___ 13. "Bad" ozone
___ 14. "Good" ozone
___ 15. A polynuclear aromatic hydrocarbon capable of causing cancer
___ 16. One of the principal producers of SO_2
___ 17. Cause rain to have an abnormally low pH
___ 18. Contribute to the destruction of the ozone layer
___ 19. An alternative to certain ozone-depleting chemicals
___ 20. Ozone-depleting chemicals that contain a carbon–bromine bond
___ 21. The principal greenhouse gas
___ 22. An air pollutant that bonds to hemoglobin

a. Oxygen
b. NO_2 and SO_2
c. Nitrogen
d. Halons
e. Aerosol
f. Chlorofluorocarbons
g. CO_2
h. Secondary pollutant
i. SO_2
j. HCFC-134a
k. Smog
l. NO
m. Fossil fuel electricity generators
n. Primary pollutant
o. Thermal inversion
p. CO
q. Ozone in the air we breathe
r. N_2O
s. Benzo(α)pyrene
t. Free radical
u. Ozone molecule
v. Stratospheric ozone

QUESTIONS FOR REVIEW AND THOUGHT

1. Define the following terms:
   (a) Polluted air (b) Aerosol
   (c) Photochemical smog (d) Secondary pollutant
2. Define the following terms:
   (a) Photodissociation (b) Particulate
   (c) Thermal inversion (d) Reducing type smog
   (e) Free radical
3. Define the following terms:
   (a) Acid rain (b) CFC
   (c) Ozone hole (d) Greenhouse gas
   (e) Global warming (f) Cryogen
4. Name three major air pollutants and give a source for each.
5. What federal legislation has the abbreviation CAA? Describe how this legislation has changed over the years.
6. Describe what a thermal inversion is and explain how it can aggravate the problems caused by smog.
7. How is an industrial-type smog different from a photochemical smog?
8. Explain how ozone is formed in the troposphere.
9. Explain how ozone is formed in the stratosphere.
10. What is the difference between "good" ozone and "bad" ozone?
11. What are the main sources of nitrogen oxides in the atmosphere?
12. What is an aerosol and how does it play a role in air pollution?
13. Describe how a volcanic eruption can contribute to global cooling.
14. Name three ingredients necessary for the formation of a chemically oxidizing type of smog and explain how they interact.
15. Name two sources of hydrocarbons in the atmosphere. Which one is more readily controlled?
16. Write the equation for the reaction that occurs when lightning causes nitrogen to react with oxygen.
17. Nitrogen dioxide plays a role in the formation of what secondary pollutant?
18. What are the names of the two regions of the atmosphere discussed in this chapter? Which one is closer to the surface of Earth?
19. What happens when a photon of light strikes a molecule of NO_2? Write the reaction.
20. Describe how ozone can be harmful when it is present in the air we breathe.
21. How are the harmful effects of SO_2-containing aerosols, NO_2, and ozone similar?
22. Describe two sources of carbon monoxide.
23. Describe how pure oxygen can be obtained from air.
24. How is the control of ozone in the lower atmosphere connected to the control of NO_x emissions?
25. What health risks are posed by breathing air contaminated with particulates?
26. What two air pollutants does an automotive catalytic converter help control?
27. Describe the trends in automobile-related air pollution during the past two decades. Name some factors that have contributed to these trends.
28. What is the product of the combustion of sulfur or sulfur-containing compounds in air? Write the reaction.
29. Pick a source of SO_2 emissions and describe two ways SO_2 emissions from that source can be controlled.
30. Assume you have at your disposal some kind of SO_2 control that produced a mole of CO_2 for every mole of SO_2 removed. Explain how this might have both good and adverse effects on the environment.
31. Explain how rain can have a pH slightly below 7, the pH of pure water. In your explanation be sure to include both natural as well as other sources of compounds that can acidify rain.
32. Briefly describe the adverse effects acid rain produces. How do these effects sometimes become international issues?
33. Write the chemical equation for the reaction between carbon and oxygen when an insufficient amount of oxygen is present. What is the name of the product of this reaction?
34. What are CFCs? Describe some of the uses of this class of compounds.
35. Someone asks you, "What's all the fuss about CFCs?" Explain in the simplest terms you can why CFCs are "in the news."
36. Describe briefly the history of the CFC problem and steps that have been taken to address the problem.
37. Write the equation for the chemical reaction that occurs when UV light strikes an oxygen molecule. Then write the equation for the reaction that produces ozone in the upper atmosphere.
38. Why is a small concentration of ozone in the upper atmosphere important?
39. Name the CFC that was commonly used in air conditioners in new model cars until recently. Why is this CFC no longer used in automotive air conditioners?
40. Explain how a single chlorine atom can cause the destruction of up to 100,000 ozone molecules.

CHAPTER 17

Feeding the World

When hunting and gathering from nature's bounty were the primary means of obtaining food, only catastrophic events could void the fundamental relationship between the effort and a satisfied stomach. As time has passed, population increases and the concentration of people in larger and larger towns and cities have forced the development of basic agriculture, which has mostly succeeded but sometimes dramatically failed. To help grow the enormous amount of food needed to feed the world population, currently 5.9 billion and growing at the rate of 1.6% per year, the chemical industry supplies modern scientific agriculture with a large assortment of chemicals—the **agrichemicals**. These include fertilizers, medicine and food supplements for livestock, and chemicals to destroy unwanted pests and plant diseases. However, even with the use of new agrichemicals, sound agricultural practices, and biotechnology advances, the world's food supply cannot keep pace with the rate of growth of the world's human population. To have sustained food security for the world's population, global cooperation is essential in slowing the rate of population growth and reducing both environmental problems that threaten food productivity and problems of food distribution.

- What are current projections for feeding the growing world population?
- What factors affect the productivity of soil?
- What agricultural practices are essential to sustaining soil productivity?
- How dependent is modern agriculture on insecticides and herbicides?
- What biotechnology advances in agriculture are helping to increase food production?
- What is sustainable agriculture?

Will there always be enough farms to feed our expanding population? *(Richard H. Smith/Dubinsky Photo Associates)*

Through the 8000 to 10,000 years of recorded human history, food production techniques have developed enormously. It is generally estimated that 90% of the U.S. population worked to provide food and fiber during most of the

19th century. Although more efficient agricultural methods now allow one U.S. farm worker to feed 100 people, the world's farmers are falling behind in their ability to feed the growing world population. More than 1 billion people in the world depend on agricultural lands that are not productive enough to support them adequately.

A report released by the United Nations Environment Program in 1992 states that 4.84 billion acres of soil—an area the size of China and India combined—have been degraded to the point where it will be difficult or impossible to reclaim them. The main causes of soil degradation are soil erosion, overgrazing, and deforestation. However, urbanization and industrialization also contribute to cropland loss. One of the main reasons that the world grain harvest tripled between 1950 and 1990 was a 2.5-fold increase in irrigated land that produces about 40% of the world's food. However, farmers of irrigated land are now facing the prospect of water scarcity as competition with urban areas for shrinking water supplies increases.

17.1 World Population Growth

Global pollution, extinction of wildlife, degradation and loss of natural resources, and depletion of energy reserves in today's world are all related to the increasing world population. How close are we to reaching the earth's capacity to support the world population? What is the maximum sustainable population level? What factors determine the earth's capacity? These questions have been studied by organizations, such as the Worldwatch Institute; and at world conferences such as the United Nations Conference on Environment and Development, commonly known as Earth Summit, in Rio de Janeiro in 1992 and the United Nations Internal Conference on Population and Development in Cairo in September 1994. All such studies recognize that control of the rate of growth of the world's population is essential in maintaining adequate food supplies to feed the world. The present world population of 5.9 billion is growing at the rate of about 80 million people per year and is projected to reach about 8 billion by the year 2030 and level off at 11 billion around 2050 (Figure 17.1). Even now, 20% of the world's population (1 out of 3 persons in developing countries) are living in extreme poverty, with 700 million of them malnourished.

Another factor that affects the earth's carrying capacity is food-use efficiency. Currently, 38% of the world's grain is fed to animals each year. A kilogram of feedlot-produced beef requires 7 kg of grain compared to 4 kg for 1 kg of pork, and 2 kg for 1 kg of poultry or 1 kg of fish. Just reducing the world's consumption of beef would free up grain either for direct consumption or as feed for pork, poultry, and fish, which are more efficient meat producers than beef. At present, the average American requires 800 kg of grain per year, the great bulk of it consumed indirectly in the form of beef, pork, poultry, eggs, milk, cheese, yogurt, and ice cream. If Americans cut their annual grain intake in half, to the level that Italians consume each year, 105 million tons of grain would be saved, which is enough to feed two thirds of the population of India for one year. The good news is that the greater awareness among Americans for lower fat diets (Section 12.6) during the past several years has already resulted in a decrease in beef consumption and a growing preference for pork, poultry, and fish.

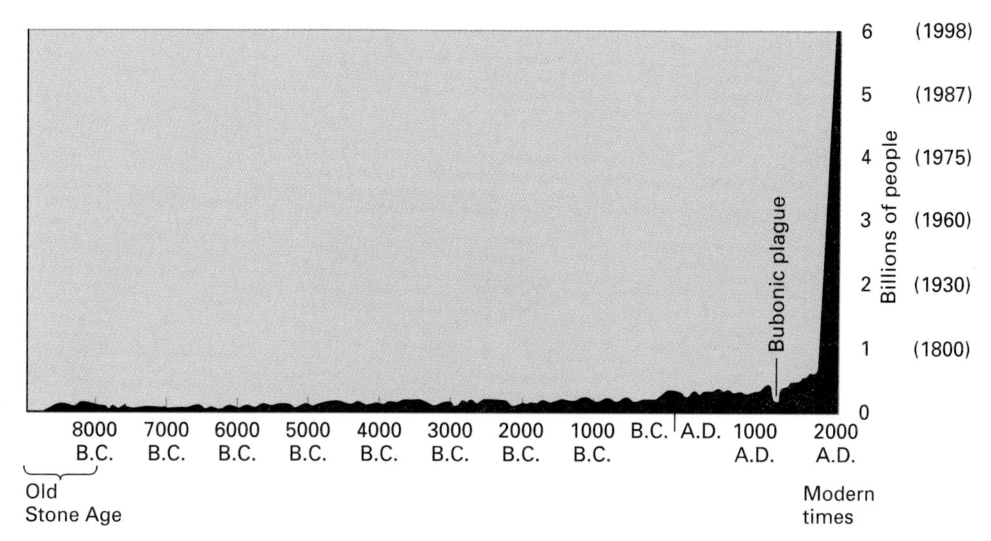

Figure 17.1 Human population growth. The plot shows population growth for the 10,000 years since the beginning of agriculture. Notice the decreasing time it has taken to add each additional billion. It took 130 years to grow from 1 to 2 billion; we have grown from 5 to 6 billion in the 11 years between 1987 and 1998.

Our fundamental food is plants. Through photosynthesis powered by energy from the Sun, Earth's green plants provide the food to sustain life on Earth. Either we eat plants or animals that eat plants. To grow, plants require the proper temperature, nutrients, air, water, and freedom from disease, weeds, and harmful pests. Agricultural productivity for the past 50 years has relied on the use of chemicals to assist nature in giving plants the proper nutrients and freedom from disease and competitive life forms. However, the use of agrichemicals involves risks to the environment and human health; it is important to measure the risks versus benefits in the use of these chemicals.

The World of Chemistry

Program 23, *Proteins: Structure and Function*

Plants and animals, sustained by photosynthesis. Powered by energy from the Sun, Earth's green plant population provides the food to sustain life on Earth.

17.2 What Is Soil?

Soil is a mixture of four components—mineral particles, organic matter, water, and air (Figure 17.2, p. 456). Weathering processes in nature over thousands of years break rock into small mineral particles found in soil. Organic matter in soil is a mixture that includes leaves, twigs, plant and animal parts in various stages of decomposition, and microorganisms. **Humus**, the dark-colored decomposed organic material, is important to a good soil structure. As a source of nutrients for plants, humus is almost like a time-release capsule, slowly releasing its contents.

■ Humus releases its nutrients to plants slowly.

Maintaining humus in the soil is of major concern to the agriculturist. Humus such as peat moss or organic fertilizer can be added. However, there is no real substitute for natural plant growth that is returned to the ground for humus formation. Clover is often grown for this purpose and plowed under at the point of its maximum growth. The compost pile of the gardener is another effort to maintain humus for a productive soil.

Figure 17.2 Soil formation. Weather, plants and their litter, earthworms and other organisms, and topography interact to produce soil.

Evaporation and transpiration
Precipitation
Plant litter
Earthworms
Runoff and erosion affected by topography
Organic additions (underground) from soil organisms
Mineral nutrients
Soil microorganisms
Fine particles and solubles are washed downward
Capillarity and evaporation cause some materials to rise
Water table
Loss of soluble materials in groundwater

■ Friable material crumbles easily under slight pressure.

In addition to being a source of plant nutrients, humus is important in maintaining good soil structure, often keeping it *friable.* Soil rich in humus may contain as much as 5% organic matter. Soils in the grasslands of North America have humus to a considerable depth, in contrast to rain-forest regions, where there is only a thin film of humus on the ground surface.

A handful of humus, which is partially decomposed organic material from plants and animals. Where soil contains ample humus, it is spongy, holds water well, and is a healthy environment for plants and organisms that live in the soil. *(USDA/Soil Conservation Service)*

Soil Profile

Layers within the soil are called **horizons** (Figure 17.3). The **topsoil** contains most of the presently living material and humus from dead organisms. Topsoil is usually several inches thick, and in some locations more than 3 ft of topsoil can be found. The **subsoil**, up to several feet in thickness, contains the inorganic materials from the parent rocks as well as organic matter, salts, and clay particles washed out of the topsoil.

Because healthy topsoil has abundant life forms, it must contain an abundant supply of oxygen. Soil that supports vegetative growth and serves as a host for insects, worms, and microbes is typically full of pores; such soil is likely to have as much as 25% of its volume occupied by air. The ability of soil to hold air depends on soil particle size and how well the particles pack and cling together to form a solid mass. The particle size groups in soils, called **separates**, vary from clays (the finest) through silt and sand to gravel (the coarsest). The particle size of a clay is 0.005 mm or less. The small particles

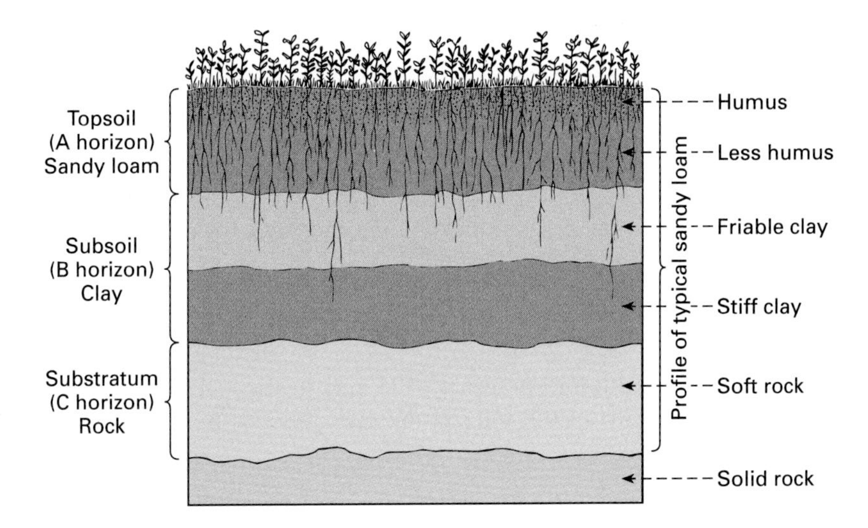

Figure 17.3 Structure of a sandy soil. The layers, known as *horizons*, are built up through weathering of rock and interaction with water, air, plants, and animals as shown in Figure 17.2.

in a clay deposit pack closely together to eliminate essentially all air and thus support little or no life. A typical soil horizon is composed of several separates. A **loam**, for example, is a soil consisting of a friable mixture of varying proportions of clay, sand, and organic matter; a loam has a high air content.

Air in soil has a different composition from the air we breathe. Normal dry air at sea level contains about 21% oxygen (O_2) and 0.03% carbon dioxide (CO_2). In soil the percentage of oxygen may drop to as low as 15%, and the percentage of carbon dioxide may rise above 5%. This results from the partial oxidation of organic matter in the closed space. The carbon in the organic

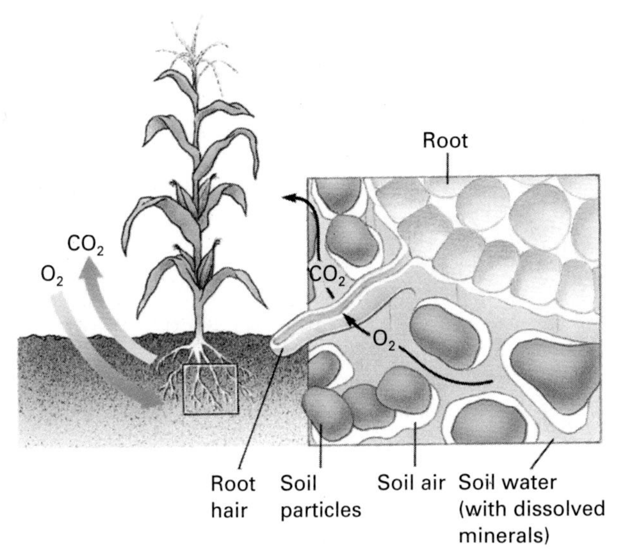

Exchanges between water and air in soil pores. Plant roots absorb oxygen and water, and release carbon dioxide.

Spreading "lime" on the lawn. "Lawn lime" is powdered limestone, which is $CaCO_3$. It increases the pH, that is, makes the soil more alkaline, or "sweeter." Calcium oxide (CaO) and calcium hydroxide ($Ca(OH)_2$, which is made by adding water to CaO) are also both often referred to as "lime." They are too strongly basic for lawns, but can be used in commercial agriculture.
(C. D. Winters)

material combines with oxygen to form carbon dioxide. This increased concentration of carbon dioxide tends to cause groundwater to become acidic; acidic soils are described as *sour* soils because of the presence of aqueous acids.

$$CO_2(g) + H_2O(\ell) \longrightarrow H^+(aq) + HCO_3^-(aq)$$

Crushed limestone ($CaCO_3$) applied to soil combines with hydrogen ions to form bicarbonate ions, thus raising the pH.

$$H^+(aq) + CO_3^{2-}(aq) \longrightarrow HCO_3^-(aq)$$

■ A slightly basic soil is a "sweet" soil.

If enough limestone is added to neutralize the acid in the soil and leave an excess of limestone, the pH of the soil becomes alkaline (basic).

Water in the Soil: Too Much, Too Little, or Just Right

Water can be held in soil in three ways: it can be *absorbed* into the structure of the particulate material, it can be *adsorbed* onto the surface of the soil particles, and it can occupy the pores ordinarily filled with air.

Water is removed from soil in four ways: plants transpire water while carrying on life processes, soil surfaces evaporate water, water is carried away in plant products, and water moves through the subsoil and rock formations below in a process called **percolation**. Soils with good percolation drain water from all but the small pores in the natural flow of the water.

■ It takes several hundred pounds of water for the typical food crop to make 1 lb of food.

The percolation of a soil depends on the soil particle size and its chemical composition. Because of their small particle sizes, clays, and to a lesser degree silts, tend to pack together in an impervious mass with little or no percolation. Of course, sand, gravel, and rock pass water readily. Waterlogged soils that do not percolate support few crops because of their lack of air and oxygen. Rice is an important exception. A negative aspect of the massive flow of water through soil is the **leaching effect**. Water, known as the universal solvent because of its ability to dissolve so many different materials, dissolves away, or leaches, many of the chemicals needed to make a soil productive. If the leached material is not replaced, the soil becomes increasingly unproductive.

Soils become acidic, or sour, not only because of the oxidation of organic matter but also because of *selective leaching* by the passing groundwater. Salts of Group IA and IIA metals are more soluble than salts of the Group IIIA and transition metals. For example, a soil containing calcium, magnesium, iron, and aluminum ions is likely to be slightly alkaline, or sweet, before leaching with water. After the selective removal of calcium and magnesium salts, the soil becomes acidic because the iron and aluminum ions each tie up hydroxide ions from water and release hydrogen ions.

$$Fe^{3+} + H_2O \longrightarrow FeOH^{2+} + H^+$$
$$Al^{3+} + H_2O \longrightarrow AlOH^{2+} + H^+$$

17.3 Nutrients

At least 18 known elemental nutrients are required for normal green plant growth (Table 17.1). Three of these, the **nonmineral nutrients**—carbon, hy-

TABLE 17.1 ■ Essential Plant Nutrients

Nonmineral	Primary	Secondary	Micronutrients
Carbon	Nitrogen	Calcium	Boron
Hydrogen	Phosphorus	Magnesium	Chlorine
Oxygen	Potassium	Sulfur	Copper
			Iron
			Manganese
			Molybdenum
			Sodium
			Vanadium
			Zinc

drogen, and oxygen—are obtained from air and water. The **mineral nutrients** must be absorbed through the plant root system as solutes in water. The 15 known mineral nutrients fall into three groups: **primary nutrients**, **secondary nutrients**, and **micronutrients**, depending on the amounts necessary for healthy plant growth.

A natural method of nitrogen fixation. The energy in a bolt of lightning is sufficient to disrupt the very stable triple bond in a nitrogen molecule (N_2). The result is the reaction with oxygen in the air to produce NO. *(Gordon Garrado/SPL/Photo Researchers)*

Primary Nutrients

The primary nutrients are nitrogen, phosphorus, and potassium. Although bathed in an atmosphere of nitrogen, most plants are unable to use the air as a supply of this vital element. **Nitrogen fixation** is the process of changing

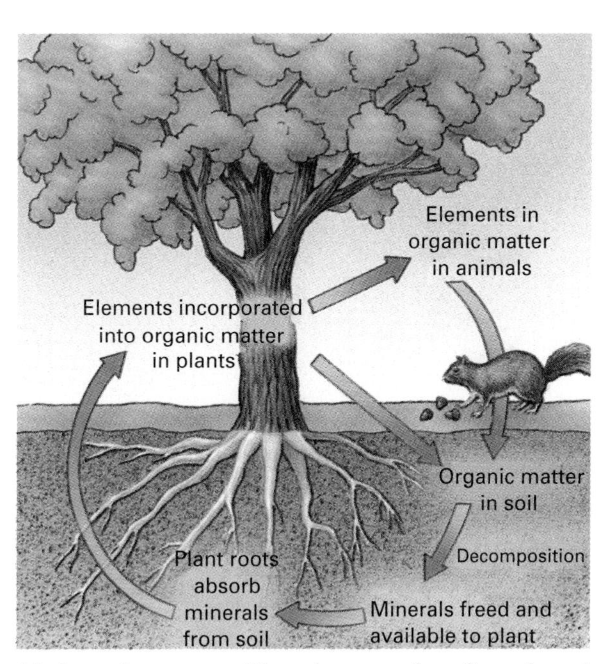

A balanced ecosystem. There is a smooth cycling of nutrients from the soil to plants and animals, and then back to the soil.

atmospheric nitrogen into water-soluble compounds that can be absorbed through the plant's roots and assimilated by the plant.

Nitrogen fixation requires breaking the strong triple bond connecting the two atoms in the nitrogen molecule N_2 (:N≡N:), as described in Section 5.3.

Nature fixes nitrogen in two ways. In the first method, bolts of lightning provide the high energy needed to oxidize nitrogen to nitric oxide (NO). The NO is further oxidized to NO_2 (Section 16.5) which reacts with water to form nitric acid (HNO_3). This method is estimated to provide less than 10% of the nitrogen fixed by nature. Nitric acid is readily soluble in rain, clouds, or ground moisture and thus increases nitrate concentration in soil.

A German, H. Hellriegel, showed in 1886 that leguminous plants such as alfalfa, beans, and soybeans "fix" nitrogen.

In the most important method, nitrogen-fixing bacteria that live in the roots of legumes, such as soybeans and alfalfa, use an enzyme, **nitrogenase**, to catalyze a complex series of reactions that convert atmospheric nitrogen into ammonia under normal atmospheric conditions. Legume nitrogen fixation can add more than 100 lb of nitrogen per acre of soil in one growing season.

Another major source of nitrogen replenishment in soil is dead organisms and animal wastes. Even in the absence of legumes, this can be an adequate source of nitrogen.

Like nitrogen, phosphorus must be in a soluble mineral or inorganic form before it can be used by plants. Unlike nitrogen, phosphorus comes totally from the mineral content of the soil. Salts of the dihydrogen phosphate ion ($H_2PO_4^-$) and monohydrogen phosphate ion (HPO_4^{2-}) are the dominant phosphate ions in soils of normal pH (Figure 17.4). Because of the great concentration of electric charge associated with the trivalent phosphate ion (PO_4^{3-}), phosphates are more tightly held to positive ions such as Ca^{2+} and Fe^{3+} and are not as easily leached by groundwater as are nitrate salts, which are generally soluble in water.

Potassium is absorbed as the free ion, $K^+(aq)$.

Potassium in the form of K^+ ion is a key element in the enzymatic control of the interchange of sugars, starches, and cellulose. Although potassium is the seventh most abundant element in Earth's crust, soil used heavily in crop production can be depleted of this important metabolic element, especially if the soil is regularly fertilized with nitrate, with no regard to potassium content.

Secondary Nutrients

Calcium and magnesium are available in small amounts as Ca^{2+} and Mg^{2+} ions as well as in complex ions and crystalline formations. These abundant elements are bound tightly enough by soil so they are not readily leached yet are held loosely enough to be available to plants. When in the soil as sulfate (SO_4^{2-}), sulfur is readily available to plants.

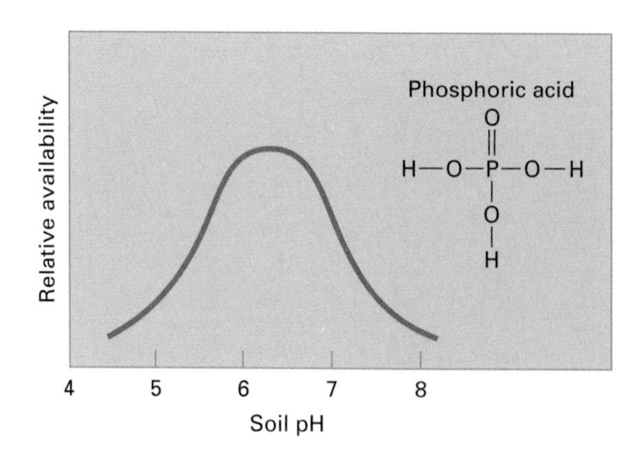

Figure 17.4 Availability of phosphate in the soil as a function of pH. The ions present vary with the pH. At pH 5 to pH 8, $H_2PO_4^-$ and HPO_4^{2-} predominate. At very low pH values, phosphorus is in the form of the nonionized acid H_3PO_4. At a very high pH, all three protons are removed and the phosphorus is in the form of the phosphate ion (PO_4^{3-}). Low soil temperatures in temperate regions significantly reduce phosphorus uptake by plants.

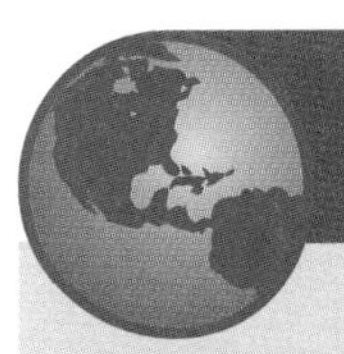

THE WORLD OF CHEMISTRY

Nitrogen Fixation

Program 8, *Chemical Bonds*

There is nitrogen in all living things. Muscles, hair, and DNA all contain nitrogen bonded to other elements. But 80% of the atmosphere is composed of nitrogen molecules held together by strong triple bonds. How do living things get the form of nitrogen they need? Lightning helps. The electrical flash in the sky has enough energy to break apart nitrogen molecules, which then react with oxygen in the air, eventually forming nitric acid. The natural acid dissolves in rain and falls to the earth as a dilute solution. There it is absorbed and metabolized by plants.

Some plants, though, convert molecular nitrogen in a different way. Soybeans and other legumes, such as peas and peanuts, host a unique bacterium in their roots. This bacterium converts the nitrogen molecule into a nitrogen compound, ammonia, which the plant can then use to make amino acids.

Exactly how the bacterium works is the subject of vigorous research. Don Keister, of the U.S. Department of Agriculture, claims,

Soybean roots. The nodules are the location of nitrogen fixation.

"This is one of the very unique enzymes in all of nature, because it is the only solution that nature has evolved for biologically reducing nitrogen."

The soybean and the bacterium have a symbiotic relationship. The plant houses and feeds the bacterium and, in turn, it receives the nitrogen it needs. But not all plants can host these nitrogen fixers. They have to rely on rain and natural fertilizers, as well as expensive manufactured fertilizers, such as ammonium nitrate.

Keister goes on to point out,

"We are currently using something like 300 million barrels of oil per year in the United States alone to produce nitrogen fertilizers. We forget sometimes that we're going to need to double the food supply over the next 20 years."

He further asks the unanswered questions,

"Where is that energy going to come from? Where is the fertilizer going to come from?"

For feeding the world, there are two basic options: We can either produce more fertilizer at greater cost and some risk to the environment, or we can create new varieties of nitrogen-fixing plants. Both options are being pursued worldwide.

Micronutrients

Only very small amounts of micronutrients are required by plants; therefore, unless extensive cropping or other factors deplete the soil of these nutrients, sufficient quantities are usually available.

Iron is also an essential component of the enzyme involved in the formation of chlorophyll. When the soil is iron deficient or when too much lime, $Ca(OH)_2$, is present in the soil, iron availability decreases. Often a gardener or lawn worker will apply phosphate and lime to adjust soil acidity, only to see green plants turn yellow because of **chlorosis**. What happens in such cases is that both phosphate and the hydroxide from the lime tie up the iron and make iron unavailable to the plants.

$$Fe^{3+}(aq) + 2\ \underset{\text{Phosphate}}{PO_4^{3-}(aq)} \longrightarrow \underset{\text{Tightly bound complex}}{Fe(PO_4)_2^{3-}(aq)}$$

$$Fe^{3+}(aq) + 3\ OH^-(aq) \longrightarrow \underset{\text{Insoluble hydroxide}}{Fe(OH)_3(s)}$$

A leaf on a blackberry plant suffering from chlorosis. Chlorophyll, the green plant pigment, requires nitrogen, magnesium, and iron from the soil. Deficiencies of any of these nutrients cause chlorosis, a condition of low chlorophyll content indicated by yellowing of the leaves. *(C. D. Winters)*

FRONTIERS IN THE WORLD OF CHEMISTRY

Trying to Mimic Nature

The process for making ammonia from the reaction of nitrogen and hydrogen at high temperatures and pressures in the presence of a catalyst was invented by Haber in 1908 and is still the major industrial process for the synthesis of ammonia. However, this is far from doing the job of fixing nitrogen as well as nature does. Fixation of nitrogen by bacteria on root nodules of legume crops occurs at atmospheric temperature and pressure. The function of nitrogenase, the biological catalyst for this process, is still not fully understood, but iron and molybdenum are known to be important to the process by which the high-energy triple bond in N_2 is broken during the reduction of N_2 to NH_3 in biological nitrogen fixation. For years, chemists have been studying reactions of transition metal compounds with nitrogen in an attempt to understand the function of metal ions in nitrogenase. Although several compounds were isolated that contained N_2 bonded to the metal center, none involved breaking the triple bond, a necessary step in nitrogen fixation. The recent discovery of a molybdenum compound that breaks the triple bond of N_2 at room temperature and atmospheric pressure to give another molybdenum compound with a single nitrogen atom bonded directly to molybdenum is a major breakthrough in synthetic chemistry. The discovery could lead to both lower energy requirements for industrial nitrogen fixation and a better understanding of the role of molybdenum in the function of nitrogenase in biological nitrogen fixation. H. Hellriegel showed in 1886 that leguminous plants can fix nitrogen; and more than 100 years later, scientists appear to be coming closer to mimicking nature's efficient process for nitrogen fixation.

Reference C. E. Laplaza and C. C. Cummins: Dinitrogen cleavage by a three-coordinate molybdenum(III) complex, *Science*, Vol. 268, pp. 861–863, May 12, 1995.

SELF-TEST 17A

1. The present world population of ____________ is growing at a rate of ____________ per year.
2. Rank the following types of soils from those with the smallest soil particles to those with the largest: silts, sandy soils, loams, clays.
3. Carbon dioxide causes soils to be (a) acidic, (b) basic. Limestone ($CaCO_3$) causes soils to be (a) acidic, (b) basic.
4. The two factors that determine the percolation of a soil are ____________ and ____________.
5. Which is more acidic, a monovalent ion such as Na^+ or a trivalent ion such as Fe^{3+}?
6. A well-decomposed, dark-colored plant residue that is relatively resistant to further decomposition is known as ____________.
7. The primary elemental plant nutrients necessary in the soil for healthy plant growth are ____________, ____________, and ____________.
8. The secondary elemental plant nutrients are ____________, ____________, and ____________.
9. Nitrogen fixation by lightning discharges involves breaking a nitrogen–nitrogen triple bond and combining nitrogen with ____________.

17.4 Fertilizers Supplement Natural Soils

Primitive people raised crops on a cultivated plot until the land lost its fertility; then they moved to a virgin piece of ground where they cut down natural vegetation ("slash") and burned off the stubble to clear the land. In many cases, the slash–burn–cultivate cycle was no more than a year in length, and few found a piece of ground anywhere that could support successful cropping for more than five years without fertilization. In farming villages, developed in ancient times and prevalent throughout the Middle Ages, innovation in fertilization was demanded, because the same land had to be used for many years. With the use of legumes in crop rotations, manures, dead fish, or almost any organic matter available, the land was kept in production.

■ Vast amounts of acres have been torched each year in the Amazon basin. Current efforts have curbed this deforestation but have not halted it.

An estimated 4 billion acres are used worldwide in the cultivation of crops for food, less than 0.8 acre per person. This acreage would probably be sufficient if modern chemical fertilization were employed on all of it. If about $40 were spent on fertilizer for each cultivated acre, world crop production would increase by 50%, the equivalent of having 2 billion more acres under cultivation. However, the cost to produce this additional food would approach a prohibitive $160 billion, and the environmental impact of such a large dispersion of fertilizer chemicals would probably be massive. For example, at the present time, aquifer contamination with nitrates due to corn crop fertilization renders well water from these natural underground basins unfit for drinking in large areas of the U.S. corn belt.

■ Examples of crop yield explosions include (1) U.S. corn—25 bushels per acre in 1800, 110 bushels per acre in the 1980s, 130 bushels per acre in the 1990s; (2) English wheat—less than 10 bushels per acre from 800 to 1600 A.D., more than 75 bushels per acre in the 1980s; and (3) rice in Japan, Korea, and Taiwan—fourfold increase in the last 40 years.

Fertilizers that contain only one nutrient are called **straight fertilizers.** Potassium chloride for potassium is an example of a straight fertilizer. Those containing a mixture of the three primary nutrients are called **complete**, or **mixed, fertilizers**. The macronutrients are absorbed by plant roots as simple inorganic ions: nitrogen in the form of nitrates (NO_3^-), phosphorus as phosphates ($H_2PO_4^-$ or HPO_4^{2-}), and potassium as the K^+ ion. Organic fertilizers can supply these ions, but only when used in large quantities over a long time. For example, a manure might be a 0.5–0.24–0.5 fertilizer, in contrast to a typical chemical fertilizer, which might carry the numbers 6–12–6. These numbers indicate the **grade**, or **analysis**, in order of the percentage of nitrogen as N, phosphorus as P_2O_5, and potassium as K_2O (commonly known as potash) in the fertilizer (Figure 17.5). In addition to containing the desired ions, chemical fertilizers place the ions in the soil in a form that can be absorbed directly by plants. The problem is that these inorganic ions are relatively easily leached from the soil and may pose pollution problems if not contained. The much slower organic fertilizer tends to stay put. **Quick-release fertilizers** are water soluble, as opposed to **slow-release fertilizers**, which require days or weeks for the material to dissolve completely. Table 17.2 (p. 464) lists the necessary plant nutrients and suitable chemical sources of each.

■ Organic fertilizers contain mostly organic material and very little mineral content. For example, manure as a fertilizer is graded less than 1–2–1.

■ The first commercial ammonia plant in the United States began production in 1921.

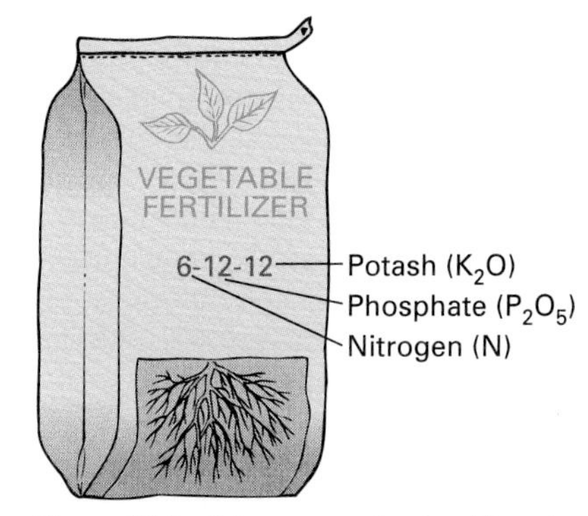

Figure 17.5 How to read a fertilizer label. The three numbers, in order, refer to the percentage by weight of N (nitrogen), P_2O_5 (phosphate), and K_2O (potash). Following the lead of J. von Liebig, his German teacher and the first to suggest adding nutrients to soils, Samuel William Johnson, an American, burned plants and analyzed their ashes. He expressed the nutrient concentrations as the oxides present in the ashes, a practice that has continued to this day.

Nitrogen Fertilizers

The cheapest source of nitrogen is the air, but it must be combined with relatively expensive hydrogen, obtained from petroleum, to form ammonia in the Haber process.

The complex interactions among social, economic, and political necessities are well illustrated by the development of the industrial synthesis of ammonia. The need for an industrial process for nitrogen fixation was recognized as early as 1890. Scientists in England noted that the world's future food supply would be determined by the amount of nitrogen compounds available for fertilizers. At the time, the sources of such compounds were

TABLE 17.2 ■ Some Chemical Sources of Plant Nutrients

Element	Source Compound(s)
	Nonmineral Nutrients
C	CO_2 (carbon dioxide)
H	H_2O (water)
O	H_2O (water)
	Primary Nutrients
N	NH_3 (ammonia), NH_4NO_3 (ammonium nitrate), H_2NCONH_2 (urea)
P	$Ca(H_2PO_4)_2$ (calcium dihydrogen phosphate)
K	KCl (potassium chloride)
	Secondary Nutrients
Ca	$Ca(OH)_2$ (calcium hydroxide, slaked lime), $CaCO_3$ (calcium carbonate, limestone), $CaSO_4$ (calcium sulfate, gypsum)
Mg	$MgCO_3$ (magnesium carbonate), $MgSO_4$ (magnesium sulfate, epsom salts)
S	Elemental sulfur, metallic sulfates
	Micronutrients
B	$Na_2B_4O_7 \cdot 10H_2O$ (borax)
Cl	KCl (potassium chloride)
Cu	$CuSO_4 \cdot 5H_2O$ (copper sulfate pentahydrate)
Fe	$FeSO_4$ (iron(II) sulfate, iron chelates)
Mn	$MnSO_4$ (manganese(II) sulfate, manganese chelates)
Mo	$(NH_4)_2MoO_4$ (ammonium molybdate)
Na	NaCl (sodium chloride)
V	V_2O_5, VO_2 (vanadium oxides)
Zn	$ZnSO_4$ (zinc sulfate, zinc chelates)

Anhydrous ammonia as a fertilizer. At normal temperatures, anhydrous ammonia (ammonia gas containing no water) can be held as a liquid under tank pressure. On release from pressure, the ammonia returns to the gaseous state and is injected into the soil by an "ammonia knife." In slightly acid soil, the ammonia is immediately converted to the water-soluble ammonium ion and enters the natural nitrogen pathways of the soil. *(Grant Heilman/Grant Heilman Photography)*

■ The Haber process is based on a delicate balancing act among the reaction energy, the reaction rate, and the reaction equilibrium. The combination of nitrogen with hydrogen is exothermic, but slow. The rate is increased by a higher temperature, but the yield of ammonia decreases with increasing temperature (*because* the reaction is exothermic). Therefore, a catalyst that acts at a moderate temperature is necessary. Also, because 4 mol of reactant give 2 mol of product, the extent of the forward reaction can be increased by increasing the pressure (Le Chatelier's principle; Section 6.4).

limited to sodium nitrate from Chile and rapidly depleting supplies of guano (bird droppings) from Peru.

Shortly after 1900, many researchers in England and Germany began investigating methods of preparing ammonia, and in 1908 the German chemist Fritz Haber developed a feasible process for the direct synthesis of ammonia from its elements.

$$N_2(g) + 3\ H_2(g) \rightleftharpoons 2\ NH_3(g)$$

The first industrial plant based on the **Haber process** (see Figure 17.6) for ammonia synthesis began operation in Germany in 1911.

Subsequently, the Haber process has provided ammonia for fertilizers on a huge scale and is now largely responsible for our ability to feed a world population of more than 5.9 billion. The Haber process has been so well developed that ammonia is very inexpensive (about $150 per ton) and is usu-

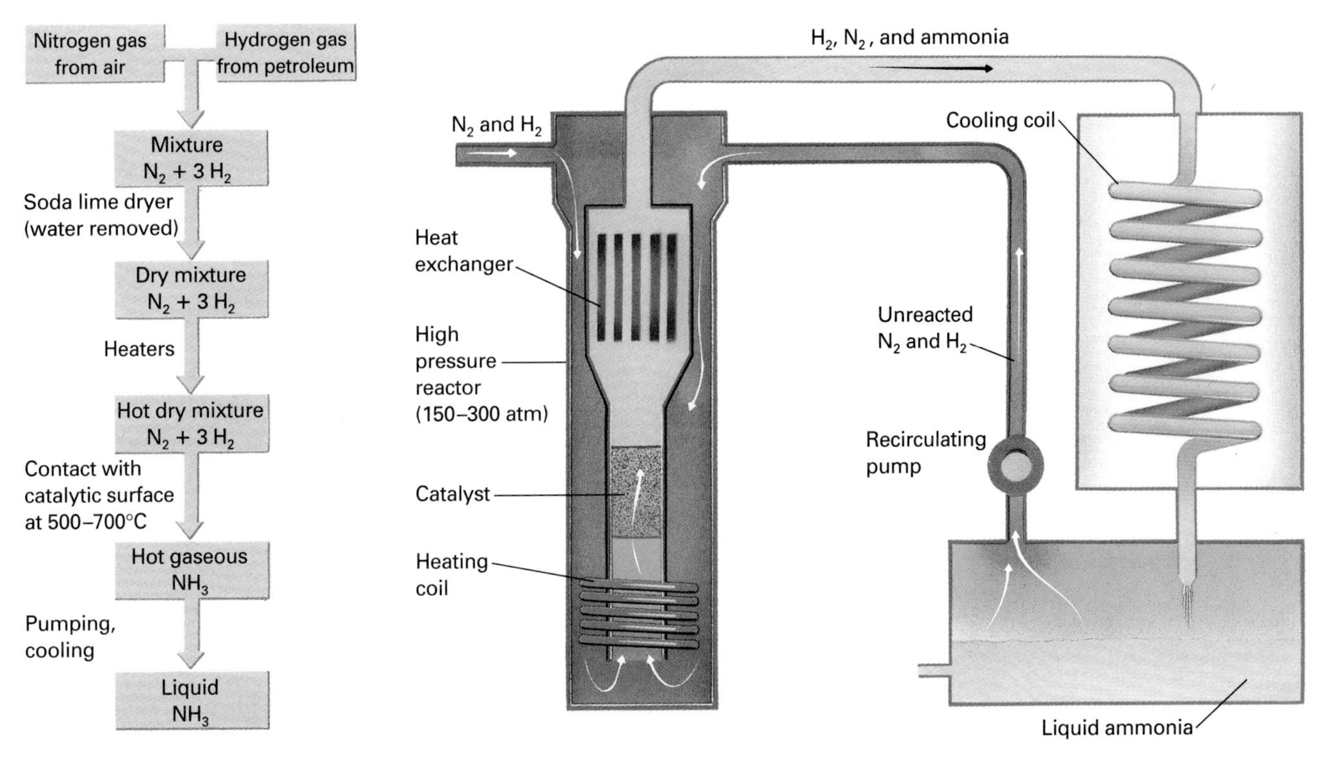

Figure 17.6 The Haber process for ammonia production. A mixture of N_2 and H_2 in the proper proportions for reaction is heated and passed under pressure over the catalyst. The ammonia is collected as a liquid and unreacted gases are recycled.

ally among the top ten industrial chemicals in terms of quantity produced each year (number three in 1996).

Ammonia can be applied directly to the soil or converted into numerous agrichemicals (Figure 17.7, p. 466). Two common straight fertilizers for ni-

■ Ammonia is a gas at normal temperatures and pressures. It is very soluble in water.

THE PERSONAL SIDE

Fritz Haber (1868–1934)

Although Fritz Haber was a fine scientist and the success of his ammonia synthesis made him a rich man, he ultimately had a tragic life. At the start of World War I he joined the German Chemical Warfare Service, where he supervised the use of chlorine as a chemical weapon during the battle of Ypres in France. This first use of a chemical weapon led to further tragic developments in chemical warfare and also to personal tragedy for Haber. His wife pleaded with him to stop his work in this area; and when he refused, she committed suicide. In 1918, he was awarded the Nobel Prize for the ammonia synthesis, but the choice was criticized because of his role in developing chemical warfare. After World War I, Haber continued his academic research in Germany until, shortly before his death in 1934, he left the country to avoid danger because of his Jewish background.

Program 14, *Molecules in Action*

Fritz Haber.

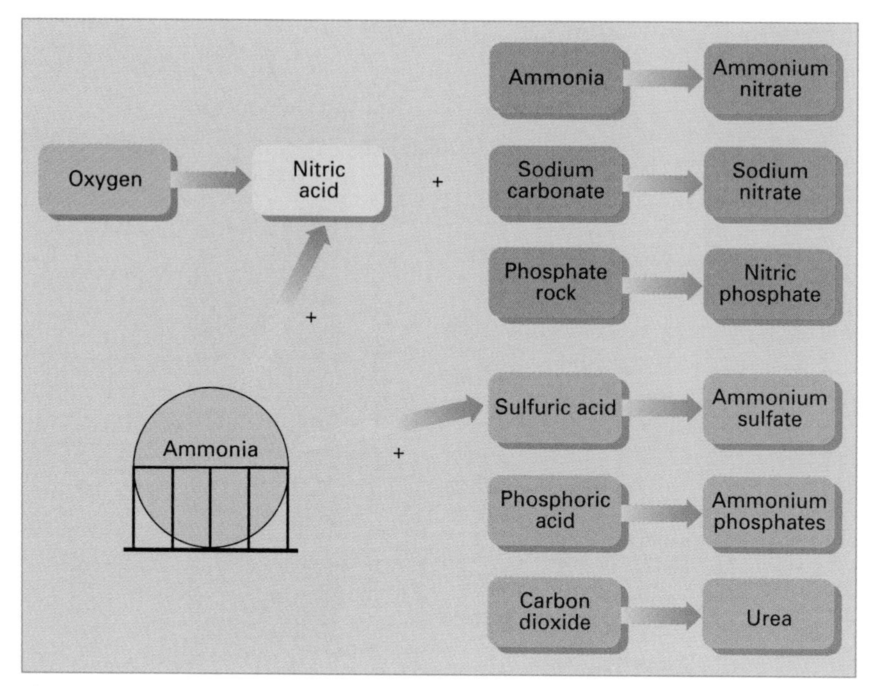

Figure 17.7 Nitrogen fertilizers produced from anhydrous ammonia.

■ The structural formula of urea is

$$H_2N-\overset{\overset{\displaystyle O}{\|}}{C}-NH_2$$

trogen are ammonium nitrate (NH_4NO_3) and urea, numbers eight and nine in U.S. chemical production in 1996. A slurry of water, urea, and ammonium nitrate is often applied to crops under the name of "liquid nitrogen." Such a solution can contain up to 30% nitrogen and is easy to store and apply.

When applied to the surface of the ground around plants, urea is subject to considerable nitrogen loss unless it is washed into the soil by rain or irrigation. When urea decomposes, ammonia is formed, some of which is lost to the air and some is absorbed by moist soil particles. As much as half of the nitrogen applied to the soil can be lost in this way.

A phosphate mine in Florida. In the background is a part of this mine that has been restored after mining was completed and is now used as a pasture. *(William Felger/Grant Heilman)*

Phosphate Fertilizers

Phosphorus is readily available in the form of phosphate rock, which can be transformed into the needed fertilizers (Figure 17.8). World deposits of phosphate rock are limited, and costs for supplying phosphorus fertilizers will increase as deposits are depleted. The phosphate rock, $Ca_3(PO_4)_2$, is not very useful because of its exceedingly low solubility. When treated with sulfuric acid, however, phosphate rock becomes more soluble; and the resulting mixture of phosphate and sulfate salts of calcium is called "superphosphate."

$$\underset{\text{Phosphate rock}}{Ca_3(PO_4)_2} + 2\,H_2SO_4 \longrightarrow \underset{\text{Superphosphate}}{Ca(H_2PO_4)_2 + 2\,CaSO_4}$$

SCIENCE AND SOCIETY

Ammonium Nitrate: A Dilemma

Ammonium nitrate is the number 8 chemical produced in the United States because of its use as both a fertilizer and an explosive. The Oklahoma City bombing on April 19, 1995, that killed 169 people has resulted in a call for regulations to eliminate the ready availability of ammonium nitrate. Regulations on the use of ammonium nitrate as an explosive are feasible, but restrictions on the purchase of ammonium nitrate fertilizer present a major problem. Fifty-pound bags of ammonium nitrate can be purchased at any farm supply store. One approach that is being discussed is to require the addition of a substance that would make fertilizer-grade ammonium nitrate unexplodable.

Samuel J. Porter, a hazardous chemicals consultant, received a patent in 1968 for rendering ammonium nitrate unexplodable by adding 5% to 10% diammonium phosphate to molten ammonium nitrate before the formation of the final product, small spherical pellets. At the time, none of the chemical companies approached by Porter felt that the increased cost of adding a substance to make ammonium nitrate inert was justified. The patent has expired and is now in the public domain.

In May 1995, Representative W. J. Tauzin from Louisiana called on the fertilizer industry to test Porter's process. Gary Myers, president of the Fertilizer Institute says the fertilizer industry supports studies on how to make fertilizers useless as explosives. The question of whether additives such as diammonium phosphate really work cannot be answered without further testing. Testing is also necessary to determine if additives could be separated from ammonium nitrate by terrorists. In Northern Ireland and the Republic of Ireland, regulations require the addition of calcium carbonate to ammonium nitrate before it can be sold as fertilizer. The salts in this mixture, referred to as calcium ammonium nitrate, can be separated; however, available information indicates that terrorists do not separate the ammonium nitrate but instead make an explosive mixture by grinding the calcium ammonium nitrate into very fine particles and mixing it with other substances to make a bomb.

The dilemma: How do we retain the availability of ammonium nitrate as an economical fertilizer and eliminate its use by terrorists to make bombs?

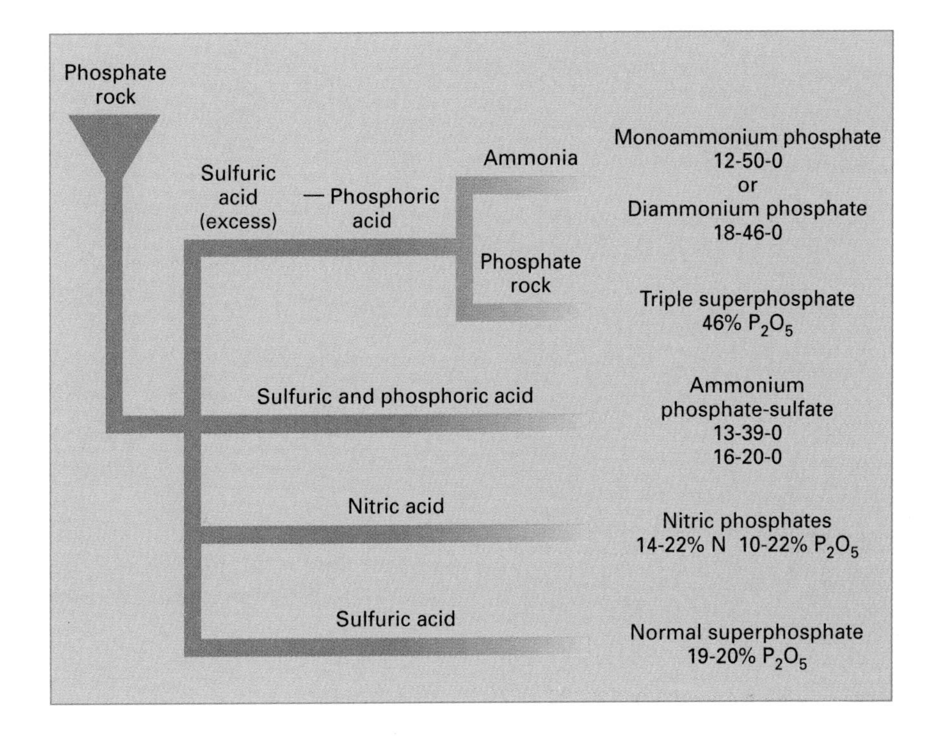

Figure 17.8 Fertilizers produced from phosphate rock.

EXAMPLE 17.1 *Fertilizer Labels*

The label on a fertilizer bag lists the numbers 22–6–8. What do these numbers mean? Calculate the pounds of each nutrient present in a 50.0-lb bag of fertilizer.

SOLUTION

The numbers mean that the fertilizer contains 22.0% nitrogen (N) in the form of some nitrogen-containing compound, 6.00% of a phosphorus-containing compound (calculated as a percentage of P_2O_5), and 8.00% of a potassium-containing compound (calculated as a percentage of K_2O). The pounds of each nutrient present are calculated as follows:

Nitrogen: The pounds of nitrogen present are calculated by multiplying the percentage of nitrogen times 50.0 lb. Since the unit in this problem is pounds, 22.0% of N can be represented as 22.0 lb of N per 100 lb of fertilizer.

$$\frac{22.0 \text{ lb N}}{100 \cancel{\text{lb fertilizer}}} \times 50.0 \cancel{\text{lb fertilizer}} = 11.0 \text{ lb N}$$

Phosphorus: Since the amount of phosphorus present is represented as 6.00% P_2O_5, the actual percentage present as elemental phosphorus is smaller. To calculate the fraction of phosphorus in P_2O_5 requires using the atomic masses of P and O in a conversion factor

$$\frac{2 \times \text{atomic mass P}}{(2 \times \text{atomic mass P}) + (5 \times \text{atomic mass O})}$$

$$2 \text{ P} = 2 \times 31.0 \text{ amu} = 62.0 \text{ amu}$$

$$5 \text{ O} = 5 \times 16.0 \text{ amu} = 80.0 \text{ amu}$$

$$\text{Total for } P_2O_5 = 142.0 \text{ amu, so factor is}$$

$$\frac{62.0 \text{ amu P}}{142 \text{ amu } P_2O_5}$$

The relative masses are the same no matter what units are used. Since the unit in this problem is pounds, the conversion factor can be written as

$$\frac{62.0 \text{ lb P}}{142 \text{ lb } P_2O_5}$$

$$\frac{6.00 \cancel{\text{lb } P_2O_5}}{100 \cancel{\text{lb fertilizer}}} \times \frac{62.0 \text{ lb P}}{142 \cancel{\text{lb } P_2O_5}} \times 50.0 \cancel{\text{lb fertilizer}} = 1.31 \text{ lb P}$$

Potassium: Since the amount of potassium present is represented as 8.00% K_2O, calculating the pounds of potassium present requires using the atomic masses of K and O in a conversion factor

$$\frac{2 \times \text{atomic mass K}}{(2 \times \text{atomic mass K}) + (\text{atomic mass O})}$$

2 K = 2 × 39.0 amu = 78.0 amu; 1 O = 16.0 amu; total of 2 K and 1 O = 94.0 amu so the conversion factor showing the ratios in pounds is

$$\frac{78.0 \text{ lb K}}{94.0 \text{ lb } K_2O}$$

$$\frac{8.00 \cancel{\text{lb } K_2O}}{100 \cancel{\text{lb fertilizer}}} \times \frac{78.0 \text{ lb K}}{94.0 \cancel{\text{lb } K_2O}} \times 50.00 \cancel{\text{lb fertilizer}} = 3.32 \text{ lb K}$$

Exercise 17.1 *Fertilizer*

Ammonium nitrate (NH_4NO_3) is widely used as a fertilizer. A bag of ammonium nitrate sold as fertilizer lists the numbers 35–0–0. Use the formula weight for NH_4NO_3 to verify that this is the correct number.

17.5 Protecting Food Crops

The natural enemies of crops include more than 80,000 diseases brought on by viruses, bacteria, fungi, algae, and similar organisms; 30,000 species of weeds; 3000 species of nematodes; and about 10,000 species of plant-eating insects. About one third of the food crops in the world are lost to pests each year, with the loss going above 40% in some developing countries. Crop losses to pests amount to $20 billion per year. "Pest" is any organism that in some way reduces crop yields or human health. **Pesticides**, chemicals used to control pests, are classified according to the pests they control: **insecticides** kill insects, **herbicides** kill weeds, and **fungicides** kill fungi.

The World of Chemistry

Program 25, *Chemistry and the Environment*

Effect of DDT on eggshells. The massive use of DDT for insect control nearly wiped out several species of birds due to interference in the formation of eggshells.

Pesticides

Pesticides are the chemical answer to pest control. Eighteen common classes of pesticides are fortified with more than 2600 active ingredients to fight the battle with pests. More than 5 billion pounds of pesticides are produced worldwide each year. In 1993, pesticide sales in the United States totaled $6.8 billion. Although the dollar cost was up to $7.5 billion in 1995, the actual poundage use began a decline in 1987. There are three reasons for a decrease in the demand for pesticides in the United States: (1) Cropland planted is less than 330 million acres, down from a high of 383 million acres in 1982. (2) Farming is becoming more cost-effective as farmers learn to use a minimum of pesticide for the desired effect. (3) Farmers are becoming more concerned with environmental and health issues related to the use of pesticides.

■ Paul Müller was awarded the Nobel Prize in medicine and physiology in 1948 for his discovery of the insecticidal properties of DDT.

DDT

Insecticides

Before World War II, the list of insecticides included only a few compounds of arsenic, petroleum oils, nicotine, pyrethrum (obtained from dried chrysanthemum flowers), rotenone (obtained from roots of derris vines), sulfur, hydrogen cyanide gas, and cryolite (the mineral Na_3AlF_6). Dichlorodiphenyltrichloroethane (DDT), the first of the chlorinated organic insecticides, was originally prepared in 1873, but it was not until 1939 that Paul Müller of Geigy

Pharmaceutical in Switzerland discovered the effectiveness of DDT as an insecticide.

The use of DDT increased enormously on a worldwide basis after World War II, primarily because of its effectiveness against mosquitoes that spread malaria and lice that carry typhus. The World Health Organization estimates that approximately 25 million lives have been saved through the use of DDT. DDT seemed to be the ideal insecticide—it is cheap and has relatively low toxicity to mammals (oral LD_{50} is 300–500 mg/kg). However, problems related to extensive use of DDT began to appear in the late 1940s. Many species of insects developed resistance to DDT, and DDT was also discovered to have a high toxicity toward fish. The chemical stability of DDT and its fat solubility compounded the problem.

■ LD_{50} is defined in Section 13.6.

DDT is not metabolized very rapidly by animals; instead, it is deposited and stored in the fatty tissues. The biological half-life of DDT is about eight years; that is, half the amount of DDT an animal assimilates today will be metabolized in eight years. If ingestion continues at a steady rate, DDT builds up within the animal over time.

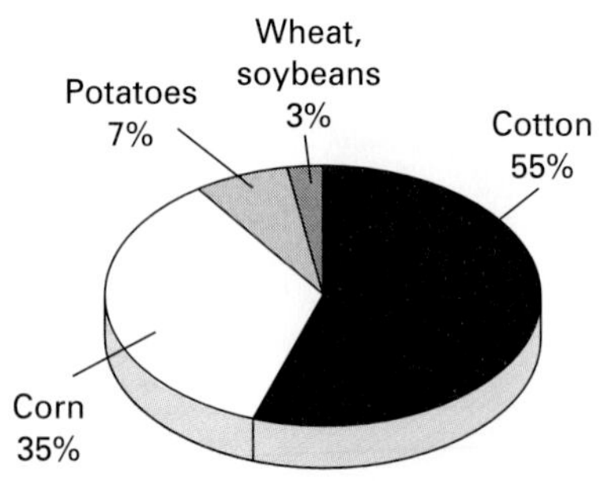

Cotton and corn get 90% of the 265 million pounds of insecticides used in 1995. *(Source: Department of Agriculture)*

The use of DDT was banned in the United States in 1973, although it is still in use in some other parts of the world. The buildup of DDT in natural waters is a reversible process; the EPA reported a 90% reduction of DDT in Lake Michigan fish by 1978 as a result of the ban on the use of the insecticide in the United States.

The most important insecticides can be grouped in three structure classes—chlorinated hydrocarbons, organophosphorus compounds, and carbamates. Chlorinated hydrocarbons such as DDT are referred to as **persistent pesticides** because they persist in the environment for years after their use.

Malathion, a widely used organophosphorus pesticide, is both biodegradable and less toxic than DDT (LD_{50} for female rats is 1000 mg/kg).

The carbamate insecticides are derivatives of carbamic acid

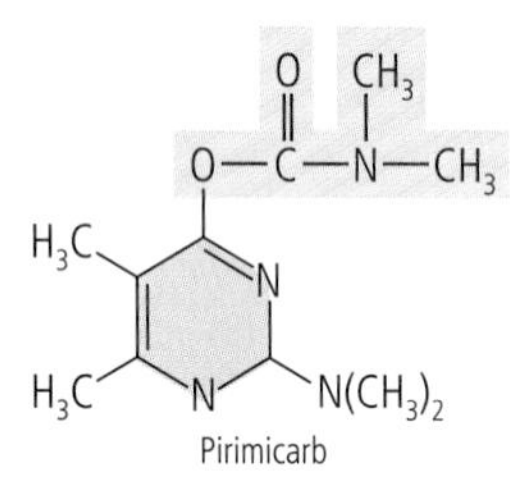

Pirimicarb

Carbamic acid

Carbaryl

They are also nonpersistent insecticides. The most widely used carbamate is carbaryl, a general-purpose insecticide with a relatively low mammalian toxicity (oral LD_{50} for rats is 250 mg/kg). A serious drawback to the use of carbaryl for spraying crops is its high toxicity toward honeybees. Another carbamate of interest is pirimicarb (oral dose for female rats is 147 mg/kg), which is a selective insecticide for aphids and has the advantage of controlling strains that have developed resistance to organic phosphates. Pirimicarb has a low toxicity to predators such as bees. It is rapidly metabolized and leaves no lasting residues in plant materials. Pirimicarb is expensive, however; its preparation requires five or six steps and special handling procedures.

A corn earworm at work. *(Gilbert S. Grant/Photo Researchers)*

The choice of solutions to the problems of insecticide use is not an easy one. Their use introduces trace amounts into our environment and our water supplies. If we fail to use them, we must tolerate malaria, plague, sleeping sickness, and the consumption of a large part of our food supply by insects. Continuing research in the development of more effective and safer insecticides is intense, and new products are introduced each year.

■ The goal of the insecticide quest is a selectively toxic chemical that is quickly biodegradable.

Herbicides

Herbicides kill plants. They may be **selective** and kill only a particular group of plants, such as the broad-leaved plants or the grasses; or they may be **nonselective**, making the ground barren of all plant life.

Selective herbicides act like hormones, very selective biochemicals that control a particular chemical change in a particular type of organism at a particular stage in its development. Most selective herbicides in use today are growth hormones; they cause cells to swell, so that leaves become too thick for chemicals to be transported through them, and roots become too thick to absorb needed water and nutrients. Nonselective herbicides usually interfere with photosynthesis and thereby starve the plant to death. On application, the plant quickly loses its green color, withers, and dies.

The most widely used herbicide is 2,4-dichlorophenoxyacetic acid (2,4-D). The corresponding trichloro- compound (common name: 2,4,5-T) has also been shown to be highly effective, but it was banned by the EPA because of a number of health problems associated with its use.

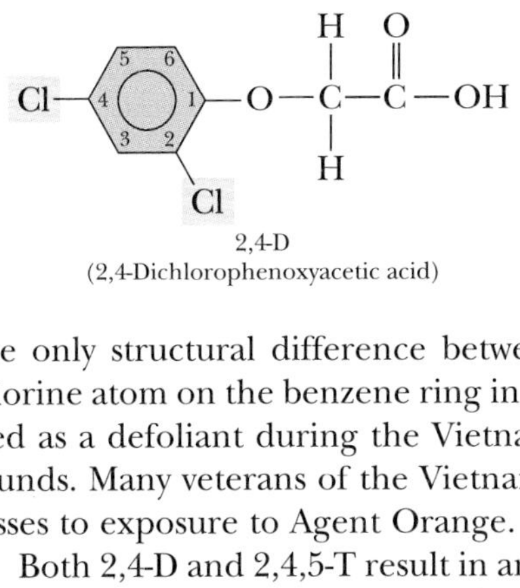

2,4-D
(2,4-Dichlorophenoxyacetic acid)

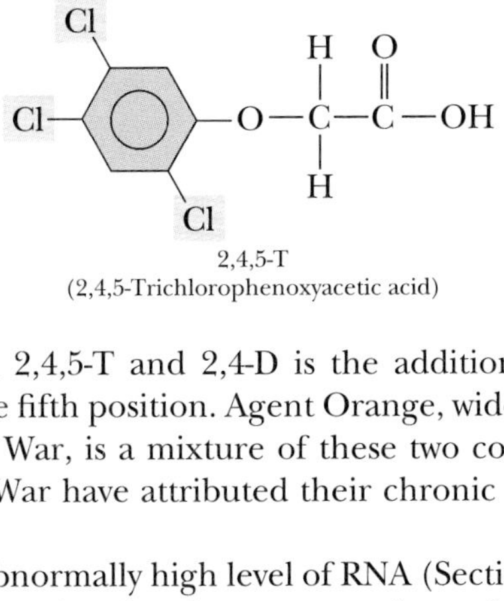

2,4,5-T
(2,4,5-Trichlorophenoxyacetic acid)

The only structural difference between 2,4,5-T and 2,4-D is the additional chlorine atom on the benzene ring in the fifth position. Agent Orange, widely used as a defoliant during the Vietnam War, is a mixture of these two compounds. Many veterans of the Vietnam War have attributed their chronic illnesses to exposure to Agent Orange.

Both 2,4-D and 2,4,5-T result in an abnormally high level of RNA (Section 11.11) in the cells of the affected plants, causing the plants to grow themselves to death.

Triazines have been found to be effective as herbicides; the most important one is atrazine.

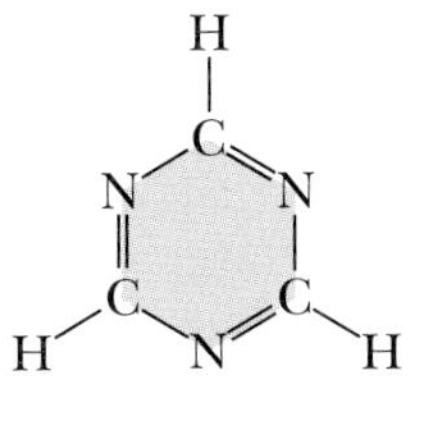

1,3,5-Triazine

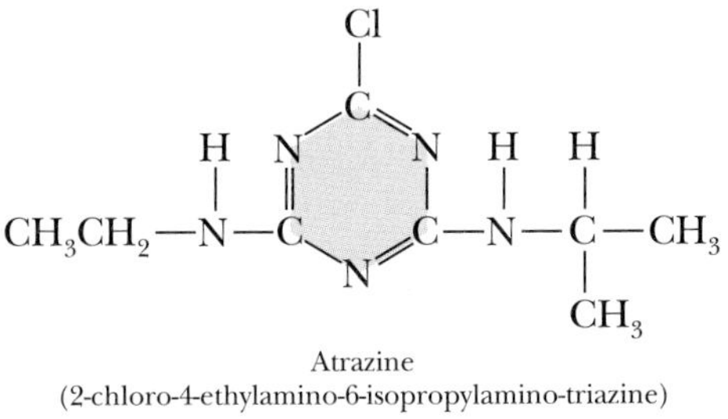

Atrazine
(2-chloro-4-ethylamino-6-isopropylamino-triazine)

■ Tillage weed control requires tilling or plowing soil between furrows to uproot weeds.

Atrazine is widely used in no-till corn production or for weed control in minimum tillage. Atrazine is a poison to any green plant if it is not quickly metabolized. Corn and certain other crops have the ability to render atrazine harmless, which weeds cannot do. Hence, the weeds die, and the corn shows no ill effect.

Several herbicides work by inhibiting plant enzymes that catalyze the synthesis of amino acids in plants that are essential amino acids in animals (Section 11.6). Unlike animals, plants can synthesize all 20 amino acids. Since animals do not synthesize essential amino acids, developing inhibitors for amino acids that are synthesized by plants but not by animals would produce safer herbicides. Glyphosate and sulfonylureas are herbicides with this mechanism of action. Glyphosate, the active ingredient in Round-up, is a phosphate derivative of glycine that inhibits the synthesis of the essential amino acids tyrosine and phenylalanine, and is used to control perennial grasses.

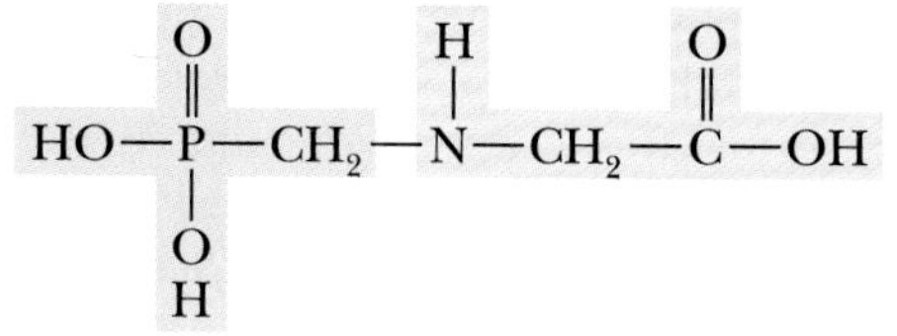

$$HO-P(=O)(OH)-CH_2-N(H)-CH_2-C(=O)-OH$$

Glyphosate

Sulfometuron methyl, the active ingredient of Oust, is a sulfonylurea that inhibits the synthesis of the essential amino acids valine, leucine, and isoleucine.

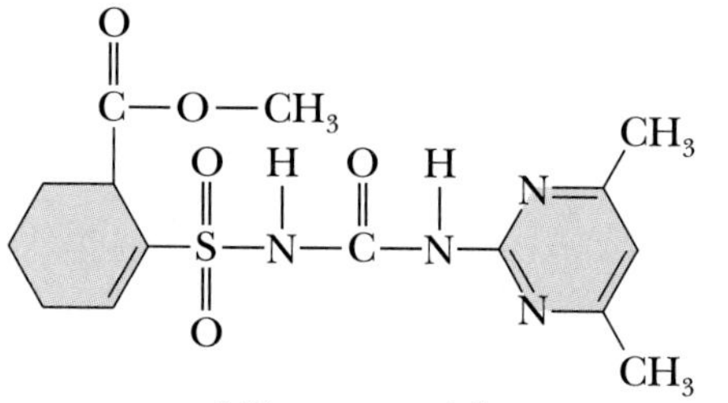

Sulfometuron methyl

Paraquat is a herbicide and can be used to kill weeds before the crop sprouts. Such herbicides are called "pre-emergent" herbicides. When applied directly to susceptible plants, paraquat quickly causes a frostbitten appearance and death. Paraquat has a nitrogen atom in each aromatic ring of the two-ring system:

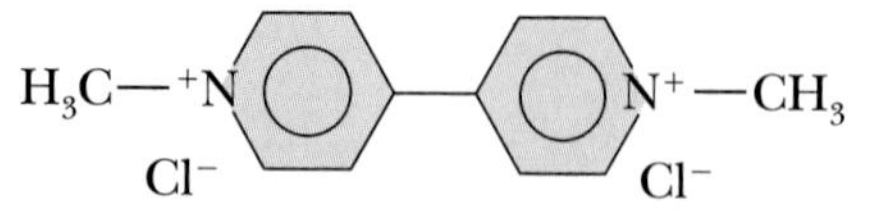

Paraquat
(1,1′-dimethyl-4,4′-bipyridinium dichloride)

Paraquat has received considerable attention because it was used to spray illegal poppy and marijuana fields in Mexico and elsewhere, which caused drug users to suffer lung damage from residual paraquat.

The amount of energy saved by herbicides used in no-till farming is enormous. The saving of topsoil is also considerable, because the cover from the previous crop holds the soil against wind and water runoff. However, agriculturists who use herbicides are highly dependent on agricultural research institutions for the selection of herbicides that will do the desired job without harmful side effects. Such selections depend on considerable research, much of which is carried out on a trial-and-error basis on test plots. A procedure that is recommended today may be outdated by the next growing season.

Fungicides

About 200 of the 100,000 classified fungal species are known to cause serious plant disease. Agricultural fungicide application accounts for about 20% of all pesticide use. Most fungicides are applied to the seed or foliage of the growing plant or to harvested produce to prevent storage losses. The earliest fungicides were inorganic substances such as elemental sulfur and compounds of copper and mercury. In the 19th century, sulfur was used to control mildew on fruit and grapes.

■ Sulfur is a solid at room temperature. Powdered sulfur is used to control fungi.

In recent times, different types of organic compounds have been used. An important class of fungicides is the dithiocarbamates and their derivatives, which are widely used on many crops such as fruits and field vegetables.

$$(CH_3)_2N-\overset{\overset{\displaystyle S}{\|}}{C}-S-S-\overset{\overset{\displaystyle S}{\|}}{C}-N(CH_3)_2$$

Thiram, a dithiocarbamate

17.6 Sustainable Agriculture

Extensive use of fertilizers and pesticides, and development of higher yielding varieties of crops have made it possible to increase crop yields dramatically. However, poor farm practices have resulted in soil erosion and loss of soil fertility. The annual amount of topsoil lost to wind and water erosion is from 2 to 4 tons per acre in Africa, Europe, and Australia; 4 to 8 tons per acre in North, Central, and South America; and nearly 12 tons in Asia. These amounts are about 20 times the rate of replenishing the topsoil. Compaction of soil, which decreases soil fertility, is caused by (1) growing shallow-rooted crops year after year; (2) not incorporating enough organic matter into the soil; (3) practices that alter soil microbial and earthworm populations; and (4) using heavy machinery, particularly on wet soils. Compacted soil, when dry, looks like brick. It restricts root growth and has soil oxygen levels below those necessary for optimum uptake of nutrients.

Soil erosion caused by water runoff, which carries soil, fertilizer, and pesticides with it. *(T. McCabe/Visuals Unlimited)*

Another major problem is the increasing resistance of pests to pesticides. For example, even though there has been a 33-fold increase in the amount of pesticides used since 1945, crop losses from insects, diseases, and weeds have increased from 31% in 1945 to 37% today. About 500 different insect and mite species, 80 fungus species, and 80 weed species are now resistant to commonly used pesticides.

These problems have led to consideration of alternatives such as **organic farming** and **alternative** or **sustainable agriculture** methods.

"Organic" produce, which is grown without pesticides. Consumer demand for such products is growing. *(Steve Feld)*

■ New FDA standards are being developed for labelling "organic" food products.

Organic farming is farming without chemical fertilizers and pesticides. Organic farming uses only about 40% of the energy required for modern farming with synthetic chemicals and produces about 90% of the yield. The costs of energy saved in organic farming are offset by the costs of human labor required by the use of natural fertilizers. Many claims are made that organic farming produces a better product for human consumption. However, there is no real evidence that these claims are generally true. Organic farming does have one clear advantage, however; it is definitely less of a threat to the environment than regular farming if agrichemicals are not very carefully controlled.

Alternative farming, a term that was popularized by a 1989 special report of the National Research Council, is an effort to stake out middle ground between organic farming and the heavy use of agrichemicals. The goals of alternative agriculture are to improve profits, limit the use of agrichemicals, and increase the use of environmentally friendly procedures to fight pests and produce food and fiber. Examples of significant changes in alternative farming include (1) expanding crop rotations because the same pests do not afflict every crop, (2) using multiple crops in alternate plantings within a given planting field (Figure 17.9), (3) using as much natural fertilizer as possible before using agrichemicals, (4) increasing the use of biological pest controls, (5) employing renewed efforts at soil and water conservation, and (6) having a diversification in livestock as well as field crops on the same farm.

The main sources of plant nutrients in alternative farming systems are animal and green manures. A green manure crop is a grass or legume that is

Figure 17.9 Strip farming, in which different crops are grown side by side. Often, some strips contain legumes that produce nitrogen and reduce the need for chemical fertilizers. Soil is preserved by contouring the strips according to the topography of the land. *(USDA)*

plowed into the soil or surface-mulched at the end of a growing season to enhance soil productivity.

One broad approach to limiting use of pesticides is **integrated pest management** (IPM), which relies more on disease-resistant crop varieties and biological controls such as natural predators or parasites that control pest populations than on agrichemicals. Farmers can select tillage methods, planting times, crop rotations, and plant-residue management practices to optimize the environment for beneficial insects that control pest species. If pesticides are used as a last resort, they are applied when pests are more vulnerable or when any beneficial species and natural predators are least likely to be harmed.

IPM programs have been most effective for cotton, sorghum, peanuts, and fruit orchards, but less effective for corn and soybeans. Some of the biological controls include release of sterilized insect pests, use of insect pheromones to disrupt mating, release of natural predator pests, and use of natural insecticides.

Chrysanthemum flowers being harvested in Rwanda, where they are grown as a source of pyrethrum. *(Robert E. Ford/Terraphotographics)*

Natural Insecticides

Many plant species contain natural protection against insects. For example, nicotine protects tobacco plants from sucking insects. However, its commercial use is limited by its high mammalian toxicity (oral LD_{50} for rats is 50 mg/kg).

One of the oldest and best known natural insecticides is pyrethrum, a contact insecticide obtained from dried chrysanthemum flowers by extraction with hydrocarbon solvents. Pyrethrum is very effective in killing flying insects and is relatively nontoxic toward mammals (oral LD_{50} for rats is 129 mg/kg). Pyrethrum aerosol sprays are excellent home insecticides because of their safety for humans and rapid action on pests. However, pyrethrum lacks persistence against agricultural insects because of its instability to air and light, so it must be mixed with small amounts of other insecticides to kill insects that might recover from sublethal doses of pyrethrum.

The insecticidal properties of pyrethrum result from six esters that are collectively called pyrethrins. After the isolation and identification of these compounds, a number of derivatives have been synthesized that are more effective than the natural pyrethrins. For example, dimethrin is effective against mosquito larva and is safe to use (oral LD_{50} for rats is greater than 10,000 mg/kg). The basic pyrethrin structure is shown in black, and the groups that vary in different pyrethrins are shown in yellow.

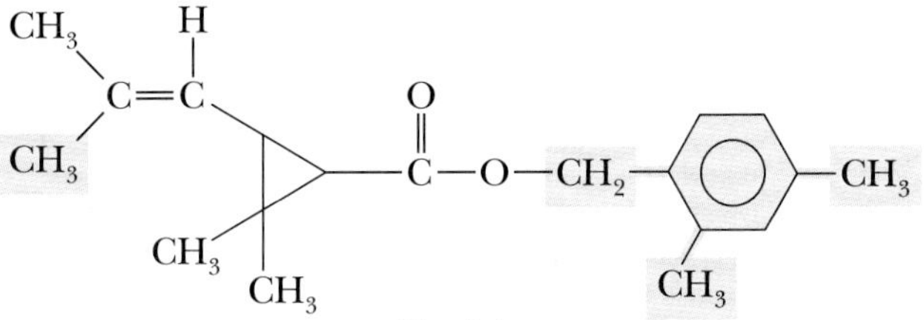

Dimethrin

A class of natural insecticides based on naturally occurring bacterial toxins of *Bacillus thuringiensis* (*Bt*), a bacterium found in the soil, has been used widely in the United States in forestry to control gypsy moths and other insects. Since different *Bt* strains are highly selective in killing targeted insects, aerial spraying of *Bt* toxins can be used to control moths or mosquitoes without being hazardous to other beneficial insects, or to animals, humans, or plants. The *Bt* toxins also degrade after a few days so they do not contaminate the environment. Two commercial products containing *Bt* are Sharpshooter for weed control in landscape management and recreational turf, and DeMoss for controlling moss and algae growth on roofs, buildings, sidewalks, and decks.

However, there is evidence that insects are developing a resistance to *Bt* toxins. Farmers using *Bt* for control of moths in fields of watercress reported a loss of effectiveness. Research has shown that when insects were exposed to high doses that kill 60% to 90% of the insects, their offspring were more resistant to *Bt*. Rotating crops and using *Bt* biopesticides sparingly are recommended ways to impede the development of resistance.

■ Natural insecticides are commonly called *biopesticides.*

Another source of effective natural pesticides is the neem tree that grows widely in Africa and Asia. For centuries, people of India have known of the insect-fighting ability of the neem tree. The oil extracted from neem tree seeds has been found to be effective against more than 200 species of insects, including locusts, gypsy moths, cockroaches, California medflies, and aphids. Azadirachtin, an active ingredient of oil from neem tree seeds, interferes with

THE WORLD OF CHEMISTRY

Pheromones

Program 10, *Signals from Within*

One important area in insect research is the development of pheromones, the sex attractants used by female insects to entice males. They offer a safe, nontoxic way to lure insects into traps and avoid the use of pesticides. Dr. Meyer Schwartz makes such synthetic insect pheromones in his laboratory.

An important part of Dr. Schwartz's work is confirming that he has made exact copies of the natural pheromones. He uses infrared spectroscopy to verify the structure of the molecules he's tried to duplicate in the lab.

Once an accurate copy is made, it's tried out to see if it works. A miniature wind tunnel lets Agriculture Department scientists see just how tantalizing their creation is. The pheromone is placed at one end of the tunnel, and a love-struck male gypsy moth flies against the wind to get it. It's bad enough that he's going to all this trouble for a synthetic chemical, but the Agriculture Department scientists have another trick up their sleeve. They can control the speed of a striped conveyor belt on the wind tunnel floor. The distracted moth sees the belt moving by and thinks he's flying full tilt toward a female. By carefully adjusting the belt speed, the researchers can stop the moth's forward progress. The speed required to stop the moth gives them a good measure of how strong a sex lure their pheromone is. When he finally is allowed to reach the end, all he finds are synthetic pheromones in a cold steel cage.

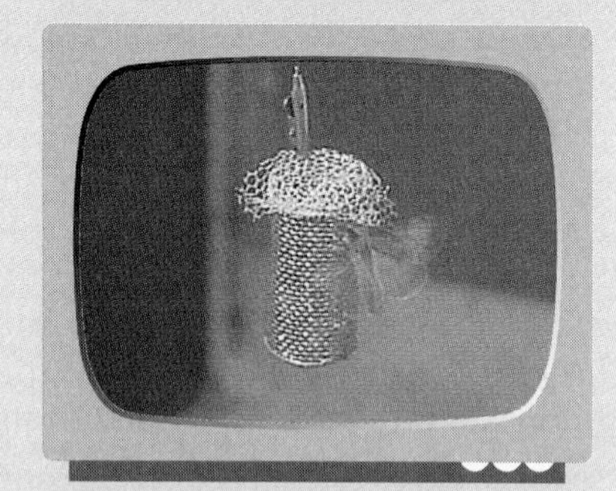

A pheromone trap. A male moth is attracted to the trap during pheromone testing. By placing the bait in a wind tunnel with a moving floor, the effort of the moth to reach the "female" can be measured numerically, giving quantitative data that can be used to evaluate the success of the pheromone.

insect molting, reproduction, and digestion. Tests have shown it is specific to insects without affecting pest predators. For example, use of azadirachtin on an aphid-infested field killed the aphids without harming ladybugs and lacewings, which are aphid predators. Margosan-O is the first commercial neem-based biopesticide, but a large market for neem-based insecticides seems likely in view of their apparently ideal characteristics.

17.7 Agricultural Genetic Engineering

Armed with the ability to insert genes into organisms (Section 11.11), it follows that we might be able to introduce genes into food plants and animals to fight pests, control diseases, and improve the quality of food being produced.

The number of **transgenic crops** (genetically altered crops) undergoing testing is expanding rapidly. The U.S. Department of Agriculture has authorized tests of more than 1000 genetically engineered fruits, vegetables, and grains; and 486 field tests were in progress in 1994 compared to five in 1987. Figure 17.10 illustrates the variety of tests involved. There are concerns that weeds might pick up resistance to herbicides, viruses, and pests from these transgenic crops, but it is unlikely that data based on small test plots will yield any information on hazards associated with transgenic crops. Genetic uniformity is a worry because all the plants would have the same strengths and vulnerabilities. A blight or disease could wipe out an entire crop. Proponents of transgenic crops feel that safety concerns can be tested in closely monitored large-scale field tests while opponents want transgenic crops tested for their ability to transfer genes to weeds before large-scale trials are conducted. Data from large-scale field trials already underway in China for transgenic tobacco, tomatoes, and rice may provide answers to the question of transgenic crop safety.

Natural breeding methods require up to ten years to produce plants suitable for field testing that show resistance to a particular virus. Genetic engi-

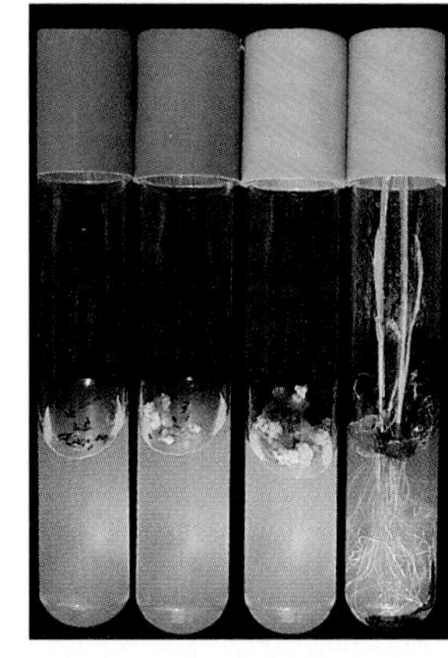

Growth of a transgenic plant. From left to right, you can see progress from a few cells that contain the altered DNA to a young plant with leaves and roots. *(Dan McCoy/Rainbow)*

The World of Chemistry

Program 2, 25, and 26, *Color, Genetic Code, and Futures*

Figure 17.10 The sequence of tests in evaluating a gene-modified plant. (a) After the laboratory work of gene modification, the test plants must be evaluated in (b) greenhouse and (c) field tests to verify that the desired characteristic has been obtained.

Experimental tomatoes, genetically engineered for long shelf life. Ordinarily, tomatoes left on the vine until they become this ripe do not last long enough to reach the supermarket. *(© Richard Nowitz/Phototake NYC)*

neering requires one year. One of the first examples of a genetically engineered plant was the insertion of DNA segments from the tobacco mosaic virus into the genetic code of tomato plants (see Figure 17.11). The implants caused the tomato plants to be strongly resistant to attack by this virus.

The first genetically engineered food to be sold to consumers is the Flavr Savr tomato developed by Calgene. Ordinary tomatoes are picked green and ripened with ethylene gas to avoid excess softening and spoilage before the tomatoes reach the consumer. The Flavr Savr tomato ripens more slowly so it can be picked at a ripened stage, enhancing its flavor and increasing its shelf life. This property was accomplished by inserting a reverse version of a tomato gene for *polygalacturonase* (PG), an enzyme that breaks down cell walls.

Several more genetically engineered plants have now received approval for marketing (Table 17.3), and transgenic seed sales are projected to reach more than $6 billion by 2005. The common method for developing crops with insect resistance is to use recombinant DNA techniques to insert a *Bt* gene. For example, Monsanto's genetically engineered Bollgard cotton seed was used for about 1.8 million of the estimated 14 million acres of cotton plantings during its first year. Bollgard uses a *Bt* gene to produce proteins for controlling bollworm and budworm. However, there are concerns that insects will develop resistance to *Bt* toxins produced by transgenic plants, and the Environmental Protection Agency requires companies who are developing transgenic crops to submit and implement pest resistance management plans as a condition of product registration. Genetic engineering has also been used to develop transgenic crops that are resistant to certain herbicides. This allows the control of weeds without damaging the plant. For example, cotton, corn, and soybeans have been genetically engineered with resistance to glyphosate, the active ingredient in Roundup herbicide.

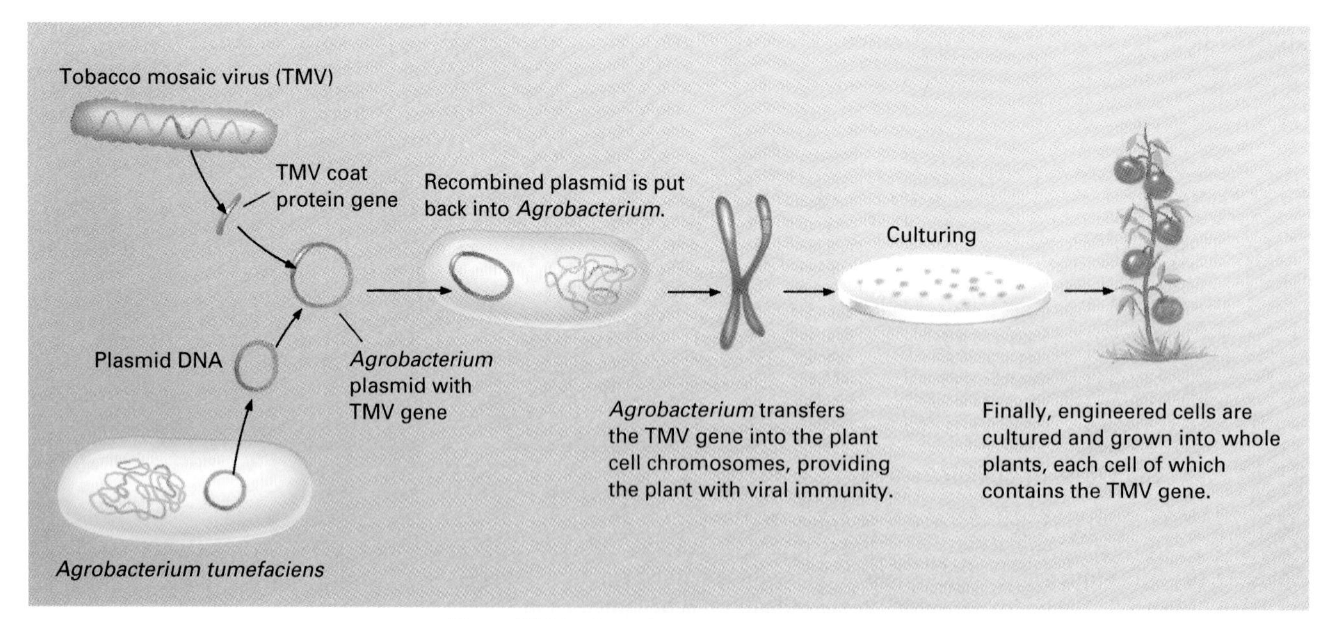

Figure 17.11 Steps necessary to insert a gene from tobacco mosaic virus into a tomato plant to make the plant resistant to a viral disease.

TABLE 17.3 ■ Genetically Engineered Crops

Genetic Trait	Crop
Insect resistance	Corn
	Cotton
	Potato
	Rice
Herbicide resistance	Canola, soybean
	Cotton
	Corn
Virus resistance	Squash
Specialty oils	Canola
Slower ripening	Tomato

A different type of application is to use genetic engineering to increase the amount of commercially useful substance in a plant. Transgenic canola plants have been developed using a gene from the California bay tree that shuts off fatty acid synthesis at 12 carbons instead of 18 carbons. The resulting transgenic plant contains up to 40% lauric acid (Table 11.2), a saturated fatty acid used to make soaps, detergents, and shampoos (Section 11.4). However, one risk is that canola seeds from these transgenic canola plants will get mixed up with seeds from the unmodified plant whose oil receives high nutritional marks for its low amount of saturated fatty acids.

■ SELF-TEST 17B

1. Most virgin soils can support crop production for a decade or more before fertilization is needed. (a) True, (b) False.
2. What do the numbers 6–12–8 on a fertilizer mean? The 6 is the percentage of __________, the 12 is the percentage of __________, and the 8 is the percentage of __________ in the fertilizer.
3. Would the nitrogen in ammonia be considered "fixed" nitrogen?
4. Pure ammonia under ordinary conditions is (a) a solid, (b) a liquid, (c) a gas.
5. Approximately what percentage of the food crops of the world is lost to pests each year?
6. The first chlorinated organic insecticide was __________.
7. Which of the following is a persistent insecticide? (a) DDT, (b) Malathion, (c) Carbaryl, (d) Pirimicarb.
8. __________ is a natural insecticide extracted from chrysanthemum flowers.
9. Which is more likely to be a hormone, a selective or a nonselective herbicide?
10. The most widely used herbicide today is __________.
11. Three common classes of pesticides are __________, __________, and __________.

MATCHING SET

___ 1. Legume crops	a. Most common herbicide
___ 2. Chlorosis	b. *Bacillus thuringiensis*
___ 3. Ammonium nitrate	c. Caused by nutrient deficiency in soil
___ 4. 2,4-D	d. First genetically engineered food on the market
___ 5. Glyphosate	e. Causes basic soil to become acidic
___ 6. Natural insecticide	f. Roots contain nitrogen-fixing bacteria
___ 7. Pheromones	g. Solid added to neutralize acidic soil
___ 8. Flavr Savr tomato	h. Nitrogen fertilizer also used as an explosive
___ 9. Fixed nitrogen	i. Organophosphorus pesticide
___ 10. Malathion	j. Herbicide inhibits amino acid synthesis in plants
___ 11. Selective leaching	k. Sex attractants used in integrated pest management
___ 12. Limestone	l. DDT
___ 13. Chlorinated hydrocarbon pesticide	m. Ammonia

QUESTIONS FOR REVIEW AND THOUGHT

1. What factors affect the earth's carrying capacity?
2. What are the likely consequences if the present rate of human population growth continues?
3. What is an operational definition of "soil"?
4. Give definitions and descriptions for the following:
   (a) Topsoil (b) Subsoil
   (c) Loam (d) Separates
5. What is the structure of a typical soil?
6. Name what causes soil to be (a) sour (b) sweet
7. If crushed limestone is spread on soil, will it raise or lower the pH of the soil? Explain.
8. Give definitions for the following:
   (a) Leaching (b) Selective leaching
9. What are the three nonmineral nutrients obtained from water and air required for normal plant growth?
10. Which is more easily leached from soils, nitrates or phosphates? Why?
11. (a) Which groups of elements are first leached from soils, the Group IA and IIA metals or the Group IIIA and transition metals? (b) What is the effect of this selective leaching on soil pH?
12. (a) What are two important roles of humus in the soil? (b) Do leaves turned into the soil to produce humus raise or lower the soil pH?
13. Give definitions and/or descriptions for the following:
   (a) Nutrients (b) Nonmineral nutrients
   (c) Mineral nutrients
14. What are the three primary mineral plant nutrients that are considered in fertilizer formulations?
15. Which is more likely to be a problem in farming, a soil shortage of N, P, and K, or a shortage of Ca, Mg, and S? Give a reason for your answer.
16. (a) What does the term "nitrogen fixation" mean? (b) Give two natural ways that this process occurs.
17. Write the balanced equation for the Haber process. Why is this process important in the production of fertilizers?
18. What are legumes? What is nitrogenase?
19. What is chlorosis? What causes it? What are the symptoms of chlorosis?
20. An inorganic fertilizer has a grade of 10–0–0. Is this a complete fertilizer? Explain.
21. Inorganic fertilizers are graded on their content of nitrogen, phosphorus, and potassium. Are any of these actually present as the element? Are any of the reference substances actually in a bag of fertilizer?
22. Phosphate rock is treated with sulfuric acid to make superphosphate. Why isn't the phosphate rock simply used as a phosphorus-containing fertilizer?
23. Urea (NH_2CONH_2) is a common straight fertilizer supplying nitrogen. What happens to urea when it is applied to soil? What nitrogen compound is formed when urea decomposes?
24. What is a straight fertilizer? What is a complete fertilizer?
25. Give definitions or descriptions for the following:
   (a) Quick-release fertilizers (b) Slow-release fertilizers
26. Give definitions for the following:
   (a) Pesticide (b) Insecticide
   (c) Herbicide (d) Fungicide

27. Ammonium nitrate ($NH_4NO_3(s)$) is a common straight fertilizer. It decomposes at high temperature to produce N_2, H_2O, and O_2. Handling ammonium nitrate has led to accidental explosions. Give one reason for continued use of ammonium nitrate as a fertilizer and one reason why it should be under greater control.
28. What is the reason why the use of DDT is banned in the United States? Why do you suppose that DDT is still used in other parts of the world?
29. Why is DDT fat soluble and not water soluble? What two properties make DDT a problem?
30. Trace the rise and fall of the use of DDT in agriculture. Debate whether it has been more good than bad for the human race.
31. What is the approximate percentage of food crops that is lost to pests every year? What is the estimated dollar value of these lost crops?
32. What is a persistent pesticide? Give an example.
33. What is biodegradable pesticide? Give an example.
34. (a) What two herbicides were formulated to produce Agent Orange? (b) Which of these herbicides is currently banned in the United States for agricultural use?
35. (a) Discuss the benefits and possible harms in using pesticides in agriculture. (b) What conclusions can you draw on this controversial issue?
36. Organic farming saves energy in one area but loses it in another. Explain.
37. What is sustainable agriculture?
38. If you grew a garden, would you use chemical fertilizers and pesticides? Why or why not?
39. What is integrated pest management?
40. Two effective natural insecticides are *Bt* toxins and neem tree seeds. Explain what these are and why they are becoming so popular.
41. (a) What are transgenic crops? (b) Give some examples.
42. What mechanism of action is followed by glyphosate and sulfonylureas? What is the relationship between "essential" amino acids and these herbicides?
43. Why are pheromones interesting compounds for insect control? How are they used?
44. Assume it is possible to use genetic engineering to improve the storage life of a ripe harvested food like oranges. Give two questions you would want answered before you would buy and eat these oranges.
45. What do you think about the connection between population growth and limitations on the amount of cultivable land? Do you see any conflict? Explain.
46. A 1992 United Nations study indicated that 4.84 billion acres of farmland soil were so degraded that it would be impossible to reclaim them. What are the main causes of this soil degradation?
47. Order the following insecticides in increasing order of toxicity based on the LD_{50} values for rats given in parentheses.
    DDT (LD_{50} = 100 mg/kg)
    malathion (LD_{50} = 1000 mg/kg)
    carbaryl (LD_{50} = 250 mg/kg)
    pirimicarb (LD_{50} = 150 mg/kg)
    dimethrin (LD_{50} = 10,000 mg/kg)

PROBLEMS

1. How many acres of land are estimated to be under cultivation? What is the approximate number of acres cultivated per person? Assume Earth's population is 5.9 billion people.
2. An inorganic fertilizer has an analysis of 10–20–20. What is the percentage of N in this fertilizer? How many pounds of N would be in a 20-lb bag?
3. A fertilizer bag label displays the following grade numbers, 20-10-5. Is this a complete fertilizer? Explain. What is the percentage of N, the percentage of P as P_2O_5, and the percentage of K as K_2O in this fertilizer? How many pounds of N are in a 50-lb bag?

APPENDIX

A

Significant Figures

A temperature of 52.7°C can be read from the thermometer pictured in the margin by estimating the 0.7 part of the measurement. The thermometer can only measure this temperature to three **significant figures**, which are the number of digits that can be measured with certainty, plus one more that is estimated. Suppose the temperature of 52.7°C is measured during a chemical experiment and it is decided to try the next experiment at a lower temperature, one third of 52.7°C. Dividing 52.7°C by 3 on a calculator gives an answer of 17.56666667°C. But since 52.7°C has three significant figures, only three digits should be used in the answer to this calculation. The 17.56666667 on the calculator must be rounded off to give a temperature for the new experiment of 17.6°C.

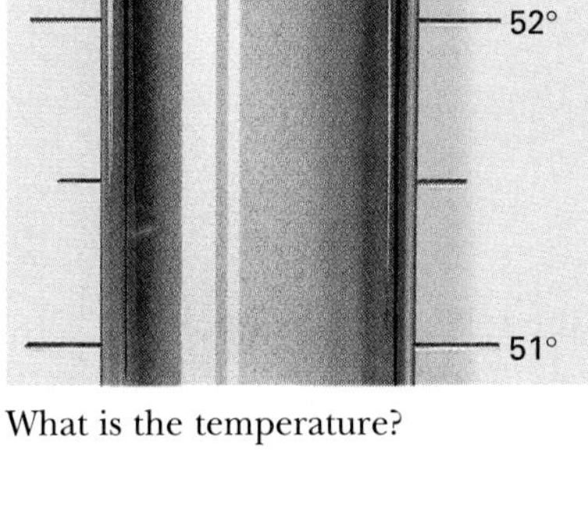

What is the temperature?

In scientific calculations, it is important to report the results of measurements with only the number of digits that are significant. When these measurements are then used in calculations, the number of digits in the answer must be neither more nor less than are significant. Keeping track of significant figures in calculations requires two skills: (1) counting digits that are significant and (2) applying rules for rounding off the answers to calculations.

Counting Significant Figures

1. **All nonzero digits are significant.**

1 2 23	1 2 3 4 5 15,699
Two significant figures	Five significant figures

2. **Zeroes to the left of the first nonzero digit are not significant.**

1 2 0.23	1 2 0.00023
Two significant figures	Two significant figures

3. **Zeroes between nonzero digits are significant.**

1 2 3 4 7077	1 2 3 50.2
Four significant figures	Three significant figures

4. Zeroes at the end of a number that includes a decimal point are significant.

These zeroes have been added to represent the correct number of significant figures.

12 345 50.020	1 23 7.00
Five significant figures	Three significant figures

For a number with zeros at the end and *no* decimal point, it is impossible to know how many of the digits are significant. Does 99,000 have two, three, or four significant figures? To represent the significance of digits in numbers like this, scientists rely on scientific notation, which is discussed in Appendix B. Sometimes a decimal point is placed after a final zero to show that the zero is significant. For example, 20. has *two* significant figures.

Counting Significant Figures

Measured Value	Number of Significant Figures
1 3 cm	1
12 34 45.87 mL	4
123 0.223 g	3
1234 4529 m	4
12 0.70 kg	2
12 3 30.6 cm	3
12 34 65.00 m^3	4
1 0.005 mg	1
1 2 3.8×10^3 L	2
1 23 1.05×10^{-10} mm	3

EXAMPLE A.1 *Significant Figures*

How many significant figures are there in each of the following measurements?

(a) 0.007 m (b) 99.00 kg (c) 2.0004°C

SOLUTION

Always start counting with the first nonzero digit. In (a) this is the 7 and there is only one significant figure. In (b), you continue to count past the decimal point—the zeros have been added to show that this measurement has four significant figures. In (c) the zeros in the middle of the number are significant, so this measurement has five significant figures.

(a) 0.007 m (1) (b) 99.00 kg (12 34) (c) 2.0004°C (1 2345)

Exercise A.1

How many significant figures are in each of the following quantities?

(a) 63,332 kg (b) 0.06 mm (c) 407 L (d) 0.0600 s (e) 0.0101 g

Answers

(a) 5 (b) 1 (c) 3 (d) 3 (e) 3

■ For calculations that have several parts, the best approach is to do the entire calculation first, and then round off the final answer. Rounding off at every separate step can cause errors.

Calculations with Significant Figures

To find the correct number of significant figures for the answer to a calculation requires three steps: (1) Do the arithmetic. (2) Decide how many significant figures the answer can have. (3) Round off the answer.

In **addition and subtraction** an answer should have no more *decimal* places than the number in the calculation with the fewest decimal places. Consider the following addition.

13.673	Three decimal places
4.0	One decimal place
[17.673]	Three decimal places (unrounded answer)

The unrounded answer has two more decimal places than 4.0 and must therefore be rounded off to include only one decimal place as in 4.0. The rules for **rounding off** the answer are simple:

If the numbers to be dropped are 5 or greater than 5, then 1 is added to the final digit that is retained.

If the numbers to be dropped are less than 5, they are dropped and the final retained digit is unchanged.

In the preceding addition, 17.673 must be rounded off to have one decimal place, which means dropping the 73 and increasing the 6 by 1 to give 17.7.

An exception to the need to round off answers occurs with numbers described as *exact.* One source of exact numbers is counting. The number of students in a class or the number of objects in one dozen are exact numbers. Such numbers do not limit the number of digits in the answer to a calculation. Another source of exact numbers is definition. For example, 1 kg is defined as *exactly* equal to 1000 g.

■ Note that in addition and subtraction, rounding off isn't needed if all of the numbers in the calculation are whole numbers.

Rounding Off Numbers

Original Number*	Number of Significant Figures Desired	Rounded-off Number
1.67	2	1.7
521.1	3	521
4.565	3	4.57
351.53	3	352
1.568	1	2
16.043	3	16.0
8.235×10^3	2	8.2×10^3
7540.916	4	7541

*The first number being dropped is underlined.

EXAMPLE A.2 *Significant Figures in Addition*

What is the total weight of three jars of chemicals weighing 14.57 g, 1835.5 g, and 0.875 g?

SOLUTION

By aligning the numbers on their decimal points for the addition and mentally drawing a vertical line after the number with the smallest number of decimal places, it is easy to see that the answer must be rounded off to one decimal place.

14.5¦7 g	
1835.5¦ g	
0.8¦75 g	
[1850.9¦45 g]	Rounded-off answer: 1850.9 g

The total weight of the jars is 1850.9 g.

EXAMPLE A.3 *Significant Figures in Subtraction*

An African Goliath beetle, one of the world's heaviest insects, weighs 99.790 g. What is the difference in weight between this beetle and a hummingbird with a weight of 10.6 g?

SOLUTION

In subtraction as in addition, align the numbers on the decimal point and use a vertical line to find the place to which the answer should be rounded.

$$\begin{array}{r} 99.7|90\text{ g} \\ -\ 10.6|\text{ g} \\ \hline [89.1|90\text{ g}] \end{array} \qquad \text{Rounded-off answer: 89.2 g}$$

The African Goliath beetle is 89.2 g heavier than a hummingbird.

Exercise A.2

Round off each of the following numbers to three significant figures.

(a) 187.6 (b) 5.281 (c) 1.0334 (d) 0.9265 (e) 30.099

Answers

(a) 188 (b) 5.28 (c) 1.03 (d) 0.927 (e) 30.1

Exercise A.3

Perform the following calculations and give the answers with the proper numbers of significant figures.

(a) 77.2 + 0.531 + 13.27 (b) 0.815 − 0.2346
(c) 62 + 49.1 (d) 1.386 − 1.2

Answers

(a) 91.0 (b) 0.580 (c) 111 (d) 0.2

Calculations with Significant Figures

Computation	Result
18.2 + 5.35 + 20.	44
15.02 + 0.003 + 700.1	715.1
7109.3 − 40.352	7068.9
1.521 − 0.81	0.71
38.2 × 0.95	36
17.32 × 1.66	28.8
182/32.800	5.55
0.881/5.2	0.17
$\frac{39.3 + 17.21}{190.}$	0.297
(58.1 × 0.82) − 1.19	46

For **multiplication and division** the answer is limited to the number of digits in the number with the fewest significant figures. When, for example, 75 is multiplied by 0.05, the answer can have only one significant figure (as in 0.05); when 0.084 is divided by 0.00298, the answer can have only two significant figures (as in 0.084).

$$75 \times 0.05 = [3.75] \qquad \text{Rounded-off answer: 4}$$

$$\frac{0.084}{0.00298} = [28.18791946] \qquad \text{Rounded-off answer: 28}$$

EXAMPLE A.4 *Significant Figures in Multiplication*

A sample of a metal ore analyzed in the laboratory is found to contain 11.00% of titanium. How much titanium is present in 55 kg of this ore?

SOLUTION

To find the answer requires multiplying 55 kg by 0.1100

$$0.1100 \times 55 \text{ kg} = [6.05 \text{ kg}] \qquad \text{Rounded-off answer: } 6.1 \text{ kg}$$

The answer is limited to two significant figures by 55 kg. The ore contains 6.1 kg of titanium.

■ The number of significant figures in the answer is determined by *decimal places* in addition and subtraction, and by total *significant figures* in multiplication and division.

■ To use a percent in a calculation, you must move the decimal point two places to the left:

$$10\% = 0.10 \qquad 59.33\% = 0.5933$$
$$230\% = 2.30$$

EXAMPLE A.5 *Significant Figures in Division*

The Pentagon building in Washington, D.C. has an area of 117,355 square meters (m^2), and a football field has an area of 5358 m^2. How many football fields would fit inside the Pentagon?

SOLUTION

To find the answer requires dividing the area of the Pentagon by the area of the football field. On a calculator the answer has ten digits. It must be rounded off to four digits as in 5358.

$$\frac{117{,}355 \text{ m}^2}{5358 \text{ m}^2} = [21.90276222] \qquad \text{Rounded-off answer: } 21.90$$

The Pentagon could hold 21.90, or just about 22 football fields.

Exercise A.5

Perform the following calculations and give the answers with the proper numbers of significant figures.

(a) 68.3×0.92 (b) 3.01×20.225 (c) $1.594/6.23$

(d) $\dfrac{5.0}{25.0}$ (e) $\dfrac{(2.151 - 1.30)}{0.591}$

Answers

(a) 63 (b) 60.9 (c) 0.256 (d) 0.20 (e) 1.4

APPENDIX

B

Scientific Notation

Scientific notation, also known as **exponential notation**, is a way of representing large and small numbers as the product of two terms. The first term, the coefficient, is a number between 1 and 10. The second term, the exponential term, is 10 raised to a power—the **exponent**. For example,

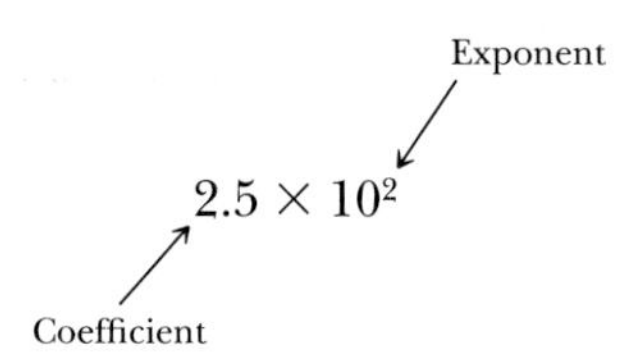

The exponent indicates the number of 10's by which the coefficient is multiplied to give the number represented in scientific notation

$$2.5 \times 10^2 = 2.5 \times 10 \times 10 = 250$$

There are two great advantages to scientific notation. The first, as illustrated in Table B.1, is that the very large and very small numbers often dealt with in the sciences are much less cumbersome in scientific notation. The second is that it removes the ambiguity in the number of significant figures in a number ending with zeroes. We can, for example, write

Two significant figures	2.5×10^2
Three significant figures	2.50×10^2
Four significant figures	2.500×10^2

All the digits in the coefficient are always considered significant.

For numbers larger than 1, the exponent in scientific notation is a positive whole number, as illustrated above. For numbers less than 1, the exponent is

TABLE B.1 ■ Large and Small Numbers in Scientific Notation

Distance from Earth to the Sun	150,000,000,000 m	1.5×10^{11} m
Diameter of Earth	13,000,000 m	1.3×10^{7} m
Height of Mt. Everest	10,000 m	1×10^{4} m
Height of Empire State Building	400 m	4×10^{2} m
Human height	1.7 m	1.7×10^{0} m
Length of a cockroach	0.040 m	4.0×10^{-2} m
Size of a grain of sand	0.00010 m	1.0×10^{-4} m
Size of a red blood cell	0.0000065 m	6.5×10^{-6} m
Size of a polio virus	0.000000025 m	2.5×10^{-8} m

a negative whole number that indicates the number of times the coefficient must be divided by 10 (or multiplied by 0.1) to give the number represented in scientific notation

■ Numbers in scientific notation are read as follows:

2×10^4 "Two times ten to the fourth power"

2×10^{-4} "Two times ten to the minus fourth power"

$$1.2 \times 10^{-2} = 1.2 \times \frac{1}{10} \times \frac{1}{10} = 0.012$$

There are two points to remember in converting a number into or out of scientific notation: (1) The value of the exponent is the number of places by which the decimal is shifted. (2) The coefficient should have only one digit before the decimal point. Look through the following examples and count the number of places the decimal points have been moved.

$$10000 = 1 \times 10^4 \qquad 12345 = 1.2345 \times 10^4$$

$$1000 = 1 \times 10^3 \qquad 1234 = 1.234 \times 10^3$$

$$100 = 1 \times 10^2 \qquad 123 = 1.23 \times 10^2$$

$$10 = 1 \times 10^1 \qquad 12 = 1.2 \times 10^1$$

$$1 = 1 \times 10^0 \qquad \text{(Any number} \times 10^0 = \text{the number itself.)}$$

$$1/10 = 1 \times 10^{-1} \qquad 0.12 = 1.2 \times 10^{-1}$$

$$1/100 = 1 \times 10^{-2} \qquad 0.012 = 1.2 \times 10^{-2}$$

$$1/1000 = 1 \times 10^{-3} \qquad 0.0012 = 1.2 \times 10^{-3}$$

$$1/10000 = 1 \times 10^{-4} \qquad 0.00012 = 1.2 \times 10^{-4}$$

Some electronic calculators allow you to easily convert numbers to scientific notation. If you have such a calculator, you can change a number shown in the usual form to scientific notation simply by pressing the EE or EXP key and then the "=" key (or in some cases the "=" key and then the exponent key).

■ Do not use $\times 10$ when entering exponential numbers on a calculator; this multiplies your answer by 10.

EXAMPLE B.1 *Converting into and out of Scientific Notation*

(a) The Sun is about 93 million miles from Earth. Express this number in scientific notation. (b) A blue whale has a weight of 1.36×10^5 kg. Convert this number from scientific notation into a conventional number.

SOLUTION

(a) Writing out the zeros in 93 million gives

$$93{,}000{,}000 \text{ mi}$$

To express this distance in scientific notation requires moving the decimal point 7 places to the left. Because the number is larger than 1, the exponent is positive.

$$\underbrace{93{,}000{,}000}_{7\,6\,5\,4\,3\,2\,1} \text{ miles} = 9.3 \times 10^7 \text{ mi}$$

(b) To go the other way, the positive exponent indicates a number greater than 1, so that the decimal point must be moved to the right, in this case by 5 places.

$$1.36 \times 10^5 \text{ kg} = \underbrace{136{,}000}_{1\,2\,3\,4\,5} \text{ kg}$$

EXAMPLE B.2 *Converting into and out of Scientific Notation*

(a) Convert the length of a virus from 0.00000002 m to scientific notation. (b) Convert the diameter of a protein molecule from 6.5×10^{-6} m to a decimal number.

SOLUTION

(a) For a number less than 1, the exponent is the negative of the number of places the decimal is moved to the *right*

$$\underbrace{0.00000002}_{1\,2\,3\,4\,5\,6\,7\,8} \text{ m} = 2 \times 10^{-8} \text{ m}$$

(b) The negative exponent shows that the number is less than 1, so the decimal point must be moved to the left, by 6 places in this case, to convert out of scientific notation.

$$6.5 \times 10^{-6} \text{ m} = \underbrace{0.0000065}_{6\,5\,4\,3\,2\,1} \text{ m}$$

Exercise B1

Write the following numbers in scientific notation.

(a) 3785 (b) 0.093 (c) 19.2
(d) 552,000 (e) 0.00075

Answers
(a) 3.785×10^3 (b) 9.3×10^{-2} (c) 1.92×10^1
(d) 5.52×10^5 (e) 7.5×10^{-4}

Exercise B.2
Write the following numbers in conventional form.
(a) 4.42×10^2
(b) 8.9×10^{-3}
(c) 3.01×10^{-1}
(d) 6.9×10^4
(e) 7.7×10^0

Answers
(a) 442 (b) 0.0089 (c) 0.301 (d) 69,000 (e) 7.7

Calculations with Numbers in Scientific Notation

For **addition and subtraction** of numbers in scientific notation, the numbers must first be converted to the same powers of 10. The coefficients are then added or subtracted as usual and the exponential term remains the same in the sum or difference.

$$\begin{aligned}(1.234 \times 10^{-3}) + (5.623 \times 10^{-2}) &= (0.1234 \times 10^{-2}) + (5.623 \times 10^{-2}) \\ &= 5.746 \times 10^{-2}\end{aligned}$$

$$\begin{aligned}(6.52 \times 10^{2}) - (1.56 \times 10^{3}) &= (6.52 \times 10^{2}) - (15.6 \times 10^{2}) \\ &= -9.1 \times 10^{2}\end{aligned}$$

For **multiplication** of numbers in scientific notation, the coefficients are multiplied in the usual manner and the exponents are added. If both exponents have the same sign, their values are added and the exponent in the answer has the same sign. If the exponents have different signs, they must be added algebraically, taking into account the + and − signs. The final answer is given with one nonzero digit to the left of the decimal and the correct number of significant figures.

■ To algebraically add a number with a + sign to a number with a − sign, drop the signs, subtract the smaller number from the larger number, and give the answer the sign of the larger number.

$$\begin{aligned}(1.23 \times 10^{3})(7.60 \times 10^{2}) &= (1.23)(7.60) \times 10^{3+2} \\ &= 9.348 \times 10^{5} \\ &= 9.35 \times 10^{5} \text{ (3 significant figures)}\end{aligned}$$

$$\begin{aligned}(6.02 \times 10^{23})(2.32 \times 10^{-2}) &= (6.02)(2.32) \times 10^{23-2} \\ &= 13.966 \times 10^{21} \\ &= 1.40 \times 10^{22} \text{ (3 significant figures)}\end{aligned}$$

$$\begin{aligned}(5.2 \times 10^{-4})(6.1 \times 10^{2}) &= (5.2)(6.1) \times 10^{-4+2} \\ &= 31.72 \times 10^{-2} \\ &= 3.2 \times 10^{-1} \text{ (2 significant figures)}\end{aligned}$$

■ To algebraically subtract numbers when they have different signs, reverse the sign of the number that is being subtracted, and then add the numbers algebraically.

For **division** of numbers in scientific notation, the coefficients are divided in the usual fashion. The exponent of the divisor is subtracted algebraically from the exponent of the number that is being divided.

$$\frac{7.60 \times 10^3}{1.23 \times 10^2} = \frac{7.60}{1.23} \times 10^{3-2} = 6.18 \times 10^1$$

$$\frac{6.02 \times 10^{23}}{9.10 \times 10^{-2}} = \frac{6.02}{9.10} \times 10^{(23)-(-2)} = 0.662 \times 10^{25} = 6.62 \times 10^{24}$$

EXAMPLE B.3 *Calculations with Numbers in Scientific Notation*

The distance around the globe from New York to London is 5.6×10^3 km and the distance from London to New Zealand is 1.9×10^4 km. What is the sum of these distances? What is the difference between these distances?

SOLUTION

$$\begin{aligned}(1.9 \times 10^4 \text{ km}) + (5.6 \times 10^3 \text{ km}) &= (1.9 \times 10^4 \text{ km}) + (0.56 \times 10^4) \\ &= 2.5 \times 10^4 \text{ km}\end{aligned}$$

$$\begin{aligned}(1.9 \times 10^4 \text{ km}) - (5.6 \times 10^3 \text{ km}) &= (1.9 \times 10^4 \text{ km}) - (0.56 \times 10^4) \\ &= 1.3 \times 10^4 \text{ km}\end{aligned}$$

It is 2.5×10^4 km from New York to New Zealand via London, and it is 1.3×10^3 km farther from London to New Zealand than it is from New York to London.

EXAMPLE B.4 *Calculations with Numbers in Scientific Notation*

To compare their sizes, divide the size of a grain of sand, 1.0×10^{-4} m, by the size of a red blood cell, 6.5×10^{-6}.

SOLUTION

$$\begin{aligned}\frac{1.0 \times 10^{-4} \text{ m}}{6.5 \times 10^{-6} \text{ m}} &= \frac{1.0}{6.5} \times 10^{-4-(-6)} \\ &= 0.15 \times 10^2 \\ &= 1.5 \times 10^1 \text{ or } 15\end{aligned}$$

A grain of sand is 15 times larger than a red blood cell.

EXAMPLE B.5 *Calculations with Numbers in Scientific Notation*

In 1 mol of oxygen gas there are 6.02×10^{23} O_2 molecules. How many O_2 molecules are there in 1.4×10^3 mol of oxygen gas?

$$(1.4 \times 10^3 \text{ mol } O_2) \times \left(\frac{6.02 \times 10^{23} \text{ } O_2 \text{ molecules}}{\text{mol } O_2}\right)$$

$$= (1.4 \times 6.02) \times 10^{3+23} \text{ molecules}$$
$$= 8.4 \times 10^{26} \text{ } O_2 \text{ molecules}$$

Exercise B.3

Perform the following calculations.

(a) $(8.91 \times 10^{-2}) \times (1.2 \times 10^5)$

(b) $(1.15 \times 10^8) + (6.2 \times 10^7)$

(c) $\dfrac{5.09 \times 10^5}{8.2 \times 10^{-2}}$

(d) $(9.82 \times 10^4) - (1.35 \times 10^3)$

(e) $\dfrac{(2.731 \times 10^{-2}) \times (1.52 \times 10^9)}{8.3 \times 10^{14}}$

Answers

(a) 1.1×10^4 (b) 1.77×10^8 (c) 6.2×10^6 (d) 9.69×10^4
(e) 5.0×10^{-8}

APPENDIX

C

Units of Measure, Unit Conversion, and Problem Solving

Units of the SI System

The metric system was begun by the French National Assembly in 1790 and has undergone many modifications. The International System of Units or *Système International* (SI), which represents an extension of the metric system, was adopted by the 11th General Conference of Weights and Measures in 1960. It is constructed from seven base units, each of which represents a particular physical quantity (Table C.1).

The first five units listed in Table C.1 are those most useful in chemistry. They are defined as follows:

1. The *meter* is the length of the path traveled by light in vacuum during a time interval of 1/299792458 of a second.
2. The *kilogram* represents the mass of a platinum–iridium block kept at the International Bureau of Weights and Measures of Sèvres, France.
3. The *second* was redefined in 1967 as the duration of 9,192,631,770 periods of a certain line in the microwave spectrum of cesium-133.
4. The *kelvin* is 1/273.16 of the temperature interval between absolute zero and the triple point of water (the temperature at which liquid water, ice, and water vapor coexist).
5. The *mole* is the amount of substance that contains as many entities as there are atoms in exactly 0.012 kg of carbon-12 (12 g of ^{12}C atoms).

Decimal fractions and multiples of metric and SI units are designated by using the ***prefixes*** listed in Table C.2. The prefix *kilo*, for example, means a unit is multiplied by 10^3

$$1 \text{ kilogram} = 1 \times 10^3 \text{ grams} = 1000 \text{ grams}$$

TABLE C.1 ■ SI Fundamental Units

Physical Quantity	Name of Unit	Symbol
Length	Meter	m
Mass	Kilogram	kg
Time	Second	s
Temperature	Kelvin	K
Amount of substance	Mole	mol
Electric current	Ampere	A
Luminous intensity	Candela	cd

and the prefix *centi* means the unit is multiplied by the factor 10^{-2}

$$1 \text{ centigram} = 1 \times 10^{-2} \text{ gram} = 0.01 \text{ gram}$$

The prefixes are added to give units of a magnitude appropriate to what is being measured. The distance from New York to London (5.6×10^3 km = 5,600 km) is much easier to comprehend measured in kilometers than in meters (5.6×10^6 m = 5,600,000 m). The following is a list of units for measuring very small and very large distances:

nanometer (nm)	0.000000001 meter	$= 1 \times 10^{-9}$ meter
micrometer (μm)	0.000001 meter	$= 1 \times 10^{-6}$ meter
millimeter (mm)	0.001 meter	$= 1 \times 10^{-3}$ meter
centimeter (cm)	0.01 meter	$= 1 \times 10^{-2}$ meter
decimeter (dm)	0.1 meter	$= 1 \times 10^{-1}$ meter
meter (m)	1 meter	
dekameter (dam)	10 meters	$= 1 \times 10$ meters
hectometer (hm)	100 meters	$= 1 \times 10^2$ meters
kilometer (km)	1,000 meters	$= 1 \times 10^3$ meters
megameter (Mm)	1,000,000 meters	$= 1 \times 10^6$ meters

TABLE C.2 ■ Prefixes for Metric and SI Units*

Factor	Prefix	Symbol	Factor	Prefix	Symbol
10^{12}	tera	T	10^{-1}	*deci*	d
10^9	giga	G	10^{-2}	*centi*	c
10^6	mega	M	10^{-3}	*milli*	m
10^3	*kilo*	k	10^{-6}	micro	μ
10^2	hecto	h	10^{-9}	*nano*	n
10^1	deka	da	10^{-12}	*pico*	p
			10^{-15}	femto	f
			10^{-18}	atto	a

*The most common prefixes are shown in italics.

TABLE C.3 ■ Derived SI Units

Physical Quantity	Name of Unit	Symbol	Definition	Symbol
Area	Square meter	m^2	—	
Volume	Cubic meter	m^3	—	
Density	Kilogram per cubic meter	kg/m^3	—	
Force	Newton	N	(kilogram) (meter)/(second)^2	$kg\ m/s^2$
Pressure	Pascal	Pa	Newton/square meter	N/m^2
Energy	Joule	J	(kilogram) (square meter)/(second)^2	$kg\ m^2/s^2$
Electric charge	Coulomb	C	(ampere) (second)	A s
Electric potential difference	Volt	V	joule/(ampere) (second)	J/(A s)

In the International System of Units, all physical quantities are represented by appropriate combinations of the base units listed in Table C.1. The result is a derived unit for each kind of measured quantity. The most common derived units are listed in Table C.3. It is easy to see that the derived unit for area is length × length = meter × meter = square meter (m^2) or that the derived unit for volume is length × length × length = meter × meter × meter = cubic meter (m^3). The more complex derivations are arrived at by the same kind of combination of units. Units such as the one for *force* have been given simple names that represent the units by which they are defined.

Conversion of Units for Physical Quantities

The result of a measurement is a **physical quantity**, which consists of a number plus a unit. To convert a physical quantity from one unit of measure to another requires a conversion factor based on equivalences between units of measure like those given in Table C.4. Each equivalence provides two conversion factors that are the inverse of each other. For example, the equivalence between a quart and a liter, 1 quart = 0.9463 liter (the 1 quart is exact), gives

$$\frac{1\ \text{quart}}{0.9463\ \text{liter}} \qquad \text{There is 1 quart per 0.9463 liter.}$$

$$\frac{0.9463\ \text{liter}}{1\ \text{quart}} \qquad \text{There is 0.9463 liter per 1 quart.}$$

The cancellation of units provides the basis for choosing which conversion factor is needed: It is always the one that allows the unit being converted to be canceled and leaves the new unit uncanceled.

To convert 2 quarts to liters:

$$2\ \cancel{\text{quarts}} \times \frac{0.9463\ \text{liter}}{1\ \cancel{\text{quart}}} = 1.893\ \text{liters}$$

TABLE C.4 ■ Common Units of Measure

Mass and Weight
1 pound = 453.59 grams = 0.45359 kilogram
1 kilogram = 1000 grams = 2.205 pounds
1 gram = 10 decigrams = 100 centigrams = 1000 milligrams
1 gram = 6.022×10^{23} atomic mass units
1 atomic mass unit = 1.6605×10^{-24} gram
1 short ton = 2000 pounds = 907.2 kilograms
1 long ton = 2240 pounds
1 metric tonne = 1000 kilograms = 2205 pounds
Length
1 inch = 2.54 centimeters (exactly)
1 mile = 5280 feet = 1.609 kilometers
1 yard = 36 inches = 0.9144 meter
1 meter = 100 centimeters = 39.37 inches = 3.281 feet = 1.094 yards
1 kilometer = 1000 meters = 1094 yards = 0.6215 mile
1 Ångstrom = 1×10^{-8} centimeter = 0.1 nanometer = 100 picometers
= 1×10^{-10} meter = 3.937×10^{-9} inch
Volume
1 quart = 0.9463 liter
1 liter = 1.0567 quarts
1 liter = 1 cubic decimeter = 1000 cubic centimeters = 0.001 cubic meter
1 milliliter = 1 cubic centimeter = 0.001 liter = 1.056×10^{-3} quart
1 cubic foot = 28.316 liters = 29.924 quarts = 7.481 gallons
Force and Pressure
1 atmosphere = 760 millimeters of mercury (exactly) = 1.013×10^{5} pascals
= 14.70 pounds per square inch
1 bar = 10^{5} pascals
1 torr = 1 millimeter of mercury
1 pascal = 1 $kg/m\ s^2$ = 1 N/m^2
Energy
1 joule = 1×10^{7} ergs
1 calorie = 4.184 joules (exactly) = 4.184×10^{7} ergs
= 4.129×10^{-2} liter-atmospheres
= 2.612×10^{19} electron volts
1 Calorie (food) = 1 kilocalorie
1 erg = 1×10^{-7} joule = 2.3901×10^{-8} calorie
1 electron volt = 1.6022×10^{-19} joule = 1.6022×10^{-12} erg
1 liter-atmosphere = 24.217 calories = 101.32 joules = 1.0132×10^{9} ergs
1 British thermal unit = 1055.06 joules = 1.05506×10^{10} ergs = 252.2 calories

To convert 2 liters to quarts:

$$2 \text{ liters} \times \frac{1 \text{ quart}}{0.9463 \text{ liter}} = 2.113 \text{ quarts}$$

EXAMPLE C.1 *Unit Conversion*

Find the weight in (a) kilograms, (b) grams, and (c) decigrams of a person who weighs 115 pounds.

SOLUTION

(a) Table C.4 gives the equivalence between pounds and kilograms in two ways: 1 pound = 0.454 kilogram (rounded off), or 1 kilogram = 2.205 pounds. Either equivalence can be used to convert 115 pounds to kilograms by setting up the problem so that *pounds* cancels out

$$115 \text{ pounds} \times \frac{0.454 \text{ kilogram}}{1 \text{ pound}} = 52.2 \text{ kilograms}$$

$$115 \text{ pounds} \times \frac{1 \text{ kilogram}}{2.205 \text{ pounds}} = 52.2 \text{ kilograms}$$

(b) Using the equivalence 1 kilogram = 1000 grams gives

$$52.2 \text{ kilograms} \times \frac{1000 \text{ grams}}{1 \text{ kilogram}} = 52{,}200 \text{ grams or } 5.22 \times 10^4 \text{ grams}$$

(Remember that equivalences for multiples of units, e.g., 1 kilogram = 1000 grams are exact conversion factors and do not limit significant figures.)
(c) Using the equivalence 1 gram = 10 decigrams gives

$$5.22 \times 10^4 \text{ grams} \times \frac{10 \text{ decigrams}}{1 \text{ gram}} = 52.2 \times 10^4 \text{ decigrams or } 5.22 \times 10^5 \text{ decigrams}$$

EXAMPLE C.2 *Unit Conversion*

What is the volume of a 5.00×10^2 milliliter flask in (a) liters, (b) quarts, and (c) ounces (1 quart = 32 ounces, exactly)?

(a)

$$5.00 \times 10^2 \text{ milliliters} \times \frac{1 \text{ liter}}{1000 \text{ milliliters}} = 0.500 \text{ liter}$$

(b)

$$0.500 \cancel{\text{liter}} \times \frac{1.0567 \text{ quart}}{1 \cancel{\text{liter}}} = 0.528 \text{ quart}$$

(c)

$$0.528 \cancel{\text{quart}} \times \frac{32 \text{ ounces}}{1 \cancel{\text{quart}}} = 16.9 \text{ ounces}$$

Exercise C.1

Use the information in Table C.4 to write the equivalences and two conversion factors for each of the following pairs of units:

(a) grams and milligrams (b) kilometers and miles
(c) calories and joules

Answers

(a) 1 gram = 1000 milligrams

$$\frac{1 \text{ gram}}{1000 \text{ milligrams}} \qquad \frac{1000 \text{ milligrams}}{1 \text{ gram}}$$

(b) 1 kilometer = 0.6215 mile

$$\frac{1 \text{ kilometer}}{0.6215 \text{ mile}} \qquad \frac{0.6215 \text{ mile}}{1 \text{ kilometer}}$$

(c) 1 calorie = 4.184 joules

$$\frac{1 \text{ calorie}}{4.184 \text{ joules}} \qquad \frac{4.184 \text{ joules}}{1 \text{ calorie}}$$

Exercise C.2

Carry out the following unit conversions:

(a) 15 milligrams to grams (b) 453 grams to milligrams
(c) 95 kilometers to miles (d) 5 miles to kilometers
(e) 325 calories to joules (f) 325 joules to calories

Answers

(a) 0.015 g (b) 4.53×10^5 milligrams (c) 59 miles
(d) 8 kilometers (e) 1360 joules (f) 77.7 calories

Sometimes it is convenient or necessary to use more than one conversion factor to convert a physical quantity into some other unit. Instead of doing one conversion at a time, a string of conversion factors can be put together and unit cancellation checked before the calculation is done. For example, atmospheric pressure is usually given in the weather report in inches of mercury. What is the atmospheric pressure in millimeters of mercury (a customary unit in scientific measurement) on a day when the weatherman reports it to be 29.2 inches of mercury? Since we have Table C.4 handy, it will be most convenient to use conversion factors based on that table. There isn't an equivalence there for millimeters and inches, but the table does give 1 inch = 2.54 centimeters. Since we know from the prefixes that 100 *centi*meters = 1 meter and 1000 *milli*meters = 1 meter, we can apply these equivalences to go from centimeters to millimeters. Combining the factors we know and checking cancellation of units gives

$$29.2\ \cancel{\text{inches}} \times \frac{2.54\ \cancel{\text{centimeters}}}{1\ \cancel{\text{inch}}} \times \frac{1\ \cancel{\text{meter}}}{100\ \cancel{\text{centimeters}}} \times \frac{1000\ \text{millimeters}}{1\ \cancel{\text{meter}}}$$

The cancellations leave only the unit we're looking for—millimeters—so the setup is correct and the answer is

$$29.2\ \cancel{\text{inches}} \times \frac{2.54\ \cancel{\text{centimeters}}}{1\ \cancel{\text{inch}}} \times \frac{1\ \cancel{\text{meter}}}{100\ \cancel{\text{centimeters}}} \times \frac{1000\ \text{millimeters}}{1\ \cancel{\text{meter}}} = 742\ \text{millimeters}$$

Exercise C.3

(a) Is the following setup correct for converting milligrams to pounds?

$$225\ \text{milligrams} \times \frac{1\ \text{gram}}{1000\ \text{milligrams}} \times \frac{1\ \text{pound}}{454\ \text{grams}}$$

(b) Is the following setup correct for converting yards to kilometers?

$$25\ \text{yards} \times \frac{36\ \text{inches}}{1\ \text{yard}} \times \frac{1\ \text{yard}}{3\ \text{feet}} \times \frac{5280\ \text{feet}}{1\ \text{mile}} \times \frac{1\ \text{mile}}{1.609\ \text{kilometers}}$$

Answers

(a) Yes (b) No

Exercise C.4

Carry out the following calculations by using multiple conversion factors (1 quart = 2 pints; 1 pound = 16 ounces)

(a) What is the volume in milliliters of 1.00 pint of milk?
(b) What is the mass in grams of 15.0 ounces of cereal?
(c) What is the distance in centimeters of the very small distance of 30.5 picometers measured by using X rays?
(d) What is the volume in cubic meters of a 250-gallon tank?

Answers

(a) 473 milliliters (b) 425 grams (c) 3.05×10^{-9} centimeters
(d) 0.95 cubic meter

Problem Solving by Cancellation of Units

Many kinds of numerical problems in everyday life as well as in science can be solved by extending the use of conversion factors and cancellation of units beyond unit conversion. The use of density expressed as mass per unit volume is a simple example of such an extension. Density provides the *connection* between mass and volume. Given that the density of lead is 11.4 g/cm^3, you can find the mass in grams of a piece of lead of known volume or the volume of a piece of lead of known mass. If the known information is that a piece of lead has a volume of 25.0 cm^3 and the unknown information is its mass, the problem is set up and solved as follows

$$25.0\ \cancel{cm^3} \times \frac{11.4\ \text{g lead}}{1\ \cancel{cm^3}} = 285\ \text{g lead}$$

As an example of an everyday problem, consider assigning "units" to the number of people served by a gallon of ice cream. Suppose you know that 1 gallon of ice cream will serve 16 people—16 servings per gallon. To calculate an unknown (the number of gallons needed to serve 50 people) setting up and solving the problem gives

$$50\ \cancel{\text{servings}} \times \frac{1\ \text{gallon of ice cream}}{16\ \cancel{\text{servings}}} = 3.1\ \text{gallons of ice cream}$$

Three gallons should do the job. Applying the method again to calculate the cost, if the ice cream is $4.35 a gallon, gives

$$3\ \cancel{\text{gallons}} \times \frac{\$4.35}{1\ \cancel{\text{gallon}}} = \$13.05$$

Notice how the "conversion factors" in the two ice cream calculations are arranged so that the unwanted "units" cancel. That is the secret of what is sometimes called the *dimensional method* of problem solving.

For a more complex example, suppose a manufacturer must figure out how many people are needed to assemble 6000 Snivies each week (6000 Snivies/week). He knows that it takes a person 1 hour to assemble 8 Snivies (8 Snivies/hour) and that each person will put in 30 hours a week on the assembly line (30 hours/person). How many people must be working on the assembly line each week? Using the dimensional method, the "conversion factors" must be arranged so that the "units" cancel to give an answer of "persons per week." In single setup, the calculation can be done as follows

$$\frac{6000\ \cancel{\text{Snivies}}}{1\ \text{week}} \times \frac{1\ \cancel{\text{hour}}}{8\ \cancel{\text{Snivies}}} \times \frac{1\ \text{person}}{30\ \cancel{\text{hours}}} = 25\ \text{persons/week}$$

The manufacturer will need 25 persons on the assembly line each week.

In Section 6.2, we illustrated dimensional problem solving for calculating the numbers of moles and the masses of reactants and products in chemical reactions. The coefficients in chemical equations and the molar masses of reactants and products provide the factors that must be combined to solve problems of this kind. Consider the equation for the reaction that occurs when copper(II) sulfide is heated in oxygen

$$Cu_2S(s) + O_2(g) \longrightarrow 2\ Cu(s) + SO_2(g)$$

The molar masses of the reactants and products are

$$159 \text{ g } Cu_2S/1 \text{ mol } Cu_2S \qquad 32 \text{ g } O_2/1 \text{ mol } O_2$$
$$63.5 \text{ g Cu}/1 \text{ mol Cu} \qquad 64 \text{ g } SO_2/1 \text{ mol } SO_2$$

How much copper can be made from 2.50 mol of Cu_2S by this reaction? The connection between Cu_2S and Cu is given by the mole ratio 1 mol Cu_2S/2 mol Cu. This ratio and the molar masses must be combined so that the answer has the units of grams of copper

$$2.50 \text{ } \cancel{\text{mol } Cu_2S} \times \frac{2 \text{ } \cancel{\text{mol Cu}}}{1 \text{ } \cancel{\text{mol } Cu_2S}} \times \frac{63.5 \text{ g Cu}}{1 \text{ } \cancel{\text{mol Cu}}} = 318 \text{ g Cu}$$

The mass of copper that can be produced by roasting 2.50 mol of copper sulfide is 318 g.

EXAMPLE C.4 *Problem Solving*

The Snivie manufacturer has one truck available to deliver his Snivies. The Snivies are packed 50 in a carton and the truck can carry 10 cartons per trip. How many trips will the truck have to make to deliver the 6000 Snivies produced each week?

SOLUTION

Expressing the known information as "conversion factors" gives 50 Snivies per carton and 10 cartons per trip. The information must be set up to yield the answer in trips per week.

$$\frac{6000 \text{ Snivies}}{1 \text{ week}} \times \frac{1 \text{ carton}}{50 \text{ Snivies}} \times \frac{1 \text{ trip}}{10 \text{ cartons}} = 12 \text{ trips/week}$$

EXAMPLE C.5 *Problem Solving*

Using the information before Example C.4, find the number of moles of oxygen that will be consumed in the conversion of 683 g of copper sulfide to copper metal.

SOLUTION

The connection here is supplied by the mole ratio, 1 mol O_2/1 mol Cu_2S, and the molar mass of copper sulfide.

$$683 \text{ g } Cu_2S \times \frac{1 \text{ mol } Cu_2S}{159 \text{ g } Cu_2S} \times \frac{1 \text{ mol } O_2}{1 \text{ mol } Cu_2S} = 4.30 \text{ mol } O_2$$

Exercise C.5

(a) If the Snivie manufacturer decides to increase his production to 20,000 Snivies per week, how many persons will he need working on the assembly line each week? (b) If the most trips a truck can make in a week is 20, how many trucks will the manufacturer need to deliver 20,000 Snivies per week?

Answers

(a) 83 persons per week (b) 2 trucks

Exercise C.6

Use the information about the reaction of copper sulfide with oxygen to answer the following questions:

(a) How many grams of copper can be made from 683 g of copper sulfide?
(b) How many moles of sulfur dioxide will be released during the production of the copper from 683 g of copper sulfide?

Answers

(a) 546 g Cu (b) 4.30 mol SO_2

Naming Organic Compounds

Chapters 9 and 10 include both common names and systematic names for organic compounds representing the various classes of hydrocarbons and functional groups. This appendix focuses on the systematic nomenclature for organic compounds, as proposed by the International Union of Pure and Applied Chemistry (IUPAC).

Hydrocarbons

The name of each of the members of the hydrocarbon classes has two parts. The first part, the prefix—*meth-*, *eth-*, *prop-*, *but-*, and so on—reflects the number of carbon atoms. When more than four carbons are present, the Greek or Latin number prefixes are used: *pent-*, *hex-*, *hept-*, *oct-*, *non-*, and *dec-*. The second part of the name, or the suffix, tells the class of hydrocarbon. Alkanes have carbon–carbon single bonds, alkenes have carbon–carbon double bonds, and alkynes have carbon–carbon triple bonds and are indicated by the suffixes *-ane*, *-ene*, and *-yne*, respectively.

Unbranched Alkanes and Alkyl Groups

The names of the first eight unbranched (straight-chain) alkanes are given in Table 9.3. Alkyl groups are named by dropping "-ane" from the parent alkane and adding "-yl." See Table 9.4.

Branched-Chain Alkanes

The rules for naming branched-chain alkanes are as follows:

1. *Find the longest continuous chain of carbon atoms: this chain determines the parent name for the compound.* For example, the following compound has two methyl groups attached to a *heptane* parent.

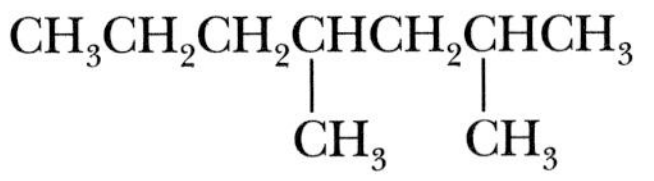

$$\underset{}{CH_3CH_2CH_2}\underset{|\atop CH_3}{C}HCH_2\underset{|\atop CH_3}{C}HCH_3$$

The longest continuous chain may not be obvious from the way the formula is written, especially for the straight-line format that is commonly used. For example, the longest continuous chain of carbon atoms in the following chain is *eight,* not *four* or *six.*

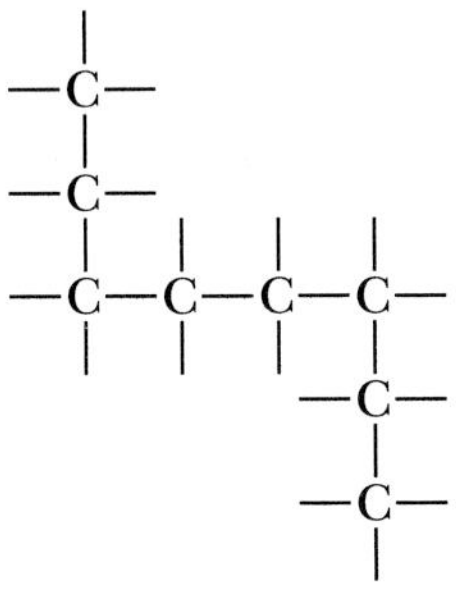

```
  |
 —C—
  |
 —C—
  |   |   |   |
 —C—C—C—C—        is equivalent to    —C   C   C   C
  |   |   |   |                         |\ /|\ /|\ /|\ |
           —C—                           C   C   C   C—
            |                            |   |   |   |
           —C—
            |
```

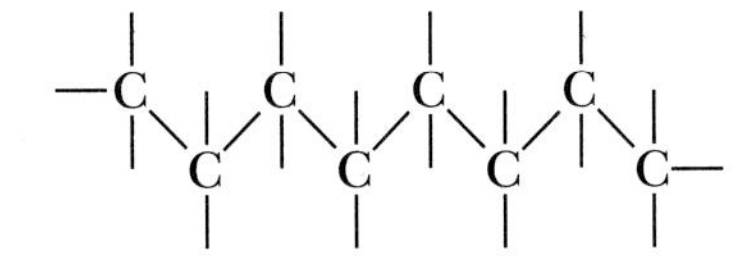

2. *Number the longest chain beginning with the end of the chain nearest the branching. Use these numbers to designate the location of the attached group. When two or more groups are attached to the parent, give each group a number corresponding to its location on the parent chain.* For example, the name of

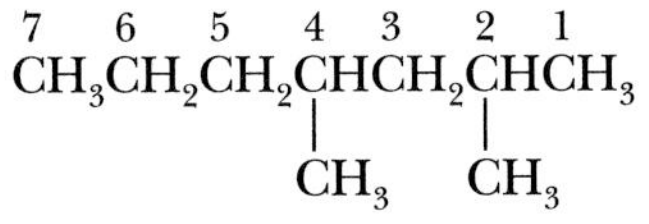

$$\overset{7}{C}H_3\overset{6}{C}H_2\overset{5}{C}H_2\underset{|\atop CH_3}{\overset{4}{C}}H\overset{3}{C}H_2\underset{|\atop CH_3}{\overset{2}{C}}H\overset{1}{C}H_3$$

is 2,4-dimethylheptane. The name of the following compound is 3-methylheptane, not 5-methylheptane or 2-ethylhexane.

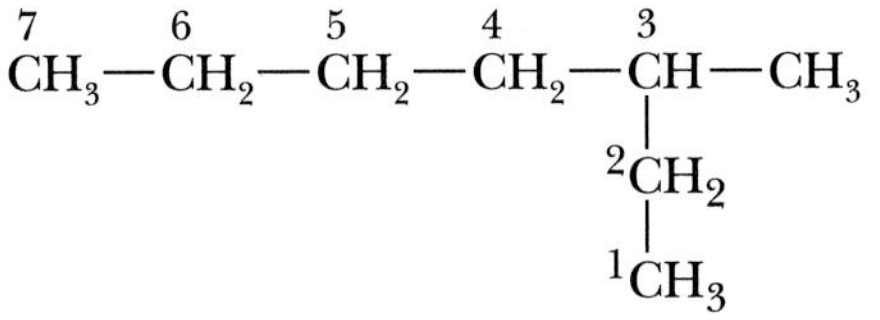

$$\overset{7}{C}H_3-\overset{6}{C}H_2-\overset{5}{C}H_2-\overset{4}{C}H_2-\underset{|\atop {}^{2}CH_2 \atop | \atop {}^{1}CH_3}{\overset{3}{C}}H-CH_3$$

3-Methylheptane

3. *When two or more substituents are identical, indicate this by the use of the prefixes di-, tri-, tetra-, and so on. Positional numbers of the substituents should have the smallest possible sum.*

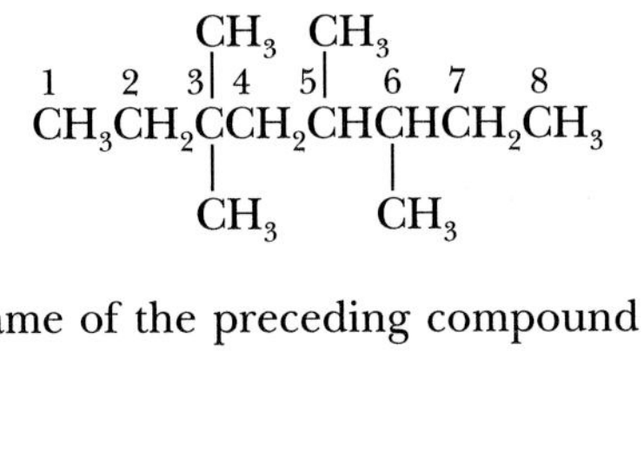

$$\overset{1}{C}H_3\overset{2}{C}H_2\underset{|\atop CH_3}{\overset{CH_3 \atop |\,3}{C}}\overset{4}{C}H_2\overset{CH_3\atop |\,5}{C}H\underset{|\atop CH_3}{\overset{6}{C}}H\overset{7}{C}H_2\overset{8}{C}H_3$$

The correct name of the preceding compound is 3,3,5,6-tetramethyloctane.

4. *If there are two or more different groups, the groups are listed alphabetically.*

$$\begin{array}{ccccccc} & & CH_3 & & & & \\ & 1 & 2\,| & 3 & 4 & 5 & 6 \\ & CH_3 & C & CH_2 & CH & CH_2 & CH_3 \\ & & | & & | & & \\ & & CH_3 & & CH_2 & & \\ & & & & | & & \\ & & & & CH_3 & & \end{array}$$

The correct name of the preceding compound is 4-ethyl-2,2-dimethylhexane. Note that the prefix "di" is ignored in determining alphabetical order.

Alkenes

Alkenes are named by using the prefix to indicate the number of carbon atoms and the suffix *-ene* to indicate one or more double bonds. The systematic names for the first two members of the alkene series are *ethene* and *propene*.

$$CH_2{=}CH_2 \qquad CH_3CH{=}CH_2$$

When groups, such as methyl or ethyl, are attached to carbon atoms in an alkene, the longest hydrocarbon chain is numbered from the end that will give the double bond the lowest number and then numbers are assigned to the attached group. For example, the name of

$$\begin{array}{cccccc} & CH_3 & & & & \\ 5 & 4\,| & 3 & & 2 & 1 \\ CH_3 & CH & CH & = & CH & CH_3 \end{array}$$

is 4-methyl-2-pentene. See Section 9.3 for a discussion of *cis–trans* isomers of alkenes.

Alkynes

The naming of alkynes is similar to that of alkenes, with the lowest number possible being used to locate the triple bond. For example, the name of

$$\begin{array}{cccccc} & & & & CH_3 & \\ 1 & 2 & & 3 & 4\,| & 5 \\ CH_3 & C & \equiv & C & CH & CH_3 \end{array}$$

is 4-methyl-2-pentyne.

Benzene Derivatives

Monosubstituted benzene derivatives are named by using a prefix for the substituent. Some examples are

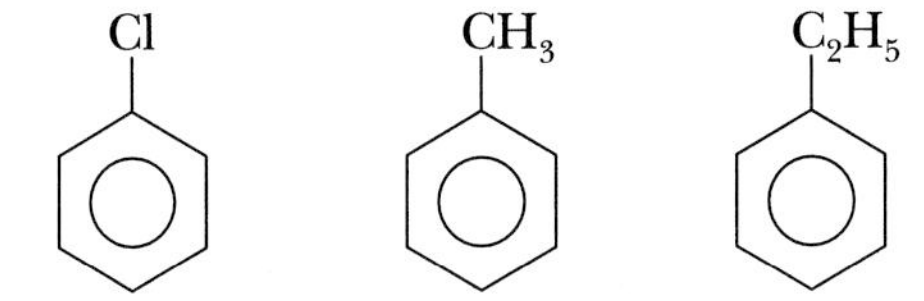

Cl CH$_3$ C$_2$H$_5$

Chlorobenzene Methylbenzene (toluene) Ethylbenzene

Three isomers are possible when two groups are substituted for hydrogen atoms on the benzene ring. The relative positions of the substituents are indicated by either the prefixes *ortho-*, *meta-*, *para-* (abbreviated *o-*, *m-*, *p-*), or by numbers. For example,

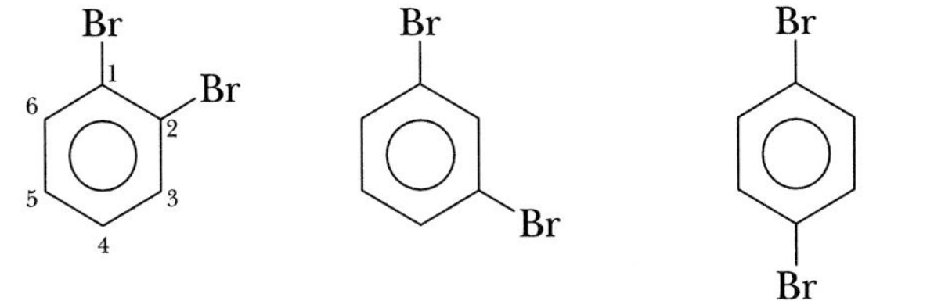

Br Br Br Br Br Br

1,2-Dibromobenzene (*o*-dibromobenzene) 1,3-Dibromobenzene (*m*-dibromobenzene) 1,4-Dibromobenzene (*p*-dibromobenzene)

The dimethylbenzenes are called *xylenes* (*o*-xylene, *m*-xylene, *p*-xylene).

If more than two groups are attached to the benzene ring, numbers must be used to identify the positions. The benzene ring is numbered to give the lowest possible numbers to the substituents.

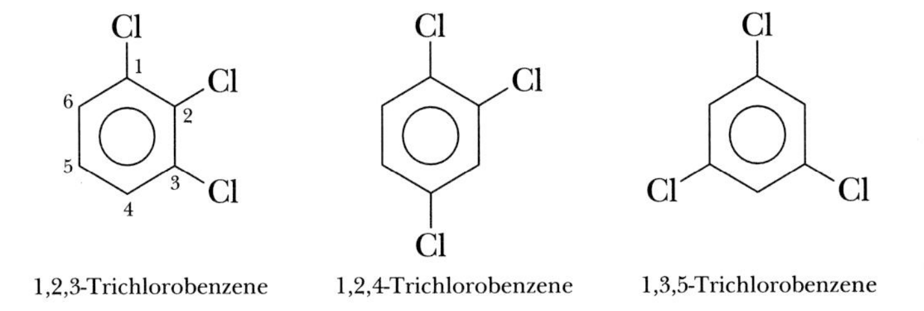

1,2,3-Trichlorobenzene 1,2,4-Trichlorobenzene 1,3,5-Trichlorobenzene

Functional Groups

An atom or group of atoms that defines the structure of a specific class of organic compounds and determines their properties is called a **functional group**. The millions of organic compounds include classes of compounds that are obtained by replacing hydrogen atoms of hydrocarbons with functional groups (Chapter 10). The important functional groups are shown in Table D.1.

The "R" attached to the functional group represents the nonfunctional, hydrocarbon framework with one hydrogen removed for each functional group added. The IUPAC system provides a systematic method for naming all members of a given class. For example, alcohols end in *-ol* (methan*ol*); aldehydes end in *-al* (methan*al*); carboxylic acids end in *-oic* (ethan*oic* acid); and ketones end in *-one* (propan*one*).

TABLE D.1 ■ Classes of Organic Compounds Based on Functional Groups*

General Formulas of Class Members	Class Name	Typical Compound	Compound Name	Common Use of Sample Compound
R—X	Halide	CH_2Cl_2 (H—C(H)(Cl)—Cl)	Dichloromethane (methylene chloride)	Solvent
R—OH	Alcohol	CH_3—OH	Methanol (wood alcohol)	Solvent
R—C(=O)—H	Aldehyde†	H—C(=O)—H	Methanal (formaldehyde)	Preservative
R—C(=O)—OH	Carboxylic acid†	CH_3—C(=O)—OH	Ethanoic acid (acetic acid)	Vinegar
R—C(=O)—R′	Ketone	CH_3—C(=O)—CH_3	Propanone (acetone)	Solvent
R—O—R′	Ether	C_2H_5—O—C_2H_5	Diethyl ether (ethyl ether)	Anesthetic
R—C(=O)—O—R′	Ester	CH_3—C(=O)—O—C_2H_5	Ethyl ethanoate (ethyl acetate)	Solvent in fingernail polish
R—N(H)—H	Amine‡	CH_3—N(H)—H	Methylamine	Tanning (foul odor)
R—C(=O)—N(H)—H	Amide‡	CH_3—C(=O)—N(H)—H	Acetamide	Plasticizer

*R stands for a hydrocarbon group such as —CH_3 or —C_2H_5. R′ could be a different group from R.
†R can be an H atom or a hydrocarbon group.
‡The H atoms in amines and amides can also be replaced by R groups.

Alcohols

Isomers are also possible for molecules containing functional groups. For example, three different alcohols are obtained when a hydrogen atom in pentane is replaced by —OH, depending on which hydrogen atom is replaced. The rules for naming the "R" or hydrocarbon framework are the same as those for hydrocarbon compounds.

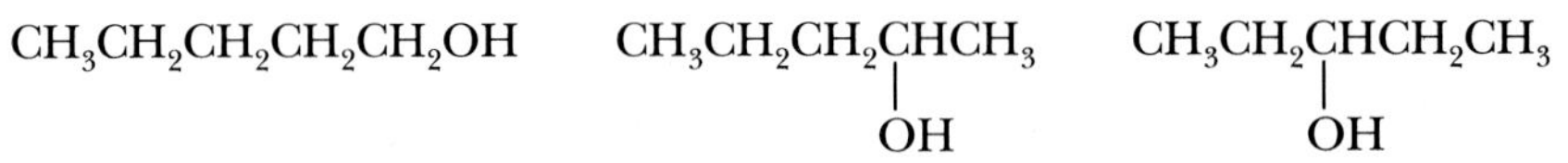

$$CH_3CH_2CH_2CH_2CH_2OH$$

1-Pentanol

$$\begin{array}{l} CH_3CH_2CH_2\underset{\textstyle OH}{\underset{|}{C}}HCH_3 \end{array}$$

2-Pentanol

$$\begin{array}{l} CH_3CH_2\underset{\textstyle OH}{\underset{|}{C}}HCH_2CH_3 \end{array}$$

3-Pentanol

Compounds with one or more functional groups and alkyl substituents are named so as to give the functional groups the lowest number. For example, the correct name of

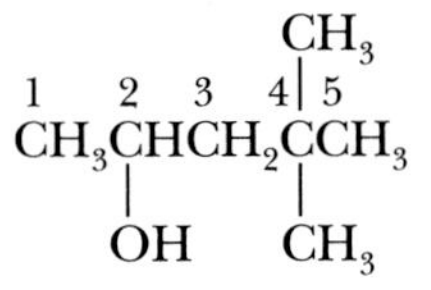

$$\begin{array}{l} \quad\quad\quad\quad\quad\; CH_3 \\ 1 \quad 2 \quad 3 \quad 4| \;5 \\ CH_3\underset{\textstyle OH}{\underset{|}{C}}HCH_2\underset{\textstyle CH_3}{\underset{|}{C}}CH_3 \end{array}$$

is 4,4-dimethyl-2-pentanol.

Aldehydes and Ketones

The systematic names of the first three aldehydes are methanal, ethanal, and propanal.

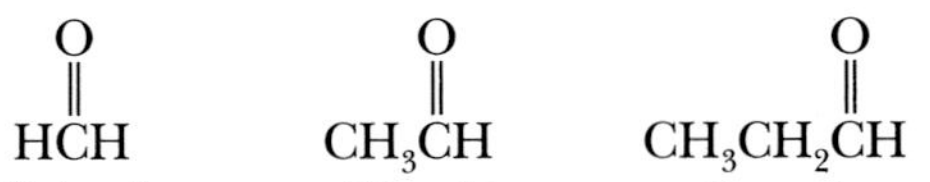

$$H\overset{\textstyle O}{\overset{\|}{C}}H$$

Methanal
(formaldehyde)

$$CH_3\overset{\textstyle O}{\overset{\|}{C}}H$$

Ethanal
(acetaldehyde)

$$CH_3CH_2\overset{\textstyle O}{\overset{\|}{C}}H$$

Propanal
(propionaldehyde)

For ketones, a number is used to designate the position of the carbonyl group, and the chain is numbered in a way that gives the carbonyl carbon the smallest number.

$$CH_3\overset{\textstyle O}{\overset{\|}{C}}CH_3$$

2-Propanone
(acetone)

$$CH_3CH_2\overset{\textstyle O}{\overset{\|}{C}}CH_3$$

2-Butanone
(methyl ethyl ketone)

$$CH_3\overset{\textstyle O}{\overset{\|}{C}}CH_2CH{=}CH_2$$

4-Penten-2-one

Carboxylic Acids

The systematic names of carboxylic acids are obtained by dropping the final *e* of the name of the corresponding alkane and adding *oic acid.* For example, the name of

$$CH_3CH_2CH_2CH_2CH_2COOH$$

is hexanoic acid. The systematic names of the first five carboxylic acids are given in Table 10.6. Other examples are

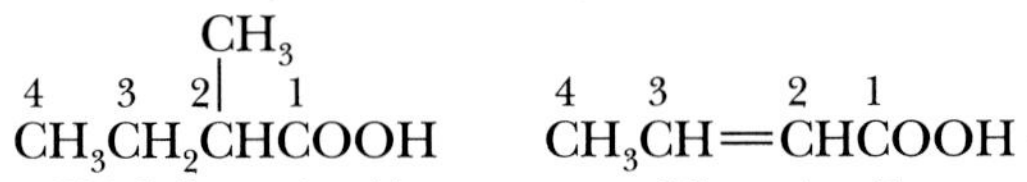

$$\begin{array}{l} \quad\quad\quad\quad\; CH_3 \\ 4 \quad 3 \quad 2| \quad 1 \\ CH_3CH_2CHCOOH \end{array}$$

2-Methylbutanoic acid

$$\begin{array}{l} 4 \quad\; 3 \quad\quad\; 2 \quad 1 \\ CH_3CH{=}CHCOOH \end{array}$$

2-Butenoic acid

Esters

The systematic names of esters are derived from the names of the alcohol and the acid used to prepare the ester. The general formula for esters is

$$R-\overset{\displaystyle O}{\overset{\|}{C}}-OR'$$

The $R-\overset{\displaystyle O}{\overset{\|}{C}}$ comes from the acid and the R′O comes from the alcohol. The alcohol part is named first, followed by the name of the acid changed to end in *-ate.* For example,

$$CH_3CH_2\overset{\displaystyle O}{\overset{\|}{C}}-OCH_3$$

is named methyl propanoate, and

$$CH_3\overset{\displaystyle O}{\overset{\|}{C}}-OCH=CH_2$$

is named ethenyl ethanoate.

APPENDIX

E

Answers to Self-Tests and Matching Sets

CHAPTER 2

SELF-TEST 2A

1. (a), (b)
2. an element; cannot be broken down
3. elements and chemical compounds
4. homogeneous
5. (a) True
6. chemical change; changed in identity
7. (a), (b), (c), (d), (e)
8. (a) sugar and water
9. (a) methane and oxygen (b) hydrogen peroxide
10. (a) chemical property (b) physical property (c) chemical property
11. (a) True

SELF-TEST 2B

1. structure of matter
2. solids: carbon, sulfur, phosphorus, iodine, all metals except mercury; liquids; mercury and bromine; gases: hydrogen, oxygen, nitrogen, fluorine, chlorine (also neon, argon, krypton, xenon, radon)
3. C, S, P, I, metal symbols in Table 2.2 and inside front cover, Hg, Br; H, O, N, F, Cl, Ne, Ar, Kr, Xe, Rn
4. Elements: potassium, hydrogen, phosphorus, oxygen; 8 atoms
5. (a) no (b) yes (c) no (d) yes
6. (a) oxygen atom (in a compound)
   (b) two molecules of oxygen
   (c) one molecule of methane (d) yields
   (e) one water molecule (f) two water molecules
7. nonmetal
8. (a) and (d)

SELF-TEST 2C

1. a number, a unit
2. prefixes, multiples of ten
3. cubic centimeter
4. kilo-, milli-
5. mm
6. (b)
7. (a) quantitative (b) qualitative

MATCHING SET

1. a	7. k	12. l
2. d (or o)	8. f	13. p
3. c	9. g	14. j
4. b	10. i	15. d (or o)
5. e	11. m	16. n
6. h		

CHAPTER 3

SELF-TEST 3A

1. (b) by early Greek philosophers
2. (b) philosophy (use of logic)
3. gained, chemical
4. 50%
5. 5.88%
6. (a) the same (b) atoms
7. (a) Yes. Although the total mass of Ag_2S may change, the percent mass of Ag in Ag_2S remains constant because the elements in a chemical compound are always present in a definite proportion by mass.
8. (a) Yes. Pure methane will have the same composition regardless of source because the percent mass of hydrogen and carbon in CH_4 is constant.

9. (b) +1
10. (b) electron
11. (a) True
12. (a) nucleus
13. 1800

SELF-TEST 3B
1. 10 protons, 10 electrons, 11 neutrons
2. atomic number, mass number
3. protons, neutrons
4. electrons, protons, neutrons
5. protons and electrons
6. atomic
7. 33 electrons, 33 protons, 42 neutrons
8. 100,000
9. ${}^{1}_{1}H$ (P) Nucleus ${}^{2}_{1}H$ (P, n) Nucleus ${}^{3}_{1}H$ (P, n, n) Nucleus
10. (a) True
11. (b) False

SELF-TEST 3C
1. (d) Infrared light
2. (a) Blue
3. farther from, closer to
4. 18
5. 2-8-8-1, 1 valence electron
6. 2-8-7, 7 valence electrons

SELF-TEST 3D
1. (a) 1, (b) 2, (c) 3, (d) 3, (e) 8, (f) 7
2. (a) True
3. (b) False
4. (a) True
5. (a) True
6. (b) and (d)
7. $SrCl_2$
8. Cs
9. (a) oxygen (b) calcium (c) fluorine (d) nitrogen (e) sulfur (f) neon
10. (a) True
11. (b) False
12. (a) cesium (b) fluorine (c) calcium (d) oxygen (e) sodium (f) xenon

MATCHING SET I

1. d	5. c	9. j
2. a	6. h	10. e
3. i	7. b	11. g
4. k	8. f	

MATCHING SET II

1. d	5. c	9. f
2. e	6. j	10. b
3. a	7. i	
4. g	8. h	

CHAPTER 4

SELF-TEST 4A
1. (a) True
2. (c) Gamma
3. (b) False
4. (b) False
5. ${}^{212}_{82}Pb$
6. (a) True
7. 1 μCi, or 37,000 dps

SELF-TEST 4B
1. ${}^{239}_{93}Np$
2. Neptunium
3. (a) True
4. (a) True
5. Radium
6. 4 pCi/L

SELF-TEST 4C
1. Food irradiation and medical imaging
2. Imaging
3. (c) metastable
4. (a) 1/8 of the original dose—(1/2)(1/2)(1/2) or $(1/2)^3$
5. The 6h half-life isotope, because it would decay more quickly producing less exposure.

SELF-TEST 4D
1. (b) Uranium-235
2. ${}^{236}_{92}U$
3. ${}^{134}_{52}Te$
4. smaller
5. (a) Iron
6. (b) 23%
7. (a) True
8. (b) 24,000 years

MATCHING SET

1. l	7. c	12. g
2. h	8. p	13. j
3. k	9. d	14. a
4. i	10. e	15. m
5. n	11. f	16. b
6. o		

CHAPTER 5

SELF-TEST 5A
1. ionic
2. (a) losing electrons
3. (a) +1, Li^+
   (b) +3, Al^{3+}
   (c) −2, S^{2-}
   (d) −1, Br^-

4. (a) potassium, K^+
   (d) sodium, Na^+
5. (b) gaining electrons
6. NaCl, sodium chloride

SELF-TEST 5B

1. Sulfur dioxide
2. SO_3
3. Hydrogen bromide
4. Cl_2O
5. SCl_2
6. Silicon tetrachloride
7. (a) four electrons, (b) six electrons
8. (c) $C{\equiv}C$

SELF-TEST 5C

1. (a) hydrogen molecule, H_2 (b) hydrogen chloride molecule, HCl
2. 109.5°, the tetrahedral angle
3. (a) H—F
   (b) C—O
   (c) H—N
4. (a) H_2O
5. (d) CCl_4

SELF-TEST 5D

1. gas, liquid, solid
2. increases
3. liquid and solid
4. liquid and gas
5. Gases are compressible while solids and liquids are noncompressible.
6. Decreases
7. (a) True
8. hydrogen bonding

SELF-TEST 5E

1. (a) True
2. (a) True
3. solute, solvent
4. (b) corn oil
5. (b) False
6. increases
7. (a) True
8. melting
9. sublimation

MATCHING SET

1. b	7. a	13. p
2. i	8. f	14. n
3. e	9. k	15. m
4. d	10. q	16. l
5. h	11. j	17. c
6. g	12. o	

CHAPTER 6

SELF-TEST 6A

Note: When the coefficient is "1" in balanced equations, the "1" is not written. In questions that relate to balancing equations, coefficients of "1" will be placed in parentheses to provide answers for the blanks given in the question.

1. law of conservation of mass
2. mole, dozen
3. (a) (1) Si + 2 Cl_2
   (b) 4 Al + 3 $O_2 \rightarrow$ 2 Al_2O_3
   (c) (1) $(NH_4)_2CO_3$ + (1) $Cu(NO_3)_2 \rightarrow$ (1) $CuCO_3$ + 2 NH_4NO_3
4. atomic weight
5. (a) 6 CO_2 + 6 $H_2O \rightarrow$ (1) $C_6H_{12}O_6$ + 6 O_2
   (b) 6 molecules (c) 6 mol (d) 180 g (e) 264 g
6. (c)

SELF-TEST 6B

1. reactant converted to product
2. energy
3. concentration, temperature
4. catalyst
5. activation energy
6. no
7. no
8. yes
9. Le Chatelier's principle

SELF-TEST 6C

1. released
2. absorbed
3. (a) True
4. energy
5. (b)
6. decreases
7. energy, entropy
8. the first law of thermodynamics
9. the second law of thermodynamics
10. using energy and increasing entropy

MATCHING SET

1. h	5. b	9. g
2. d	6. c	10. j
3. a	7. e	11. i
4. k	8. f	

CHAPTER 7

SELF-TEST 7A

1. donor, acceptor
2. acid
3. a basic substance
4. $H_3O^+(aq) + OH^-(aq) \rightarrow 2\,H_2O(\ell)$
5. (a) salt, (b) acid, (c) salt, (d) base

6. weak base, establishes an equilibrium, OH^- ions are formed.
7. hydronium ion, H_3O^+
8. sulfuric acid, H_2SO_4
9. $HCl(aq) + KOH(aq) \rightarrow KCl(aq) + H_2O(\ell)$
10. weak acids

SELF-TEST 7B

1. pH of 2
2. (b) low hydronium ion concentration
3. (a) high hydronium ion concentration
4. H_2CO_3 neutralizes added base, HCO_3^- neutralizes added acid
5. 3
6. 11
7. 1.0×10^{-11} M, 1.0×10^{-3} M
8. 1.0×10^{-4} M
9. 1×10^{-2} M
10. acidic, basic
11. (a) sodium bicarbonate (c) $MgCO_3$

MATCHING SET

1. b	5. h	8. g
2. a	6. c	9. d
3. j	7. e	10. f
4. i		

CHAPTER 8

SELF-TEST 8A

1. (b)
2. reducing agent
3. (a)
4. (b)
5. (b)
6. oxidizing agent
7. (a)
8. (a)
9. (b)
10. reducing agent
11. reducing agent
12. oxidizing agent

SELF-TEST 8B

1. anode, cathode
2. all the reactants are used up
3. electrons to flow, charge balance is maintained.
4. oxidation, anode
5. reduction, cathode
6. electrical energy is converted to chemical energy
7. electrolysis, aluminum
8. Because the rust falls away from the surface to expose a fresh surface for further corrosion.
9. oxygen, water, iron

MATCHING SET

1. a	5. g	9. k
2. j	6. d	10. b
3. i	7. e	11. h
4. c	8. f	

CHAPTER 9

SELF-TEST 9A

1. (c) natural gas
2. (b) hydrogen
3. (a) True
4. oxygen, water
5. tetrahedral
6. ethene or ethylene
7. ethyne or acetylene
8. $—C_2H_5$
9. structural
10. ethylene
11. *cis* and *trans*

SELF-TEST 9B

1. hydrogen
2. 12
3. CO and H_2
4. 3
5. 3
6. (a) ethanol
7. R—O—R′

SELF-TEST 9C

1. fractional distillation
2. methane, CH_4
3. methyl-*tertiary*-butyl ether
4. catalytic re-forming
5. catalytic cracking
6. (c)

MATCHING SET

1. d	5. j	8. b
2. f	6. a	9. e
3. i	7. c	10. g
4. h		

CHAPTER 10

SELF-TEST 10A

1. $CH_3CH_2CH_2—$
2. aromatic hydrocarbons
3. acetaldehyde
4. rubbing alcohol
5. ethylene glycol
6. 42%

7. denatured
8. 3
9. glycerol
10. CH_3OH

SELF-TEST 10B
1. formaldehyde
2. acetic acid
3. ethanol with acetic acid
4. methyl acetate
5. 3
6. acetone
7. (b) esters

SELF-TEST 10C
1. monomers
2. free radicals, organic peroxides
3. double
4. single
5. (a) True
6. CH_2CHCl
7. *cis*
8. (a) low-density polyethylene (b) high-density polyethylene (c) cross-linked polyethylene

SELF-TEST 10D
1. adipic acid, hexamethylenediamine, nylon 66, water
2. condensation
3. water
4. (b) condensation reactions
5. (a) True
6. glass fibers
7. polymer

MATCHING SET

1. e	5. l	9. c
2. d	6. h	10. i
3. f	7. j	11. b
4. g	8. k	12. a

CHAPTER 11

SELF-TEST 11A
1. 4
2. monosaccharides
3. glucose and fructose
4. D-glucose
5. Hydrogen
6. D-glucose
7. glycerol and fatty acids
8. double bonds. A saturated fat contains only single carbon-carbon bonds while an unsaturated fat contains one or more carbon-carbon double bonds and hence fewer hydrogen atoms than a saturated fat.
9. (a) steroid
10. (a) True
11. long-chain fatty acids and long-chain alcohols

SELF-TEST 11B
1. aqueous base
2. hydrophobic, hydrophilic
3. (a) traditional soaps
4. emollient
5. emulsion
6. shaving cream
7. smoke
8. fog
9. emulsion

SELF-TEST 11C
1. glycine
2. (a) True
3. essential
4.
```
  O  H
  ‖  |
—C—N—
```
5. (a) 27 (b) 6
6. (a) amino acid sequence
   (b) hydrogen bonded structures to form helices or sheets
   (c) how the protein molecule is folded
   (d) shape assumed by all chains in a protein with two or more chains
7. hydrogen
8. active site
9. (a) catalyst
10. enzyme
11. hydrogen, ionic
12. hydrogen

SELF-TEST 11D
1. CO_2 and H_2O, energy
2. ATP
3. ADP, energy
4. DNA
5. ribose, deoxyribose
6. phosphoric acid group, a sugar (ribose or deoxyribose), a nitrogen heterocyclic base (adenine, guanine, thymine, cytosine, or uracil)
7. double helix
8. hydrogen
9. T, C, A, G
10. three billion

MATCHING SET

1. d	5. g	9. f
2. c	6. i	10. h
3. a	7. j	11. k
4. l	8. e	12. b

CHAPTER 12

SELF-TEST 12A

1. simple sugars, amino acids, glycerol and fatty acids, energy
2. basal metabolic rate, weight in pounds
3. 1 kilocalorie
4. carbohydrates, fats, proteins
5. (a) True
6. (b) glucose
7. ATP

SELF-TEST 12B

1. dietary fiber
2. triglycerides
3. high-density lipoproteins (HDLs), transport cholesterol to liver for excretion
4. low-density lipoproteins (LDLs), transport cholesterol to arteries
5. nitrogen
6. fats
7. fruits and vegetables
8. 30%

SELF-TEST 12C

1. (b) False
2. fat-soluble, water-soluble
3. vitamin A
4. antioxidants
5. enzymes
6. enriched
7. ions
8. electrolyte balance, acid-base balance (also fluid balance), nerve
9. anemia, osteoporosis

SELF-TEST 12D

1. drying, salt or sugar
2. Generally Recognized as Safe
3. (b) False
4. antioxidants
5. EDTA
6. (b) False
7. polysaccharides
8. 100 kcal divided by 230 kcal
9. 0.30×2000 kcal

MATCHING SET

1. d	7. o	13. l
2. i	8. k	14. c
3. n	9. p	15. j
4. g	10. h	16. e
5. a	11. m	
6. b	12. f	

CHAPTER 13

SELF-TEST 13A

1. generic, trade name
2. pneumonia, influenza, HIV infection
3. microorganisms
4. S (sulfur) and N (nitrogen), four
5. cross-linking
6. an $—NH_2$ group and an —OH group
7. inside, outside
8. retrovirus
9. AIDS, DNA
10. chemotherapy
11. protease inhibitors, enzymes

SELF-TEST 13B

1. hormones and neurotransmitters
2. receptors
3. proteins or steroids
4. (a) and (d)
5. dopamine
6. epinephrine
7. glands
8. increasing

SELF-TEST 13C

1. heroin
2. does
3. (a) ii (b) i (c) iii
4. acetylsalicylic acid
5. carboxylic acid, anti-inflammatory
6. addictive (an abused drug), medical
7. depressants
8. cocaine, addictive
9. (c) allergic reaction
10. (b) False
11. decongestants, antitussives, expectorants, analgesics, antihistamines
12. ultraviolet (uv) light
13. sun protection factor
14. (a) iii (b) iii (c) i (d) ii
15. dose

SELF-TEST 13D

1. cardiovascular disease
2. oxygen
3. (a) faster
4. enzymes
5. (a), (b), (c)
6. (a) True
7. DNA
8. kill

MATCHING SET

1. f
2. e
3. c
4. m
5. b
6. h
7. o
8. k
9. l
10. d
11. g
12. a
13. p
14. j
15. i
16. n

CHAPTER 14

SELF-TEST 14A

1. Oxygen
2. aluminum
3. (a) chlorine, (i) hydrosphere
   (b) sodium chloride, (i) hydrosphere
   (c) magnesium, (i) hydrosphere
   (d) sand, (ii) earth's crust
   (e) marble, (ii) earth's crust
4. magnesium
5. reduced

SELF-TEST 14B

1. iron
2. carbon
3. carbon in the form called coke
4. (a) True
5. sulfur
6. (a) True

SELF-TEST 14C

1. n- and p-
2. superconductor
3. increases
4. n-type
5. p-type
6. (b) transistors
7. the one whose resistance dropped to zero at the boiling point of nitrogen

SELF-TEST 14D

1. oxygen and silicon
2. (a) SiO_2, (i) silica
   (b) quartz, (i) silica
   (c) contain metal ions, (ii) silicates
   (d) clay, (ii) silicates
   (e) built up of tetrahedral units, both (i) and (ii)
3. Silica is pure SiO_2 while glass is a mixture containing metal ions, oxide ions, and SiO_2
4. clay, sand, and aluminosilicates
5. temperatures, shock
6. cement, sand, gravel

MATCHING SET

1. d
2. j
3. n
4. h
5. g
6. a
7. e
8. o
9. i
10. m
11. b
12. c
13. k
14. f
15. l

CHAPTER 15

SELF-TEST 15A

1. 70%
2. the oceans
3. groundwater
4. surface water
5. 168 gal
6. 2 gal
7. pathogens, toxic metals, organic compounds
8. aquifer
9. it is returned to the sea or it evaporates

SELF-TEST 15B

1. Superfund
2. (b) False
3.

Household waste	Harmful Chemical
automobile battery	sulfuric acid and lead
oil-based paint	organic solvents
batteries	lead, mercury, cadmium

4. plastics, unused paint, used motor oil, used auto batteries
5. biochemical oxygen demand, BOD
6. The manufacture of auto batteries can introduce lead into the groundwater. The manufacture of fluorescent lamps can introduce mercury into the groundwater.
7. The manufacture of plastics can introduce chlorinated organics into the groundwater. The manufacture of textiles can introduce chlorinated organics into the groundwater.
8. Mercury and lead

SELF-TEST 15C

1. (d) reverse osmosis
2. (b) persistent
3. Aeration and filtration
4. aerobic
5. Anaerobic
6. (b) calcium, (c) magnesium
7. chlorination and ozonation
8. Na^+, Mg^{2+}, Ca^{2+}, K^+
9. Osmosis

MATCHING SET

1. l	7. k	13. g
2. o	8. q	14. n
3. p	9. d	15. i
4. a	10. m	16. h
5. c	11. j	17. b
6. e	12. f	

CHAPTER 16

SELF-TEST 16A

1. (b) decrease
2. ozone
3. adsorb
4. absorb
5. warm, cool
6. (a) coal burning
7. oxides
8. nitric oxide and free oxygen atoms
9. free oxygen atoms

SELF-TEST 16B

1. NO_x and SO_2
2. SO_3, H_2SO_4
3. (b) CaO, a basic oxide
4. H_2SO_4 and HNO_3
5. 5.6, CO_2
6. 1872
7. The C—Cl bond
8. $O_2 + h\nu \rightarrow 2\ O$
   $O + O_2 \rightarrow O_3$
9. ClO (chlorine oxide free radical)
10. Antarctica
11. Ultraviolet light

SELF-TEST 16C

1. respiration, fossil fuel burning for transportation, fossil fuel burning in electricity generation
2. photosynthesis, dissolving in the world's oceans
3. carbon dioxide, nitrogen dioxide, and water
4. carbon dioxide
5. (a) True
6. 370 ppm
7. Louisiana
8. methanol
9. Cigarette smoking

MATCHING SET

1. c	9. i	16. m
2. e	10. u	17. b
3. o	11. r	18. f
4. k	12. l	19. j
5. h	13. q	20. d
6. a	14. v	21. g
7. n	15. s	22. p
8. t		

CHAPTER 17

SELF-TEST 17A

1. 5.9 million, 80 million
2. clays, silts, sandy soils, loams
3. (a) acidic, (b) basic
4. particle size and chemical composition
5. a trivalent ion like Fe^{3+} (hydrolyzes more than Na^+)
6. humus
7. nitrogen, phosphorus, potassium
8. calcium, magnesium, sulfur
9. oxygen

SELF-TEST 17B

1. (b) False
2. nitrogen, phosphorus (calculated as P_2O_5), potassium (calculated as K_2O)
3. Yes
4. (c) a gas
5. about 33%
6. DDT
7. (a) DDT
8. Pyrethrum
9. Selective herbicide
10. 2,4-D
11. insecticides, herbicides, fungicides

MATCHING SET

1. f	6. b	11. e
2. c	7. k	12. g
3. h	8. d	13. l
4. a	9. m	
5. j	10. i	

APPENDIX F

Answers to Exercises

CHAPTER 2

2.1. (a) Pb (b) P (c) HCl (d) $AlBr_3$ (e) F_2

2.2. (a) Hydrogen gas reacts with chlorine gas to form hydrogen chloride gas.
(b) *Reactants:* 2 H atoms in one H_2 molecule, 2 Cl atoms in one Cl_2 molecule; *products:* 2 HCl molecules, which contain 2 H atoms and 2 Cl atoms

2.3. *mega* = 1 million, 20 megabucks = 20 million dollars ($20,000,000)

CHAPTER 3

3.1. 28 protons, 28 electrons, and 31 neutrons

3.2. $^{107}_{47}Ag$ $^{109}_{47}Ag$

3.3. (a) Group IIA is the only main group that has all metals. Group IA is often regarded as having only metals since hydrogen is a nonmetal. Figure 3.9 shows hydrogen separated from the rest of the members of Group IA for this reason.
(b) Groups VIIA and VIIIA have only nonmetals.
(c) Groups IIIA, IVA, VA, and VIA include metalloids, nonmetals, and metals. The number of valence electrons are: (a) Group IA, 1; Group IIA, 2. (b) Group VIIA, 7; Group VIIIA, 8. (c) Group IIIA, 3; Group IVA, 4; Group VA, 5; Group VIA, 6.

3.4A. (a) Ba (b) Se (c) Si (d) Ga

3.4B. (a) Sr (b) Cl (c) Cs

CHAPTER 4

4.1. $^{237}_{93}Np \rightarrow {}^{4}_{2}He + {}^{233}_{91}Pa$

4.2. $^{234}_{91}Pa \rightarrow {}^{0}_{-1}e + {}^{234}_{92}U$

CHAPTER 5

5.1. CaF_2

5.2. (a) Rubidium chloride (b) Gallium oxide (c) Calcium dibromide (d) Iron(II) nitride

5.3. (a) CoS (b) MgF_2 (c) KI

5.4. (a) $MgCO_3$ (b) NaH_2PO_4

5.5. Tetraphosphorus trisulfide

CHAPTER 6

6.1A. $2\,Al(s) + 3\,Cl_2(g) \rightarrow 2\,AlCl_3(s)$

6.1B. (a) yes (b) no

6.2. $50.\text{ mol } Ba(NO_3)_2 \times \dfrac{261\text{ g } Ba(NO_3)_2}{1\text{ mol } Ba(NO_3)_2} =$
$13{,}000\text{ g } Ba(NO_3)_2$

6.3.

$CO(g)$	+	$2\,H_2(g)$	$\rightarrow CH_3OH(\ell)$
1 mol CO		2 mol H_2	1 mol CH_3OH
28 g CO		2×2 g/mol = 4 g H_2	32 g CH_3OH

CHAPTER 7

7.1. $\dfrac{4.0\text{ mol NaOH}}{1\text{ L}} \times \dfrac{40.\text{ g NaOH}}{1\text{ mol NaOH}} = \dfrac{160\text{ g NaOH}}{1\text{ L}}$

7.2. For $[H_3O^+] = 1 \times 10^{-4}$ M, pH = 4. A pH of 4 is less acidic than a pH of 3; the tomatoes are less acidic than the cola.

CHAPTER 8

8.1. Li is oxidized (converted to Li^+); O_2 is reduced (converted to O^{2-})

8.2. (a) Cu is oxidized (addition of oxygen)
(b) $CH_3C{\equiv}N$ is reduced (addition of hydrogen)
(c) SnO is reduced (removal of oxygen)

CHAPTER 9

9.1. (a) 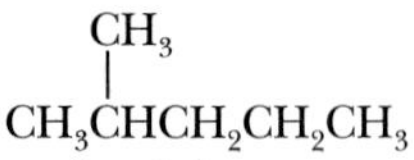 $CH_3CH(CH_3)CH_2CH_2CH_3$
2-methyl pentane

(b) 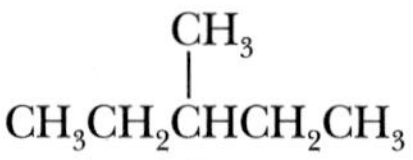 $CH_3CH_2CH(CH_3)CH_2CH_3$
3-methyl pentane

(c) 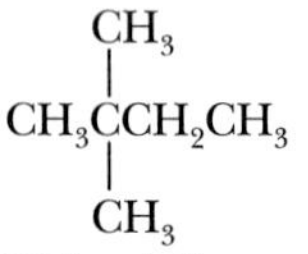 $CH_3C(CH_3)_2CH_2CH_3$
2,2-dimethylbutane

(d) 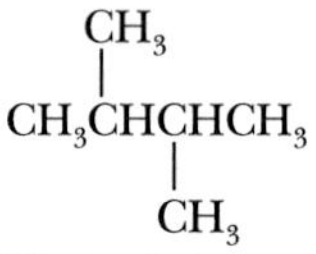 $CH_3CH(CH_3)CH(CH_3)CH_3$
2,3-dimethylbutane

9.2. $H{-}C(H){=}C(H){-}C(H)(CH_3){-}C(H)(H){-}H$
3-methyl-1-butene

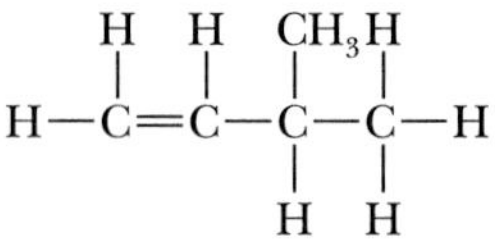

CHAPTER 10

10.1. (a) 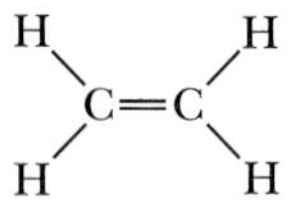 $H_2C{=}CH_2$ (H, H on each carbon)

(b) 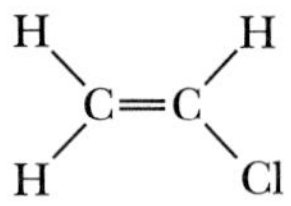 $H_2C{=}CHCl$

(c) 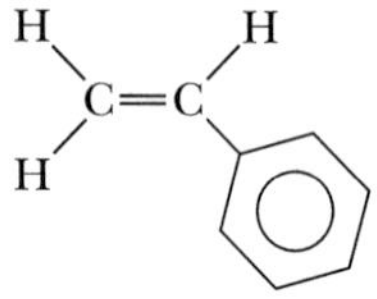 $H_2C{=}CH{-}C_6H_5$

10.2. $\left(\overset{O}{\overset{\|}{C}}{-}C_6H_4{-}\overset{O}{\overset{\|}{C}}{-}O{-}CH_2{-}CH_2{-}O \right)_n$

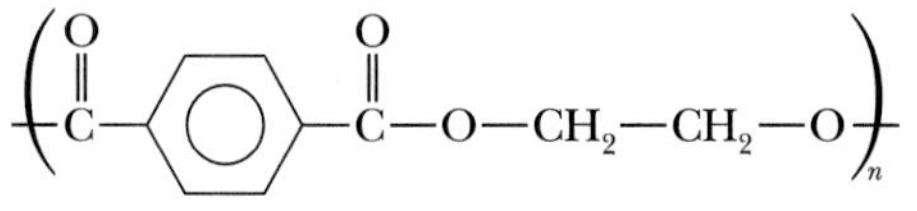

CHAPTER 11

11.1. (b) is chiral.

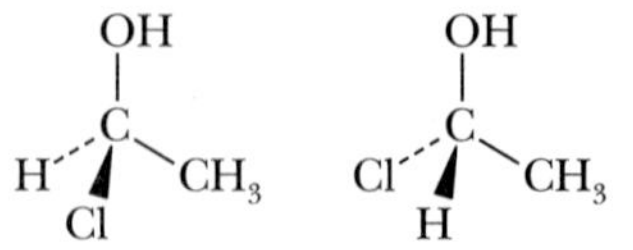

OH, H, CH_3, Cl on C (two mirror-image forms)

11.2.

$H_2N{-}CH(CH_2SH){-}\overset{O}{\overset{\|}{C}}{-}NH{-}CH(CH_2C_6H_5){-}\overset{O}{\overset{\|}{C}}{-}NH{-}CH(CH_2OH){-}\overset{O}{\overset{\|}{C}}{-}NH{-}CH(CH_3){-}\overset{O}{\overset{\|}{C}}{-}OH$

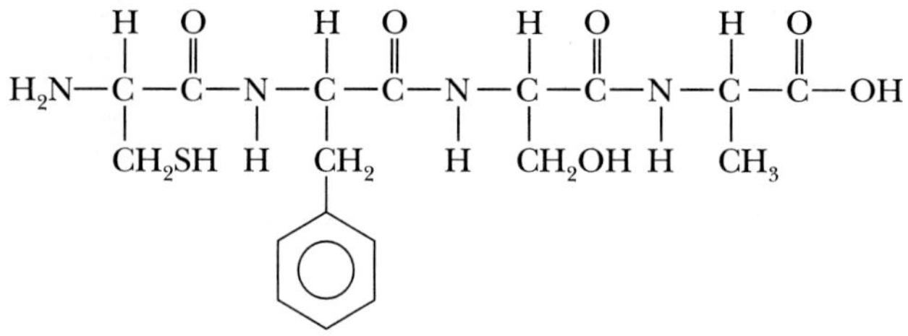

11.3. AGGCTA

CHAPTER 12

12.1. 160. lb × 10 kcal/lb = 1600 kcal
1600 kcal × 1.3 = 2080 kcal

12.2. Healthy Chicken:

$$15 \text{ cal fat} \times \frac{1 \text{ g}}{9 \text{ cal}} = 2 \text{ g fat}$$

$$\frac{2 \text{ g fat}}{65 \text{ g fat}} \times 100\% = 3\% \text{ of daily fat allowance}$$

Cream of Mushroom:

$$80 \text{ cal fat} \times \frac{1 \text{ g}}{9 \text{ cal}} = 9 \text{ g fat}$$

$$\frac{9 \text{ g fat}}{65 \text{ g fat}} \times 100\% = 14\% \text{ of daily fat allowance}$$

12.3.

$$3 \text{ g protein} \times \frac{4 \text{ kcal}}{\text{g protein}} = 12 \text{ kcal}$$

$$8 \text{ g fat} \times \frac{9 \text{ kcal}}{\text{g fat}} = 72 \text{ kcal}$$

$$40 \text{ g carbohydrate} \times \frac{4 \text{ kcal}}{\text{g carbohydrate}} = 160 \text{ kcal}$$

Total = 244 kcal per slice

CHAPTER 17

17.1. Formula weight: 2 N × 14 = 28, 4 H × 1 = 4,
3 O × 16 = 48. Total = 80.

$$\text{Percentage N} = \frac{28}{80} \times 100\% = 35\%$$

There is no K or P in ammonium nitrate, so their values are zero.

Appendix G

Answers to Selected Questions and Problems

CHAPTER 2

1. Answer will depend on each person's experience.
3. No, they would be the same substance.
5. A spark provides energy to ignite the air and fuel mixture. The oxygen in the air and the hydrocarbon fuel are chemically changed to heat , water vapor, and carbon dioxide (new substances).
7. a. physical property b. physical property
   c. chemical property d. physical property
   e. chemical property f. chemical property
9. a. element, contains only Hg atoms
   b. mixture of water, minerals, proteins, fats
   c. compound, contains only one kind of molecule (H_2O)
   d. mixture of cellulose, water. Wood changes weight when dried.
   e. mixture of dye and solvent
   f. mixture of water, caffeine, tea extract
   g. compound, solid pure water containing only one kind of molecule
   h. element, contains only C atoms
   i. element, contains only Sb atoms
11. Properties of iron do not change because all particles in iron are atoms of iron. Steel is a mixture of iron and other atoms. The type of steel depends on what is added to the iron.

13.

	Major Source	Compound
nitrogen	air	ammonia, NH_3
sulfur	underground deposits	sulfuric acid, H_2SO_4
chlorine	sea water	sodium chloride, NaCl
magnesium	sea water	Milk of Magnesia, $Mg(OH)_2$
cobalt	mineral deposits	cyanocobalamin, Vitamin B_{12}

15. Cytoxan, $C_7H_{15}O_2N_2PCl_2$
   a. twenty nine, 29 atoms total
   b. carbon, hydrogen, oxygen, nitrogen, phosphorus, chlorine
   c. 15 hydrogens/2 nitrogen
   d. yes, it is organic

17.

	solid	liquid	gas
pure substances	dry ice, CO_2 iron, Fe copper, Cu	mercury, Hg octane, C_8H_{18}	helium, He nitrogen, N_2 methane, CH_4
mixtures	butter steel 14 K "gold"	sea water homogenized milk coffee	natural gas fuel (hydrocarbons mixed with mercaptans) a person's exhaled breath (CO_2, O_2, H_2O)

19. Yes, a mixture of H_2 and O_2 can exist at room temperature. This mixture will be stable as long as no spark or activation energy is added. A reaction produces water, H_2O, which contains both elements.
21. a. Two sodium atoms react with 1 chlorine molecule to form two formula units of sodium chloride solid.
   b. One nitrogen molecule reacts with three chlorine molecules to produce two molecules of nitrogen trichloride.
   c. One molecule of carbon dioxide reacts with one molecule of water to produce one molecule of carbonic acid.
   d. Two molecules of hydrogen peroxide react to produce one molecule of oxygen gas and two molecules of water liquid.

23. b. $\underbrace{N_{2(g)} + 3\ Cl_{2(g)}}_{\text{reactants}} \rightarrow \underbrace{2\ NCl_{3(g)}}_{\text{products}}$

d. $\underbrace{2H_2O_2(aq)}_{\text{reactants}} \rightarrow \underbrace{2\ H_2O(\ell) + O_{2(g)}}_{\text{products}}$

25. The tea in tea bags is a mixture. It can be partially separated by dissolving some water-soluble substances with hot water. Instant tea is a mixture of the water-soluble substances in tea.
27. a. 1 gram = 1000 milligrams
    b. 1 kilometer = 1000 meters
    c. 1 gram = 100 centigrams
29. a. 9 cal/gram; no.
    b. 100 cm/meter; no.
    c. 1.5 g/mL; yes, grams/milliliter is mass/volume.
    d. 454 g/lb.; no.

SELECTED PROBLEMS

1. 5.5 acres/55 cows
2. 0.200 g, 200,000 μg
3. 10,000 meters; ? meters = 10 km $\times \frac{1000\text{ m}}{1\text{ km}} =$ 10,000 meters
4. 3000 mg protein/1 oz. cereal; $\frac{?\text{ mg}}{1\text{ oz}} = \frac{3.00\text{ g protein}}{1\text{ oz}} \times \frac{1000\text{ mg}}{1\text{ g}} = \frac{3000\text{ mg protein}}{1\text{ oz cereal}}$
5. a. 0.04 m
   b. 43 mg
   c. 15500 mm
   d. 0.328 L
   e. 980 g
6. 163 kg $\times \frac{1000\text{ g}}{1\text{ kg}} = 163000$ g
7. 70 kg $\times \frac{1000\text{ g}}{1\text{ kg}} \times \frac{1000\text{ mg}}{1\text{ g}} = 70{,}000{,}000$ mg

CHAPTER 3

1. Matter is neither created nor destroyed. Examples: flash bulb before and after use, hard boiling an egg, yarn knitted into clothing, melting ice cubes in a glass. There is no change in mass during the chemical changes (first two) or the physical changes.
3. Experiments indicated that matter was conserved. Elements had been identified. Compounds had been shown to be composed of definite amounts of specific elements. The composition of a compound had been shown to always be the same regardless of its source. The composition of a combination of elements could be predicted.
5. a. Rutherford used a beam of alpha particles, ${}^4_2\alpha$, from a radioactive source to bombard a thin gold foil. The alpha particles could be detected because they produce a pulse of light when they strike a luminescent screen.
   b. Alpha particles usually passed through the foil, but occasionally they were reflected back towards the source.
   c. Rutherford proposed a planetary or nuclear model for the atom. The alphas that passed through encountered electrons and the reflected alphas were reflected back by a heavy, positively-charged nucleus.
7. a. The number of protons in the nucleus of an atom.
   b. The total count of protons *plus* neutrons in the nucleus of an atom.
   c. The mass of an average atom of an element compared to an atom of ${}^{12}C$ which is assigned a mass of exactly 12 atomic mass units.
   d. Atoms with the same number of protons but with different numbers of neutrons.
   e. On Earth, the % that this isotope is of all the atoms of that element
   f. 1 amu is one-twelfth of the mass of the carbon-12 atom. See also 13.c.
9. Elements of the same element all have the same atomic number. This means they have the same number of protons. For example these two atoms both have 10 protons ${}^{21}_{10}Ne$, ${}^{22}_{10}Ne$.
11. Lithium has 3 protons; the mass number 7 means that the sum of the numbers of protons and neutrons = 7, so it has 4 neutrons. The number of neutrons is determined by subtracting the atomic number from the mass number.
    A = mass number; Z = atomic number
    A − Z = number of neutrons; 7 − 3 = 4
13. Atoms with the same number of protons and different numbers of neutrons are isotopes like these isotopes of neon, ${}^{21}_{10}Ne$ and ${}^{22}_{10}Ne$. A and D both have 25 protons but different numbers of neutrons so they are isotopes. B and C both have 24 protons but different numbers of neutrons so they are isotopes.
15. The atomic number has the symbol Z. The number of protons equals the atomic number.
    a. Germanium, Z = 32. or 32 protons
    b. Silicon, Z = 14 or 14 protons
    c. Nickel, Z = 28 or 28 protons
    d. Cadmium, Z = 48 or 48 protons
    e. Iridium, Z = 77 or 77 protons
17. Yes, the statement is correct. The atomic mass of 24.305 is a weighted average based on the masses of individual Mg isotopes and their abundances.
19. Choice "c" is correct. The ratio by weight of the elements in a compound.
21.

Bohr arrangement	Number of electrons	Element
2-4	6	Carbon
2-8	10	Neon
2-8-3	13	Aluminum
2-8-8-2	20	Calcium

23. Na, 1; Mg, 2; Al, 3; Si, 4; P, 5; S, 6; Cl, 7; Ar, 8
25. When elements are arranged in order of their Atomic Numbers, their chemical and physical properties show repeatable trends.
27. The known elements were arranged in order and gaps between elements occurred. These gaps suggested the existence of elements not yet isolated.
29. Metals are good conductors of heat and electricity. Metals are malleable and ductile. Nonmetals are the opposite, they are not ductile, not malleable and are poor conductors of heat and electricity.
31. a. There are 7.
    b. There are 8 representative groups.
    c. There are two representative groups that are all metals if you ignore H in IA.
    d. The elements in Group VIIA and VIIIA are all nonmetals.
    e. Yes, all the elements in period 7 are metals.
33. Choices "a. 2,1" and "g. 2, 8, 8, 1" both have filled inner levels and one electron in the outermost level.
    Choices "b. 2,6" and "d. 2,8,6" both have filled inner levels and have 6 electrons in the outermost level.
35. a. B, Al, Ga, In, Tl each have 3 valence electrons
    b. C, Si, Ge, Sn, Pb all have 4 valence electrons
    c. F, Cl, Br, I, At all have 7 valence electrons
    d. H, Li, Na, K, Rb, Cs, Fr all have 1 valence electron
37. a. Be is more metallic than B
    b. Ca is more metallic than Be
    c. K is more metallic than Ca
    d. Ge is more metallic than As
    e. Bi is more metallic than As
39. a completed outer shell or set of eight electrons in the outermost energy level
41.

Atomic number	Element name	Number of valence electrons	Period	Metal, M or Nonmetal, NM
6	carbon, C	4	2	NM
12	magnesium, Mg	2	3	M
17	chlorine, Cl	7	3	NM
37	rubidium, Rb	1	5	M
42	molybdenum, Mo	6	5	M
54	xenon, Xe	8 or 0	5	NM

43. The outer or last electron in Cs is in a higher energy level, $n = 6$ and Li has outer electron in $n = 2$. The higher the level the greater the average distance from the nucleus.
45. The smaller the radius of the nonmetal atom, the more reactive it is.
47. There is a stepwise change in proton number across a period. The attraction between the nucleus and the electrons in the atom gradually increases because of this change. This is duplicated in each period so trends in properties appear for each period that match these electronic structure changes.
49. K (largest), Al, P, S, Cl (smallest)
51. If element 36 is a noble gas, then element 35 is a halogen and element 37 is an alkali metal.
53. Sulfur, S, and tellurium, Te, are above and below selenium, Se, in the same group.

CHAPTER 4

1. gamma rays
3. The more hazardous radioisotope is $^{222}_{86}Rn$, more ionizing radiation is emitted over a shorter time period.
5. It is a gas and can be inhaled. The high density makes it difficult to exhale and emitted alpha particles produce cell damage. Radioactive daughters can cause more damage.
7. The mass number on the phosphorus atom should be 30 not 49.
9. Seaborg proposed that Th and the transuranium elements were a group under the rare earth elements Ce, etc.
11. a sequence of radioactive decay steps that starts with radioactive U-238 and ends with stable Pb-206
13. diagnosis and treatment of diseases
15.

	relative charge	relative mass
a.	-1	0
b.	$+2$	4
c.	0	0
d.	$+1$	0
e.	0	1

17. Gamma rays can damage the DNA sequence needed for duplication of the genetic code in the rapidly multiplying cancer cells which duplicate more often than normal cells.
19. natural radiation in food, water and air
21. four neutrons
23. a. discovered ionizing radiation was emitted by uranium compounds
    b. demonstrated that radioactivity was an atomic property
    c. proposed that radioactive decay was a natural change of an isotope of one element into an isotope of another element
25. no effect on the mass number of the daughter
27. catastrophic accident and release of radioactive material, plutonium production, disposal of radioactive wastes
29. a. separation of useful plutonium and uranium from other nuclides in spent fuel
    b. plutonium may be diverted for nuclear weapons, radioactive material may escape into the environment

31.

	Fuels	Benefits	Problems	Current Status
Fission	Uranium Plutonium	renewable from breeder reactors	waste storage and disposal Pu toxicity	commercially available, proven method
Fusion	Deuterium Tritium	unlimited fuel supply	radioactive hardware from power plants, no containment vessel available	research only

33. a. $^{64}_{29}Cu \rightarrow {}^{64}_{30}Zn + {}^{0}_{-1}e$ b. $^{69}_{30}Zn \rightarrow {}^{69}_{31}Ga + {}^{0}_{-1}e$ c. $^{131}_{53}I \rightarrow {}^{131}_{54}Xe + {}^{0}_{-1}e$
35. a. 24,000 years.
 b. it does not dissipate quickly and plutonium is chemically extremely toxic.
37. The "magnetic bottle" uses magnetic fields to contain a plasma by interacting with the magnetic fields generated by the ions in the plasma.

SELECTED PROBLEMS

1. 18,750 tritium atoms
2. 100,000 alpha particles

CHAPTER 5

1. a. A cation is an ion with a positive charge.
 b. An anion is an ion with a negative charge.
 c. Atoms react to acquire an electron configuration with 8 electrons in the outermost shell to match the configuration of the nearest noble gas.
 d. The formula unit is the simplest element ratio for an ionic compound like NaCl instead of Na_2Cl_2.
3. a. a pair of electrons shared between two atoms
 b. four electrons (2 pairs) shared by two atoms
 c. six electrons (3 pairs) shared between two atoms
 d. pair of valence electrons on an atom that are not shared with another atom.
 e. a pair of electrons shared between two atoms
 f. either a double or a triple bond.
5. a. bond between two atoms that share electrons equally.
 b. bond between two atoms that do not attract shared electrons equally.
7. a. A hydrocarbon is a compound consisting only of carbon and hydrogen.
 b. a compound consisting only of carbon and hydrogen with only single bonds
 c. a compound consisting only of carbon and hydrogen with one or more multiple bonds between carbon atoms
 d. Alkenes are hydrocarbons with one or more carbon-carbon double bonds.
9. a. Br^{1-}
 b. Al^{3+}.
 c. Na^{1+}.
 d. Ba^{2+}.
 e. Ca^{2+}.
 f. Ga^{3+}.
 g. I^{1-}.
 h. S^{2-}.
 i. Group IA atoms lose one valence electron to form a +1 ion, see 9c above.
 j. Group VIIA atoms have seven valence electrons and gain one to form a 1− ion, see 9g.
11. a. AlI_3 aluminum iodide
 b. $SrCl_2$ strontium chloride
 c. Ca_3N_2 calcium nitride
 d. K_2S potassium sulfide
 e. Al_2S_3 aluminum sulfide
 f. Li_3N lithium nitride
13. Answer will depend on individual search. Suggested categories: soft drinks, antacids, pickled items, breakfast cereals, preserved and smoked meats, breads, canned soups, jams and jellies, cleaning products.
15. a. Lithium is in Group Ia so it forms the Li^{1+} ion. Tellurium is in Group VIa and forms the Te^{2-} ion. The formula for the neutral ionic combination is Li_2Te.

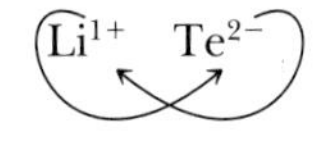

 b. $MgBr_2$
 c. Ga_2S_3
17. a. ionic
 b. covalent
 c. ionic
 d. covalent
 e. covalent
19. a. NO, nitrogen monoxide b. SO_3, sulfur trioxide
 c. N_2O, dinitrogen oxide d. NO_2, nitrogen dioxide
21. Ionic bonding exists between oppositely-charged ions. The polar covalent bond exists between atoms that share electrons unequally. The nonpolar bonding exists between atoms that share electrons equally (atoms with similar attractions for electrons).
23. a. The electron attracting power differences determine the polarity of the bonds. The farther apart two elements are in the periodic table the more polar the bond. The C—Cl bond is more polar because they are farther apart in the periodic table.
 b. The C—F bond is more polar than the C—Cl bond.

25. a. :C:::O: (arrow ⟶) Polar; O is the negative end and C is the positive end. The arrow indicates the direction electrons shift in molecule.
b. H:Ge:H with H above and below, and the tetrahedral structure Ge bonded to four H. Nonpolar; the bonds are nonpolar. The molecule is tetrahedral and symmetric.
c. Cl, Cl, Cl around B (Lewis structure). Nonpolar; the bonds are polar but the planar triangular shape puts the center of positive change and the center of negative charge in the middle of the boron atom.
d. H:F: (arrow ⟶) Polar; F is the negative end and H is the positive end.
27. a. The gaseous state has no definite shape or volume; a gas takes the shape and volume of its container. Gases are compressible.
b. A liquid has definite volume and assumes the shape of the container. The liquid molecules are free to move past each other, allowing the liquid to flow and assume the shape of its container. Liquids are relatively incompressible.
c. The solid state has definite shape and definite volume. The molecules or ions making up the solid are in direct contact and located at fixed positions. The solid is incompressible due to direct contact between the particles.
29. solid
31. The number of gas molecules in a given volume of gas decreases as you move away from the earth's surface. The fewer number of molecules collide with any object less often and exert less push or pressure. Additionally, the force of earth's gravity decreases with altitude. This results in less weight or force per unit area at greater altitudes.
33. Gas molecules are constantly moving. The gas molecules of the perfume vapor diffuse away from the person wearing the perfume.
35. Water has an unusually high normal boiling point, and surface tension. Additionally it expands on freezing.
37. Evaporation of water from one's skin draws energy from the skin. This produces a resulting "chilling" sensation. We are cooled when the energy flows into the evaporating liquid.
39. Without hydrogen bonding the boiling point for ammonia would be -100 °C. See Figure 5.9
41. The increased pressure creates more collisions of gas molecules with the solvent surface. These more frequent collisions create more opportunities for the gas to dissolve in the solvent. Champagne, sparkling waters, Coca Cola® and Pepsi® are examples.
43. The air in the frost-free refrigerator reaches an equilibrium between water molecules in the gas state and water molecules as frost. This equilibrium gas mixture is "wet" or "moist." Dry air from the outside is blown into the refrigerator and the wet air is forced out. Then equilibrium is reestablished and the amount of frost decreases. This process is repeated over and over again leading to the removal of frost.

CHAPTER 6

1. a. 6 and 6
b. 7 and 7
c. 1, 3, 2, 3
d. The number of atoms of each element is the same on both reactant and product side.
3. a. $2\ Al + 3\ Cl_2 \rightarrow 2\ AlCl_3$.
b. $3\ Mg + N_2 \rightarrow Mg_3N_2$
c. $2\ NO + O_2 \rightarrow 2\ NO_2$
d. $2\ SO_2 + O_2 \rightarrow 2\ SO_3$
e. $3\ H_2 + N_2 \rightarrow 2\ NH_3$
5. a. $Ba(s) + 2\ H_2O(\ell) \rightarrow Ba(OH)_2(aq) + H_2(g)$
b. $3\ Fe(s) + 4\ H_2O(\ell) \rightarrow Fe_3O_4(s) + 4\ H_2(g)$
c. $2\ Na(s) + 2\ H_2O(\ell) \rightarrow 2\ NaOH(aq) + H_2(g)$
d. $2\ Li(s) + 2\ H_2O(\ell) \rightarrow 2\ LiOH(aq) + H_2(g)$
7. a. $Sn(s) + 2\ HBr(aq) \rightarrow SnBr_2(aq) + H_2(g)$
b. $Mg(s) + 2\ HCl(aq) \rightarrow MgCl_2(aq) + H_2(g)$
c. $Ca(s) + 2\ H_2O(\ell) \rightarrow Ca(OH)_2(aq) + H_2(g)$
d. $Zn(s) + 2\ HNO_3(aq) \rightarrow Zn(NO_3)_2(aq) + H_2(g)$
e. $2\ Cs(s) + 2\ H_2O(\ell) \rightarrow 2\ CsOH(aq) + H_2(g)$
9. The carbon-12 isotope with a *defined* mass of *exactly* 12 amu is the basis.
11. Each is 0.500 mol of that element.
13. Avogadro's number is 6.022×10^{23} to 4 significant digits or 6.02×10^{23} to 3 s.d.
15. The mole is defined as the number of atoms in exactly 12 g of carbon-12, this equals roughly 6.02×10^{23} atoms.
17. Some reactions are fast because reactants have weak bonds. These reactions have low activation energies; the reactants have a "low hill to climb" to form products. Other reactions are slow because reactants have strong bonds and a high activation energy; these reactants have a high "energy hill to climb" to form products.
19. Slows reaction rates. Freezing slows reaction rates, slowing spoilage and slowing growth of mold or bacteria.
21. Reactions with high or large activation energies are slower than reactions with low or small activation energies.
23. The activation energy is high so there is no reaction at room temperature. Once started by the spark (source of activation energy), the reaction continues with produc-

tion of energy because the products have lower potential energy than the reactants.
25. Faster because contact with oxygen molecules would be better and more frequent.
27. Reversible reactions occur in both forward and reverse directions.
29. a. shift to form $CaCO_3$
b. shift to form CaO and CO_2
31. Reactants are consumed to make more products until a new balance between reactants and products is reached.
33. Unreacted reactants are essentially zero; all are converted to products.
35. a. stored energy
b. a process that gives off energy
c. a process that consumes energy
d. a measure of disorder
e. a process that favors products at the expense of reactants
37. Every spontaneous process occurs with an increase in entropy for the universe. This entropy increase consumes energy that cannot be used for other purposes. Every spontaneous process has a built-in amount of this "wasted" energy.
39. exothermic
41. No, a continuous input of energy is required to keep the process going; without sunlight energy the process will stop.
43. No, because recycling may organize the recycled materials, but the steps in the process create entropy in other matter.

SELECTED PROBLEMS

1. a. 18 g b. 254 g c. 56 g
d. 17 g e. 44 g f. 28 g
2. a. 180.0 g $C_6H_{12}O_6$
b. 98.0 g H_2SO_4
c. 142.0 g Na_2HPO_4
d. 164 g $Ca(NO_3)_2$
e. 884.0 g $C_{57}H_{104}O_6$
3. b. Pb, is the largest.
4. The correct answer is 12 moles of hydrogen has the largest mass of 24.0 g H_2
5. a. 0.5 mol copper b. 1.5 mol $CuCl_2$
6. a. $6 \text{ mols } C_6H_{12}O_6 \times \dfrac{2 \text{ mol } CH_3CH_2OH}{1 \text{ mol } C_6H_{12}O_6} =$
12 mols CH_3CH_2OH
b. $10.5 \text{ mols } CO_2 \times \dfrac{2 \text{ mol } CH_3CH_2OH}{2 \text{ mol } CO_2} =$
10.5 mols CH_3CH_2OH

CHAPTER 7

1. a. Excess hydronium ion, $[H_3O^{1+}] > [OH^{1-}]$, pH < 7
b. Excess hydroxide ion, $[H_3O^{1+}] < [OH^{1-}]$, pH > 7.
c. Equal amounts of hydronium and hydroxide, pH = 7 and $[H_3O^{1+}] = [OH^{1-}]$
3. A salt is the substance formed between the anion of an acid and the cation of a base.
5. a. measures the fraction of molecules that ionize
b. ionizes 100%
c. ionizes 100%
d. only partially ionized, majority of acid molecules are still intact
e. partially ionized, the majority of base molecules or ions are still intact
7. soap, baking soda, household ammonia and milk of magnesia.
9. a. acidic b. acidic c. basic
d. basic e. acidic f. basic
11. a. An indicator changes color with changes in acidity or pH.
b. Molarity is a concentration measure equal to the moles of solute in one liter of solution.
c. measure of the relative amount of solute in a definite amount of solution
13. An acid buffer is an aqueous mixture of both a weak acid and its anion that will maintain a stable pH when either acid or base is added. A basic buffer is a mixture of a weak base and its cation that will maintain a stable pH.
15. a. $CH_3COOH(aq) + KOH(aq) \rightarrow H_2O(\ell) + CH_3COOK(aq)$
b. $H_2SO_4(aq) + Ca(OH)_2(aq) \rightarrow CaSO_4(aq) + 2\ H_2O(\ell)$
c. $H_2SO_4(aq) + 2\ NaOH(aq) \rightarrow Na_2SO_4(aq) + 2\ H_2O(\ell)$
17. $2\ HCl(aq) + CaCO_3(s) \rightarrow CaCl_2(s) + CO_2(g) + H_2O(\ell)$
19. black coffee with the pH = 5.0 The lower the pH the more acidic the solution.
21. Gastric juice pH = 1
23. b. A mixture of CH_3COOH and $Na^+CH_3COO^-$ would be a good combination to prepare a buffer. Any added base would be neutralized by the acidic CH_3COOH (aq). Added acid would be neutralized by the basic proton acceptor, $CH_3COO^-(aq)$.
25. B with pH = 1.1 will be the stronger acid.

SELECTED PROBLEMS

1. a. 0.14 M HCl
b. 1.0 M NaOH
c. 1.00 M NaCl
d. 0.017 M NaCl
2. 87.66 g NaCl; rounded off to 88. g
3. 9.808 g H_2SO_4; rounded off to 9.8 g H_2SO_4.
4. a. pH = 2
b. pH = 3
c. pH = 13.
d. pH = 1
5. a. pH = 11.
b. pH = 3

c. pH = 12.
d. pH = 7

6. a. $10^{-1} = 1 \times 10^{-1}$ or 0.1
b. $10^{-0} = 1 \times 10^{-0}$ or 1
c. $10^{-5} = 1 \times 10^{-5}$ or 0.00001
d. $10^{-3} = 1 \times 10^{-3}$ or 0.001

CHAPTER 8

1. a. Oxidation is the loss of electrons
b. Reduction is the gain of electrons
c. Oxidation is the gain of oxygen
d. Reduction is the gain of hydrogen
e. Oxidizing agent is a substance that accepts electrons
f. Reducing agent is a substance that gives up electrons
3. a. $CO_2 > CO$
b. $NO_2 > NO$
c. $SO_3 > SO_2$
d. $CrO_3 > CrO$
e. $CaO > Ca$
f. $N_2 > NH_3$
5. water H_2O
7. carbon monoxide, CO
9. a. oxidation
b. oxidation
c. reduction
d. reduction
e. reduction
f. oxidation
11. reducing because there is a ready supply of hydrogen to act as a reducing agent
13. zinc is oxidized.
15. Rusting is less of a problem in Arizona because there is less water in the air.
17. a. true
19. Both produce electrical energy. The reactants are added to a fuel cell and products are removed. A battery is usually sealed or self-contained.
21. See answer to question 19. The reactants and products are still inside the battery. No material was added or removed.

CHAPTER 9

1. fuel formed by decomposition of plant and animal matter
3. compound containing only carbon and hydrogen
5. the energy released when a compound reacts completely with oxygen
7. a. molecules with the same general formula but different molecular structures
b. compounds with a linear chain of CH_2-units and terminal CH_3 units
c. compounds with a backbone like a straight-chain hydrocarbon, but with some, C_nH_{2n+1}, alkyl groups replacing H's along the chain
9. n = 1 methane CH_4
n = 2 ethane C_2H_6
n = 3 propane C_3H_8
n = 4 butane C_4H_{10}
11. a. 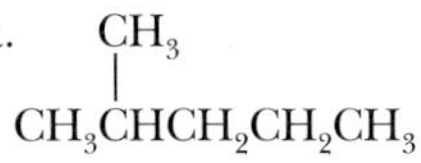
```
CH3
|
CH3CHCH2CH2CH3
```
b. 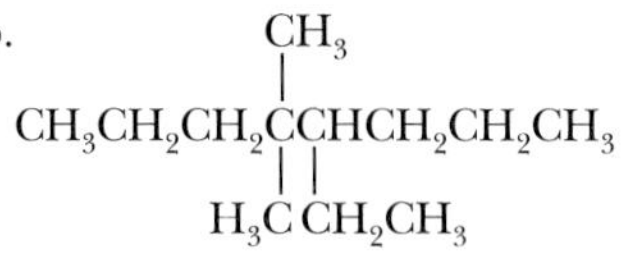
```
              CH3
              |
CH3CH2CH2CCHCH2CH2CH3
         | |
       H3C CH2CH3
```
c. 
```
   CH3
   |
CH3C=CHCH2CH2CH3
```
13. pentane
$CH_3CH_2CH_2CH_2CH_3$

2-methylbutane 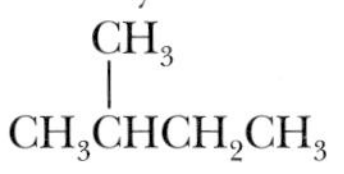
```
   CH3
   |
CH3CHCH2CH3
```
2,2-dimethylpropane 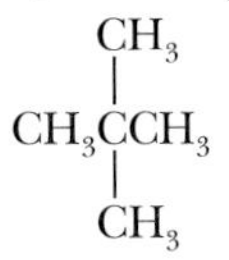
```
   CH3
   |
CH3CCH3
   |
   CH3
```
15. para-dimethylbenzene (benzene ring with CH_3 at top and CH_3 at bottom)

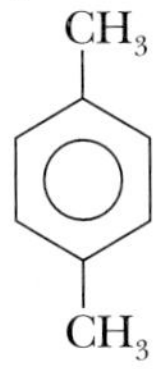

meta-dimethylbenzene (benzene ring with CH_3 at top and CH_3 at lower right)

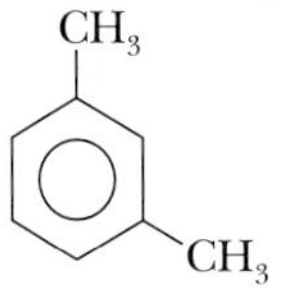

ortho-dimethylbenzene (benzene ring with adjacent CH_3 groups)

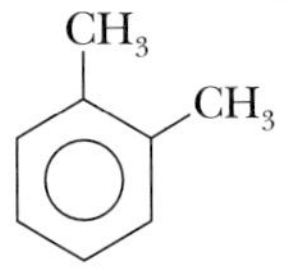

17. There are *cis*- and *trans*- isomers for 2-butene because there are carbon chains on both double-bonded carbon atoms. The 1-butene has only hydrogens on the #1 carbon so there is no group that can be cis or trans to the groups on the #2 carbon.
19. Hot petroleum at about 400°C is distilled in a fractionation tower. There are trays at various levels in the tower, each at a unique temperature. Each volatile component will vaporize. These will condense at a temperature range of a specific tray. The compounds that are most volatile are collected near the top. The compounds like tars that are least volatile with the highest boiling point are collected at the bottom.
21. Aromatic hydrocarbons, branched-chain hydrocarbons, 2-methyl-2-propanol, methanol, ethanol, or methyl-tertiary-butyl ether

23. Synthesis gas, a mixture of CO and H_2, is produced by passing super-heated steam over pulverized coal.

$$C + H_2O + 31 \text{ kcal} \longrightarrow CO + H_2$$

Coal gasification using a catalyst, crushed coal and synthesis gas will produce methane.

$$2\,C + 2\,H_2O + 2 \text{ kcal} \longrightarrow CH_4 + CO_2$$

25. Methanol-to-gasoline is possible through a multi-step catalyzed process.

$$2\,CH_3OH \xrightarrow{\text{ZSM-5 catalyst}} (CH_3)_2O + H_2O$$

$$2\,(CH_3)_2O \xrightarrow{\text{ZSM-5 catalyst}} 2\,C_2H_4 + 2\,H_2O$$

$$2\,C_2H_4 \xrightarrow{\text{ZSM-5 catalyst}} \text{gasoline, a mixture of } C_5\text{-}C_{12} \text{ hydrocarbons}$$

27. Legal requirements for enhanced oxygenated gasoline. A need for cleaner burning gasoline. Higher prices for gasoline from crude oil, petroleum.
29. They burn more completely.
31. Treating pulverized coal with superheated steam forms either carbon monoxide and hydrogen or, using a catalyst, carbon dioxide and methane.
33. M100 means 100% methanol.
E100 means 100% ethanol.
M85 means 85% methanol, 15% gasoline.
E85 means 85% ethanol, 15% gasoline.
FFV means flexible-fueled vehicle which can operate on gasoline or other fuels.
35. a. methane, alkane
b. benzene, aromatic
c. 1-butene, alkene
d. acetylene, alkyne

CHAPTER 10

1. Carbon atoms can bond to other carbon atoms in almost unlimited numbers and in a variety of ways. Introducing other elements allows different molecules due to different atom sequences (functional groups and isomers).
2. There are four structural isomers for the alcohols $C_4H_{10}O$.

3.
a. carboxylic acid, pentanoic acid

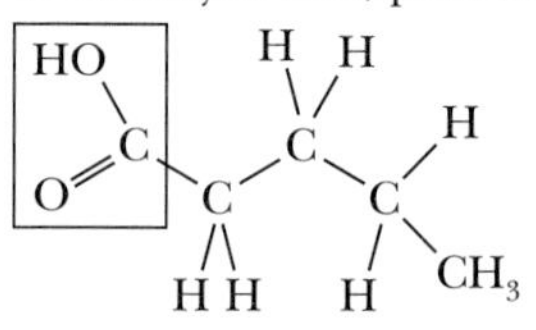

b. alkene, 2-pentene

c. symmetric ether, diethyl ether

d. secondary alcohol, 3-pentanol

e. tertiary alcohol, 2-methyl-2-propanol

f. aldehyde, ethanal

7. 90 proof $= \frac{90}{2}\% = 45\%$

9.

ethanol	ethylene glycol	glycerol
CH_3CH_2OH	CH_2OHCH_2OH	$CH_2OHCHOHCH_2OH$
one —OH group	two —OH groups	three —OH groups
use: solvent	use: automobile antifreeze	use: cosmetics

11.
a. aldehyde

b. alcohol

c. ketone

d. carboxylic acid

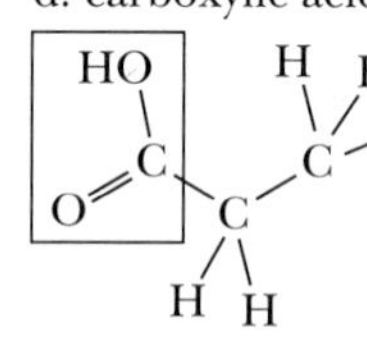

e. ester

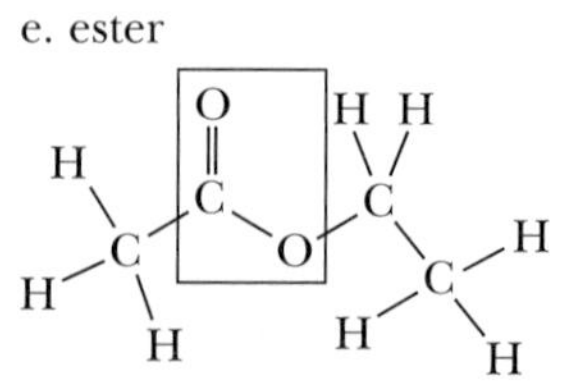

f. ether

g. aldehyde

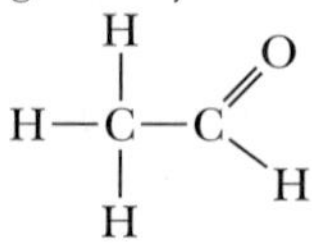

13. Primary alcohols have the —OH group bonded to a carbon atom attached to one other carbon. Secondary alcohols have the —OH group bonded to a carbon atom attached to two other carbon atoms. Tertiary alcohols have the —OH group attached to a carbon atom that is bonded to three other carbon atoms.

Primary: $H_3C{-}CH_2{-}OH$ Secondary: $H_3C{-}CH(OH){-}CH_3$ Tertiary: $H_3C{-}CH_2{-}CH_2{-}C(H_3C)(CH_3){-}OH$

15. Ethanol is oxidized to acetaldehyde which is oxidized to acetic acid. This is further oxidized to CO_2 and H_2O.
17. A railroad train has a series of repeating, identical, linked units, the railroad cars. Polystyrene is a long chain of repeating, identical, linked units, the styrene molecules.
19. A macromolecule is a molecule with very high molecular mass.
21. Monomers need to contain a multiple bond in order to form addition polymers.
23. Natural rubber is poly-*cis*-isoprene. The individual monomer is isoprene.

isoprene: $H_2C{=}C(CH_3){-}CH{=}CH_2$

poly-*cis*-isoprene: $-CH_2{-}C(H_3C){=}CH{-}CH_2{-}[CH_2{-}C(H_3C){=}CH{-}CH_2{-}CH_2{-}C(H_3C){=}CH{-}CH_2]_n$

25. Sulfur is reacted with rubber to align rubber polymers and bond the polymer chains to one another.
27. Cross linking makes the polymer more rigid.
29. a. Yes, styrene can undergo an addition reaction because it can add to double bond in the ethene group.
    b. Yes, propene can undergo an addition reaction because it can add to the double bond.
    c. No, there are only single bonds so nothing can add to any multiple bonds in ethane.
31. a. vinyl chloride $H_2C{=}CHCl$ b. styrene $C_6H_5{-}CH{=}CH_2$
    c. butadiene $H_2C{=}CH{-}CH{=}CH_2$ d. propylene $H_2C{=}CH{-}CH_3$

33. a. Polyester is a polymer built from ester monomers.
    b. Polyamide is a polymer built from amide monomers.
    c. Nylon 66 is a polymer built from 6-carbon monomers. The condensation of hexamethylene diamine and adipic acid yields a polymer with 6 carbon atoms between the nitrogen atoms.

$$\left[-NH{-}(CH_2)_6{-}NH{-}C(=O){-}(CH_2)_4{-}C(=O)-\right]_n$$

d. A diacid is a molecule containing two carboxylic acid groups.

$$HO{-}C(=O){-}CH_2{-}CH_2{-}CH_2{-}CH_2{-}C(=O){-}OH$$

e. A diamine is a molecule containing two amine groups.

$$H_2N{-}CH_2{-}CH_2{-}CH_2{-}CH_2{-}CH_2{-}CH_2{-}NH_2$$

f. A peptide linkage has a carbonyl group, $>C{=}O$, adjacent to an amine —NH— group.

$$-CH_2{-}CH_2{-}C(=O){-}NH-$$ (peptide linkage)

35. Condensation reactions occur between two molecules to produce a larger molecule from the smaller reactant molecules and a separate small molecule like water which is eliminated.
37. Petroleum. Yes, this will change. Plant material sources are renewable and relatively cheaper so they will probably become the major source eventually. In addition the earth's petroleum reserves are limited, petroleum will become more expensive and will not last indefinitely.
39. The four parts to successful recycling are: collection, sorting, reclamation, end-use. Unfortunately, collected materials will not be sorted or reclaimed if the supporting infrastructure for processing doesn't exist.
41. a. aldehyde, alcohol
    b. ketone, alkene, alcohol
    c. aromatic, aldehyde, ether, alcohol
    d. alcohol, carboxylic acid
43. Polymer composite materials are substances that have a polymer matrix with reinforcing fibers. Glass-fiber-reinforced polyesters are used for boat hulls and car body panels. Graphite fiber-epoxy composites are used in fishing rods and tennis racquets.

SOLUTIONS FOR SELECTED PROBLEMS

1. $\frac{100{,}000 \text{ ethylene monomers}}{1 \text{ polyethylene polymer}}$
2. $\frac{84{,}000 \text{ g}}{42 \text{ g}} = 2{,}000$

CHAPTER 11

1. It describes objects that are non-superimposeable mirror images of one another.
3. a 50-50 mixture of enantiomers with no net optical activity
5. a nonsugar molecule that stimulates the sweetness receptors in human taste buds.
7. a naturally occurring organic substance in living systems that is soluble in nonpolar organic solvents, but insoluble in water.
9. the conversion of C=C double bonds to single bonds by adding hydrogen
11. The charged end of the soap and detergent molecule is hydrophilic (attracted to water). The hydrocarbon end of the molecule is attracted to nonpolar solutes but not to water (hydrophobic). A layer of soap or detergent molecules forms around the greasy solute. This envelops the solute particles so the solute particles "float away" in the water.
13. Nonionic detergents do not form insoluble calcium or magnesium salts; hence, less is needed and they are more effective.
15. The listing of water after the names of the oils indicates that the mixture is a water-in-oil emulsion.
17. The peptide bond is the bond formed between the amino group of one amino acid and the carboxyl carbon in another amino acid.
19. An amino acid contains the carboxyl , —COOH, and amino, $—NH_2$, functional groups
21. Order of amino acid residues in a protein or peptide
23. The protein primary structure is the sequence of amino acid residues.
    The secondary structure of proteins is usually either the regular repetitive pattern of the α-helix or the β-pleated sheet. The tertiary structure is the three dimensional structure resulting from the folding of the protein molecule.
25. a. the spiral secondary protein structure
    b. a common secondary protein structure where protein chains form parallel strands held together by hydrogen bonds
27. An active site is a part or region of an enzyme where the substrate sits down to interact with the enzyme.
29. a. 24 possible combinations.
    b. The 24 possible combinations for the amino acids glycine, Gly; alanine, Ala; serine, Ser: and cystine, Cys, are shown in the following table.

Gly-Ala-Ser-Cys	Ala-Ser-Cys-Gly	Ser-Cys-Gly-Ala	Cys-Gly-Ala-Ser
Gly-Ala-Cys-Ser	Ala-Ser-Gly-Cys	Ser-Cys-Ala-Gly	Cys-Gly-Ser-Ala
Gly-Ser-Cys-Ala	Ala-Cys-Gly-Ser	Ser-Gly-Ala-Cys	Cys-Ala-Ser-Gly
Gly-Ser-Ala-Cys	Ala-Cys-Ser-Gly	Ser-Gly-Cys-Ala	Cys-Ala-Gly-Ser
Gly-Cys-Ala-Ser	Ala-Gly-Ser-Cys	Ser-Ala-Gly-Cys	Cys-Ser-Ala-Gly
Gly-Cys-Ser-Ala	Ala-Gly-Cys-Ser	Ser-Ala-Cys-Gly	Cys-Ser-Gly-Ala

31. Valylcystyltyrosine.
33. The D-glucose fits its active site in an enzyme, but not the L-glucose.
35. Protein chains in hair are held together by hydrogen bonds, ionic bonds and disulfide bonds. Hydrogen bonds form between H atoms attached to very electronegative atoms (N and O) and the electron rich atoms with lone pairs. Ionic bonds result when carboxylate ions ($RCOO^-$) and the protonated amino groups $—NH_3^+$; are attracted to one another. Disulfide bonds form between sulfur atoms in cysteine fragments in adjacent strands.
37. The disulfide bond is a single bond between two sulfur atoms. The disulfide bonds between cysteine amino acid units hold parallel strands of hair protein in place.
39. DNA and RNA differ structurally because each contains a different sugar for the polymer chain and they contain different bases. DNA contains the sugar, α-2-deoxy-D-ribose, $C_5H_{10}O_4$, while RNA contains α-2-D-ribose, $C_5H_{10}O_5$. Both DNA and RNA contain adenine, guanine, and cytosine. The base thymine occurs only in DNA. The base uracil occurs only in RNA.
41. The monomers that polymerize to form DNA and RNA are nucleotides. The nucleotide monomers contain a phosphoric acid unit, a ribose unit and a nitrogenous base. See Figure 11.19 in text.
43. bases like adenine-thymine and guanine-cytosine. There are two hydrogen bond sites between the A-T pair and three hydrogen bond sites between the G-C pair.
45. to map the sequence of base pairs in human DNA.
47. a.
    DNA sequence T G T C A G T G G G C C G C T
    mRNA sequence A C A G U C A C C C G G C G A.
    b.
    mRNA sequence A C A G U C A C C C G G C G A
    tRNA sequence U G U C A G U G G G C C G C U
    c.
    Cysteine Glutamine Tryptophan Alanine Alanine.
49. hydrogen bonding between specific bases that have complementary hydrogen bonding.
51. DNA sequence . . . G T A G C . . .
    mRNA sequence C A U C G . . .
53. mRNA sequence C U G U

CHAPTER 12

1. Digestion is the process of breaking large molecules in food into substances small enough to be absorbed by the body from the digestive tract.

3. a. 9 kcal/ g
   b. 4 kcal/g
   c. 4 kcal/ g
5. The basal metabolic rate, BMR, is the rate at which the body uses energy to support the normal maintenance operations of the body. It is the minimum energy required for a person to stay alive.
7. A triglyceride is a triester formed from glycerol and three fatty acids.
9. a. a disease that is associated with the buildup of fatty deposits on the inner walls of arteries
   b. a lipid steroidal alcohol that contributes to the development of artherosclerosis.
   c. the yellowish deposit of cholesterol and lipid-containing material on artery walls that is a symptom of artherosclerosis
   d. complex assemblages of lipids, cholesterol and proteins that serve to transport water-soluble lipids in the blood stream through the body.
   e. lipoproteins richer in lipid than in protein with a corresponding lower density
   f. lipoproteins richer in protein than in lipid with a corresponding higher density
11. a. nutrients the body needs in large amounts.
    b. nutrients needed by the body in small amounts.
13. a. plants; they are found in foods like pasta, rice, potatoes, corn.
    b. produced in the liver and limited amounts are stored in liver tissue and muscle.
    c. plants.
15. an organic compound essential to health; it is needed in small amounts in the diet.
17. a substance that dissolves in water to produce ions. The electrolyte balance refers to a condition of proper transfer of material through osmosis, normal nerve impulse transmission, normal acid-base balance and extracellular volume.
19. phosphorus, calcium, magnesium, sodium, potassium, chlorine, and sulfur
21. to tie up metal ions to keep them from catalyzing the decomposition of food
23. 30% of daily calories
25. Milk should be promoted as a healthy food for the general public because it is an excellent source of calcium and protein. Milk should not be promoted as a healthy food because people who are lactose intolerant may experience serious adverse reactions and become ill when they consume milk.
27. Cholesterol is water insoluble because water is a small polar molecule that will have very weak attractions to the large, organic nonpolar cholesterol molecule. High levels of cholesterol are linked to atherosclerosis or hardening of the arteries. There are 27 carbon atoms in cholesterol, $C_{27}H_{45}OH$.
29. Women produce less estrogen after menopause. Estrogen inhibits calcium loss and bone erosion. Reduced estrogen levels result in more rapid calcium and bone loss. Osteoporosis can be minimized if adequate calcium intake occurs, especially from adolescence through young adulthood, and if estrogen replacements are taken after menopause.
31. Iodine deficiency leads to an enlargement of the thyroid. This condition is called "goiter." Potassium iodide, KI, is added to table salt to make iodized salt in order to prevent iodine deficiency.
33. There is no right answer for this question. Each answer will depend on the specific product.

SELECTED PROBLEMS

1. 1100 Calories (estimated daily caloric need = 1760 Calories)
2. 1450 BMR
3. 20 minutes
4. 20.7 minutes
5. 77 minutes for 100 g bread
6. 11400 mg cholesterol
7. 14758 mg = 15000 mg This is 15 grams with only two significant digits.
8. pecan pie: 39.2% calories from fat
   apple pie: 35.4% calories from fat
   The apple pie is a better low fat choice.
9. 598 Calories total
10. 726 Calories total
11. 2750 Calories total, grams of fat = 92 g;
    grams of carbohydrate = 413 g;
    grams of protein = 69 g.
12. 20% of Calories from fat
    5.6% of daily Calorie needed
    This food would not be a good product for someone on a low salt (low sodium) diet. This product contains 730 mg sodium or 30% of someone's daily amount in each serving.

CHAPTER 13

1. Malaria, pneumonia, bone infections, gonorrhea, gangrene, tuberculosis, and typhoid fever are bacterial diseases. Polio, AIDS, and Rubella (German measles) are viral diseases.
3. The United States Food and Drug Administration, FDA, has the responsibility for classifying drugs as either over-the-counter drugs or prescription drugs.
5. Three major classes of antibiotics are penicillins, cephalosporins, and tetracyclines.
7. Chemotherapy is the treatment of disease with chemical agents.
9. A retrovirus is a virus that uses RNA directed synthesis of DNA instead of the usual DNA-directed synthesis of RNA.
11. a. Vasodilators are used to treat heart disease and asthma.

The vasodilator relaxes the walls of blood vessels and creates a wider passage for blood flow. This lowers blood pressure and reduces the amount of work that the heart must do to pump blood. It also enables a person to breathe more deeply and easily by dilating the bronchial tubes.
b. Alkylating agents are used to treat cancer. Alkylating agents react with the nitrogen bases of DNA in cancer cells and normal cells. Alkyl groups are added to the nitrogens in the bases. This has an effect on both cancer cells and normal cells, but cancer cells are usually dividing and duplicating DNA more often, so the cancer cells are impacted more.
c. Beta blockers are used to treat heart disease. Propranolol (Inderol) is used to treat angina, cardiac arrhythmia, and hypertension. Beta blockers act to keep epinephrine and norepinephrine from stimulating the heart.

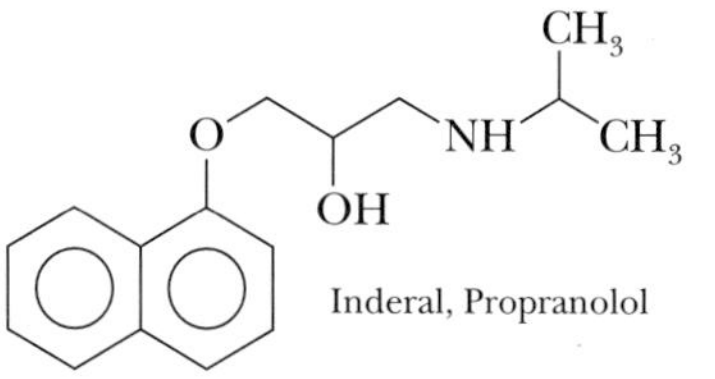

Inderal, Propranolol

d. Antimetabolites are used to treat cancer. Antimetabolites interfere with DNA synthesis, and cancer cells are more susceptible than normal cells because the cancer cells are generally replicating DNA more frequently than normal cells.

13. Histamines and neurotransmitters bind to receptor sites.
15. a. Lovastatin is a cholesterol lowering drug.
b. Methotrexate is an antimetabolite used to treat cancer.
c. Chlorphenirimine (in Chlortrimeton) is an antihistamine
d. Pseudoephedrine is a decongestant.
17. Codeine is
a. an analgesic b. an opioid d. a controlled substance, a Schedule 2 drug.
19. a. Angina results from heart disease with the symptom of chest pain on exertion.
b. Arrhythmia is a heart disease; its symptom is an abnormal heart rhythm.
21. A barbiturate such as phenobarbital, is a depressant. Barbiturates bind to a receptor for GABA. This keeps channels for chloride ion, Cl^{1-}, transmission open. This inhibits transmission of nerve impulses. The physiological effects are a progression from sedation or relaxation, to sleep, to general anesthesia, to coma and death.
23. Heart muscle contracts more often and heart rate increases.
25. Estradiol has an alcohol group and a hydrogen on a carbon in the 5 membered ring while ethynyl estradiol has both an alcohol group and an ethynyl group on that carbon.
27. a. Dopamine helps control memory, emotion and regulate fine muscle movement.
b. Epinephrine is a neurotransmitter that produces increased blood pressure, dilation of blood vessels and widening of the pupils of the eye.
29. Over-the-counter, prescription, unregulated nonmedical drugs, controlled substances.
31. a. hallucinogen b. hallucinogen c. hallucinogen d. hallucinogen e. depressant f. antidepressant or stimulant
33. a. heart disease — Drugs like cholestyramine and lovastatin act to reduce cholesterol blood levels. Cholestyramine accelerates cholesterol excretion while lovastatin interferes with cholesterol synthesis in the liver.
b. angina — Vasodilators work to expand or dilate veins. This decreases resistance to blood flow and the blood pressure goes down. This reduces the work the heart must do to circulate the blood.
c. hypertension — Diuretics reduce blood pressure and hypertension by stimulating excretion of Na^{1+} and increasing urine production. Blood volume and blood pressure decrease when urine output increases.
35. Surgery, irradiation, chemotherapy
37. Hallucinogens cause a person to experience vivid illusions, fantasies, and hallucinations. Examples of hallucinogens are mescaline, PCP, and LSD.
39. a. Morphine is a more effective pain killer than heroin and codeine.
b. Heroin is not a natural alkaloid.
c. Heroin is so addictive that it is not legal to sell or use it in the United States.
41. Nitroglycerine is a heart muscle relaxant. It is used to treat angina which is a symptom of heart disease.
43. The blood-brain boundary is defined physically by small openings in capillaries and the astrocyte cell membranes that prevent passage of large polar molecules. The barrier keeps large toxic polar molecules from passing into the brain. Drugs must be soluble in blood and soluble in the lipid layer of the membrane in order to reach the brain.

CHAPTER 14

1. a. the layer of fresh water and salt water above and below the Earth's surface
b. rock formed by solidification of molten rock
c. rock formed by deposition of dissolved or suspended substances
3. a. Annealing is the process of heating and then slowly cooling a substance to make the substance less brittle and reduce strain in the solid.
b. Amorphous solids have no regular crystalline order.

c. Ceramics are materials generally made from clays and then hardened by heat.
d. Cement is a substance able to bond mineral fragments into a solid mass.
e. A glass is a hard, noncrystalline substance with random (liquid-like) structure.
5. Magnesium is extracted from sea water. It is used for alloys for auto and aircraft parts.
7. Gold, copper, and platinum all can be found in the pure metallic element in nature.
9. The slag is lower density than the molten iron.
11. Carbon is the main impurity in pig iron. Pig iron is brittle partly because it contains Fe_3C.
13. a. $2\ Cu_2S(s) + 3\ O_2(g) \rightarrow 2\ Cu_2O(s) + 2\ SO_2(g)$
b. $2\ FeS(s) + 3\ O_2(g) \rightarrow 2\ FeO(s) + 2\ SO_2(g)$
15.

ductile	malleable	electrical conductor
copper wire	steel sheet metal	gold connectors
steel tubing	aluminum foil	mercury switches

heat conductor	lustrous
copper pots	silver mirrors
aluminum pans	gold jewelry

17. a. The p-type semiconductors have a shortage of electrons. The p-type semiconductors are created by using dopants like B, boron, or Ga, gallium.
b. The n-type semiconductors have excess electrons. The n-type semiconductors are created by using dopants like As, arsenic, or P, phosphorus.
19. The structure of an SiO_4 unit is tetrahedral.

(structure: Si bonded tetrahedrally to four O atoms)

21. Superconductivity is the behavior displayed by a material at temperatures at or below the superconducting transition temperature. The material has no electrical resistance at these conditions. Superconductivity research is aimed at minimizing energy losses when transferring electricity.
23. Iron(III) compounds give glass a yellow color.
25.

Property	Metal	Ceramic
Hardness	variable	yes
Strength	yes	yes
Ductility	yes	no
Electrical Conductivity	yes	no
Brittleness	no	yes

27. Lime, CaO, reacts with CO_2 to form $CaCO_3$
29. No, the silicates will crystallize first because of their higher melting points.
31. Yes. If elements were distributed evenly separation of useful pure elements would require processing more material. This would cause greater environmental disturbance.

CHAPTER 15

1. a. water available in rivers, lakes and streams.
b. water beneath the earth's surface.
c. a layer of water-bearing porous rock.
d. fresh water contaminated by salty sea water.
e. any condition that causes the natural usefulness of water, air, or soil to be diminished.
f. pumping water into aquifers to maintain water volume
g. drinkable water; water that is safe to drink
h. process of reducing concentration by addition of pure water
i. waste substances that have the potential to harm the environment as defined by the EPA
3. Most of the rain water that falls on the United States each day returns to the atmosphere by evaporation or transpiration from plants.
5. Groundwater can be contaminated with pollutants when rainwater runs over or filters through materials, dissolves the pollutants, then percolates into the ground water.
7. Agriculture is the largest single user of water in the United States.
9. Water use per day differs from person to person.
11. Ground water recharge describes the process of pumping water into aquifers to maintain water volume. Purified recycled water from sewage effluent is used as recharge water.
13. The user was responsible for water quality prior to the 1977 Clean Water Act. The discharger is now responsible for maintaining water quality.
15. Two methods for disposal of solid wastes from industry and households are landfills and incineration. Landfills have the greater potential for water contamination.
17. A landfill can be made more secure with respect to water quality protection by controlling the materials placed in the landfill. The materials can be immobilized.
19. batteries, heavy metals
furniture polish, organic solvents
bathroom cleaners, acids or caustics
paint, organic polymers
oven cleaners, caustics
21. Some of the solder used in joints in older copper pipe plumbing contains lead. This lead will dissolve in acidic water passing through these pipes, so the lead can be picked up by the water.
23. Water is vaporized from one container and condensed in another while dissolved, nonvolatile substances remain behind.
25. Settling separates high-density suspended solids such as sand, while filtration removes suspended low-density matter such as algae.
27. The concentration of these chlorinated hydrocarbons and branched chain hydrocarbons rises because they accumulate in the environment. They are fat soluble and stored

by living organisms. They accumulate in organisms at the top of the food chain.

29. Ammonia, $NH_3(aq)$, and ammonium ion, $NH_4^+(aq)$, can be removed from wastewater by using denitrifying bacteria. These bacteria convert the ammonia and ammonium ion to nitrogen, $N_2(g)$. The unbalanced reaction is:

$$NH_3(aq) \text{ or } NH_4^+(aq) \xrightarrow[\text{bacteria}]{\text{denitrifying}} N_2(g)$$

31. Both kill bacteria by oxidizing organic compounds. Aeration depends on O_2 as the oxidizing agent while chlorination uses Cl_2.
33. Mechanical pressure is used to force water molecules through a semi-permeable membrane from the salty aqueous side to the pure water side of the membrane.

CHAPTER 16

1. a. Polluted air is air that contains unwanted and harmful substances.
   b. Aerosols are mixtures of water droplets and particulates
   c. smog containing compounds formed in reactions initiated by sunlight
   d. substances formed by reactions of emitted compounds with components in the air
3. a. Acid rain is rainwater with a pH lower than 5.6.
   b. CFC is the abbreviation for chlorofluorocarbons such as CFC-11, CCl_3F.
   c. The ozone hole is a region in the stratosphere that has lower than normal ozone concentration.
   d. A greenhouse gas is a molecule that absorbs infrared light and radiates it to the atmosphere.
   e. Global warming refers to a worldwide increase in atmospheric temperature.
5. The abbreviation CAA refers to the Clean Air Act. The first act was passed in 1970 and originally controlled air pollution from cars and industry. Amendments in 1977 imposed stricter auto emission standards. The 1990 CAA extends to manufacturing and commercial activity. This version of the CAA regulates particulates, ozone, carbon monoxide, oxides of nitrogen and sulfur, carbon dioxide, and substances that would deplete stratospheric ozone. A newer version was passed in 1996.
7. Industrial smog is chemically reducing while photochemical smog is chemically oxidizing. Industrial smog contains sulfur dioxide mixed with soot, fly ash, smoke, and partially oxidized organic compounds. Photochemical smog is essentially free of SO_2, but contains ozone, ozonated hydrocarbons, organic peroxide compounds, nitrogen oxides, and unreacted hydrocarbons.
9. Ozone is formed in the stratosphere when UV light breaks up O_2 to produce oxygen atoms; these react with additional O_2 to form O_3, ozone.
11. Of the nitrogen oxides "NO_x" in the atmosphere, 97% come from natural sources. Lightning strikes during electrical storms produce NO. Nitric oxide, NO, is so reactive that it combines with O_2 in the atmosphere to produce NO_2. Some bacteria produce N_2O. About 3% of the atmospheric nitrogen oxides come from human activity such as combustion in automobile engines.
13. Volcanic eruptions can contribute to global cooling because the eruption can throw dust particles into the air. These dust particles scatter and reflect sunlight into space, so the solar energy never reaches the earth's surface. This will decrease the amount of energy striking the earth and decrease the atmospheric temperature.
15. Hydrocarbons are released into the atmosphere by natural sources and human sources. Hydrocarbons are put into the atmosphere by living plants such as deciduous trees, by the decay of dead plants and animals, and by excrement from insects and animals. Human activity introduces hydrocarbons when organic solvents are used, for example in the handling of petroleum products, etc. Generally only human activities are within our control.
17. Nitrogen dioxide plays a role in the formation of ozone in the troposphere.
19. Nitrogen dioxide (NO_2) will dissociate to form an oxygen atom and nitric oxide, if it is excited by a sufficiently energetic photon of UV light. The O atom forms O_3 by reaction with O_2.

$$NO_2(g) + h\nu \longrightarrow NO(g) + O(g)$$
$$O + O_2 \longrightarrow O_3$$

21. All three are oxidizing agents. Ozone, sulfur dioxide and nitrogen dioxide cause lung damage.
23. Air is precooled below 0°C = 273°K to remove water vapor as ice. The temperature is decreased to less than −78°C or 194°K to remove CO_2. The dry air, free of CO_2, is compressed to more than 100 atmospheres. This compression heats the air because energy is added to compress it. The air is cooled to room temperature. The room temperature air is allowed to expand and cool. This compression expansion cycle is repeated until all of the air is liquefied. The liquid air is allowed to warm and each gas in the mixture can be collected when it "boils" off at its own boiling point.
25. Lung diseases, cancer and mutagenic effects.
27. All three categories of emissions decreased on a per mile basis because of catalytic converters. Lead emissions decreased because unleaded gasolines are now used. These data show what happens per mile of automobile use. What these do not show is how the totals compare because of changes in the number of miles driven. The cars are cleaner, but increases in total miles driven could increase actual pollution.

1960 (no catalytic converters)	g/mile	1993 (catalytic converters)	g/mile
HC	10.6	HC	0.41
CO	84.0	CO	3.4
NO_x	4.1	NO_x	1.0

1995 (estimated)	g/mile
HC	0.41
CO	3.4
NO_x	1.0

29. Oil burning electric fuel plants generate SO_2. The amount of SO_2 emitted can be reduced by either using low sulfur fuel oil or by passing plant exhaust gas through molten sodium carbonate to form sodium sulfite.
31. Rain tends to be acidic because it dissolves CO_2 to make H_2CO_3. This normally creates a solution with a pH of 5.6. Human sources put SO_2 and NO_x into the atmosphere. They react with rainfall to yield mixtures with pH lower than the natural 5.6. This precipitation is called acid rain.
33. The reaction between carbon and oxygen when there is insufficient oxygen is

$$2\,C + O_2 \longrightarrow 2\,CO$$

The product is carbon monoxide.

35. CFCs are linked to depletion of ozone concentrations in the stratosphere. These regions of lowered ozone concentration are relative ozone "holes." These ozone holes allow increased UV light levels at sea level causing greater rates of skin cancer. Some unexpected effects on algae, plants and animals that could upset the ecological balance are also causes for worry.
37. Ultraviolet light can cause the break up of oxygen molecules.

$$O_2 + h\nu \longrightarrow 2\,O$$

The oxygen atoms formed by the dissociation of O_2 can form ozone in the following reaction.

$$O_2 + O \longrightarrow O_3$$

39. Automobile air conditioners primarily used CFC-11, CCl_3F. CFCs are now longer used because when they reach the stratosphere UV light can break the carbon chlorine bond to produce Cl atoms. These Cl atoms can react to deplete ozone concentrations in the stratosphere. The dots indicate free radicals.

$$\mathrm{F{-}\overset{\displaystyle Cl}{\underset{\displaystyle F}{C}}{-}F} + h\nu \longrightarrow \mathrm{F{-}\underset{\displaystyle F}{\dot{C}}{-}F} + \mathrm{Cl\cdot}$$

CHAPTER 17

1. The earth's carrying capacity depends on the amount of productive land, agricultural practices, biotechnology advances, the rate of degradation of farmland global pollution, the amount of water for irrigation, and the level of an acceptable quality of life.
3. Soil is a mixture of mineral particles, organic matter, water, and air.
5. Soil has a structure made up of a series of layers called horizons which are loosely packed and permeable near the surface and gradually change to impermeable solid rock.
7. Limestone will neutralize acids in the soil and raise the pH.
9. carbon, hydrogen and oxygen
11. a. Metal ions from Group IA and Group IIA
    b. The result is a more acid soil, and the pH goes down.
13. a. substances needed by green plants for healthy growth
    b. Nonmineral nutrients are carbon, oxygen and hydrogen. They are available from water and atmospheric carbon dioxide.
    c. Mineral elemental nutrients are water soluble substances that plants can only absorb as solutes through their roots.
15. A soil shortage of N, P and K is more likely. The nutrients N, P, and K are primary elemental nutrients and are needed in greater amounts. They are more easily leached from the soil.
17. $N_2(g) + 3\,H_2(g) \rightleftharpoons 2\,NH_3(g)$. It is the synthetic source of ammonia fertilizer.
19. Chlorosis is a plant condition of low chlorophyll. It is caused by a deficiency of any one of the three nutrients: magnesium, nitrogen or iron. A symptom of chlorosis is the presence of leaves that are pale yellow instead of green.
21. The fertilizer labeling system indicates the percentage by weight of nitrogen (N), phosphorus (P_2O_5) (phosphate) and potassium as K_2O (potash). None of the pure elements are present in the fertilizer. Each element is present in some compound. None of the reference substances are present in the fertilizer bag either.
23. Urea, H_2NCONH_2, decomposes in the soil to form ammonia, NH_3.
25. a. A quick-release fertilizer dissolves readily in water. It dissolves easily and can be picked up quickly by the plant roots.
    b. A slow-release fertilizer is not as water soluble. It dissolves very slowly in water and it is available to the plant more slowly, so it is taken up slowly by the plant.
27. Ammonium nitrate is inexpensive and a good source of water soluble nitrogen. Farmers benefit because cheap fertilizers keep food production costs down. Consumers benefit because food prices are low. The dark side of ammonium nitrate is that it is an explosive. Handling it can lead to industrial accidents. It can be used as an explosive by terrorists as was done in 1995 when approximately a

ton of ammonium nitrate was detonated next to the Federal Building in Oklahoma City.

29. DDT is fat soluble because "like dissolves like" and both are nonpolar. DDT poses problems because it is a carcinogen and does not degrade quickly. It is stored in the fat tissue of animals. This accumulation problem is exaggerated when one organism high in the food chain eats many other ones lower in the chain.
31. About 33% to 40% of food crop production is lost to pests each year worldwide. The monetary value of these losses is estimated at $20 billion per year.
33. A biodegradable pesticide is one that is quickly converted to harmless products by microorganisms and natural processes. Pyrethrins are an example of a class of biodegradable insecticides. An example of a pyrethrin is shown here.

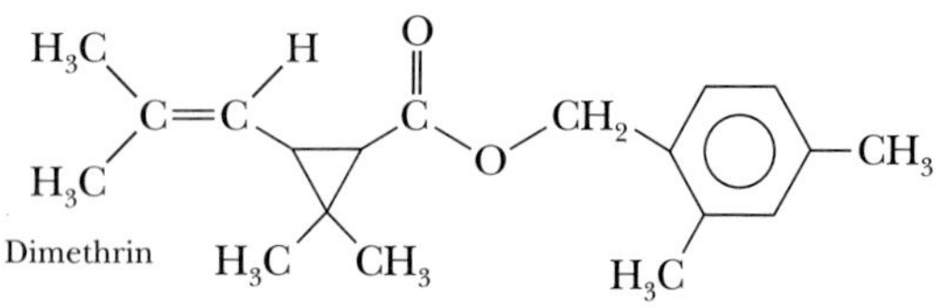

Dimethrin

35. a. Pesticides can increase crop yields and protect them when stored. They reduce the loss of crops to pests. Every bit of food saved is food that need not be replaced. Food supplies increase and food is more plentiful at lower costs. Diseases carried by pests can be controlled or even eliminated. Pesticides can cause problems of water and soil contamination if they are misused. Pesticide residues can contaminate crops and create long term poisoning problems. Resistant strains of pests can develop and require higher levels of pesticides.
    b. Pesticides should be used early enough to require smaller amounts. They should be used only when absolutely necessary and when alternative methods do not exist.
37. Sustainable agriculture is a set of practices intended to improve profits, limit use of agrichemicals and increase the use of environmentally friendly farming procedures.
39. Integrated pest management involves the use of disease resistant plant varieties and biological controls such as predators or parasites to control pests.
41. a. Genetically altered crops with specific genes inserted to produce plants with desirable properties.
    b. Flavr Savr tomato, a weevil resistant garden pea and Bollgard cotton seed.
43. Pheromones are insect sex attractants. They are used to lure insects into traps so insecticide spraying is not needed.
45. Yes, there is a conflict. Population growth will decrease the number of productive acres per person. The present 0.82 acre per person will decrease and the efficiency of farming methods will need to improve in order to avoid famine and food shortages. This problem will be compounded because population centers usually are in the middle of prime agricultural land. The growth of urban areas will remove such land from production.
47.

Agrichemical	LD_{50}	
DDT	100 mg/kg	most toxic
primicarb	150 mg/kg	
carbaryl	250 mg/kg	
malathion	1,000 mg/kg	
dimethrin	10,000 mg/kg	least toxic

SELECTED PROBLEMS

1. $\dfrac{\text{acres of land under cultivation}}{\text{number of people}} = \dfrac{4.84 \text{ billion}}{5.9 \text{ billion}} =$

   $\dfrac{4.84 \times 10^9 \text{ acres}}{5.9 \times 10^9 \text{ people}} = 0.82 \text{ acres/person}$
2. $\dfrac{10 \text{ pounds of N}}{100 \text{ pounds total}} \times 20 \text{ pound bag} = 2 \text{ pounds of N}$
3. A fertilizer with 20-10-5 label is a complete fertilizer containing all three mineral plant nutrients listed in the order: nitrogen, phosphorus, potassium. On a 50 pound bag the 20-10-5 label means that 20% of the weight comes from nitrogen as N; 10% comes from phosphorus as P_2O_5; and 5% comes from potassium as K_2O. Therefore, 20% of the 50 pounds comes from nitrogen:

$$50 \text{ pounds fertilizer} \times \frac{20 \text{ pounds N}}{100 \text{ pounds fertilizer}} = 10 \text{ pounds N}$$

Glossary/Index

Note: The letters i, t, and s following page numbers indicate the entry refers to an illustration, a table, or a structure, respectively.

A

B

C

D

G

H

I

J

K

N

O

P

Protease inhibitors, 337
Protein Biomolecule that is a polymer of amino acids and may also include nonamino acid parts, 281
dietary, 312–313
digestion of, 305–306, 306i
in foods, 307t
hair, 289–291
natural synthesis, 296–297, 298i
structure and function, 285–289, 285i
Protein-energy malnutrition, 313
Proton acceptors, 160
Proton donors, 160
Proton Positively charged subatomic particle found in the nucleus, 40, 160, 40t
Proust, Joseph Louis, 39
Prozac (fluoxetine), 331t, 342s
Pseudoephedrine, 351t
p-type semiconductors, 377, 379
Public perception, of risk, 11
Pure substance Matter with a uniform and fixed composition at submicroscopic level, 16. *See also* Compounds; Element
studying, reasons for, 20–22
PVA (poly(vinyl acetate), 242t
PVC (poly(vinyl chloride), 242t, 243, 367
Pyrethins, 475
Pyrethrum, 475
Pyrite, 431
Pyrolysis, 225
Pyroxenes, 383, 383i, 384t

Q

Qualitative Describes information or experiments that are not numerical, 31
Quanta Discrete packets of energy, 51
Quantitative Describes information or experiments that are numerical, 31
Quantized, 51
Quantum Minimum energy change in an atom emitting or absorbing radiation, 51
Quartz, 365t, 383
Quaternary protein structure Shape assumed by a protein with more than one polypeptide, 288
Quick-release fertilizers Water-soluble combinations of plant nutrients, 463

R

R groups, 227
Racemic mixture Mixture of equal amounts of enantiomers, 262
Radiators, smog-eating, 430
Radioactive decay, 75–76, 76i, 77t
natural series, 79, 80f
rates of, 77
Radioactive isotopes
artificial production, 81
half-lives of, 78–79, 78t, 79i
Radioactivity Spontaneous decomposition of unstable atomic nuclei; produces alpha, beta, and gamma radiation, 25, 41, 70
discovery of, 71, 71i
natural, 41–42, 41i
useful applications of, 85–87, 85i, 86t
Radiation exposure, 83i
Radioisotopes
for medical imaging, 86–87, 86t
in medicine, 77
Radon, 83i, 84–85, 85i
Radon daughters, 84
Rainwater, 405
Rantidine (Zantac), 331t
RCRA (Resource Conservation and Recovery Act), 398
Reactants Substances that undergo change in a chemical reaction, 18, 136–138
Reaction pathways, 138–143, 140i–143i
Reaction rate The amount of reactant converted to product in a specific unit of time, 141–143, 141i
Reactivity, of element, 62
Receptor A molecule or portion of a molecule that interacts with another to cause a change in biochemical activity, 339
Recombinant DNA technology Process of splicing and recombining genes, 300
Recycling, 151–154, 152i, 414
of plastics, 253, 255, 254i
of tires, 9–10
Red blood cells (erythrocytes), 287
Redox reactions Reactions in which electrons are transferred between reactants so that oxidation and reduction occur; oxidation-reduction reactions, 176
Reducing agent(s) Anything that causes the loss of oxygen, gain of hydrogen, or gain of electrons by another reactant in a chemical reaction, 177t, 178–181
Reduction The loss of oxygen, gain of hydrogen, or gain of electrons, 175, 180
Reference daily values, 313, 313t
Reformulated gasoline Gasoline whose composition has been changed to reduce the percentage of olefins, aromatics, and sulfur and with additional oxygenated compounds such as MTBE, 217, 218
Refractory, 373
Refrigerants, fluorine-containing, 6–7
Reinforced plastics Plastics with fibers of a substance, such as graphite, imbedded in a polymer matrix, 252
Representative elements Elements in A groups of the periodic table, 58
Reprocessing, of nuclear fuel, 94
Research, basic. *See* Basic science
Resource Conservation and Recovery Act (RCRA), 398
Retinol, 315
Retrovirus, 336
Reverse osmosis The application of pressure on a solution to cause water molecules to flow through a semipermeable membrane from a more-concentrated to a less-concentrated solution, 412, 412i, 413, 413i
Reverse transcriptase, 336, 337
Reversible Describes a chemical reaction able to proceed in the reverse direction and to come to equilibrium, 144
Rhodium, 145
Riboflavin, 317s
Ribonucleic acid (RNA) The nucleic acid that transmits genetic information and directs protein synthesis, 293, 296, 297
D-Ribose, 293s
Rifampin, 335
Risk assessment, 10–11, 11t
Risk management, 12
Ritonovir, 338i
RNA (ribonucleic acid) The nucleic acid that transmits genetic information and directs protein synthesis, 293, 296, 297
Roasting Heating an ore in air, usually to produce an oxide, 375
Rocks, types of, 365
Rounding off Reducing the number of significant digits in the result of a calculation, 485
Rowland, F. S., 439
RPIvory, 253
Rubber, 9
Rubella virus, 333
Rum, 230t
Rust, 146, 175, 191
Rutherford, Ernest, 41–42, 43–44, 71
artificial nuclear reactions and, 80–81
gold foil experiment of, 42, 44, 42i, 43i

S

Saccharin, 266, 266t
Safe Drinking Water Act, 11
Safe Drinking Water and Toxic Environment Act of 1986, 400
Salicylic acid, 346s
Salt bridge, 184, 185
Salt Compound made up of positive ions from a base and negative ions from an acid, 101, 160
separation from seawater, 366, 367–368, 367i
Saponification Hydrolysis in basic solution of fats and oils to produce glycerol and salts of fatty acids, 274
Saturated hydrocarbons Another name fro alkanes, 110
Saturated solutions Solutions that have dissolved all the solute in a quantity of solvent that it can at a given temperature, 125
Scanning tunneling microscope (STM), 46, 46i, 47i
Scarton, Henry, 253
Schockley, William, 378
Science, and technology, 8i
Scientific fact, 2
Scientific knowledge, use of, 8
Scientific notation Representation of a number by a number between 1 and 10 times 10 raised to a power; exponential notation, 165, 488–493
calculations with, 491–493
conversions, 489–491
small numbers in, 489t
Scrubbing, 423i
Scurvy, 318
Seaburg, Glenn T., 83
Seawater
desalinization of, 411–412, 411t, 412i
magnesium recovery from, 368–369, 368i
separation of salt from, 366, 367–368, 367i
Second, 494
Second law of thermodynamics The total entropy of the universe is constantly increasing, 150–151
Secondary battery A battery that can be recharged by reversing the flow of electricity, 185–187, 186i